Lecture Notes in Computer Science 6457

Commenced Publication in 1973
Founding and Former Series Editors:
Gerhard Goos, Juris Hartmanis, and Jan van Leeuwen

Kalyanmoy Deb   Arnab Bhattacharya
Nirupam Chakraborti   Partha Chakroborty
Swagatam Das   Joydeep Dutta
Santosh K. Gupta   Ashu Jain
Varun Aggarwal   Jürgen Branke
Sushil J. Louis   Kay Chen Tan (Eds.)

Simulated Evolution and Learning

8th International Conference, SEAL 2010
Kanpur, India, December 1-4, 2010
Proceedings

 Springer

Volume Editors

Kalyanmoy Deb
Arnab Bhattacharya
Partha Chakroborty
Joydeep Dutta
Santosh K. Gupta
Ashu Jain
Indian Institute of Technology Kanpur, Kanpur, Uttar Pradesh 208016, India
{deb, arnabb, partha, jdutta, skgupta, ashujain}@iitk.ac.in

Nirupam Chakraborti
Indian Institute of Technology Kharagpur, India, nchakrab@gmail.com

Swagatam Das
Jadavpur University, Kolkata, India, swagatamdas19@yahoo.co.in

Varun Aggarwal
Aspiring Minds, New Delhi, India, varun@aspiringminds.in

Jürgen Branke
University of Warwick, Coventry, UK, juergen.branke@wbs.ac.uk

Sushil J. Louis
University of Nevada, Reno, USA, sushil@cse.unr.edu

Kay Chen Tan
National University of Singapore, eletankc@nus.edu.sg

Library of Congress Control Number: 2010939055

CR Subject Classification (1998): F.1.1, I.2.6, I.6, G.1.6, H.3, D.2.2, J.3-4

LNCS Sublibrary: SL 1 – Theoretical Computer Science and General Issues

ISSN   0302-9743
ISBN-10   3-642-17297-0 Springer Berlin Heidelberg New York
ISBN-13   978-3-642-17297-7 Springer Berlin Heidelberg New York

springer.com

© Springer-Verlag Berlin Heidelberg 2010
Printed in Germany

Typesetting: Camera-ready by author, data conversion by Scientific Publishing Services, Chennai, India
Printed on acid-free paper   06/3180

Preface

This LNCS volume contains the papers presented at the 8th Simulated Evolution and Learning (SEAL 2010) Conference held during December 1–4, 2010 at the Indian Institute of Technology Kanpur in India. SEAL is a prestigious international conference series in evolutionary optimization and machine learning. This biennial event started in Seoul, South Korea in 1996 and was thereafter held in Canberra, Australia in 1998, Nagoya, Japan in 2000, Singapore in 2002, Busan, South Korea in 2004, Hefei, China in 2006 and Melbourne, Australia in 2008.

SEAL 2010 received 141 paper submissions in total from 30 countries. After a rigorous peer-review process involving 431 reviews in total (averaging a little more than 3 reviews per paper), 60 full-length and 19 short papers were accepted for presentation (both oral and poster) at the conference. The full-length papers alone correspond to a 42.6% acceptance rate and short papers add another 13.5%.

The papers included in this LNCS volume cover a wide range of topics in simulated evolution and learning. The accepted papers have been classified into the following main categories: (a) theoretical developments, (b) evolutionary algorithms and applications, (c) learning methodologies, (d) multi-objective evolutionary algorithms and applications, (e) hybrid algorithms and (f) industrial applications.

The conference featured three distinguished keynote speakers. Narendra Karmarkar's talk on "Beyond Convexity: New Perspectives in Computational Optimization" focused on providing new theoretical concepts for non-convex optimization and indicated a rich connection between optimization and mathematical physics and also showed a deep significance of advanced geometry to optimization. The advancement of optimization theory for non-convex problems is beneficial for meta-heuristic optimization algorithms such as evolutionary algorithms. Manindra Agrawal's talk on "PRIMES is in P" provided a much-improved version of his celebrated and ground-breaking 2002 work on polynomial time algorithm for testing prime numbers. The theoretical computation work presented in this keynote lecture should be motivating for the evolutionary optimization and machine learning community at large. Toshio Fukuda's talk on "Intelligent Robot for Multi-mode Locomotion" showcased how multiple locomotions adopted by animals can be mimicked in developing highly robust robots for performing different tasks. The learning behaviors portrayed in the talk should be motivating to the researchers in evolutionary learning and robotics alike.

SEAL 2010 also included two tutorials, which were free to all conference participants. Tutorial topics were chosen from two complementary areas of evolutionary computing. B. Yegnanarayana's tutorial on "Artificial Neural Networks and Applications in Optimization" systematically introduced the principles of artificial neural networks (ANN) and their applications in various problems. ANN has been extensively used in learning and modeling problems; however, an efficient use of ANN requires knowledge and understanding of the intricacies of its fundamental principles. This tutorial inspired both novices and experts of ANN to a greater understanding of the very fundamentals of its working principles. The tutorial by Debabrata Goswami on "Quantum Computing" introduced the fast-growing methodologies of quantum computing techniques. The ideas portrayed in the tutorial motivated evolutionary computing researchers to pay more attention to the collaborative activities between the two fields. These two tutorials made an excellent start to the four-day conference.

We take this opportunity to thank authors of all submitted papers for their hard work, adherence to the deadlines and patience with the review process. The quality of a refereed volume depends mainly on the expertise and dedication of the reviewers. We are indebted to the Program Committee members, who not only produced excellent reviews but also did these in the short time frames that they were given. The review process would not have been possible without the tireless dedication of Amit Saha and Shivam Gupta and their coherent core team at KanGAL (Sunith Bandaru, Rituparna Dutta, Soumil Srivastava, Rupesh Srivastava, and Rupesh Tulshyan) in interacting with authors, Program Committee members and simultaneously with the Chairs.

We would also like to thank our sponsors for providing all the support and financial assistance. First, we are indebted to the IIT Kanpur administration team (the director, the deputy director, deans, and faculty colleagues and administrative personnel) for supporting our cause and encouraging us to organize the conference at IIT Kanpur. We would like to thank Xin Yao, Xiaodong Li, Mengjie Zhang for showing confidence in us in organizing the SEAL 2010 conference in India. The financial assistance from the Golden Jubilee Committee, IIT Kanpur, Department of Science and Technology (DST), New Delhi, and the Council of Scientific and Industrial Research (CSIR), New Delhi in meeting a major portion of the expenses is highly appreciated. Contributions including financial support from Esteco, Italy and USA, General Electric, Bangalore, General Motors, Bangalore, and TCS Innovation Lab, Delhi, were extremely helpful in arranging the conference. We would also like to thank the participants of this conference who, despite the difficulties in getting to Kanpur, considered attending the conference above all hardships. Finally, we would like to thank all the volunteers whose tireless efforts included meeting the deadlines and arranging every detail to make sure that the conference ran smoothly. We hope the readers of these proceedings find the papers inspiring and enjoyable.

December 2010

Kalyanmoy Deb
Varun Aggarwal
Arnab Bhattacharya
Nirupam Chakraborti
Partha Chakroborty
Swagatam Das
Joydeep Dutta
Santosh K. Gupta
Ashu Jain

Organization

SEAL 2010 was organized by the Kanpur Genetic Algorithms Laboratory (KanGAL), Indian Institute of Technology Kanpur, India.

Executive Commitee

General Chair:	Kalyanmoy Deb, India
Program Chairs:	Arnab Bhattacharya, India
	Nirupam Chakraborti, India
	Partha Chakroborty, India
	Swagatam Das, India
	Joydeep Dutta, India
	Santosh K. Gupta, India
	Ashu Jain, India
Industrial Session Chair:	Varun Aggarwal, India
Technical Co-chairs:	Jürgen Branke, UK
	Sushil J. Louis, USA
	Kay Chen Tan, Singapore
International Advisory Committee:	Thomas Bäck, The Netherlands
	Vijay Chandru, India
	Peter Fleming, UK
	Eric Goodman, USA
	Jong-Hwan Kim, Korea
	Zbigniew Michalewicz, Australia
	Una-May O'Reilly, USA
	Asim K. Pal, India
	Hans-Paul Schwefel, Germany
	Lothar Thiele, Switzerland
	Xin Yao, UK

Tutorials

Title: Quantum Computing
Speaker: Debabrata Goswami

Title: Artificial Neural Networks and Applications in Optimization
Speaker: B. Yegnanarayana

Keynote Lectures

Title: PRIMES is in P
Speaker: Manindra Agrawal

Title: Intelligent Robot for Multi-mode Locomotion
Speaker: Toshio Fukuda

Title: Beyond Convexity: Towards Mathematical Foundation for Non-convex
          Optimization
Speaker: Narendra Karmarkar

Sponsoring Agencies

Kanpur Genetic Algorithms Laboratory (KanGAL)
Indian Institute of Technology Kanpur, India
Golden Jubilee Committee, IIT Kanpur, India
Department of Science and Technology (DST), New Delhi
Council of Scientific and Industrial Research (CSIR), New Delhi
ESTECO, Italy and USA
TCS Innovation Laboratory, Delhi, India
General Electric, Bangalore, India
General Motors, Bangalore, India

Program Committee

Aboubekeur Hamdi-Cherif	Bishakh Bhattacharya	Fransisco Ruiz
Adam Berry	Bob Mckay	G. Saravana Kumar
Adam Ghandar	B.V. Babu	G.N. Sashi Kumar
Ah King Robert	Byoung-Tak Zhang	Gary Lamont
Ali Riza Yildiz	Carlos Coello Coello	Gaspar Cunha
Amit Saha	Carlos Fonseca	Gustavo Recio
Amos Ng	Christie Myburgh	Hans-Georg Beyer
Andrew Lewis	Clarisse Dhaenens	Helio J.C. Barbosa
Andries Engelbrecht	Dario Landa-Silva	Henrik Saxén
Andy Tyrrell	David Corne	Hisao Ishibuchi
Aniruddha Basak	Deepak Sharma	Hussein Abbass
Ankit Palliwal	Dhanesh Padmanabhan	Jesper Genri Hattel
Ankur Sinha	Dhish Saxena	Jin-Kao Hao
Anna Piwonska	Dilip Datta	João Vasconcelos
Arnab Bhattacharya	Dilip Pratihar	Jong-Hwan Kim
Arjun Chandra	Dimo Brockhoff	Julian Molina
Arnob Ghosh	Dipankar Dasgupta	Kalyan Veeramachaneni
Arup Nandi	Efrén Mezura-Montes	Kalyanmoy Deb
Ashutosh Tiwari	Enrico Rigoni	Kareti V.R.B. Prasad
	Enrique Alba	Karthik Sindhya

Table of Contents

Learning Methodologies

Multi-Objective Evolutionary Algorithms and Applications

Hybrid Algorithms

Industrial Applications

Beyond Convexity: New Perspectives in Computational Optimization

Narendra Karmarkar

Laboratory for Computational Mathematics
narendrakarmarkar@yahoo.com

Abstract. For computational solutions of convex optimization problems, a rich body of knowledge including theory, algorithms, and computational experience is now available. In contrast, nothing of comparable depth and completeness can be offered at the present time, for non-convex problems. The field of convex optimization benefited immensely from pre-existing body of concepts and knowledge from pure mathematics, while non-convex problems seems to require formulation and exploration of entirely new mathematical concepts, as well as new models of computation. The intent of this paper is to describe our efforts in this direction, at a philosophical or conceptual level, without going into specific applications or implementation in software. We also point out connections with other areas, particularly mathematical physics.

1 Introduction

Theory of convex optimization successfully exploits geometric and topological properties of the underlying space. Any extension to non-convex problems requires critical examination of these properties, extracting what is essential for success, and formulating more general conceptual framework. In this paper, we describe our investigation in this direction, organized along the following topics:

- Graded Connectivity: Moduli Spaces for Connected Sets
- Space Curvature and Generalized Parallelism
- Locally Exact Models: Linear and Quadratic
- Linear Models
- Quadratic Models
- Relativized Concept of Algebraic Curves
- Space Extension for Connectivity
- Continuous Models of Computation

2 Graded Connectivity

2.1 Introduction

Let $f(\mathbf{x})$, $\mathbf{x} \in \mathbb{R}^n$ be the function to be minimized. The downward level set of f corresponding to level α is defined as: $L_f(\alpha) = \{\mathbf{x} \in \mathbb{R}^n | f(\mathbf{x}) \leq \alpha\}$.

K. Deb et al. (Eds.): SEAL 2010, LNCS 6457, pp. 1–23, 2010.
© Springer-Verlag Berlin Heidelberg 2010

For maximization problems, upward level sets are defined similarly, reversing the inequality above. Unless otherwise stated, the level sets are assumed to be connected. If not, a remedy is given in the section on *space extension*.

A *convex* set is defined by requiring that every pair of points in the set can be joined by a straight line segment lying entirely in the set. However, if the function we wish to minimize is not convex, but has connected level sets, one should, in principle be able to construct a path leading to the solution [20]. The topological definition of *connected* set only requires that we can connect any pair of points in the set by a continuous path, lying entirely in the set, a much less stringent requirement than the straight line segment required for convexity. In *differential topology*, one requires more – the path should be *smooth*, while these notions serve the goals of topology very well, in the discipline of *computational mathematics*, one needs to have two further properties – possibility of

finite representation, and
efficient operations.

Therefore, we introduce the following concept of *graded connectivity*:

Definition 2.1. *A connected subset of $\mathbb{R}^n$ is $(k,l) - connected$ if any pair of points in the set can be joined by segment of a curve of degree not exceeding k and dimension not exceeding l, lying entirely in the set.*

Parameters k and l will be referred to as *connectivity indices*. As per this def., a *convex* set has the simplest type of connectivity – it is $(1,1) - connected$. Let us consider a few more examples. Suppose any pair of points in the set can be joined by a segment of plane cubic curve in the set. Then it would be $(3,2) - connected$. If any pair of points in the set can be joined by segment of a planar curve lying in the set, it would be $(\infty, 2) - connected$.

Definition 2.2. *A function defined on open connected subset of $\mathbb{R}^n$ has connectivity indices k and l for minimization if the downward level sets of the function, are $(k,l) - connected$.*

As per this definition, a convex or pseudo-convex function has lowest values for connectivity indices – $(1,1)$. Let us consider examples in the next *layer*.

2.2 Example

Consider the following two subsets of $\mathbb{R}^2$:

$$C = \{(x,y) \in \mathbb{R}^2 \mid x^2 + y^2 \le 1\} \tag{1}$$
$$H = \{(x,y) \in \mathbb{R}^2 \mid x^2 - y^2 \le 1\} \tag{2}$$

From a computational point of view, the first set, bounded by a circle, would be regarded as "easy", since it is *convex*, while the second set, bounded by a pair of hyperbolas, would be regarded as "difficult", since it is not convex. But both are connected, the first one being $(1,1) - connected$. What can we say about the connectivity indices of the second? It is well known that a circle and hyperbola are both *conic sections*, and projectively equivalent. Hence if we give a *projectively invariant* construction for joining a pair of points in one of these sets, it will be equally applicable to the other.

We give such a construction, for not just one curve but an entire *family of curves parameterized by a parameter* λ. Suppose we are given, in addition to the parameter λ, a pair of points to be joined, say P and Q on the boundary of the circle. (Boundary points are the extreme case, joining interior points is simpler.) Here is the construction:

1. Draw tangents to the boundary at points P and Q.
2. Let R be their point of intersection.
3. Consider a *pencil* of lines passing through R.
4. A generic line l in the pencil will intersect the circle in two points, say S and T (they may have complex co-ordinates).
5. Take an unknown point X on the line l so that the *cross-ratio of the four points R, S, T, X denoted by [R,S,T,X]* satisfies the following equation for the curve parameter λ

$$[R, S, T, X] = \lambda. \tag{3}$$

6. As we vary the line l in the pencil, the points X trace out a curve.
7. For each value of the parameter λ, we get one such curve segment connecting P and Q.

The result of this construction for the hyperbola is shown in Figures 1 and 2. The upper and lower portions of the ellipse in Figure 1 actually are two different joining paths, corresponding to two different values of λ.

Note that each step of the construction has a *projectively invariant* meaning – concept of tangents, intersection of the two tangents, intersection of the members of the pencil with the curve and the solution of cross-ratio equation defining the final point X. Hence the construction itself is projectively invariant. If we apply a *projective transformation* to the original curve and carry out the construction

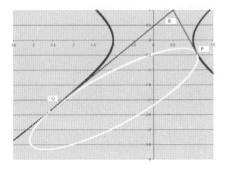

Fig. 1. Joining P and Q on two different branches of the hyperbola

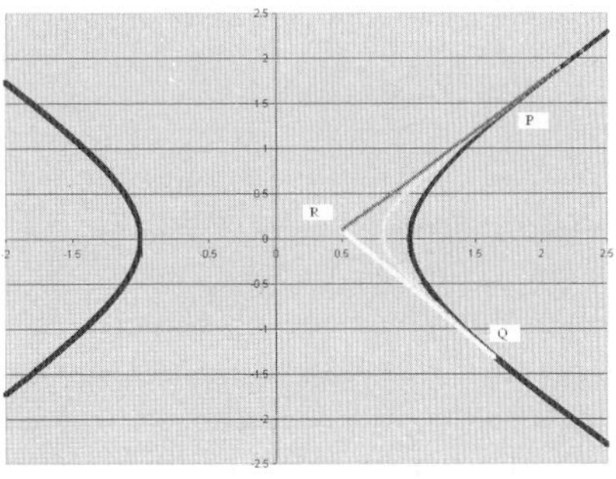

Fig. 2. Joining P and Q on the same branch of the hyperbola

in the transformed space, the joining path we get is the image of the joining path in the original space under the same transformation. It turns out that this construction applied to conic sections always leads to segments of conic sections connecting the two points. Hence the set H is $(2,2) - connected$.

The same construction also works for the complement of the set under consideration, provided one works in the projective space. In that case, the outside of circle and H^c are also $(2,2) - connected$.

Some further observations regarding the family of curves obtained by this method as the parameter λ is varied:

1. All curves in the family pass through the same two points P and Q and also share the same two tangents at those two points.
2. The family of curves forms what is known as a *linear system* [34] in algebraic geometry.

The family of curves obtained by this construction also shows that the straight line segment connecting P and Q can be alternatively viewed in an interesting form – as a limiting case of an ellipse degenerating into double cover of the straight line segment as the length of the minor axis of the ellipse tends to zero. It can also be obtained directly for $\lambda = \frac{1}{2}$.

This example shows that a set such as H defined above, while not convex, is only moderately more difficult to work with than a convex set. The intent here is to rectify the current "All or Nothing" situation in optimization. For convex optimization we have a great theory and efficient algorithms, otherwise, nothing much can be said. By introducing gradation in the nature of connectivity, we have the possibility of more efficient methods for smaller values of connectivity indices. In Section 6, we will analyze $s_{02}(\mathbf{x})$, related to potential function in [20], in n dimension. For now, we mention that for $n = 2$ level sets of s_{02} correspond to symmetrized version of H:

$$\{\mathbf{x} \in \mathbb{R}^n | s_{02}(\mathbf{x}) \leq 1\} = \{(x,y) \in \mathbb{R}^2 \mid x^2 - y^2 \leq 1, y^2 - x^2 \leq 1\} \qquad (4)$$

3   Space Curvature and Generalized Parallelism

3.1   Introduction

Note that the definition of graded connectivity in the last section implicitly used the *Euclidean Geometry* of $\mathbb{R}^n$. The notions of dimension and degree depend on the geometry of the underlying space in a very fundamental way, since they depend on the concept of *parallelism* itself. Evolution of foundational aspects of geometry in last two centuries have clarified that the notion of parallelism is not *absolute*, but a matter of choice. In Riemannian geometry, parallel transport is formalized in terms of Levi-Civita connection [28]. The general definition of connectivity indices also includes specification of a connection.

Interior point advances in optimization are based on Riemannian geometry [21]. The main features are:

Simplex as a Lie Group. Consider interior of simplex S_n in $\mathbb{R}^n$ defined by $\sum_i \mathbf{x}^i = 1$, $\mathbf{x}^i > 0$. This is a Lie Group with the following definition for multiplication:

$$\mathbf{x}, \mathbf{y} \in S_n, (\mathbf{x} \circ \mathbf{y})^i = \frac{\mathbf{x}^i \mathbf{y}^i}{\sum_i \mathbf{x}^i \mathbf{y}^i} \tag{5}$$

Projectively Invariant Metric. This is given by:

$$g_{ij}(\mathbf{x}) = \frac{1}{\mathbf{x}^i \mathbf{x}^j} \left(\delta_{ij} - \frac{1}{n} \right) \tag{6}$$

This metric has an interesting property that an observer at any point in the interior of the simplex will think that he is at the "center" of the universe! Note that, in dimension four, Gödel's [17] p.168 model of universe had similar homogeneity. In the model above, transitivity holds for any dimension.

Performance proportional to Curvature. The algorithm follows a continuous trajectory given by a differential equation, and it's performance is shown to be proportional to integrated curvature of the curve measured in the Riemannian metric.

Averaging based on Fibre Bundle. A set of linear programming problems are stacked into a fibre bundle for the purpose of defining an ensemble over which average performance is measured. A tight bound proved on this performance in terms of curvature corroborates well with performance observed in practice.

For other treatments of applications of Riemannian Geometry to linear and nonlinear optimization, see [49,38]. Earlier, there were also some attempts, in the context of convex functions to take the hessian of the function as metric, but this choice is somewhat ad-hoc and the corresponding affine connection does not satisfy the PDE described in next section.

Riemannian Geometry, which is based on a positive definite metric, is not adequate for more general non-convex situations. Differential Geometry has undergone major broadening of the concepts laid down by Riemann, often motivated by requirements of major scientific application, and in the same vein, the field of optimization is making further demands for it's broadening.

From the point of view adopted here, the major conceptual streams are:

Riemannian Geometry. This is based on positive definite differential quadratic form, which gives the metric, and geodesics based on this metric. These developments were initiated by Gauss [13], Lubachevsky [29], Riemann [41], Christoffel [7], Ricci [40], and extended further in many directions (e.g Finsler [12]).

Indefinite Metric, Lorentzian. This is a major step, that allows the metric to be indefinite (but still non-singular). In particular, signature of the Lorentz metric is $[+, -, -, -]$. This was developed and applied by Minkowski [31] and Einstein [30], to provide mathematical foundation for the theory of general relativity. Global structure of Lorentzian geometry is quite different from Riemannian case. There is no analog of Hopf-Rinow [18] theorem.

Also, Lorentzian manifolds encountered in applications can be be geodesically disconnected, i.e. not (1,1)-connected as per above definition. But if not (1,1)-connected, could they be connected for higher values of k and l? The concepts introduced here raise an interesting question of connectivity indices for Lorentzian universe with various mass distributions.

Theory of Connections. While a Riemannian metric gives a unique compatible *affine connection*, one can take the connection as the main quantity defining parallelism and geodesics. This more general point of view was developed by Weyl [30], Eddington [8], Einstein [30], Cartan [5], Schordinger [43], Eisenhart [10], Veblen [50], Thomas [50], Ehresmann [9]. In this paper similar approach helps in more general framework for optimization.

Multi-Cellular Differential Geometry. Insisting that the metric is non-singular everywhere seems too restrictive, from the point of view of applications. A more general view is that it should be non-singular *almost everywhere*, but one should allow for the possibility of lower dimensional sets where the metric is singular. In case the singular hyper-surface has co-dimension one, it can partition the original connected manifold into multiple connected components called "cells" or "zones". In case the metric is semi-definite, interior of each cell still has a Riemannian geometry, but in the more general case, the signature of the metric may change when crossing the zone boundary.

We came across examples (see section on quadratic models) of interior-point trajectory crossing from Riemannian to Lorentzian cell, in our "synthetic" universe of non-convex optimization. Similar "signature changing" transition has been hypothesized in the "natural" universe by Hartle and Hawking [16], and significant efforts directed towards it's analysis.

In the example we give, the transition point can be controlled to some extent by choice of parameters, constrained by a quantity called "reciprocal distance to closest non-solution", but the transition is essential and distinguishes satisfiable instances from non-satisfiable ones.

In the context of classical General Relativity, geodesics crossing cell boundaries have been analyzed by Kossowski [25], Larsen [26] and many others. In multi-dimensional non-convex optimization there is greater diversity of cell-crossing behaviors. This is to be expected in view of effects like refraction or total internal reflection encountered in classical ray optics based on Fermat's variational principle and more modern negative-index meta-materials.

It seems that a more general development of such geometry has not yet been undertaken by pure mathematicians. We are referring to this as "multi-cellular" geometry. This may be done in "top-down" fashion by decomposing a given connected manifold by the hyper-surface defined by $det(g) = 0$ or by gluing together cells as done in some approaches to quantum gravity. In the gluing approach, any projectively invariant metric as mentioned in Section 3.1 has the natural advantage that in case of two simplices intersecting in a lower dimensional face (which has a metric given by the same formula), the metrics induced on that face from both, agree.

3.2   Summary of Notation

Terminology and notation used in differential geometry and its applications has been changing considerably, over the last one and half century. Modern notation emphasizes independence from co-ordinates, while the older classical notation sometimes looks cumbersome, since it conveys not only the operations on tensors, but how to carry them out in one particular co-ordinate system. Therefore, it still has a merit when it comes to actual software implementation, since an explicit representation is needed anyway. The notation is also expressive enough to permit automated parsing and conversion into computer programs. Since efficient implementation is a big part of our research program, we chose to follow the older notation and terminology. Our experience is corroborated by those who implement codes for numerical General Relativity. For help in interpretation and inter-conversion of various notations and terminology, we recommend Misner [32], written for physicists and Spivak [46], which is written in the spirit of a friendly tour guide to differential geometry. For more formal treatment, see Kobayashi [24], Chern [6]. We follow Einstein summation convention for repeated covariant and contra-variant tensor indices. If a particular subscript or superscript does not have a significance in the tensor notation, but is simply a label to identify members of set, we enclose it in (), e.g. $f_i^{(\alpha)}, \alpha \in S$ refers to i^{th} component of the first covariant derivative of function f from a set S, superscripted with membership label α.

The metric will denoted by g_{ij} The corresponding affine connection is denoted by Γ_{ij}^k, and is given in terms of the metric by means of eqs. (7-9). Some times only the connection is defined without any metric.In this paper, connection is always symmetric. Eq. (10) gives geodesics in terms of the connection.

Christoffel's symbols of the first kind

$$[ij, k] = \frac{1}{2} \left[\frac{\partial g_{jk}}{\partial x^i} + \frac{\partial g_{ik}}{\partial x^j} - \frac{\partial g_{ij}}{\partial x^k} \right] \tag{7}$$

Christoffel's symbols of the second kind

$$\{_{ij}^l\} = g^{lk}[ij, k] \tag{8}$$

Coefficients of Affine Connection

$$\Gamma_{ij}^k = \{_{ij}^k\} \tag{9}$$

Differential Equation for the geodesics

$$\ddot{\mathbf{x}}^{\mathbf{k}} + \mathbf{\Gamma_{ij}^k} \dot{\mathbf{x}}^{\mathbf{i}} \dot{\mathbf{x}}^{\mathbf{j}} = \mathbf{0} \tag{10}$$

A variation of this equation is obtained by multiplying by $det(g)$. It is particularly advantageous when computing geodesic crossing cell boundary.

$$det(g)\ddot{\mathbf{x}}^{\mathbf{l}} + adj(g)^{lk}[ij, k]\dot{\mathbf{x}}^{\mathbf{i}}\dot{\mathbf{x}}^{\mathbf{j}} = 0 \tag{11}$$

4   Locally Exact Models: Linear and Quadratic

4.1   Introduction

A number of computational methods in optimization and solving equations make use of *locally approximate* linear or quadratic models based on Taylor expansion of the functions involved at a point. This includes Newton's method, homotopy methods, interior-point methods, SQP [14], trust region methods [33], etc.

- Strength of this approach lies in the availability of efficient linear algebraic methods for various computations on the linear of quadratic models.
- Limitations of the approach results from the fact the models are *local* and *approximate*.

Much of the analysis of such methods is aimed at bounding the effect of such approximations. e.g. For Newton's method, see $\alpha-theory$ given in Smale [45]. For interior point methods for convex problems, considerable unification is obtained by means of the concept *self-concordant* functions Nesterov [35].

Given the success of linear and quadratic models, can we improve on their limitations? In this paper, we give a methodology that still continues to be *local* but can be made *exact*, by defining appropriate *space curvature* adapted to the function(s) under consideration. Note that in many applications of differential geometry, a manifold with a *metric* or *affine connection* is given first, and then one wants study behavior of functions and their *covariant derivatives* relative to the given connection. Here we are reversing the roles. Functions of interest, defining objective(s) or constraints are given first, and we want to explore if we can set up a suitably curved space so that the functions behave in a particularly simple way, as linear or quadratic, along the geodesics in the space. This is similar to the situation in the theory of *General Relativity*. Matter acts like source of the curvature of space.Given a mass distribution, one asks if there is some curvature that can be assigned to the space so that behavior of test particles in response would follow from the space curvature alone – along the generalized straight lines, i.e. geodesics. Here, the given functions are the sources of curvature.

4.2   Partial Differential Equation for the Affine Connection

Let $f(\mathbf{x})$ be a function of interest. If we are to find a symmetric affine connection Γ_{ij}^k so that f is linear or quadratic w.r.t. that connection, then second or third covariant derivative of f based on the connection should be zero, respectively. This gives a P.D.E. to be satisfied by the connection. Expressions for first three covariant derivatives are:

First Derivative

$$f_i = \frac{\partial f}{\partial x^i} \tag{12}$$

Second Derivative

$$f_{ij} = \frac{\partial^2 f}{\partial x^i \partial x^j} - \Gamma_{ij}^k \frac{\partial f}{\partial x^k} \tag{13}$$

Third Derivative

$$
f_{ijk} = \frac{\partial^3 f}{\partial x^i \partial x^j \partial x^k} - \left\{ \Gamma^l_{ij} \frac{\partial^2 f}{\partial x^l \partial x^k} + \Gamma^l_{jk} \frac{\partial^2 f}{\partial x^l \partial x^i} + \Gamma^l_{ki} \frac{\partial^2 f}{\partial x^l \partial x^j} \right\}
$$

$$
+ \{ \Gamma^l_{ij} \Gamma^m_{lk} + \Gamma^l_{jk} \Gamma^m_{li} + \Gamma^l_{ki} \Gamma^m_{lj} \} \frac{\partial f}{\partial x^m}
$$

$$
- \left\{ \frac{\partial}{\partial x^k}(\Gamma^l_{ij}) + \Gamma^m_{ij} \Gamma^l_{mk} \right\} \frac{\partial f}{\partial x^l} \tag{14}
$$

If we set the third covariant derivative to zero, we get a first order, non-linear (just quadratic) PDE in the unknown Γ^l_{ij}, and we also have the right number of unknown functions to try to solve it. But solving this numerically in higher dimensional space can be computationally expensive. While there may be situations where it's important enough to get a solution computationally, the approach here is different. We have identified situations where under certain conditions we can get a closed form analytical solution by symbolic means for a *class* of problems. The closed form solutions can then be deployed in particular numerical instances of the class.

5   Linear Models

Suppose we are given a set of m functions, on a manifold of dimension n,

$$
f^{(i)}, i = 1, .., m, m \le n \tag{15}
$$

which are functionally independent in the given co-ordinate patch. Since we are describing a local model, there is no loss of generality in working with an open set $U \subseteq \mathbb{R}^n$, to keep the exposition simple. Using the co-ordinates $x^j, j = 1, \ldots, n$ the jacobian matrix J, in this patch is given by

$$
J^i_j = \frac{\partial f^{(i)}}{\partial x^j}, \ rank(J) = m \tag{16}
$$

Define a set of $m \times n$ functions

$$
h^j_{(k)}(\mathbf{x}), k = 1, .., m, j = 1, .., n \text{ so that } \sum_{j=1}^{n} h^j_{(k)} \frac{\partial f^{(l)}}{\partial x^j} = \delta^l_k. \tag{17}
$$

In general there are many ways to do this, e.g. one can use inverse of any $m \times m$ non-singular sub-matrix of J. If $m < n$, there are many choices possible. one can use randomization or the following least-squares type equation, for positive definite $Q(\mathbf{x})$:

$$
h^j_{(k)} = ((JQJ^T)^{-1}JQ)^j_k \tag{18}
$$

Now we can define the desired connection as

$$
\Gamma^l_{ij} = \sum_{k=1}^{m} \frac{\partial^2 f^{(k)}}{\partial x^i \partial x^j} h^l_{(k)} \tag{19}
$$

It is easy to check that the second covariant derivative of each of the functions $f^{(i)}, i = 1, .., m$, as given in eq. (13) is zero. Now we mention many interesting consequences and observations related to this construction:

Behavior along Geodesics: There is a geodesic in each direction going through every point in the set U satisfying eq. (10). Along each such geodesic, every function in our set varies linearly w.r.t. affine parameter of the geodesic. In effect we have linearized all these functions over the set U. This is much stronger than taking linear terms in the Taylor expansion at a point and ignoring higher order terms. If the functions in the set were used for defining non-linear inequality constraints, they will now become linear inequalities.

Generalized Polynomials: If we construct new functions by addition, multiplication of the functions in the set, augmented by constant functions, we get functions which behave like polynomials along the geodesics. We refer to these functions as *generalized polynomials*. Note that we could have taken transcendental functions subject to the functional independence condition and carried out the construction. The set of all such functions is denoted by $R[f^{(1)}, f^{(2)}, \ldots, f^{(m)}]$ and forms a commutative algebra.

Complex case: Similar construction works for complex manifolds and complex analytic functions. The construction can be carried out even if the jacobian condition is violated at some points, but is satisfied generically. In this case, the points where the condition is violated are singularities. The main difference between the real and complex case is that in the complex case, complement of the singular set still remains connected and singular set does not necessarily "obstruct" interior point algorithms.

Linear Systems: Starting with a set of positive functions and applying the construction to their logarithms,original functions vary exponentially along the geodesics. In the simplest case of ordinary linear differential equations with constant coefficients, the solution trajectories are given by exponential functions. Ability of the construction to linearize such transcendental functions is a major advantage. Linear systems are rather important in engineering applications. The methods is also applicable to non-linear systems. In many higher order numerical methods for solving ordinary differential equations, a polynomial basis is implicitly used for local expansion of the solution trajectories. Instead, using generalized polynomials adapted to the problem, longer prediction steps are possible.

Jacobian Conjecture: The main reason we refer to the construction as "local linearization" is that the independence condition can be satisfied globally or in the "large", only in rather exceptional situations. One such situation is the collection of polynomials satisfying the hypothesis of the jacobian conjecture [1,11]. This situation is almost "too perfect", for the jacobian is constant. More generally, the construction of affine connection given in the paper becomes global in case of analytic automorphisms of $\mathbb{C}^n$.

Convex case and Legendre Transform: Although the emphasis of the present paper is on non-convex problems, the linearization can be applied to the n derivatives $\frac{\partial f}{\partial x^i}$ of a convex function. The independence condition is assured for strictly convex function by non-singularity of the Hessian of f. There is a relation to the Legendre transform, since geodesics w.r.t. to the connection constructed in the original space correspond to ordinary Euclidean geometry in transformed space. For application of Legendre transform to interior point methods, see Lagarias [2].

Relation to Convex Homotopy Methods: In convex homotopy methods for zero finding, one creates a single curve so that each of the given functions varies linearly w.r.t. homotopy parameter along the curve. The construction we have given creates an entire family of curves in a region along which the functions vary linearly.

6   Quadratic Models

6.1   Introduction

Here we give a construction of metric as well as affine connection applicable to homogeneous function of n variables. In earlier paper on Tensor Optimization [22], which dealt with third order tensors, we had already described technique for homogenization. The present construction is applicable to fourth (and higher) order tensors, which have many applications. Let $f(\mathbf{x})$ be an homogenous function of degree d on n real variables

$$f(\lambda \mathbf{x}) = \lambda^d f(\mathbf{x}) \tag{20}$$

We would like to find a metric and the corresponding affine connection which leads to vanishing of third covariant derivative of this function. It turns out that such a metric is not unique, and permits linear superposition of terms we call "null metric" terms. In general, signature of the metric also depends on the choice of such terms, i.e. the same function may be simultaneously quadratic w.r.t. a Riemannian metric and a Lorentzian metric.

6.2   Construction

Definition 6.1. *A **null metric term** is any homogenous $g_{ij}^{(0)}(\mathbf{x})$ such that*

$$g_{ij}^{(0)}(\mathbf{x})\mathbf{x}^i \mathbf{x}^j = 0 \tag{21}$$

The set of null metric terms form a linear space, and set of all homogeneous null metric terms of a given degree a linear subspace. A simple example of such term is given below:

$$g_{ij}^{(0)}(\mathbf{x}) = \delta_{ij} - \frac{1}{n}\left(\frac{\mathbf{x}_i \mathbf{x}_j}{r^2}\right) \text{ where } r^2 = \sum_{k=1}^{n} x_k^2 \tag{22}$$

As an illustration of the effect of null metric terms on the geodescis, let us superpose the above term and standard Euclidean metric on $\mathbb{R}^n$ to get:

$$g_{ij}(\mathbf{x}) = \delta_{ij} - \frac{\gamma}{n}\left(\frac{\mathbf{x}_i\mathbf{x}_j}{r^2}\right), \text{ where } \gamma > 1 \tag{23}$$

Note that this is Lorentzian, while the original Euclidean metric was Riemannian.

By solving eq. (10) for this metric, geodesics are easily shown to be logarithmic spirals. They share a well known property of straight lines and circles of possessing a continuous transformation group of *self-similarity* [52].

A simple way to construct a homogeneous null metric term of degree d is to take two homogeneous functions, $u(\mathbf{x})$ and $v(\mathbf{x})$ of degrees $d+1$ and 1 respectively, and construct the following product:

$$g_{ij}^{(0)}(\mathbf{x}) = u(\mathbf{x})\frac{\partial^2 v}{\partial x^i \partial x^j} \tag{24}$$

We can also take $v(\mathbf{x})$ to be any Minkowski metric i.e. a positive homogeneous function. Null metric property can be proved easily using Euler's identity.

A solution to the main problem is then obtained by superposing second derivative of f with a homogeneous null metric term of the same degree:

$$g_{ij}(\mathbf{x}) = \frac{\partial^2 f}{\partial x^i \partial x^j} + g_{ij}^{(0)}(\mathbf{x}) \tag{25}$$

Using this metric and corresponding affine connection, The second covariant derivative of f is just a constant multiple of the metric, hence the third covariant derivative of f vanishes. In case the manifold is a direct product of two manifolds $M_a \bigotimes M_b$ we can apply the above construction separately to M_a and M_b and make any linear superposition of the two metrics with constant coefficients.

To construct some interesting examples, consider power sums which form a basis for symmetric polynomials: $p^k(\mathbf{x}) = \sum_{k=1}^{n} x^k$. For completeness define constant $p^0(\mathbf{x}) = n$ where $dim(\mathbf{x}) = n$. This allows many interesting polynomial equations and inequalities involving symmetric polynomials to be expressed so that the *same* expressions hold for any n, and $\mathbf{x}$ does not appear explicitly.

For $\mathbf{x} \in \mathbb{R}^n$, we define a family of non-negative symmetric polynomials for any non-negative integers $0 \le k \le l, k+l,$ *even*:

$$s_{kl}(\mathbf{x}) = p^{2k}(\mathbf{x})p^{2l}(\mathbf{x}) - (p^{k+l}(\mathbf{x}))^2 \ge 0 \tag{26}$$

Equality holds for rays generated by

$$\mathbf{x}^i = \pm 1 \tag{27}$$

If we impose a system of homogeneous linear equations $A\mathbf{x} = 0$, the above two properties continue to hold. Consider the potential function with connected level sets, for ± 1 integer programming problems introduced in Karmarkar [20]. It is a real analytic function in the interior of polytope. If it is separated in it's homogeneous components, the leading term is of degree 4, a multiple of $s_{02}(\mathbf{x})$, the first non-negative symmetric polynomial defined above. Now the reader should browse through the next section.

6.3   Example 1: Same Curve, Two Connections

We want to give example of curves in multi-cellular geometry which are algebraic relative to two different affine connections, and having in each cell:

- Degree 2^n w.r.t. Euclidean connection
- Degree 1 (i.e. geodesic), w.r.t. metric constructed as per recipe above.
- Geodesics given by eq. (11) crossing cell boundary are not smooth but C^1. However, they are given by component functions belonging to:

$$\sqrt{C^\omega} = \{f(\mathbf{x}) | \exists k \in \mathbb{Z}^+ \ni f^k \in C^\omega\} \tag{28}$$

It is easy for interior point methods to follow such curve, since we maintain representation of f^k as well.

We first came across such situation in analysis of potential function with connected level sets [20]. Now we can offer much simpler example without any reference to NP-complete problems:

1. Apply either the quadratic method above to $p^4(\mathbf{x})$ without any null metric terms or the linear method to $x_i^2, i = 1, \ldots, n$ Then $g_{ij} = x_i x_j \delta_{ij}$ (up to constant factors) Cells are orthants in $\mathbb{R}^n$.
2. Let $s(x) = \int_0^x |t| dt, x \in \mathbb{R}$ clearly, $s(x) \in \sqrt{C^\omega}$
3.

$$\text{define} \quad \phi(\mathbf{x}) : \mathbb{R}^n \to \mathbb{R}^n$$
$$\phi_i(\mathbf{x}) = s(x_i).$$

4. Within a cell, geodesics are pullbacks of straight lines, by applying ϕ^{-1}.

6.4   Example 2: Fourth Degree, Non-convex Function

In the above example, $p^4(\mathbf{x})$ was convex in Euclidean space, and we reduced it from 4th degree to quadratic. Now, we take $s_{02}(\mathbf{x})$, which is 4th degree and non-convex, and create curved space so that it becomes quadratic and has positive semi-definite second covariant derivative, by following the quadratic recipe and choosing appropriate null metric term. The resulting metric is

$$g_{ij}(\mathbf{x}) = x_i x_j \left(\delta_{ij} - \frac{1}{n} \right) \tag{29}$$

The second covariant derivative remains positive semi-definite after imposing linear constraints $A\mathbf{x} = 0$, To see this, choose a basis for the null space of A and let $\mathbf{x} = B\mathbf{y}, dim(\mathbf{y}) = co - rank(A)$. In matrix notation, denoting the 2nd covariant derivatives in X and Y spaces by H_x and H_y, we get, $H_y = B^T H_x B$.

6.5   Example 3: Eighth Degree, Non-Convex Function

Assuming $rank(A) > 0$ define:

Null Space: $\Omega(A) = \{\mathbf{x} | A\mathbf{x} = 0\}$
Non-solutions: $NS(A) = \{-1, +1\}^n \cap \Omega^c(A)$
Closest Non-solution: $r_{min}(A) = \min\{dist(u, \Omega) | u \in NS(A)\}$
Condition Number: $C_{\#}(A) = \frac{1}{r_{min}(A)}$

As a final example, consider the following 8th degree function, along with constraints $A\mathbf{x} = 0$, and control parameter ε

$$f(\mathbf{x}) = p^0(\mathbf{x})^2 p^4(\mathbf{x})^2 - (1 + \varepsilon)p^2(\mathbf{x})^4 \qquad (30)$$

The metric obtained by the procedure:

$$g_{ij}(\mathbf{x}) = p^4(\mathbf{x})(x_i x_j (\delta_{ij} - (1 + \varepsilon)\frac{\rho(\mathbf{x})}{n})) \qquad (31)$$

$$\text{where } \rho(\mathbf{x}) = \frac{p^2(\mathbf{x})^2}{p^0(\mathbf{x})p^4(\mathbf{x})} \qquad (32)$$

Note that $\rho(\mathbf{x}) \leq 1$ with equality on rays generated by $\mathbf{x}^i = \pm 1$.
For any $r_{min}(A), \exists \varepsilon > 0$, such that the following holds:

Non-satisfiable case
1. The space defined with the above metric is Riemannian.
2. The function $f(\mathbf{x})$ is geodesically convex, and strictly positive.
3. Expansion of f, using eigenfunctions of the second covariant derivative, gives proof of non-satisfiability as sums of squares with n terms.

Satisfiable case
1. The space defined has a **Riemannian** cell and a **Lorentzian** cell.
2. The rays corresponding to satisfiable solutions are in **Lorentzian** cell.
3. The starting point is (generally) in **Riemannian** cell, any continuous trajectory, or evolution process, deterministic, non-deterministic, or random needs to cross from **Riemannian** cell to **Lorentzian** cell.
4. The function $f(\mathbf{x})$ is geodesically convex in the interior of **Riemannian** cell.

7   Relativized Concept of Algebraic Curves

While polynomials are normally defined algebraically, there is another way to define them in $\mathbb{R}^n$ or $\mathbb{C}^n$ as functions whose all derivatives, except for a finite number, vanish. In this form they generalize to manifolds with a connection immediately, by using covariant derivatives w.r.t. the connection, instead of ordinary derivatives. Such functions behave as polynomial functions along geodesics. Algebraic curves and higher dimensional algebraic varieties on the manifold can be defined as intersection of zero sets of a set of such "generalized polynomials".

As per this view point, being *algebraic* is not an *absolute* property of a curve, but it's relative to the notion of parallelism, and thus a joint property of an entire "eco-system" curves, a connection, set of geodesics, and a collection of functions constituting generalized polynomials. In case there is more than one affine connection defined on the manifold, a curve may be algebraic in the above sense w.r.t. one connection and not so w.r.t. the others. More interesting situation is when a curve may be algebraic w.r.t. two different affine connections. In this case, it may have a different degree w.r.t. the two connections. It is important to recognize that if an interior point trajectory has high degree w.r.t. one metric, say Euclidean, it does not necessarily imply that the curve is difficult to follow, it may be much simpler when viewed in another metric or connection.

Even in case of Linear Programming, the simplest and well understood semi-algebraic problem, it is noteworthy that algebraic methods such as Fourier-Motzkin elimination method, or combinatorial method such as the Simplex method are not polynomial, where as the polynomial-time interior point method uses Riemannian metric. Euclidean space is quite adequate for defining the linear programming problem, but not for its best solution algorithm. This effect is even more pronounced in case of more difficult problems.

Optimization problems involve not just finding optimal solution but also constructing proofs of their optimality. Proving that f_{min} is the minimum value of $f(\mathbf{x})$ is equivalent to showing that $f(\mathbf{x}) - f_{min} \geq 0$ Current approach to this can be found in Stengle [47], Shor [44], Schmüdgen [42], Karmarkar [22], Putinar [37], Resnick [39], Lasserre [27], Parrilo [36], Bachnak [4], Todd [48]. These approaches can be generalized using ideas in this paper, and made more effective by working in appropriate commutative algebra $R[f^1, f^2, \ldots, f^m]$ defined on the manifold of interest. This is a subject of a separate paper.

8 Space Extension for Connectivity

8.1 Introduction

A classic example of *space extension* is *projectivization* of *affine space* into *projective space* including *points at infinity*. Resulting benefits in algebraic geometry are well known. In the subject of interior point methods, space extension was used [19] to create projectively invariant polynomial-time algorithm. Here it is used to transform optimization problem so that it has connected level sets.

8.2 Generalization of Linear Interpolation on Simplex

Let Δ be a simplex in $\mathbb{R}^n$ with barycentric co-ordinates $x^1, x^2, \ldots, x^{n+1}$ with $\sum_i \mathbf{x^i} = 1$, If values $f_{(1)}, f_{(2)}, \ldots, f_{(n+1)}$ are specified on the vertices, the corresponding *linearly interpolated* function on Δ is given by $f(\mathbf{x}) = \sum_i f_{(i)} x^i$. This could be rewritten as:

$$f(\mathbf{x}) = \frac{\sum_i f_{(i)} \mathbf{x}^i}{\sum_i \mathbf{x}^i} \tag{33}$$

In this form, it can be generalized easily to degree d as follows:

$$f_d(\mathbf{x}) = \frac{\sum_i f_{(i)}(\mathbf{x}^i)^d}{\sum_i (\mathbf{x}^i)^d} \tag{34}$$

This is valid in all of $\mathbb{R}^n$. For even d the field of values of the extrapolated function remains the same as that of linearly interpolated one:

$$\min_i(f_{(i)}) \leq f_d(\mathbf{x}) \leq \max_i(f_{(i)}) \quad \forall \mathbf{x} \in \mathbb{R}^n \tag{35}$$

8.3   Extension to Achieve Connectivity

While projectivization involves taking one point outside the space and taking union of all rays going through that point and points in the original space, we do similar extension with two outside points P_m and P_M to achieve connectivity for both downward and upward level sets. Extension of the function f and manifold M are denoted by $\widetilde{f}$ and $\widetilde{M}$ $\widetilde{f}$ agrees with f on M. Then assign values on P_m and P_M as:

$$\widetilde{f}(P_m) = \inf_{\mathbf{x} \in M} f(\mathbf{x}) \text{ and } \widetilde{f}(P_M) = \sup_{\mathbf{x} \in M} f(\mathbf{x}) \tag{36}$$

(We assume that both these values are finite.)

For each $\mathbf{x} \in M$ Let $\Omega(\mathbf{x}) = span\langle \mathbf{x}, P_m, P_M \rangle$ be the $2 - dimensional$ affine space created by the three points, and $\Delta(\mathbf{x})$ the simplex spanned by the same. Union of these spaces $\Omega(\mathbf{x})$ makes the extended manifold $\widetilde{M}$, and $\widetilde{f}(\mathbf{x})$ is defined on $\Omega(\mathbf{x})$ by eq. (34), with even d. For any two points P_a and P_b in a downward level set, the line segments $P_a P_m$ and $P_m P_b$ also lie in the same level set, hence it is connected. Similarly for upward level sets. If connectivity is to be achieved for only one of these, extension generated by one extra point suffices.

9   Continuous Models of Computation

9.1   Introduction

One of the goals of the work reported here is to understand computational difficulty of various optimization problems and classify them accordingly. Since non-convex optimization can be applied to problems such as satisfiability, there is partial overlap with the goals of computational complexity theory based on Turing machines. At the same time, there are important differences in methodology and overall approach. Each time we presented the ideas in this paper along with computational results on finding satisfying truth assignments or computing proofs of non-satisfiability, one of the questions from the audience is "Does this mean that P might be equal to NP?" To avoid such misinterpretations, and to point out that the ideas presented here actually suggest a new path for proving $P \neq NP$, we are clarifying some basic differences in the two approaches. The approach based on classical complexity theory has four components:

Pre-classification. Problem instance are sorted according to the number of bits required for their description and all instances requiring the same number of bits are grouped into one bin or one layer.

Model of Computation. This is usually based on the Turing machine with its tape and finite state machine.

Post-classification. This is based on the number of time steps taken by the particular Turing machine or algorithm, to solve the problem instance. All instance requiring same number of steps are put in one layer.

Complexity of the Algorithm. If the post-classification agrees with the pre-classification upto a polynomial function, it is considered to be a "good" algorithm or polynomial-time algorithm.

9.2   A Different Approach

Our approach to classification of computational difficulty is based on the view that the conceptual basis for all the above aspects should be developed hand-in-hand, and whatever is learned from progress in one aspect should be used to refine the other aspects. In *Linear Programming*, which is one of the best understood sub-domains in optimization, it is now clear that the computational difficulty depends primarily on the condition number and dimension and its dependence on number of bits required for input description is rather weak (except that it's still polynomial) Layering of the input instances based on condition number and dimension leads to more accurate classification of their difficulty, since instances in the same layer require similar computing effort, while grouping on the basis of same input length leads to rather large variance in the computational effort required for instances in the same layer. With the latter type of grouping as the first step of the complexity theory, only statements (true or false) that can be made about all problems in the same bin as a group are admissible statements. Any prior restrictions on admissible statements can create a situation in which there are statements which are true but not provable, since the set of all true statements may be disconnected in the implication graph, restricted to such class of statements. Note that the alternative pre-classification, even for the simpler problem of linear programming indicated above, is two dimensional and does not impose a simple ordering on the input instances. In the context of more difficult non-convex optimization, concepts of connectivity indices and appropriate notion of condition number are likely to play a major role in arranging input instance into layers. In any event, our approach is to not impose any particular pre-conceived classification or layering on input instances,but to allow it to evolve along with our understanding of the problem domain itself.

The other major difference in the two approaches is the computational model, described briefly in the next section.

9.3   Towards Abstract Continuum Computing Model AC^2M

A major intellectual step beyond the Turing machine model is based on real or complex number computing BCSS [3]. While this model has not received the attention from the complexity theory community that it should have received (in our opinion),it is clearly more appropriate in the context of optimization. However this model has carried over the discrete time steps from the Turing machine model, thereby inheriting some of its limitations.

Integrating concept of continuous time into the computational model is necessary for further development of interior-point algorithms, which have turned out be extremely powerful. It is well known that fundamental theories in physics could make a major leap in the last century, only after earlier concepts of separate space and time were replaced by a concept of a unified *space-time continuum*. A similar situation prevails in development of interior point algorithms today.

This is incorporated in the following model called *Abstract Continuum Computing Model* or $\mathbf{AC^2M}$. It is called "abstract", since we don't go into various possibilities of mapping it on real physical computing devices in this paper. There are many kinds of issues involved in the mapping, some of philosophical import and some of engineering type. The former kind involves possibility of *direct mapping* of $\mathbf{AC^2M}$ into laws of nature. The latter type involves indirect mapping or effective simulation of the abstract model on concrete or discrete computing devices, based on classical or quantum behavior. In a separate research program, we also have proposals (see Karmarkar [23]) for implementation by combining traditional localized devices such as transistors with new type of distributed devices called "electron transporter".

In any case, measure of computing "cost" or effort depends on details of the mapping. The abstract model is only concerned with trajectories for solving various computing problems, in same spirit as theory of differential equations in the "large", and any measure of "cost" at this level is limited to quantities such as integrated curvature or length of trajectories. In case the mapping is done on conventional digital computers, the mapping process belongs to the discipline of numerical analysis or numerical engineering. Adaptive time-steps and adaptive grids, which are rather common in modern numerical solvers show the value of treating the time variable as continuous and integrated into a space-time continuum in the first abstract phase of the process and giving freedom to design most effective discretization to the second step of the process. Consideration of appropriate precision level also belongs to the second step.

Such "separation of concern" allows one to focus on the deeper aspects that are not yet well understood, such as relation between determinism and non-determinism and whether "transcendental" concepts are necessary for fuller understanding of mathematical logic.Much of the contemporary complexity theory is inspired by Gödelś work. The least one expects from faithful followers of Gödel is that they study his work in its *entirety*. It is noteworthy that almost two decades after his famous work on foundational questions raised by Hilbert, Gödel [15] turned to investigation of "Lorentzian" geometry. Could it be that solutions of many intellectual puzzles that his work has led to, actually *require* such methods?. Our investigation so far seems to suggest so.

The main features of the model, called *Abstract Continuum Computing Model*, implicit in the present work, are as follows:

- Retain real or complex number computing from [3].
- Replace discrete time by continuous time.
- Replace Turing machine tape by a multi-dimensional curve in $\mathbb{R}^n$ or $\mathbb{C}^n$.
- Retain the three main principles of Finite State Machine in the Turing model: The rule to be applied uses only *local* information, it's a *fixed* or *uniform rule* repeated over and over, and it's *easy to compute*. The difference is that the main computational step uses real or complex numbers,instead of 0'a and 1's. The input and/or output may be restricted to special subsets such as integer, rational, algebraic integers or bounded in their magnitude. Such restrictions, if any, constitute important part of problem specification.

In the abstract phase of the model, we are only concerned about the existence of the trajectory to solve the problem deterministically or non-deterministically.

9.4   Examples of Deterministic Computing in AC^2M

Interior Point Algorithms. Such methods for linear programming, convex programming, tensor optimization and its particular variant known more commonly as "semi-definite programming" are all deterministic methods in AC^2M.

Autonomous Dynamical Systems. Consider a dynamical system in $\mathbb{R}^n$:
$\dot{\mathbf{x}} = f(\mathbf{x})$ If $f(\mathbf{x})$ is easy to compute from $\mathbf{x}$ as per cost model in [3], this is an example of deterministic algorithm.
   In analysis of dynamical systems f is given, and the aim is to understand behavior of the solution to the differential equation. However, when a dynamical system is proposed to be used as a means of getting a solution to a decision or computational problem, there is again a "reversal" of roles described earlier. f is to be chosen or *designed* to achieve the solution. Such designed f can be rather special and very different from a *generic* dynamical system of similar dimension or structure.

9.5   Examples of Non-deterministic Computing in AC^2M

Differential Linear Programming (DLP). Suppose M is a real manifold and we are given a set S of covariant fields $a_i^{(\alpha)}(\mathbf{x})$ and scalar fields, $b^{(\alpha)}$, $\alpha \in S, \mathbf{x} \in M$. We use them to define a linear programming problem at each point $\mathbf{x}$ on the manifold, which is required to be satisfied by the tangent vector $\dot{\mathbf{x}}$ of the curve, at each point on the curve.:

$$a_i^{(\alpha)}\dot{\mathbf{x}}^i \leq b^{(\alpha)}, \forall \alpha \in S \tag{37}$$

A curve satisfying the **DLP** is not unique (except in certain degenerate family of LP's, where the feasible region at each point is one-dimensional). The set of all possible curves satisfying the **DLP** define all points reachable "non-deterministically" (Note that the fibre bundle mentioned in Section 3.1, differs from the one implicit in DLP, as the former one had multiple *non-interacting* linear programs.)

Riemann Surface as a tape replacement. The "multi-dimensional" curve mentioned in features of AC^2M has a double meaning in the complex case: Not only is the curve embedded in a multidimensional space, the curve itself is a two-dimensional object, e.g. Riemann surface. The set of all possible real one-dimensional curves that can be drawn on the Riemann surface, from the starting point create a non-deterministic capability.

Even before the conception Riemann surface,the non-determinism in the complex case was present in Weierstrass's [51] method of analytic continuation. Suppose we are given a complex analytic function at a point A and wish to continue it analytically to point B, we choose a curve connecting A to B "non-deterministically". After this choice is made, analytic continuation from A to B is uniquely defined (if it can be done at all), without further free choices. Different non-deterministic choices of curves may lead to different values at B. If one of them gives solution to the original decision problem, we say that the problem has been solved non-deterministically.

9.6   Relation between Determinism and Non-determinism

Forms of non-determinism mentioned above, may turn out to be more amenable to mathematical reasoning as compared to the non-determinism in the Turing model. Truth and provability are not the same, and we find greater focus on provability desirable, leading us to AC^2M. The non-determinism mentioned above is more constrained: If the point B in the example above was contained in the disk of convergence of the power series at A, non-deterministic and deterministic results would be the same.

An example of how constrained form of non-determinism can be exploited is hidden in the first proof of polynomial-time interior point algorithm for linear programming [19]. The proof is based on showing that a *polynomially weakened* non-deterministic TM for solving linear programming can be effectively simulated by a deterministic one: In the proof, the current interior point is transformed to the center of the simplex by means of a projective transformation. Then it is joined by a straight-line to the optimal solution point. How does one know the optimal point? A NDTM can guess that and solve the problem in "one iteration", which is not particularly interesting. But we *weaken* the NDTM so that it is allowed to proceed only up to the boundary of the inscribed ball, along the line joining the current point to the optimum. This is weakening only by a polynomial amount,(actually just by factor of n) due to the theorem on ratio of radii of circumscribed and inscribed balls, but now something interesting happens: A deterministic TM can match the performance of the weakened NDTM,by optimizing a linear function over a ball, resulting in a polynomial time bound. This kind of proof technique is expected to play similar role in AC^2M, with the role of straight line joining current point to the optimum, replaced by curves with higher values of connectivity indices.

A potential approach to $P \neq NP$ in classical Turing Model using properties of AC^2M consists of proving the following steps:

- No DTM can *consistently* beat certain class of methods in AC^2M by more than a polynomial factor
- A tight upper and lower bound on the latter methods in terms of the layer number in its pre-classification. This is the hardest step and requires that the the pre-classification method is well designed, otherwise lower bounds become too low and upper bounds become too high, which are two sides of the same coin, and only show weakness of the pre-classification method.
- Since the two models have distinct layering methods in the pre-classification step, transferring all (infinite) input instances from one pre-classification to another involves a permutation of infinite number of objects. Proving that this permutation involves changing the layer number by super-polynomial amount in either direction.

References

1. Abhyankar, S., Brenner, H., Campillo, A., Cutkosky, S., Gaffney, T., Ghezzi, L.: Some Thoughts on the Jacobian Conjecture. In: Conference on Valuation Theory and Integral Closures in Commutative Algebra (2006)
2. Bayer, D., Lagarias, J.: The Nonlinear Geometry of Linear Programming. Affine and Projective Scaling Trajectories. Trans. Am. Math. Soc. 314(2), 499–526 (1989)
3. Blum, L.: Complexity and real computation. Springer, Heidelberg (1998)
4. Bochnak, J., Coste, M., Roy, M.: Real algebraic geometry. Springer, Heidelberg (1998)
5. Cartan, É.: On manifolds with an affine connection and the theory of general relativity. Humanities Press (1922)
6. Chern, S.: Complex manifolds without potential theory: With an appendix on the geometry of characteristic classes. Springer, Heidelberg (1979)
7. Christoffel, E.: Üeber die Transformation der homogenen Differentialausdrücke zweiten Grades. J. für die reine und angewandte Math. 1869(70), 46–70 (1869)
8. Eddington, A.: The mathematical theory of relativity. Cambridge Univ. Press, Cambridge (1963)
9. Ehresmann, C.: Les connexions infinitésimales dans un espace fibré différentiable, Colloque de topologie de Bruxelles. Georges Thone, Liège, pp. 29–55 (1950)
10. Eisenhart, L.: Non-Riemannian Geometry. Am. Math. Soc. Colloq. Pub. (1990)
11. Van den Essen, A.: Polynomial automorphisms and the Jacobian conjecture. Birkhäuser, Basel (2000)
12. Finsler, P.: Üeber Kurven und Flächen in allgemeinen Räumen. Ph.D. thesis, Georg August Universitat Gottingen (1918)
13. Gauss, C.: Disquisitiones circa superficies curvas. Comm. Soc. Gottingen 6 (1828)
14. Gill, P., Murray, W., Saunders, M., Wright, M.: Some issues in implementing a sequential quadratic programming algorithm. ACM Signum Newsletter 20(2) (1985)
15. Gödel, K.: An example of a new type of cosmological solutions of Einstein's field equations of gravitation. Reviews of Modern Physics 21(3), 447–450 (1949)
16. Hartle, J.B., Hawking, S.W.: Wave function of the Universe. Phys. Rev. D 28(12), 2960–2975 (1983)
17. Hawking, S., Ellis, G.: The large scale structure of space-time. Cambridge University Press, Cambridge (1973)
18. Hopf, H., Rinow, W.: Ueber den Begriff der vollständigen differentialgeometrischen Fläche. Commentarii Mathematici Helvetici 3(1), 209–225 (1931)

19. Karmarkar, N.: A new polynomial-time algorithm for linear programming. In: Proc. of the 16th Annual ACM Symposium on Theory of Computing. ACM, New York (1984)
20. Karmarkar, N.: An Interior Point Approach to NP Complete Problems: Part I. AMS Contemporary Mathematics 114, 297–308 (1988)
21. Karmarkar, N.: Riemannian Geometry Underlying Interior Point Methods for LP. AMS Contemporary Mathematics 114, 51–75 (1988)
22. Karmarkar, N., Thakur, S.: An interior-point approach to a tensor optimization problem with application to upper bounds in integer quadratic optimization problems. In: Integer Programming and Combinatorial Optimization, pp. 406–419 (1992)
23. Karmarkar, N.: A novel approach to overcome bandwidth limitations of parallel computers based on CMOS, Part-1: General concepts. In: Electron Devices and Semiconductor Technology, IEDST 2009, pp. 1–7 (2009)
24. Kobayashi, S., Nomizu, K.: Foundations of differential geometry, vol. I. Wiley Interscience, Hoboken (1963)
25. Kossowski, M., Kriele, M.: Transverse, Type Changing, Pseudo Riemannian Metrics and the Extendability of Geodesics. Proceedings: Mathematical and Physical Sciences 444(1994), 297–306 (1921)
26. Larsen, J.: Geodesics and Jacobi fields in singular semi-Riemannian geometry. Proceedings of the Royal Society of London. Series A: Mathematical and Physical Sciences 446(1928) (1994)
27. Lasserre, J.: Global Optimization with Polynomials and the Problem of Moments. SIAM Journal on Optimization 11, 796–817 (2001)
28. Levi-Civita, T.: The Absolute Differential Calculus, London and Glasgow (1947)
29. Lobachevsky, N.: On the foundations of geometry. Kazan Messenger 25 (1829)
30. Lorentz, H., Einstein, A., Minkowski, H., Sommerfeld, A., Weyl, H.: The principle of relativity. Dover, New York (1952)
31. Minkowski, H.: Raum und Zeit. Physik. Zeits. 10(104) (1909)
32. Misner, C., Thorne, K., Wheeler, J.: Gravitation. WH Freeman & Co., New York (1973)
33. Moré, J., Sorensen, D.: Computing a trust region step. SIAM Journal on Scientific and Statistical Computing 4, 553–572 (1983)
34. Mumford, D.: Algebraic geometry I: Complex projective varieties. Springer, Heidelberg (1995)
35. Nesterov, Y., Nemirovsky, A.: Self-concordant functions and polynomial-time methods in convex programming. USSR Academy of Sciences, Central Economic & Mathematic Institute, Moscow (1989)
36. Parrilo, P., Sturmfels, B.: Minimizing polynomial functions. In: Algorithmic and quantitative real algebraic geometry. DIMACS Series in Discrete Mathematics and Theoretical Computer Science, vol. 60, pp. 83–99 (2001)
37. Putinar, M.: Positive polynomials on compact semi-algebraic sets. Indiana University Mathematics Journal 42(3), 969–984 (1993)
38. Rapcsák, T.: Smooth nonlinear optimization in R^n. Kluwer Academic Pub., Boston (1997)
39. Reznick, B.: Some concrete aspects of Hilbert's 17th problem. Real algebraic geometry and ordered structures: AMS Special Session 253 (2000)
40. Ricci, M., Levi-Civita, T.: Méthodes de calcul différentiel absolu et leurs applications. Mathematische Annalen 54(1), 125–201 (1900)
41. Riemann, B.: On the Hypotheses which lie at the Bases of Geometry. Nature 8(183), 14–17 (1873)

42. Schmüdgen, K.: The K-moment problem for compact semi-algebraic sets. Mathematische Annalen 289(1), 203–206 (1991)
43. Schrödinger, E.: Space-time structure. Cambridge University Press, Cambridge (1929)
44. Shor, N.: Class of global minimum bounds of polynomial functions. Cybernetics and Systems Analysis 23(6), 731–734 (1987)
45. Smale, S.: Newton's method estimates from data at one point. In: The Merging of Disciplines: New Directions in Pure, Applied, and Computational Mathematics, pp. 185–196. Springer, Heidelberg (1986)
46. Spivak, M.: A Comprehensive Introduction to Differential Geometry. Publish or Perish, Berkeley, CA 1 (1979)
47. Stengle, G.: A Nullstellensatz and a Positivstellensatz in semi-algebraic geometry. Mathematische Annalen 207(2), 87–97 (1973)
48. Todd, M.: Semidefinite optimization. Acta Numerica 10, 515–560 (2001)
49. Udrişte, C.: Convex functions and optimization methods on Riemannian manifolds. Springer, Heidelberg (1994)
50. Veblen, O., Thomas, T.: The geometry of paths. Transactions of the American Mathematical Society 25, 551–608 (1923)
51. Weierstrass, K.: Definition analytischer funktionen einer Veränderlichen vermittelst algebraischer Differentialgleichungen. Werke I, 75–84 (1842)
52. Weyl, H.: Symmetry. Princeton University Press, Princeton (1952)

Optimal μ-Distributions for the Hypervolume Indicator for Problems with Linear Bi-objective Fronts: Exact and Exhaustive Results

Dimo Brockhoff*

TAO Team, INRIA Saclay, LRI, Paris Sud University, 91405 Orsay Cedex, France
`dimo.brockhoff@inria.fr`

Abstract. To simultaneously optimize multiple objective functions, several evolutionary multiobjective optimization (EMO) algorithms have been proposed. Nowadays, often set quality indicators are used when comparing the performance of those algorithms or when selecting "good" solutions during the algorithm run. Hence, characterizing the solution sets that maximize a certain indicator is crucial—complying with the optimization goal of many indicator-based EMO algorithms. If these optimal solution sets are upper bounded in size, e.g., by the population size μ, we call them *optimal μ-distributions*. Recently, optimal μ-distributions for the well-known hypervolume indicator have been theoretically analyzed, in particular, for bi-objective problems with a linear Pareto front. Although the exact optimal μ-distributions have been characterized in this case, not all possible choices of the hypervolume's reference point have been investigated. In this paper, we revisit the previous results and rigorously characterize the optimal μ-distributions also for all other reference point choices. In this sense, our characterization is now exhaustive as the result holds for any linear Pareto front and for any choice of the reference point and the optimal μ-distributions turn out to be always unique in those cases. We also prove a tight lower bound (depending on μ) such that choosing the reference point above this bound ensures the extremes of the Pareto front to be always included in optimal μ-distributions.

Keywords: multiobjective optimization, hypervolume indicator, optimal μ-distributions, theory.

1 Introduction

Many evolutionary multiobjective optimization (EMO) algorithms have been proposed to tackle optimization problems with multiple objectives. The most recent ones employ quality indicators within their selection in order to (i) directly incorporate user preferences into the search [1, 16] and/or to (ii) avoid cyclic behavior of the current population [15, 18]. In particular the hypervolume indicator [17] is of interest here and due to its refinement property [18] employed in several EMO algorithms [4, 6, 14]. The hypervolume indicator assigns a set of solutions the "size of the objective value space which is covered" and at the same time is bounded by the indicator's reference point [17].

---

* Currently at LIX, École Polytechnique, Palaiseau, France.

K. Deb et al. (Eds.): SEAL 2010, LNCS 6457, pp. 24–34, 2010.

Although maximizing the hypervolume indicator, according to its refinement property, results in finding Pareto-optimal solutions only [10], the question arises which of these points are favored by hypervolume-based algorithms. In other words, we are interested in the optimization goal of hypervolume-based algorithms with a fixed population size μ, i.e., in finding a set of μ solutions with the highest hypervolume indicator value among all sets with μ solutions. Also in performance assessment, the hypervolume is used quite frequently [19]. Here, knowing the set of points maximizing the hypervolume is crucial as well. On the one hand, it allows to evaluate whether hypervolume-based algorithms really converge towards their optimization goal on certain test functions. On the other hand, only the knowledge of the best hypervolume value achievable with μ solutions allows to compare algorithms in an absolute manner similar to the state-of-the-art approach of benchmarking single-objective continuous optimization algorithms in the horizontal-cut view scenario, see [12, appendix] for details.

Theoretical investigations of the sets of μ points maximizing the hypervolume indicator—also known under the term of *optimal μ-distributions* [2]—have been started only recently. Although quite strong, i.e., very general, results on optimal μ-distributions are known [2, 7], most of them are approximation or limit results in order to study a wide range of problem classes. The only exact results consider problems with very specific Pareto fronts, namely linear fronts that can be described by a function $f : x \in [x_{\min}, x_{\max}] \mapsto \alpha x + \beta$ where $\alpha < 0$ and $\beta \in \mathbb{R}$ in the bi-objective case [2, 5, 9] or fronts that can be expressed as $f : x \in [1, c] \mapsto c/x$ with $c > 1$ [11].

The main scope of this paper is to revisit the results on optimal μ-distributions for bi-objective problems with linear Pareto fronts and to consider all conditions under which the exact optimal μ-distributions have not been characterized yet. The result is both exact and exhaustive, in the sense that a single formula is proven that characterizes the unique optimal μ-distribution for any choice of the hypervolume indicator's reference point and for any $\mu \geq 2$, covering also the previously known cases. It turns out that the specific case of $\mu = 2$ complies with a previous results of [2] and that for all linear front shapes, the optimal μ-distributions are always unique.

Before we present our results in Sec. 5–7, we introduce basic notations and definitions in Sec. 2, define and discuss the problem of finding optimal μ-distributions in Sec. 3 in more detail, and give an extensive overview of the known results in Sec. 4.

2 Preliminaries

Without loss of generality (w.l.o.g.), we consider bi-objective minimization problems where a vector-valued function $\mathcal{F} : X \to \mathbb{R}^2$ has to be minimized with respect to the weak Pareto dominance relation $\preceq$. We say a solution $x \in X$ is weakly dominating another solution $y \in X$ ($x \preceq y$) iff $\mathcal{F}_1(x) \leq \mathcal{F}_1(y)$ and $\mathcal{F}_2(x) \leq \mathcal{F}_2(y)$ where $\mathcal{F} = (\mathcal{F}_1, \mathcal{F}_2)$. We also say $x \in X$ is dominating $y \in X$ ($x \prec y$) if $x \preceq y$ but $y \npreceq x$. The set of nondominated solutions is the so-called Pareto set $\mathcal{P}_s = \{x \in X \mid \nexists y \in X : y \prec x\}$ and its image $\mathcal{F}(\mathcal{P}_s)$ in objective space is called Pareto front. Note that, to keep things simple, we make an abuse of terminology throughout the paper and use the term *solution* both for a point x in the decision space X and for its corresponding objective vector $\mathcal{F}(x) \in \mathbb{R}^2$. Moreover, we also define the orders $\preceq$ and $\prec$ on objective vectors.

In order to optimize multiobjective optimization problems like the bi-objective ones considered here, several recent EMO algorithms aim at optimizing the *hypervolume indicator* [17], a set quality indicator $I_H(A, r)$ that assigns a set A the Lebesgue measure λ of the set of solutions that are weakly dominated by solutions in A but that at the same time weakly dominate a given reference point $r \in \mathbb{R}^2$, see Fig. 1:

$$I_H(A, r) = \lambda \left(\{ z \in \mathbb{R}^2 \mid \exists a \in A : f(a) \preceq z \preceq r \} \right) \tag{1}$$

The hypervolume indicator has the nice property of being a refinement of the Pareto dominance relation [18]. This means that maximizing the hypervolume indicator is equivalent to obtaining solutions in the Pareto set only [10]. However, it is more interesting to know *where* the solutions maximizing the hypervolume lie on the Pareto front if we restrict the size of the sets A to let us say, the population size μ. This set of μ points maximizing the hypervolume indicator among all sets of μ points is known under the term *optimal μ-distribution* [2] and finding an optimal μ-distribution coincides with the optimization goal of hypervolume-based algorithms with fixed population size.

To investigate optimal μ-distributions in this paper, we assume the Pareto front to be given by a function $f : \mathbb{R} \rightarrow \mathbb{R}$ and two values $x_{\min}, x_{\max} \in \mathbb{R}$ such that all points on the Pareto front have the form $(x, f(x))$ with $x \in [x_{\min}, x_{\max}]$. In case of a linear Pareto front, $f(x) = \alpha x + \beta$ for $\alpha, \beta \in \mathbb{R}$, see Fig. 1 for an example. W.l.o.g, we assume that $x_{\min} = 0$ and $\beta > 0$ in the remainder of the paper—otherwise, a simple linear transformation brings us back to this case. Moreover, $\alpha < 0$ follows from minimization. Note also that under not too strong assumptions on the Pareto front, and in particular for linear fronts, optimal μ-distributions always exist, see [2].

3   Problem Statement

In case of a linear Pareto front described by the function $f(x) = \alpha x + \beta$ ($\alpha < 0, \beta \in \mathbb{R}$), finding the optimal μ-distribution for the hypervolume indicator with reference point $r = (r_1, r_2)$ can be written as finding the minimum of the function

$$I_H(x_1, \ldots, x_\mu) = \sum_{i=1}^{\mu} (x_{i+1} - x_i)(f(x_0) - f(x_i)) = \sum_{i=1}^{\mu} (x_{i+1} - x_i)(\alpha x_0 - \alpha x_i)$$

$$= \alpha \sum_{i=1}^{\mu} \left[(x_i)^2 + x_0 x_{i+1} - x_0 x_i - x_i x_{i+1} \right] \tag{2}$$

with $x_{\min} \leq x_i \leq x_{\max}$ for all $1 \leq i \leq \mu$

where we define $x_{\mu+1} = r_1$ and $x_0 = f^{-1}(r_2)$ [2], Fig. 1. According to [2], we denote the x-values of the optimal μ-distribution, maximizing (2), as $x_1^\mu \ldots x_\mu^\mu$. Although the term in (2) is quadratic in the variables $x_0, \ldots, x_{\mu+1}$, and therefore, in principle, solvable analytically, the restrictions of the variables to the interval $[x_{\min}, x_{\max}]$ makes it difficult to solve the problem. In the following, we therefore investigate the minima of (2) depending on the choice of r_1 and r_2 with another approach: we use the necessary condition for optimal μ-distributions of [2, Proposition 1] and apply it to linear fronts while the restriction of the variables to $[x_{\min}, x_{\max}]$ are handled "by hand".

4   Overview of Recent and New Results

Characterizing optimal μ-distributions for the hypervolume indicator has been started only recently but the number of results is already quite extensive, see for example [1–5, 7, 9–11]. Here, we restate, to the best of our knowledge, all previous results that relate to linear Pareto fronts and point out which problems are still open.

Besides the proof that maximizing the hypervolume indicator yields Pareto-optimal solutions [10], the authors of [5] and [9] were the first to investigate optimal μ-distributions for linear fronts. Under the assumption that the extreme points $(0, \beta)$ and $(x_{\max}, 0)$ are included in the optimal μ-distribution, it was shown for linear fronts with $\alpha = -1$ that neighbored points within a set maximizing the hypervolume are equally spaced. However, the result does not state where the leftmost and rightmost point of the optimal μ-distribution have to be placed in order to maximize the hypervolume and it has been shown later [2] that the assumption about the extreme points does not always hold.

The first results without assuming the positions of the leftmost and rightmost point have been proven in [2] where the result is based on a more general necessary condition about optimal μ-distributions for the hypervolume indicator. In particular, [2] presents the exact distribution of μ points maximizing the hypervolume indicator when the reference point is chosen close to the Pareto front (region I in Fig. 1, cp. [2, Theorem 5]) or far away from the front (region IX in Fig. 1, cp. [2, Theorem 6]). In the former case, both extreme points of the front do not dominate the reference point and the (in this case unique) optimal μ-distribution reads

$$x_i^\mu = f^{-1}(r_2) + \frac{i}{\mu + 1} \cdot (r_1 - f^{-1}(r_2)) \ . \tag{3}$$

In the latter case, the reference point is chosen far enough such that—independent of the reference point and μ—both extreme points are included in an optimal μ-distribution[1] and the (again unique) optimal μ-distributions can be expressed as

$$x_i^\mu = x_{\min} + \frac{i - 1}{\mu - 1} (x_{\max} - x_{\min}) \ . \tag{4}$$

Note that the region IX, corresponding to choices of the reference point within Theorem 6 of [2] does not depend on μ but on a lower bound on the reference point to ensure that both extremes are included in the optimal μ-distribution. Recently, a limit result has been proven [3] which shows that the lower bound of [2, Theorem 6] converges to the nadir point[2] if μ goes to infinity but the result does not state how fast (in μ) the nadir point is approached. Clearly, choosing the reference point within the other regions II–VIII in Fig. 1 is possible and the question arises how the reference point influences the optimal μ-distributions in these uninvestigated cases as well. The answer to this question is the main focus of this paper.

---

[1] Which is proven to be true for $r_1 > 2x_{\max}$ and $r_2 > 2\beta$ in another general theorem [2].

[2] In case of a linear front as defined above, the nadir point equals $n = (x_{\max}, f(x_{\min}))$.

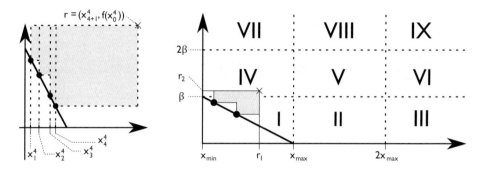

Fig. 1. Left: Illustration of the hypervolume indicator $I_H(A, r)$ (gray area). **Right:** Optimal μ-distributions and the choice of the reference point for linear fronts of shape $y = \alpha x + \beta$. Up-to-now, theoretical results are only known if the reference point is chosen within the regions I and IX [2]. Exemplary, the optimal 2-distribution (circles) is shown when choosing the reference point (cross) within region IV.

5   If the Reference Point Is Dominated by Only the Right Extreme

As a first new result, we consider choosing the reference point within the regions II or III of Fig. 1. Here, the left extreme cannot be included in an optimal μ-distribution as it is never dominating the reference point and thereby always has a zero hypervolume contribution. Thus, the proof of the optimal μ-distribution has to consider only the restrictions of the μ points at the right extreme. Moreover, the uniqueness of the optimal μ-distribution in the cases II and III follows directly from case I.

Theorem 1. *Given $\mu \in \mathbb{N}_{\geq 2}$, $\alpha \in \mathbb{R}_{<0}$, $\beta \in \mathbb{R}_{>0}$, and a linear Pareto front $f(x) = \alpha x + \beta$ within $[0, x_{\max} = -\frac{\beta}{\alpha}]$. If $r_2 \leq \beta$ and $r_1 \geq x_{\max}$ (cases II and III), the unique optimal μ-distribution $(x_0^\mu, \ldots, x_\mu^\mu)$ for the hypervolume indicator I_H with reference point (r_1, r_2) can be described by*

$$x_i^\mu = f^{-1}(r_2) + \frac{i}{\mu+1}\left(\min\left\{r_1, \frac{\mu+1}{\mu}x_{\max} - \frac{f^{-1}(r_2)}{\mu}\right\} - f^{-1}(r_2)\right) . \quad (5)$$

Proof. According to (3) and assuming no restrictions of the solutions on the linear front $\alpha x + \beta$ with $x \in \mathbb{R}$, the optimal μ-distribution would be given by $x_i^\mu = f^{-1}(r_2) + \frac{i}{\mu+1} \cdot (r_1 - f^{-1}(r_2))$ where the x_i^μ are possibly lying outside the interval $[0, x_{\max}]$. However, as long as r_1 is chosen such that $x_\mu^\mu \leq x_{\max}$, we can use (3) for describing the optimal μ-distributions, i.e., in the case that

$$x_\mu^\mu = f^{-1}(r_2) + \frac{\mu}{\mu+1} \cdot (r_1 - f^{-1}(r_2)) \leq x_{\max} \Leftrightarrow \frac{f^{-1}(r_2)}{\mu+1} + \frac{\mu}{\mu+1}r_1 \leq x_{\max}$$

$$\Leftrightarrow r_1 \leq \frac{\mu+1}{\mu}x_{\max} - \frac{f^{-1}(r_2)}{\mu}\left(= \frac{-r_2 - \beta\mu}{\alpha\mu}\right) . \quad (6)$$

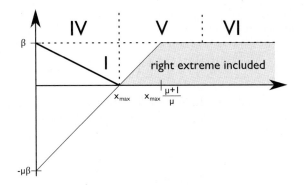

Fig. 2. When choosing the reference point within regions II and III, we prove that the right extreme is included in optimal μ-distributions if the reference point is chosen within the gray shaded area right of the line $y = -\alpha\mu x - \mu\beta$, see Corollary 1. The picture corresponds to $\mu = 2$.

With larger r_1, the optimal μ-distribution does not change any further (only the hypervolume contribution of x_μ^μ increases linearly with r_1), i.e., we can rewrite (3) as (5).   □

The previous theorem allows us also a more precise statement of when the right extreme is included in optimal μ-distributions than the statement in [2].

Corollary 1. *In case that $r_2 \leq \beta$ and $r_1 \geq \frac{\mu+1}{\mu} x_{\max} - \frac{f^{-1}(r_2)}{\mu}$, the right extreme point $(x_{\max}, 0)$ is included in all optimal μ-distributions for the front $\alpha x + \beta$.*   □

Note that the choice of r_1 to guarantee the right extreme in optimal μ-distributions depends both on μ and r_2 here whereas the (not so tight) bound for r_1 to ensure the right extreme proven in [2] equals $2x_{\max}$. This is independent of μ and coincides with the new (tighter) result if $\mu = 2$ and $r_2 = \beta$. Figure 2 illustrates the region for which, if the reference point is chosen within, the right extreme is always included in an optimal μ-distribution. Compare also to the old result of [2] which states this inclusion of the right extreme only in case the reference point is chosen in region IX of Fig. 1. The description of the line $y = -\alpha\mu x - \mu\beta$ where choosing the reference point to the right of it ensures the right extreme in the optimal μ-distribution results from writing r_2 within $r_1 = \frac{\mu+1}{\mu} x_{\max} - \frac{f^{-1}(r_2)}{\mu}$ as a function of r_1.

6   If the Reference Point Is Dominated by Only the Left Extreme

Obviously, the two cases IV and VII of Fig. 1 are symmetrical to the cases II and III where mainly the left extreme and the reference point's coordinate r_2 take the roles of the right extreme and the coordinate r_1 respectively from the previous proof.

Theorem 2. *Given $\mu \in \mathbb{N}_{\geq 2}$, $\alpha \in \mathbb{R}_{<0}$, $\beta \in \mathbb{R}_{>0}$, and a linear Pareto front $f(x) = \alpha x + \beta$ within $[0, x_{\max} = -\frac{\beta}{\alpha}]$. If $r_1 \leq x_{\max}$ and $r_2 \geq \beta$ (cases IV and VII), the unique* *optimal μ-distribution $(x_0^\mu, \ldots, x_\mu^\mu)$ for the hypervolume indicator I_H with reference point (r_1, r_2) can be described by*

$$x_i^\mu = f^{-1}\left(\min\left\{r_2, \tfrac{\mu+1}{\mu}\beta - \tfrac{f(r_1)}{\mu}\right\}\right) + \frac{i}{\mu+1}\left(r_1 - f^{-1}\left(\min\left\{r_2, \tfrac{\mu+1}{\mu}\beta - \tfrac{f(r_1)}{\mu}\right\}\right)\right). \quad (7)$$

Proof. The proof is similar to the one of Theorem 1: As in case I, we can write the optimal μ-distribution according to (3) except that we have to ensure that $x_1^\mu \geq x_{\min} = 0$. This is equivalent to $f^{-1}(r_2) + \tfrac{1}{\mu+1}\left(r_1 - f^{-1}(r_2)\right) \geq 0$ or $\tfrac{r_2-\beta}{\alpha} + \tfrac{1}{\mu+1}\left(r_1 - \tfrac{r_2-\beta}{\alpha}\right) \geq 0$ or $\tfrac{r_2-\beta}{\alpha} + \tfrac{\alpha r_1 - r_2 + \beta}{(\mu+1)\alpha} \geq 0$. With $\alpha < 0$, this gives $(\mu+1)r_2 - (\mu+1)\beta + \alpha r_1 - r_2 + \beta \leq 0$ and finally $r_2 \leq \tfrac{(\mu+1)\beta-(\alpha r_1+\beta)}{\mu} = \tfrac{\mu+1}{\mu}\beta - \tfrac{f(r_1)}{\mu}$ such that (3) becomes (7). □

7   General Result for All Cases I–IX

By combining the above results, we can now characterize the optimal μ-distributions also for the other cases V, VI, VII, and IX and give a general description of optimal μ-distributions for problems with bi-objective linear fronts, given any $\mu \geq 2$ and any meaningful choice of the reference point[3].

Theorem 3. *Given $\mu \in \mathbb{N}_{\geq 2}$, $\alpha \in \mathbb{R}_{<0}$, $\beta \in \mathbb{R}_{>0}$, and a linear Pareto front $f(x) = \alpha x + \beta$ within $[0, x_{\max} = -\tfrac{\beta}{\alpha}]$, the unique optimal μ-distribution $(x_0^\mu, \ldots, x_\mu^\mu)$ for the hypervolume indicator I_H with reference point $(r_1, r_2) \in \mathbb{R}_{>0}^2$ can be described by*

$$x_i^\mu = f^{-1}(F_l) + \frac{i}{\mu+1}\left(F_r - f^{-1}(F_l)\right) \quad (8)$$

for all $1 \leq i \leq \mu$ where

$$F_l = \min\{r_2, \frac{\mu+1}{\mu}\beta - \frac{1}{\mu}f(r_1), \frac{\mu}{\mu-1}\beta\} \text{ and}$$

$$F_r = \min\{r_1, \frac{\mu+1}{\mu}x_{\max} - \frac{1}{\mu}f^{-1}(r_2), \frac{\mu}{\mu-1}x_{\max}\} .$$

Proof. Again, the optimal μ-distribution would be given by (3) if we prolongate the front linearly outside the interval $[x_{\min}, x_{\max}]$ and therefore, no restrictions on the x_i^μ would hold. However, the points x_i^μ are restricted to $[x_{\min}, x_{\max}]$ and therefore (since we assume $x_i^\mu < x_{i+1}^\mu$) we have to ensure that both $x_1^\mu \geq x_{\min} = 0$ and $x_\mu^\mu \leq x_{\max} = -\beta/\alpha$ hold. According to the above proofs, the former is equivalent to

$$r_2 \leq \frac{\mu+1}{\mu}\beta - \frac{f(r_1)}{\mu} \quad (9)$$

---

[3] Choosing the reference point such that it weakly dominates a Pareto-optimal point does not make sense as no feasible solution would have a positive hypervolume.

and the latter is equivalent to

$$r_1 \leq \frac{\mu+1}{\mu} x_{\max} - \frac{f^{-1}(r_2)}{\mu} \tag{10}$$

however, with restrictions on r_1 ($r_1 \leq x_{\max}$) and $r_2 \leq \beta$ respectively which we do not have here. As long as both (9) and (10) hold as in the white area in Fig. 3, i.e., no constraint is violated, (3) can be used directly to describe the optimal μ-distribution as in region I. To cover all other cases, we could, at first sight, simply combine the results for the cases II, III, IV, and VII from above and use

$$F_l^* = \min \left\{ r_2, \frac{\mu+1}{\mu}\beta - \frac{f(r_1)}{\mu} \right\} \quad \text{and} \quad F_r^* = \min \left\{ r_1 \frac{\mu+1}{\mu} x_{\max} - \frac{f^{-1}(r_2)}{\mu} \right\}$$

as the extremes influencing the set $x_i^{\mu,*} = F_l^* + \frac{i}{\mu+1}(F_r^* - F_l^*)$. However, r_1 and r_2 are unrestricted and thus, F_l^* and F_r^* can become too large such that the points $x_i^{\mu,*}$ lie outside the feasible front part $[x_{\min}, x_{\max}]$. To this end, we compute where the two constraints (9) and (10) meet, i.e., what is the smallest possible reference point that results in having both extremes in the optimal μ-distribution. This point is depicted as the lower left point of the dark gray area in Fig. 3.

By combining the equalities in (9) and (10) which is equivalent to $r_2 = -\alpha\mu r_1 - \beta\mu$ (see end of Sec. 5), we obtain

$$r_2 = \frac{\mu+1}{\mu}\beta - \frac{f(r_1)}{\mu} = \frac{\mu+1}{\mu}\beta - \frac{\alpha r_1 + \beta}{\mu} = -\alpha\mu r_1 - \beta\mu \text{ or } r_1 = -\frac{\beta}{\alpha}\frac{\mu}{\mu-1} = \frac{\mu}{\mu-1}x_{\max}$$

and thus $r_2 = \frac{\mu+1}{\mu}\beta - \frac{f(\frac{\mu}{\mu-1}x_{\max})}{\mu} = \frac{\mu}{\mu-1}\beta$. Hence, if we choose the reference point $r = (r_1, r_2)$ such that $r_1 \geq \frac{\mu}{\mu-1}x_{\max}$ and $r_2 \geq \frac{\mu}{\mu-1}\beta$, both extremes will be included in the optimal μ-distribution $x_i^\mu = F_l^{\text{extr}} + \frac{i}{\mu+1}(F_r^{\text{extr}} - F_l^{\text{extr}})$ with $F_l^{\text{extr}} = \frac{\mu}{\mu-1}\beta$ and $F_r^{\text{extr}} = \frac{\mu}{\mu-1}x_{\max}$. With this result, we know that, independent of r_2, the right extreme is included if $r_1 \geq \frac{\mu}{\mu-1}x_{\max}$ (if the leftmost extreme is not included, r_2 must be smaller than $\frac{\mu}{\mu-1}\beta$ and in this case $r_1 \geq \frac{\mu+1}{\mu}x_{\max}$ ensures that it is also greater or equal to $\frac{\mu+1}{\mu}x_{\max} - \frac{f^{-1}(r_2)}{\mu}$). The same can be said for the left extreme, which is included in an optimal μ-distribution whenever $r_2 \geq \frac{\mu}{\mu-1}\beta$. The optimal μ-distribution for those cases are the same than the optimal μ-distributions if we restrict r_1 and r_2 to be at most $\min\{\frac{\mu+1}{\mu}x_{\max} - \frac{1}{\mu}f^{-1}(r_2), \frac{\mu}{\mu-1}x_{\max}\}$, and $\min\{\frac{\mu+1}{\mu}\beta - \frac{1}{\mu}f(r_1), \frac{\mu}{\mu-1}\beta\}\}$ respectively, i.e., to the cases where the reference point is lying on the boundary of the white region of Fig. 3 and having one or even both extremes included in the optimal μ-distributions. In those cases, (3) can be used again for characterizing the optimal μ-distribution as the constraints on the x_i^μ are fulfilled. Using the mentioned restrictions on r_1 and r_2 results in the theorem. $\square$

Note that the previous proof gives a tighter bound for how to choose the reference point $r = (r_1, r_2)$ in order to obtain the extremes in comparison to the old result in [2]: The former result states that whenever r_1 is chosen strictly larger than $2x_{\max}$ and r_2 is chosen strictly larger than 2β, both extremes are included in an optimal μ-distribution in the case of a linear Pareto front. This bound holds for every $\mu \geq 2$ but the previous

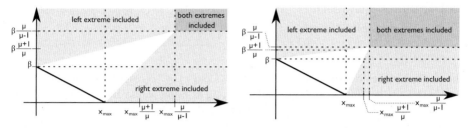

Fig. 3. How to choose the reference point to obtain the extremes in optimal μ-distributions: $\mu = 2$ (left) and $\mu = 4$ (right) for one and the same front $y = -x/2 + 1$

theorem precises this bound to $r_1 \geq \frac{\mu+1}{\mu} x_{\max}$ and $r_2 \geq \frac{\mu+1}{\mu} \beta$ for a given μ which coincides with the old bound for $\mu = 2$ but is the closer to the nadir point $(x_{\max}, \beta)$, the larger μ gets—a result that has been previously shown as a limit result for arbitrary Pareto fronts [3].

Last, we want to note that, though the two equations (8) and (4) do not look the same at first sight, Theorem 3 complies with the characterization of optimal μ-distributions given in (4) [2, Theorem 6] for the case IX which can be shown by simple algebra.

8   Conclusions

Finding optimal μ-distributions, i.e., sets of μ points that have the highest quality in-dicator value among all sets of μ solutions coincides with the optimization goal of indicator-based multiobjective optimization algorithms and it is therefore important to characterize them. Here, we rigorously analyze optimal μ-distributions for the often used hypervolume indicator and for problems with linear Pareto fronts. The results are exhaustive in a sense that a single formula covers all possible choices of the hypervol-ume's reference point, including two previously proven cases. In addition to the newly covered cases, the new results show also how the choice of μ influences the fact that the extremes of the Pareto front are included in optimal μ-distributions for the case of linear fronts—a fact that has been only shown before by a lower bound result of choos-ing the reference point and not exact as here. The proofs also show that the optimal μ-distributions for problems with linear Pareto fronts are, given a $\mu \geq 2$ and a certain choice of the reference point, always unique.

Besides being the first exhaustive theoretical investigation of optimal μ-distributions for a specific front shape, the presented results are expected to have an impact in prac-tical performance assessment as well. For the first time, it is now possible to use the exact optimal μ-distribution and its corresponding hypervolume when comparing algo-rithms on test problems with linear fronts such as DTLZ1 [8] or WFG3 [13] for any choice of the reference point[4]. It remains future work to theoretically characterize the optimal μ-distributions for test problems with other front shapes for which the optimal μ-distributions can only be approximated numerically at the moment [2].

---

[4] Theorem 3 can be applied directly with $\alpha = -1$ and $\beta = 0.5$ (DTLZ1) or $\beta = 1$ (WFG3).

Acknowledgments. The author would like to thank Anne Auger for many fruitful discussions and comments as well as the French national research agency (via the SYS-COMM project ANR-08-SYSC-017) and the LIX laboratory for financial support.

References

[1] Auger, A., Bader, J., Brockhoff, D., Zitzler, E.: Investigating and Exploiting the Bias of the Weighted Hypervolume to Articulate User Preferences. In: Genetic and Evolutionary Computation Conference (GECCO 2009), pp. 563–570. ACM, New York (2009)

[2] Auger, A., Bader, J., Brockhoff, D., Zitzler, E.: Theory of the Hypervolume Indicator: Optimal μ-Distributions and the Choice of the Reference Point. In: Foundations of Genetic Algorithms (FOGA 2009), pp. 87–102. ACM, New York (2009)

[3] Auger, A., Bader, J., Brockhoff, D., Zitzler, E.: Hypervolume-based Multiobjective Optimization: Theoretical Foundations and Practical Implications. Theor. Comput. Sci. (submitted 2010)

[4] Bader, J.: Hypervolume-Based Search For Multiobjective Optimization: Theory and Methods. PhD thesis, ETH Zurich (2010)

[5] Beume, N., Fonseca, C.M., Lopez-Ibanez, M., Paquete, L., Vahrenhold, J.: On the Complexity of Computing the Hypervolume Indicator. Technical Report CI-235/07, University of Dortmund (December 2007)

[6] Beume, N., Naujoks, B., Emmerich, M.: SMS-EMOA: Multiobjective Selection Based on Dominated Hypervolume. Eur. J. Oper. Res. 181, 1653–1669 (2007)

[7] Bringmann, K., Friedrich, T.: The Maximum Hypervolume Set Yields Near-optimal Approximation. In: Genetic and Evolutionary Computation Conference (GECCO 2010), pp. 511–518. ACM, New York (2010)

[8] Deb, K., Thiele, L., Laumanns, M., Zitzler, E.: Scalable Test Problems for Evolutionary Multi-Objective Optimization. In: Evolutionary Multiobjective Optimization: Theoretical Advances and Applications, pp. 105–145. Springer, Heidelberg (2005)

[9] Emmerich, M., Deutz, A., Beume, N.: Gradient-Based/Evolutionary Relay Hybrid for Computing Pareto Front Approximations Maximizing the S-Metric. In: Bartz-Beielstein, T., Blesa Aguilera, M.J., Blum, C., Naujoks, B., Roli, A., Rudolph, G., Sampels, M. (eds.) HCI/ICCV 2007. LNCS, vol. 4771, pp. 140–156. Springer, Heidelberg (2007)

[10] Fleischer, M.: The measure of Pareto optima. Applications to multi-objective metaheuristics. In: Fonseca, C.M., Fleming, P.J., Zitzler, E., Deb, K., Thiele, L. (eds.) EMO 2003. LNCS, vol. 2632, pp. 519–533. Springer, Heidelberg (2003)

[11] Friedrich, T., Horoba, C., Neumann, F.: Multiplicative Approximations and the Hypervolume Indicator. In: Genetic and Evolutionary Computation Conference (GECCO 2009), pp. 571–578. ACM, New York (2009)

[12] Hansen, N., Auger, A., Finck, S., Ros, R.: Real-Parameter Black-Box Optimization Benchmarking 2009: Experimental Setup. INRIA Research Report RR-6828, INRIA Saclay—Ile-de-France (May 2009)

[13] Huband, S., Hingston, P., Barone, L., While, L.: A Review of Multiobjective Test Problems and a Scalable Test Problem Toolkit. IEEE T. Evolut. Comput. 10(5), 477–506 (2006)

[14] Igel, C., Hansen, N., Roth, S.: Covariance Matrix Adaptation for Multi-objective Optimization. Evol. Comput. 15(1), 1–28 (2007)

[15] Wagner, T., Beume, N., Naujoks, B.: Pareto-, Aggregation-, and Indicator-Based Methods in Many-Objective Optimization. In: Obayashi, S., Deb, K., Poloni, C., Hiroyasu, T., Murata, T. (eds.) EMO 2007. LNCS, vol. 4403, pp. 742–756. Springer, Heidelberg (2007)

[16] Zitzler, E., Künzli, S.: Indicator-Based Selection in Multiobjective Search. In: Yao, X., Burke, E.K., Lozano, J.A., Smith, J., Merelo-Guervós, J.J., Bullinaria, J.A., Rowe, J.E., Tiňo, P., Kabán, A., Schwefel, H.-P. (eds.) PPSN 2004. LNCS, vol. 3242, pp. 832–842. Springer, Heidelberg (2004)
[17] Zitzler, E., Thiele, L.: Multiobjective Optimization Using Evolutionary Algorithms - A Comparative Case Study. In: Eiben, A.E., Bäck, T., Schoenauer, M., Schwefel, H.-P. (eds.) PPSN 1998. LNCS, vol. 1498, pp. 292–301. Springer, Heidelberg (1998)
[18] Zitzler, E., Thiele, L., Bader, J.: On Set-Based Multiobjective Optimization. IEEE T. Evolut. Comput. 14(1), 58–79 (2010)
[19] Zitzler, E., Thiele, L., Laumanns, M., Fonseca, C.M., Grunert da Fonseca, V.: Performance Assessment of Multiobjective Optimizers: An Analysis and Review. IEEE T. Evolut. Comput. 7(2), 117–132 (2003)

A Parallel Algorithm for Solving Large Convex Minimax Problems

Ramnik Arora[1], Utkarsh Upadhyay[2], Rupesh Tulshyan[3], and J. Dutta[4]

[1] The Courant Institute of Mathematical Sciences, New York University
[2] The Department of Computer Science, EPFL, Lausanne
[3] The Department of Mathematics and Statistics,
Indian Institute of Technology Kanpur
[4] Indian Institute of Technology Kanpur
Kanpur - 208016, India
ra1221@nyu.edu, utkarsh.upadhyay@epfl.ch, {tulshyan,jdutta}@iitk.ac.in

Abstract. We consider unconstrained minimax problem where the objective function is the maximum of a finite number of smooth convex functions. We present an iterative method to compute the optimal solution for the unconstrained convex finite minimax problem. The algorithm developed estimates the direction of steepest-descent rapidly and using Armijo's condition proceeds towards the solution. Owing to the highly parallel nature of the algorithm, it is highly suitable for large minimax problems. Algorithm is implemented on Nvidia Tesla C1060 graphics card using CUDA and numerical comparisons with RGA & CFSQP are presented.

Keywords: Steepest-Descent, Large Minimax Problem, CUDA.

1 Introduction

We deal with convex unconstrained finite minimax problems which maybe characterized as:

$$\min_{\Re^n} F(x)$$

where

$$F(x) = \max\{f_i(x) : i = 1, \ldots, m\}$$

and $f_i : \Re^n \to \Re$ are differentiable convex functions. We can trivially see that the function F is convex but may not be necessarily differentiable.

Constrained finite minimax programs arise in engineering design, computer-aided-design, circuit design and optimal control [1], and are often tackled by converting them into unconstrained problems using Penalty parameter approach. Many decision models can be formulated as continuous minimax problems and interested readers can find applications and examples in [2] and [3].

Finite minimax problems form an important class of nonsmooth optimization problems, which have attracted much attention from researchers. Prominent approaches include smoothing methods [1], interior-point methods [4], active-set

K. Deb et al. (Eds.): SEAL 2010, LNCS 6457, pp. 35–44, 2010.

strategy [5], trust region strategy [6], direct search methods [7], method of sub-gradients [8], semidefinite programming [2], bundle methods [9] and sequential quadratic programming [10]. However, standard algorithms often fail on large-scale problems since function evaluations are computationally expensive.

2   Algorithm - Basics

Descent algorithms are broad category of iterative algorithms which compute the next iterate (say x_{k+1}) from the current iterative (x_k) in two steps: firstly, they compute the direction of descent, say s_k and then define a step-size (say t_k). Now the next iterate is:

$$x_{k+1} = x_k + t_k s_k, \ where \ t_k > 0$$

2.1   Terminology

An index i is said to be active at a given point x iff $f_i(x) = F(x)$. The active index-set, $I(x)$ contains all the active indices at point x.

$$I(x) := \{i : f_i(x) = F(x)\}$$

Similarly, $f_i(x)$ and $\nabla f_i(x)$ for $i \in I(x)$ are called the active function and the active gradient respectively at point x.

The subdifferential, at a point x, represented by $\partial F(x)$ is the convex hull of the active gradients at point x or,

$$\partial F(x) = co\{\nabla f_i(x) : i \in I(x)\}$$

where co is the convex hull of the gradient vectors.

$$\partial F(x) = \{ \sum_{i \in I(x)} \alpha_i \nabla f_i(x) : \sum_{i \in I(x)} \alpha_i = 1, \ \alpha_i \geq 0 \ \forall \ i\}$$

It must be noted that the subdifferential (∂F) at any point x is a *compact convex polyhedron* for the finite minimax problem characterized by the convex hull of the ∇f_i where $i \in I(x)$.

2.2   Condition for Optimality

The proofs of the following results maybe found in [11].

Theorem 2.1. *A necessary and sufficient condition for $\hat{x}$ to minimize F, as defined above, is that there exist non-negative α_i's such that,*

$$\sum_{i \in I(\hat{x})} \alpha_i = 1$$

and

$$\sum_{i \in I(\hat{x})} \alpha_i \nabla f_i(\hat{x}) = 0$$

Theorem 2.2. *If the subdifferential is a compact convex polyhedron character-ized as a convex hull of $s_1, s_2, \ldots s_m$ each in $\Re^n$ such that,*

$$\partial F(x) := co\{s_1, s_2 \ldots s_m\}$$

and α^o is the solution of the convex quadratic minimization problem

$$\min_{\alpha \in \Delta^{m-1}} \frac{1}{2}\{\|\sum_{j=1}^{m} \alpha_j s_j\|^2\}$$

where Δ^{m-1} is the unit simplex[1], then the direction of steepest-descent is given as,

$$s = \sum_{j=1}^{m} \alpha_j^o s_j$$

3   Algorithm-Serial

The sequence of iterates x_k is generated by

$$x_{k+1} = x_k + t_k s_k$$

where t_k is chosen using appropriate Line search method.

Using formulation of Theorem 2.2, the direction of steepest descent (at x_k) may be found by solving the quadratic programming problem

$$\min_{\alpha \in \Delta^{l-1}} \frac{1}{2}\|\sum_{j=1}^{l} \alpha_j s_j\|^2,$$

where $|I(x)| = l$, $(s_1, \ldots, s_l)$ is the set of active gradients at the point x_k and Δ^{l-1} is the unit simplex. The direction of steepest-descent is then unique and is given by,

$$s = \sum_{j=1}^{l} \alpha_j^o s_j$$

where α^o solves the above convex quadratic programming problem. Therefore the steepest-descent is opposite to the projection of the non-optimal point x onto $\partial F(x)$.

Let x be the given point with $|I(x)| = l$, and $S = [s_1 s_2 \ldots s_l]$ where S is an $n \times l$ matrix. Thus, our problem is equivalent to:

$$\min_{\alpha \in \Delta^{l-1}} < S\alpha, S\alpha >$$

The above minimization is done iteratively. In each step of the iteration, random i and j are chosen and α_k's for $k \neq \{i, j\}$ are treated as constants. The problem then reduces to,

---

[1] $\Delta^n = \{(t_1, \ldots, t_{n+1}) \in \Re^{n+1} \mid \sum_{i=1}^{n+1} t_i = 1 \text{ and } t_i \geq 0 \ \forall \ i\}$.

$$\min_{\alpha_i} < \alpha_i s_i + (p - \alpha_i)s_j + c, \ \alpha_i s_i + (p - \alpha_i)s_j + c >$$

where $c = \sum_{k \neq \{i,j\}} \alpha_k s_k$ and $\alpha_i + \alpha_j = p$. If the problem was unconstrained the minima would be at:

$$\alpha_i^* = -\frac{< s_i - s_j, \ p s_j + c >}{< s_i - s_j, \ s_i - s_j >}$$

However, another condition needs to be kept in mind, that is $\alpha_i, \alpha_j \geq 0$ and $\alpha_i + \alpha_j = p$. Therefore, if $\alpha_i^* < 0$ then we update $\alpha_i = 0$, or if $\alpha_i^* > p$, then we update $\alpha_i = p$ and α_j is set equal to $p - \alpha_i$. The iterations are continued till we are unable to minimize objective value any further. Thereafter, we have found out the direction of steepest-descent at point x_k.

A step length, t_k is said to satisfy the Armijo condition for any descent direction $d_k \in \partial F(x)$ iff: $F(x_k + t_k d_k) \leq F(x_k) + c_1 t_k d_k^T s_k$ where c_1 is a pre-defined constant. Inexact linesearch is used to calculate a step length satisfying the Armijo condition, using which x_k is updated as $x_{k+1} = x_k + t_k s_k$, where s_k is the steepest-descent direction at x_k. The iterations are continued till $0 \in co_{i \in I(x)}\{\nabla f_i(x)\}$.

4   Numerical Results - Small Scale Problems

The algorithm was tested (using Matlab) on various small convex optimization problems ([12], [13]) and results compared with performance of algorithm as presented in [11].

4.1   Problem 1

$$f_1(x) = x_1^4 + x_2^2$$
$$f_2(x) = (2 - x_1)^2 + (2 - x_2)^2$$
$$f_3(x) = 2e^{-x_1 + x_2}$$

The convergence path can be traced in Fig.1 and numerical results are tabulated in Table 1.

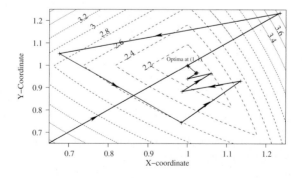

Fig. 1. Convergence path in Problem 1 for $x_{init} = [0, 0]$

Table 1. Problem 1

x_{init}	[0, 0]	[100, 100]
Minima	2.0	2.0
ArgMin	[1, 1]	[1, 1]
Function Evaluations	719	2465
Descent Iterations	29	67
Internal Iterations	33	47

4.2   Problem 2

$$f_1(x) = x_1^2 + x_2^4$$
$$f_2(x) = (2 - x_1)^2 + (2 - x_2)^2$$
$$f_3(x) = 2e^{-x_1 + x_2}$$

The results are present in Table 2.

Table 2. Problem 2

x_{init}	[0, 0]	[100, 100]
Minima	1.9522	1.9523
ArgMin	[1.1390, 0.8996]	[1.1390, 0.8996]
Function Evaluations	500	805
Descent Iterations	25	46
Internal Iterations	12	12

4.3   Problem 3

$$f_1(x) = x_1^2 + x_2^2 + 2x_3^2 + x_4^2 - 5x_1 - 5x_2 - 12x_3 + 7x_4$$
$$f_2(x) = f_1(x) - 10(-x_1^2 - x_2^2 - x_3^2 - x_4^2 - x_1 + x_2 - x_3 + x_4 + 8)$$
$$f_3(x) = f_1(x) - 10(-x_1^2 - 2x_2^2 - x_3^2 - 2x_4^2 + x_1 + x_4 + 10)$$
$$f_4(x) = f_1(x) - 10(-x_1^2 - x_2^2 - x_3^2 - 2x_1 + x_2 + x_4 + 5)$$

The results are presented in Table 3.

Table 3. Problem 3

x_{init}	[0, 0, 0, 0]	[100, 100, 100, 100]
Minima	−44	−44
ArgMin	[0, 0.999, 2, −1]	[0, 0.996, 2, −1]
Function Evaluations	375246	2081639
Descent Iterations	3782	19846
Internal Iterations	75441	403940

5   Large Minimax Problem

As we go onto increase m, the number of differentiable convex functions, the problem becomes computationally tougher. Large minimax problems have m of the order upto 10^6 and the mere task of computing the active set $I(x)$ is very heavy and the practical implementation of the serial algorithm described above is inefficient. Therefore, computing the function value at a given point x and the full subdifferential has traditionally been avoided for large minimax problems [11].

5.1   GPGPU and CUDA

Recent advancements in parallel computing have seen the usage of Graphics Processing Units for general purpose computing (GPGPU). The GPUs employ parallel multicore architecture which favours the inherently data parallel nature of graphics processing. With the launch of Nvidia's Compute Unified Device Architecture - Software Development Kit (CUDA) in 2007, using the GPU's computational powers for general purpose computing has become easy. With dynamic scheduling and fast creation and destruction of light-weight threads, CUDA is best suited for problems whereby same instruction set is to be executed on different data via multiple threads. The SIMT paradigm is suitable for tasks similar to evaluation of $F(x)$, the computationally dominant step in the serial algorithm. Now we present the parallel algorithm.

5.2   Parallel Algorithm

The parallel algorithm derives its efficiency by parallelizing three major sections of the serial code namely the function evaluation, active index set construction and steepest-descent direction search.

The evaluation of f_i is independent of f_j $(j \neq i)$ and therefore if each thread is passed the point x_k and the function f_i, it computes $f_i(x_k)$ efficiently. Parallel reduction technique is used to compute $F(x_k)$, thereby reducing the time taken in active index set construction. Furthermore, computation of the ∂F at a point x_k is similarly done in parallel using suitable finite difference method.

The number of active constraints are expected to be small and therefore the task of finding the direction of descent by choosing a pair of $\alpha_i's$ could be treated serially. However, using suitable heuristics, this task is also parallelized. Instead of choosing one pair of α_i and α_j, we run multiple threads to randomly choose many more mutually exclusive pairs and minimize with respect to these (keeping others constant). Note that if all the threads were to write their changed values to α, $\| \sum_{i \in I(x_k)} \alpha_i \nabla f_i(x_k) \|$ might actually increase. Therefore, a thread writes its modified α_i and α_j only if these changes decrease the quadratic objective function. Such atomic operations are supported in CUDA, similar to locks in other parallel programming paradigms.

5.3   CUDA Optimizations

In order to efficiently port the code on CUDA, significant code optimizations have been made:

1. The warp level optimization in function evaluation is achieved through the use of reverse Polish notation for evaluating functions step by step. The functions are sorted on the basis of their length and concurrency of arithmetic operators, so that same arithmetic operations can be performed parallely by maximum number of threads in a warp.
2. Parallel reduction is used for calculation of $F(x)$ once all $f_i(x)$ have been provided. This reduces the complexity of the operation from $O(m)$ to $O(\log m)$.
3. The gradient calculation of each active function is done in a separate thread block, with the threads in the block parallely calculating the gradient in each dimension.

6   Numerical Results - Large Scale Problems

To test the efficacy and performance of the proposed algorithm, we tested the algorithm against some of the popular algorithms for large scale minimax optimization, namely Real Coded Genetic Algorithms (RGA)[2] & Feasible Sequential Quadratic Programming (CFSQP).[3]

Experiments were run on a *AMD Athlon 64 X2 Dual Core processor 3800+*, with a *Tesla C1060* graphics card from Nvidia with 30 SM's, 240 cores and 4 GB of device memory. The convex functions were generated by taking a weighted sum of two objective functions where the weights were randomly generated in the range (0, 10). Objectives chosen were first and third convex functions specified in Prob.1.

6.1   Results

As the terminating criterion of the algorithms (our algorithm, RGA and CFSQP) differ, the number of iterations does not reflect the true performance of an algorithm. The number of function evaluations of $F(x)$ is also not a good measure because CFSQP does not necessarily evaluate all $f_i(x)$ for the calculation of $F(x)$. Hence, for the comparison, the wall-clock time taken by the algorithm to converge was chosen, using standard RGA code and CFSQP[4].

It can be seen in figure 2 that the time taken by RGA[5] is orders of magnitude higher than the time taken either by CFSQP or the proposed algorithm. A

---

[2] More descriptions of RGA can be found in [14].

[3] FSQP developed by University of Maryland, is a commercial tool for nonlinear optimization, esp. large scale convex minimax problems.

[4] RGA Code available at `www.iitk.ac.in/kangal/codes.shtml` while CFSQP at `www.aemdesign.com`. Faster evolutionary algorithms may also be tried, such as G3-PCX[15] and CMA-ES[16].

[5] As suggested in the literature, the parameters chosen are: Population Size = 30, Generations = 80, Mutation Probability = 0.5 Crossover probability=0.7 and a range of (-10, 10) for each variable.

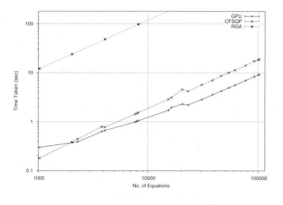

Fig. 2. Time comparison using 64 Threads

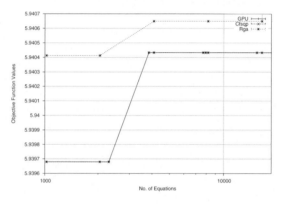

Fig. 3. Comparison of results (Objective function value)

noticeable feature of the graph is that there is a slight drop in the time taken by the gradient based methods when the number of functions increases from 20,480 to 23,040. The reason for the change is that the minima changes and both the methods take less iterations to arrive at the solution. However, RGA is unable to take advantage of the same owing to the deterministic number of generations.

The algorithms are also compared based on the results they produce, namely the minimum objective function value obtained and the accuracy of the minima. It can be seen in figure 3 that the minima found by the gradient based methods is significantly better than the minima obtained by RGA.

In figure 4, we can see the points to which the algorithms converged. CFSQP and the proposed algorithm converge to the same set of values for all function set sizes while RGA is insensitive to minute changes (notice the changing point of minima between 20,480 and 23,040 functions).

The scalabililty of the algorithm was studied using Fig.2. As expected the CFSQP and the RGA scale linearly (slope on the log-log graph is 1.0), while our

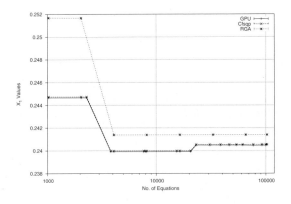

Fig. 4. x_1 at the point of minima

proposed algorithm is sublinear (slope is 0.746124), making it suitable for large scale problems.

7   Conclusion

In this study we have proposed an algorithm to solve large scale convex minimax problems. The algorithm has been developed on the Graphics Processing Unit using CUDA, achieving appreciable speed-ups on test problems. The algorithm scales sublinearly while increasing the number of functions from 1000 to 100000. Hence, it is highly suitable for large problems. In the future we would like to adapt some classical optimization algorithms for the SIMT architecture to solve large scale optimization problems.

References

[1] Xu, S.: Smoothing Method for Minimax Problems. Computational Optimization and Applications 20, 267–279 (2001)
[2] Parpas, P., Rustem, B.: An Algorithm for the Global Optimization of a Class of Continuous Minimax Problems. Journal of Optimization Theory and Applications 141, 461–473 (2009)
[3] Pillo, G.D., Grippo, L., Lucidi, S.: A Smooth Method for the Finite Minimax Problem. Math. Program. 60, 187–214 (1993)
[4] Rustem, B., Zakovic, S., Parpas, P.: An Interior Point Algorithm for Continuous Minimax. Journal of Optimization Theory and Applications 136, 87–103 (2008)
[5] Polak, E., Womersley, R.S., Yin, H.X.: An Algorithm Based on Active Sets and Smoothing for Discretized Semi-Infinite Minimax Problems. Journal of Optimization Theory and Applications 138, 311–328 (2008)
[6] Sun, W.: Non-Monotone Trust Region Method for Solving Optimization Problem. Applied Mathematics and Computation 156, 159–174 (2004)
[7] Charalamous, C., Conn, A.: An Efficient Method to Solve the Minimax Problem Directly. SIAM J. Numer. Anal. 15, 162–187 (1978)

[8] Chen, R.: Solution of Minimax Problem using Equivalent Differentiable Functions. Comp. and Maths. with Appls. 11, 1165–1169 (1985)
[9] Gaudioso, M., Monaco, M.F.: A Bundle Type Approach to the Unconstrained Minimization of Convex Non-Smooth Functions. Mathematical Programming 23, 216–226 (1982)
[10] Zhua, Z., Cai, X., Jian, J.: An Improved SQP Algorithm for Solving Minimax Problems. Applied Mathematical Letters 22, 464–469 (2009)
[11] Hiriart-Urruty, J.B., Lemarechal, C.: Convex Analysis and Minimization Algorithms I: Fundamentals. Springer, Heidelberg (1996)
[12] Vardi, A.: New Minimax Algorithm. Journal of Optimization Theory and Applications (1992)
[13] Wang, F., Zhang, K.: A Hybrid Algorithm for Nonlinear Minimax Problems. Annals of Operation Research (2008)
[14] Goldberg, D.E.: Genetic Algorithms for Search, Optimization, and Machine Learning. Addison-Wesley, Reading (1989)
[15] Deb, K., et al.: A Computationally Efficient Evolutionary Algorithm for Real-Parameter Optimization. Evol. Comput. 10(4), 371–395 (2002)
[16] Hansen, N., Muller, S., Koumoutsakos, P.: Reducing the Time Complexity of the Derandomized Evolution Strategy with Covariance Matrix Adaptation (CMA-ES). Evolutionary Computation 11(1), 1–18 (2003)

Towards Efficient and Effective Negative Selection Algorithm: A Convex Hull Representation Scheme

Mahshid Majd, Farzaneh Shoeleh, Ali Hamzeh, and Sattar Hashemi

Computer Science and Engineering Dept., Shiraz University, Mollasadra Avn., Shiraz, Iran
{majd,shoeleh,ali}@cse.shirazu.ac.ir, s_hashemi@shirazu.ac.ir

Abstract. Negative Selection Algorithm (NSA) is one of several algorithms inspired by the principles of natural immune system. The algorithm received the researchers attention due to its applicability in various research areas and a number of valuable efforts are made to increase the effectiveness and efficiency of it. The heart of NSA is to somehow find rules called detectors to discriminate self and anomaly areas. Each detector in NSA defines a subspace of problem space where no self data is located. One of the major issues in NSA is detector's shape or representation of detectors which can affect the detection performance significantly. This paper for the first time proposes a new representation for detectors based on convex hull. Since convex hull is a general form of other geometric shapes, it retains the benefits of other shapes meanwhile it provides some new features like the asymmetric shape. Experimental results show a significant enhancement in the accuracy of negative selection algorithm compared to other common representation shapes.

Keywords: Negative Selection Algorithm, Convex hull Representation, Detector Representation.

1 Introduction

The natural immune system is a self organized learning system which employs a multilevel defense against invaders. Negative Selection Algorithm (NSA) is one of the several algorithms inspired by the natural immune system. It has potential applications in various research areas such as anomaly detection where the problem is discriminating the incoming data as normal or anomaly data, given a collection of normal data as the training samples. NSA has been widely used for such problems because of its ability to model the problem space using just normal (self) samples.

NSA was firstly proposed by Forrest [1] and soon later D'haeseler et al. [2] presented an efficient implementation of it named greedy algorithm. Generally, during NSA, a group of detectors are generated by some random process to cover those parts of problem space where anomaly (non-self) samples might appear. Each detector should cover some portion of non-self space and any incoming data which is laid in this region would be classified as non-self data. In other words, each detector tries to model a subspace of non-self space and the ultimate rule set would model the whole non-self space. Consequently, NSA would be effective if its detectors cover the

K. Deb et al. (Eds.): SEAL 2010, LNCS 6457, pp. 45–54, 2010.

non-self space completely. Also the algorithm would be efficient if it covers the non-self space with minimum number of detectors. Therefore, detector representation (i.e. detector shape) has effective role in NSA.

A traditional representation of NSA detectors was binary representation due to its simplicity [1,2]. In [3] it is mentioned that although almost all representations can theoretically be translated into binary form, it imposes some limitations such as binary string size, accessible distance measures and comprehensibility. Hence, real valued representation has received great attention in the recent decades.

Subsequent works [4,5,6,7] used real valued representation to characterize the self/non-self space. Different shapes have been used as detector representation such as hyper rectangle [4], hyper sphere [5] and hyper ellipse [6]. Also in [7], a frame work for multi-shape detector representation was presented. In addition to these different shapes, the idea of having individual specific properties for detectors have been used either implicitly [4] or explicitly [8]. This approach is discussed extensively in [9] where the algorithm is named as *V-detector* algorithm.

This paper introduces a new representation of detector definition and matching mechanism for NSA which is based on convex hull shape. Since convex hull provides asymmetric shape, it can partition the problem space into complex regions and model complicated boundaries. Indeed, convex hull is general form of other geometric shapes such as hyper rectangle and hyper sphere. Therefore each detector would have its own shape and size.

The remainder of this paper is organized as follows. Section 2 describes the main principles of NSA and convex hulls. Section 3 introduces convex hull representation for NSA. Experiments are given in Section 4. The last section includes summary and conclusion.

2 Preliminaries

In this section the basic elements needed to describe the proposed method are presented. First, a detailed description of negative selection algorithm is provided and second, the definition of convex hull as well as the algorithms of creating a convex hull is briefly presented.

2.1 Negative Selection, Principle and Algorithm

One of the best known types of natural immune cells is T cell. Negative Selection process (NS) is applied during the production of T cells. The surface of these cells is covered by a number of receptors. Any object that could be matched with those receptors is known as malicious organisms. Therefore the receptors must be designed in a way that they would never detect self cell as a non-self one. NS is a process in which the cells that react against self cells are filtered out. This would happen during the generation process of T cells. A huge number of T cells are made through a pseudo random genetic process. Then they undergo the NS process which eliminates those that recognize self cells. The remaining ones would be distributed in body to recognize invaders. For more information, interested reader can refer to [10].

This maturation process is modeled in computer science as the Negative Selection Algorithm (NSA). The algorithm has two phases, training and detection phase. During the training phase, a group of detectors are produced by some generation mechanism, usually a pseudo random process. These detectors play the role of T cells and should cover the non-self region of problem space. The self region can be recognized using the self data instances given as the training set. As in the generation process of T cells, the detectors are checked through the training data and those detectors who match any self data would be eliminated. Next, in detection phase, these detectors are used to categorize the incoming data instance as self or non self. The NSA mechanism is summarized in following diagram.

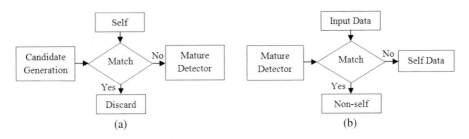

Fig. 1. NSA flow diagram (a) Training phase; (b) Detection phase

2.2 Convex Hull, Definition and Algorithms

Given a set P of limited number of points, n, convex hull of P, $CH(P)$, is defined as the largest convex polygon whose vertices are from P, or unique convex polygon that contains all points in P and whose vertices are in P. In the plane the convex hull takes a polygonal shape. The extreme points in the set are the vertices of the convex polygon.

There are many algorithms for finding the convex hull of a point set in the plane, 3-D and higher dimension space. For the convex hull problem in the plane, the Graham scan algorithm is the well known one [11]. The algorithm consists of 2 simple steps to construct a convex hull from $n>3$ points. First, an external point is chosen, for example a point with smallest y coordinate, and is labeled as p_0 and then the remaining n-1 points are sorted radially, using p_0 as origin. Now a simple polygon with n vertices is produced. Next, to change this polygon into convex hull, each point goes through the Three Penny Algorithm which determines whether the point along with its two adjacent are in clockwise order, counterclockwise order or on a line. Those points which are in counterclockwise order with respect to their adjacent, are known to be the convex hull vertices. For example, consider a sorted set of points $\{P_0$ $P_1...P_n\}$. First, P_i's two adjacent points in set P, are P_{i-1} and P_{i+1}. Note that P_0 is an adjacent of P_n. Next, it is determined whether P_{i-1} ; P_i ; P_{i+1} is in counterclockwise order. If so, the point P_i is supposed to be a vertex of the convex hull.

To find out whether three points are in counterclockwise order or vice versa, we can simply calculate the cotangent of their angle since it is a monotonically decreasing function. For three points (x_1,y_1), (x_2,y_2) and (x_3,y_3), this is formulated as bellow:

$$(x_2 - x_1)(y_3 - y_1) - (y_2 - y_1)(x_3 - x_1) : \begin{cases} = 0 & Collinear \\ > 0 & Counterclockwise \\ < 0 & Clockwise \end{cases} \qquad (1)$$

3   Proposed Method

As mentioned before, there had been some valuable efforts to achieve a proper representation for the detectors in terms of shape and size. It is shown that detectors with individual specific properties lead to more effective and efficient results [7,8,9]. Meanwhile, convex hull offers asymmetric shape which means that two convex hulls with different or even the same number of vertices may have different shapes and consequently different sizes. Therefore detectors with convex hull representation can form a wide variety of shapes and sizes compared to spherical and even the multi-shaped representation. Also it is worth mentioning that all the other geometric shapes can be approximated through convex hulls with sufficient number of points. The recent property suggests convex hull as the shape of choice by offering more general representation than other geometric shapes.

Convex hull representation is not a complete stranger in knowledge representation field. In fact, it has been used previously in learning classifier systems literature [12] where representation showed to be a challenging issue. The reported experimental results of this representation showed a significant improvement at a 99.99% confidence level. This result and the studied characteristics of convex hulls motivated us to test this representation in NSA as well.

As previously indicated, NSA has two phases, training and detection phase. In what follows, we explain NSA's phases for convex hull representations.

3.1   Training Phase

A detector with convex hull representation can be defined as a set of points by their Cartesian coordinates when they define the convex hull's vertices $\{P_0, P_1,...,P_n\}$. Another choice is to represent the detector in polar coordinates originated in a random point which can be considered as center of the detector. Although the former case is a straightforward schema, the latter one provides more control on the covered area of convex hull. This control mechanism has been used in spherical representation to keep detectors away from the self space.

The proposed algorithm uses five steps to generate a detector with convex hull representation. Next, these steps are explained in details.

Step 1. Produce the center point of convex hull. A point is generated randomly and is checked to be out of the self space. Also as an optimization mechanism to keep number of generated detectors under control, the point is examined to find the number of already generated detectors it matches with. If it is more than a predefined ($\theta_{overlap}$) percentage of the whole number of generated detectors by now, the point is eliminated and a new one is produced. Otherwise, it is accepted as center point.

Step 2. Define the number of convex hull vertices, n. It can be done in two ways; first, a fixed number of points for all detectors and second, a variable number of points chosen randomly within a lower and upper bound for each detector. In the former case, the larger the number of points is fixed on, more complex regions is covered. However, generating such a complex detector would be harder. In the latter case, which we preferred, the lower bound should obviously set to three. The upper bound has a direct effect on the complexity of detector. Since a complex convex hull is less feasible to generate, smaller amount of such detectors would be generated. Because a convex hull detector is represented by its points, the number of convex hull's vertices is considered as the detector size.

Step 3. Generate n points to be the convex hull vertices. The points are generated in polar coordinate originated in the accepted center point. Again, to reduce possible self coverage by the detector, it is tried to choose the points out of the self space. Thus, the points are generated as the following. The angular space around the center point is divided into n parts. For each part, a random angle is generated, and then the radius corresponding to each angle is defined equal to distance between the center point and nearest self data point located close to the corresponding angle. As usual [6,7,8], if one of the radiuses is less than a predefined value (θ_{self}), the detector is ignored and a new center would be generated.

Step 4. Make convex hull from generated points. In this step, one of the convex hull algorithms as like as the one explained in the previous section, Graham scan algorithm, should be used to generate convex hull of the generated points. Note that the number of vertices in the resulted convex hull may be less than the number of points, n.

Step 5. Check the detector against training data. As usual, the detector which includes any self data instance is eliminated. The matching rule component is responsible to define whether or not a data instance is covered by a detector. Convex hull detectors match all the points located in their corresponding convex hull. Therefore, matching rule will be the algorithm which defines whether a point is in the convex hull or not. A simple algorithm for this purpose in plane is Three Penny Algorithm which has been explained in Section 2.2. Figure 2 summarizes the above steps in a flow diagram.

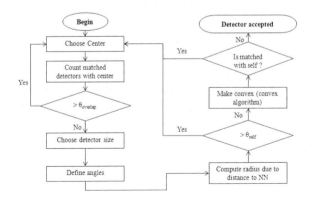

Fig. 2. Flow diagram of the proposed method

3.2 Detection Phase

In this phase (i.e. test phase), the incoming data would be examined whether it is matched with any detectors generated in the previous phase. As a rule, if the data is matched with even one detector, it is known to be non-self. The matching rule is the same as the one used in the training phase.

4 Experimental Results

This section presents the experimental results in three subsections. First, the tested datasets are described and second, the experimental set up for control parameters of proposed method is given in detail. The next subsection presents the achieved results of proposed method in comparison with the well known V-detector algorithm.

4.1 Experiments Design

To verify the basic behavior of the proposed algorithm, experiments were carried out using three Synthetic data which have been used in literature as benchmark problems [7,8,9]. These datasets are selected due to their various shapes, small size and few dimensions which would lead to better visual understanding of the algorithm behavior. The different shapes are designed as the self space which are normalized in the unit square $[0,1]^2$. These shapes are shown in Figure 3. The datasets contains certain known characteristics of common problem spaces such as sharp corners, curvilinear spaces and disjoint parts.

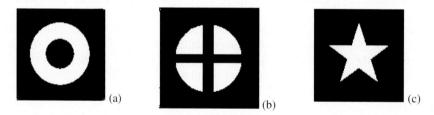

(a) (b) (c)

Fig. 3. The synthetic data, drawn from [9]; (a) Ring, (b) Intersection, (c) Pentagram

In Figure 3, the white area presents the self space and the remaining is considered as non-self space. The training data is composed of 1000 random points sampled from self space and the test data consists of 1000 random points from both non-self and self space. Each instance of the training data has two real valued attribute according to the position of corresponding sample pixel. In addition to these two attributes, test data contains a class label for each instance which determines self/non-self status of the sample.

4.2 Experimental Setting

To obtain performance of the proposed algorithm, the average and standard deviation of the accuracy over 50 independent runs of the algorithm is calculated. The algorithm has four control parameters; self radius, θ_{self}, maximum number of allowed overlapped detectors, $\theta_{overlap}$, the number of convex hull vertices which is called the detector size, $|d|$, and the number of detectors $|S|$.

To compare performance of the well known *V-detector* algorithm and the proposed algorithm, the parameters are configured as follows; θ_{self} is set to 0.05 as it is common in the literature, [7,8,9]. $\theta_{overlap}$ is set to 1/15 experimentally, $|d|$ is randomly chosen among 4,5 and 6 and the number of detectors , $|S|$, is set to 50 the same as in *V-detector* to have a fair comparison. Also, the one-tailed pairwise t-test is applied to check whether the difference between results of two methods over the datasets is statistically significant or not. The null hypothesis of this test is that the proposed method and *V-detector* algorithm perform the same on average over 50 independent runs against the alternative that the average performances are not equal.

4.3 Results

Table 1 highlights the difference between the proposed method and *V-detector* algorithm. *V-detector* algorithm is implemented as in [9] and is justified by results mentioned there.

Table 1. The results of applying the proposed method and *V-detector* algorithm in selected datasets

Data Set	Proposed method		V-detector algorithm		p-value
	TP	FP	TP	FP	
Pentagram	99.6±0.2	1.7	96.4 ± 1.7	0.5	2.8e-021
Intersection	98.6±0.4	3.7	90.6±2.7	4.3	1.4e-035
Ring	98.7±0.5	5	93.2±3.2	6.3	1.1e-018

The table contains the average of true positive rate (TP), false positive rate (FP) and their standard deviation over 50 independent runs. Also, the obtained p-value of pairwaise t-test with 5% significance level is presented. The results show that with the same number of detectors, the proposed method can achieve better TP and well FP rate. The results of pairwise t-test verified this performance improvement over 95% confidence level. In addition, the variances of obtained rates are less. Therefore it can be concluded that a more stable and higher performance is acquired by the proposed method.

It was stated earlier that the results presented in Table 1 is obtained by a fixed number of detectors, 50. Nevertheless, using proposed method, acceptable results can be achieved with a very small number of detectors. This is shown in Figure 4 where, results of the proposed method and V-detector algorithm with different values for $|S|$ on pentagram, intersection, and ring benchmarks are illustrated. It is not surprising that increasing the number of detectors results in a higher detection performance. Comparison of the two methods shows that the proposed method results in slightly few detectors to reach 90 percentage of true positive rate.

The visual results of proposed method is presented in Figure 5 where the training data is illustrated as black dots and the gray area shows the detectors coverage.

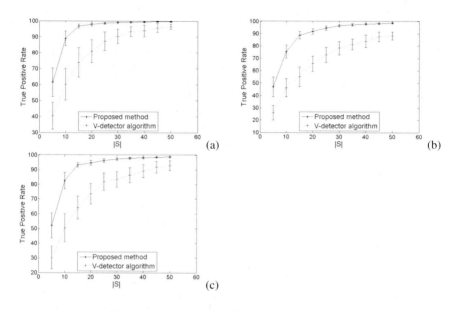

Fig. 4. Comparison of proposed method and V-detector algorithm in different values of |S| on, (a) Pentagram, (b) Intersection, (c) Ring

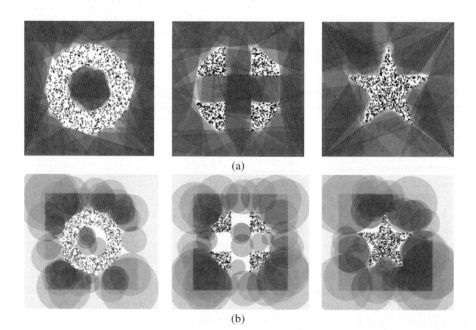

Fig. 5. Visual results by (a) the proposed method (b) *v-detector* algorithm. The training data is illustrated as black dots and the dark gray area shows the detectors coverage.

As shown in the Figure 5 part b, it seems that the spherical shape can effectively cover the curvilinear boundaries but it can estimate lines and corners only if a large number of them are used. In contrast, the convex hull representation has better ability to match with straight lines and corners. Also, the symmetric shape of convex hull brings it to match the curvilinear boundaries as well.

5 Conclusion

In this work a new extension of NSA with convex hull based representation is presented. The proposed method is designed to enhance the performance of NSA by partitioning the problem space into convex regions with maximum coverage and minimum overlap between detectors. Convex hull representation offers our model the ability to cover complex area with more flexibility. To explore the advantages of our method over the well known NSA, named *V-detector* algorithm, in terms of true positive rate, standard deviations and number of detectors, a number of experiments on known benchmark are conducted. As comparison results show, convex hull based representation outperforms the variable size spherical representation due to asymmetric shape and higher flexibility of convex hulls in covering complex areas. The flexibility of convex hull representation brings the benefit to the proposed method in the sense that it provides less number of detectors with more efficient detection of new sample, and also, promising true positive rate.

The detector generation mechanism used in this algorithm is a pseudo random process which considers the uncovered area of problem space. Hence, it seems that a good topic as a future work can be designing a detector generation mechanism which can distinguish the best detectors in terms of its covered area. Also, the influence of higher dimensional data sets needs further study including more experiments and analysis.

Acknowledgments. This work was supported by the Iran Tele communication Research Center.

References

1. Forrest, S., Perelson, A.S., Allen, L., Cherukuri, R.: Self-Nonself Discrimination in a Computer. In: The 1994 IEEE Symposium on Security and Privacy, pp. 202–212. IEEE Computer Society, Washington (1994)
2. D'haeseleer, P., Forrest, S., Helman, P.: Immunological Approach to Change Detection: Algorithms, Analysis and Implications. In: Proceedings of the 1996 IEEE Symposium on Security and Privacy, SP 1996. IEEE Computer Society, Washington (1996)
3. Ji, Z., Dasgupta, D.: Revisiting Negative Selection Algorithms. Evolutionary Computation 15(2), 223–251 (2007)
4. Dasgupta, D., Gonzalez, F.: An Immunity-based Technique to Characterize Intrusion In Computer Networks. IEEE Transactions on Evolutionary Computation 6(3), 1081–1088 (2002)
5. Gonzalez, F., Dasgupta, D.: Anomaly Detection Using Real-valued Negative Selection. Proceedings of Genetic Programming and Evolvable Machines 4, 383–403 (2003)

6. Shapiro, J.M., Lamont, G.B., Peterson, G.L.: An Evolutionary Algorithm to Generate Hyper-Ellipse Detectors for Negative Selection. In: Proceedings of Genetic and Evolutionary Computation Conference (GECCO), Washington, D.C, June 25-29 (2005)

7. Balachandran, S., Dasgupta, D., Nino, F., Garrett, D.: A framework for evolving multi-shaped detectors in negative selection. In: Proceedings of IEEE Symposium Series on Computational Intelligence, Honolulu, pp. 401–408. IEEE Press, Los Alamitos (2007)

8. Ji, Z., Dasgupta, D.: Real-valued negative selection algorithm with variable-sized detectors. In: Deb, K., et al. (eds.) GECCO 2004. LNCS, vol. 3102, pp. 287–298. Springer, Heidelberg (2004)

9. Ji, Z., Dasgupta, D.: V-detector: An efficient negative selection algorithm with "probably adequate" detector coverage. Information Science 179(10), 1390–1406 (2009)

10. Jerne, N.K.: Towards a Network Theory of The Immune System. Annales d'immunologie 125c(1), 373–389 (1974)

11. Graham, R.L.: An efficient algorithm for determining the convex hull of a finite planar set. Information Processing Letters 1, 132–133 (1972)

12. Lanzi, P.L., Wilson, S.W.: Using Convex Hulls to Represent Classifier Conditions. In: Proceedings of GECCO (2006)

To Handle Real Valued Input in XCS: Using Fuzzy Hyper-trapezoidal Membership in Classifier Condition

Farzaneh Shoeleh, Ali Hamzeh, and Sattar Hashemi

Department of Computer Science and Engineering and Information Tehnology
School of Electrical Computer, Shiraz University, Mollasadra Ave., Shiraz, Iran
{shoeleh,ali}@cse.shirazu.ac.ir, s_hashemi@shirazu.ac.ir

Abstract. Learning classifier systems (LCSs) are evolutionary learning mechanisms that combine genetic algorithms (GAs) with the power of the reinforcement learning paradigm. XCS, eXtended Classifier System, is currently considered as state of the art learning classifier systems due to its effectiveness in data analysis and its success in applying to varieties of learning problems. Generalization and the size of evolved population in XCS is one of the most challenging issues in XCS. This paper suggests that rule representation in XCS is not matured to the point where the condition parts of classifiers are covering the problem space in an effective manner with minimum redundancy. The key idea in the proposed system, named XCS-HT, is to use a novel representation scheme based on presenting a subspace of problem space with two kinds of regions named certain and vague regions. Using such mechanism significantly boils the number of evolved XCS-HT's classifiers down, while changing the performance only marginally. The experimental results show that XCS-HT has a better ability to solve problems having indistinguishable class boundaries comparing to XCSR, a common extension of XCS standing for handling real valued data.

Keywords: Learning Classifier System, XCS, Representation, Fuzzy Membership Function.

1 Introduction

The accuracy based learning classifier system (XCS) [1] is a Learning Classifier System (LCS) that was introduced by S. W. Wilson. The Learning Classifier system term is used generally for a system which tries to propose a classification for a set of environmental states. The aim of learning classifier systems is to evolve a population of classifiers using an internal evolutionary algorithm. This population is expected to enable the system to interact with the environment to achieve maximum available payoff. In Lanzi's view [2], the effectiveness of XCS as a machine learning technique to solve a learning problems is that *"XCS was the first classifier system to be both general enough to allow applications to several domains and simple enough to allow duplication of the presented results"*.

A rule based system like XCS is able to solve a problem efficiently if its rule set can cover the whole problem space properly and also each rule makes an effective

K. Deb et al. (Eds.): SEAL 2010, LNCS 6457, pp. 55–64, 2010.

decision. Therefore, it can be concluded that rule representation, i.e. knowledge representation, has an essential role in rule based systems. In addition, one of the most important goals in XCSs is to achieve a compact and an accurate solution for a given problem. These features depend on the representation method used to cover the problem space. For example, to handle real valued problems, many representation methods have been proposed in XCS realm, such as; interval [3-6], ellipsoidal [7], and convex hull [8]. It can be concluded that many of proposed representations are designed to describe the shape of subspace in the problem space. This shape certainly identifies which environmental states belong and which do not. Our main contribution in this paper is to propose a new representation scheme for XCS. In this scheme, each classifier presents two kinds of regions; 1) Certain region, identifying which environmental states can be matched by the corresponding classifier certainly, 2) Vague region, identifying which environmental states might be matched by the corresponding classifier.

The rest of this paper is organized as follows; Section 2 summarizes the important works that have been done to enhance the knowledge representation component of XCS. Our proposed method is described in Section 3. In Section 4, the test problems are represented and the experimental results are demonstrated and discussed.

2   Related Work

For the sake of continuous-valued input, the traditional "Ternary" representation including {0,1,#} alphabets has been substituted to the interval-based representation. For each dimension, there is an interval in condition part of classifiers. The disjunction of intervals in condition part of each classifier identifies the subspace of problem space which can be covered with this classifier. According to definition of these intervals four different representation techniques have been introduced; Center Spread Representation (CSR) [3], Lower-Upper Bound Representation or Min-Max Representation (MMR) [4], Unordered-Bound Representation (UBR) [5], and Min-Percentage Representation (MPR) [6]. Besides, in order to handle real valued inputs, several invaluable efforts have been done to define a geometric shape in each classifier to cover the problem space. Some of these efforts are done in [7] and [8] where each classifier presents an ellipsoidal and a convex hull respectively. Other representations with general purpose have also been proposed to use in XCS to make it more effective; such as, fuzzy [9-13], GP-like conditions [14], and neural network [15].

There are a number of approaches to use fuzzy logic as a technique for representing rules in Michigan-style LCS [9-13]. The main goal behind such efforts is combining the generalization capabilities of XCS with the fine interpretability of fuzzy rules to achieve an online learning system with more accurate, general and well understandable rule set. A general framework of LCS using fuzzy logic was introduced by Bonarini et al. in [10, 11]. Recently, a system named Fuzzy-XCS [12] was proposed where the rules are expressed in fuzzy format. Soon later, this approach is expanded to be useable in UCS [13].

3   XCS-HT

The aim of this paper is to propose a mechanism for handling real valued input and also covering the problem space with classifiers containing certain and vague regions. Here, an incoming state x will be certainly matched by a classifier cl if and only if it falls in the certain region presented by cl and x might be matched by cl if and only if it locates in the vague region of cl. In former case, it can be said that cl matches x with 100% matching degree and in latter case x might be matched by cl with less than 100% matching degree. So, in XCS-HT, each classifier has a parameter to identify its matching degree when faces the current incoming state. In following, we describe the modification of XCS's components in XCS-HT in detail. The other components of XCS-HT are the same as those in XCS.

Representation. Each classifier covers a subspace of the problem space and makes a decision in this subspace regarding to its experiences. In XCS-HT, the condition part of each classifier presents two hyper rectangles. One of these hyper rectangles embraces the other one completely. The inner hyper rectangle is indeed the certain region and the space between these two hyper rectangles is a vague region where instances can be matched by less than 100% matching degree. In other words, in each classifier, we use an asymmetric hyper trapezoidal membership function defined as Formula 1. This function identifies the matching degree of corresponding classifier while it is checked with current incoming instances in forming match set, [M], process. Therefore, the condition part of each classifier in XCS-HT consists of four genes for each dimension to present such trapezoidal membership function, that is, $cl.C=\{a_1,b_1,c_1,d_1,a_2,b_2,c_2,d_2, ..., a_n,b_n,c_n,d_n\}$ where n identifies the dimensions of given problem and $\{a_i,b_i,c_i,d_i\}$ are the parameters of the defined trapezoidal membership function in i'th dimension.

$$T(x,a,b,c,d) = \begin{cases} 0 & x \le a \\ \dfrac{x-a}{b-a} & a < x < b \\ 1 & b \le x \le c \\ \dfrac{d-x}{d-c} & c < x < d \\ 0 & d \le x \end{cases} \tag{1}$$

Matching. To match a classifier condition against the incoming instance the trapezoidal membership functions defined in condition for each dimension are checked. If the outputs of these functions are greater than zero, this classifier can match the current input. As mentioned before, each classifier has a μ identifying its matching degree when faces the current incoming instance. This parameter is calculated by multiplying the output of defined trapezoidal membership function for each dimension in the condition part of classifier, that is, $cl.\mu = \prod_{i=1}^{n} T(cl.C_{a_i}, cl.C_{b_i}, cl.C_{c_i}, cl.C_{d_i})$.

Covering. In XCS-HT, the matching degree parameter of each classifier, i.e. $cl.\mu$, has an essential. In XCS-HT, two mechanisms can be considered to use such parameter in forming [M] process; 1) the probability of a classifier being a part of [M] is equal to its matching degree parameter. In this case, a random number between (0,1] is generated. If this random number is less than the classifier's matching degree parameter, the corresponding classifier will be a member of [M]. This mechanism is named *Ordinary mechanism*. 2) Another approach called *RW mechanism* is using roulette wheel selection mechanism to select classifiers for forming [M]. So, [M] consists of classifiers that its matching degree is equal to one and the classifiers who win in roulette wheel selection. In this mechanism, a predefined threshold is used to identify the minimum sum of candidate classifiers' matching degree.

If no classifier was found to match the current instance, the system would generate a sequence of random numbers as $a_1, b_1, c_1, d_1,\ a_2, b_2, c_2, d_2,\ ...\ ,\ a_n, b_n, c_n, d_n$ in condition part of a new classifier as if it could match the current instance certainly.

Subsumption. Two subsumption procedures named *GA subsumption* and *Action Set Subsumption* were introduced in XCS to improve its generalization capability. In each subsumption procedure, we must determine whether the condition part of a classifier $cl_1.C$ can cover the other one $cl_2.C$ or not. It must be mentioned that the action part of both classifiers cl_1 and cl_2 must be the same. In the case of interval based representation, a classifier cl_1 subsumes cl_2 if the interval presented in each dimension of cl_1 covers the interval defined in the corresponding dimension of cl_2. In other words, cl_1 can subsume cl_2 if $cl_1.C_{l_i} \leq cl_2.C_{l_i}$ and $cl_1.C_{u_i} \geq cl_2.C_{u_i}$ for all dimensions. But in XCS-HT we use another mechanism to determine the covering classifiers in both subsumption procedures. Here, to identify whether cl_1 can cover cl_2 or not, for each dimension the overlapping area of the trapezoidal membership functions defined in the condition parts of two classifiers is calculated. If in all problem dimensions the overlapping area ($S_{overlap}$) is greater than a specific percentage of total area which is covered by more specific classifier, i.e. cl_2, ($S_{overlap} > \theta_{overlap}*S_{cl2}$), we can say that cl_2 can be covered by the more general classifier, cl_1. To illustrate such mechanism, consider an example which is shown in Figure 2. According to our approach, cl_1 can subsume cl_2 in subsumption procedures whereas cl_1 cannot subsume cl_3.

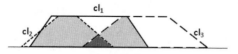

Fig. 1. An example of our subsumption mechanism; let consider $\theta_{overlap}$ =0.9, cl_1 can subsume cl_2 ($S_{overlap} > 0.9\ *S_{cl2}$) whereas it cannot subsume cl_3 ($S_{overlap} < 0.9*S_{cl3}$)

Algorithm 1 describes our *IS MORE GENERAL (cl_{gen}, cl_{spec})* function which is used in both *GA subsumption* and *Action Set Subsumption* procedures to identify which classifier is more general than others.

Algorithm 1. The pseudo code of our *IS MORE GENERAL (cl_{gen}, cl_{spec})* function

```
IS MORE GENERAL (cl  , cl    ):
                   gen    spec
  i=0
  do{
    S       =OverlapArea(Trap(cl  [i]),Trap(cl    [i]))
     overlap                    gen           spec
    S    =(|cl   [i].d-cl   [i].a|+|cl   [i].c-cl   [i].b|)/2
     spec     gen        gen          gen        gen
    If(S       <S    *θ       )
        overlap  spec  overlap
      Return false
    i++
  }while(i<length of Condition part of cl    )
                                         spec
  Return true

OverlapArea(Trap ,Trap ):
                1     2
  max_a=max(Trap .a, Trap .a)
                1        2
  min_d=min(Trap .d, Trap .d)
                1        2
  if(min_d<=max_a)
    retrun 0
  Δx=constant less than 0.1
  i=0
  S       =0
   overlap
  do{
    p1= max_a+Δx*i
    p1_value= min(T(p1,Trap .a,Trap .b,Trap .c,Trap .d)
                           1       1       1       1
                 ,T(p1,Trap .a,Trap .b,Trap .c,Trap .d))
                          2       2       2       2
    p2=max_a+Δx *(i+1)
    p2_value= min(T(p2,Trap .a,Trap .b,Trap .c,Trap .d)
                           1       1       1       1
                 ,T(p2,Trap .a,Trap .b,Trap .c,Trap .d))
                          2       2       2       2
    S       =S       +|p1_value + p2_value|/2
     overlap  overlap
  } while(p2<min_d)
  retrun S
          overlap
```

4   Experimental Results and Discussion

This section precisely covers our observation and analysis. At first, the experimental setup and used data sets are introduced. Then, the obtained results are presented and analyzed. Finally, we introduce a measurement to approve our hypothesis that XCS-HT can cover the problem space in an effective manner with minimum redundancy.

4.1   Experimental Set Up

To verify the basic behavior of our proposed representation in XCS-HT, the experiments were carried out using four synthetic data sets named: "Zigzag Boundary", "Vague Boundary", "Tao" [16] and "Pentagram" [17]. As shown in Figure 2, we use these data sets due to their various class boundary shapes and few dimensions which would lead to better visual understanding of the resulting classifiers. In Figure 2, the white area presents the space of the positive class and the black area is the negative class space. The instances of these data sets are obtained by sampling from their figures. Each instance has two real valued attributes regarding the position of

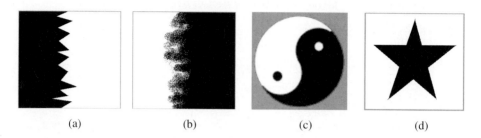

Fig. 2. Different shapes used as a synthetic data set, (a) Zigzag Boundary, (b)Vague Boundary, (c) Tao, (d) Pentagram

corresponding sample pixel and the associated class label which is the color (black or white) of the sample pixel. All sampled instances are normalized in the range of [0,1].

We compare the XCS-HT's behavior with XCSR's one. We selected XCSR [] due to its commonness in real valued problems. The XCS-HT and XCSR parameters are set as follows: $N=1000$, $\eta = 0.2$, $\beta = 0.2$, $\alpha = 0.1$, $v = 5$, $\chi = 0.8$, $\mu = 0.5$, $\tau=0.4$, $\varepsilon_0 = 10$, $\theta_{del} = 20$, $\theta_{GA} = 12$, $\delta = 0.1$, $\theta_{sub} = 20$, $m_0 =0.1$ and $r_0 = 0.04$ as mentioned in [3]. Each experiment consists of two phases with limited number of trials; exploration and exploitation phases. In exploration trials, the winner action is chosen at random but in exploitation trials the winner action is chosen with respect its prediction. Here, for the synthetic data set we use 50000 trials in exploration phase and 50000 in exploitation phase.

4.2 Results and Discussion

Table 1 shows the obtained results of XCS-HT using *Ordinary mechanism* to form [M] whereas Table 2 presents the results obtained using *RW mechanism*. Note that, these tables contain the average performance and the number of evolved macro classifiers over 50 independent runs.

With respect to Table 1 and 2, the performance of XCS-HT with new representation scheme containing both certain and vague approaches that of XCSR. But the size of evolved rule set, i.e. the number of macro classifiers, is so less than ones in XCSR. So, it can be concluded that XCS-HT has a more generality capacity. Note that, in general when we use higher level of $\theta_{overlap}$, the obtained performance will be better while the size of evolved rule set is increased.

It is worth to note that the obtained XCS-HT results in "Zigzag Boundary" and "Vague Boundary" data sets justify our hypothesis that the proposed representation has a better ability to solve problems having indistinguishable class boundaries. The proposed representation is good in these problems because it produces classifiers which present two types of regions, i.e. certain and vague regions, simultaneously.

Also, comparing the results presented in Table 1 with the results in Table 2 indicates that *RW mechanism* is better than the other one, i.e. *Ordinary mechanism* to form [M]. *RW mechanism* causes higher selection pressure than *Ordinary mechanism* in forming [M] process. It means that the classifiers with lower matching degree have higher chance to be a member of [M] so the covering procedure is called less

Table 1. Comparison XCSR using MMR representation against XCS-HT using *Ordinary mechanism* to form match set, [M]. The results are achieved at different level of $\theta_{overlap}$. The performance (Perf.) and the number of evolved macro classifiers (#Rules) are averaged over 50 independent runs.

Data set		XCSR	XCS-HT $\theta_{overlap}=1.0$	XCS-HT $\theta_{overlap}=0.95$	XCS-HT $\theta_{overlap}=0.9$	XCS-HT $\theta_{overlap}=0.85$	XCS-HT $\theta_{overlap}=0.8$
Zigzag Boundary	Perf.	95.95 ($\pm$1.33)	**96.27** ($\pm$0.63)	**96.46** ($\pm$0.67)	**96.39** ($\pm$0.66)	**96.33** ($\pm$0.81)	95.92 ($\pm$0.78)
	#Rules	231.62 ($\pm$40.1)	247.74 ($\pm$34.9)	257.76 ($\pm$54.4)	**191.28** ($\pm$39.5)	**169.56** ($\pm$20.5)	**179.48** ($\pm$26.8)
Vague Boundary	Perf.	93.71 ($\pm$1.36)	**94.66** ($\pm$0.76)	**94.25** ($\pm$0.81)	**94.34** ($\pm$0.85)	**94.14** ($\pm$0.82)	93.99 ($\pm$0.88)
	#Rules	234.36 ($\pm$35.6)	303.34 ($\pm$46.9)	283.02 ($\pm$47.1)	**213.62** ($\pm$40.5)	**187.92** ($\pm$28.2)	**190.78** ($\pm$27.0)
Tao	Perf.	92.85 ($\pm$1.26)	**93.99** ($\pm$1.00)	**93.24** ($\pm$0.88)	91.9 ($\pm$1.62)	92.09 ($\pm$1.44)	91.54 ($\pm$1.49)
	#Rules	416.58 ($\pm$47.0)	673.5 ($\pm$43.5)	443.7 ($\pm$53.4)	**296.44** ($\pm$43.2)	**230.10** ($\pm$37.8)	**115.44** ($\pm$30.3)
Pentagram	Perf.	94.34 ($\pm$1.14)	93.44 ($\pm$0.84)	92.70 ($\pm$1.02)	91.73 ($\pm$1.21)	90.94 ($\pm$1.23)	90.60 ($\pm$1.60)
	#Rules	576.34 ($\pm$36.2)	689.72 ($\pm$34.5)	564.78 ($\pm$36.0)	**484.4** ($\pm$43.7)	**416.48** ($\pm$59.5)	**359** ($\pm$49.8)

Table 2. Comparison XCSR using MMR representation against XCS-HT using *RW mechanism* to form match set, [M]. The results are achieved at different level of $\theta_{overlap}$. The performance (Perf.) and the number of evolved macro classifiers (#Rules) are averaged over 50 independent runs.

Data set		XCSR	XCS-HT $\theta_{overlap}=1.0$	XCS-HT $\theta_{overlap}=0.95$	XCS-HT $\theta_{overlap}=0.9$	XCS-HT $\theta_{overlap}=0.85$	XCS-HT $\theta_{overlap}=0.8$
Zigzag Boundary	Perf.	95.95 ($\pm$1.33)	**96.56** ($\pm$0.72)	**96.46** ($\pm$0.67)	**96.48** ($\pm$0.68)	**96.47** ($\pm$0.61)	95.92 ($\pm$0.79)
	#Rules	231.62 ($\pm$40.1)	**214.60** ($\pm$39.5)	**204.50** ($\pm$36.4)	**167.56** ($\pm$29.3)	**145.8** ($\pm$24.1)	**153.62** ($\pm$23.0)
Vague Boundary	Perf.	93.71 ($\pm$1.36)	**94.40** ($\pm$0.78)	**94.46** ($\pm$0.76)	**94.58** ($\pm$0.60)	**94.45** ($\pm$0.75)	93.92 ($\pm$0.95)
	#Rules	234.36 ($\pm$35.6)	256.36 ($\pm$51.1)	259.10 ($\pm$50.6)	**193.24** ($\pm$32.5)	**162.74** ($\pm$28.3)	**154.28** ($\pm$24.6)
Tao	Perf.	92.85 ($\pm$1.26)	**93.95** ($\pm$1.02)	**93.22** ($\pm$0.84)	92.52 ($\pm$1.27)	92.15 ($\pm$1.18)	92.10 ($\pm$1.36)
	#Rules	416.58 ($\pm$47.0)	611.50 ($\pm$46.7)	**407.94** ($\pm$52.2)	**267.56** ($\pm$37.9)	**217.28** ($\pm$40.2)	**205.34** ($\pm$41.2)
Pentagram	Perf.	94.34 ($\pm$1.14)	93.91 ($\pm$1.04)	93.2 ($\pm$1.08)	92.89 ($\pm$0.96)	91.65 ($\pm$1.22)	90.88 ($\pm$1.20)
	#Rules	576.34 ($\pm$36.2)	653.78 ($\pm$30.4)	**512.28** ($\pm$33.6)	**424.86** ($\pm$39.1)	**379.74** ($\pm$44.1)	**312** ($\pm$43.9)

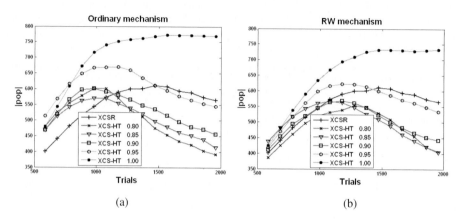

Fig. 3. Difference between applying *Ordinary mechanism* (a) or *RW mechanism* (b) to form match set, [M], in XCSR and XCS-HT with different values of $\theta_{overlap}$.

frequently in comparison to using *Ordinary mechanism*. Figure 3 shows the difference of applying these two mechanisms for forming [M] in "Tao" data set. Note that, the horizontal axis is the first 5000 exploitation trials and the vertical axis is the mean size of evolved population (|pop|). As shown in this figure, Using *RW mechanism* causes lower increasing of |pop| in the beginning and also after the subsumption is applied the size of evolved classifiers will be decreased more rapidly.

Minimum Redundancy. If the covered subspaces of classifiers in XCS have low overlap with each other, XCS can cover whole problem space with less number of classifiers. In other words, producing and evolving classifiers with minimum redundancy would solve the problem space in an effective manner. To investigate whether the classifiers of XCS-HT have the minimum redundancy property or not, a measure named $\overline{cl}_{CoveringArea}$ is defined. This measurement shows the average of covering space for each classifier in corresponding system. Figure 4 illustrates the obtained $\overline{cl}_{CoveringArea}$ measurement in XCS-HT and XCSR which is computed as Formula 2 and 3 respectively.

$$\overline{cl}_{CoveringArea} = \frac{\sum_{cl \in [P]} \prod_{i=1}^{n}(cl.u_i - cl.l_i)}{[P].size} \qquad (2)$$

$$\overline{cl}_{CoveringArea} = \frac{\sum_{cl \in [P]} \prod_{i=1}^{n} 0.5 * (cl.b_i - cl.a_i) + (cl.c_i - cl.b_i) + 0.5 * (cl.d_i - cl.c_i)}{[P].size} \qquad (3)$$

As mentioned before, the performance of XCS-HT can approach that of XCSR but it produces smaller evolved rule set. So, if XCS-HT has a lower $\overline{cl}_{CoveringArea}$ in comparison to XCSR, it can be claimed that the evolved classifiers in XCS-HT have a lower overlap with each other. As the results shown in Figure 4 approve our claim in most of data sets especially in ones having distinguished class boundaries, it is can be

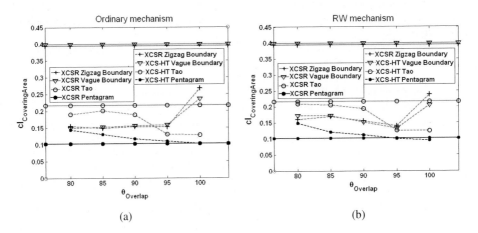

Fig. 4. Difference between applying *Ordinary mechanism* (a) or *RW mechanism* (b) to form match set, [M], in XCSR and XCS-HT with different values of $\theta_{overlap}$.

said that the proposed representation scheme can producing classifiers covering the problem space in an effective manner with minimum redundancy.

5   Conclusion

In this work, we propose an accuracy based classifier system named XCS-HT with new representation scheme, which is based on presenting two kinds of regions, i.e. certain and vague regions. In this scheme, in each dimension a trapezoidal membership function is defined. Some main components of XCS are modified in XCS-HT to be compatible with the proposed representation scheme. XCS-HT is applied on some real valued benchmark problems and it compared with the original XCS with interval based representation named XCSR. As experimental results showed, in comparison with XCSR, XCS-HT using the proposed representation scheme leads significant improvement in term of the number of evolved classifiers without noticeable loss of performance. To conclude, our experimental observations show that the obtained performances of XCS-HT on the problems with no distinguishable class boundaries (like Vague Boundary and Zigzag Boundary) are promising. As a feature work, we decide to improve this system to be able apply on multi step problems.

References

1. Wilson, S.W.: Classifier fitness based on accuracy. Evol. Comput. 3(2), 149–175 (1995)
2. Lanzi, P.L.: Learning classifier systems: then and now. Evol. Intel. 1, 63–82 (2008)
3. Wilson, S.W.: Get real! XCS with continuous-valued inputs. In: Lanzi, P.L., Stolzmann, W., Wilson, S.W. (eds.) IWLCS 1999. LNCS (LNAI), vol. 1813, pp. 209–222. Springer, Heidelberg (2000)

4. Wilson, S.W.: Mining oblique data with XCS. In: Lanzi, P.L., Stolzmann, W., Wilson, S.W. (eds.) IWLCS 2000. LNCS (LNAI), vol. 1996, pp. 158–176. Springer, Heidelberg (2001)

5. Stone, C., Bull, L.: For real! XCS with continuous-valued inputs. Evolutionary Computation 11(3), 298–336 (2003)

6. Dam, H., Abbass, H., Lokan, C.: Be real! XCS with continuous-valued inputs. In: Proceedings of the 2005 Workshops on Genetic and Evolutionary Computation, GECCO 2005, Washington, D.C, June 25-26, ACM, New York (2005)

7. Butz, M.V., Lanzi, P.L., Wilson, S.W.: Hyper-ellipsoidal conditions in XCS: rotation, linear approximation, and solution structure. In: Cattolico, M. (ed.) Proceedings of the 8th Annual Conference on Genetic and Evolutionary Computation, GECCO 2006, Seattle, Washington, July 8–12, pp. 1457–1464. ACM, New York (2006)

8. Lanzi, P.L., Wilson, S.W.: Using convex hulls to represent classifier conditions. In: Cattolico, M. (ed.) Proceedings of Genetic and Evolutionary Computation Conference, GECCO 2006, July 8–12, pp. 1481–1488. ACM Press, New York (2006)

9. Valenzuela-Rendoń, M.: The fuzzy classifier system: a classifier system for continuously varying variables. In: Booker, L.B., Belew, R.K. (eds.) Proceedings of the 4th International Conference on Genetic Algorithms (ICGA 1991), pp. 346–353. Morgan Kaufmann, San Mateo (1991)

10. Bonarini, A., Bonacina, C., Matteucci, M.: Fuzzy and crisp representations of real valued input for learning classifier systems. In: Lanzi, P.L., Stolzmann, W., Wilson, S.W. (eds.) IWLCS 1999. LNCS (LNAI), vol. 1813, pp. 107–124. Springer, Heidelberg (2000)

11. Bonarini, A.: An introduction to learning fuzzy classifier systems. In: Lanzi, P.L., Stolzmann, W., Wilson, S.W. (eds.) IWLCS 1999. LNCS (LNAI), vol. 1813, pp. 83–106. Springer, Heidelberg (2000)

12. Casillas, J., Carse, B., Bull, L.: Fuzzy-XCS: A Michigan genetic fuzzy system. IEEE Transaction Fuzzy Systems 15, 536–550 (2007)

13. Orriols-Puig, A., Casillas, J., Bernadó-Mansilla, E.: Fuzzy-UCS: preliminary results. In: Proceedings of the Genetic and Evolutionary Computation Conference (GECCO 2007), pp. 2871–2874 (2007)

14. Wilson, S.W.: Classifier conditions using gene expression programming. Technical Report 2008001, Illinois Genetic Algorithms Laboratory, University of Illinois at Urbana-Champaign (2008)

15. Bull, L., O'Hara, T.: Accuracy-based neuro and neuro-fuzzy classifier systems. In: Langdon, W.B., Cantu'-Paz, E., Mathias, K.E., Roy, R., Davis, D., Poli, R., Balakrishnan, K., Honavar, V., Rudolph, G., Wegener, J., Bull, L., Potter, M.A., Schultz, A.C., Miller, J.F., Burke, E.K., Jonoska, N. (eds.) Proceedings of the Genetic and Evolutionary Computation Conference, GECCO 2002, New York, pp. 905–911. Morgan Kaufmann, San Francisco (July 2002)

16. Llora, X., Guiu, J.M.G.: Inducing partially-defined instances with evolutionary algorithms. In: Brodley, C.E., Danyluk, A.P. (eds.) ICML, pp. 337–344. Morgan Kaufmann, San Francisco (2001)

17. Ji, Z., Dasgupta, D.: Real-valued negative selection algorithm with variable-sized detectors. In: Deb, K., et al. (eds.) GECCO 2004. LNCS, vol. 3102, pp. 287–298. Springer, Heidelberg (2004)

Development of Optimal Control System for Safe Distance of Platooning Using Model Predictive Control

Xin Zhao, Dongmei Wu, Yichun Yeh, and Harutoshi Ogai

Graduate School of Information, Production and Systems, Waseda University,
808-0135, Kitakyushu, Fukuoka, Japan
zhaoxin_19870121@yahoo.co.jp, ogai@waseda.jp

Abstract. Platooning technology is becoming a future task which suggests as a way of reducing carbon dioxide emissions and realizing safe driving at a high velocity. This paper presents a unique optimal control method of velocity and distance for platooning using model predictive control. The vehicle-platoon's distance model which is based on the road condition and weather condition is used in this rigorous approach of deriving the control input. A combination of Continuation and Generalized Minimum Residual Methods is used to optimize the sequence of vehicle control commands which is required in the prediction horizon aiming at minimizing the relative velocity and keeping safe distance of the vehicle-platoon while the vehicle-platoon is on a high velocity driving.

Keywords: Platooning, model predictive control, safe distance.

1 Introduction

Platooning is considered as one of the innovations in the automotive industry that aim to improve the safety, efficiency, mileage, and time of travel of vehicles while relieving traffic congestion, decreasing pollution and reducing stress for passengers [1]. Also, platooning makes it possible for vehicles to travel together closely and safely. This leads to a reduction in the amount of space used by a number of vehicles on a highway. Thus more vehicles can use the highway without traffic congestion [2].

Model predictive control is a potential control technique for non-linear systems that suits vehicle driving [3], [4]. This paper presents a unique approach for the control of vehicle-platoon's relative velocity and safe distance that measures the conditions of vehicles in the vehicle-platoon, determining the optimal control input based on minimization of relative velocity and difference between safe distances of the vehicle-platoon.

The optimum control model of the vehicle-platoon, used in this paper, expresses the dynamic relationship between leading vehicle and following vehicles at any instant. The optimization of control inputs in the prediction horizon is conducted using Continuation method combined with generalized minimum residual method known as C/GMRES method [5], [6].

The contribution of this paper is two-fold. Firstly, develops an optimum control system to minimizing the relative velocity and keeping safe distance of vehicle-platoon

K. Deb et al. (Eds.): SEAL 2010, LNCS 6457, pp. 65–74, 2010.

for platooning using model predictive control. Secondly, a safe distance adjustment model according to road condition and weather condition is designed.

2 Overview of Optimal Control System

The proposed optimal control system is composed of vehicle-platoon model and model predictive control block, which is shown in Fig. 1. Vehicle-platoon model describe dynamics of the leading vehicle and following vehicles. The optimizer calculates a group of control inputs over the whole horizon, and feed the optimal commands to the vehicle-platoon. This process is repeated when control vector is updated.

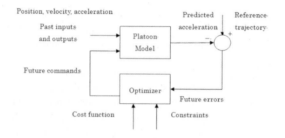

Fig. 1. Optimal controller for platooning

3 Model Predictive Control

In the model predictive control, dynamic models are usually described as [7]

$$\dot{x} = f(x_t, u_t) \tag{1}$$

Performance is evaluated by a cost function

$$J(u(\tau)) = \int_t^{t+T} L(x(\tau), u(\tau), \tau) d\tau \tag{2}$$

where, L is the function applied to evaluate automobile performance, τ is the virtual time, t is the time when control vector is updated, T is the horizon length, x is a state variable vector of the host vehicle and the surrounding vehicles, its initial value is determined by data from vehicle sensors, and u is an optimal control input vector. The problem is how to optimize control vector u to minimize the value of J. in this paper, GMRES method is applied [8], [9].

Requirements for solving this problem are defined as

$$\dot{x} = H_\lambda, x(t): given, \dot{\lambda} = H_x, \lambda(t+T) = 0 \tag{3}$$

$$H_u = 0 \tag{4}$$

H is a Hamiltonian defined as

$$H(x, u, \lambda, \tau) = L(x, u, \tau) + \lambda^T \cdot f(x, u) \tag{5}$$

where, H_λ, H_x and H_u are partial derivatives of H, and λ is a costate vector with the same dimension of x. Prediction horizon $[t, \ t + T]$ is divided into N sampling times, and control input vector is defined as

$$U(t) = \left(u(\tau_0), u(\tau_1), \cdots u(\tau_{N-1}) \right) \ \tau_i = t + \frac{i}{N} T, i = 0, 1, \cdots N - 1 \tag{6}$$

$x(\tau_i)$ and $\lambda(\tau_i)$ can be calculated by integral of equations (3).

A function of u and x is constructed as

$$F(U(t), x(t), t) = (H_u^0, H_u^1, \cdots, H_u^{N-1}) \tag{7}$$

$$H_u^i = \frac{\partial H(x(\tau_i), u(\tau_i), \lambda(\tau_i), \tau_i)}{\partial u} \tag{8}$$

The optimal condition can be derived from

$$F(U(t), x(t), t) = 0 \tag{9}$$

With differentiation of equation (10), equation of $\dot{U}$ is derived as,

$$F_U \dot{U}(t) + F_x \dot{x}(t) + F_i = 0 \tag{10}$$

Equation (11) is solved by GMRES method. Therefore optimal control vector $U(t)$ can be obtained. Optimal control vector after Δt is represented as $U(t + \Delta t)$, and calculated by

$$U(t + \Delta t) = U(t) + \dot{U}(t) \cdot \Delta t \tag{11}$$

4 Modeling and Control

4.1 Platooning Model

For the optimum control system for safe distance of platooning, the most important element is the formulation of the vehicle control problem. A simplified platooning model considering only the longitudinal motion of three vehicles is taken into account Fig. 2.

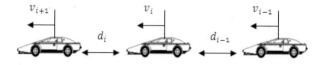

Fig. 2. Vehicle-platoon in a string

For each vehicle, we define, x_i is position, v_i is the velocity, u_i is acceleration, d_i is range signal measured, r_i is relative velocity measured by i^{th} vehicle, i shows the ith vehicle.

Here we assume that each vehicle can transmit its status to the others. In this modeling, the state equation of the platooning:

$$\dot{x} = f(x, u) = [\dot{x}_1, u_1, \dot{x}_2, u_2, \dot{x}_3, u_L]^T \qquad (12)$$

Where, $x = [x_1, \dot{x}_1, x_2, \dot{x}_2, x_3, \dot{x}_3]^T$ denotes state vector representing position and velocity of the vehicle-platoon including the leading vehicle. The control input of this equation u_L is the acceleration/deceleration of the leading vehicle, which is bounded by an inequality constraint as $-u_{max} \leq u_L \leq u_{max}$, to meet physical limits of actuators. The time dependent parameter u_1 and u_2 represent the model of following vehicles in term of these acceleration/deceleration by which movement of the following vehicles is anticipated in the prediction horizon.

For leading vehicle, the motion equation is defined as

$$\dot{x}_L = v_L, \dot{v}_L = u_L \qquad (13)$$

and the motion equations of following vehicles is defined as

$$\dot{x}_i = v_i, \dot{v}_i = k_1^i e_r^i + k_2^i e_{rv}^i + k_3^i e_v^i \qquad (14)$$

$$e_r^i = x_{i+1} - x_i - h_i v_i, e_{rv}^i = v_{i+1} - v_i, e_v^i = v_t^d - v_i \qquad (15)$$

where, h_i is the desired time headway, v_t^d is desired velocity of vehicle i.

4.2 Safe Distance Model

As we know, when the vehicle is driving on the road, the road condition and weather condition usually changes, so a fixed distance is not safe. We should consider a safe distance adjustment model according to different road condition and weather condition.

Road condition for platooning is about the frictional factor of different road types in Table 1.

Table 1. Road frictional factor

Road type	Dry road	Rainy road	Snow road	Freeze road
Frictional factor	0.6	0.4	0.28	0.18

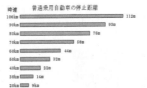

Fig. 3. Actual stopping distance

There is a difference between actual stopping distance and ideal stopping distance. Fig. 3 [10] shows the actual stopping distance according to different velocities.

Here we assume the human driving by velocity of 60 *km/h* on dry road, and the ideal stopping distance is calculated as 44 *m* by the equation as

$$d_{ideal} = v^2/2\mu g \qquad (16)$$

Where, μ is the frictional factor, $g = 9.8m/s^2$.

For the difference between ideal stopping distance and real stopping distance, we consider two parts, which are driving reaction time and correction factor.

$$d_{real} = d_{ideal} + vt_{rt}, \mu = \mu_{road}k \qquad (17)$$

Where, t_{rt} is driving reaction time which means the reaction delay time to carry out commands, k is the influence between road and shoes.

We use the data of human driving stopping distance by the velocity of 60 *km/h* and we can calculate $k = 0.865$. The reaction time of human is 1*s*, while the reaction time of platooning is 0.1*s* for signal transmission. So the safe distance of vehicle-platoon can be designed as

$$d = v^2/2k\mu g + 0.1v \qquad (18)$$

Here I consider a method for vehicle adjusting the distance of vehicle-platoon automatically. For leading vehicle, we assume that it loads a camera to discriminate the road every mixed time period. Fig 4 shows the dry road and wet road which comes from Kitakyushu, Japan (pixel 400×400).

Here we use the HSV method which is the two most common cylindrical-coordinate representations of points in an RGB color model, which rearrange the geometry of RGB in an attempt to be more perceptually relevant than the cartesian representation. For the HSV method, we can pick up the hue, saturation, and lightness of the picture and compare with other pictures, finding the most obvious part. After simulation, the biggest difference appears in the saturation which is shown in Fig 5.

Fig. 4. Road (right-dry road and left-wet road)

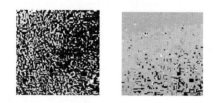

Fig. 5. Saturation (right-dry road and left-wet road)

We select the first value of saturation of the pictures and show compared results in Fig 6.

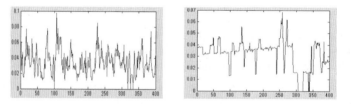

Fig. 6. Saturation value (right-dry road and left-wet road)

The average value of two pictures is almost the same, but the variance of dry road data is 0.000246 while variance of wet road data is 0.000145. We can see that colour comparison is stronger in dry road from Fig 5.

Here we assume ideal condition that variance of wet road data is 0 which means it is full of gray, and variance of dry road data is 0.00025, so we can give the frictional factor forecast shown in Table 2. So we can calculate the variance of the picture of road condition at present and estimate the road frictional factor, adjusting the distance of vehicle-platoon automatically.

Table 2. Frictional factor forecast

Variance	0.00025	$0.00025<x<0$	0
Frictional factor	0.6	$(0.6-0.4)x/0.00025=\mu-0.4$	0.4

4.3 Evaluation Function

The vehicle performance is evaluated from the viewpoint of safe and smooth driving, by the following criteria:

(1) Vehicle acceleration should be as small as possible, which is evaluated with an equation of longitudinal acceleration,

$$L_1 = \frac{1}{2}u_i^2 \tag{19}$$

(2) The relative velocity of the vehicle-platoon is maintained as small as possible,

$$L_2 = \frac{1}{2}(v_{i+1} - v_i)^2 \tag{20}$$

(3) Keep the safe distance of vehicle-platoon as long as possible,

$$L_3 = \frac{1}{2}(x_{i+1} - x_i - h_i v_i)^2 \tag{21}$$

(4) The desired velocity of the vehicle is maintained as long as possible,

$$L_4 = \frac{1}{2}(v_i^d - v_i)^2 \tag{22}$$

The basic form of function L is defined as

$$L = w_1 L_1 + w_2 L_2 + w_3 L_3 + w_4 L_4 \tag{23}$$

Where, L_1, L_2, L_3 and L_4 are evaluation indexes, w_1, w_2, w_3 and w_4 are their weightings.

5 Simulation Results

5.1 Simulation Result with Fixed Distance

In this part, the simulation conditions are shown in Fig. 7, the figure of left is for the first case, while right is for the second case. For the first case, weightings for evaluation indexes are set to $[w_1, w_2, w_3, w_4] = [0.1, 0.35, 0.35, 0.2]$ and $[k_1, k_2, k_3] = [0.2, 0.5, 0.3]$, headway time $h=1s$, distance is fixed as $10m$ and simulation time is $30s$.

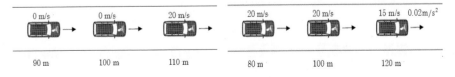

Fig. 7. Initial simulation conditions

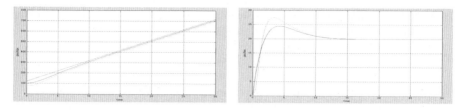

Fig. 8. Simulation results (left- position of platooning and right-velocity of platooning)

Simulation results are shown in Fig. 8. Following vehicles achieve the same velocity with the leading vehicle and keep the 10m distance after 15s, and then keep the same status all the time.

For the second case, weightings for evaluation indexes are set to $[w_1, w_2, w_3, w_4]$ = [0.1, 0.35, 0.35, 0.2] and $[k_1, k_2, k_3]$ = [0.2, 0.5, 0.3], headway time h=1s, distance is fixed as 10m and simulation time is 90s. In the second case, it is assumed that vehicle-platoon drive on the road of plane, uphill and downhill, shown as Table 2, and the angle of roll is set as 10/π.

Table 3. Road condition

Time	0-30 s	30-40 s	40-60 s	70-80 s	80-90 s
Road condition	plane	uphill	plane	downhill	plane

Simulation results are shown in Fig. 9. After 10s, three vehicles keep the same velocity and the distance achieves to 10m.

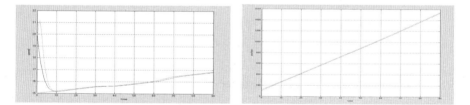

Fig. 9. Simulation results (left- position of platooning and right-velocity of platooning

5.2 Simulation Result with Safe Distance Model

In this part, the simulation conditions are shown in Fig. 10, the figure of left is for the first case, while right is for the second case.

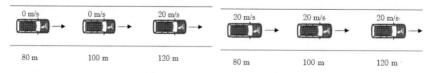

Fig. 10. Initial simulation conditions

For the first case, weightings for evaluation indexes are set to $[w_1, w_2, w_3, w_4] =$ [0.1, 0.35, 0.35, 0.2] and $[k_1, k_2, k_3] = [0.2, 0.5, 0.3]$, headway time $h=1s$, frictional factor $\mu = 0.6$ and simulation time is $30s$.

Simulation results are shown in Fig. 11.The safe distance of vehicle-platoon is changed according to velocity of vehicle-platoon. We can find that the safe distance of $5s$ is longer than the safe distance of $25s$.

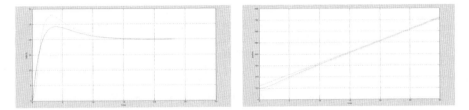

Fig. 11. Simulation results (left-position of platooning and right-velocity of platooning)

For the second case, weightings for evaluation indexes are set to $[w_1, w_2, w_3, w_4] =$ [0.1, 0.35, 0.35, 0.2] and $[k_1, k_2, k_3] = [0.2, 0.5, 0.3]$, headway time $h=1s$, frictional factor is changed from $\mu = 0.6$ to $\mu = 0.28$, the velocity of vehicle-platoon is fixed as 20m/s and simulation time is 30s.

Simulation results are shown in Fig. 12. The safe distance of vehicle-platoon is changed according to different road conditions. We can see that there is longer safe distance of snow road than dry road.

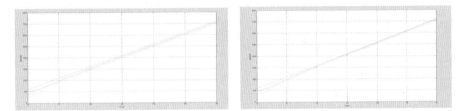

Fig. 12. Simulation results (left-dry road and right-snow road)

6 Conclusion

In this paper, an optimal driving control system calculating optimal driving mode based on model predictive control is presented. The information of road condition and statement of vehicle-platoon is formulated and transformed to a more intuitive and acceptable form, an optimal driving commands for vehicle-platoon. By the road condition information, the optimal control system for platooning adjusts the safe distance of vehicle-platoon.

In the future research, weightings such as w_1, w_2, w_3, w_4 would be optimized. Safe distance model would be improved for more precise prediction. Besides, the road condition automatic measurement by the leading vehicle using image discrimination would be considered. The optimal driving control system is expected to be applied in automatic driving systems in the future.

References

1. Using ITS to cut CO2 emissions in half by 2050, Development of Energy-saving ITS Technologies (2009)
2. Yamamura, Y., Seto, Y., Nagai, M.: Study on a String-stable ACC Using Vehicle-to-Vehicle Communication. Japan Society of Mechanical Engineers, No.06-7044
3. Kamal, M.A.S., Mukai, M., Murata, J., Kawabe, T.: Development of Ecological Driving System Using Model Predictive Control. In: ICROS-SICE International Joint Conference 2009, pp. 3549–3554 (2009)
4. Kamal, M.A.S., Raisuddin, K.M., Wahyudi, Muhida, R.: Comprehensive Driving Behavior Model for Intelligent Transportation Systems. In: Proc. of the Int. Conf. on Computer and Communication Engineering (ICCCE 2008), May13-15, pp. 1233–1236 (2008)
5. Kelley, C.T.: Iterative Methods for Linear and Nonlinear Equations. In: Frontiers in applied mathematics, vol. 16. SIAM, Philadelphia (1995)
6. Ohtsuka, T.: A Continuation/GMRES Method for Fast Computation of Nonlinear Receding Horizon Control. Automatica 40(4), 563–574 (2004)
7. Kawabe, T., Nishira, H., Ohtsuka, T.: Optimal Path Generator for Automobiles using Receding Horizon control. In: Proceeding of SICE 3rd Annual Conference of Control Systems, Kobe, May 28-30, pp. 405–408 (2003)
8. Xia, Y.: Automobile Optimal Driving Control based on Model Prediction Control, Master thesis, Waseda University (2009)
9. Ohtsuka, T.: A continuation/GMRES method for fast computation of nonlinear receding horizon control. Automatica, 563–574 (November 2003)
10. Society of Automotive Engineers of Japan: Automobile technology handbook 1, 136 (2004)

A Comparative Study on Theoretical and Empirical Evolution of Population Variance of Differential Evolution Variants

G. Jeyakumar and C. Shunmuga Velayutham

Amrita School of Engineering
Amrita Vishwa Vidyapeetham, Coimbatore, Tamil Nadu, India
g_jeyakumar@cb.amrita.edu, cs_velayutham@cb.amrita.edu

Abstract. In this paper we derive theoretical expressions to compute expected population variance for Differential Evolution (DE) variants – *DE/best/1/bin, DE/rand/2/bin* and *DE/best/2/*bin by directly extending Zaharie's work on *DE/rand/1/bin*. The study includes comparing the theoretical and empirical evolution of population variance of three DE variants. This work provides insight about the explorative power of the variants and explains their behavior.

Keywords: differential evolution, population variance, theoretical and empirical analysis, convergence rate, explorative power.

1 Introduction

Differential Evolution (DE), proposed by Storn and Price [1,2], is a very simple yet very powerful stochastic global optimizer for continuous search domain [3,4,5]. DE has some unique characteristics for it uses a differential mutation operation, coupled with recombination operation to generate a trial vector (offspring) followed by a one-to-one greedy selection scheme between the trial vector and the parent. Depending on the way the parent solutions are perturbed, there exist many DE variants.

In [6], we performed empirical comparative analyses of fourteen variants of DE on fourteen numerical benchmark optimization problems. In a preliminary effort to understand the explorative power of DE variants on theoretical grounds, in this paper, we have directly extended Zaharie's theoretical measure of population diversity of the DE variant *DE/rand/1/bin* [8] to three other closely related, commonly used, variants viz. *DE/best/1/bin, DE/rand/2/bin* and *DE/best/2/bin.*

This paper is organized as follows. Section 2 provides the expressions to compute the expected population variance of three above said DE variants. While Section 3 briefs the design of experiments to compare the theoretical and empirical evolution of population variance, Section 4 provides results and discussion. Finally, Section 5 concludes the paper.

K. Deb et al. (Eds.): SEAL 2010, LNCS 6457, pp. 75–79, 2010.
© Springer-Verlag Berlin Heidelberg 2010

2 Computation of Expected Population Variance

As the evolution of population variance is a measure of the explorative power of an evolutionary algorithm (EA) [7], Zaharie in [8], derived a theoretical relationship between the expected population variance after mutation and crossover and the initial population variance of *DE/rand/1/bin*. The main result of [8], is the following theorem.

Let X={x₁,...,xₘ} be the current population, Y={Y₁,...,Yₘ} the intermediate population obtained after applying the mutation and Z={Z₁,...,Zₘ} the population obtained by crossing over the population X and Y. If F is the parameter of the mutation step and Pc is the parameter of the crossover step then

$$E(Var(Z)) = \left(2F^2 P_c + 1 - \frac{2p_c}{m} + p_c^2\right) Var(X) \tag{1}$$

In this paper we have directly extended the Zaharie's work to the three closely related, commonly used, DE variants viz. *DE/best/1/bin, DE/rand/2/bin* and *DE/best/2/bin*. In [6], we observed that strategies relying on best solutions, in spite of fast convergence, get stuck at local optimum and those relying on random solution demonstrated stronger exploration capability. The choice of DE variants, in this paper, is to gain insight, on theoretical basis, the above observations.

Following [8], the extension of theoretical relationship for the chosen DE variants is presented in two steps viz. expected population variance after differential mutation i.e $E(Var(Y))$ and expected population variance after binomial crossover i.e, $E(Var(Z))$.

In case of the variant *DE/best/1/bin*, the expected population variance after mutation and crossover has been derived respectively as

$$E(Var(Y)) = 2F^2 Var(X) \tag{2}$$

$$E(Var(Z)) = \left(2F^2 p_c + \frac{(1-p_c)^2}{m} + \frac{m-1}{m}(1-p_c)\right) Var(X) + \left(\frac{m-1}{m}\right) p_c(1-p_c)(\bar{x} - x_{best})^2 \tag{3}$$

For *DE/best/2/bin*, the derived population variances after mutation and crossover are respectively

$$E(Var(Y)) = 4F^2 Var(X) \tag{4}$$

$$E(Var(Z)) = \left(4F^2 p_c + \frac{(1-p_c)^2}{m} + \frac{m-1}{m}(1-p_c)\right) Var(X) + \left(\frac{m-1}{m}\right) p_c(1-p_c)(\bar{x} - x_{best})^2 \tag{5}$$

The expected population variances of *DE/rand/2/bin* are as follows

$$E(Var(Y)) = 4F^2 Var(X) \tag{6}$$

$$E(Var(Z)) = \left(4F^2 p_c + \frac{(1-p_c)^2}{m} + \frac{m-1}{m}(1-p_c)\right) Var(X) + \left(\frac{m-1}{m}\right) p_c(1-p_c)(\bar{x} - x_{best})^2 \tag{7}$$

In the following section, experiments have been designed to compare $E\left(Var(X^{(g)})\right)$ and the empirical expected population variance$\langle Var(X^{(g)})\rangle$, where $X^{(g)}$ is the population obtained after g generations through successive mutation and crossovers.

3 Design of Experiment

Having derived the theoretical expression to compute the expected population variance, we set out to compare the theoretical and empirical evolution of population variance of chosen DE variants. For the purpose of comparison we have used Sphere model [9,10], with dimension 30, as the benchmark function.

The parameters of the DE variants were: population size NP = 50, crossover rate P_c = 0.5, scaling factor F = 0.1, 0.12, 0.2 and 0.3 and the maximum number of generations GMax = 30. With this experimental set up we compare the theoretical and empirical evolution of the population variance for the three DE variants.

For further analysis and comparison, we extend our experiment with a different parameter setup and by considering the following variants: *DE/rand/1/bin* (considered from Zaharie's work [8]) and the chosen three variants - *DE/best/1/bin, DE/rand/2/bin* and *DE/best/2/bin*. The new parameters were : NP=60, CR={0.5,0.2,0.2,0.2} (a bootstrap test yielded these values for the above said four variants), following [10,11], F ∈ [0.3, 0.9], GMax = 3000 (or if tolerance error 1 x 10^{-12} is obtained). As EA's are stochastic in nature, 100 independent runs per variant were performed and mean values are presented.

For the extended experiment we report convergence rate [10] and vanish point for all four variants. Convergence rate is used to detect which variant is most competitive. It is calculated as the mean percentage out of the total 1,80,000 function evaluations required by each of the variant to reach its best objective function value, for all the 100 independent runs. The vanish point shows the generation at which the population diversity vanished (ie. population variance = 0).

4   Results and Discussion

The expected population variance and empirical variances for the variants *DE/best/1/bin, DE/rand/2/bin* and *DE/best/2/bin* are displayed in Table 1.

Table 1. Empirical and Theoretical Variance measured for *DE/best/1/bin, DE/rand/2/bin* and *DE/best/2/bin*

DE/best/1/bin's Empirical / Theoretical Variance				DE/rand/2/bin's Empirical / Theoretical Variance				DE/best/2/bin's Empirical / Theoretical Variance						
G	F=0.1	F=0.12	F=0.2	F=0.3	G	F=0.1	F=0.12	F=0.2	F=0.3	G	F=0.1	F=0.12	F=0.2	F=0.3
1	8.52/8.64	8.52/8.72	8.30/8.34	8.71/8.85	1	8.75/8.92	8.81/9.14	8.46/9.65	8.44/11.31	1	8.44/8.41	8.48/8.90	8.68/8.58	8.39/8.09
4	1.36/0.57	1.46/0.59	1.79/0.71	2.48/1.05	4	9.11/9.76	9.52/10.70	11.34/16.44	15.51/36.72	4	1.55/0.62	1.59/0.35	2.35/0.98	3.66/1.81
7	0.33/0.07	0.36/0.08	0.52/0.11	0.91/0.21	7	9.18/10.36	9.94/11.95	13.82/24.36	24.13/88.36	7	0.38/0.09	0.42/0.05	0.83/0.19	1.97/0.56
10	0.073/0.01	0.08/0.01	0.16/0.02	0.33/0.05	10	9.37/10.98	10.29/13.35	16.54/36.09	37.87/212.61	10	0.09/0.01	0.11/0.01	0.29/0.04	1.05/0.18
13	0.02/0.00	0.02/0.00	0.05/0.00	0.13/0.00	13	9.53/11/66	10.71/14.91	20.29/54.47	60.24/511.55	13	0.02/0.00	0.02/0.00	0.10/0.01	0.54/0.06
16	0.00/0.00	0.01/0.00	0.02/0.00	0.05/0.00	16	9.64/12.38	11.26/16.67	24.34/79.21	93.68/1230.85	16	0.01/0.00	0.01/0.00	0.04/0.00	0.29/0.02
19	0.00/0.00	0.00/0.00	0.00/0.00	0.02/0.00	19	9.74/13.13	11.66/18.61	29.54/117.36	145.21/2961.55	19	0.00/0.00	0.00/0.00	0.01/0.00	0.15/0.01
22	0.00/0.00	0.00/0.00	0.00/0.00	0.01/0.00	22	9.93/13.04	11.99/20.79	35.78/173.86	227.65/7125.80	22	0.00/0.00	0.00/0.00	0.01/0.00	0.09/0.00
25	0.00/0.00	0.00/0.00	0.00/0.00	0.00/0.00	25	10.11/14.79	12.57/23.23	43.36/257.59	354.89/17145.41	25	0.00/0.00	0.00/0.00	0.00/0.00	0.05/0.00
28	0.00/0.00	0.00/0.00	0.00/0.00	0.00/0.00	28	10.27/15.70	13.09/25.96	52.87/381.63	550.62/41253.64	28	0.00/0.00	0.00/0.00	0.00/0.00	0.03/0.00
30	0.00/0.00	0.00/0.00	0.00/0.00	0.00/0.00	30	10.39/16.33	13.33/27.93	59.18/495.97	749.29/74075.03	30	0.00/0.00	0.00/0.00	0.00/0.00	0.01/0.00

As can be seen from the table, the decrease and increase in theoretical expected variance is matched by the empirical variance. Figure 1 reiterates this similar evolution pattern of theoretical and empirical population variance of *DE/rand/2/bin* and *DE/best/2/bin*. However there is large difference between the theoretical and empirical variances, which may be attributed to the fact that the theoretical derivation ignores the restriction that the indices of chosen solution vectors should not be equal to the parent vector index.

It is worth noting that *DE/rand/2/bin* variant displays increase in population variance (theoretical and empirical) as evolution proceeds, a measure of strong explorative capability. On the other hand *DE/best/*/bin* variants lose population diversity, as is evident from decreasing variances.

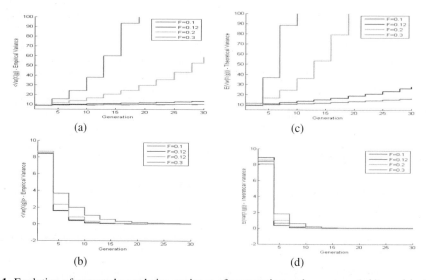

Fig. 1. Evolution of expected population variance after mutation and crossover (a,b) empirical evolution and (c,d) theoretical expected evolution, for *DE/rand/2/bin* and *DE/best/2/bin*.

Table 2. Mean Objective Function Value, Convergence Rate and Vanish Point measured

Variant	G	MOV	CoRate %	VanishPoint
DE/rand/1/bin	1227	0	40.94	1200
DE/best/1/bin	3000	457.24	100	600
DE/rand/2/bin	2100	0	70.05	1359
DE/best/2/bin	1278	0	42.67	768

Table 2 displays the mean objective function value (MOV), Convergence rate and Vanish point for the four DE variants considered. As can be seen from the table, for the variant *DE/best/1/bin* the population variance vanishes soon. In the case of *DE/best/2/bin*, even though the variance vanished rapidly, the optimum is also reached equally fast. This substantiates the observation that strategies relying on best solution have fast convergence but equally vulnerable to premature convergence. Relatively the *DE/rand/*/bin* variants retain diversity as is evident from the Table 2.

From the experiments, it has been observed that the theoretical expected population variance match with empirical variance evolution behavior for the chosen DE variants and substantiates our observation made elsewhere [6].

5 Conclusion

In this paper, a theoretical relationship between expected population variance after mutation, crossover and initial population variance for *DE/best/1/bin, DE/rand/2/bin* and *DE/best/2/bin* has been derived by directly extending Zaharie's work on *DE/rand/1/bin*. Simulation results show that the theoretical expected population variance evolve similar to empirical variance. In future we will extend this theoretical analysis to other DE variants such as *DE/current-to-rand/1/bin, DE/current-to-best/1/bin* and *DE/rand-to-best/1/bin* to get further insight about the variants.

References

1. Storn., R., Price, K.: Differential Evolution – A Simple and Efficient Adaptive Scheme for Global Optimization over Continuous Spaces. Technical Report TR-95-012, ICSI (1995)
2. Storn., R., Price, K.: Differential Evolution – A Simple and Efficient Heuristic Strategy for Global Optimization and Continuous Spaces. Journal of Global Optimization 11, 341–359 (1997)
3. Price, K.V.: An Introduction to differential evolution. In: Corne, D., Dorigo, M., Glover, F. (eds.) New Ideas in Optimization, pp. 79–108. Mc Graw-Hill, UK (1999)
4. Price, K., Storn, R.M., Lampinen, J.A.: Differential Evolution: A practical Approach to Global Optimization. Springer, Heidelberg (2005)
5. Vesterstrom, J., Thomsen, R.: A Comparative Study of Differential Evolution Particle Swarm Optimization and Evolutionary Algorithm on Numerical Benchmark Problems. In: Proceedings of the IEEE Congress on Evolutionary Computation (CEC 2004), vol. 3, pp. 1980–1987 (June 2004)
6. Jeyakumar, G., Shunmuga Velayutham, C.: An Empirical Performance Analysis of Differential Evolution Variants on Unconstrained Global Optimization Problems. International Journal of Computer Information Systems and Industrial Management Applications (IJCISIM) 2, 77–86 (2010)
7. Bayer, H.G., Deb, K.: On the Analysis of Self Adaptive Evolutionary Algorithms, Technical Report, CI-69/99, University of Dortmund (1999)
8. Zaharie, D.: On the Explorative Power of Differential Evolution Algorithms. In: 3rd Int. Workshop "Symbolic and Numeric Algorithms of Scientific Computing"-SYNASC 2001, Romania (October 2001)
9. Yao, X., Liu, Y., Liang, K.H., Lin, G.: Fast Evolutionary Algorithms. In: Rozenberg, G., Back, T., Eiben, A. (eds.) Advances in Evolutionary Computing: Theory and Applications, pp. 45–94. Springer, New York (2003)
10. Mezura-Montes, E., Velazquez-Reyes, J., Coello Coello, C.A.: A Comparative Study on Differential Evolution Variants for Global Optimization. In: GECCO 2006, July 8-12 (2006)
11. Mezura-Montes, E.: Personal Communication (unpublished)

Generating Sequential Space-Filling Designs Using Genetic Algorithms and Monte Carlo Methods

Karel Crombecq[1] and Tom Dhaene[2]

[1] University of Antwerp, 2020 Antwerp, Belgium
Karel.Crombecq@ua.ac.be
[2] Ghent University - IBBT, 9050 Ghent, Belgium

Abstract. In this paper, the authors compare a Monte Carlo method and an optimization-based approach using genetic algorithms for sequentially generating space-filling experimental designs. It is shown that Monte Carlo methods perform better than genetic algorithms for this specific problem.

Keywords: surrogate modelling, active learning, sequential design, Monte Carlo, genetic algorithm, space-filling.

1 Introduction

For many modern engineering problems, accurate high fidelity simulations are often used instead of controlled real-life experiments, in order to reduce the overall time, cost and/or risk. These simulations are used by the engineer to understand and interpret the behaviour of the system under study and to identify interesting regions in the design space. They are also used to understand the relationships between the different input parameters and how they affect the outputs.

However, the simulation of one single instance of a complex system with multiple inputs (also called factors or variables) and outputs (also called responses) can be a very time-consuming process. For example, Ford Motor Company reported on a crash simulation for a full passenger car that takes 36 to 160 hours to compute [2]. Because of this long computational time, using this simulation directly is still impractical for engineers who want to explore, optimize or gain insight into the system.

The goal of *global* surrogate modelling is to find a function that mimics the original system, but can be computed much faster. This function is constructed by performing multiple simulations (called samples) at key points in the design space, analyzing the results, and selecting a model that approximates the samples and the overall system behavior quite well.

It is clear that the choice of the data points (or samples) is of paramount importance to the success of the surrogate modelling task. Intuitively, the data points must be spread out in such a way as to convey a maximum amount of

K. Deb et al. (Eds.): SEAL 2010, LNCS 6457, pp. 80–84, 2010.

information about the behaviour of simulator. This is a non-trivial challenge, since little or nothing is known about this (black-box) simulator in advance. From now on, we will refer to the entire set of samples that were selected for evaluation as the experimental design.

Sequential design (which is also known as adaptive sampling [7] or active learning [10]) methods generate an experimental design by selecting samples one by one, allowing for an integrated approach with the surrogate modelling task. After each simulation, new models are built, and the accuracy of these models is estimated. If the target accuracy is reached, the algorithm is halted. Otherwise, another sample is selected and the process starts all over again.

In this paper, we study the sequential generation of space-filling designs. Space-filling designs attempt to spread out the samples as evenly as possible, in order to get as much information about the entire design space as possible. We will investigate why it is difficult to sequentially generate good space-filling designs, and we will compare two approaches to tackling this problem: optimization using genetic algorithms, and Monte Carlo methods.

2 Space-Filling Experimental Design Criteria

From now on, we will consider the d-dimensional experimental design $P = \{\mathbf{p}_1, \mathbf{p}_2, \ldots, \mathbf{p}_n\}$ containing n samples $\mathbf{p}_i = (p_i^1, p_i^2, \ldots, p_i^d)$ in the (hyper)cube $[-1, 1]^d$. This experimental design P will be constructed by selecting samples one by one, without knowing the total number of samples n at any point during the construction process. In order to evaluate the space-filling qualities of the final design, we consider two criteria.

First and foremost, the generated design should be space-filling. Intuitively, a space-filling design is an experimental design in which the points are spread out evenly over the design space. However, there are several ways do define this property mathematically. Over the course of the years, many different space-filling criteria have been proposed. Depending on the criterion, the optimal design P will look differently. A popular and intuitive choice is the maximin criterion or *intersite distance* [1,5,6,8,9,12]. The intersite distance of an experimental design P is the smallest distance between two points in the design, and is defined as follows:

$$\text{idist}(P) = \min_{\mathbf{p_i}, \mathbf{p_j} \in P} \sqrt{\sum_{k=1}^{d} \left\| p_i^k - p_j^k \right\|^2} \qquad (1)$$

Secondly, a good space-filling design should also have good projective properties. This is also called the non-collapsing property by some authors [1]. An experimental design P has good projective properties if, for every point $\mathbf{p}_i$, each value p_i^j is strictly unique, and as different from the other values as possible. This property also means that, when the experimental design is projected from d-dimensional space to $(d-1)$-dimensional space along one of the axes, no two points are ever projected onto the same location. The quality of a design in terms of its projective properties can be defined as the minimum *projected distance* of points from each other:

$$\text{pdist}(P) = min_{\mathbf{p_i},\mathbf{p_j}\in P} \min_{1\leq k\leq d}\left|p_i^k - p_j^k\right|$$

$$= min_{\mathbf{p_i},\mathbf{p_j}\in P}\left\|\mathbf{p_i} - \mathbf{p_j}\right\|_{-\infty} \qquad (2)$$

3   Results

Generating an experimental design that satisfies the two criteria mentioned in the previous section is a multi-objective optimization problem. Starting with two initial points (for example, opposing corner points), a new point is selected by finding a location in the design space that maximizes both the intersite and projective distance. Many different methods have been proposed to solve such multi-objective optimization problems efficiently. The simplest approach is to combine the different objectives in a single aggregate objective function. This solution is only acceptable if the scale of both objectives is known, so that they can be combined into a formula that gives each objective equal weight. Fortunately, in the case of the intersite and projective distance, this is indeed the case.

The final objective function, which scores a new candidate point $\mathbf{p}$ when it is added to an existing design P, is defined as:

$$\text{dist}(P,\mathbf{p}) = \frac{(n+1)^{\frac{1}{d}-1}}{2}min_{\mathbf{p_i}\in P}\sqrt{\sum_{k=1}^{d}\left|p_i^k - p^k\right|^2} + \frac{n+1}{2}min_{\mathbf{p_i}\in P}\left\|\mathbf{p_i} - \mathbf{p}\right\|_{-\infty}$$

$$(3)$$

In this paper, we will compare two approaches to finding the best location for the next point at each iteration. The first one is a Monte Carlo method. In this method a large number of uniformly distributed random points is generated, and for each candidate $\mathbf{p}$, the objective function $\text{dist}(P,\mathbf{p})$ is calculated and the best candidate is selected as the new point to be added to P. In the second method, the genetic algorithm toolbox from Matlab will be used to optimize the objective function and find the best next candidate. Both methods will be compared for different settings, and the final designs be evaluated on the idist and pdist criteria to compare both approaches.

At each iteration, the Monte Carlo method will generate kn random points, where n is the number of samples evaluated thus far, and k is an algorithm parameter. It is expected that, for larger k, the quality of the design will improve. In this study, the following values for k were considered: $50, 250, 2000, 10000, 50000$.

For the genetic algorithm, the implementation from the Matlab Genetic Algorithm and Direct Search Toolbox (version 3.0) was used. Most of the options were kept at their default values, but some were changed in order to improve the performance. The default mutation function (which offsets each input by a value drawn from a gaussuan distribution) wasn't usable, because it did not respect the boundary constraints (each input must lie in $[-1, 1]$). It was changed to a mutation function that changes each input with a chance of 0.01 to a random values in the $[-1, 1]$ interval. Preliminary results have shown that playing with the crossover/mutation fraction settings, changing the elite behaviour etc does not affect the outcome much, so these settings were kept at their default values. This experiment was repeated for different numbers of generations: $50, 100, 250, 1000, 2000$.

All these experiments were carried out using the SUMO Toolbox research platform [3,4]. This freely available Matlab toolbox, designed for adaptive surrogate modelling and sampling, has excellent extensibility, making it possible for the user to add, customize and replace any component of the sampling and modelling process. Because of this, SUMO was the ideal choice for conducting this experiment[1].

In order to compare the two methods, an experimental design of 144 points was generated in a $2D$ input space, and the intersite and projective distance of the final design generated by both methods were compared. Because both the Monte Carlo and genetic algorithms use random numbers, each experiment was repeated 10 times to get a good average of the performance of the methods. This demonstrated a clear trend: even the Monte Carlo method with $k = 50$ produced better results than the genetic algorithm with 2000 generations. And even though the final designs generated by the genetic algorithm are considerably worse than the ones generated by the Monte Carlo method, the genetic algorithm requires much more time to generate them. The genetic algorithm with 2000 generations takes 3 times longer than the Monte Carlo method with $k = 50$, but still produces worse results.

It is also noticeable that the difference between 50 generations and 2000 generations is smaller than the difference between $k = 50$ and $k = 50000$, while the difference in elapsed time is larger for the genetic algorithm. This indicates that the rate at which the genetic algorithm improves is actually lower than the improvement rate for the Monte Carlo method. So no matter how many generations are computed, there will always be a Monte Carlo alternative that requires less time to get the same result.

4   Conclusion

This study shows that, when sequentially generating space-filling experimental designs, Monte Carlo methods are prefered above genetic algorithms or other optimization methods. This can be explained by the extremely complex and multimodal optimization surface obtained by adding the intersite and projective distance. It is possible that other optimization methods might perform better than the genetic algorithm implementation from the Matlab toolbox used in this study. However, the authors find it unlikely that any optimization method

---

[1] The SUMO Toolbox v7.0 can be downloaded from
http://www.sumo.intec.ugent.be

will do better than the Monte Carlo approach, considering the nature of the optimization surface.

Preliminary experiments have shown that the results published in this paper also hold in higher dimensions, and for different criteria, such as the ϕ_p criterion proposed by [8]. In subsequent publications, these preliminary results will be examined and expanded upon.

The ultimate goal of this experiment is to develop highly efficient sequential space-filling algorithms that can compete with proven and popular one-shot experimental design techniques such as the optimized Latin hypercube [1,11]. These methods will use Monte Carlo methods as the optimization method of choice, as opposed to global optimization methods. Finally, hybrid methods will also be investigated. Hybrid methods use Monte Carlo to find promising locations, and then perform a local optimization to further improve the initial result.

References

1. van Dam, E.R., Husslage, B., den Hertog, D., Melissen, H.: Maximin latin hypercube design in two dimensions. Operations Research 55(1), 158–169 (2007)
2. Gorissen, D., Crombecq, K., Hendrickx, W., Dhaene, T.: Adaptive distributed metamodeling. In: Daydé, M., Palma, J.M.L.M., Coutinho, Á.L.G.A., Pacitti, E., Lopes, J.C. (eds.) VECPAR 2006. LNCS, vol. 4395, pp. 579–588. Springer, Heidelberg (2007)
3. Gorissen, D., Tommasi, L.D., Crombecq, K., Dhaene, T.: Sequential modeling of a low noise amplifier with neural networks and active learning. Neural Computation & Applications 18(5), 485–494 (2009)
4. Gorissen, D., Turck, F.D., Dhaene, T.: Evolutionary model type selection for global surrogate modeling. Journal of Machine Learning Research 10(1), 2039–2078 (2009)
5. Johnson, M., Moore, L., Ylvisaker, D.: Minimax and maximin distance designs. Journal of Statistical Planning and Inference 26, 131–148 (1990)
6. Joseph, V.R., Hung, Y.: Orthogonal-maximin latin hypercube designs. Statistica Sinica 18, 171–186 (2008)
7. Lehmensiek, R., Meyer, P., Müller, M.: Adaptive sampling applied to multivariate, multiple output rational interpolation models with application to microwave circuits. International Journal of RF and Microwave Computer-Aided Engineering 12(4), 332–340 (2002)
8. Morris, M.D., Mitchell, T.J.: Exploratory designs for computer experiments. Journal of Statistical Planning and Inference 43, 381–402 (1995)
9. Rennen, G., Husslage, B., Dam, E.V., Hertog, D.D.: Nested maximin latin hypercube designs. Tech. Rep., Tilburg University (2009)
10. Sugiyama, M.: Active learning in approximately linear regression based on conditional expectation of generalization error. Journal of Machine Learning Research 7, 141–166 (2006)
11. Viana, F.A.C., Venter, G., Balabanov, V.: An algorithm for fast optimal latin hypercube design of experiments. International Journal for Numerical Methods in Engineering (2009)
12. Ye, K.Q., Li, W., Sidjianto, A.: Algorithmic construction of optimal symmetric latin hypercube designs. Journal of Statistical Planning and Inference 90(1), 145–159 (2000)

MP-EDA: A Robust Estimation of Distribution Algorithm with Multiple Probabilistic Models for Global Continuous Optimization[*]

Jing-hui Zhong[1], Jun Zhang[1,**], and Zhun Fan[2]

[1] Dept. of Computer Science, SUN yat-sen University
Key Laboratory of Digital Life (Sun Yat-Sen University), Ministry of Education
Key Laboratory of Software Technology, Education Department of Guangdong
Province, P.R. China
junzhang@ieee.org
[2] Denmark Technical University

Abstract. Extending Estimation of distribution algorithms (EDAs) to the continuous field is a promising and challenging task. With a single probabilistic model, most existing continuous EDAs usually suffer from the local stagnation or a low convergence speed. This paper presents an enhanced continuous EDA with multiple probabilistic models (MP-EDA). In the MP-EDA, the population is divided into two subpopulations. The one involved by histogram model is used to roughly capture the global optima, whereas the other involved by Gaussian model is aimed at finding highly accurate solutions. During the evolution, a migration operation is periodically carried out to exchange some best individuals of the two subpopulations. Besides, the MP-EDA adaptively adjusts the offspring size of each subpopulation to improve the searching efficiency. The effectiveness of the MP-EDA is investigated by testing ten benchmark functions. Compared with several state-of-the-art evolutionary computations, the proposed algorithm can obtain better results in most test cases.

Keywords: Estimation of Distribution Algorithm, Evolutionary Computation, Histogram, Multivariate Gaussian Distribution, Global Optimization.

1 Introduction

The estimation of distribution algorithms (EDAs) are a new class of evolutionary computation algorithms [1] [2]. They generate new individuals by sampling a probabilistic model, which is estimated based on the current promising solutions. As the probabilistic model can capture promising areas in a statistically sound manner and can explicitly express the interactions among variables, EDAs usually can outperform traditional EAs on a number of complex problems .

---

[*] This work was supported in part by the National Natural Science Foundation of China No. U0835002 and No.61070004, by the National High-Technology Research and Development Program ("863" Program) of China No. 2009AA01Z208.
[**] Corresponding author.

K. Deb et al. (Eds.): SEAL 2010, LNCS 6457, pp. 85–94, 2010.

The EDAs are first proposed to solve discrete problems with binary representation. For the last few years, various efforts have been made on extending them to the continuous optimization [3] - [11]. Most existing continuous EDAs use the Gaussian model or the histogram model to estimate the distribution of promising areas. The marginal Gaussian probabilistic model was used to guide the search in early continuous EDAs, such as the PBILc [3] and the UMDAc [4]. Lately, the multivariate Gaussian models also appeared in continuous EDAs [5] [6]. EDAs with a single Gaussian model are excellent in finding the global optima for unimodal optimization. However, for multimodal problems, they usually suffer from a slow convergence speed or even a local stagnation. Though clustering techniques have been utilized to address some of these issues [7], more efficient methods are greatly desirable.

Meanwhile, the histogram has multimodal density and can capture multiple local optima at the same time. Hence the histogram-based EDAs (HEDAs) are less likely to get trapped in local optima [8]. However, the HEDAs usually need a heavy computational cost to search for a highly accurate solution. To tackle this problem, the local search techniques [9], the shrink strategy [10], and the sub-divided method [11], have been proposed in recent few years. Nevertheless, since the histogram ignores interactions among variables, the HEDAs are not efficient enough to solve problems containing variable dependency.

In this paper, we present an enhanced EDA with multiple probabilistic models (named MP-EDA) for the global continuous optimization. The proposed MP-EDA has made two major improvements as below. First, the robustness to find a highly accurate solution is guaranteed by adopting two different types of probabilistic models: the multivariate Gaussian model and the fixed-height histogram (FHH) model. Second, the efficiency of the search is improved by utilizing an adaptive control strategy. In the MP-EDA, the population is initially divided into two subpopulations, with each involved by a probabilistic model. The one involved by the FHH is used to roughly capture the global optima, and the other involved by the multivariate Gaussian model is aimed at finding highly accurate solutions. During the evolution, a migration operation is carried out periodically to exchange some best individuals between the two subpopulations. By sharing information of better individuals, both subpopulations can converge faster. Besides, in order to improve the search efficiency, an adaptive strategy is used to adjust the offspring sizes of the two subpopulations. The performance of a probabilistic model is measured by the average fitness of new individuals generated by it during recent generations. The better probabilistic model is allowed to generate more offspring individuals, and vice versa.We investigate the effectiveness of the proposed MP-EDA by solving ten test functions with different characteristics. Several state-of-the-art EAs (i.e PLSO [12], DE [13], FEP [14] and CMA-ES [15]) are used for the comparison. Experimental results show that the proposed MP-EDA can achieve better results in most test cases.

The rest of the paper is organized as follows. Section 2 briefly describes the EDAs' general framework and two classical continuous EDAs. Section 3 illustrates the detailed implementations of the MP-EDA. The experimental studies on the MP-EDA are presented in Section 4. At last, Section 5 draws the conclusions.

2 Estimation of Distribution Algorithms

2.1 Framework of EDAs

The general framework of EDAs is similar to that of the GA. However, there is neither crossover nor mutation in EDAs. Instead, they use a probabilistic model to generate new individuals. Specifically, the common outline of EDAs consists of the following four steps.

Step 1: Initialize the algorithm parameters and the initial population.

Step 2: Select a certain number of excellent individuals. There are several selection strategies, such as the truncation selection and the tournament selection.

Step 3: Construct probabilistic model by analyzing information of the selected individuals.

Step 4: Create new population by sampling new individuals from the constructed probabilistic model.

There is a repetition from *Step2* to *Step4* until the algorithm meets the termination condition.

2.2 Multivariate Gaussian Model

The multivariate Gaussian model is most commonly used in continuous EDAs. It characterizes the distribution of data points by two parameters μ and Σ. The former gives the mean vector of data points, while the latter describes the covariance information. In particular, given a set of data points $X = \{x_1, x_2, ..., x_Z\}$, the multivariate Gaussian model of X can be expressed by

$$N(\mu, \Sigma) = \frac{1}{\sqrt{2\pi\Sigma}} \cdot \exp(-\frac{1}{2} \cdot (x - \mu) \cdot \Sigma^{-1} \cdot (x - \mu)) \tag{1}$$

where μ is the mean vector of X, and Σ is the covariance matrix of X.

The multivariate Gaussian model is able to capture the interaction between variables, owing to the covariance matrix in the density function. However, it is not effective to estimate the distribution of multimodal data points, due to its unimodal density.

2.3 Histogram Model

There are usually two types of histograms, namely the fixed-width histogram (FWH) and the FHH. The FWH consists of bins with the same width, whereas the FHH contains bins with the same height. In this paper, we choose the FHH as one probabilistic model of the proposed method, because it has shown to perform better than the FWH. Given a set of data points $X = \{x_1, x_2, ..., x_Z\}$ within an interval $\Omega = [x_{min}, x_{max}]$. The FHH of X consists of n bins, with each bin B_i (i=1, 2, ..., n) containing the same number of data points. Hypothesizing that the data points in each bin are uniformly distributed, the estimated density $\hat{f}_H(x)$ of the underlying probability density at any point x can be computed by

$$\hat{f}_H(x) = \frac{1}{w_k \cdot n} \tag{2}$$

where w_k is the width of the bin $\mathbf{B}_k$ that contains x. According to the definition of the FHH, bins in the dense regions would have a narrower width than those in the sparse regions. As the population of an evolutionary algorithm evolves, individuals will gradually gather around promising areas. By giving more bins to these regions, the FHH is able to capture multiple peaks at the same time. Nevertheless, one drawback of the FHH based EDA is that it may get stuck when the population converges. It this case, a heavy computational cost is usually required to sample a highly accurate solution.

3 MP-EDA for Multimodal Continous Optimization

3.1 Algorithm Framework

There are two subpopulations (POP_{fhh} and POP_{gm}) involved by two probabilistic models in the MP-EDA. The POP_{fhh} is involved by the FHH model, whereas the POP_{gm} is involved by the Gaussian model. Specifically, the framework of the proposed MP-EDA contains following six steps.

1) Step 1 - Initialization
This step initializes parameters of the algorithm, such as the size of the population N, the number of excellent individuals S, the migration cycle T and the number of bins n. Besides, new individuals of POP_{fhh} and POP_{gm} are randomly generated in the search space, with their offspring sizes respectively set as $N_{fhh} = N$ and $N_{gm} = N$.

2) Step 2 – Constructing probabilistic models
This step aims at constructing two different probabilistic models for the two subpopulations in parallel. For the POP_{fhh}, S best individuals are firstly selected. Then the marginal histogram for all variables is updated. Let $X=\{x_0, x_1,..., x_S\}$ with $x_0 \leq x_1 \leq ... \leq x_S$, be the i-th variable values of these selected individuals, then the lower bound and the upper bound of each bins is updated by

$$l_{j,i} = \begin{cases} \text{lower bound of the } i\text{-th variable,} & \text{if } j = 1 \\ (x_{\lfloor (j-1)\cdot \Delta x \rfloor} + x_{\lfloor (j-1)\cdot \Delta x \rfloor+1})/2, & \text{otherwise} \end{cases} \tag{3}$$

$$u_{j,i} = \begin{cases} \text{upper bound of the } i\text{-th variable,} & \text{if } j = n \\ l_{j+1,i}, & \text{otherwise} \end{cases} \tag{4}$$

where $\Delta x = S/n$ is the average number of data points in each bin.

As for the multivariate Gaussian model, the constructing process aims to update values of two parameters: the mean μ and the covariance matrix $\sum$. Firstly, S promising individuals are selected in POP_{gm}, and then the mean and the covariance of these S individuals are computed and set as the values of μ and $\sum$.

3) *Step 3 – Sampling new population*
The sampling process of the FHH consists of two sub-steps. Firstly, a bin is randomly selected according to the constructed marginal histogram. Then a random value uniformly distributed in the interval of the selected bin is generated.

As for the multivariate Gaussian model, the sampling process contains three sub-steps.

Sub-step1: Use the Cholesky decomposition to generate a lower triangular matrix S, with

$$\sum = S \cdot S^T \tag{5}$$

Sub-step2: obtain a 1-by-D matrix Z with random elements sampled from a standard Normal distribution N (0,1).

Sub-step3: generate a individual P by

$$P = \mu + S \cdot Z \tag{6}$$

In order to maintain higher population diversity, a scale parameter $\delta > 1$ is suggested to add in (6) [16] and expressed as

$$P = \mu + \delta \cdot S \cdot Z \tag{7}$$

Note that the parameter δ should be set appropriate, for a trade-off between the diversity and the convergence of the population.

4) *Step 4 –Evaluation and Replacement*
This step aims to evaluate the fitness values of all new individuals and replace some worst individuals of the current population by better new individuals. For example, let $C = \{C_1, C_1, ..., C_N\}$ be the current subpopulation POP_{fhh}, $C^* = \{C^*_1, C^*_2, ..., C^*_N\}$ be the newly generated individuals by the FHH and $C^\# = C \cup C^*$. Then the best N individuals in $C^\#$ are selected as the new POP_{fhh}. The similar process can be done to generate the new POP_{gm}.

5) *Step 5 –Migration*
During the evolution, a migration operation is carried out every T generations. In the migration, the best M individuals in one subpopulation are migrated to the other. By sharing the best individuals of two subpopulations, the algorithm can utilize the advantages of both probabilistic models, which makes the algorithm less likely to be trapped into local optima and more efficient to find highly accurate solutions.

6) *Step 6 – Adaptive control operation*
This step aims to adjust the size of new offspring for each subpopulation, so as to save the computational cost. The key idea is to make the better probabilistic model generate more new individuals in each generation. Specifically, F_{fhh} and F_{gm} are supposed to be the average fitness of new individuals generated by the FHH and the GM respectively during the recent T generations. If F_{fhh} is better than F_{gm}, the value of N_{fhh} would be enlarged by a small step Δ, whereas the value of N_{gm} would be reduced

by Δ. Otherwise, the reverse operation can be done. The maximum and the minimum of N_{fhh} and N_{gm} are both set as P_{min} and P_{max}.

There is a repetition from *Step2* to *Step6* and the evolution processes iteratively until reaching the maximum number of evaluations.

4 Experiments and Comparisons

4.1 Test Functions and Parameter Settings

Ten benchmark functions are contained in the experimental study, as listed in Table 1. These functions are chosen in the literature [17] and [18]. Parameter settings of all comparison algorithms are listed in Table 2. The parameter settings of the DE, the CLPSO and the CMA-ES are set as those in the reference papers. The dimension of all test functions is 30 and the maximum evaluation number for each function is 3×10^5. In order to make fair comparisons, 30 independent runs are performed on each algorithm.

Table 1. Test functions

Test Function	Domain
Unimodal	
$f_1(x) = \sum_{i=1}^{n} z_i^2 + f\_bias, z = x - o$	$[-100,100]$
$f_2(x) = \sum_{i=1}^{n}(\sum_{j=1}^{i} z_j^2) + f\_bias, z = x - o$	$[-100,100]$
$f_3(x) = (\sum_{i=1}^{n}(\sum_{j=1}^{i} z_j^2)) \cdot (1 + 0.4|N(0,1)|) + f\_bias, z = x - o$	$[-100,100]$
Multimodal	
$f_4(x) = \sum_{i=1}^{n}(100(z_i^2 - z_{i+1})^2 + (z_i - 1)^2) + f\_bias, z = x - o$	$[-100,100]$
$f_5(x) = \sum_{i=1}^{n}(-z_i \sin(\sqrt{|z_i|})) + f\_bias, z = x - o$	$[-500,500]$
$f_6(x) = \sum_{i=1}^{n}(z_i^2 - 10\cos(2\pi z_i) + 10) + f\_bias, z = x - o$	$[-5,5]$
$f_7(x) = \sum_{j=1}^{n}\sum_{i=1}^{n}(\frac{y_{i,j}^2}{4000} - \cos(y_{i,j}) + 1) + f\_bias,$ where $y_{i,j} = 100(z_j - z_i^2)^2 + (1 - z_i)^2, z = x - o$	$[-100,100]$
$f_8(x) = \sum_{i=1}^{n}(\frac{z_i^2}{4000}) - \prod_{i=1}^{n}\cos(\frac{z_i}{\sqrt{i}}) + 1 + f\_bias, z = x - o$	$[-600,600]$
$f_9(x) = -20 \cdot \exp(-0.2\sqrt{\frac{1}{n}\sum_{i=1}^{n} z_i^2}) - \exp(\frac{1}{n}\sum_{i=1}^{n}\cos(2\pi z_i)) + 20 + e + f\_bias, z = x - o$	$[-32,32]$
$f_{10}(x) = \sum_{i=1}^{n}(z_i^2 - 10\cos(2\pi z_i) + 10) + f\_bias, z = (x - o) * M$	$[-5,5]$

Table 2. Parameter Settings

Algorithm	Parameter Settings
GM-EDA	population size $N = 200$, $S=100$, $\alpha=0.2$, $Q=2$, $\delta=1.2$
FHH-EDA	population size $N = 200$, $S=100$, $n=50$
MP-EDA	population size $T=5$, $N_{min}=20$, $N_{max}=200$, $\Delta=10$, $M=5$
CLPSO	particle number $ps = 40$, $w_0 = 0.9$, $w_1 = 0.4$, $c = 1.49445$, $m = 7$
DE	population size $= 100$, $F = 0.5$, $CR = 0.9$
CMA-ES	population size $= 14$ as default

Table 3. Experimental Results. 'a' stands for the average best values, 's' stands for the standard deviations and 'r' stands for the performance ranking of the corresponding algorithm.

			MP-EDA	FHH-EDA	GM-EDA	CLPSO	DE	CMA-ES
Unimodal	f_1	a	**0**	6.99×10^{-1}	2.08×10^{-14}	5.68×10^{-14}	**0**	3.79×10^{-15}
		s	**0**	3.37×10^{-1}	2.74×10^{-14}	0	**0**	1.44×10^{-14}
		r	**1**	6	4	5	**1**	3
	f_2	a	1.98×10^{-9}	3.40×10^{3}	3.51×10^{-10}	6.21×10^{2}	4.81×10^{-5}	**3.22×10^{-14}**
		s	2.23×10^{-9}	9.65×10^{3}	2.91×10^{-10}	1.52×10^{2}	4.00×10^{-5}	**2.86×10^{-14}**
		r	3	6	2	5	4	1
	f_3	a	2.46×10^{-5}	5.22×10^{3}	**1.40×10^{-6}**	6.29×10^{3}	1.39×10^{-2}	2.69×10^{4}
		s	4.56×10^{-5}	1.53×10^{3}	**5.21×10^{-6}**	1.38×10^{3}	1.38×10^{-2}	8.53×10^{3}
		r	2	4	1	5	3	6
	f_4	a	4.74×10^{-5}	2.97×10^{3}	3.46×10^{-8}	5.14×10^{0}	2.87×10^{0}	**5.12×10^{-14}**
		s	1.03×10^{-4}	3.82×10^{3}	9.21×10^{-8}	4.62×10^{0}	1.98×10^{0}	**1.73×10^{-14}**
		r	3	6	2	5	4	1
Multimodal	f_5	a	**-12569.5**	-12569.3	-6.03×10^{3}	-12557.6	-8290.5	-9555.6
		s	**1.82×10^{-12}**	6.75×10^{-2}	1.11×10^{3}	3.55×10^{1}	8.32×10^{2}	4.40×10^{2}
		r	1	2	6	3	5	4
	f_6	a	**0**	2.72×10^{-1}	1.85×10^{2}	3.69×10^{-13}	1.28×10^{2}	2.81×10^{2}
		s	**0**	1.79×10^{-1}	9.27×10^{0}	3.55×10^{-13}	2.08×10^{1}	3.27×10^{1}
		r	1	3	5	2	4	6
	f_7	a	3.51×10^{2}	5.46×10^{2}	6.79×10^{2}	2.53×10^{2}	4.95×10^{2}	**6.88×10^{1}**
		s	3.06×10^{1}	1.11×10^{2}	1.61×10^{1}	5.29×10^{1}	5.98×10^{1}	**2.44×10^{1}**
		r	3	5	6	2	4	1
	f_8	a	**0**	4.69×10^{-1}	1.52×10^{-14}	5.68×10^{-14}	**0**	2.47×10^{-2}
		s	**0**	1.67×10^{-1}	2.51×10^{-14}	0	**0**	0
		r	1	6	3	4	1	5
	f_9	a	**5.68×10^{-14}**	1.99×10^{-1}	1.06×10^{-13}	1.27×10^{-13}	6.06×10^{-14}	1.63×10^{1}
		s	**0**	7.99×10^{-2}	1.93×10^{-14}	2.40×10^{-14}	1.42×10^{-14}	7.41×10^{0}
		r	1	5	3	4	2	6
	f_{10}	a	**4.13×10^{1}**	1.58×10^{2}	1.90×10^{2}	1.03×10^{2}	1.78×10^{2}	5.13×10^{2}
		s	**1.13×10^{1}**	7.51×10^{1}	8.08×10^{0}	1.68×10^{1}	9.69×10^{1}	2.36×10^{2}
		r	1	3	5	2	4	6
Total r			**17**	46	37	37	32	39
Average r			**1.7**	4.6	3.7	3.7	3.2	3.9

4.2   Experimental Results

The comparison results are summarized in Table 3. Firstly, we investigate the performance of the FHH-EDA, GM-EDA and the MP-EDA. Results of f1 show that, among these three algorithms, the MP-EDA performs the best, while the FHH-EDA performs the worst. The GM-EDA can obtain very highly accurate solutions, but it gets stuck near the global optima. This is because the eigenvectors of the covariate matrix in the GM-EDA has dropped to zero when the mean vector was still a bit distant from the global optimum. However, the MP-EDA can overcome this drawback, owing to the diversity provided by the histogram. Results of f2 and f3 indicate that, the FHH-EDA is not suitable to solve these kinds of problems, since it contains no mechanism to deal with variable interactions. However, the GM-EDA is very effective to solve these problems, and the proposed MP-EDA can also give competitive performance. As for multimodal functions, the MP-EDA performs much better than the GM-EDA and the FHH-EDA in all test cases except for the shift rosenbrock function. It can be observed that the Gaussian-based EDA can find satisfying solutions for multimodal functions with a big-valley structure (such as f8 and f9), but it performs

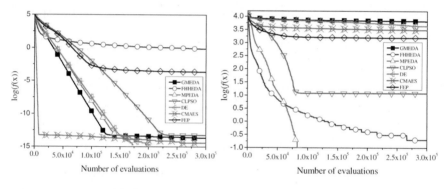

(a) f_1 (b) f_5

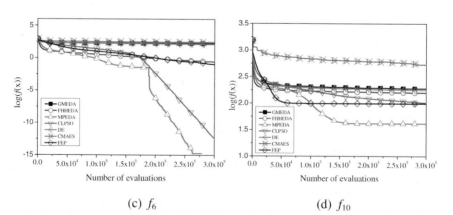

(c) f_6 (d) f_{10}

Fig. 1. Convergence graphs of the MP-EDA

very badly for those with local optima scattering in the search space. Meanwhile, the FHH-EDA can only find roughly accurate solutions for separable-multimodal functions (such as f5 and f6), but it performs badly for those containing variable interactions (such as f4). By combining two different probabilistic models, the MP-EDA performs better and more stable.

Results in table 3 indicate that the proposed MP-EDA performs the best in six test functions, and it can also give competitive performance on other test functions. According to the average rank value, the MP-EDA performs averagely much better than the other comparison EAs.

The convergence graphs of the MP-EDA and the comparison algorithms are shown in Fig. 1. It can be observed that the proposed algorithm can not only search a highly accurate solution, but also has a very fast convergence speed. Besides, the proposed MP-EDA can give good performance on both unimodal functions and multimodal functions.

5   Conclusions

This paper has proposed an enhanced EDA with multiple probabilistic models for the global continuous optimization. The proposed MP-EDA adopts a histogram and a multivariate Gaussian model to involve two subpopulations. During the evolution, a migration strategy is used to exchange some best individuals of the two subpopulations. Besides, the offspring size of each subpopulation is adaptively adjusted to reduce the computational cost. The effectiveness of the proposed MP-EDA has been investigated by testing ten benchmark functions. Compared with several state-of-the-art evolutionary algorithms, the proposed MP-EDA can obtain better results in most test cases.

This study has shown that, EDAs with multiple probabilistic models usually can work more effectively and efficiently than those with a single probabilistic model. As for future work, we will further extent the framework by hybridizing more probabilistic models. Besides, applying the proposed algorithm framework to solve real application is another promising research topic.

References

1. Baluja, S.: Population-based incremental learning: a method for integrating genetic search based function optimization and competitive learning. Technical report CMU-CS-94-163, Carnegie Mellon University (1994)
2. Mühlenbein, H., Paaß, G.: From recombination of genes to the estimation of distributions I. binary parameters. In: Ebeling, W., Rechenberg, I., Voigt, H.-M., Schwefel, H.-P. (eds.) PPSN 1996. LNCS, vol. 1141, pp. 178–187. Springer, Heidelberg (1996)
3. Sebag, M., Ducoulombier, A.: Extending population-based incremental learning to continuous search spaces. In: Eiben, A.E., Bäck, T., Schoenauer, M., Schwefel, H.-P. (eds.) PPSN 1998. LNCS, vol. 1498, pp. 418–427. Springer, Heidelberg (1998)
4. Larrañaga, P., Etxeberria, R., Lozano, J.A., Pe a, J.M.: Optimization in continuous domains by learning and simulation of Gaussian networks. In: Proceedings of the Genetic and Evolutionary Computation Conference 2000, Las Vegas, Nevada (July 2000)

5. Paul, T.K., Iba, H.: Real-Coded Estimation of Distribution Algorithm. In: Proceedings of The Fifth Metaheuristics International Conference (2003)
6. Yuan, B., Gallagher, M.: Experimental results for the special session on real-parameter optimization at CEC 2005: a simple, continuous EDA. In: Proc. of Congress on Evolutionary Computation (CEC 2005), vol. 2, pp. 1792–1799 (2005)
7. Lu, Q., Yao, X.: Clustering and learning Gaussian distribution for continuous optimization. IEEE Transactions on Systems, Man, and Cybernetics, Part C: Applications and Reviews 35(2), 195–204 (2005)
8. Tsutsui, S., Pelikan, M., Goldberg, D.E.: Evolutionary algorithm using marginal histogram models in continuous domain. IlliGAL Report No. 2001019, UIUC (2001)
9. Zhang, Q., Sun, J., Tsang, E., Ford, J.: Hybrid Estimation of Distribution Algorithm for Global Optimsation. Engineering Computations 21(1), 91–107 (2004)
10. Ding, N., Zhou, S., Sun, Z.: Histogram-based estimation of distribution algorithm: a competent method for continuous optimization. Computer Science and Technology 23(1), 35–42 (2008)
11. Xiao, J., Yan, Y.P., Zhang, J.: HPBILc: A histogram-based EDA for continuous optimization. Applied Mathematics and Computation 215(3), 973–982 (2009)
12. Liang, J.J., Qin, A.K., Suganthan, P.N., Baskar, S.: Comprehensive learning particle swarm optimizer for global optimization of multimodal functions. IEEE Trans. on Evolutionary Computation 10(3), 280–295 (2006)
13. Storn, R.M., Price, K.V.: Differential evolution – A simple and efficient heuristic for global optimization over continuous spaces. J. Global Optimization 11, 341–359 (1997)
14. Yao, X., Liu, Y., Lin, G.: Evolutionary programming made faster. IEEE Trans. On Evolutionary Computation 3(2), 82–102 (1999)
15. Auger, A., Hansen, N.: Performance evaluation of an advanced local search evolutionary algorithm. In: Proc. of IEEE Congress on Evolutionary Computation (CEC 2005), vol. 2, pp. 1777–1784 (September 2005)
16. yuan, B., Gallagher, M.: On the importance of diversity maintenance in estimation of distribution algorithms. In: Proc. of the Genetic and Evolutionary Computation Conference-GECCO-2005, pp. 719–726. ACM, New York (2005)
17. Suganthan, P.N., Hansen, N., Liang, J.J., Deb, K., Chen, Y.-P., Auger, A., Tiwari, S.: Problem Definitions and Evaluation Criteria for the CEC 2005 Special Session on Real-Parameter Optimization. Nanyang Technol. Univ., Singapore, IIT Kanpur, India, KanGAL Rep. 2005005 (May 2005)
18. Noman, N., Iba, H.: Accelerating differential evolution using an adaptive local search. IEEE Transactions on Evolutionary Computation 12(1), 107–125 (2008)

A Bi-criterion Approach to Multimodal Optimization: Self-adaptive Approach

Amit Saha and Kalyanmoy Deb

Kanpur Genetic Algorithms Laboratory (KanGAL)
Indian Institute of Technology Kanpur
PIN 208016, India
amitsaha.in@gmail.com, deb@iitk.ac.in
http://www.iitk.ac.in/kangal

Abstract. In a multimodal optimization task, the main purpose is to find multiple optimal solutions, so that the user can have a better knowledge about different optimal solutions in the search space and as and when needed, the current solution may be replaced by another optimum solution. Recently, we proposed a novel and successful evolutionary multi-objective approach to multimodal optimization. Our work however made use of three different parameters which had to be set properly for the optimal performance of the proposed algorithm. In this paper, we have eliminated one of the parameters and made the other two self-adaptive. This makes the proposed multimodal optimization procedure devoid of user specified parameters (other than the parameters required for the evolutionary algorithm). We present successful results on a number of different multimodal optimization problems of upto 16 variables to demonstrate the generic applicability of the proposed algorithm.

Keywords: Multimodal optimization, Multi-objective optimization, Self-adaptive algorithm, Hooke-Jeeves search.

1 Introduction

Single-objective optimization problems are usually solved for finding a single optimal solution, despite the existence of multiple optima in the search space. In the presence of multiple global and local optimal solutions in a problem, an algorithm is usually preferred if it is able to avoid locally optimal solutions and locate the true global optimum.

However, in many practical optimization problems having multiple optima, it is wise to find as many optimum points as possible for a number of reasons. First, an optimal solution currently favorable (say, due to unavailability of some critical resources or satisfaction of some codal principles, or others) may not remain to be so in the future. This would then demand the user to operate at a different solution when such a predicament occurs. With the knowledge of another optimal solution for the problem which is favorable to the changed scenario, the user can simply switch to this new optimal solution. Second, the sheer knowledge of

K. Deb et al. (Eds.): SEAL 2010, LNCS 6457, pp. 95–104, 2010.

multiple optimal solutions in the search space may provide useful insights into the properties of high-performing optimal solutions of the problem.

All existing methods of multimodal optimization in evolutionary computing (EC) literature use an additional *niching* operation in some form to maintain multiple optimum solutions from one generation to next. Recently we suggested a very different approach using principles of Evolutionary Multi-objective Optimization (EMO). We showed a couple of ways to select a second objective for an otherwise single-objective multimodal optimization problem, and hence solve the resulting multi-objective optimization problem using a modified EMO procedure, the NSGA-II [2]. One of the two methods that we found scalable and more generic was the Hooke-Jeeves (H-J) search based neighborhood count algorithm. The proposed algorithm, although found to be performing excellently suffered from the drawback of preselecting the correct value for three parameters. In the current work, we successfully eliminate one of the parameters and estimate the other two from the population itself. This work hence contributes a generic procedure for multimodal optimization to the research community without the additional burden of more user specified parameters.

In the remainder of the paper, we provide a brief description of past multimodal EC studies in Section 2. The concept of multi-objective optimization for solving multimodal problems is described in Section 3. We explain our self-adaptive parameter calculation in Section 4 and then present the results of the approach on a number of different test problems in Section 5. Finally, conclusions and possible extensions to this study are highlighted in Section 6.

2   Evolutionary Multimodal Optimization

As opposed to single-objective optimization, multimodal optimization entails finding the multiple solutions, rather than the best solution. Population based optimization algorithms, such as Evolutionary algorithms (EAs) have been found to be a generic tool for such problems. The *intrinsically parallel* nature of their search enables discovery of multiple solutions simultaneously. However, all EAs have the tendency to lose diversity and converge to the globally best solution due to genetic drift. The main challenge is to maintain an adequate diversity among population members such that multiple optimum solutions can be found and then a preservation mechanism to maintain the discovered solutions from one generation to another. For this purpose, *niching* methodologies are employed, in which crowded solutions in the population (usually in the decision variable space) are degraded either by directly reducing the fitness value of neighboring solutions (such as the sharing function approach [8,9]) or by directly ignoring crowded neighbors such as in the clearing approach [10] or clustering approach [11].

In [1], we proposed an EMO based approach, where a couple of suggestions for a suitable second objective for the multimodal optimization problem is put forward and their performance demonstrated. The H-J based approach, which we put forward as the better method and improve upon here is described in the next section.

3 Multimodal Optimization Using a Bi-objective Approach

In our two-objective approach, we demonstrated the high performance of H-J based neighborhood count approach, where the first objective is the given function, $f(\mathbf{x})$ and the second objective is the number of neighboring points better (we restrict ourself to minimization problems here, and hence a point is better than another, if its function value is lesser) than $\mathbf{x}$. We select the sample of neighboring points judiciously by using the way an exploratory search is performed in the Hooke-Jeeves classical optimization algorithm. The procedure starts with the first variable $(i = 1)$ dimension and creates two extra points $x_i^c \pm \delta_{hj}$ around the current solution $\mathbf{x}^c = \mathbf{x}$. Thereafter, three solutions $(\mathbf{x}^c - \delta_{hj}\mathbf{e}_i, \mathbf{x}^c, \mathbf{x}^c + \delta_{hj}\mathbf{e}_i)$ (where $\mathbf{e}_i$ is the unit vector along i-th variable axis in the n-dimensional variable space) are compared with their function values and the best is chosen. The current solution $\mathbf{x}^c$ is then moved to the best solution. Similar operations are done for $i = 2$ and continued for the remaining variables. Every time a solution having a better objective value than the original objective value $(f(\mathbf{x}))$ is encountered, the second objective value $f_2(\mathbf{x})$ is incremented by one. It is obvious that for an optimum solution $f_2(\mathbf{x}) = 0$. Multiple optimum solutions, hence possess an invariant second objective and thus form a weakly Pareto-optimal set. Some basic changes to NSGA-II were sufficient to obtain multiple optimal solutions:

1. We introduced a parameter, δ_f to prevent the convergence to only one (or some) of the optima thus giving a chance to the population of solutions to discover other optima. For any two solutions having equal $f_2(\mathbf{x})$ and function values within δ_f of each other, the solution with the lower function value *dominates* the other, else the solutions are assigned an identical non-dominated rank

2. It is interesting to observe that the above change will prevent multiple optima of equal value to be discovered. To be able to obtain multiple optimal solutions having identical function value, we introduce another parameter δ_x. We check the normalized Euclidean distance (in the variable space) of any two solutions having function values within δ_f (and equal values of $f_2(\mathbf{x})$). If the normalized distance is greater than the value of δ_x, we assign both solutions an identical non-dominated rank, otherwise both solutions are considered arising from the same optimal basin and we assign a large dominated rank to the solution having the worse objective value $f(\mathbf{x})$.

The above changes ensured that NSGA-II would be able to find a weak Pareto-optimal set (in our case the multiple optimal solutions) of the given bi-objective problem. It is clear from the above discussion, that the algorithm required the values of three parameters – δ_{hj}, δ_f and δ_x to be preset correctly for a successful operation. This is a limitation, in an otherwise novel work. We seek to remove this shortcoming by calculating the values of δ_{hj} and δ_x from the population of solutions and eliminating the need for δ_f in our algorithm. In the next section, we describe these new developments.

4   Self-adaptive Parameter Estimation

Niching is a mandatory task in multimodal optimization. Most of the niching methods use at least one Euclidean distance measure, usually called the *niching radius* to be able to maintain multiple favorable areas in the solution space. Like any other parameter based approach, niching suffers from the problem of setting the value of *niching radius*. Hence, researchers in the EC community have targeted to make it self-adaptive such as demonstrated by [3] in the realm of Particle Swarm Optimization (PSO), [4] in Genetic Algorithms (GA), and [5] in Evolution Strategies (ES). Others ([6] and [7]) have also tried to make certain aspect(s) of a GA self-adaptive to assist multimodal optimization.

In the next sub-sections, we describe how we were able to eliminate δ_f and calculate the values of δ_{hj} and δ_x from the population of solutions.

4.1   Elimination of the Objective Space Niching Parameter – δ_f

As earlier described, δ_f was introduced to allow two *almost* identical valued solutions to be eligible for the variable space niching check. We did a basic parametric study with different values of δ_f on one of the problems, the MMP(8) (introduced in our earlier work and later described in Section 5). We fixed the values of δ_{hj} and δ_x as per our earlier work. The results are show in Table 1. It is apparent that even though we keep increasing the value of δ_f, the performance of the algorithm doesn't get worse. In fact, the number of function evaluations to find all the desired optima tend to decrease as we increase the value of δ_f. (Increasing the value of δ_f implies that more solutions qualify for the variable space niching check, thus increasing the importance of the variable space niching check as the primary mode of maintaining multiple solutions)

Observing the above behavior of the algorithm, we eliminate δ_f – thus making every pair of solutions (irrespective of their relative function value) eligible for a variable space niching check. (This is equivalent to setting a very high value for δ_f). Our algorithm performed equivalently well without δ_f on our test problems – MMP(4), MMP(8) and MMP(16) (described in Section 5) introduced in the

Table 1. Performance of the H-J based procedure with respect to different values of δ_f for the MMP(8) problem

Parameter	Value	Succ. runs (out of 30)	Func. Evals. (Median, Std.dev)
δ_f ($\delta_x = 0.2$ $\delta_{hj} = 0.04$)	0.1	23	(5.1250e+05, 7.0612e+05)
	0.2	26	(5.0620e+05, 9.3033e+05)
	0.5	25	(5.0000e+05, 6.6298e+05)
	0.7	28	(4.8125e+05, 6.1725e+05)
	1.5	26	(4.3750e+05, 5.9616e+05)
	20	27	(4.2500e+05, 9.1617e+05)

earlier work. Based on this study, we conclude that the δ_f based check can be eliminated without compromising on the performance of the algorithm.

4.2   Estimation of the HJ Parameter – δ_{hj}

The value of δ_{hj} determines "how far" is the neighborhood of a solution searched for better solutions around it. One simple restriction is that if x_i is one of the vector-components of a possible optimum solution, then δ_{hj} should be such that the new point with $x_i \pm \delta_{hj}$ does not overshoot nearby optima. This is to prevent encroachment into another optima and hence get a correct estimation of $f_2(\mathbf{x})$. We consider this factor and calculate δ_{hj} for *a* solution, $\mathbf{x}$, as follows:

1. Find the closest member (and the corresponding Euclidean distance, say $d_x^{\min}$) in the population of solutions to the current solution $\mathbf{x}$.
2. Set $\delta_{hj} = 0.5 d_x^{\min}$.
3. The H-J procedure is now invoked with δ_{hj}.

The above procedure ensures that the value of δ_{hj} is such that *new* points are evaluated around the incoming solution during the neighborhood search. This not only prevents overshooting another possible optima, but also makes sure that the search does not evaluate another population member for calculating $f_2(\mathbf{x})$. Note that for every $\mathbf{x}$, the estimated value of δ_{hj} may be different.

4.3   Estimation of the Variable-Space Niching Parameter – δ_x

A multimodal problem may contain equal valued local and global minima. We introduced a simple variable space *niching* using a parameter δ_x which preserved multiple solutions having the same function value. The value of δ_x thus determines the least distance (in the variable space) that two solutions having the same function value should be to consider them as different solutions, and hence preserved.

Since we are solving for the minima, as the modified EMO progresses, the maxima will start occupying the higher ranked fronts (higher domination level) and the minima will be occupying the lower ranked fronts (higher non-domination level). On the assumption that for continuous functions, between any two minima lies a maxima, we can obtain a fair estimate of the distance of a point from its closest maxima by the calculating the minimum distance between the population members of the highest ranked front and the lowest ranked front. This is done every generation and we use this distance as δ_x. We set the lower limit on the value of δ_x to be 0.05. Note that δ_x is estimated to have a single value for the entire population.

Figure 1 is a schematic description of the calculation of δ_x. It is intuitive that if two solutions having the same function value are farther than δ_x in the variable space, then they can be safely considered to arise from different optimal basins, and hence non-dominated to each other in EMO parlance.

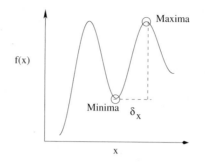

Fig. 1. Estimation of δ_x

5   Performance of the Algorithm

In this section, we present results of applying the new self-adaptive multimodal optimization algorithm to different test problems of varying dimensions and number of optima. In all cases, we use a population size, $N \leq 15 \max(n, M)$, where n is the dimension of the problem and M is the number of optima, SBX probability = 0.9, SBX index = 15, polynomial mutation probability = $\frac{1}{n}$, and mutation index = 20 (unless otherwise mentioned). The results are summarized in Table 3.

5.1   Modified Rastrigin Function

In our earlier work, we defined a scalable n-variable unconstrained multimodal test-problem, as follows:

$$\text{MMP}(n): \quad \begin{array}{l} \text{min. } f(\mathbf{x}) = \sum_{i=1}^{n} 10(1 + \cos(2\pi k_i x_i)) + 2k_i x_i^2, \\ \text{subject to } 0 \leq x_i \leq 1, \quad i = 1, 2, \ldots, n. \end{array} \tag{1}$$

Here, the total number of global and local minima are $M = \Pi_{i=1}^{n} k_i$. We use the following k_i values for the two MMP problems, each having 48 minimum points:

$$\begin{array}{c} \text{MMP(8): } k_2 = 2, \ k_4 = 2, \ k_6 = 3, \ k_8 = 4, \\ k_i = 1, \quad \text{for } i = 1, 3, 5, 7. \\ \text{MMP(16): } k_4 = 2, \ k_8 = 2, \ k_{12} = 3, \ k_{16} = 4, \\ k_i = 1, \quad \text{for } i = \text{1-3, 5-7, 9-11, 13-15.} \end{array}$$

Both these problems have one global minimum and 47 local minimum points.

Figures 2 and 3 summarize the result obtained by this modified procedure for the MMP(8) and MMP(16) problems. All 48 minimum points are discovered by the modified procedure. All 10 runs starting from different initial random populations find 48 minima in each case. For MMP(8), we use $N = 500$ and a maximum of 300 generations, and for MMP(16), we use $N = 750$ with a maximum of $1,000$ generations.

It would be interesting to see how the current self-adaptive algorithm with one less parameter fares compared to the earlier fixed-parameter approach. Table 2

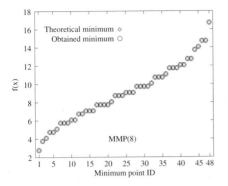

Fig. 2. All 48 minimum points are found for the eight-variable MMP(8) problem

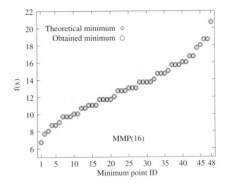

Fig. 3. All 48 minimum points are found for the 16-variable MMP(16) problem

Table 2. Results (over 10 runs) of the H-J based procedure on MMP(16) with fixed parameters and self-adaptive parameters

		Fixed Param.	Adaptive Param.
Success rate		86.67	100.0
Function Evaluations	Best	5.3802e+06	4.9500e+06
	Median	1.0628e+07	7.2150e+06
	Worst	2.5181e+07	9.5400e+06
	Standard Deviation	5.5476e+06	1.1437e+06
Average Error	Minimum	9.3035e-05	1.6803e-06
	Median	1.2113e-03	9.3366e-06
	Maximum	6.5139e-02	5.4722e-05
	Standard Deviation	1.64e-02	1.3195e-05

shows the comparative success rate, function evaluations taken to find all the 48-optima and the average error in the found optima for the MMP(16) problem.

It appears that overall the self-adaptive approach have increased the efficiency and accuracy of the algorithm.

48 Global Optimal Solutions: Next, we modify the objective function as follows:

$$f(\mathbf{x}) = \sum_{i=1}^{n} 10 + 9\cos(2\pi k_i x_i).$$

In this problem, we use $n = 16$ and k_i values are chosen as in MMP(16) above, so that there are 48 minima having an identical objective value of $f = 16$. Interestingly, there is no locally minimum solution in this problem. We believe this problem will provide a stiff challenge to the adaptive variable-space niching procedure, as all 48 global minimum points will lie on an identical point on the

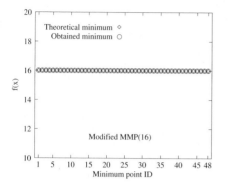

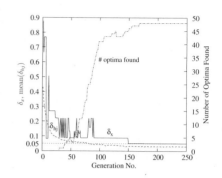

Fig. 4. All 48 global minimum points are found for the modified MMP(16) problem

Fig. 5. Representative plot showing the variation of δ_x and $\bar{\delta}_{hj}$ for the modified MMP(16) problem along with the number of optima found

objective space and the only way to distinguish them from each other would be to investigate the variable space and emphasize distant solutions.

Figure 4 shows that the proposed niching-based NSGA-II procedure is able to find all 48 minimum points for the above 16-variable modified problem. Figure 5 shows the variation of δ_x and population-average $\bar{\delta}_{hj}$ with generation. Starting with a large value, both these parameters get stabilized to a small value after a substantial number of optima are found.

5.2   Six-Hump Camel Back

The Six-Hump camel back function [3] has two global optima and 4 local optima. It is defined as:

$$\text{min. } f(\mathbf{x}) = 4[(4 - 2.1x^2 + \tfrac{x^4}{3})x^2 + xy + (-4 + 4y^2)y^2],$$
$$\text{subject to } -1.9 \leq x \leq 1.9, \qquad (2)$$
$$-1.1 \leq y \leq 1.1.$$

We report our results in Table 3 with $N = 90$ and a maximum of 100 generations. All the six optima are found.

5.3   Schwefel Function

The Schwefel function is defined as follows:

$$\text{min. } f(\mathbf{x}) = 418.9828872724339n - \sum_{i=1}^{n} \sin(\sqrt{|x_i|}),$$
$$\text{subject to } -500 \leq x_i \leq 500, \quad i = 1, 2, \ldots, n. \qquad (3)$$

This function has 7^n optima, with 1 global optima. We test our algorithm on $n = 2$ and $n = 5$. For $n = 2$, we report the results of our algorithm with $N = 750$ and a maximum of 1,000 generations. For $n = 5$, we report the results of our algorithm with $N = 1,000$ and a maximum of 5,000 generations. The global optima is found in all runs along with many other local optima.

5.4   Vincent Function

The Vincent function [5] is defined as follows:

$$\text{min. } f(\mathbf{x}) = -\frac{1}{n} \sum_{i=1}^{n} \sin(10 \log(x_i)), \\ \text{subject to } 0.25 \leq x_i \leq 10, \quad i = 1, 2, \ldots, n. \tag{4}$$

This function has 6^n global optima, with no local optima. We test our algorithm on $n = 2$ and $n = 5$. For $n = 2$, our algorithm finds all 36 optima with $N = 500$ and a maximum of 500 generations For $n = 5$, the algorithm finds a large number of global optima with $N = 1,000$ and a maximum of 5,000 generations.

Table 3. Test functions and Results (over 10 runs). A optimum solution is said to be found when the normalized Euclidean distance between the expected and the obtained optima is within 0.05. The std. dev. of the number of function evaluations have been reported when all the optima have been found within the maximum allowed number of generations.

Function	n	Range	# Optima	# Avg. Opt. obt. (Med., Std.dev)	Func. Evals. (Med., Std.dev.)
Mod. Rast.	4	$0 \leq x_i \leq 1$	48	(48, 0)	(2.9750e+05, 5.1508e+04)
Mod. Rast.	8	$0 \leq x_i \leq 1$	48	(48, 0)	(1.0920e+06, 2.1203e+05)
Mod. Rast.	16	$0 \leq x_i \leq 1$	48	(48, 0)	(7.2150e+06, 1.1437e+06)
Six-hump Camel back	2	$-1.9 \leq x_1 \leq 1.9$ $-1.1 \leq x_2 \leq 1.1$	6	(6, 0)	(7.2000e+04, 0.0000e+00)
Schwefel	2	$-500 \leq x_i \leq 500$	49	(30, 5)	(6.0060e+06, -)
Schwefel	5	$-500 \leq x_i \leq 500$	7^5	(82, 42)	(8.5000e+07, -)
Vincent	2	$0.25 \leq x_i \leq 10$	36	(36, 0)	(2.2080e+05, 7.3618e+04)
Vincent	5	$0.25 \leq x_i \leq 10$	6^5	(907, 11)	(8.5000e+07, -)

6   Conclusions

In this work we have successfully improved our earlier bi-objective multimodal optimization algorithm with a self-adaptive approach of calculating the niching parameters. The results obtained on different multimodal test problems with varying dimensions reinforce its generic applicability. An important future work is to test the self-adaptive algorithm on constrained multimodal test problems suggested earlier in [1]. An interesting area to explore would be to use the ideas of the proposed approach on multimodal problems in domains such as control systems, evolutionary neural networks and combinatorial optimization.

References

1. Deb, K., Saha, A.: Finding Multiple Solutions for Multimodal Optimization Problems Using a Multi-Objective Evolutionary Approach. In: Proceedings of the 12th Annual Conference on Genetic and Evolutionary Computation, GECCO 2010, pp. 447–454. ACM, New York (2010)
2. Deb, K., Pratap, A., Agarwal, S., Meyarivan, T.: A fast and elitist multi-objective genetic algorithm: NSGA-II. IEEE Transactions on Evolutionary Computation 6(2), 182–197 (2002)
3. Li, X.: Niching Without Niching Parameters: Particle Swarm Optimization Using a Ring Topology. IEEE Transactions on Evolutionary Computation 14(1), 150–169 (2010)
4. Goldberg, D.E., Wang, L.: Adaptive niching via coevolutionary sharing. In: Genetic Algorithms and Evolution Strategy in Engineering and Computer Science: Recent Advances and Industrial Applications, pp. 21–38. John Wiley & Son Ltd., Chichester (1997)
5. Shir, O., Back, T.: Niche radius adaptation in the CMA-ES niching algorithm. In: Proceedings of the Parallel Problem Solving from Nature-PPSN IX, pp. 142–151. Springer, Heidelberg (2006)
6. Leung, K.S., Liang, Y.: Adaptive elitist-population based genetic algorithm for multimodal function optimization. In: Cantú-Paz, E., Foster, J.A., Deb, K., Davis, L., Roy, R., O'Reilly, U.-M., Beyer, H.-G., Kendall, G., Wilson, S.W., Harman, M., Wegener, J., Dasgupta, D., Potter, M.A., Schultz, A., Dowsland, K.A., Jonoska, N., Miller, J., Standish, R.K. (eds.) GECCO 2003. LNCS, vol. 2724, pp. 1160–1171. Springer, Heidelberg (2003)
7. Srinivas, M., Patnaik, L.M.: Adaptive probabilities of crossover and mutation in genetic algorithms. IEEE Transactions on Systems, Man and Cybernetics 24(4), 656–667 (1994)
8. Goldberg, D.E., Richardson, J.: Genetic algorithms with sharing for multimodal function optimization. In: Proceedings of the Second International Conference on Genetic Algorithms and Their Applications, pp. 41–49. L. Erlbaum Associates Inc., NJ (1987)
9. Deb, K., Goldberg, D.E.: An investigation of niche and species formation in genetic function optimization. In: Proceedings of the Third International Conference on Genetic Algorithms, pp. 42–50. Morgan Kaufmann Publishers Inc., USA (1989)
10. Pétrowski, A.: A clearing procedure as a niching method for genetic algorithms. In: Proceedings of the IEEE 3rd International Conference on Evolutionary Computation (ICEC 1996), pp. 798–803. IEEE Press, Los Alamitos (1996)
11. Streichert, F., Stein, G., Ulmer, H., Zell, A.: A clustering based niching EA for multimodal search spaces. In: Liardet, P., Collet, P., Fonlupt, C., Lutton, E., Schoenauer, M. (eds.) EA 2003. LNCS, vol. 2936, pp. 293–304. Springer, Heidelberg (2004)

On the Flexible Applied Boundary and Support Conditions of Compliant Mechanisms Using Customized Evolutionary Algorithm

Deepak Sharma

Department of Mechanical Engineering
Indian Institute of Technology Kanpur, India
dsharma@iitk.ac.in

Abstract. In structure topology optimization, the applied boundary and support conditions are often fixed in a-priori. These conditions can affect the behavior and the properties of single-piece elastic structures known as compliant mechanisms. In this paper, the same aspect is explored for path generating compliant mechanisms by considering them as design variables and their values are evolved using customized NSGA-II algorithm. Three examples are solved and the innovative facts among the applied boundary and support conditions are presented. The elastic structures are also presented in this paper.

1 Introduction

Structural topology optimization is a fast growing field that is finding numerous applications in automotive, aerospace and mechanical design processes. It optimizes the material distribution or layout within a given design-domain under the applied boundary and support conditions [1].

Quite often, it has been observed in the literature of structural topology optimization that the applied boundary and support conditions are fixed a-priori. Sometimes, these conditions are known or the design constraints and variables limit them. However, the applied boundary and support conditions can affect the behavior and optimal properties of structures. It has been shown elsewhere [2] that the optimum set of support and loading positions generated the improved compliant mechanisms in their objective values. It can also influence the final shape of elastic structures [3]. However, various design principals and facts can be discovered on the basis of design goals and variables [4]. In this paper, an attempt is made to explore the innovative facts by considering the applied and boundary conditions as design variables for three examples of path generating compliant mechanisms. Unique facts of these conditions for compliant mechanisms are explored that can be beneficial to the designers. The elastic structures of three examples of path generating complaint mechanisms are also presented. In the remaining part of the paper, section 2 described the methodology followed in this paper. The experimental results are discussed and optimum elastic structures are presented in section 3. The paper is concluded in section 4.

K. Deb et al. (Eds.): SEAL 2010, LNCS 6457, pp. 105–114, 2010.

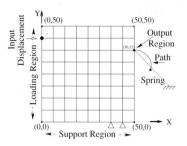

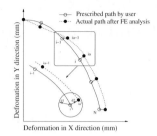

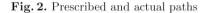

Fig. 1. A design-domain

Fig. 2. Prescribed and actual paths

2 Methodology

2.1 Formulation

The formulation is designed for path generating compliant mechanism (PGCMs) that trace-out the prescribed path by undergo through elastic deformation. In this paper, the design-domain for PGCMs is categorized into three regions called support, loading and output regions (cf. Fig. 1). The elastic structures are supported in support region whereas the load is applied at the loading region. The elastic structures trace-out the prescribed path at the output region. As Fig. 2 shows, the constraints are imposed at precision points of prescribed path so that the actual path generates the similar path ($d_2 \leq d_1$) [5].

In this paper, the compliant mechanisms are designed using two bi-objective sets. In both sets, the primary objective is to minimize the weight of elastic structures. The another objective for first bi-objective set is the minimization of supplied input energy [5] that is calculated with respect to the stress and the strain developed during the large deformation of the elastic structures. For second bi-objective set, the maximization of geometrical diversity [2] is chosen that is calculated by comparing the dissimilarity in the bits of binary strings of the *reference design* and the elastic structure of GA population. In this paper, the compliant mechanism evolved by single-objective optimization is chosen as a *reference design*. The single and bi-objective sets and the constraints are given in appendix A.

2.2 Customized Evolutionary Algorithm

Among the multi-objective evolutionary algorithms, NSGA-II [6] is the fastest and has shown to have a good convergence property to the global 'Pareto-optimal' front for various two objective test and engineering problems [7]. Thus, NSGA-II is used as a global search and optimizer in this paper. However, there is a need to modify the existing NSGA-II for structure topology optimization. A local search method is also used which acts as a post-processing method to refine the non-dominated compliant mechanisms evolved by the modified NSGA-II. The flow chart of the customized NSGA-II algorithm is shown in Fig. 3.

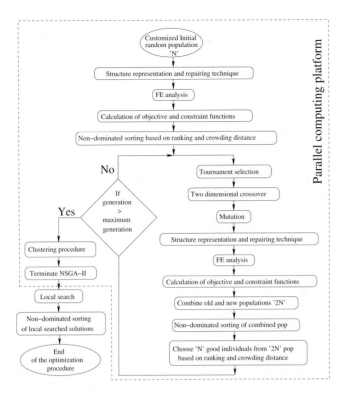

Fig. 3. A flow chart of customized NSGA-II algorithm

Start with NSGA-II parameters which are given in Table 1. A binary string of 12 bits is used to evolve the applied boundary and support conditions in their respective regions of the given design domain (cf. Fig. 1). To calculate the values, 12 bits are divided into three groups of five, three and four bits respectively as given in Table 2. The decoded value of first five bits indicates the location of an element from the origin where the elastic structure is to be supported. The decoded value of subsequent three bits helps to determine the loading position, that is, a node where the input load is applied. The decoded value of last four bits are used to evaluate the magnitude of input displacement which can vary from 1 mm to 16 mm at step of 1 mm.

Table 1. GA parameters

Population	240	Generation	100
Crossover probability	0.95	Mutation probability	1/string length
String length for a structure	625	String length for applied boundary & support conditions	12

Table 2. A binary string of a GA population member

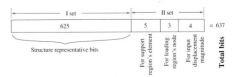

For structure representation, a binary string of 625 bits is used to represent the material distribution for the elastic structure. A binary string is copied to two dimensional array followed by the material assignment as shown in Fig. 4. The bit value '1' signifies that material is present whereas, '0' represents the void. This scheme divides a design domain of structure into 25×25 ($= 625$) grids in x and y directions, respectively.

In this paper, the domain specific initial population strategy is used which has shown its advantage over random initialization of material in the design domain [3]. The initial population strategy is described by showing the material connectivity between the support and loading regions in Fig. 5. The intermediate points (between 1 to 5) are randomly generating within the design domain. In Fig. 5, four points (P1, P2, P3 and P4) are generated and they connect the support (S1) and the loading (L1) positions by straight lines. Thereafter, a material is assigned to those elements where these straight lines pass as shown in Fig. 5. Similarly, a set of piece-wise linear line segments between the support and output regions and another set between the loading and output regions are explained. Here, the element positions of support and loading regions are calculated after decoding the binary string of 12 bits (cf. Table 2). The location of output region is fixed in this study because this point will trace-out the user-defined path. This initial population strategy ensures the geometrically feasible structures in the initial population.

As two bi-objective sets are used to capture the facts between applied boundary and support conditions, the two crossover operators are also used in this paper. For each example of PGCMs, NSGA-II is coupled with both operators individually and the conditions are evolved on different optimization platforms. The first two-dimensional crossover has shown its successful application in shape optimization [8,9] and compliant mechanisms [5,2,3]. It works on exchanging the rows or column (refer Fig. 6) with equal probability. The size and location of common patch are found randomly and it is swapped between the two parents. Another crossover operator is a domain specific crossover that divides the given two-dimensional design domain into four sub-regions. Points P1, P2 and P3 of Fig. 8 are chosen randomly on their respective edges and are joined by straight lines. With an equal probability, two sub-regions out of four are swapped between the two parents. For the crossover of 12 bit binary string, a standard

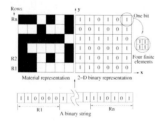

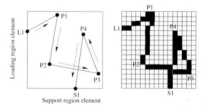

Fig. 4. A representation of structure using binary string

Fig. 5. Connectivity between support and loading regions

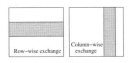

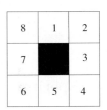

Fig. 6. Row/Column-wise crossover operator

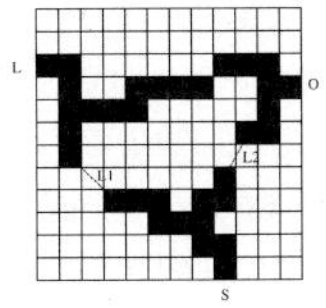

Fig. 7. Disconnected topology: a hypothetical case

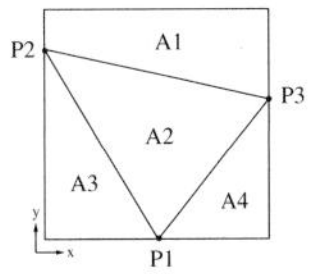

Fig. 8. Domain specific crossover operator

single point crossover operator is used. In this paper, the mutation of each bit of binary string representing the structure is done with a probability of 1/string length.

Because of the crossover and mutation operators, the new solutions can suffer from disconnected topology problem. As shown in Fig. 7, the support region (**S**) is not connected to the loading (**L**) and the output (**O**) regions. In this disconnected scenario, the individual distances are calculated from the centroid of each grid of material of **S** to the centroid of each grid of material of **L** and **O**. Then, the straight lines (**L1, L2**) are drawn from the centroid of those two grids which show minimum distances between **S-L** and **S-O**. In this way, the connectivity among **S**, **L** and **O** regions of a structure is checked.

The point singularity between the two material element's can arise due to the developed initial population strategy, GA operators and after the connectivity technique. An ad-hoc repairing technique motivated from the image processing concept [10] is employed in this paper. In Fig. 9, the material at positions 2, 4, 6 and 8 can create point singularity. If position 2 creates point connectivity, then an extra material can be filled at 1 or 3 with equal probability. In this way, the point singularity for each element of material is eliminated. Due to the mutation operator, any floating material element can appear which is not connected to the seed elements. In this case, this isolated element is changed to void by assigning value '0'.

8	1	2
7	■	3
6	5	4

Fig. 9. Eight neighborhood connectivity

After above steps, the elastic structures are now undergo for finite element analysis (FEA). In this study, one grid of a structure is further discretized into four finite elements with same Boolean variable value as shown in Fig. 4. In the present process, the structure is discretized with 4×625 (= 2500) 4-node rectangular finite elements and analyzed through a non-linear large deformation FE analysis using *ANSYS* package. However, the GA operations are performed on 625 bits representing the same structure.

The function evaluations and FE simulations are performed on the parallel computing platform. Master-slave architecture is used in the present paper in which the population members are evaluated on the slave processors and rest of the operations are done on the master processor. A MPI based Linux cluster with 24 processors is used in the present study.

When the non-dominated solutions are evolved by the customized NSGA-II, these solutions are refined by local search method. The weighted-sum approach is used in this paper to reduce the multi-objective problem into single-objective. Weights are calculated according to the positions of non-dominated solutions evolved by NSGA-II in the objective space (refer Eqn. 1) [7].

$$\overline{w}_j^x = \frac{(f_{j_{max}}^x - f_j^x) \setminus (f_{j_{max}}^x - f_{j_{min}}^x)}{\sum_{k=0}^M (f_{k_{max}}^x - f_k^x) \setminus (f_{k_{max}}^x - f_{k_{min}}^x)}, \tag{1}$$

where $\overline{w}_j^x$ is the corresponding weight to the jth objective function, f_j^x is j^{th} objective function, $f_{j_{min}}^x$ and $f_{j_{max}}^x$ are minimum and maximum values of jth objective function of non-dominated front, M is the number of non-dominated solutions.

In the local search method, the weighted sum of scaled fitness of a selected representative solution is evaluated. Thereafter, the two-dimensional array of solution is checked for the grids having a material. For each material's grid, there are maximum of eight possible neighborhoods as shown in Fig. 9. One by one, all neighboring bits including its own bit, are mutated. The new elastic structure is now extracted on which FEA is performed for objective function and constraints values. The new elastic structure will discard if it is infeasible. If it is feasible, then the changes are accepted when the weighted sum of scaled fitness of new elastic structure is better. When the scaled fitness of elastic structure before checking the material's grid is same as after mutating all bits having material and their neighborhood, the local search method is terminated. In the same way, all representative solutions are mutated.

3   Experimental Results

In this section, three examples of compliant mechanisms tracing (i) curvilinear path, (ii) straight line path and (iii) upward curvilinear path are solved with different objective sets using customized NSGA-II with two different crossover operators. The applied boundary and support conditions are evolved for the wide-variety of optimization frame-works and the innovative facts are discovered.

Deb and Srinivasan [11] introduced a new design methodology called "innovization" in which the new and innovative design principles are developed by means of optimization techniques. In this paper, an attempt is made to find such principals or facts that are based on the applied boundary and support conditions of various PGCMs. It can help the designers and decision makers to get more insight into the topology optimization of compliant mechanism tracing user-defined path.

The optimum set of applied boundary and support conditions of all single and bi-objective studies are given in Table 3. In this table, 'OX' is referred for the row/column crossover-wise operator based studies and similarly, 'NX' is used for the domain specific crossover operator based studies.

Table 3. Applied boundary and support conditions for different optimization frameworks

Example: Curvilinear Path Tracing Compliant Mechanisms (CPTCM)						
Conditions	Single-objective study		I^{st} bi-objective study		IInd bi-objective study	
	OX	NX	OX	NX	OX	NX
Support position	20	2	2	2	16	18
Loading position	32	24	32	40	24	32
Input displacement magnitude	7	5	7	9	5	7
Example: Straight Line Path Tracing Compliant Mechanisms (SLPTCM)						
Conditions	Single-objective study		I^{st} bi-objective study		IInd bi-objective study	
	OX	NX	OX	NX	OX	NX
Support position	46	46	46	46	46	46
Loading position	40	28	20	20	20	28
Input displacement magnitude	8	5	4	4	4	5
Example: Upward Non-Linear Path Tracing Compliant Mechanisms (UNPTCM)						
Conditions	Single-objective study		I^{st} bi-objective study		IInd bi-objective study	
	OX	NX	OX	NX	OX	NX
Support position	44	46	44	46	46	44
Loading position	44	48	44	44	48	44
Input displacement magnitude	5	6	5	5	6	5

Let us first identify the common support positions in Table 3. For the curvilinear path tracing compliant mechanisms (CPTCM), the support position of 2 mm is common in single-objective 'NX', I^{st} bi-objective 'OX', and I^{st} bi-objective 'NX' studies. The corresponding loading positions are at 24 mm, 32 mm and 40 mm, respectively. The required input displacement magnitudes of single-objective 'NX', I^{st} bi-objective 'OX', and I^{st} bi-objective 'NX' studies are 5 mm, 7 mm and 9 mm, respectively.

Similar information can also be unfold from the examples of straight-line path tracing compliant mechanisms (SLPTCM) and upward non-linear path tracing compliant mechanisms (UNPTCM). In case of SLPTCM, the identical support position at 46 mm is evolved for all studies. The corresponding loading positions are at 40 mm, 28 mm and 20 mm from the origin. The respective magnitudes of input displacement are 8 mm, 5 mm and 4 mm to trace the straight line prescribed path. Similarly, an example of UNPTCM shows the common support position at 44 mm for single-objective 'OX', I^{st} bi-objective 'OX' and IInd bi-objective 'NX' studies and another support position at 46 mm for single-objective 'NX', I^{st} bi-objective 'NX' and IInd bi-objective 'OX' studies. When the topologies are supported at 44 mm, then they are loaded at 44 mm and required 5 mm of input displacement magnitude. On the other hand, the elastic structures that are supported at 46 mm, require 6 mm and 5 mm of input displacement magnitudes for the loading positions at 48 mm and 44 mm respectively to trace the upward non-linear path.

For identical support positioned compliant mechanisms, we observe that the magnitude of input displacement required to trace the prescribed path increases

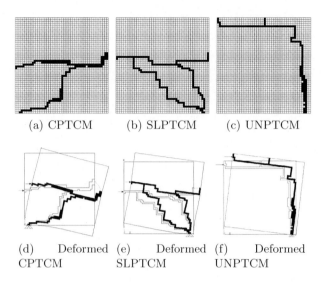

(a) CPTCM   (b) SLPTCM   (c) UNPTCM

(d)   Deformed   (e)   Deformed   (f)   Deformed
CPTCM   SLPTCM   UNPTCM

Fig. 10. Topologies

as the load is applied away from the origin. This common feature is independent of the nature of compliant mechanisms tracing variety of paths.

Another interesting information can be drawn out when we observe the common loading positions of each study mentioned in Table 3. In the example of CPTCM, the first identical loading position is at 32 mm for which the elastic structures are supported at 20 mm, 2 mm and 18 mm from the origin and require 7 mm of input displacement. The elastic structures with second common loading position of 24 mm are supported at 2 mm and 16 mm and require 5 mm of input displacement. The example SLPTCM indicates 28 mm and 20 mm common loading positions for which all compliant mechanisms are supported at same position and require 5 mm and 4 mm values of input displacement, respectively. Similarly in the example of UNPTCM, the topologies which are loaded at 44 mm, are supported at 44 mm and 46 mm from the origin and require 5 mm of input displacement whereas, the topologies with common loading position of 48 mm are supported at 46 mm and require 6 mm of input displacement. The observation reveals that if the loading position of compliant mechanisms is same, then these mechanisms require same magnitude of input displacement to trace the prescribed path irrespective of the different support positions.

The topologies of single-objective optimization using NSGA-II with domain-specific crossover operator are shown in Fig. 10 for three examples of PGCMs. According to the nature of generating prescribed paths, the topologies and shapes of CPTCM, SLPTCM and UNPTCM are evolved. Although the applied boundary and support conditions of CPTCM and UNPTCM and their prescribed paths are different but they have same topology. If we look at the support positions of above three examples, CPTCM is supported on the bottom left-hand side (cf. Fig. 10(d)) and SLPTCM is on the bottom right-hand side (refer Fig. 10(e)). It is because the elastic structure supported in left-hand side generates higher downward curvilinear paths compared to right-hand side supported PGCMs. In case of UNPTCM, the output region is positioned at the middle of top edge of the given design domain. In this scenario, the output point can only trace the upward non-linear path when the support position lies on the right hand side of the output point (refer Fig. 10(f)). Such behaviors of the elastic structures are expected and the evolved support conditions also abide the same principal [2,3].

Another important fact of considering the applied boundary and support conditions as design variables can observe when these conditions are unknown. The designer does not have to define these conditions a-priori. Moreover, the optimum set of evolved conditions can explore the possibilities of non-optimum applied boundary conditions that might be considered in the previous practice of designers [3].

4   Conclusions

This paper has explored the innovative facts of the applied boundary and support conditions for variety of PGCMs, irrespective of the different optimization frameworks. Moreover, the optimum sets of these conditions were evolved without any priori information. The possibility of non-optimum conditions was also explored which might be considered in the previous practices. Beside all facts, the evolved support positions by the customized NSGA-II for three examples of PGCMs abided the expected rule of elastic deformation of structures. These unfold facts and information can be beneficial to the designers to get deeper insight into the problem. In the future work, the concept of flexible applied boundary and support conditions can be used for variety of structure topology optimization problems.

Acknowledgement

Author would like to thank Prof. Kalyanmoy Deb and Prof. N. N. Kishore for their valuable suggestions to explore the significance of applied boundary and support conditions.

References

1. Hassani, B., Hinton, E.: Homogenization and Structural Topology Optimization: Theory, Practice and Software. Springer, UK (1998)
2. Sharma, D., Deb, K., Kishore, N.N.: Towards generating diverse topologies of path tracing compliant mechanisms using a local search based multi-objective genetic algorithm procedure. In: IEEE Congress on Evolutionary Computation (CEC 2008), June 1-6, pp. 2004–2011 (2008)
3. Sharma, D., Deb, K., Kishore, N.N.: An improved initial population strategy for compliant mechanism designs using evolutionary optimization. In: ASME International Design Engineering Technical Conferences (IDETC) and Computers and Information in Engineering Conference (CIE), New York, USA, Paper No. DETC2008-49187, August 3-6 (2008)

4. Deb, K., Chaudhauri, S., Jain, P., Gupta, N., Maji, H., Reddy, A.R.: Discovering innovative design principles using multi-conflicting objectives. In: Proceedings of Evolutionary Method for Design, Optimization and Control with Applications to Industrial Problems, EUROGEN 2003, pp. 118–119 (2003)

5. Sharma, D., Deb, K., Kishore, N.N.: Evolving path generation compliant mechanisms (PGCM) using local-search based multi-objective genetic algorithm. In: International Conference on Trends in Product Life Cycle, Modeling, Simulation and Synthesis (PLMSS), pp. 227–238 (December 2006)

6. Deb, K., Agrawal, S., Pratap, A., Meyarivan, T.: A fast and elitist multi-objective genetic algorithm: NSGA-II. IEEE Transactions on Evolutionary Computation 6(2), 182–197 (2002)

7. Deb, K.: Multi-Objective Optimization using Evolutionary Algorithms. Wiley, Chichester (2001)

8. Deb, K., Goel, T.: A hybrid multi-objective evolutionary approach to engineering shape design. In: Zitzler, E., Deb, K., Thiele, L., Coello Coello, C.A., Corne, D.W. (eds.) EMO 2001. LNCS, vol. 1993, pp. 385–399. Springer, Heidelberg (2001)

9. Deb, K., Chaudhuri, S.: Automated discovery of innovative designs of mechanical components using evolutionary multi-objective algorithms. In: Nedjah, N., de Macedo, M.L. (eds.) Evolutionary Machine Design: Methodology and Applications, pp. 143–168. Nova Science Publishers, Inc., New York (2005)

10. Sigmund, O.: On the design of compliant mechanisms using topology optimization. Mechanics Based Design of Structures and Machines 25(4), 493–524 (1997)

11. Deb, K., Srinivasan, A.: Innovization: Innovating design principles through optimization. In: Proceedings of the Genetic and Evolutionary Computation Conference (GECCO 2006), pp. 1629–1636. ACM, New York (2006)

A   PGCM Formulation

Single-objective optimization:
Minimize: Weight of structure
I^{st} Bi-objective set:
Minimize: Weight of structure (primary objective),
Minimize: Supplied Input energy (secondary objective),
IInd Bi-objective set:
Minimize: Weight of structure (primary objective),
Maximize: Geometrical Diversity of structure (helper objective),
Optimization problems are subjected to:

$$1 - \frac{\sqrt{(x_{ia}-x_i)^2+(y_{ia}-y_i)^2}}{\eta \times \sqrt{(x_i-x_{i-1})^2+(y_i-y_{i-1})^2}} \geq 0, \quad i = 1, 2, ..., N$$

$\sigma_{flexural} - \sigma \geq 0,$

where $\eta = 15\%$ is the permissible deviation and N is number of precision points representing the prescribed path. $\sigma_{flexural}$ and σ are flexural yield strength of material and maximum stress developed in the elastic structure, respectively. x_i and y_i are the coordinates of precision points whereas the coordinates of x_{ia} and y_{ia} are the corresponding points on the actual path traced by elastic structure. Note that the points on actual path are found from the non-linear finite element analysis of elastics structures based on equal load steps.

Intensification Strategies for Extremal Optimisation

Marcus Randall[1] and Andrew Lewis[2]

[1] School of Information Technology, Bond University
Queensland, Australia
mrandall@bond.edu.au
[2] Institute for Integrated and Intelligent Systems, Griffith University
Queensland, Australia
a.lewis@griffith.edu.au

Abstract. It is only relatively recently that extremal optimisation (EO) has been applied to combinatorial optimisation problems. As such, there have been only a few attempts to extend the paradigm to include standard search mechanisms that are routinely used by other techniques such as genetic algorithms, tabu search and ant colony optimisation. The key way to begin this process is to augment EO with attributes that it naturally lacks. While EO does not get confounded by local optima and is able to move through search space unencumbered, one of the major issues is to provide it with better search intensification strategies. In this paper, two strategies that compliment EO's mechanics are introduced and are used to augment an existing solver framework. Results, for single and population versions of the algorithm, demonstrate that intensification aids the performance of EO.

1 Introduction

Extremal optimisation (EO) [4] has two main attractions for the researcher and practitioner. The first is that it is very easy to understand and implement. The base algorithm, in comparison to other meta-heuristics, is very simple. This is described in detail in the next section. Perhaps more important is the fact that, unlike its counterparts, it does not converge on a single, or set of, locally optimal solution(s). This avoids the problem of premature convergence. Instead it moves continually through search space, largely being repelled by poor regions of the search space, rather than being specifically attracted to good areas.

The natural, continual diversification of EO's search suggests that work on the algorithm should be devoted to improving its ability to focus on good and promising areas of the search space. This increase in search focus is traditionally referred to as *intensification*, and is present in varying degrees in all meta-heuristic techniques. There exists a reasonably large body of literature for explicit intensification strategies applied to a range of these meta-heuristics, rather than to EO itself. According to Glover and Laguna [10], these approaches can be grouped into two broad classes. The first set of techniques gathers elite solutions found during

K. Deb et al. (Eds.): SEAL 2010, LNCS 6457, pp. 115–124, 2010.
© Springer-Verlag Berlin Heidelberg 2010

the search and allows the meta-heuristic to revisit them when in intensification mode. This is characterised by works of Gambardella, Taillard and Dorigo [9] and Blum [3]. The former propose a hybrid ant system for the quadratic assignment problem (QAP) that is known as HAS-QAP. In an intensification trial phase, the initial solution it uses is the best solution found to date, so it thus corresponds to using one elite solution. Blum [3] follows a similar idea to this, except that the search is allowed to intensify around a set of elite solutions rather than just one.

The other main approach is based on the frequency of incorporation of solution component values[1] into solutions, and their association with quality. If solution components are generally associated with better solutions, it would be prudent to intensify search around a combination of these values. This is exemplified by Randall [11] and Beausoleil [2]. The former work, based on ant colony systems (ACS) (an ant colony optimisation meta-heuristic), analysed the frequency of incorporating components into the colony's solutions. In an intensification phase, frequently incorporated components were highly weighted in the probability selection equations. This is because, according to the ACS rules, they have been associated with good quality solutions in the past. Three search phases (normal, intensification and diversification) were triggered on the basis of the progress of the search. Intensification is performed if the search frequently receives good solutions above some threshold rate, which suggested that the search was in a promising area of the space. Overall, improved solution costs were obtained for larger travelling salesman problem (TSP) instances over a standard ACS strategy. In the latter work [2], parts of the solution structure were kept constant to allow the tabu search mechanism to concentrate on regions containing high quality solution components (as identified through the frequency memory). Applying this strategy, along with frequency-based diversification and path relinking, to the one-machine problem with a weighted tardiness objective produced very good performance in terms of solution quality and time.

The remainder of the paper is organised as follows. Section 2 gives an overview of the mechanics of EO and a description of a framework to help it process constrained problems. Section 3 describes two different intensification methods and how they are integrated with, and augment EO. Using a standard set of generalised assignment problems (GAPs), the intensification-enhanced methods are compared to canonical EO (as well as other algorithms) in Section 4. Finally, the conclusions and future research directions are given in Section 5.

2   Extremal Optimisation

Boettcher and Percus [4, 6, 7] describe the general tenets of EO. In many ways, it operates counter to other meta-heuristic search algorithms. Instead of actively seeking good solutions through either incremental improvement (such as tabu

---

[1] A solution component simply refers to each separate value in the solution vector. In essence, a component represents a building block of the problem. For example, in the generalised assignment problem [8], the assignment of an agent to a job represents a single component.

search and genetic algorithms) or construction (such as ant colony optimisation), bad solutions are actively discouraged. The main advantage of this, as previously mentioned, is that EO will not prematurely converge on a locally optimal solution or a set thereof.

At each iteration of the algorithm, a poor solution component value (as defined by some incremental cost measure) has its value simply changed to a random value (which, of course, is different from the initial value). In the original version of EO, this chosen solution component was always the worst. However, the performance of EO in this form was not favourable. Allowing the component to be chosen probabilistically, according to Equation 1, helped to increase EO's performance and competitiveness with other meta-heuristics.

$$P_i = i^{-\tau} \qquad 1 \leq i \leq n \qquad (1)$$

Where: i is the rank of the component, τ is a parameter that alters the probability distribution, P_i is the probability ($P_i = [0, 1]$) that component i is chosen and n is the number of components.

All components are ranked from worst (rank 1) to best (rank n). The set of probabilities are calculated from the ranks and the parameter τ. Essentially, a τ value of 0 gives a random search whereas τ approaching ∞ corresponds to greedy search.

From the above, the general EO (Algorithm 1) is formed.

Algorithm 1. General EO algorithm for a minimisation problem

1: $best\_cost =$ Form an initial solution, x, to the problem
2: Calculate the probability, P, for each rank based on the value of τ and n (using Equation 1)
3: **while** the stopping criterion is not met **do**
4:     Rank the n solution components from worst to best, based on x and the objective function
5:     $i =$ Choose the component value to change using roulette wheel selection (or similar)
6:     Assign $x(i)$ a different, random, value
7:     $cost =$ Evaluate the cost of x according to the objective function
8:     **if** $cost < best\_cost$ **then**
9:         $best\_cost = cost$
10:    **end if**
11: **end while**
12: Report $best\_cost$
13: **end**

2.1   A Framework for Constrained Problem Solution

The following describes three elements that have previously been used [12, 13] to enable EO to solve constrained optimisation problems. The discussion is cast in terms of the problems solved in this paper, the GAP. However, they are sufficiently general to be able to be applied to other problems.

1. *Feasibility Management* – While EO only makes a small change at each iteration - allowing many transitions to be made in a computationally reasonable time - their randomness allows transitions that potentially result in infeasible solutions. Overall, many feasible solutions will likely be produced depending on the difficulty of the constraints. Increasing the proportion of feasible solutions may be accomplished by the use of the partial feasibility restoration algorithm. This is a simple, non-degenerative, parameter-free heuristic that reduces the amount of infeasibility of a solution. Effectively it helps speed EO back to feasible space, but it does not guarantee that a feasible solution will be produced. As such, it subsequently has a low computational complexity of $O(MN)$ for the GAP, where M is the number of agents and N is the number of jobs.

2. *Population Model* – Unlike many other meta-heuristics, canonical EO only manipulates a single solution. Techniques such as genetic algorithms, ant colony optimisation and particle swarm optimisation derive much of their search power from co-operating and interacting solutions. A simple way to add this extension to EO is to employ some aspects of the Bak-Sneppen model of evolution [1]. Put simply, the weakest member of a population and its two closest neighbours die and are replaced by new members. These new members are two generated at random and one that is a copy of the best found solution. The population interactions are triggered probabilistically according to a divergence measure.

3. *Local Search* – Standard greedy descent search can be applied each time a feasible solution is generated. For the GAP, two operators are possible. "Move" moves a job from one agent to another and "Swap" swaps two jobs (from different agents). This is a variable length search that stops when an improving move cannot be found. These used in combination are very effective and add relatively little to the computational time [12, 13].

Another, new component that can be added to this standard framework is a general method to introduce search intensification. This is described in the next section.

3   Intensification Methods

Intensification schemes that take advantage of the native EO algorithm can be developed using frequency and historic information. Two separate methods that integrate into, and extend, the previously described EO framework are thus developed and outlined below.

3.1   Intensification Based on Frequency Information

Over a number of iterations of EO, an association between solution component values and quality of solutions can be built. The solution component values that are generally present in good quality solutions may be worthy of further investigation. Specifically, if these values are fixed temporarily within the search process, it allows EO to better explore the (smaller) region around those values. This may in turn reveal improved solutions.

The association of solution component values and overall solution quality can be kept in a matrix indexed by component number and value. Each cell of the matrix holds the average solution quality that that combination of component and value has received to date. This can be used to calculate the probability of temporarily "locking in" that value. There are numerous, sensible, ways that such a probability can be calculated, with one such given in Algorithm 2.

Algorithm 2. Calculation of the probabilities to lock in solution component values. This assumes a minimisation problem

---
1: **for** each solution component c **do**
2:     v_c = Choose the lowest/best from the association matrix
3:     $p_c = \frac{best\_cost}{v_c}$
4: **end for**
5: Rescale the probability vector, p, between its minimum and maximum value

---

After an initial period of iterations (typically a few hundred iterations), to allow the association matrix to receive sensible values, EO has the possibility of temporarily locking in solution component values. At each iteration that a feasible solution is produced, each of the probability values are compared against uniformly distributed random numbers to determine which combinations of component and values should be locked in. This works with EO's ranking system in a novel way. If a component is locked, it is given the highest available EO rank. In reality this is a "soft" lock, as EO may, in an unlikely case, choose this component value to change. In the event that all components become locked, this protects the system from not being able to make a transition (see step 5 of Algorithm 1).

The search, of course, will quickly stagnate if components are locked for an excessive number of iterations. Conversely, if the period is too short, EO will not have sufficient opportunity to explore the search region bounded by the locked values. The system allows this choice to be a user-defined parameter. Once this number of iterations has expired, the components are then freed. The alternative to fixing this period would be to explore self-adaptation governed by sufficient solution improvement.

3.2   Intensification Based on Historic Information

During the course of EO's search, it will typically find many solutions that have objective costs that are the best to date. These solutions, apart from being comparatively good, may indicate that the surrounding space may also contain other solutions of similar, and potentially superior, quality. However, canonical EO will continue through the space without adequately exploring such regions.

The best solutions found during the course of the search can be saved in a specially created archive. EO can revisit these solutions by selecting them from the archive to substitute for poor solutions to be discarded. It is extremely unlikely that EO will follow the same search trajectory after one of the archived solutions is reinstated, since further probabilistically-chosen bad solutions will be replaced with random values. Thus, the region around an archived solution may be better explored by further sampling. The questions that naturally arise in terms of its implementation are: a) when should this intensi cation be activated in the search process; b) how many solutions should constitute the archive and c) how should they be chosen to be used?

Intensification need only be activated when the search process appears to be unable to find improved solutions for an extended period. The current region of search space likely does not contain better solutions than received previously; thus, returning to one of the archived solutions would be more profitable. There are many possible models that could provide this intensification "trigger" mechanism. One such interesting way, given in Equation 2, calculates a probability based on the number of iterations since the last solution improvement was received. As the number of iterations since the production of this solution increases, the probability of activating intensification geometrically increases.

$$p = N \times \frac{I}{T} \tag{2}$$

Where: p is the probability that intensification will be activated, N is the number of non-improving moves since the last improved solution was received, I is the number of (approximate) intensification periods in the search, and is a user definable parameter and T is the total number of iterations.

A fixed-size archive is a simple and practical way of storing the solutions to which search may be returned. There is no way of accurately determining this size *a-priori*, thus it becomes another user defined parameter. As new improved solutions are obtained, and the archive is at capacity, solutions in it will need to be replaced. An effective way to do this is to replace the current worst solution in the archive. As only ever improving solutions are added to the archive, the replaced solution will also be the oldest member.

4   Computational Experiments

The computer used to perform the experiments was a 3 GHz Pentium 4-based PC. Each problem instance was run across ten random seeds. The experimental programs were coded in the C language and compiled with gcc. The only two standard EO parameters were τ and the number of iterations for which it was to be run. τ was set as 1.5 (a value consistent with Boettcher and Percus [5]), and the latter as 500,000. This value is the same as used by Randall [12] and Randall et al. [13]. Results are reported as relative percentage deviations (RPDs) from the known best solution cost for each problem instance. Formally this is given as $\frac{a-b}{b} \times 100\%$ where a is the obtained cost and b is the best known cost.

To test the practical performance of the two intensification schemes, instances of the generalised assignment problem were used. This is because it has now been extensively trialled with EO [12, 13] and as such, new results can be sensibly compared. In addition, it is an $\mathcal{NP}$ hard problem that also has constraints. The test suite of problems was the large-sized set of Chu and Beasley [8]. Further explanations of these may also be found in Randall [12] and Randall et al. [13]. The B, C and D type problem instances were considered here.

4.1   Experiments and Results

In the development of any new algorithm, there will always be design choices and new parameters introduced. It is not possible to test all combinations of algorithm variations and parameter values to determine the best. However, some investigation and analysis is required.

For both of the intensification-enhanced EO variants, the trigger points could be infinitely varied. This is perhaps matter for a more extensive study of the subject than this paper provides. There is one parameter to vary for the Locked version and two for the Archive version. These are as follows:

- *Locked* – A substantial number of iterations should be used to give EO time to explore the region around a solution that has one or more of its components locked. As such, values from the set $\{100, 200, 500, 1000, 5000\}$ iterations were tested.
- *Archive* – The two parameters are size of the archive and the approximate number of intensification periods (see Equation 2). Both were tested with values drawn from the set $\{5, 10, 20, 50, 100\}$.

To test the effect of these parameters, the instances D5-100 and D10-200 were chosen. Both of these problems are relatively hard, yet representative of this assignment problem. Table 1 shows the results of varying the Locked parameter on the canonical and population variants respectively. Due to the sheer size of the Archive results only those for the canonical algorithm for D5-100 are shown here. The Kruskal-Wallis statistical procedure was used to determine if there were any substantive differences between the parameter values. The following was found:

- *Locked* – The single and population versions of the algorithm yielded different results as is evident from the tables. Neither Kruskal-Wallis test detected significant differences amongst the locked period values. Looking at Table 1 shows that, in terms of the median and maximum values, 500 and 5000 iterations were best for the single and population versions of the algorithm respectively.
- *Archive* – Given the combination of two parameters, there are 25 separate cases. These results (for D5-100) are shown in Table 2. D10-200 shows a

Table 1. Parameter variation results for the Locked version

Instance	Locked Iterations	Single			Population		
		Min	Med	Max	Min	Med	Max
D5-100	100	0.85	1.26	1.33	0.28	0.75	1.26
	200	0.55	1.18	1.38	0.41	0.86	1.22
	500	0.89	1.2	1.29	0.27	0.77	1.22
	1000	0.93	1.26	1.47	0.5	0.73	1.27
	5000	0.86	1.26	1.41	0.31	0.62	1.22
D10-100	100	2.45	2.93	3.12	1.38	1.83	2.49
	200	2.4	2.97	3.19	1.62	2.04	2.71
	500	2.49	2.93	3.08	1.54	2.03	2.71
	1000	2.53	3.1	3.22	1.51	1.84	2.71
	5000	2.57	3	3.2	1.55	1.83	2.71

similar pattern. In the canonical/single version of EO, a significant difference
was detected. The best results were achieved using a small archive size (5)
and a large number of intensification periods which is confirmed by inspection
of the table. This was not the case for the population version. There was no
significant difference between the Kruskal-Wallis ranks also verified by the
contents of the table. However, a larger archive size (10 members) was better
and was used subsequently.

Table 2. Parameter variation results for the archive version using D5-100

Descriptor	(Size, Periods)									
	Single					Population				
	(5,5)	(5,10)	(5,20)	(5,50)	(5,100)	(5,5)	(5,10)	(5,20)	(5,50)	(5,100)
Min	0.82	0.44	0.63	0.58	0.44	0.08	0.24	0.19	0.05	0.05
Med	1.04	0.82	0.81	0.81	0.71	0.35	0.33	0.42	0.29	0.29
Max	1.21	1.11	1.1	1.13	0.82	0.63	0.75	0.67	0.77	0.77
	(10,5)	(10,10)	(10,20)	(10,50)	(10,100)	(10,5)	(10,10)	(10,20)	(10,50)	(10,100)
Min	0.82	0.63	0.63	0.77	0.71	0.09	0.25	0.13	0.16	0.16
Med	1.08	1	0.91	0.86	0.85	0.39	0.43	0.36	0.28	0.28
Max	1.27	1.46	1.37	0.97	1.02	0.78	0.58	0.56	0.52	0.52
	(20,5)	(20,10)	(20,20)	(20,50)	(20,100)	(20,5)	(20,10)	(20,20)	(20,50)	(20,100)
Min	0.77	0.77	0.88	0.72	0.72	0.17	0.08	0.16	0.19	0.19
Med	1.28	1.25	1.11	1.15	0.97	0.29	0.34	0.42	0.34	0.34
Max	1.46	1.43	1.38	1.4	1.05	0.55	0.64	0.74	0.67	0.67
	(50,5)	(50,10)	(50,20)	(50,50)	(50,100)	(50,5)	(50,10)	(50,20)	(50,50)	(50,100)
Min	0.77	0.77	1.05	0.86	0.83	0.02	0.22	0.09	0.27	0.27
Med	1.27	1.25	1.17	1.17	0.93	0.28	0.33	0.29	0.38	0.38
Max	1.46	1.43	1.43	1.4	1.11	0.53	0.55	0.5	0.55	0.55
	(100,5)	(100,10)	(100,20)	(100,50)	(100,100)	(100,5)	(100,10)	(100,20)	(100,50)	(100,100)
Min	0.77	0.77	1.05	0.86	0.83	0.19	0.22	0.28	0.16	0.16
Med	1.27	1.25	1.17	1.17	0.93	0.31	0.35	0.41	0.38	0.38
Max	1.46	1.43	1.43	1.4	1.11	0.94	0.49	0.61	0.63	0.63

Table 3 shows the results for the Locked and Archive versions of the algorithm
(in single and population mode). Visual inspection indicates that the Archive
version clearly outperforms the Locked version. On all problem instances, for
both single and population versions, the former yields solution values with lower

Table 3. The cost results received by the Locked and Archive

Instance	Unintensified Single Med	Pop. Med	Locked Single Min	Med	Max	Population Min	Med	Max	Archive Single Min	Med	Max	Population Min	Med	Max
B5-100	1.03	**0.3**	0.81	1.47	1.74	0.54	1.33	1.84	0.27	0.33	0.65	0.22	0.33	0.54
B5-200	0.48	**0.07**	0.17	0.37	0.62	0.2	0.45	0.65	0.17	0.24	0.34	0.03	0.08	0.2
B10-100	0	0	0	0	0.14	0	0	0.07	0	0	0	0	0	0
B10-200	0.76	**0.11**	1.41	1.59	2.23	0	0.49	0.64	0.21	0.37	0.53	0	0.12	0.32
B20-100	0.17	**0**	0.17	0.17	0.43	0.09	0.17	0.43	0	0.04	0.17	0	**0**	0.17
B20-200	0.21	0.11	0.09	0.26	0.47	0.09	0.19	0.34	0.04	0.11	0.17	0	**0.06**	0.13
C5-100	0.6	0.05	0.05	0.47	0.62	0.62	0.88	1.19	0.05	0.21	0.47	0	**0**	0.1
C5-200	0.52	0.04	0.23	0.36	0.69	0.29	0.45	0.52	0.17	0.26	0.32	0	**0.03**	0.2
C10-100	1.1	0.14	0.64	1.46	2.35	0.29	0.96	1.21	0.07	0.36	1.14	0	**0.07**	0.21
C10-200	0.82	0	1.28	1.6	1.85	0.18	0.44	0.82	0.14	0.25	0.57	0	**0**	0.11
C20-100	1.05	**0.08**	1.45	2.65	3.7	0.48	0.76	1.13	0.08	0.24	1.05	0.08	0.12	0.48
C20-200	1.17	**0**	2.29	2.84	3.5	0.13	0.58	1	0.21	0.5	0.92	0	0.08	0.33
D5-100	1.54	0.45	0.89	1.2	1.29	0.89	1.3	1.68	0.44	0.71	0.82	0.16	**0.28**	0.52
D5-200	1.63	0.27	1.33	1.61	1.79	0.69	0.91	1.36	0.42	0.7	1.01	0.12	**0.24**	0.42
D10-100	2.56	**0.74**	2.4	2.84	2.98	2.01	2.42	2.62	1.07	1.43	1.91	0.66	0.94	1.3
D10-200	1.54	0.04	1.75	2.08	2.31	0.66	1.12	1.72	0.18	0.5	0.96	0	**0**	0.22
D20-100	2.47	**0.45**	3.01	3.4	3.45	1.72	2.15	2.63	0.72	1.45	1.9	0.37	0.68	1.23
D20-200	1.69	0.17	2.63	2.91	3.12	1.25	1.74	2.05	0.52	0.8	1.29	0.06	**0.13**	0.4

RPDs. In more than half the test cases, the population implementation of the Archive version finds the best known solution at least once. On several test cases it finds the best known solution in more than half of the trials, a result unmatched by any other version.

Consistent with the results of previous studies [12, 13], the population model is a more efficient way of running EO. Comparison to the work of Randall et al. [13] shows that (using the Archive version), intensification helps the single version of EO dramatically. On all test cases, it gives improved or better solution costs in terms of minimum, median and maximum values. In the majority of cases, the RPDs achieved have been at least halved. Interestingly, the performance of intensification in the population version of Archive is very similar to the results of Randall et al. [13]. This simply confirms that the population mechanism is a very powerful way of managing EO searches. However, it may be noted that, on the majority of test cases, intensification has yielded lower median RPDs than the use of the population mechanism alone.

5   Conclusions

From the results presented in this paper it is evident that explicit intensification techniques for EO can improve its ability to find good quality solutions. Two relatively simple forms of intensification, based on the notions of locking in solution components with good values, and revisiting previously known good solutions were investigated. These strategies were enhanced with heuristics that made them more suitable for use with, and contextually applicable to, EO. In particular, the Archive version produces very good solutions. We plan to investigate further refinements on these schemes for a wider range of problems in the near future.

References

[1] Bak, P.: How Nature Works. Springer, New York (1996)
[2] Beausoleil, R.: Intensification and diversification strategies with tabu search: One-machine problem with weighted tardiness objective. In: Cairó, O., Cantú, F.J. (eds.) MICAI 2000. LNCS, vol. 1793, pp. 52–62. Springer, Heidelberg (2000)
[3] Blum, C.: ACO applied to group shop scheduling: A case study on intensification and diversification. In: Dorigo, M., Di Caro, G., Sampels, M. (eds.) Ant Algorithms 2002. LNCS, vol. 2463, pp. 14–27. Springer, Heidelberg (2002)
[4] Boettcher, S., Percus, A.: Extremal optimization: Methods derived from co-evolution. In: Proceedings of the Genetic and Evolutionary and Computation Conference, pp. 825–832. Moran Kaufmann, San Francisco (1999)
[5] Boettcher, S., Percus, A.: Combining local search with co-evolution in a remarkably simple way. In: Proceedings of the Congress on Evolutionary Computation, pp. 1578–1584. IEEE Service Center, Piscataway (2000)
[6] Boettcher, S., Percus, A.: Nature's way of optimizing. Artificial Intelligence 119, 275–286 (2000)
[7] Boettcher, S., Percus, A.: Extremal optimization: An evolutionary local search algorithm. In: Bhargava, H., Ye, N. (eds.) Computational Modeling and Problem Solving in the Networked World, Interfaces in Computer Science and Operations Research, pp. 61–77. Kluwer Academic Publishers, Dordrecht (2003)
[8] Chu, P., Beasley, J.: A genetic algorithm for the generalised assignment problem. Computers and Operations Research 24, 17–23 (1997)
[9] Gambardella, L., Taillard, E., Dorigo, M.: Ant colonies for the quadratic assignment problem. Journal of the Operational Research Society 50, 167–176 (1999)
[10] Glover, F., Laguna, M.: Tabu Search. Kluwer Academic Publishers, Boston (1997)
[11] Randall, M.: A systematic strategy to incorporate intensification and diversification into ant colony optimisation. In: Abbass, H., Wiles, J. (eds.) Proceedings of the Australian Conference on Artificial Life, Canberra, Australia, pp. 199–208 (2003)
[12] Randall, M.: Enhancements to extremal optimisation for generalised assignment. In: Randall, M., Abbass, H.A., Wiles, J. (eds.) ACAL 2007. LNCS (LNAI), vol. 4828, pp. 369–380. Springer, Heidelberg (2007)
[13] Randall, M., Hendtlass, T., Lewis, A.: Extremal optimisation for assignment type problems. In: Lewis, A., Mostaghim, S., Randall, M. (eds.) Biologically-inspired Optimisation Methods: Parallel Algorithms, Systems and Applications. SCI, vol. 210, pp. 139–164. Springer, Heidelberg (2009)

Comparing Two Constraint Handling Techniques in a Binary-Coded Genetic Algorithm for Optimization Problems

Helio J.C. Barbosa[1], Afonso C.C. Lemonge[2],
Leonardo G. Fonseca[3], and Heder S. Bernardino[1]

[1] Laboratório Nacional de Computação Científica, Petrópolis, RJ, Brazil
[2] Universidade Federal de Juiz de Fora, Juiz de Fora, MG, Brazil
[3] Universidade Federal do Espírito Santo, São Mateus, ES, Brazil
{hcbm,hedersb}@lncc.br, afonso.lemonge@ufjf.edu.br, goliatt@gmail.com

Abstract. In this paper the relative performance of two constraint handling techniques, namely a parameter-less adaptive penalty method (APM) and the stochastic ranking method (SR), is studied in the context of continuous parameter constrained optimization problems. Both techniques are used within the same search engine, a binary-coded genetic algorithm.

Keywords: constrained optimization, stochastic ranking, adaptive penalty.

1 Introduction

Nature-inspired meta-heuristics in general, and genetic algorithms (GAs) in particular, can be readily applied to unconstrained optimization problems. However, when the solution must satisfy a set of constraints, such techniques must be equipped with a constraint handling procedure which can be classified either as direct (interior), when only feasible elements are considered, or as indirect (exterior), when both feasible and infeasible elements are used during the search.

Direct techniques comprise: a) special (closed) genetic operators, b) special decoders, c) repair techniques, and d) "death penalty". With the exception of the "death penalty", they are strongly problem dependent and of reduced practical applicability, specially when constraints are not known as explicit functions of the decision variables. Indirect techniques include: a) the use of Lagrange multipliers, b) the use of fitness as well as constraint violation values in a multi-objective optimization setting, c) the use of special selection techniques, and d) penalty techniques. Ensemble of such techniques have also been considered [7]. Further references for constraint handling methods in evolutionary computation can be found in the on-line bibliography [3].

It is only then natural that several performance comparisons among such techniques have been published using benchmark problems. For the case of continuous parameter optimization problems, one of the best performing technique

K. Deb et al. (Eds.): SEAL 2010, LNCS 6457, pp. 125–134, 2010.

has been shown to be that due to Runarsson & Yao [9], where a balance between the objective and the penalty function values is sought by means of a stochastic ranking (SR) procedure. However, it should be noted that the final results of a constrained optimization problem depend not only on the constraint handling technique but also on the search engine adopted. The superior results obtained by Runarsson & Yao [9] correspond to the use of an evolution strategy search technique (ES) augmented with the stochastic ranking (SR) constraint handling procedure. The question then arises as to the relative contribution of each of these components, ES and SR.

In [2] a binary-coded GA using an adaptive penalty method (GA+APM) was not able to outperform the results obtained using the ES+SR technique in [9]. However, as both search engine and constraint handling technique were different, and the (real-coded) ES technique is usually more efficient for continuous parameter optimization than a binary-coded GA, the question whether SR is in fact superior to APM as a constraint handling technique could not be satisfactorily answered.

In an attempt to clarify this issue, the objective of this paper is to compare the relative performance of the two constraint handling techniques (SR and APM) when the same search engine is adopted with SR and APM, namely a standard binary-coded GA.

2   Constrained Optimization

Optimization problems appear naturally in many areas as one is always interested in minimizing or maximizing quantities such as cost or profit, respectively.

A continuous constrained optimization problem can be stated as the minimization (or maximization) of a given objective function $f(x)$, where $x \in R^n$ is the vector of design/decision variables, subject to inequality constraints $g_p(x) \leq 0$, $p = 1, 2, \ldots, \bar{p}$ as well as equality constraints $h_q(x) = 0$, $q = 1, 2, \ldots, \bar{q}$. Additionally, the variables are usually subject to bounds $x_i^L \leq x_i \leq x_i^U$. For convenience, this formulation is retained in complex real world situations where an explicit mathematical expression for g_p or h_q as a function of the vector of decision variables x is not available. The constraints are in fact a complex implicit function of x, and the check for feasibility may require an expensive computational simulation. Furthermore, derivatives of the objective function and/or constraints with respect to the design variables may be undefined, noisy, expensive or unavailable. Constraint handling techniques which neither require the explicit form of the constraints nor additional evaluations of the objective function are thus well suited for real-world applications.

In the following sections, the Adaptive Penalty Method and the Stochastic Ranking technique, which satisfy those requirements, are summarized.

3   The Adaptive Penalty Method - APM

Due to its generality and simple intuitive basis, penalty techniques, in spite of their shortcomings, are perhaps the most popular constraint handling techniques. The fitness function value of an unfeasible solution is penalized by means of a term that grows with the magnitude of constraint violation. Usually, the performance of the technique depends strongly on the penalty parameter(s) that must be set by the user for a given problem.

In [2,6] an adaptive penalty scheme which uses feedback from the population and does not require any user defined parameter was proposed and tested.

Defining the amount of violation of the j-th constraint by the candidate solution $x \in R^n$ by

$$v_j(x) = \begin{cases} |h_j(x)|, & \text{for an equality constraint,} \\ \max\{0, -g_j(x)\} & \text{otherwise} \end{cases}$$

where $j = 1, \ldots, m$ and $m = \bar{p} + \bar{q}$, the modified fitness function is written as:

$$F(x) = \begin{cases} f(x), & \text{if } x \text{ is feasible,} \\ \overline{f}(x) + \sum_{j=1}^{m} k_j v_j(x) & \text{otherwise} \end{cases} \qquad \overline{f}(x) = \begin{cases} f(x), & \text{if } f(x) < \langle f(x) \rangle, \\ \langle f(x) \rangle & \text{otherwise} \end{cases}$$

and $\langle f(x) \rangle$ is the mean of the objective function values in the current population.

Denoting by $\langle v_l(x) \rangle$ the violation of the l-th constraint averaged over the current population, the penalty parameter is then defined at each generation by:

$$k_j = |\langle f(x) \rangle| \frac{\langle v_j(x) \rangle}{\sum_{l=1}^{m} [\langle v_l(x) \rangle]^2}$$

The idea is that the values of the penalty coefficients should be distributed in a way that those constraints which are more difficult to be satisfied should have a relatively higher penalty coefficient.

By automatically defining a penalty parameter for each constraint, APM relieves the user from the burden of having to determine sensitive parameter(s) when dealing with every new constrained optimization problem.

4   The Stochastic Ranking Method - SR

In the stochastic ranking (SR) technique [9] the balance between the objective and penalty functions is achieved through a ranking procedure based on a stochastic version of the bubble-sort algorithm. In this approach a probability p_f of using only the objective function for comparing solutions in the infeasible region of the search space is introduced. Given any pair of two adjacent candidate solutions, the probability of comparing them according to the objective function is 1 if both solutions are feasible, and p_f otherwise. The procedure (see Figure 1) is halted when no change occurs in the rank ordering within a complete sweep.

When $p_f = 0$ the ranking induces an over-penalization, as all feasible solutions are ranked highest, according to their objective value, followed by the infeasible ones. Two infeasible solutions are then compared based on their amount of constraint violation. On the other extreme, when $p_f = 1$, all solutions would always be compared according to their objective function values.

```
1: procedure SR(I, f, φ, pop, p_f)
2:     I_j = {j, j = 1 : pop}
3:     for j = 1 : pop do
4:         swap ← false
5:         for j = 1 : pop − 1 do
6:             u = RANDOM(0, 1)
7:             if φ_{I_j} = φ_{I_{j+1}} = 0 or u < p_f then
8:                 if f_{I_j} > f_{I_{j+1}}  then
9:                     tmp = I_{j+1}; I_{j+1} = I_j; I_j = tmp; swap ← true
10:                 end if
11:             else
12:                 if φ_{I_j} > φ_{I_{j+1}}  then
13:                     tmp = I_{j+1}; I_{j+1} = I_j; I_j = tmp; swap ← true
14:                 end if
15:             end if
16:         end for
17:         if not swap then BREAK
18:     end for
19: end procedure
```

Fig. 1. Stochastic ranking (SR) procedure. I is a list of ranked solutions, f denotes the objective function, and ϕ is the sum of the squares of the constraint violations.

However, as one is interested in finding feasible solutions in the end, p_f should be less than 0.5, so that there is a selective pressure against infeasible solutions. The parameter p_f can thus be used to adjust such selection pressure [8].

5   Numerical Experiments

The constraint handling techniques considered here are tested against a well known suite of 24 test-problems [10], where three levels of number of fitness function evaluations (ffe) were considered, namely, 5,000, 50,000, and 500,000.

The same generational Gray-coded GA was adopted as the search engine for both constraint handling techniques. It uses elitism (the best and one copy with one bit flipped are saved to the next generation), uniform crossover with 0.8 probability, mutation rate equal to 0.03, a population size of 100, and 25 bits for each real variable. The GA+APM technique uses linear ranking selection, which is replaced by the SR technique in the GA+SR context.

Four variants of the GA+SR technique were considered here by setting SR's user-defined parameter to $p_f = 0, 0.25, 0.35$, and 0.45. These variants were denoted by sr_{000}, sr_{025}, sr_{035}, and sr_{045}, respectively. The results for 500,000 ffe obtained by another SR variant, a (60,400)-ES published in [8] (denoted here by ES+SR), are also presented.

Twenty-five independent runs were performed for each test-problem, and the average, best, and worst results were recorded. The results are presented using

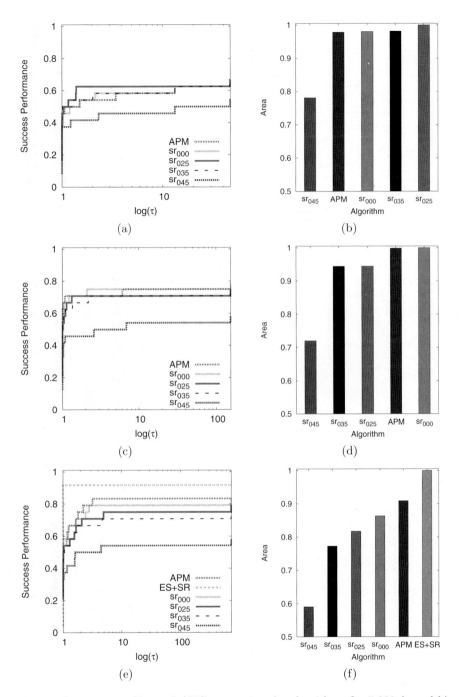

Fig. 2. Performance profiles and AUC comparing the algorithms for 5,000 (a and b), 50,000 (c and d), and 500,000 (e and f) ffe considering the average results

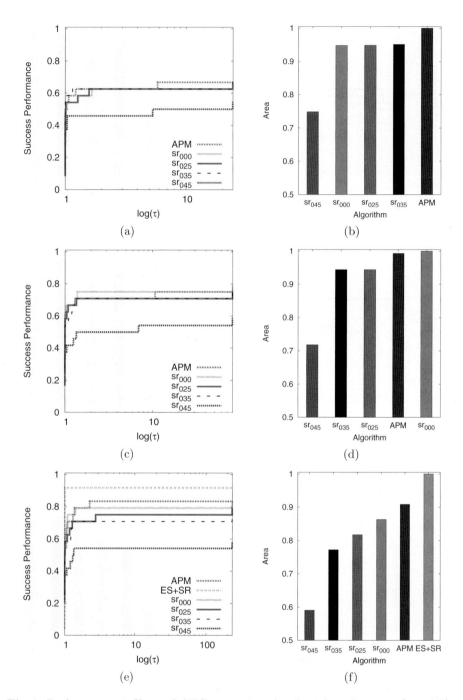

Fig. 3. Performance profiles and AUC comparing the algorithms for 5,000 (a and b), 50,000 (c and d), and 500,000 (e and f) ffe considering the best results

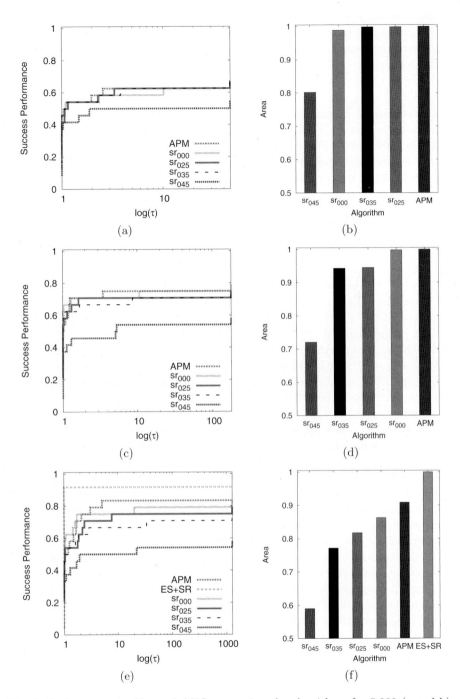

Fig. 4. Performance profiles and AUC comparing the algorithms for 5,000 (a and b), 50,000 (c and d), and 500,000 (e and f) ffe considering the worst results

performance profiles [5,1]. By testing all variants against all problems and measuring the performance $t_{p,v}$ of variant $v \in V$ when applied to problem $p \in P$,
a performance ratio can be defined with respect the best performing variant in
each problem:

$$r_{p,v} = \frac{t_{p,v}}{\min\{t_{p,v} : v \in V\}} \tag{1}$$

The performance indicators (larger values are better) considered here are the
inverse of the average, best, and worst of the minimum objective function value
found by variant v in problem p.

The relative performance of the variants in V on the whole set of problems P
can be displayed in a compact graphical form by defining [5]

$$\rho_v(\tau) = |\{p \in P : r_{p,v} \leq \tau\}| / n_p$$

where $|.|$ denotes the cardinality of a set. Then $\rho_v(\tau)$ is the probability that the
performance ratio $r_{p,v}$ of variant $v \in S$ is within a factor $\tau \geq 1$ of the best
possible ratio. If the set P is representative of problems yet to be tackled, then
variants with larger $\rho_s(\tau)$ are to be preferred. The performance profiles thus
defined have a number of useful properties [5,1]. The first one is that $\rho_v(1)$ is
the probability that variant v will provide the best performance in P among all
variants in V. If $\rho_{V1}(1) > \rho_{V2}(1)$ then variant $V1$ was the winner in a larger
number of problems in P than variant $V2$. The second one is that the area
under the ρ_v curve (AUC) is an overall performance measure for variant v in
the problem set P: the larger the AUC, the higher the variant efficiency. Finally,
a measure of the reliability of variant v is its performance ratio in the problem
where it performed worst. The most reliable variant is the one with minimum
$R_v = \sup\{\tau : \rho_v(\tau) < 1\}$.

Performance profiles for the measure defined by equation (1) are displayed in
Figures 2, 3, and 4 (a, c, and e). Each curve corresponds to one variant, and,
at the left, $\rho(1)$ indicates the fraction of test-problems where the algorithm was
the best performer. Also, these figures display the area under each performance
profile curve in the interval $[1, \tau_{max}]$ (b, d, and f). Another important information can be obtained at the right extreme of the plots, $\rho(\tau_{max})$, in which the
fraction of problems that were eventually solved by each algorithm within the
computational resources alloted can be observed. In the experiments conducted,
not a single variant was able to solve all problems in the benchmark since the
$\rho(\tau^*)$ curves never reach the maximum value of one in Figures 2–4.

Figure 2 shows performance profiles and area under their curves for the mean
of the results over the 25 independent runs. It can be seen that, for 5,000 ffe (a
and b), sr_{025} performed better than the other algorithms. APM, sr_{000}, and sr_{035}
present similar behavior but their results are slightly worse than those obtained
by sr_{025}. The worst results are found by the sr_{045} variant.

APM and sr_{000} present the best mean results for 50,000 ffe (c and d) although sr_{000} is slightly better than APM. Similarly to the case of 5,000 ffe
sr_{045} presents the worst results.

When 500,000 ffe are available to the algorithms the ES+SR variant finds
the best average results (e and f). APM is the second best variant, and the first

one if only binary-coded search engines are considered. This points to the higher performance of APM over SR when the same search mechanism is used by both algorithms, suggesting that the ES component is in fact the one responsible for the good results found by the original ES+SR variant, and not the constraint handling (SR) component. Also, it is important to notice that, when considering the average case and more than 50,000 fitness function evaluations, higher values of the p_f parameter lead to lower performance of the GA+SR variants.

Performance profiles for the best results and the area under their curves can be shown in Figure 3. For this case, using 5,000 ffe, APM presents the best performance (a and b). sr_{000}, sr_{025}, and sr_{035} variants present similar results, and all variants substantially outperform sr_{045}.

The sr_{000} variant presents the best results when 50,000 ffe are available (c and d). APM is the second one as it performs around 10 times worse than sr_{000} for one problem (c).

For 500,000 ffe (e and f) APM is outperformed only by ES+SR. Also, as for the average results (see Figure 2), with more than 50,000 fitness function evaluations, higher values of the p_f parameter lead to lower performance of the GA+SR variants.

APM presents the best worst results among the GA based algorithms. Only when 500,000 ffe are available it is outperformed by ES+SR. Again, with more than 50,000 ffe, higher values of the p_f parameter lead to lower performance of the GA+SR variants.

It is important to notice that, although $p_f = 0.45$ is the recommended value in the ES context [9], the (GA+SR) sr_{045} variant finds the worst results in all comparisons. Using 5,000 fitness function evaluations the sr_{025} and sr_{035} variants present good results showing the importance of having both feasible and infeasible candidate solutions at the start of the search process. APM performs best for 5,000 ffe when the best and worst cases are considered, but is the fourth in the average case, where the sr_{025} variant presents the best results.

When 50,000 fitness function evaluations are available, a similar performance ranking (sr_{000}, APM, sr_{025}, sr_{035}, and sr_{045}) can be observed in all cases, except for the case of worst results, in which APM presents the best performance.

The same performance ranking is observed when 500,000 ffe are allowed, in which case APM is the best GA algorithm. This result is important as it shows that SR leads to better results than APM only when embedded in the ES search context. If the same search mechanism is used, here a binary GA, then SR is no longer able to outperform APM.

6   Conclusions

Among nature-inspired techniques for continuous parameter optimization problems, superior results were obtained by Runarsson & Yao [9] using an evolution strategy technique (ES) augmented with the stochastic ranking (SR) constraint handling procedure (ES+SR). A genetic algorithm equipped with an adaptive penalty method (GA+APM) [6] was not able to outperform them. However, as both search engine and constraint handling technique were different, and the (real-coded) ES technique is usually more efficient for continuous parameter optimization than a binary-coded GA, the question whether SR is in fact superior to APM as a constraint handling technique could not be satisfactorily answered. In this paper numerical experiments were performed comparing the two constraint handling techniques (SR and APM) when the same search engine is adopted, namely a standard binary-coded GA.

The experiments performed comparing GA+SR and GA+APM in a standard test-problem suite have shown that (i) using the GA search engine, SR is no longer able to outperform APM as a constraint handling technique; (ii) in fact, for a larger number of fitness function evaluations, GA+APM outperforms GA+SR; (iii) if the SR parameter is set at its standard value $p_f = 0.45$ then GA+APM clearly outperforms GA+SR; (iv) lower values of p_f are required in the GA case due to its lower selection pressure when compared to the ES; (v) GA+SR obtains better results with $p_f = 0$, which recovers Deb's constraint handling technique [4]; and (vi) all variants are greatly outperformed by the original ES+SR technique, indicating the superiority of the ES search engine in this class of problems.

The next question is to investigate whether introducing APM into ES would provide better results than those obtained by ES+SR [9].

Acknowledgments. The authors would like to thank the support from CNPq (grants 308317/2009-2 and 301527/2008-3), FAPEMIG (TEC PPM 425/09), and FAPERJ (grants E-26/102.825/2008 and E-26/100.308/2010).

References

1. Barbosa, H.J., Bernardino, H.S., Barreto, A.M.: Using performance profiles to analyze the results of the 2006 CEC constrained optimization competition. In: IEEE World Congress on Computational Intelligence, Barcelona, Spain (2010)
2. Barbosa, H.J., Lemonge, A.C.: An adaptive penalty scheme in genetic algorithms for constrained optimization problems. In: Proc. of the Genetic and Evolutionary Computation Conference, New York, pp. 287–294 (2002)
3. Coello, C.A.C.: List of references on constraint-handling techniques used with evolutionary algorithms, http://www.cs.cinvestav.mx/~constraint/
4. Deb, K.: An efficient constraint handling method for genetic algorithms. Computer Methods in Applied Mechanics and Engineering 186(2-4), 311–338 (2000)
5. Dolan, E., Moré, J.J.: Benchmarcking optimization software with performance profiles. Math. Programming 91(2), 201–213 (2002)
6. Lemonge, A.C., Barbosa, H.J.: An adaptive penalty scheme for genetic algorithms in structural optimization. Intl. J. Num. Meth. Eng. 59(5), 703–736 (2004)
7. Mallipeddi, R., Suganthan, P.: Ensemble of constraint handling techniques. IEEE Trans. Evo. Comp. 14(4), 561–579 (2010)
8. Runarsson, T.: Approximate evolution strategy using stochastic ranking. In: Proc. of the IEEE Congress on Evolutionary Computation, pp. 745–752 (2006)
9. Runarsson, T., Yao, X.: Stochastic ranking for constrained evolutionary optimization. IEEE Trans. Evo. Comp. 4(3), 284–294 (2000)
10. Suganthan, P.N.: Special session on constrained real-parameter optimization, http://www3.ntu.edu.sg/home/epnsugan/index_files/CEC-06/CEC06.htm

Evolving Stories: Tree Adjoining Grammar Guided Genetic Programming for Complex Plot Generation

Kun Wang, Vinh Q. Bui, and Hussein A. Abbass

School of Engineering and Information Technology, University of New South Wales at ADFA
(UNSW@ADFA), Canberra 2600, Australia
Kun.Wang@student.adfa.edu.au, {V.Bui, H.Abbass}@adfa.edu.au

Abstract. In this paper, we develop a tree adjoining grammar (TAG) to capture semantics of a story with long-distance causal dependency, and present a computational framework for story plot generation. Under this framework, TAG is derived and a story plot is represented by a derivation tree of TAG. The generated plots are then evolved using grammar guided genetic programming (GGGP) to generate creative, interesting and complex story plots. To evaluate these newly generated plots, a human-in-the-loop approach is used. An experimental study was carried out, in which this framework was used to produce creative, interesting and complex plots from a predesigned fabula based on a story known as "The magpie and the water bottle". The experimental study demonstrated that TAG and GGGP can potentially contribute significantly to complex automatic story plot generation.

Keywords: Automatic storytelling, story formalism, complex plot generation, long-distance dependency, Tree Adjoining Grammar, GGGP.

1 Introduction

Automatic storytelling and story generation has drawn significant attention from researchers in Artificial Intelligence, computational linguistics, and the entertainment industry such as computer Role Playing Games [1]. Moreover, it plays an important role in planning and scenario generation [2]. Three major story generation approaches have been observed in the literature: case-based reasoning [13-16], simulation-based [17] and multi-agent-based planning [1, 5, 18-19]. In the first approach, stories generated possess limited creativity and variability, so generating stories with surprising plots is a challenge. The second approach hardly produces good story by simply recording the simulated events [20]. Stories generated by the third approach may be subject to semantic incoherence [21].

The lack of a story formalism – we conjecture – is a cause for these problems. We hold that using a formal grammar to guide story generation, the generated stories can enjoy semantic coherence and the quality of the generated stories can be further improved using already mature search techniques, such as evolutionary algorithm (EA). In a recent paper [10], regular grammar is used to generate and evolve plots with Grammar Evolution (GE). It has been shown that the syntactic constrains imposed by the story grammar is an important factor for the generated stories to enjoy desirable

K. Deb et al. (Eds.): SEAL 2010, LNCS 6457, pp. 135–145, 2010.

coherence. Through the evolutionary process, biased by human assessment of some desirable story features (e.g. interestingness, creativity), the generated stories can be improved in quality to demonstrate these features to some degree. However, only simple stories with single-character can be generated because long-distance causal dependency is a challenge for regular grammar.

In order to deal with the problem that regular grammar is incapable of capturing long-distance dependency, we propose to use Tree Adjoining Grammar (TAG) to generate stories. Two excellent properties of TAG are responsible for this: the extended domain of locality (ELD) and factoring recursion from the domain of dependencies (FRD) [22].

In this paper, we firstly develop a TAG for story plot generation, which can capture story semantics with long-distance causal dependency. This TAG is based on an existing fabula story generation model [11-12]. Subsequently, we present a computational framework for story plot generation. Under this framework, TAG is derived and a story skeleton (i.e. an uninstantiated story plot) is represented by a derivation tree of TAG. The generated story skeletons are then evolved using grammar guided genetic programming (GGGP) and a human-in-the-loop fitness evaluation. After instantiating these story skeletons with a predefined fabula, creative, interesting and complex story plots are generated. The contributions of this paper are the proposal of a TAG for story plot generation that can capture story semantics with long-distance causal dependency, and the computational framework to generate creative and interesting complex story plots using GGGP.

The rest of the paper is organized as follows. In the following part of section 1, we give a brief introduction of the fabula story generation model which is the basis of our work. In section 2, we elaborate on our proposed TAG and the computational framework for constructing creative and interesting complex plots. In section 3, an experimental study is presented and the results are discussed. Finally, some concluding remarks are made, and future work is discussed in Section 4.

Before going into the details of the TAG and the computational framework, we first present the fabula story generation model which is the basis of our framework.

According to the fabula model adapted from Bal [11] and Rimmon-Kenan [12], a story comprises three information layers: fabula, plot, and presentation. The lowest layer is called the fabula layer. It is a series of logically and chronologically related events that are caused or experienced by characters in a story world. In [24] and [25], a fabula is defined as a network of causal relationships of six type of elements: goal (g), actions (a), outcomes (o), events (e), perceptions (p) and internal elements (i), which are connected by four types of relationships: Physical Causality (φ), Motivation (μ), Psychological Causality(ψ), and Enablement(ε) depicted in Fig.1. The middle layer is called the plot layer, which is a relevant set of events taken from the fabula to form a consistent and coherent whole semantically. Semantic measures of a story such as coherence, creativity and interestingness are determined by the story at this level. In this paper, we focus our work on story generation on the plot layer. The top layer is the presentation layer where story plot unfolds in an understandable form for the audience, e.g. narration, movies and so on.

The three-layer story structure of the fabula model divides story generation work into three independent functional levels. First, we define a TAG to capture the semantics in

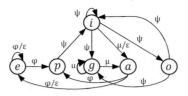

Fig. 1. Swartjes's model of fibula

the fabula layer. To build the plot layer, we first derive the corresponding TAG to obtain un-instantiated plots which we call story skeletons. These story skeletons are then instantiated with instances retrieved from the fabula; after which, plots are generated.

2   A Computational Framework for Plot Generation

In this section, a tree adjoining grammar (TAG) formalism based on the fabula model is first given. The proposed TAG is derived and a story skeleton is represented by a derivation tree of TAG. The generated story skeletons are then evolved using grammar guided genetic programming (GGGP) to generate creative, interesting and complex plots from a given fabula. We use the following notations: "$\wedge$" is a set notation; "$\mapsto$" is the valuation notation; symbols starting with capital letters or x, y, z are variables; other symbols denote constants.

2.1   Tree Adjoining Grammar (TAG) for Plot Generation

Tree adjoining grammar (TAG) is a tree rewriting and analysis system, first proposed by Joshi et al in [27] to enhance the expressive and generative power of context free grammar (CFG). TAG in its original form can not represent multi-character stories. We overcome this problem through grammar parameterization. In order to denote different characters in a story, we introduce parameters P, x, x_0 and n_m. P is used to indicate the global causal dependency, which we define to mean a causal relationship among characters. It is introduced to represent the interaction between characters which is necessary to generate complex story plots with multiple characters and branches. Our TAG for plot generation is defined as a quintuple $(\widehat{\Sigma}, \hat{n}, \hat{i}, \hat{a}, s)$, where:

(1) $\widehat{\Sigma}$ is a finite set of terminal symbols, defined as $\widehat{\Sigma} = \hat{t} (\widehat{X})$ $=\{ t_{ep}(n_m), t_{ep}(x_0), t_{ep}(P,x), t_{ee}(P,x), t_{ae}(X), t_{ia}(X), t_{ig}(X), t_{gg}(X), t_{ga}(X), t_{aa}(X), t_{ap}(X), t_{io}(X), t_{ii}(X), t_{og}(X), t_{oi}(X), \varepsilon(n_m)\}$, where X denotes n_m, x_0 or x. $\varepsilon(n_m)$ is a placeholder, and each of the other terminals is a symbol representing a causal relationship between two elements. For example, $t_{ap}(x)$ means an action done by x causes a corresponding perception of x.

(2) $\hat{n}$ is a finite set of non-terminal symbols: $\hat{n} \cap \widehat{\Sigma} = \emptyset$. In our grammar, $\hat{n} = \widehat{nt}(\widehat{Y})$ $=\{s(n_m), g(Y), a(Y), e(n_m), e(x_0), i(Y), o(Y) \}$ where Y denotes n_m, x_0 or x. g, a, e, i, o correspond to the story elements in the fabula model.

(3)   s is a starting symbol: $s \in \hat{n}$. We define $s =s(n_m)$.

(4) î is a finite set of finite trees, called initial trees (or α trees). Each of the α trees corresponds to the structure of the simplest story plot without any recursion. We define $î = \hat{\alpha}(\hat{Z}) = \hat{\alpha}(n_m)$. A typical α tree $\alpha_4(n_m)$ is illustrated in Fig.2. The initial trees and auxiliary trees in this TAG for plot generation are transformed from our existing regular grammar rules [10] using Schabe's algorithm[22, 28].

(5) â is a finite set of finite trees, called auxiliary trees (or β trees). Each auxiliary tree corresponds to a minimal recursive structure that is brought into the plot derivation when we recur on the non-terminal using adjunction to expand the main story line expressed in an α tree. Therefore, the label of the foot node (i.e. a node on the frontier annotated by an asterisk *) must be identical to the label of the root node. We define $â = \hat{\beta}(\hat{Z'}) = \hat{\beta}\ (\{x_0, P, x\})$, where x_0 represents the character of the non-terminal who calls the adjunction operation, P is to record global causal dependency in the plot derivation process, x can be assigned to represent a branch line character during the plot derivation. There exists two types of β trees in this grammar: $\beta_e(x_0, P, x)$ and $\beta_{i/o/g/a}(x_0)$, see examples $\beta_{e0}(x_0, P, x)$ and $\beta_{p1}(x_0)$ in Fig.2. The trees in $î \cup â$ are called elementary trees.

2.2   Plot Derivation

To generate a plot, we first generate a story skeleton (i.e. an un-instantiated plot) by deriving the TAG defined above using adjunction and substitution operations on elementary trees, and then instantiate it with a predefined fabula to obtain a plot. Since substitution is not an indispensable operation for TAG to maintain generative power [22], we only use adjunction operation

Definition of Adjunction

An adjunction is done by expanding a non-terminal in an elementary tree (e.g. $e(n_m)$ in $\alpha_4(n_m)$ and $p(n_1)$ in $\beta_{e0}("0", n_1)$ in Fig.2) with an auxiliary tree (e.g. $\beta_{e0}("0", n_1)$ and $\beta_{p1}(n_1)$ in Fig.2) whose root and foot node are labeled by the same non-terminal. After adjunction, the expanded non-terminal is substituted by the auxiliary tree with its original sub-tree excised and inserted below the foot node of the auxiliary tree. Because parameters have been included in our TAG, parameter valuations must be taken into account during TAG derivation: x_0 needs to be assigned to the character parameter of the non-terminal who calls the adjunction operation. As a result, we can ignore x_0 in the derivation tree. For example, in Fig. 2, when adjoining $\beta_{e0}("0", n_1)$ to the non-terminal $e(n_m)$ in $\alpha_4(n_m)$, the valuation of x_0 is $x_0 \mapsto n_m$. P and x are assigned as follows: Based on the two types of auxiliary trees: $\beta_{i/o/g/a}(x_0)$ and $\beta_e(x_0, P, x)$, we define "local" and "global" adjunction. On the one hand, the role of the local adjunction is to expand the main story or a branch line to obtain a wandering plot (see the local adjunction of $\beta_{p1}(n_1)$ on the non-terminal $p(n_1)$ in Fig.2). It can employ the first type β tree or the second type by assigning $\{P \mapsto null, x \mapsto x_0\}$. On the other hand, the global adjunction is to introduce a branch line (see the global adjunction of $\beta_{e0}("0", n_1)$ on the non-terminal $e(n_m)$ in Fig.2) which can itself be expanded by local adjunction (see the following local adjunction in Fig.2), so that long-distance causal dependency can be realized by allowing some-thing about another character (i.e. a branch line character, see n_1 in Fig.2) happening in between a causal relationship in the main story. More importantly, a branch line is not introduced arbitrarily. It will only emerge when some events (i.e. the result part of a causal relationship couple $t_{ee}(P,x)$ or $t_{ae}(x)$) have happened in the existing un-folded story, which endows the generated stories with an interaction mechanism. Parameter valuation will be explained in the story skeleton generation procedure.

In TAG, the derived tree and derivation tree [29] are used to represent the deriva-tion result. Each of them can be equivalently transformed to another.

Definition of Derived Tree and Derivation Tree

The derived tree is the resultant tree after grammar derivation. It contains the detailed information about a story skeleton, which we need to resort to when deriving TAG to generate a story skeleton and when instantiating the story skeleton to obtain a plot.

The derivation tree [29] encodes the history of adjunctions to obtain a story skele-ton. Compared to the derived tree, it enjoys conciseness in presentation. So it is used as the representation of the story skeleton during plot generation and evolutionary process. An example is given in Fig.2.

Story Skeleton Generation Procedure

By deriving TAG using the adjunction operation defined above, a story skeleton can be generated in the following procedure: First, choose one α tree from the initial tree set $\widehat{\alpha}(n_m)$ and expand it using random times of local adjunction operation within the main story length limit to obtain a complete main story; Second, for every un-adjoined non-terminal $e(n_m)$ in the existing derived tree, adjoin a randomly selected $\beta_e(x_0,P,x)$ using global adjunction to obtain a primitive branch line. Then apply random times of local adjunction on this branch line within the length limit. All the $e(n_m)$ non-terminals are adjoined in their emergence order in the derived tree. Here comes the parameter setup of x and P: There are three choices of x: n_m(main cha-racter), newly introduced branch line character n_{i+1}, or one of the existing branch line characters. That is: $x \mapsto n_m/ n_{i+1}/ n(n \in \{n_1 ... n_j\})$. P can be assigned to the ad-dress of the node (in the derivation tree representing a β tree) which has t_{ee} or t_{ae} terminals prior to the to-be-adjoined non-terminal $e(n_m)$. During the derivation process, all the addresses of this type of nodes must be recorded in a set $\widehat{P}$, so that when introducing a branch line, we can randomly select one P_j from $\widehat{P}$ and do $\{P \mapsto P_j\}$.

After the story skeleton generation, a plot can be generated by instantiating the sto-ry skeleton with the predefined fabula. It involves the following procedure: First of all, transform a story skeleton in the form of a derivation tree to a derived tree. Then, record all the terminal symbols from left to right. Finally, select an instance for each of the terminal symbol while bearing in mind the instance consistency between adja-cent terminals in the same character or the ones requiring global causal links. For example, in Fig.2, a fragment of the plot is: $\varepsilon\text{-}t_{ga^\#}(n_m)\text{-}t_{a^\#e'}(n_m)\text{-}t_{e'p^*}(0,n_1)\text{-}t_{p^*i}(n_1)$ where the symbols "# ′ *" are used to mark the instance consistency.

TAG for plot generation has the following main advantages:

(1) The length of the plots is controllable, which credits for the derivation length controllability of TAG [23].
(2) It can generate complex multi-character and multi-branch story plots with long-distance dependency owing to the global adjunction and parameterization;
(3) Using derivation tree as the story skeleton presentation, a complicated plot can be expressed in a concise form, which will simplify the evolutionary process;
(4) Sufficient evolution is facilitated. This owes to the non-fixed arity and locality property of TAG discussed in Hoai [23], which allows diversified and fine-grained genetic and search operators.

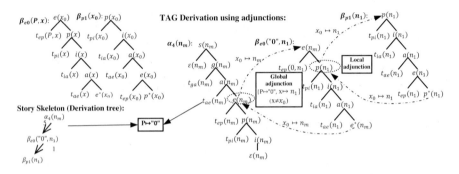

Fig. 2. Story Skeleton Generation by TAG Derivation

2.3   Plot Evolution Using GGGP

So far, using our TAG for plot generation, coherent and complex multi-character and multi-branch story plots can be generated. However, the quality of the plots may vary and become far from satisfactory. To overcome this problem, a plot evolution based on GGGP is used to gather story skeletons with preferable properties in the hope of magnifying the chance to obtain "good" story plots once instantiating them. Our plot evolution falls into three parts:

Representation: The derivation tree is used as the story skeleton representation.

Fitness Evaluation: A human-in-the-loop fitness evaluation approach is applied. In order to reduce the risk of instability of human evaluation, we use two fitness scores—creativity and interestingness. The final fitness is their linear combination.

Genetic and Search Operators: Diversified fine-grained genetic and search operators can be applied to our story skeleton evolution, which makes the evolutionary process effective and efficient. Literarily, all the operators Hoai [23] has introduced can be used. The only extra work is to reassign P and x during evolution to preserve the validity of story skeletons.

(1) Reassigning x is to unify character names within one newly-recombined story line (either the main story or a branch line).

(2) Reassigning P is to repair the global causal relationship broken during the application of genetic and search operators. This must be done after all crossovers and mutations. It requires exactly the same work to the parameter setup of P when the story skeleton is first generated.

The applied genetic and search operators we use are: sub-tree crossover; mutation operators that can make small changes [23] on a story skeleton, which are referred to sub-tree relocation, leaf node insertion and deletion, and leaf node substitution[23].

3   Experimental Study

In this section, we present some of the initial results we have obtained. In particular, we derive our proposed TAG and evolve the derivation trees to obtain different story plots from a small fabula we design called "The magpie and the water bottle". In the conducted study, a population of 20 individuals was randomly initialized: The length of each of the story lines was controlled by randomly choosing adjunction times applied on its root node between 1 and 3. The initial population was then evolved for 50 generations. During each successive generation, a binary tournament selection was used to select story skeleton individuals with higher fitness for reproducing a new generation. Here, a sub-tree crossover rate of 0.7, a sub-tree relocation rate of 0.6, a leaf node insertion and deletion rate of 0.6, and a leaf node substitution rate of 0.8 were used.

High mutation rates were used to help TAG guided genetic programming converge to solutions more quickly even with a small population [23] and to modulate the diversity of sub-structures. Diversity has shown to potentially contribute to creativity and interestingness in the experiment.

The fitness evaluation was implemented using a human-in-the-loop approach. It requires the instantiation of each story skeleton. The instance of each terminal was also randomly chosen from a set of available instances of the fabula. The fitness of each individual was computed based on two criteria with equal weights: creativity and interestingness. Each criterion was evaluated on a scale between 1 and 10. Fig.3. illustrates the evolutionary process of the total fitness and the fitness of each criterion. The average fitness value of the whole population at each generation was bounded by the fitness values of the weakest and the strongest individuals of the population. As depicted, the evolutionary process became stable around generation 40, from where onwards no significant improvement in the two fitness values was observed.

A closer look at the evolution of the fitness of each criterion offers a number of interesting insights. First, the best individual in the initial population has enjoyed desirable fitness value. That can be probably attributed to the capacity of our proposed TAG based formalism to generate complex multi-character and multi-branch story plots with long-distance dependency. Second, the gap between the weakest and the fittest individual gradually becomes narrower as evolution comes to an end. This indicates that the search starts widely and becomes more focused on potential areas along the evolutionary process. Furthermore, it is interesting to notice that even though the fitness value converges at the end of the evolution, individuals surviving through the evolutionary process are still with diversified skeletons. This can still be owed to the power of our proposed TAG. However, all of the individual stories in the final population either possess diversity in their constituents (e.g. non-terminals, elementary trees, characters, instances of terminals…) or interwoven story lines. This finding implies that diversity and interweaving lines may play important role in creativity and interestingness.

An individual (a story skeleton) with its instantiated presented story in the final population is given in Fig.4. We can notice that after an event "That breaks the bottle on the stone" occurring in the main story about "Minnie", a branch line about the character "Tom" is introduced and independently unfolded. This is regarded as an interaction between characters. Then the main story unfolds again from where it is broken by the branch line. It is at the breaking point that a long-distance causal relationship emerges in the main story. It is also interesting to note that what happened in the branch line about Tom shadows the main character Minie's death after drinking the water in that bottle. This may be another reason why this individual enjoys high fitness.

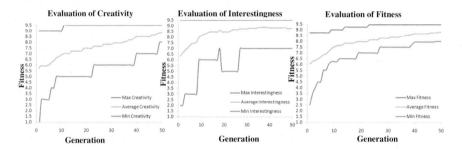

Fig. 3. The evolutionary process of the fitness and the two criteria

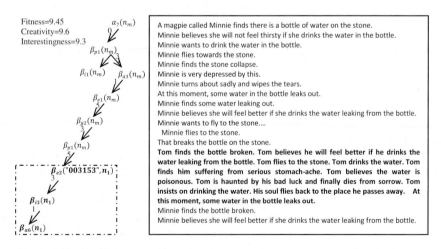

Fig. 4. An individual with high fitness in the final population

From the results and analysis, we can conclude that the proposed computational framework for plot generation can generate complex plots with multiple characters and independently unfolded branch lines. The emergence of each branch is to accommodate long-distance dependency as well as strengthen the interaction between characters.

However, the plot generation TAG we propose in this paper is a linear TAG with only right auxiliary trees. It has been demonstrated that linear TAG possesses weaker generative capacity than TAG [30-32] in the sense of limited string wrapping. Fortunately, this situation is relieved by the introduction of global adjunction which enables the wrapping of main story and branch lines. Moreover, our linear version of story generation TAG can be easily modified to full TAG by separating each of the causal relationship couples (i.e. each terminal in this grammar) into corresponding cause and result parts (two terminals), and relocating each of the non-terminals between them. The reason why this work is suspended in this paper is because of the unknown effect of overmuch wrapping of story elements on the coherence and understandability characteristics of a story, which requires further research.

4   Conclusion and Future Work

In this paper, we developed a Tree Adjoining Grammar (TAG) for plot generation that captures story semantics with long-distance causal dependency. This owes to the extended domain of locality (ELD) and factoring recursion from the domain of dependencies (FRD) properties of TAG. Consequently, we presented a computational framework for story plot generation. Under this framework, TAG is derived and a story plot is represented by a derivation tree of TAG. The generated plots are then evolved using grammar guided genetic programming (GGGP). To verify the approach, an experiment was conducted based on a designed fabula called "The magpie and the water bottle". The result obtained is promising. The system was able to produce coherent, creative, interesting and complex plots with multiple characters, branches, and long-distance dependency.

The current plot grammar we use cannot represent cause or result requiring more than one character. Future work includes designing new formalism of plot generation based on TAG. The substitution operation is still required by our TAG for plot generation to reduce redundancy in existing elementary trees and allow independently and completely unfolded branch lines. Quantitative metrics of story semantics such as creativity and interestingness are needed to enable computational story evaluation thus overcome the shortcomings of human-in-the-loop evaluation.

References

1. Hsueh-Min, C., Von-Wun, S.: Planning-Based Narrative Generation in Simulated Game Universes. IEEE Transactions on Comp. Intel. & AI in Games 1(3), 200–213 (2009)
2. McKeever, W., et al.: Scenario management and automated scenario generation. In: Modeling and Simulation for Military Applications, pp. 62281A.1–62281A.12. SPIE, Florida (2006)
3. Sebastiani, F.: Machine Learning in Automated Text Categorization. ACM Computing Surveys 34(1), 1–47 (2002)
4. Fairclough, C., Cunningham, P.: An interactive story engine. In: O'Neill, M., Sutcliffe, R.F.E., Ryan, C., Eaton, M., Griffith, N.J.L. (eds.) AICS 2002. LNCS (LNAI), vol. 2464, pp. 171–176. Springer, Heidelberg (2002)

5. Shim, Y., Kim, M.: Automatic short story generator based on autonomous agents. In: Ku-wabara, K., Lee, J. (eds.) PRIMA 2002. LNCS (LNAI), vol. 2413, pp. 151–162. Springer, Heidelberg (2002)
6. Riedl, M., Young, R.: Open-world planning for story generation. In: 9th International Joint Conference on Artificial Intelligence, pp. 1719–1720. Professional Book Center, Edin-burgh (2005)
7. Gervás, P., et al.: Story plot generation based on CBR. KBS 18(4-5), 235–242 (2005)
8. Swartjes, I., Theune, M.: The virtual storyteller: Story generation by simulation. In: 20th Belgian-Netherlands Conference on Artificial Intelligence, Enschede, pp. 257–265 (2008)
9. Lee, M.G.: A model of story generation. University of Manchester, Manchester (1994)
10. Bui, V., Abbass, H., Bender, A.: Evolving Stories: Grammar Evolution for Automatic Plot Generation. In: IEEE World Congress on Computational Intellegence, Barcelona (2010)
11. Bal, M.: Narratology: Introduction to the theory of narrative. University of Toronto Press, Toronto (1997)
12. Rimmon-Kenan, S.: Narrative fiction: Contemporary poetics. Routledge, London (2002)
13. Díaz-Agudo, B., Gervás, P., Peinado, F.: A case based reasoning approach to story plot generation. In: Funk, P., González Calero, P.A. (eds.) ECCBR 2004. LNCS (LNAI), vol. 3155, pp. 142–156. Springer, Heidelberg (2004)
14. Turner, S.: MINSTREL: a computer model of creativity and storytelling. University of California, Los Angeles (1993)
15. Pérez y Pérez, R., Sharples, M.: MEXICA: A computer model of a cognitive account of creative writing. Journal of Experimental and Theoretical Artificial Intelligence 13(2), 119–139 (2001)
16. Bringsjord, S., Ferrucci, D.: Artificial Intelligence and Literary Creativity: Inside the Mind of Brutus a Storytelling Machine. Routledge, London (2000)
17. Peinado, F.: Interactive digital storytelling: Automatic direction of virtual environments. Upgrade. Monograph:Virtual Environments 7(2), 42–46 (2006)
18. Mateas, M., Stern, A.: Writing Façade: A Case Study in Procedural Authorship. In: Second Person: Role-Playing and Story in Games and Playable Media, 183–208 (2004)
19. Cheong, Y., Young, R.: A computational model of narrative generation for suspense. In: 21st National Conference on Artificial intelligence, vol. 2, pp. 1906–1907. AAAI Press, Boston (2006)
20. Theune, M., et al.: The virtual storyteller: Story creation by intelligent agents. In: Tech-nologies for Interactive Digital Storytelling and Entertainment Conference, pp. 204–215. Citeseer, Darmstadt (2003)
21. Riedl, M., Young, R.: From linear story generation to branching story graphs. IEEE Com-puter Graphics and Applications 26(3), 23–31 (2006)
22. Joshi, A., Schabes, Y.: Tree-adjoining grammars. Handbook of Formal Languages, Beyond Words, pp. 69–123. Springer, Heidelberg (1997)
23. Hoai, N.X.: A Flexible Representation for Genetic Programming from Natural Language Processing. University of New South Wales, Canberra (2004)
24. Trabasso, T., Broek, P., Suh, S.: Logical necessity and transitivity of causal relations in stories. Discourse Processes 12(1), 1–25 (1989)
25. Swartjes, I., Theune, M.: A Fabula model for emergent narrative. Technologies for Inter-active Digital Storytelling and Entertainment, 49–60 (2006)
26. Theune, M., et al.: The virtual storyteller. ACM SigGroup Bulletin 23(2), 20–21 (2002)
27. Joshi, A., Levy, L., Takahashi, M.: Tree adjunct grammars. Journal of Computer and Sys-tem Sciences 10(1), 136–163 (1975)

28. Schabes, Y.: Mathematical and computational aspects of lexicalized grammars. University of Pennsylvania, Philadelphia (1990)
29. Weir, D.J.: Characterizing mildly context-sensitive grammar formalisms. University of Pennsylvania, Philadelphia (1988)
30. Schabes, Y., Waters, R.C.: Lexicalized Context-Free Grammars. In: 31st Annual Meeting of the ACL, pp. 121–129. ACL, Columbus (1993)
31. Chiang, D.: The weak generative capacity of linear tree-adjoining grammars. In: 8th International Workshop on TAG & Related Formalisms, pp. 25–32. ACL, Sydney (2006)
32. Rogers, J.: Capturing CFLS with Tree Adjoining Grammars. In: 32nd Annual Meeting of the ACL, pp. 155–162. ACL, Columbus (1994)

Improving Differential Evolution by Altering Steps in EC

Nikhil Padhye[1], Piyush Bhardawaj[2], and Kalyanmoy Deb[2]

[1] Department of Mechanical Engineering
Massachusetts Institute of Technology (MIT), MA 02139, USA
npdhye@mit.edu
[2] Department of Mechanical Engineering
Indian Institute of Technology Kanpur
PIN 208016, Uttar Pradesh, India
{piyush,deb}@iitk.ac.in

Abstract. In past, only a few attempts have been made in adopting a unified outlook towards different paradigms in Evolutionary Computation. The underlying motivation of these studies was aimed at gaining better understanding of evolutionary methods, both at the level of theory as well as application, in order to design efficient evolutionary algorithms for solving wide-range complex problems. One such attempt is made in this paper, where we reinstate 'Unified Theory Of Evolutionary Computation', drawn from past studies, and investigate four steps – *Initialization, Selection, Generation* and *Replacement*, which are sufficient to describe common *Evolutionary Optimization Systems* such as Genetic Algorithms, Evolutionary Strategies, Evolutionary Programming, Particle Swarm Optimization and Differential Evolution. As a next step we consider Differential Evolution, a relatively new evolutionary paradigm, and discover its inability to efficiently solve unimodal problems when compared against a benchmark Genetic Algorithm. Targeted towards enhancing DE's performance, several modifications are successfully proposed and validated through simulation results. The *Unified Approach* is found helpful in understanding the role and re-modeling of DE steps to efficiently solve unimodal problems.

1 Introduction

Much of the early research and development in evolutionary based computational methods occurred independently without any interaction(s) among various groups [3]. It was around late 1980s and early 1990s when the confluence of these paradigms began, which eventually led to the agreement on the term "Evolutionary Computation".

In-spite of advances in different EA paradigms there has been a lukewarm interest in investigating a framework which is capable of explaining the overall behavior of an EA. A plausible approach could be to decompose an EA into key standard components. Then, by understanding the role of each component individually and interaction between the components, insights into the performance of an EA could be obtained.

K. Deb et al. (Eds.): SEAL 2010, LNCS 6457, pp. 146–155, 2010.

This paper focuses on the performance of standard DE algorithm on class of unimodal problems, compared against a benchmark genetic algorithm named G3-PCX [1]. After discovering inefficient DE performance, *Unified Approach* is adopted in analyzing major DE steps. The DE steps are modified by borrowing ideas from G3-PCX and gradual improvement in performance is noted. Through a series of seamless modifications the DE performance is enhanced to an extent where it is comparable to the benchmark results, and the resulting algorithm is found to be equivalent to G3-PCX. Thus, this study highlights that how one can traverse from modifying one algorithm into the other by altering the major steps of an algorithm on the basis of functional requirements, and stresses on the importance of similarities and differences in terms of key steps of an algorithm which give rise to a difference in performance.

The rest of the paper is structured as follows: Section 2, presents an *Unified Framework* for evolutionary optimization algorithms. Section 3, provides details on chosen test problems and describes experimental methodology. Sections 4 and 5, look at the the performance of standard DE and its variants for the sake of improvement. Section 6, concludes the paper and hints on the direction for the future work.

2 Unified Framework for Evolutionary Algorithms

The most notable breaking new ground attempt in adopting *Unified Approach* towards Evolutionary Computation is made in [3], serving the goal of presenting an integrated view of Evolutionary Computation. This paper takes a step forward in demonstrating – How the *Unified Approach* can be utilized in better understanding (and thereby improving) an EA paradigm? In [3] author outlines a general *Evolutionary Optimization System(EOS)*, which is based on Darwinian evolutionary system. Such an *EOS* can be assumed to be constant in population size and with the optimization task of minimizing. The key steps pointed out in this *EOS* are: (1)*Initialization* – of the population randmomly, (2) *Selection* – of the individual(s) from the population to act as parent(s), (3) *Generation* – Creation of offspring(s) from the selected parent(s), and (4)*Replacement* – Selection of individuals(s) to survive for the next generation. After *Initialization, Selection, Generation* and *Replacement* are iteratively repeated till some termination criterion is met. Although detailed descriptions on each step are required before *EOS* can be simulated. The reader is refrred to [3] to see how major EA paradigms are studied in the above framework.

EOS described above requires an additional elaboration on population management i.e. how do offsprings compete for survival. Two popular ways are: (a) *Steady State* – or *incremental model*, implying that offsprings are produced one at a time and immediately compete for the survival i.e. if the fitness of child is better than the parent selected, the child survives or vice-versa, or (b) *Generational* – or *batch model*, implying that entire batch of child population is created and then there is competition for survival. We adopt the notation from [3] while representing evolutionary systems as follows – two populations are maintained:

one of size m for parents and second of size n for offsprings (now the system being represented as $EOS(m,n)$). In $EOS(m,n)$, n offsprings are created from the parent population of size m and then each child competes for space in the parent population. For, a special case, $n = 1$ we arrive at steady state model, and any value of $n > 1$ symbolizes generational model. EOSs, in this paper, are associated with real parameter optimization and solutions shall be represented as vectors of real parameter decision variables. The initialization of population for EOSs shall be done randomly. Next, we discuss DE as an instance of $EOS(m,n)$.

2.1   Differential Evolution as EOS(m,m)

Differential Evolution (DE) algorithm has emerged as a very competitive form of evolutionary computing more than a decade ago. The main goal of this study is to develop a thorough understanding of DE algorithm as an EOS and then systematically exploit this understanding in improving DE's performance. As majority of simulations presented in this study are based either on standard DE, its pseudo code is presented in Figure 1.

Selection, Generation and *Replacement* steps in DE here are same as those in "DE/best/1/exp" [4]. The population is scanned serially and for creation of a child, corresponding to any individual, four parents are selected (i.e. the individual itself, also referred to as base or index parent, *best fitness* individual from

---

Input DE parameters: scale factor *(F)*, crossover-rate *(Cr)* and population size *(M)*
Randomly Generate the initial population of M individuals in the defined region
and compute the fitness of each individual.
Set Generation Counter $t = 1$
Do until a defined stopping criterion is met:
   <u>*For*</u> $i = 1$ to M
   Selection
   – Choose, i^{th} individual $(X_{i,t})$, two random individuals $(X_{r1,t}, X_{r2,t})$, and best member
   in the population at previous $(t-1)$ generation (X_{best}) as parents
   Generation
     (a)*Create i^{th} Donor Vector:*
     $V_{i,t} = X_{best,t-1} + F \cdot (X_{r1,t} - X_{r2,t})$
     (b)*Create i^{th} Trial Vector:*
     $U_{i,t}$=CombineElements$(X_{i,t}, V_{i,t})$ // with probability CR
   Replacement
   If (Fitness$(U_{i,t}) \leq$Fitness$(X_{i,t})$)
   Then $X_{i,t+1} = U_{i,t}$
   Else $X_{i,t+1} = X_{i,t}$
   End <u>For</u>
   Update(X_{best})
   $t = t + 1$
   Update(P_{t+1})
   End **Do**
Return the individual with *best fitness value.*

Fig. 1.   Standard Differential Evolution Algorithm *(DE/best/1/exp)*, borrowed from [4]

the previous generation and any two population members chosen at random). First a donor vector $(V_{i,t})$ is created (step a) and then a trial vector $(U_{i,t})$ is created (step b) by stochastically combining elements from $X_{i,t}$ and $V_{i,t}$. This combination is commonly done using an exponential distribution with crossover factor of CR. If the newly created child $U_{i,t}$ is *better* compared to $X_{i,t}$ then $U_{i,t}$ is stored for updating $X_{i,t+1}$. It should be noted carefully that $X_{i,t}$s are updated to $X_{i,t+1}$s after entire set of $U_{i,t}$s are created. Once the population is updated, the generation counter is incremented and termination criteria is checked.

Following properties of this DE should be noted: (i) There is 'elitism' at an individual level i.e. if the newly created trial vector $U_{i,t}$ is inferior compared to the individual then individual is preserved as a child for the next generation and $V_{i,t}$ is ignored. (ii) The algorithm follows a generational model i.e. the current population is updated only after the entire offspring population is created.

3   Test Suite

We consider unimodal problems (having one optimum solution) or problems having a few optimal solutions, so as to test an algorithm's ability to progress towards the optimal region and then to focus to find the optimum with a specified precision. A previous study considered a number of evolutionary algorithms like, generalized generation gap (G3) model using a parent-centric crossover (PCX) operator, differential evolution, evolution strategies (ESs), CMA-ES, and a classical method on following test problems [1]:

$$F_{elp} = \sum_{i=1}^{n} i x_i^2 \quad \text{(Ellipsoidal function)} \tag{1}$$

$$F_{sch} = \sum_{i=1}^{n} \left(\sum_{j=1}^{i} x_j \right)^2 \quad \text{(Schwefel's function)} \tag{2}$$

$$F_{ros} = \sum_{i=1}^{n-1} \left(100(x_i^2 - x_{i+1})^2 + (x_i - 1)^2 \right) \quad \text{(Generalized Rosenbrock's function)} \tag{3}$$

In all these problems we use $n = 20$ The first two problems have their minimum at $x_i^* = 0$ with $F^* = 0$ and the third function has its minimum at $x_i^* = 1$ with $F^* = 0$. We initialize the population away from the known optima while restricting $x_i \in [-10, -5]$ for all i, in all problems. In subsequent generations we do not confine solutions to lie in the above range. After initialization, we count the number of function evaluations needed for the algorithm to find a solution close to the optimal solution and call this our first evaluation criterion S_1. We choose a value of 0.1 for this purpose. This criterion will denote how fast an algorithm is able to reach the optimal region. The second evaluation criterion (S_2) involves the overall number of function evaluations needed to find a solution having a function value very close to the optimal function value. We choose a value of 10^{-20} for this purpose.

Table 1. G3-PCX results, as reported in [1]

	F_{elp}			F_{sch}			F_{ros}		
	Best	Median	Worst	Best	Median	Worst	Best	Median	Worst
S_2	5,744	6,624	7,372	14,643	16,326	17,712	14,847 (38)	22,368	25,797

The earlier extensive study on G3-PCX algorithm reported the best, median and worst number of function evaluations needed based on 50 different runs on the three problems with the S_2 criterion. Table 1 presents those results. G3-PCX outperformed other state-of-the-art algorithms [1] and is treated as a benchmark for this study.

4   Performance Analysis of Standard DE

DE algorithm presented in section 2.1 belongs to the DE family of Storn and Price [4]. The family comprises of 10 different *Generation* strategies. Based on preliminary experiments we found *strategy* 1 to yield overall best performance. According to this strategy new solutions are created around the previous generation's *best solution* (step a in *Generation*, Figure 1). This feature of generating solutions around the best population member is desirable in solving unimodal problems and also done in G3-PCX [1], hence it is no surprise that *strategy* 1 was the best performer. In remainder of this paper, DE with strategy 1 is employed for simulations and referred to as standard DE. We performed a parametric study on standard DE for M, CR and F, and found $M = 50$, $CR = 0.95$ and $F = 0.7$, as optimal values with respect to all the three test problems. The results of standard DE are reported in Table 2.

Table 2. Standard DE, "DE/best/1/exp" [4], $F = 0.7$, $CR = 0.95$, $M = 50$

	F_{elp}			F_{sch}			F_{ros}		
	Best	Median	Worst	Best	Median	Worst	Best	Median	Worst
S_1 with 10^{-1}	6,100	6,600	7,200	7,600	9,000	11,200	21,050(43)	29,500	33,850
S_2 with 10^{-20}	31,700	33,550	35,100	48,050	51,100	55,200	55,400(43)	63,350	69,350

5   Functional Analysis and Modification of Standard DE

One of the noticeable features of standard DE is *elitism* at the individual level i.e. a child is compared with its *base parent* (i.e. the individual at the index corresponding to which child has been created), and only the better of the two survives in the next generation. We modified this *Replacement* scheme by always accepting the newly created child i.e. without carrying out the parent-child comparison. This resulted in a significant performance degradation in all three test problems with respect to both the metrics, indicating that elitism in DE by parent-child comparison is key to its performance.

Next, we try two *Selection* schemes, *Tournament* and *Random*, instead of the usual serial parent selection. The results are shown in Table 3 indicate that the alternate selection schemes perform poorly compared to serial selection, and we conclude that deterministic serial approach works most appropriately for DE.

Table 3. Standard DE with *Random* and *Tournament* selection, $CR = 0.95$, $F = 0.7$, $M = 50$

	F_{elp}			F_{sch}			F_{ros}		
	Best	Median	Worst	Best	Median	Worst	Best	Median	Worst
Random Selection									
S_1 with 10^{-1}	14,250	16,700	20,050	15,950	20,400	24,600	50,800(38)	56,900	64,500
S_2 with 10^{-20}	58,350	64,900	72,100	116,000	122,000	132,700	136,750(38)	147,050	158,900
Tournament Selection									
S_1 with 10^{-1}	14,600	16,850	19,400	17,700	21,250	25,300	39,500(44)	52,950	66,650
S_2 with 10^{-20}	86,350	92,950	97,650	114,800	124,700	134,900	100,050(44)	114,850	132,300

The *Generation* scheme (Step a) in standard DE involves creation of a child around $X_{best,t-1}$. This approach of creating solutions around the *best* is particularly useful in solving problems exibhiting unimodality, and the benchmark algorithm G3-PCX successfully exploits this property. A major difference between G3-PCX and the standard DE arises from the fact that former uses the *current best* location in the population, whereas DE utilizes the *previous generation's best*. We incorporate this feature in standard DE by using X_{best} instead of using $X_{best,t-1}$, where X_{best} indicates the best known location so far. This is achieved by checking and updating X_{best} after every child creation. The results shown in Table 4 reflect an improved performance in all cases. Thus, we conclude that creating solutions around X_{best} is an effective strategy for standard DE while solving unimodal problems.

Next we observe a basic difference in steady state and generational models of G3-PCX and standard DE. In standard DE a newly created child has to wait till next generation before it can be selected as the index parent. We test the steady state version of standard DE in which as soon as a child is created it is compared with its index parent. The index parent is replaced if the child is better. Since the created child is compared with the index parent itself, we refer this as *Serial Parent Replacement*. In another steady state version of standard DE, we compare the created child with a randomly selected member of the population and carry

152   N. Padhye, P. Bhardawaj, and K. Deb

Table 4. Standard DE + *Best Update*, $F = 0.7$, $CR = 0.95$, $M = 50$

	F_{elp}			F_{sch}			F_{ros}		
	Best	Median	Worst	Best	Median	Worst	Best	Median	Worst
S_1 with 10^{-1}	5,500	6,250	6,900	7,450	8,800	10,800	18,550(44)	24,350	29,000
S_2 with 10^{-20}	31,400	32,650	34,500	43,600	48,600	52,100	51,400(44)	56,750	62,400

out the replacement. In this version even if the newly created child is inferior to the index parent, it has a chance of surviving while being compared against a randomly chosen individual. The results for both the steady state versions are shown in Tables 5. Both the steady state versions show an improvement over standard DE (Table 2). The steady state versions are also an improvement over the DE with *Best Update* except for F_{ros}. Between the two versions, the *Random Parent Replacement* performs better compared to the *Serial Parent Replacement*. Thus, we conclude that the steady state model is useful over the generational model, and in particular *Random Parent Replacement* is a preferred strategy.

Now, we combine *Best Update* and *Steady State (Random Parent Replacement)* with standard DE, results shown in Table 6. The performance of this modified DE turns out to be best so far.

Table 5. Standard DE + *SteadyStateVersions*, $CR = 0.95$, $F = 0.7$, $M = 50$

	F_{elp}			F_{sch}			F_{ros}		
	Best	Median	Worst	Best	Median	Worst	Best	Median	Worst
Standard DE + *Steady State(Serial Parent Replacement)*									
S_1 with 10^{-1}	5,300	6,000	6,850	7,200	9,050	12,600	23,600(36)	28,950	35,850
S_2 with 10^{-20}	28,750	29,650	32,200	46,300	50,100	55,250	54,550(36)	60,950	70,850
Standard DE + *Steady State(Random Parent Replacement)*									
S_1 with 10^{-1}	4,000	5,200	6,850	6,050	8,500	11,450	20,750(40)	29,750	37,200
S_2 with 10^{-20}	23,150	25,200	27,150	42,100	47,700	54,350	53,050(40)	66,300	76,900

Till now we have been successful in improving the performance of standard DE by borrowing ideas, particularly *Steady State* and *Best Update*, from G3-PCX algorithm. This emphasizes the fact that a better understanding of *EOSs* at the level of operators can be highly useful in developing and enhancing other *EOSs*. At this stage, we also identify a mutation operator proposed by [2] in context to

Table 6. Standard DE + *Best Update* + *Steady State(RPR)*, $F = 0.7$, $CR = 0.95$, $M = 50$

	F_{elp}			F_{sch}			F_{ros}		
	Best	Median	Worst	Best	Median	Worst	Best	Median	Worst
S_1 with 10^{-1}	3,800	4,350	5,150	5,500	7,150	9,600	15,050(38)	20,050	24,800
S_2 with 10^{-20}	20,350	21,700	24,200	36,350	40,550	44,350	40,850(38)	46,600	51,500

Table 7. Standard DE + *Best Update* + *Steady State(RPR)* + *Mutation*, $CR = 0.95$, $F = 0.7$, $P_m = 0.25$

	F_{elp}			F_{sch}			F_{ros}		
	Best	Median	Worst	Best	Median	Worst	Best	Median	Worst
S_1 with 10^{-1}	2,450	3,050	3,850	5,350	7,400	9,300	11,550(39)	19,450	24,600
S_2 with 10^{-20}	13,200	14,700	15,700	37,250	41,700	46,250	43,300(39)	52,900	61,150

development of efficient PSO for solving unimodal problems. In short, the goal of this mutation operator is to probabilistically (P_m indicating the mutation probability) perturb a newly created child randomly around the *Best* solution. This serves for following two purposes: (a) to explicitly promote the diversity in the population, and (b) aid search around the *Best* region. We combine the mutation operator with the best DE so far and show the results in Table 7. P_m is chosen as 0.25 as done in [2]. The results show a definitive improvement on F_{elp} and a mixed improvement on F_{sch} and F_{ros}. Such trends reconcile with those presented in [2]. The possible explanation for the improved performance on F_{elp} lies in the variable-separable and unimodal properties of this problem.

5.1   PCX Based DE

The overall best performance from all the modified DEs is compared against G3-PCX in Table 8, and the DE performances are unable to match-up with those of G3-PCX. While facing a similar predicament with PSO, in [2], authors successfully introduced a parent-centric *Generation* mechanism based on PCX operator and enhanced the PSOs performance, which we attempt next. The step *a* of *Generation*, shown in Figure 1, is replaced by PCX operation in which child is created around the best solution. More details on PCX operator can be found in [1]. Two parameters required in PCX, σ_ζ and σ_η, were taken as 0.1.

The PCX operation with standard DE (referred to as PCX-DE) failed to give any satisfactory results. Following which we introduced *Best Update* strategy. The performance of 'PCX-DE with *Best Update*' was studied and with a population size of $M = 100$ and higher values of CR (taken here as 0.95) yielded

Table 8. G3-PCX and DE's best-so-far performance

	F_{elp}			F_{sch}			F_{ros}		
	Best	Median	Worst	Best	Median	Worst	Best	Median	Worst
				G3-PCX					
S_2 with 10^{-20}	5,744	6,624	7,372	14,643	16,326	17,712	14,847 (38)	22,368	25,797
				Best So Far in DE					
S_2 with 10^{-20}	12,500	13,550	15,250	31,000	34,550	38,850	33,400(40)	43,950	49,800

Table 9. PCX-DE + *Best Update + Steady State (RPR)*, $F = 0.7$, $CR = 0.95$, $NP = 100$

	F_{elp}			F_{sch}			F_{ros}		
	Best	Median	Worst	Best	Median	Worst	Best	Median	Worst
S_1 with 10^{-1}	1,800	2,300	2,900	3,900	5,200	6,600	21,500(39)	24,500	31,100
S_2 with 10^{-20}	9,300	10,200	11,600	28,000	32,500	35,000	45,500(39)	55,300	66,500

an overall better performance. The results were also better than best-so-far DE results. We also tried mutation operator in conjunction with DE-PCX and discovered a degradation in performance. This could be explained based on the fact that mutation brings undesirable randomness into the child creation and destroys the ellipsoidal distribution from PCX operation.

Next, we introduce the *Steady State with Random Parent Replacement* in DE-PCX with *Best Update* and observe a slight improvement in few cases, Table 9. As a next step, the index parent was selected randomly as opposed to being being selected serially. Random selection of index parent further improved the performance but still did not take it closer to G3-PCX. At this point we increased the value of CR to 1.0 and achieved a performance similar to that of G3-PCX, Table 10.

Table 10. PCX-DE + *Random Parent Selection + Best Update + Steady State (RPR)*, $F = 0.7$, $CR = 1.0$, $NP = 100$

	F_{elp}			F_{sch}			F_{ros}		
	Best	Median	Worst	Best	Median	Worst	Best	Median	Worst
S_1 with 10^{-1}	1,000	1,400	1,900	2,300	2,800	3,300	8,800(42)	11,700	14,400
S_2 with 10^{-20}	5,700	6,300	6,900	13,700	15,200	16,500	19,500(42)	23,800	27,800

6 Conclusion

Drawing concepts from existing literature, this paper makes an attempt in developing and employing a unified approach towards *Evolutionary Optimization Systems*. The key steps required for describing an *EOS* are *Initialization, Selection, Generation* and *Replacement*. The central focus of this study is then to improve the performance of standard DE on the class of unimodal problems by identifying modifying its key steps – *Selection, Generation* and *Replacement*. Drawing principles from G3-PCX, a benchmark algorithm, key steps in DE are modified one-by-one. At each stage certain degree of performance improvement is obtained. Finally, PCX operation is introduced in standard DE along with the other alterations, and the performance is comparable to G3-PCX. Although, the modified DE is algorithmically equivalent to G3-PCX, the study suggests how two seemingly different algorithms can be converted from one into another by modifying the key steps. Such a study should enable researchers in Evolutionary Computation to adopt unified approach towards evolutionary algorithms and work towards identifying the properties of key steps, useful in order to develop efficient EAs for any given task.

References

1. Deb, K., Annand, A., Joshi, D.: A computationally efficient evolutionary algorithm for real-parameter optimization. Evol. Comput. 10(4), 371–395 (2002)
2. Deb, K., Padhye, N.: Development of efficient particle swarm optimizers by using concepts from evolutionary algorithms. In: Proceedings of the 2010 GECCO Conference Companion on Genetic and Evolutionary Computation, pp. 55–62. ACM, New York (2010)
3. De Jong, K.A.: Evolutionary Computation: A Unified Approach. MIT Press, Cambridge (2006)
4. Price, K.V., Storn, R.M., Lampinen, J.A.: Differential Evolution: a practical approach to global optimization. Springer, Heidelberg (2005)

A Dynamic Island-Based Genetic Algorithms Framework

Frédéric Lardeux and Adrien Goëffon

LERIA, Université d'Angers
UFR Sciences, 2 bd Lavoisier
49045 Angers, France
frederic.lardeux@univ-angers.fr,
adrien.goeffon@univ-angers.fr

Abstract. This work presents a dynamic island model framework for helping the resolution of combinatorial optimization problems with evolutionary algorithms. In this framework, the possible migrations among islands are represented by a complete graph. The migrations probabilities associated to each edge are dynamically updated with respect to the last migrations impact. This new framework is tested on the well-known 0/1 Knapsack problem and MAX-SAT problem. Good results are obtained and several properties of this framework are studied.

1 Introduction

Genetic algorithms (GAs) [8] are widely used to tackle NP-hard problems. They are easy to implement and can provide good results on classic discrete and continuous optimization problems in term of solutions quality and robustness. Nevertheless, the efficiency of GAs mainly depends on the representation of configurations [15], the fitness function [26], the mutation and crossover operators used [18] as well as global parametrization (population size, mutation frequency, diversity control, selections, elitism, ...) [10,9,22]. Even with a good effort to adapt an efficient GA to a given problem, one quickly observes on critical problem instances some limitations in terms of general performance or scalability.

In order to make GAs more powerful, classic techniques include hybridizations with local search (memetic algorithms [16]) and/or multi-island parallelization schemes [25], which we are investigating in this paper.

Since twenty years and the first distributed evolutionary algorithms [23], island-based genetic algorithms (or island models [25]) are more and more studied in the community. The main problem is to define both the model topology and the migration policies in order to slow down the general convergence of the population while preserving the global mixing of promising individuals. Araujo *et al.* give in [2] a nice review of state-of-the-art island models, in particular concerning the question of migration policies. One can observe that an important number of topologies (Gustafson and Burke in [11] or Rucinski *et al.* in [19] cite numerous topologies models like chains, rings, hypercube and many more) and policies [3,6,24,5,7,1], greatly based on local or global diversity measures, have been defined. In all cases, migration sizes and intervals

K. Deb et al. (Eds.): SEAL 2010, LNCS 6457, pp. 156–165, 2010.

remain difficult to fix [21]. In his recent work, Skolicki [20] emphasizes the fondamental interactions between the two levels of evolution in island models: intra-islands and inter-islands. Ideally, a master intelligent evolution strategy should take advantage of these interactions and maximize the benefits of migrations. But, depending on the current intra- and inter-islands situations (traditionnally only with diversity and fitness measures), it is difficult to predict when individuals have to move, which ones and where, and for which impact.

These considerations motivate us to develop a dynamic island model framework which aims to auto-adapt topology and migration policies during the search in function of some chosen indicators (typically subpopulations properties and previous migrations effects). A particularity of our dynamic island model is to use a complete graph modeling. Nodes represent islands while edges symbolize possible migrations.

Section 2 contains both general and concrete descriptions of our dynamic island model framework. In section 3, we apply our model to two benchmark problems: 0/1 KP and MAX-SAT. Section 4 is a short discussion with additional experiments, with a view to measure the influence of migrations. The conclusion includes future investigations.

2   Dynamic Island Model Framework

2.1   General Description

As recalled by [2], several parameters specify an island model, like:

- the number of individuals undergoing migration,
- the frequency of migrations,
- the policy for selecting immigrants,
- the immigrant replacement policy,
- the topology of the communication among subpopulations, and
- the synchronous or asynchronous nature of the connection among subpopulations.

Now, let us propose an island model framework which generalizes all these parameters, while giving us the possibility to make the model dynamic.

The island model is materialised by a graph, where vertices symbolize islands (subpopulations), while edges represent the possibilities of migrations. Each edge is oriented, and valued with the probability for an individual to migrate from an island to a destination one. The auto-adaptation of this modeling is made with a reward/penalty mechanism. Migration probabilities (values of the edges) are updated after each migration cycle in function of the last migration effects. If the island which receives an individual observes any improvement (resp. deterioration) of its population, then the corresponding migration probability increases (resp. decreases). Here, the population quality is impacted by the average fitness of individuals as well as their diversity if the modeling imposes it.

The dynamic control of parameters like migration rate, can produce different size islands (unless we specifically forbid it). This mechanism prevents poor-quality subpopulations to require as many computational effort as promising ones, and manages the merging of populations. If different islands represent different mutation operators, local search effort or local parametrization, then the algorithm will dynamically provides a well-adapted repartition of individuals considering the search progression.

2.2   Practical Use of the Framework

Figure 1 is an example of our Island Model framework with three islands (i^1, i^2 and i^3). Figure 1.a represents islands with their individuals as well as the migration values (probabilities) from an island to another. In figure 1.b, the destination for each individual are chosen with respect to the values. Most of them remain in the same island (due to the probability values close to 1) but several individuals migrate to other islands (two individuals go from i^1 to i^2 and one from i^3 to i^2). After those migrations, on each island, operators like crossover, migration, or local search, are applied on the individuals. If an offspring individual (*i.e.* an individual obtained by crossover) improves (resp. deteriorates) the population, then its parents are used to update the migration values. For each parent, the values of edges between the last visited island and the current one are increased (resp. decreased) to take into account the impact of the migration. For instance, on Figure 1.c, if we only observe i^2 and a reward/penaly fixed to 5 points (± 0.05), several values are updated:

- ($i^1 \rightarrow i^2$) decreases from 0.50 to 0.40, because the two indivduals becoming from i^1 have produced individual offspring deterioring the population of i^2;
- ($i^2 \rightarrow i^2$) decreases from 0.95 to 0.85, because individuals of island i^2 do not improve the population of i^2;
- ($i^3 \rightarrow i^2$) increases from 0.20 to 0.30, because the individual becoming from i^3 has produced individual offsprings improving the population of i^2;
- due to normalization, other values in relation with i^2 have to be adjusted.

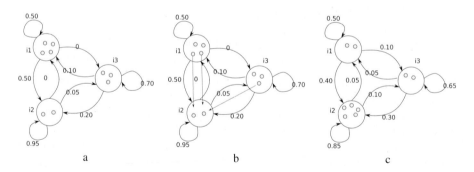

Fig. 1. Communication among subpopulations with a complete graph representation

3   Results

In this section we propose to measure the overall efficiency of the Dynamic Island Model scheme (DIM), applied to two well-known NP-hard problems: 0/1 Knapsack problem and MAX-SAT problem. For this study, the main goal is not to propose a ready-to-use algorithm which outperforms best available softwares, but to measure the global relevance of such a model. For this purpose, we compare for both problems the performance of four basic configurations of the GA:

- A classic 1-island GA ($GA_{classic}$),
- A standard DIM algorithm (DIM_{stand}),
- A specially-parametrized DIM algorithm with uni-directional ring topology DIM_{ring}, which simulates a classic island-based GA with rotative migrations (at each migration process, best individuals migrate to the following populations),
- A parallel GA ($GA/\!/$), with several islands but no migration.

Let us notice that a classic GA corresponds to an island-based GA where migration intervals are minimal, while the parallel partitioned GA ($GA/\!/$) is an island-based GA with no migrations (*i.e.* infinite migration intervals).

3.1 Genetic Algorithms Characteristics

The four configurations of the GA have two types of characteristics. All the numerical values are empirically obtained and confirmed by the REVAC method [17].

1. Intra-islands characteristics:
    - type of population management: steady state
    - elitism: yes
    - selection: tournament
    - mutation: random on offspring with probability 0.5
    - crossover: uniform crossover
2. Inter-islands characterisics:
    - islands number: 20
    - total number of individuals: 600
    - starting repartition: well-balanced (30 individuals per island)
    - total number of crossovers: 216000 (360×600 individuals)
    - initial migration probabilities: see below
    - reward: 5 points
    - penalty: 5 points

The number of crossovers in an island between two migrations is proportional to its number of individuals. This choice ensures the same crossover rate per island, whatever its size.

Initial migration probabilities. To give the same attractive power to each island, the initial migration probabilities must be symmetric. At the beginning, we fix a highest probability to stay on the same island than to move to another one, in order to exploit initial populations. For instance, the initial matrix corresponding to an island model with three islands can be the next one:

		Destination		
		i^1	i^2	i^3
	i^1	0.75	0.125	0.125
Source	i^2	0.125	0.75	0.125
	i^3	0.125	0.125	0.75

3.2   Experimental Settings

Algorithms used in our experiments are applied 10 times for each instance. To be sure that the difference of behaviours is not due to the initial populations and other stochastic factors, 10 distinct random seeds are used by each algorithm. Results presented in the tables are averages; standard deviations are not mentioned since they are very low.

3.3   0/1 Knapsack Problem

The Knapsack Problem (KP) is a well-known combinatorial problem. Given n items whose weights w_i and values v_i are known ($x_i \in \{1, \ldots, n\}$), the goal is to find a subset of items of maximal value such that the total weight is less than a given capacity W. In the most common 0/1 KP, each item can be selected only once ($x_i \in \{0, 1\}$, where x_i is the number of selected copies of object i).

More precisely, 0/1 KP is shortly formulated as an optimization problem by:

$$\text{maximize} \sum_{i=1}^{n} v_i x_i, \text{s.t.} \sum_{i=1}^{n} w_i x_i \leq W, x_i \in \{0, 1\}$$

For more information on 0/1 KP, we invite the reader to refer to [12].

Island-based algorithms $\text{DIM}_{\text{stand}}$ and DIM_{ring}, as well as edgeless topologies $\text{GA}_{\text{classic}}$ and $\text{GA}/\!/$, have been tested on five 0/1 KP instances. Instances have been generated according to the definition given by [4] and the generator proposed in [13], with the following parameters:

- number of items $\in \{100, 250, 500, 1000, 2000\}$
- range of coefficient: 10000
- type: avis subset-sum
- number of tests in series: 1000

Experiments have shown that only the three last instances (those with resp. 500, 1000 and 2000 items) are representative for comparison, the two first ones appearing too much easy to solve, with similar results for all algorithms. Consequently, we only focus on three random instances: n500, n1000 and n2000.

Table 1 shows the efficiency of each method on these instances. It is not surprising that the Dynamic Island Model outperforms traditional GAs. However, performance differences are quite important, taking into account that last improvements are particularly hard to find for knapsack problems. An interesting point is that, in this experiment, the classic rotative scheme (DIM_{ring}) is not competitive; comparatively, the classic GA works even better for the two hardest instances. The main reason is probably the relatively-small size of islands (20 individuals), which is adaptive in $\text{DIM}_{\text{stand}}$ while it remains unchanged during the entire process in DIM_{ring}.

3.4   MAX-SAT Problem

In order to test the DIM framework with an other problem, we try to handle the MAX-SAT problem. Given a Boolean formula in CNF (conjunction of clauses which are

Table 1. Comparison between DIM and classic GAs

Instance	GA	DIM$_{stand}$	DIM$_{ring}$	GA$/\!/$
n500	755 626.54	**760 620.50**	755 818.38	748 230.25
n1000	1 485 393.86	**1 502 549.22**	1 483 248.36	1 465 757.47
n2000	2 866 752.92	**2 910 891.46**	2 853 072.50	2 832 290.37

disjunctions of literals), the aim is to provide an assignment to the Boolean variables such that the number of true clauses is maximum. The formula is satisfiable iff it exists an assignment which makes true all the clauses.

Three instances are used for our experiments:

- f600: random instance with 600 variables and 2550 clauses;
- f1000: random instance with 1000 variables and 4250 clauses;
- qg1-7.suffled: latin square instance with 686 variables and 6816 clauses.

All these instances are satisfiable thus there is an assignment of the Boolean variables satisfying all the clauses.

Table 2. Comparison between DIM and classic GAs

Instance	Nb Clauses	GA	DIM$_{stand}$	DIM$_{ring}$	GA$/\!/$
f600	2550	2513.80	**2533.40**	2518.70	2357.20
f1000	4250	4174.20	**4208.90**	4174.10	3890.50
qg1-7.shuffled	6816	6756.30	**6787.00**	6776.50	6211.70

In table 2, we observe that DIM$_{stand}$ provides the best results on the three instances. GA$/\!/$ is the worst and GA obtains a little less interesting results than DIM$_{ring}$. The results for DIM$_{stand}$ and DIM$_{ring}$ are computed with the best migration frequency empirically found. In the next section, a more detailled study of this parameter is given.

4 Discussion

As seen in section 3.3 and 3.4, DIM$_{stand}$ provides very promising results with respect to the other GAs. The difference between DIM$_{stand}$ and DIM$_{ring}$ is only concerning the type of migration, whereas the difference among GA$/\!/$, GA and DIM$_{stand}$ is the migrations periodicity. This periodicity is given by a mean number of crossovers per individuals. Then, the number of crossovers between two migrations differs from an island to another and depends on their size (number of individuals).

GA can be considered as an island model with a very weak migrations periodicity and GA$/\!/$ with a very strong migrations periodicity (recall that a weak periodicity corresponds to a high frequency). Between these two algorithms, a dynamic island model can use different frequencies which provide different algorithm behaviours. In figure 2,

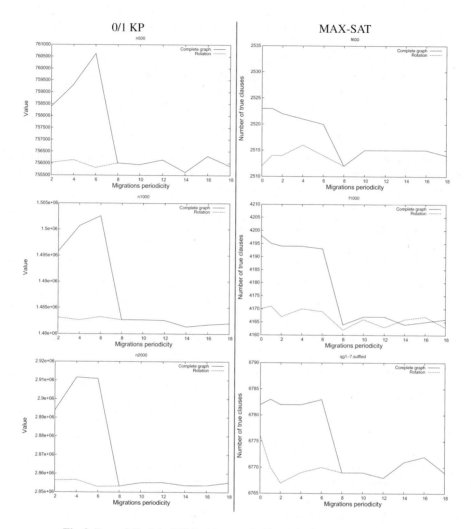

Fig. 2. Powerfull of the DIM with several values of migrations periodicity

one can observe the impact of the migrations periodicity on all the studied 0/1 KP and MAX-SAT instances in this paper.

It is interesting to see that, when the migrations periodicity is higher than 8 crossovers per individual, DIM$_{stand}$ and DIM$_{ring}$ provide equivalent results. A possible explanation of this behavior is that the population is converging before the first migration. Indeed, in our experiments, 240 crossovers are applied on the population of each island before the first migration. If each population only contains clones after these crossovers, then rotative migration provides the same effect than the complete graph migration scheme. Let us precise that in this study, we have deliberately not regulated the diversity of the population, in order to observe more precisely the behaviour of the models. That is the reason for which scores are lower than those shown in Table 2.

An other observation is the weak difference among the results obtained by DIM_{ring} for all values of migrations periodicity. Except for instance qg1-7.suffled, where results are better with a small periodicity, this parameter seems not determinant for the rotative migration scheme.

With a migrations periodicity smaller than 8, the complete graph migration scheme is clearly better than the rotative one. For the 0/1 KP instances, value 6 seems to be the best periodicity. For MAX-SAT instances, a small periodicity provides very good results. The reason is probably that a small periodicity avoid the convergence of all the populations.

It is clear that the migration periodicity is an important parameter for DIM_{stand}. A next work will be to control it with autonomous mechanism like in [14].

5   Conclusion

In this work, we have introduced a dynamic island model framework, which aims both to generalize migration toplogies and to auto-regulate migration policies. First, the complete graph modeling allows every definition of topologies, from edgeless graphs (standard sequential or parallel genetic algorithms) to well-studied island model topologies like uni-directional or bi-directional rings, lattices, hypercubes, full topologies or anything else. If the dynamic regulation is activated, then the topology is evolving during the search following the rewards and penalties due to previous migration effects. Contrary to traditional island model mechanisms, where migrations are evaluated *a priori* in measuring divergence between individuals (which is much more a guided repartition of individuals to provide diversity, but a nonsense in a nature-inspired algorithm), we encourage (resp. dissuade) moves of which previous executions and produced mixing yield good (resp. weak) offsprings, in term of fitness and/or diversity. During the search, this auto-regulation of migration probabilities makes the model more or less dynamic in terms of number of migrations, which favouring either diversification, or intensification.

Experiments realized on two major combinatorial problems like 0/1 KP and MAX-SAT show that this dynamic scheme, even with basic parametrization, provides good results, notably if we compare its performance with a classic uni-directional ring migration topology.

The most promising prospect of our ongoing and future works is to parametrize differently each island. One can imagine that different islands can work with their own rules in terms of mutation or crossover operators, selection or replacement criterions for instance. In particular, considering local search operators, if different islands working with different operators, or different parametrizations of the search (more intensive or more stochastic), are coexisting within a dynamic island model, it would be interesting to observe the evolution of each island activity during the search. We think that such a model can provide an adaptative operator selection as well as a diversity regulation due to the island topology.

References

1. Araujo, L., Guervós, J.J.M., Cotta, C., de Vega, F.F.: Multikulti algorithm: Migrating the most different genotypes in an island model. CoRR, abs/0806.2843 (2008)
2. Araujo, L., Guervós, J.J.M., Mora, A., Cotta, C.: Genotypic differences and migration policies in an island model. In: GECCO, pp. 1331–1338 (2009)
3. Cantú-Paz, E.: Migration policies, selection pressure, and parallel evolutionary algorithms. Journal of Heuristics 7(4), 311–334 (2001)
4. Chvátal, V.: Hard knapsack problems. Operations Research 28, 1402–1411 (1980)
5. Denzinger, J., Kidney, J.: Improving migration by diversity. In: IEEE Congress on Evolutionary Computation, vol. (1), pp. 700–707 (2003)
6. Ricarte, I.L.M., Yamakami, A., Noda, E., Coelho, A.L.V., Freitas, A.A.: Devising adaptive migration policies for cooperative distributed genetic algorithms. In: Proceedings of the IEEE International Conference on Systems, Man and Cybernetics, pp. 438–443 (2002)
7. Eldos, T.: A new migration model for the distributed genetic algorithms. In: Proceedings of the International Conference on Scientific Computing (CSC 2006), Las Vegas, NV, pp. 26–29 (2006)
8. Goldberg, D.E.: Genetic algorithms in search, optimization, and machine learning. Addison-Wesley, Reading (1989)
9. Goldberg, D.E., Deb, K.: A comparative analysis of selection schemes used in genetic algorithms. In: Foundations of Genetic Algorithms, pp. 69–93. Morgan Kaufmann, San Francisco (1991)
10. Goldberg, D.E., Deb, K., Clark, J.H.: Genetic algorithms, noise, and the sizing of populations. Complex Systems 6, 333–362 (1991)
11. Gustafson, S., Burke, E.K.: The speciating island model: An alternative parallel evolutionary algorithm. J. Parallel Distrib. Comput. 66(8), 1025–1036 (2006)
12. Lagoudakis, M.G.: The 0-1 knapsack problem – an introductory survey. Technical report, University of Southwestern Louisiana (1996)
13. Martello, S., Pisinger, D., Toth, P.: Dynamic programming and strong bounds for the 0-1 knapsack problem. Manage. Sci. 45(3), 414–424 (1999)
14. Maturana, J., Lardeux, F., Saubion, F.: Autonomous operator management for evolutionary algorithms. Journal of Heuristics (in press, 2010)
15. Michalewicz, Z.: Genetic algorithms + data structures = evolution programs, 3rd edn. Springer, London (1996)
16. Moscato, P.: Memetic algorithms: a short introduction, pp. 219–234 (1999)
17. Nannen, V., Smit, S.K., Eiben, A.E.: Costs and benefits of tuning parameters of evolutionary algorithms. In: Rudolph, G., Jansen, T., Lucas, S., Poloni, C., Beume, N. (eds.) PPSN 2008. LNCS, vol. 5199, pp. 528–538. Springer, Heidelberg (2008)
18. Richter, J.N.: On Mutation and Crossover in the Theory of Evoutionary Algorithms. PhD thesis, Montana State University (2010)
19. Rucinski, M., Izzo, D., Biscani, F.: On the impact of the migration topology on the island model. CoRR, abs/1004.4541 (2010)
20. Skolicki, Z.: Studies in Computational Intelligence. In: Linkage in Island Models, pp. 41–60. Springer, Heidelberg (2008)
21. Skolicki, Z., De Jong, K.A.: The influence of migration sizes and intervals on island models. In: GECCO, pp. 1295–1302 (2005)
22. Spears, W.M.: Evolutionary Algorithms: The Role of Mutation and Recombination. Springer-Verlag New York, Inc., Secaucus (2000)
23. Tanese, R.: Distributed genetic algorithms. In: Proceedings of the Third International Conference on Genetic Algorithms, pp. 434–439. Morgan Kaufmann Publishers Inc., San Francisco (1989)

24. Ursem, R.K.: Diversity-guided evolutionary algorithms. In: Guervós, J.J.M., Adamidis, P.A., Beyer, H.-G., Fernández-Villacañas, J.-L., Schwefel, H.-P. (eds.) PPSN 2002. LNCS, vol. 2439, pp. 462–474. Springer, Heidelberg (2002)
25. Whitley, D., Rana, S., Heckendorn, R.B.: The island model genetic algorithm: On separability, population size and convergence. Journal of Computing and Information Technology 7, 33–47 (1998)
26. Wolpert, D.H., Macready, W.G.: No free lunch theorems for optimization. IEEE Transactions on Evolutionary Computation 1(1), 67–82 (1997)

Solving the Optimal Coverage Problem in Wireless Sensor Networks Using Evolutionary Computation Algorithms[*]

Zhi-hui Zhan[1], Jun Zhang[1,**], and Zhun Fan[2]

[1] Dept. of Computer Science, SUN Yat-sen University
Key Laboratory of Digital Life (Sun Yat-Sen University), Ministry of Education
Key Laboratory of Software Technology,
Education Department of Guangdong Province, P.R. China
junzhang@ieee.org
[2] Denmark Technical University

Abstract. This paper formulates the optimal coverage problem (OCP) in wireless sensor network (WSN) as a 0/1 programming problem and proposes to use evolutionary computation (EC) algorithms to solve the problem. The OCP is to determine to active as few nodes as possible to monitor the area in order to save energy while at the same time meets the surveillance requirement, e.g., the full coverage. This is a fundamental problem in the WSN which is significant for the network lifetime. Even though lots of models have been proposed for the problem and variants of approaches have been designed for the solution, they are still inefficient because of the local optima. In order to solve the problem effectively and efficiently, this paper makes the contributions to the following two aspects. First, the OCP is modeled as a 0/1 programming problem where 0 means the node is turned off whilst 1 means the node is active. This model has a very natural and intuitive map from the representation to the real network. Second, by considering that the EC algorithms have strong global optimization ability and are very suitable for solving the 0/1 programming problem, this paper proposes to use the genetic algorithm (GA) and the binary particle swarm optimization (BPSO) to solve the OCP, resulting in a direct application of the EC algorithms and an efficient solution to the OCP. Simulations have been conducted to evaluate the performance of the proposed approaches. The experimental results show that our proposed GA and BPSO approaches outperform the state-of-the-art approaches in minimizing the active nodes number.

Keywords: Wireless sensor networks (WSN), optimal coverage problem, evolutionary computation (EC), genetic algorithm (GA), particle swarm optimization (PSO).

---

[*] This work was supported in part by the National Natural Science Foundation of China No.U0835002 and No.61070004, by the National High-Technology Research and Development Program ("863" Program) of China No. 2009AA01Z208.
[**] Corresponding author.

K. Deb et al. (Eds.): SEAL 2010, LNCS 6457, pp. 166–176, 2010.

1 Introduction

The wireless sensor networks (WSN) is a very new technology which has become a hottest and most challenging research topic recently [1][2]. The WSN consists lots of sensor nodes that monitor the area for specialized applications such as battlefield surveillance, habitat monitoring, environmental observation, health applications, and many others [3]. The environments of these applications are usually not friendly and it is difficult to deploy the sensors determinately. Therefore, a large amount of nodes are randomly deployed in the area, resulting in more sensors than required. The high density of sensors on the one hand compensates for the lack of exact positioning and improves the fault tolerance, while on the other hand may cause the larger energy consumption due to conflicting in accessing the communication channels, maintaining information about neighboring nodes, and some other factors [4]. Therefore, research into optimally scheduling the sensor nodes and making the redundant nodes turned off to sleep in order to save the energy to prolong the network lifetime has become one of the most significant and promising areas in the WSN [5][6].

A considerable number of researches have devoted to address the energy efficient problem in the WSN in order to prolong the network lifetime. Most of the researches transform this issue to the optimal coverage problem (OCP) [5]. The OCP is based on the fact that the WSN contains a large number of sensor nodes with many nodes sharing the same monitored regions, and some of the nodes are redundant and can be turned off to preserve the energy while the others still work to offer the full coverage. The OCP is to find out a minimal set of nodes to monitor the area, and turning off the other redundant nodes to save energy, while at the same time meeting the coverage requirement. This way, not only the nodes can reduce the energy consumption caused by the nodes confliction, but also the network lifetime can be significantly prolonged because the nodes can be scheduled to work in turn [6].

In the literature, different models and assumptions have been introduced to this problem and variant of approaches have been proposed for solution. Xing *et al.* [7] proved that the full sensing coverage of the network can guarantee the connectivity of the network when the communication range is not shorter than twice the sensing range. As many kinds of wireless sensor can meet this condition, many researches only concentrate on the coverage problem of the network. Consequently, only the coverage problem is considered in this paper. Approaches such as coverage-based off-duty eligibility rule [8], time axis dividing node working schedule [9], and probing environment and adaptive sleeping (PEAS) protocol [10] have been proposed to address the OCP in finding out a minimal set of nodes to be active. Among the above state-of-the-art approaches, it should be noted that the approaches in [8][9] can guarantee the full coverage while the one in [10] can not. Our model and approaches are designed to guarantee the full coverage.

The motivations and contributions of our work include the following four aspects.

1)   The WSN consists of lots of sensor nodes with very limited energy. It has been a promising and significant research area to solve the OCP in order to save sensor energy and prolong network lifetime.

2)   The existing models and their approaches to OCP are always not easy to understand or implement. Therefore, it is significant and promising to design a

simple OCP model to describe the problem, and at the same time propose simple but effective and efficient approach to solve the problem.

3)   The OCP is NP-complete [11] and some of the traditional approaches have their natural disadvantages of being trapped into local optimal. Therefore, we propose to use evolutionary computation (EC) algorithm [12] to solve the problem. As the EC algorithms have strong global search ability, good adaptation and robustness, it is expected that the EC algorithms can solve the OCP efficiently.

4)   Some existing approaches can not guarantee the full coverage when dealing with the OCP [10], resulting in disadvantages when used in practical applications. In our work, the model and approaches are hence designed to guarantee the full coverage.

In this paper, the OCP is modeled as a 0/1 programming problem. Given a WSN topology with a set of randomly deployed sensor nodes, our model marks the nodes with values of 0 or 1, where 0 means that the node is turned off to sleep while 1 means that the node is active to work. In this way, the OCP is transformed to optimizing a 0/1 string. That is, minimizing the number of 1 and at the same time making sure that the nodes with value 1 can provide full coverage for the area. In order to solve this problem effectively and efficiently, two EC algorithms, named the genetic algorithm (GA) [13] and the particle swarm optimization (PSO) [14], are adopted. The GA approach is used to solve the problem directly with the chromosomes coded as a 0/1 string. The standard PSO was designed for the problems in continuous domain. Therefore it is not suitable to directly use the standard PSO to solve the OCP. In our work, we use the discrete binary PSO (BPSO) [15] as the approach. The BPSO is promising in solve discrete binary problems [15], and therefore it is also suitable for the OCP. Simulations are conducted to evaluate the performance of the proposed GA and BPSO approaches. Experimental results show that our GA and BPSO approaches both win the state-of-the-art approaches in minimizing the active sensor number in the WSN. The experimental results also show that the BPSO approach is more robust and more efficient than the GA approach at minimizing the active sensor number over different sensing ranges and different deployed nodes number.

The rest of this paper is organized as follows. Section 2 gives the problem formulations of the OCP in the WSN. Section 3 proposes our methodology that uses the EC algorithm to solve the OCP. Section 4 gives the experimental results and compares our approaches with some state-of-the-art approaches. Moreover, the search behaviors and scalabilities of the GA and BPSO approaches are investigated. At last, conclusions are summarized in Section 5.

2   Optimal Coverage Problem in WSN

Given an $L{\times}W$ (Length×Width) rectangle area A for monitoring, and a great amount of N sensors are randomly deployed in the area. The OCP is to determine using only a sub-set of M sensors from the N sensors to fully cover the monitored area, supposed that the area can be fully covered by the original N sensors (as the N sensors are randomly deployed, the area may be not fully covered in the original network topology,

we do not consider this situation in our paper). The objective of the OCP is to minimize the number of M.

In order to know whether the area A is fully covered by the sensors network, we
assume that the location of the sensor is prior known. Moreover, the area is divided
into grids and the coverage issue can be transformed to check whether each of the
grids is covered by at least one active sensor [8].

All the N sensors form the sensors set $S=\{s_1, s_2, ..., s_N\}$, where each sensor node s_i
is with the location (x_i, y_i) and the sensor radius R. For any grid $g=(x, y)\in A$ in the
monitored area, the relationship between the s_i and the g is defined as:

$$P(s_i, g) = \begin{cases} 1, & \text{if } (x-x_i)^2 + (y-y_i)^2 \le R^2 \\ 0, & \text{otherwise} \end{cases} \quad (1)$$

where 1 means that the grid g is covered by the sensor s_i while 0 means the sensor s_i
does not cover the grid g. Therefore, for any grid point g, if there exists at least one
sensor $s_i (1 \le i \le N)$ that makes $P(s_i,g)=1$ follow, we say that the g is covered by the
sensor network. In this sense, the monitored area A is fully covered if any grid point g
in the area is covered by the sensor network.

In OCP, the area is monitored by an optimally selection sub-set $S^* \subseteq S$ with M sensors from the N sensors, satisfying the constraint that the area A is still fully covered
by the M sensors, and with the objective of minimizing the value of M, as:

$$f = \min M, \quad \text{where } M = \left|S^*\right|, \quad S^* \subseteq S \quad (2)$$
$$subject \ to \ (\oplus_{s_i \in S^*} P(s_i, g)) = 1, \quad \forall g \in A$$

Here, the operator $\oplus$ results in a value of 0 if all the elements are 0. Otherwise, the
result is 1 if at least one of the elements is 1.

Therefore, the OCP can be modeled as a 0/1 programming problem to determine
whether a sensor is selected (with the value 1) or is not selected (with the value 0).
Such a model has the following features and advantages.

1)   The 0/1 programming model is easy and intuitive to understand, 1 means the
     sensor is selected to active, whilst 0 means the sensor is scheduled to sleep.
2)   Unlike some other scheduling algorithms and protocols, our OCP model does
     not need the neighborhood information of the sensor nodes. This makes the
     model robust to adapt to different network topologies.
3)   The OCP model is naturally suitable to be solved by the EC algorithms such
     as the GA and the BPSO. Therefore, good performance can be expected by
     using the EC algorithms to solve the OCP.

3   Methodology: EC for OCP

In this section, the EC approaches are designed to solve the OCP. As the OCP is
modeled as a 0/1 programming problem, the GA [13] and the BPSO [15] are naturally
suitable for it. Therefore, we describe the EC approaches to the OCP based on these
two algorithms.

3.1   Solution Representation and Fitness Function

The OCP is modeled as a 0/1 programming problem and therefore the individual (e.g., the chromosome in the GA and the particle in the BPSO) is coded as a binary string of 0 and 1. The length of the binary string is the same as the sensor nodes number N, and the representation is as:

$$X = [x_1, x_2, \ldots, x_N], \text{ where } x_j=0 \text{ or } 1 \tag{3}$$

where $x_j=1$ means that the jth sensor is selected to active and $x_j=0$ means that the jth sensor is selected to sleep.

The objective of the problem is to minimize the number of active sensor nodes. Therefore, the fitness function can be simply defined as Eq. (2) as in Section 2.

3.2   Evolutionary Process

The main process of the EC algorithms consists of the evolutionary operators. For example, the selection, crossover, and mutation operators in GA, and the velocity, position updating operators in BPSO. In this section, we describe these evolutionary processes of GA and BPSO in solving the OCP.

(1) Initialization: In this phase, a check is first carried out to make sure that the area can be fully covered by the original sensor network with all the nodes active. Otherwise, the optimization process reports a failing result and terminates.

In the initialization, a population of individuals is randomly generated. For each dimension j, the value of x_j is set as 0 or 1 randomly. Moreover, the velocity in BPSO is randomly initialized as a real value within the velocity range $[-Vmax, Vmax]$.

One thing should be noted is that the initialized individual X may be infeasible because it can not provide the full coverage of the area. In this case, a repair procedure would be performed on the individual to make it feasible. The repair procedure will be discussed in the next part.

(2) GA Related Operators: In a genetic cycle, the GA performs the selection, crossover, mutation, evaluation, and elitist operations.

In the selection, a tournament selection strategy is adopted. In each selection, partial of the chromosomes (e.g., 20% of the population) are randomly chosen to compete for survival. The winner (the chromosome with the best fitness value, i.e., with fewest active sensor nodes) is selected into the next generation. Repeat the selection round until a population size of chromosomes have been selected. After selection, the survived chromosomes produce offspring via the single-point crossover and the bits flip mutation operations [13]. After the above three genetic operations, all the chromosomes are to be evaluated. The evaluation operation is to count the sensor nodes that the chromosome uses to fully cover the area. That is, the number of value 1 in the solution string. If the solution is infeasible, it will be repaired before evaluation. At last the elitist operation is used to store the best-so-far solution.

(3) BPSO Related Operators: The BPSO performs the velocity update similarly to the one used in the standard PSO. The only difference is that the value for x_{ij}, p_{ij}, and g_j is 0 or 1. Also, the velocity is to be clamped within the range of $[-Vmax, Vmax]$. The jth dimension of velocity, v_{ij}, is regarded as the probability of the jth position x_{ij} being the value of 1. Therefore, the value of v_{ij} should be mapped into the interval of

[0.0, 1.0]. In the proposal of [15], the sigmoid function is used to obtain this transformation, as:

$$p_{ij} = Sigmoid(v_{ij}) = \frac{1}{1+e^{-v_{ij}}} \tag{4}$$

With the value of p_{ij} obtained, the BPSO performs the position update as:

$$x_{ij} = \begin{cases} 1, & \text{if } rand() < p_{ij} \\ 0, & \text{otherwise} \end{cases} \tag{5}$$

where $rand()$ is a random value generated from a uniform distribution in the range [0.0, 1.0].

After the velocity and position update operations, the particle will be evaluated. However, we do not repair the particle if its current position X is infeasible. Instead, we do not evaluate the infeasible particle. As we have made sure that all the particles are initialized feasibly, all the particles will store a feasible solution in their **Pbest** vectors. Therefore, all the particles will eventually return to the search range because they are attracted by the feasible guidance **Pbest** and **Gbest**. Using such a strategy can save a lot of computational time by avoiding the repair procedure.

3.3 Repair Procedure

As mentioned above, the solution represented by the individual may be infeasible sometimes because the active sensor nodes can not fully cover the area. Therefore, it is necessary to design a repair procedure to make the individual feasible. The repair procedure is performed on all the infeasible individuals in the initialization and on all the infeasible GA chromosomes during the evolutionary process. The pseudo-code of the repair procedure is given as Fig. 1 and is described as follows.

```
01. //Input: A infeasible binary string X
02. //Output: A feasible binary string X
03. procedure Repair
04. {
05.     int k=rand()%N;
06.     while (X is infeasible){
07.         while (X_k==1){
08.             k=k+1;
09.             if (k>=N) k=k-N;
10.         }//end of while (X_k==1)
11.         X_k=1;
12.     }//end of while (X is infeasible)
13. }
```

Fig. 1. The pseudo-code of the repair procedure

Given an infeasible individual X to be repaired, it is supposed that the X is feasible if the values of all the dimensions of X are 1. First, a random integer value k in range $[1, N]$ is generated as the start dimension where N is the total sensor nodes number. Then the procedure checks the dimensions of X from k to find the first dimension k^* with the value of 0. This dimension is forced to set as 1. Then the new X is checked to see whether it is feasible. If the X is still infeasible, the procedure goes on finding another dimension k^{**} with the value of 0 and force it to be 1. The procedure terminates until the X is feasible.

4 Experiments and Comparisons

The GA and BPSO approaches for solving the OCP are implemented and evaluated. The parameters configurations of the two algorithms are described as follows.

In the GA approach, a population size of 40 chromosomes is used to solve the OCP. A tournament selection strategy is used. In each selection round, 20% of the chromosomes compete for survival. The crossover probability p_x and mutation probability p_m are 0.7 and 0.03 respectively. The algorithm terminates at the maximal generations of 200.

In the BPSO approach, the population size is also 40 and the maximal generations is 200. The algorithm parameters Vmax is 6, acceleration coefficients c_1 and c_2 are both 2.0 [15]. These parameters configurations are determined by considering the commonly recommended values in the literature and our empirical study.

4.1 Comparisons with the State-of-the-Art Approaches

We take the representative state-of-the-art work in [8] and [10] for comparisons. It is should be noticed that the approach in 8 ensures the full coverage while the one in [10] can not.

In order to make a fair comparison, we follow the network topology used in [8]. That is, with the monitored area 50m×50m, the sensing range 10m, and the original deployed nodes number 100. We carried out the simulations on 100 different random topologies and the mean results are compared.

Table 1. Comparisons on the Sleep Nodes Number Obtained by Different Approaches

Approach	Probing range	Sleep nodes	Blind points	Full coverage
[8]	N/A	53	0	Y
[10]	3	38	13	N
	4	54	26	N
	5	66	68	N
	6	71	91	N
	7	81	100	N
GA	N/A	81.74	0	Y
BPSO	N/A	81.12	0	Y

We show the average number of sleep nodes obtained by our GA approach, BPSO approach, and the approaches of [8] and [10] in Table I where the data of approaches [8] and [10] is derived directly from [8]. It can be observed from the table that our GA and BPSO approaches have the strong search ability to identify the redundant nodes and let more nodes sleep to save the energy. The performance of the PEAS in [10] is significantly relied on the probing range. The sleep node number increases as the probing range increases. However, the larger probing range results in more blind points. The approach in [8] can avoid the blind point, but it is not efficient enough to schedule the redundant nodes to sleep. Our proposed GA and BPSO approaches can not only avoid the blind point, but also are efficient in turning off the redundant nodes to save energy.

4.2 Behaviors of the EC Approaches

As the stochastic nature of the EC optimization approaches, it is interesting to investigate the search behaviors of the EC approaches in solving different network topologies with different node density and sensing range. Therefore, more experiments are conducted in this subsection.

In the investigations, the sensing range is set as 8m, 10m, and 12m, while the original deployed nodes number is set as 100, 150, 200, 250, and 300. We test the approaches in 100 random topologies for each configuration, and the average results are plotted in Fig. 2. The observations from the Fig. 2 show that increasing the sensing range and increasing the number of original deployed nodes can both result in more nodes being scheduled to sleep. This result is consistent with our expectation.

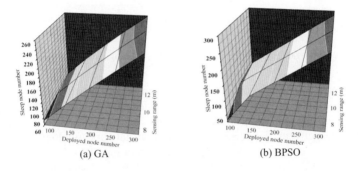

(a) GA   (b) BPSO

Fig. 2. The sleep node number over different sensing range and deployed node number

Another interesting investigation is to see whether the number of active nodes remains constant over different number of original deployed nodes when the sensing range is fixed. The data in Table 2 and the curves in Fig. 3(a) show that the behavior of the GA approach does not have the good ability to keep the number of active nodes in a dense network as small as the one in a sparse network. Instead, the number of active nodes increases as the increasing of the original deployed nodes. Such a phenomenon was also observed in the coverage-based off-duty eligibility rule which was proposed in [8]. The authors of [8] claimed that that was caused by the increasing of the edge nodes which can not be turned off by their approach. However, the GA and BPSO approaches seem to be less affected by the edge nodes. This can be demonstrated by the performance of BPSO that has very good ability in keeping the small set of active nodes over different networks. As shown in Fig. 3(b), within the same sensing range, the number of active nodes almost does not change when the original deployed node number increases.

Table 2. The Active Node Number Over Different Original Deployed Node Number on Different Sensing Ranges

Sensing range		Original Deployed Node Number				
		100	150	200	250	300
GA	8 m	27.97	31.29	37.59	45.43	54.91
	10 m	18.26	21.59	27.75	35.63	45.32
	12 m	13.26	16.27	22.48	31.14	42.25
BPSO	8 m	28.90	28.77	28.41	28.64	29.13
	10 m	18.88	19.04	19.17	19.44	19.62
	12 m	13.48	13.70	14.03	13.85	14.22

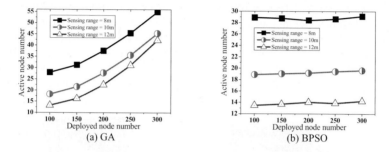

(a) GA   (b) BPSO

Fig. 3. The active node number over different sensing range and deployed node number

So what makes the GA sensitive to nodes density of the network? We conducted experiments on another two mutation probabilities, 0.005 and 0.07, while the other parameters keep the same as above. The results are plotted in Fig. 4. It can be observed that the performance of GA over different network topologies can be maintained when the p_m is small, e.g., 0.005, as shown in Fig. 4(a). However, if the p_m is large, GA can not find a solution with as few nodes as in a sparse network when dealing with a dense network. This means that GA is sensitive to the parameters, specially, the mutation probability p_m. The BPSO approach is more robust than the GA approach in dealing with different optimization environments.

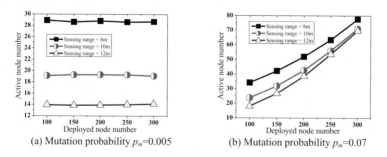

(a) Mutation probability p_m=0.005   (b) Mutation probability p_m=0.07

Fig. 4. The active node number over different sensing range and deployed node number for the GA approach with different mutation probability p_m

5 Conclusions

The optimal coverage problem in WSN has been formulated as a 0/1 programming problem and the GA and BPSO approaches have been designed to solve the problem. In the OCP model, 0 stands for sleep and 1 stands for active. Therefore the model is not only simple and easy to understand, but also has a natural map from the representation to the real network topology. The approaches to solve the OCP are based on evolutionary computation algorithms and therefore they are adaptive to the problem, with strong global search ability and good robustness. We have described the implementation details of using the GA and the BPSO to solve the problem. The performance is evaluated and compared with the state-of-the-art approaches. The experimental results have shown the effectiveness and efficiency of the proposed approaches. The experimental results also show that BPSO is more robust and more efficient than GA in reducing the active node number to save the network energy, especially in dense network.

In the future work, we will try to use the most recent adaptive strategy [16] and the orthogonal learning strategy [17] in to BPSO to design more efficient algorithm for solving the OCP in WSN.

References

1. Akyildiz, I.F., Su, W., Sankarasubramaniam, Y., Cayirci, E.: Wireless sensor networks: a survey. Computer Networks 38(4), 393–422 (2002)
2. Quintao, F.P., Nakamura, F.G., Mateus, G.R.: Evolutionary algorithm for the dynamic coverage problem applied to wireless sensor networks design. In: Proc. IEEE Int. Conf. on Evolutionary Computation, pp. 1589–1596 (2005)
3. Kuorilehto, M., Hannikainen, M., Hamalainen, T.D.: A survey of application distribution in wireless sensor networks. EURASIP Journal on Wireless Communications and Networking 5, 774–788 (2005)
4. Shih, E., Cho, S.H., Ickes, N., Min, R., Sinha, A., Wang, A., Chandrakasan, A.: Physical layer driven protocol and algorithm design for energy-efficient wireless sensor networks. In: Proc. of the 7th Annual Int. Conf. on Mobile Computing and Networking, pp. 272–287 (2001)
5. Huang, C.F., Tseng, Y.C.: A survey of solutions to the coverage problems in wireless sensor networks. Journal of Internet Technology 6(1), 1–8 (2005)
6. Cardei, M., Wu, J.: Energy-efficient coverage problems in wireless ad-hoc sensor networks. Computer Communications 29, 413–420 (2006)
7. Xing, G., Wang, X., Zhang, Y., Lu, C., Pless, R., Gill, C.: Integrated coverage and connectivity configuration for energy conservation in sensor networks. ACM Trans. on Sensor Networks 1(1), 36–72 (2005)
8. Tian, D., Georganas, N.D.: A node scheduling scheme for energy conservation in large wireless sensor networks. Wireless Commun. And Mobile Comput. 3, 271–290 (2003)
9. Yan, T., He, T., Stankovic, J.A.: Differentiated surveillance for sensor networks. In: ACM 1st International Conf. on Embedded Networked Sensor Systems (SenSys), pp. 51–62 (2003)
10. Ye, F., Zhong, G., Lu, S., Zhang, L.: PEAS: A robust energy conserving protocol for long-lived sensor networks. In: Proc. Int. Conf. on Distributed Computing Systems (ICDCS), pp. 28–37 (2003)

11. Zou, Y., Chakrabarty, K.: A distributed coverage- and connectivity-centric technique for selecting active nodes in wireless sensor networks. IEEE Trans. Computers 54(8), 978–991 (2005)
12. Back, T., Hammel, U., Schwefel, H.: Evolutionary computation: comments on the history and current state. IEEE Trans. Evol. Comput. 1(1), 15–28 (1997)
13. Michalewicz, Z.: Genetic Algorithms + Data Structure = Evolution Programs. Springer, Heidelberg (1996)
14. Kennedy, J., Eberhart, R.C.: Particle swarm optimization. In: Proc. IEEE Int. Conf. Neural Networks, Perth, Australia, pp. 1942–1948 (1995)
15. Kennedy, J., Eberhart, R.C.: A discrete binary version of the particle swarm algorithm. In: Proc. IEEE Int. Conf. on Syst., Man, and Cybern., pp. 4104–4109 (1997)
16. Zhan, Z.H., Zhang, J., Li, Y., Chung, S.H.: Adaptive particle swarm optimization. IEEE Trans. Syst., Man, and Cybern. B. 39(6), 1362–1381 (2009)
17. Zhan, Z.H., Zhang, J., Li, Y., Shi, Y.H.: Orthogonal learning particle swarm optimization. IEEE Trans. Evol. Comput. (in press)

A Comparative Study of Different Variants of Genetic Algorithms for Constrained Optimization

Saber M. Elsayed, Ruhul A. Sarker, and Daryl L. Essam

School of Engineering and Information Technology, University of New South Wales at
ADFA (UNSW@ADFA), Canberra 2600, Australia
saber.elsayed@student.adfa.edu.au,
{r.sarker,d.essam}@adfa.edu.au

Abstract. Over the last few decades, many different variants of Genetic Algorithms (GAs) have been introduced for solving Constrained Optimization Problems (COPs). However, a comparative study of their performances is rare. In this paper, our objective is to analyze different variants of GA and compare their performances by solving the 36 CEC benchmark problems by using, a new scoring scheme introduced in this paper and, a nonparametric test procedure. The insights gain in this study will help researchers and practitioners to decide which variant to use for their problems.

Keywords: Constrained Optimization, genetic algorithm, a non-parametric test.

1 Introduction

Over the last few decades, many different variants of GA have been introduced for solving COPs. These variants differ mainly in their use of crossover and mutation operators. Due to the variability of function properties in practical COPs, one such variant, even if it works well for one problem or a class of problems, does not guarantee that it will work well for another class of problems or a range of problems. As a result, finding a suitable variant is still a headache for researchers and practitioners. Keeping this in mind, in this research, we have implemented ten variants of GA using five different specialized crossover and two mutation operators. The crossovers considered in this paper are widely used in practice, such as the blend crossover (BLX-α) [7], the simulated binary crossover (SBX) [2], the simplex crossover (SPX) [13], the parent centric crossover (PCX) [3] and the triangular crossover (TC) [6]. Each of these operators has its own positives and negatives when applied to evolutionary problem solving. Interestingly, none of these crossovers is suitable for all possible scenarios of the constrained problems. The mutations considered are non-uniform and polynomial mutations.

The variants are compared based on their performances in solving a set of 36 problem instances introduced in CEC2010 (18 test problems, 2 instances each [9]). In this paper, we have introduced a new scoring scheme to compare the performance of different algorithms which is very simple to use, and easy to judge the performances of different algorithms. The scheme produces a single score for each test problem. To judge the performance over a set of problems, an overall score can be generated by

K. Deb et al. (Eds.): SEAL 2010, LNCS 6457, pp. 177–186, 2010.

simply adding the individual scores. The decision produced by the proposed scheme is consistent with a nonparametric test procedure. The analysis shows that no single operator of GA is able to reach the high quality solutions for all the test problems. In addition, no one is clearly the winner. These insights of algorithms' performances are analyzed and discussed. We believe the insights will help researchers and practitioners to decide the appropriate variant for their problems.

This paper is organized as follows. After the introduction, the search operators and the constraint handling technique used are discussed in section 2. Section 3 presents the new comparison technique. Section 4 provides the computational results and analyzes the performances. Finally the conclusions are drawn.

2   Search Operators and Constraint Handling

In this section, the search operators and constraint handling technique used in this research are briefly discussed.

As indicated earlier, we have implemented five different crossovers with two mutations. These crossover and mutation operators are briefly reviewed below.

Blend Crossover (BLX-α) has an advantage of generating diverse offspring [7], that allows GA to converge, diverge, or adapt to changing fitness landscapes without incurring extra parameters or mechanisms [8]. However, it has a disadvantage in the consideration of the epistasis problems, in which a piece of information of one variable is very dependent to another one [8]. Also, BLX-α works well for separable functions, but it does not perform well in solving optimization problems with non-separable functions [12].

Simulated Binary Crossover (SBX) is widely used in practice. The SBX operator has been found to work well in many test problems having a continuous search space when compared to other real-coded crossover implementations. The SBX operator can restrict offspring solutions to any arbitrary closeness to the parent solutions, thereby not requiring any separate mating restriction scheme for better performance. SBX is also useful in problems where the bounds of the optimum point are not known a priori and where there are multiple optima [2].

Simplex Crossover (SPX) is a multi-parent recombination operator for real-coded GAs. The SPX uses the property of a simplex in the search space. The simplex crossover works well on functions having multimodality and/or epistasis with a medium number of parents: three parents on a low dimensional function and four parents on a high dimensional function [13]. However, SPX fails on functions that consist of tightly linked sub-functions [13].

Parent Centric Crossover (PCX) allows a large probability of creating a solution near to each parent, rather than near the centroid of the parents [3]. PCX is a self adaptive type approach that has shown excellent performance in solving some test problems when implemented with a real coded GA [3]. However, GA with PCX has a difficulty in separable multimodal problems compared to other EAs such as DE [11].

Triangular Crossover (TC) is a three-parent crossover approach that concentrates on the boundary of a feasible region. TC works well where the optimal solution lies on the boundary of the feasible region of a problem, where the problem also has a single bounded feasible region in the continuous domain [6].

In Non-uniform Mutation, the step size is decreased as the generations increase, thus making a uniform search in the initial stage and very little at the later stages [10]. In contrast, in Polynomial Mutation, the probability of mutating a solution near to the parent is higher than the probability of mutating one distant from it. The shape of the probability distribution is directly controlled by an external parameter and the distribution remains unchanged throughout the entire evolution process [4]. Hence non-uniform mutation is often better at refining a solution, while polynomial mutation maintains diversity throughout a run.

In this paper, we measure the superiority of feasible points (during a tournament [5]) as follows: i) between two feasible solutions, the fittest one (according to the fitness function) is better, ii) a feasible solution is always better than an infeasible one, iii) between two infeasible solutions, the one having the smaller sum of its constraint violation is preferred. The equality constraints are transformed to inequalities of the form: $|h_j(\vec{x})| - \varepsilon \leq 0, for\ j = q + 1, ..., m$, where ε is a small number.

3   A New Comparison Technique

In this section, we propose a new technique for comparing different variants of GA. It can also be used to compare any other stochastic algorithms. To judge the quality of any variant, we assign a score of '1.0' if a variant obtains the best fitness value for a given test instance and '0.0' if a variant fails to achieve any feasible solution. If a variant achieves a feasible solution, but not the best fitness, it will receive a fractional score (between 0.0 and 1.0) as discussed below. We assume that all of the problem instances have a minimization objective function.

For a variant i and test instance j, and a total number of test problems J, we define F_{ij} as the actual fitness, and $BF_j = min_i(F_{ij})$, $WF_j = max_i(F_{ij})$ as the overall best and worst fitness value for a test instance j, respectively. The score of a variant i for instance j is then:

$$S_{ij} = \begin{cases} \left(1 - \frac{|F_{ij} - BF_j|}{a \times (|BF_j - WF_j|)}\right)^p, & \text{if } F_{ij} \text{ is feasible} \\ 0, & otherwise \end{cases} \tag{1}$$

where $a \geq 1$ and $p > 1$. A value $a > 1$ will differentiate between the worst feasible and any infeasible solution by having a small positive value for S_{ij}. A higher value of p will put a higher emphasis on good solutions. In this study, we use $a = 1.1$ and $p = 2$. In a similar way we can also calculate scores for averages. In that case, the final score for a variant i can be calculated as follows:

$$FS_{ij} = \left(\vartheta \times \sum_{j=1}^{J} S_{ij}^{best} + (1 - \vartheta \times) \sum_{j=1}^{J} S_{ij}^{average}\right) \times FR_{ij} \tag{2}$$

where, FS_{ij} is the final score of variant i for test problem j, FR_{ij} is the feasibility ratio of variant i for test problem j, S_{ij}^{best} is the score based on the best solutions, $S_{ij}^{average}$ is the score based on the average values, and ϑ is a constant $\in [0, 1]$. A higher value of ϑ (1 or close to 1), will put a higher emphasis on the best solutions, which is appropriate when we are interested in only the best fitness value, while a lower value of ϑ (0 or close to 0) will put a higher emphasis on the average solutions, which is appropriate when we are interested in a number good alternative solutions. In this study, we use $\vartheta = 0.5$ to make a balance between the best and the average results. The overall score (OS_i) for each variant i can then be calculated using the following equation.

$$OS_i = \sum_j FS_{ij} \tag{3}$$

If a variant finds the best fitness value for all test instances, then $OS_i = J$. For the worst fitness value for all test instances, $OS_i = 0$. In statistical significance testing, if a variant is significantly better in some instances and significantly worse in some other instances, it is not easy to decide the best performing variant, as the precise difference in magnitudes is not reflected in the test results. In that case, the proposed scoring scheme would provide better insights in such comparisons, as the precise difference in magnitudes is taken into account. In addition, this is not a pair-wise comparison, but rather a simultaneous comparison for any number of algorithms /samples.

4   Experimental Results and Analysis

As discussed earlier, the variants are designed using one of five crossover operators with one of two mutation operators. The crossover operators are TC, SBX, PCX, SPX and BLX-α, and the mutation operators are non-uniform (NU) and polynomial (P) mutations. The test problems details are presented in Table 1. The parameters settings are: initial population size $PS = 30$, crossover rate $(CR) = 100\%$, mutation rate $(MR) = 10\%$, tournament size $(TS) = 3$, $\alpha = 0.366$ according to [12], the index parameter $\eta = 3$, $\delta_\xi = \delta_\eta = 0.01$ according to [3]. As suggested for the number of parents in SPX, three parents are used on a low 10D function and four parents on a high 20D function [13], we have used 5 parents for 30D. For the non-uniform mutation, b = 5, and for the polynomial external parameter $\eta_m = 10$, finally ε = 0.0001. The detailed results showing best fitness (b), mean (M), standard deviation (Sd) and the average feasibility ratio (Avg. Fr) are presented in Table 2 for 10D, and in Table 3 for 30D. All results are out of 25 independent runs. Due to space and formatting limitation, we only provide the results of the best variants. However the analysis is performed using all ten variants.

Firstly, based on the feasibility ratio for 10D and 30D, we found that TC-NU is in the 1st position followed by, BLX-α-NU, PCX-NU, SBX-NU, SPX-NU, PCX-P, SBX-P, TC-P, BLX-α-P and SPX-P, with total averages of FR 84%, 83.5%,83.5%, 83%, 78.5%, 78%, 75%, 72%, 69% and 61.5%, respectively. So, it is clear that the non-uniform mutation is superior to polynomial in regards of the feasibility ratio. The number of best solutions found by each variant, TC-NU, TC-P, SBX-NU, SBX-P, SPX-NU, SPX-P, PCX-NU, PCX-P, BLX-α-NU, and BLX-α-P are 3, 0, 4, 2, 3, 0, 3, 4, 3 and 0, for 10D, respectively. Note that all variants with the non-uniform mutation were able to reach the same solution for C16, while TC-NU, SBX-NU, SPX-NU, PCX-NU, PCX-P and BLX-α-NU are able to solve C03. Further only SBX-NU, SPX-NU, PCX-NU, and BLX- α-NU are able to solve C04, while all variants except TC-NU cannot solve C11. For 30D, the numbers of best solutions found by each variant (in the same order) are: 3, 2, 2, 2, 1, 0, 1, 0, 4 and 0. No variant can solve C03, C04 and C11, and the variants with only the non-uniform mutation are able to solve C12.

Table 1. Properties of the CEC2010 test problems. D is the number of decision variables, $p = |F|/|S|$ is the estimated ratio between the feasible region and the search space, I is the number of inequality constraints, E is the number of equality constraints

Prob	Search Range	Objective Type	Number of constraints		Feasibility Region p	
			E	I	10D	30D
C01	$[0,10]^D$	Non Separable	0	2 Non Separable	0.997689	1.000000
C02	$[-5.12,5.12]^D$	Separable	1 Separable	2 Non Separable	0.000000	0.000000
C03	$[-1000,1000]^D$	Non Separable	1 Separable	0	0.000000	0.000000
C04	$[-50,50]^D$	Separable	2 Non Separable, 2 Separable	0	0.000000	0.000000
C05	$[-600,600]^D$	Separable	2 Separable	0	0.000000	0.000000
C06	$[-600,600]^D$	Separable	2 Rotated	0	0.000000	0.000000
C07	$[-140,140]^D$	Non Separable	0	1 Separable	0.505123	0.503725
C08	$[-140,140]^D$	Non Separable	0	1 Rotated	0.379512	0.375278
C09	$[-500,500]^D$	Non Separable	1 Separable	0	0.000000	0.000000
C10	$[-500,500]^D$	Non Separable	1 Rotated	0	0.000000	0.000000
C11	$[-100,100]^D$	Rotated	1 Non Separable	0	0.000000	0.000000
C12	$[-1000,1000]^D$	Separable	1 Non Separable	1 Separable	0.000000	0.000000
C13	$[-500,500]^D$	Separable	0	2 Separable, 1 Non Separable	0.000000	0.000000
C14	$[-1000,1000]^D$	Non Separable	0	3 Separable	0.003112	0.006123
C15	$[-1000,1000]^D$	Non Separable	0	3 Rotated	0.003210	0.006023
C16	$[-10,10]^D$	Non Separable	2 Separable	1 Separable, 1 Non Separable	0.000000	0.000000
C17	$[-10,10]^D$	Non Separable	1 Separable	2 Non Separable	0.000000	0.000000
C18	$[-50,50]^D$	Non Separable	1 Separable	1 Separable	0.000010	0.000000

According to the problem properties, it can be seen, that TC with the non-uniform mutation is the only variant that is able to reach a feasible solution for 10D rotated objective functions (e.g., C11). In contrast, PCX with the polynomial mutation is pre-ferred for either separable or non-separable objective function, with only rotated equality constraints, (e.g. C06 and C10). Also, PCX with the polynomial mutation is the best for the separable objective function, with separable equality constraints and non-separable inequality constraints, (e.g. C02). For the non-separable objective functions with only inequality constraints (rotated or separable), it is preferable to use the non-uniform mutation with BLX-α, or TC (e.g. C07, C08, C14 and C15). For the non-separable objective function with both equality constraints (separable) and in-equality constraints (only non-separable or only separable), SBX and PCX with the polynomial mutation reached more robust solutions (e.g. C17 and C18). In test prob-lems where the objective function is non-separable/or separable, and does have a mix of non-separable and separable inequality constraints, the non-uniform mutation is preferred with SBX (e.g. C13). Finally, for the separable objective function with only equality constraints (separable), it is preferable to use the non-uniform mutation with SBX or PCX (e.g. C05). The non-uniform mutation is preferred in the test problems where the objective function is non-separable with only inequality constraints (non-separable) (e.g. C01). Also, the same mutation performs well for test problems with a separable objective function with equality constraints (only non-separable) and in-equality constraints (only separable), (e.g. C03 and C09).To compare the variants, we have performed a non-parametric test (Wilcoxon Signed Rank Test) [1] that allows us

Table 2. Function values out of different GA variants, for 10D test problems

Pr.		TC-NU	SBX-NU	SBX-P	SPX-NU	PCX-NU	PCX-P	BLX-α-NU
1	b	-0.74726	-0.74728	-0.74641	-0.74731	-0.74729	-0.74687	-0.74729
	M	-0.72839	-0.72992	-0.69813	-0.68265	-0.71796	-0.70001	-0.71900
	Sd	0.02172	0.02331	0.04945	0.06857	0.02762	0.05356	0.04048
2	b	-2.20403	-2.20140	-2.25830	-1.60231	-2.24070	-2.27097	-2.18671
	M	-0.07300	-1.18436	-1.69085	1.53784	-0.90938	-1.73062	0.75009
	Sd	1.85484	1.31417	0.72668	1.49232	1.40541	0.65883	1.94768
3	b	1.1E+10	10028669	-	3.44E+09	40707467	3.516912	3.99E+08
	M	2.8E+13	7.54E+13	-	1.01E+14	6.49E+13	1.39E+14	5.13E+13
	Sd	3.7E+13	2.22E+14	-	2.14E+14	1.01E+14	2.83E+14	1.03E+14
4	b	-	0.004377	-	0.00045	0.014563	-	14.02194
	M	-	-	-	4.387388	8.102383	-	15.10635
	Sd	-	-	-	7.87455	11.4379	-	1.53359
5	b	-231.962	-473.5781	-237.2348	-464.6771	-400.9999	-427.6626	-265.2265
	M	-106.338	-204.7016	-105.0468	-50.0089	-244.1168	-160.4987	-150.5770
	Sd	77.2131	93.7422	105.7225	241.3993	59.3223	130.6904	100.1925
6	b	-575.789	-575.7960	-576.2588	-576.1843	-576.7053	-572.6023	-574.6536
	M	-83.7290	-160.4517	-345.7701	-68.3604	-144.7258	-87.8436	-189.3297
	Sd	321.1026	483.5311	334.8700	379.0355	392.5743	328.1714	380.5476
7	b	3.35E-21	1.22E-20	6.70E-11	9.32E-17	1.38E-21	2.97E-12	4.64E-22
	M	5.40E-19	1.70E-18	5.72E-09	1.90E-15	3.02E-20	1.88E-11	1.47E-20
	Sd	1.10E-18	2.34E-18	1.92E-08	2.38E-15	2.56E-20	2.01E-11	2.44E-20
8	b	1.20E-20	5.87E-20	3.52E-10	1.76E-16	4.95E-21	3.25E-11	1.93E-22
	M	2.17E-19	1.31E-18	4.62E-08	0.5147464	6.29E-20	1.80E-08	1.10E-20
	Sd	2.33E-19	1.93E-18	5.72E-08	0.6465979	6.96E-20	5.34E-08	1.76E-20
9	b	207.7978	375.54613	24.32602211	4.30E+08	218.49958	0.0920886	374.7197
	M	2.6E+07	4.58E+08	3.74E+09	4.88E+12	1.04E+06	5.47E+09	1.89E+09
	Sd	1.1E+08	2.24E+09	17348769574	4.77E+12	2209044.7	2.71E+10	7.33E+09
1 0	b	0.248143	60.033514	6.61064209	1.34E+10	69.104363	7.73E-06	75.291298
	M	9.6E+06	5.93E+09	2.50E+09	5.02E+12	2.97E+08	3.31E+08	5.49E+09
	Sd	3.0E+07	2.44E+10	1.05E+10	4.24E+12	1.21E+09	1.27E+09	1.16E+10
1 1	b	-0.00106	-	-	-	-	-	-
	M	-	-	-	-	-	-	-
	Sd	-	-	-	-	-	-	-
1 2	b	-115.445	-374.4726	-304.7074	-161.8349	-304.6433	-260.0960	-303.0400
	M	44.3670	20.1254	8.7836	25.7515	5.2216	-27.8116	-26.1526
	Sd	126.6006	384.7512	241.1720	135.0566	197.5852	71.4188	112.6932
1 3	b	-67.3679	-68.2885	-68.1437	-68.0511	-67.3388	-68.2287	-67.8926
	M	-61.2049	-63.2528	-64.9840	-61.7082	-63.3479	-65.4860	-63.7720
	Sd	3.4104	1.9835	2.8576	3.5873	1.9832	1.9890	2.2564
1 4	b	0.1758	0.0910	0.2467	0.0882	0.0818	0.0000	0.2258
	M	4783.407	7931.4724	2335.7943	4197.8557	5130.0824	8377.1144	10656.307
	Sd	18106.75	16877.661	11231.1540	18074.121	18711.9980	17430.0030	23138.143
1 5	b	4.3E+07	3.42E+11	4.0787E+10	5.84E+10	1.09E+11	5.33E+08	9E+08
	M	2.4E+12	7.057E+13	4.0229E+13	1.39E+13	3.2E+13	4.56E+12	2.03E+13
	Sd	5.3E+12	6.744E+13	5.2273E+13	1.65E+13	4.14E+13	6.48E+12	5.72E+13
1 6	b	0	0	4.78173E-13	1.11E-16	0	1.33E-14	0
	M	0.153187	0.0119858	0.1556867	0.633979	0.19949	0.245698	0.280637
	Sd	0.221928	0.0293591	0.15681833	0.317308	0.255503	0.248255	0.488704
1 7	b	0.111628	0.0035024	0.000678985	15.44855	0.001073	0.000924	0.106927
	M	6.632889	6.594218	3.80935626	198.1898	9.641727	3.198288	19.50653
	Sd	7.659975	8.162406	5.49315642	181.9092	11.98974	4.413743	32.84651
1 8	b	0.353893	2.488E-10	0.082019072	1.97547	0.007973	0.003178	0.075067
	M	1107.45	762.80204	104.199786	6695.769	834.5436	26.33942	1004.014
	Sd	1545.649	1441.3715	178.745095	4695.514	1723.946	51.25978	1384.054
Fr		89%	88%	81%	83%	88%	86%	89%

Table 3. Function values out of different GA variants, for 30D test problems

Pr.		TC-NU	SBX-NU	SBX-P	SPX-NU	PCX-NU	PCX-P	BLX-α-NU
1	b	-0.81714	-0.821671	-0.815216	-0.815679	-0.817997	-0.817050	-0.817855
	M	-0.79892	-0.781180	-0.794330	-0.764440	-0.799509	-0.792670	-0.783985
	Sd	0.014196	0.031671	0.014201	0.030688	0.012201	0.019100	0.029011
2	b	-1.25658	-1.126926	-1.652844	3.390203	-1.789475	-1.422927	-0.640858
	M	-0.13683	1.289422	-0.525870	4.545690	0.996184	-0.296748	1.283116
	Sd	0.84786	1.361109	0.640276	0.636066	1.383242	0.921409	0.915661
3	b	-	-	-	-	-	-	-
	M	-	-	-	-	-	-	-
	Sd	-	-	-	-	-	-	-
4	b	-	-	-	-	-	-	-
	M	-	-	-	-	-	-	-
	Sd	-	-	-	-	-	-	-
5	b	-457.322	-439.8795	-429.2540	215.0637	-431.9744	-378.6791	-435.1072
	M	-145.951	-274.4893	-162.1598	481.4305	-335.2496	-33.7101	-111.7567
	Sd	102.1364	173.4741	214.5627	97.3591	83.9306	386.8334	181.6593
6	b	-252.319	-517.8180	-436.9482	45.3549	-519.5224	-310.4290	-521.3494
	M	-147.094	-352.6657	-271.6617	339.0001	-123.7365	-214.1711	-409.0158
	Sd	57.6696	228.0035	104.7799	194.7921	231.2404	97.0451	175.9162
7	b	7.12E-19	4.43E-17	0.111546449	5.9393812	1.96E-16	0.063957987	4.70E-20
	M	0.197979	3.08E-16	1.034270613	9.5030099	5.78E-16	1.972052263	6.12E-19
	Sd	0.494948	1.60E-16	0.847198591	2.4415226	2.97E-16	0.98146394	3.65E-19
8	b	4.50E-19	6.38E-17	0.429089011	4.9494844	1.30E-16	3.072100581	1.63E-19
	M	1.629232	0.435555	3.794085142	11.129745	0.5939709	5.021856481	0.5939381
	Sd	1.643779	0.6440672	1.473428046	2.9225308	0.857331	1.392382668	1.0303173
9	b	1.8E+09	2.29E+12	7.86E+10	2.52E+13	3.97E+12	2.19E+11	4.43E+12
	M	1.4E+11	1.94E+13	2.03E+12	3.33E+13	1.84E+13	3.36E+12	1.41E+13
	Sd	1.6E+11	1.17E+13	1.86E+12	8.78E+12	1.05E+13	2.85E+12	5.73E+12
10	b	4.3E+11	2.19E+12	1.36E+11	9.26E+12	7.41E+12	4.31E+11	5.86E+10
	M	2.9E+12	1.74E+13	3.35E+12	2.87E+13	2.00E+13	3.56E+12	1.05E+13
	Sd	2.3E+12	1.09E+13	3.67E+12	1.53E+13	8.17E+12	2.02E+12	5.43E+12
11	b	-	-	-	-	-	-	-
	M	-	-	-	-	-	-	-
	Sd	-	-	-	-	-	-	-
12	b	-0.13045	-0.190433	-	-0.192083	-0.117789	-	-0.086306
	M	0.003453	3.155255	-	-0.054965	0.998620	-	8.751375
	Sd	0.148807	4.731518	-	0.130925	1.993884	-	12.498368
13	b	-60.5334	-65.79880	-62.000733	-63.25430	-64.448800	-64.213130	-64.56370
	M	-55.6431	-61.68457	-59.837132	-58.94818	-61.069500	-60.510728	-59.50287
	Sd	2.509570	2.120910	1.756991	3.094427	1.768428	2.112393	1.987751
14	b	8.47152	4.81551	161.87403	22.70745	10.92657	85.12049	0.12090
	M	2.7E+03	7.08E+03	4.09E+05	1.53E+05	6.56E+03	4.55E+03	4.14E+03
	Sd	6.6E+03	1.30E+04	1.83E+06	7.44E+05	1.83E+04	9.17E+03	1.24E+04
15	b	52.11062	6017.2691	3.43E+10	7.26E+12	225712214	6.88E+10	0.7314271
	M	9.8E+03	3.13E+12	6.76E+11	2.84E+13	3.42E+12	2.01E+12	2.61E+12
	Sd	3.8E+04	6.67E+12	6.99E+11	2.51E+13	6.67E+12	2.12E+12	5.23E+12
16	b	0.749937	0.290163	0.000841	1.017013	0.258111	0.001214	0.824704
	M	1.017990	0.958698	0.852487	1.080749	1.004627	0.777738	1.027524
	Sd	0.103935	0.188695	0.336429	0.044407	0.238504	0.397705	0.054663
17	b	0.749937	60.024590	3.598506	182.72766	49.991087	3.623220	90.666040
	M	124.8363	258.79140	93.818115	1016.4893	265.721340	86.013982	420.60479
	Sd	45.59627	155.21920	61.064014	503.07045	214.048620	72.386892	180.46487
18	b	6.2E+03	2.63E-01	5.57E+01	2.54E+03	2.44E+02	9.42E+01	1.63E+03
	M	1.1E+04	3.14E+03	1.72E+03	1.57E+04	4.70E+03	1.55E+03	5.70E+03
	Sd	2.9E+03	3.99E+03	1.94E+03	5.57E+03	7.16E+03	1.28E+03	2.06E+03
Fr		79%	78%	69%	74%	79%	70%	78%

Table 4. The Wilcoxon non-parametric test for ten variants of GA for both 10D and 30D, based on the best fitness values obtained

b \ a	10D 1	2	3	4	5	6	7	8	9	10	30D 1	2	3	4	5	6	7	8	9	10
1		≈	≈	≈	≈	+	≈	≈	≈	≈		≈	≈	≈	+	+	≈	≈	≈	≈
2			≈	≈	≈	+	≈	−	≈	≈			≈	≈	+	+	≈	≈	≈	+
3				≈	≈	+	≈	≈	≈	≈				≈	+	+	≈	≈	≈	≈
4					≈	+	≈	≈	≈	≈					+	+	≈	≈	≈	≈
5						≈	≈	≈	≈	≈						+	−	−	−	−
6							−	−	−	−							−	−	−	−
7								≈	+	≈								≈	≈	≈
8									≈	≈									≈	≈
9										≈										≈

Table 5. The Wilcoxon non-parametric test for ten variants of GA for both 10D and 30D, based on the average fitness values obtained

b \ a	10D 1	2	3	4	5	6	7	8	9	10	30D 1	2	3	4	5	6	7	8	9	10
1		≈	≈	≈	+	+	≈	≈	≈	≈		≈	≈	≈	+	+	≈	≈	≈	≈
2			≈	≈	≈	+	≈	≈	≈	≈			≈	≈	+	+	≈	≈	≈	≈
3				≈	+	+	≈	≈	≈	≈				≈	+	+	≈	≈	≈	≈
4					+	+	≈	≈	≈	≈					+	+	≈	≈	≈	≈
5						≈	≈	−	≈	−						+	−	−	−	−
6							−	−	≈	−							−	−	−	−
7								≈	≈	≈								≈	≈	≈
8									≈	≈									≈	≈
9										−										≈

to judge the difference between paired scores when it cannot make the assumption required by the paired-samples t test, such as that the populations should be normally distributed. The results based on the best and average fitness values are presented in Table 4 and 5, respectively. As a null hypothesis, it is assumed that there is no significant difference between the best/or mean values of two samples, whereas the alternative hypothesis is that there is a significant difference of the two samples at the 5% significance level. Based on the test results /rankings, we assign one of three signs (+, −, and ≈) for the comparison of any two algorithms (shown in the last column), where the " + " sign means the first algorithm (a) is significantly better than the second (b), " − " sign means that the first algorithm significantly worse, and " ≈ " sign means that there is no significant difference between the two algorithms. From the results it can be seen that there is no significant difference between most of the algorithms, with the exception of SPX with both non-uniform and polynomial in 30D, but SPX with non-uniform is a little bit competitive to the other variants in the 10D test problems. Note that the numbers in the second row of Tables 4 and 5 represent the algorithms TC-NU, TC-P, SBX- NU, SBX-P, SPX-NU, SPX-P, PCX- NU, PCX-P, BLX-α-NU and BLX-α-P, respectively.

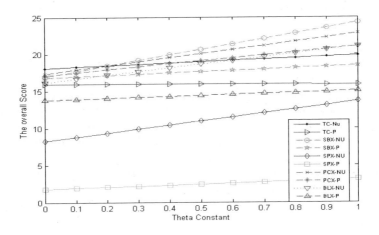

Fig. 1. The effect of changing ϑ on the overall scoring over 36 test problems

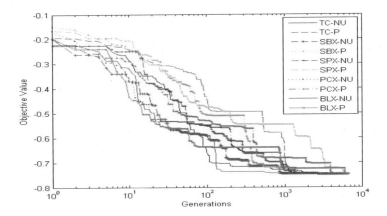

Fig. 2. The convergence pattern for 10 variants of GA, for one problem "C01". Note that, the *x*-axis is in the log scale.

Based on our proposed scoring scheme, when considered all 36 test problems, the overall scores for the variants SBX-NU, PCX-NU, PCX-P, TC-NU, BLX-α-NU, SBX-P, TC-P, BLX-α-P, SPX-NU and SPX-P are 20.62, 20.09, 19.12, 18.98, 18.66, 17.65, 15.2, 14.45, 10.98, and 2.48, respectively. Note that, the maximum possible score is 36 when a variant dominates all other variants. That means SBX-NU is the best performing variant for the 36 test problems considered in this paper. However it is not easy to make a similar decision with the Wilcoxon Signed Rank Test results.

To see the effect of the constant ϑ on the overall scorings, we have plotted ϑ versus the overall scores for all the variants in Fig. 1, which shows that TC-NU is the best when ϑ is less than 0.2, and SBX-NU is the best when ϑ is above 0.2. However, PCX-NU is always in the second place irrespective of the value of ϑ.

Finally, a sample convergence pattern for problem "C01" for all variants is shown in Fig. 2.

5 Conclusions

In this paper, ten different variants of GA were implemented and extensively tested on 36 well-known benchmark problems. The results obtained showed that there is no single variant that is able to reach the best results for all test problems. However, the non-uniform mutation is suitable in most of the test problems in both 10D and 30D, but there are also some test problems in which the polynomial mutation works better. The analysis and insights provided in this paper will help researchers and practitioners to decide the variant suitable for their problems.

References

1. Corder, G.W., Foreman, D.I.: Nonparametric Statistics for Non-Statisticians: A Step-by-Step Approach. John Wiley, Hoboken (2009)
2. Deb, K., Agrawal, R.B.: Simulated Binary Crossover for Continuous Search Space. Complex Syst. 9, 115–148 (1995)
3. Deb, K., Anand, A., Joshi, D.: A computationally efficient evolutionary algorithm for real-parameter evolution. IEEE Trans. Evol. Comput. 10(4), 371–395 (2002)
4. Deb, K., Pratap, A., Agarwal, S., Meyarivan, T.: A Fast and Elitist Multiobjective Genetic Algorithm: NSGA-II. IEEE Trans. Evol. Comput. 6(2), 182–197 (2002)
5. Deb, K.: An Efficient Constraint Handling Method for Genetic Algorithms. Computer Methods in Applied Mechanics and Engineering 186, 311–338 (2000)
6. Elfeky, E.Z., Sarker, R., Essam, D.: Analyzing the simple ranking and selection process for constrained evolutionary optimization. Journal of Computer Science and Technology 23(1), 19–34 (2008)
7. Eshelman, L.J., Schaffer, J.D.: Real-Coded Genetic Algorithms and Interval-Schemata. Foundations of Genetic Algorithms 2, 187–202 (1993)
8. Herrera, F., Lozano, M., Molina, D.: Continuous Scatter Search: An Analysis of the Integration of Some Combination Methods and Improvement Strategies. European Journal of Operational Research 169, 450–476 (2006)
9. Mallipeddi, R., Suganthan, P.N.: Problem definitions and evaluation criteria for the CEC 2010 competition and special session on single objective constrained real-parameter optimization. Tech. Rep., Nangyang Technological University, Singapore (2010)
10. Michalewicz, Z.: Genetic Algorithms + Data Structures = Evolution Programs. Springer, New York (1992)
11. Rönkkönen, J.: Multimodal Global Optimization with Differential Evolution-Based Methods. Thesis for the degree of Doctor of Science, Lappeenranta University of Technology, Lappeenranta, Finland (2009) ISBN 978-952-214-851-3
12. Takahashi, M., Kita, M.: A Crossover Operator Using Independent Component Analysis for Real-Coded Genetic Algorithms. In: IEEE Congress on Evolutionary Computation, pp. 643–649 (2002)
13. Tsutsui, S., Yamamura, M., Higuchi, T.: Multi-parent Recombination with Simplex Crossover in Real Coded Genetic Algorithms. In: Genetic Evolutionary Computation Conf. (GECCO 1999), pp. 657–664 (1999)

Evolutionary FCMAC-BYY Applied to Stream Data Analysis

D. Shi[1], M. Loomes[1], and M.N. Nguyen[2]

[1] School of Engineering and Information Sciences, Middlesex University,
London NW4 4BT, UK
[2] School of Electrical and Electronics Engineering, Nanyang Technological University,
Singapore 639798

Abstract. A data stream is an ordered sequence of instances that can be read only once or a small number of times using limited computing and storage capabilities. Stream data analysis is a critical issue in many application areas such as network fraud detection, stock market prediction, and web searches. In this research, our previously proposed FCMAC-BYY, that uses Bayesian Ying-Yang (BYY) learning in the fuzzy cerebellar model articulation controller (FCMAC), will be advanced by evolutionary computation and dynamic rule construction. The developed FCMAC-EBYY has been applied to a real-time stream data analysis problem of traffic flow prediction. The experimental results illustrate that FCMAC-EBYY is indeed capable of producing better performance than other representative neuro-fuzzy systems.

1 Introduction

The sequence of data in a time dependent manner is called a datastream. Stream data flow in and out of a computer system continuously and with varying update rates. They are temporally ordered, fast changing, massive, and potentially infinite. Applications involving stream data include telecommunications, financial markets, and satellite data processing [1]. For algorithms designed to analyze data stream, the ability to process the data in a single pass, or a small number of passes, while using little memory, is crucial [2].

Inspired by the Taoist Yin-Yang philosophy, Xu [3] proposed the concept of Bayesian Ying-Yang (BYY) learning, here the word "Ying", substituted for "Yin" in order to create symmetry with the word "Yang". The BYY learning algorithm is well-known as a harmony learning theory with a new learning mechanism that implements model selection implemented either *automatically* during parameter learning or *subsequently* after parameter learning via a new class of model selection criteria obtained from this mechanism. The implementation of the BYY learning model is done in two phases, parameter learning and cluster number selection, in order to determine all unknown parameters and to select the optimal solution for the input data, respectively.

The Cerebellar Model Articulation Controller (CMAC) is a type of associative memory neural network based on a model of the mammalian cerebellum [4]. In our

K. Deb et al. (Eds.): SEAL 2010, LNCS 6457, pp. 187–194, 2010.

previous work, the BYY was applied to the fuzzification layer of the FCMAC structure, thereafter referred to as FCMAC-BYY [5-8]. To help comparison with conventional clustering algorithms that are "one-way" clustering, BYY harmonizes the training input and the solution/clusters by considering not only forward mapping from the input data into the clusters, but also the backward path from the obtained clusters to the input data. With the introduction of the BYY learning algorithm, FCMAC-BYY has a higher generalization ability because the fuzzy rule sets are systematically optimized by BYY; it also reduces the memory requirement of the network by a significant degree compared with the original CMAC; finally, it provides intuitive fuzzy logic reasoning and has clear semantic meanings. However, the FCMAC-BYY suffers from two problems: First, when the BYY-based fuzzification is separated from the weight training of the neural network, local optima may occur. Second, the weights are trained by using gradient-based methods, which may also lead to local optima.

To address the above-mentioned problems, evolutionary computation (EC) is introduced in this research to utilize global search for both the optimal fuzzy clusters and the optimal weights of FCMAC-BYY. Traditional evolutionary approaches are limited to small populations of short binary string length, whereas neural network training involves a large search space due to the complex connections and real values. One of the solutions to applying EC to neural network training is coevolution, in which a complex solution is decomposed into pieces to be optimized in different populations/species and then reassembled [9]. Typically, the coevolutionary algorithms can be classified into two types, namely, competitive coevolution [10] and cooperative coevolution [11].

In a competitive coevolutionary algorithm the fitness of an individual is based on direct competition with individuals of other species, which, in turn, evolve separately in their own populations. Increased fitness of one of the species implies a diminution in the fitness of the other species. Such an evolutionary pressure tends to produce new strategies in the populations involved so as to maintain their chances of survival. Consequently, the capability of reaching global optima can be improved [10].

García-Pedrajas, Hervás-Martínez and Ortiz-Boyer presented a cooperative coevolutive approach for designing neural network ensembles [11]. For each network, different objectives are defined, considering not only its performance in the given problem, but also its cooperation with the rest of the networks. In addition, a population of ensembles is evolved, improving the combination of networks and obtaining subsets of networks to form ensembles that perform better than the combination of all the evolved networks. The performance has been thoroughly tested over a set of ten real-world problems with different features. However, the problem of how to define the objectives, especially the diversity objectives, still remains.

This research aims to incorporate coevolutionary computation into the FCMAC-BYY to search for the optimal fuzzy sets as well as the connection weights. The paper is organized as follows. Section 2 describes the work mechanism of coevolutionary learning incorporated with FCMAC-BYY. In Section 3, dynamic rule construction suitable for stream data analysis is introduced. Experimental results are presented and compared in Section 4, followed by our conclusion in Section 5.

2   FCMAC –EBYY Model

As depicted in Figure 1, the FCMAC-EBYY network has a hierarchical structure of 5 layers, the input layer, fuzzification layer, association layer, post association layer and output layer.

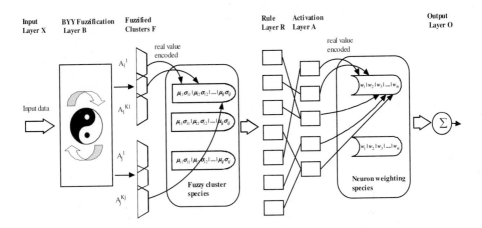

Fig. 1. The block diagram of FCMAC-EBYY

2.1   Fuzzification with BYY Learning

The input to FCMAC-BYY is a non-fuzzy data vector corresponding to a measure of the input parameter represented in the respective dimension. The fuzzification layer maps the input patterns into the fuzzy sets in the association layer through BYY learning. Thereafter, the association layer associates the fuzzy rules with the memory cell and tries to imitate a human cerebellum. The logical *AND* operation is carried out in this layer to ensure that a cell is activated only when all the inputs associated with it are fired. The association layer is then mapped to the post association layer where the logical *OR* operation will fire those cells whose connected inputs are activated. For the output layer, the defuzzification center of area (COA) method is used to compute the output of the structure.

In neural network models, input data x and solution y can be treated as random processes. The joint probability distribution $p(x,y)$ can be calculated by either of these two formulae:

$$p(x,y) = p(y|x)p(x) \tag{1}$$

$$p(x,y) = p(x|y)p(y) \tag{2}$$

However, the result of Equation (1) is not equal to that of Equation (2) unless y is the optimal solution and Ying and Yang achieve harmony.

Ying-Yang fuzzification involves two phases, namely, parameter learning and cluster selection. Parameter learning does the task of determining all the unknown parameters for a specific number of clusters in each dimension. In the cluster selection phase, the optimal cluster number is selected according to a cost function calculation. In this research, we focus on the parameter learning phase, in which the optimization of the clusters is converted to minimization of the difference between Ying and Yang by using the Kullback-Leibler (KL) divergence. Minimization of the KL divergence will produce the optimal parameter at each number of clusters [5].

2.2   FCMAC-BYY with Coevolutionary Learning

The original FCMAC-BYY suffers from two problems: First, as the BYY-based fuzzification is separated from the weight training of the neural network, local optima may occur. Second, the weights are trained using gradient-based methods, which may also lead to local optima. These two problems motivate us to introduce EC so that we can utilize global search for both the optimal fuzzy clusters and the optimal weights of FCMAC-BYY.

In traditional evolutionary computation the expected solution is sought among a population of individuals via genetic operations such as crossover and mutation. However, single-population evolutionary algorithms often perform poorly-manifesting stagnation and convergence to local optima when confronted with complex problems, or strong interdependencies among the components of the solution. Co-evolutionary learning allows simultaneous evolution of two or more species with coupled fitness. Such coupled evolution favors the discovery of complex solutions. On the other hand, the traditional binary GA has some drawbacks when applied to multidimensional and high-precision numerical problems. The situation can be improved if GA in real numbers is used. Each chromosome is coded as a vector of floating point numbers that has the same length as the solution vector. A large domain can thus be handled. Much research effort has been spent to improve the performance of real-coded GA (RCGA) [12].

In this research, coevolutionary computation and real-coded GA are applied to FCMAC-BYY, referred to as FCMAC-EBYY, in which the fuzzy sets and the weights are evolved independently and then cooperated to form the final solution. One can see from Figure 1 that co-evolutionary learning is integrated into the optimization of fuzzy clusters and weights of the FCMAC-EBYY network. There are two kinds of species in the FCMC-EBYY, namely, fuzzy cluster species and weight species. In the species of fuzzy clusters, all the chromosomes are initialized with the fuzzification results obtained by the BYY learning. In the species of connection weights, all the chromosomes are initialized with randomly generated real values. Such a treatment keeps a good mixture of the optimized outputs from BYY fuzzification and random processes, so as to speed up the searching process.

The chromosome representation has been also shown in Figure 1, where μ_{ij} and σ_{ij} represent the mean values and widths of the ith dimension in the jth cluster of fuzzy cluster species respectively, w_n represents the weights of the corresponding rules in the neuron weighting species. The following are the details of the five processes in coevolutionary learning: evaluation, selection, reproduction (crossover and mutation) and re-insertion.

Evaluation. In the fuzzy cluster species, the membership value of training data for each cluster is used as the fitness function of each chromosome, whereas the output error is used as the fitness function in the connection weights species to train the weights of the neurons in the rule layer.   An individual undergoing fitness evaluation cooperates with one or more representatives of the other species to construct a candidate FCMAC.   A selected individual from the fuzzy cluster species will cooperate with the selected individuals from the connection weights species to form a full representation of the structure.

Selection. Stochastic Universal Sampling (SUS) is employed in the FCMAC-EBYY, in which the chromosomes are mapped one-to-one into contiguous segments of a line, where the size of the segment of each chromosome relates to its fitness value. Then, equally-spaced pointers are placed along the line. The number of pointers on the line corresponds to the number of individuals to be selected.

Crossover. Intermediate recombination is suitable to fulfil crossover between two real-valued parent chromosomes [13]. Each variable in the offspring is the result of combining the variables in the parents according to the above expression with a new α chosen for each pair of parent genes. Intermediate crossover is capable of producing new variables within a slightly larger hypercube than that defined by the parents but is constrained by the range of α.

Mutation. Mutation can be accomplished by replacing the parameter values with random selection of new values. These new values, which are selected from the allowable range for the respective dimension, ensure that the mutated values are within reasonable regions. Similar to the crossover operator, the probabilistic rate of the mutation determines the frequency of mutations. In all, mutation allows for an increase in the level of possible exploration of the search space without adversely affecting the convergence characteristics.

Reinsertion. The elitist strategy is employed to select a number of old chromosomes to be reinserted into the resultant population. This strategy chooses a prespecified number of the fittest chromosomes in the old population to be inserted into the resultant population while the rest of the chromosomes are discarded. The deleted cluster parameters are replaced by the chromosomes created from the reproduction process. This is done to retain the best quality chromosomes in the population. The production of the resultant population marks the completion of one generation.

3   Dynamic Rule Construction Based on Credit Assignment

In steam data analysis, a learning system must have the ability to update its structure to fit to the new data. In this research, two parameters, *fired_frequency* and *time_fired*, are used as predefined quality creation to evaluate the current neurons in the rule layer. The *fired_frequency* evaluates the efficiency of a rule, as it indicates the historical contribution of individual neurons to the past patterns, whereas the *time_fired* records the nearest time a neuron is fired. By combination of these two parameters, the best-performing neurons will be retained while the others are deleted.

Figure 2 depicts how to determine all neighborhoods of a fired cell in two-dimensional input space. The green clusters are those activated in each input dimension, for example, the neighborhoods of rule 11 are: 01, 21, 10, and 12. In the FCMAC-EBYY, each rule is identified by a unique index that represents the connection between clusters in layer 2 according to that rule. Two rules are called neighbors

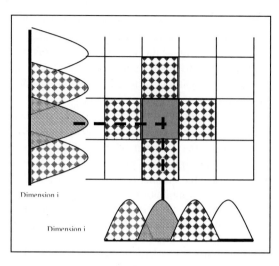

Fig. 2. Cell neighbors in two-dimensional input space

if they connect to two neighboring clusters in the same dimension while connecting to the same clusters in the other dimensions. The details can be seen in [6].

4   Experimental Results

A data stream is an ordered sequence of instances that can be read only once or a small number of times using limited computing and storage capabilities. Examples of data streams include computer network traffic, phone conversations, ATM transactions, web searches, and sensor data. In this section, the FCMAC-OBYY is applied to a real-time traffic prediction application. The data were collected from a site (Site 29) located at exit 15 along the east-bound Pan Island Expressway (PIE) in Singapore using loop detectors embedded beneath the road surface. There are a total of five lanes at the site, two exit lanes and three straight lanes for the main traffic. For this experiment, only the traffic flow data for the three straight lanes were considered. The traffic dataset has four input attributes, including the time and the traffic density of the three lanes. It was used to make predictions for traffic density of each lane at time $t + \tau$, where $\tau = 5,15,30,45,60$ minutes.

For the simulation, three cross-validation groups, CV1, CV2 and CV3, of training and test sets were used. The square of the Pearson product-moment correlation value (denoted as R^2) was used to compute the accuracy of the predicted traffic trends obtained using the FCMAC-EBYY network. The FCMAC-EBYY network is compared with the Falcon-class networks, and the GenSoFNN-CRI(S) network [14] in Table 1. The "Var" indicator (the change in Avg R^2 value from min to min expected as a percentage of the former) and the "Avg Var", the mean "Var" values across all 3 lanes, were used for the benchmarking of the various systems. These 2 indicators reflect the consistency of the predictions made by the benchmarked systems over

Table 1. Simulation results of traffic prediction

Network	Lane1 Var(%)	Lane2 Var(%)	Lane3 Var(%)	Avg Var(%)
Falcon-FCM(CL)	24.17	9.32	30.47	21.32
Falcon-MLVQ(CL)	36.41	25.94	30.21	30.85
Falcon-FKP(CL)	23.87	22.09	35.19	27.05
Falcon-PFKP(CL)	20.78	21.05	28.25	25.70
Falcon-MART	20.78	15.47	20.58	18.94
GenSoFNN-VRI(S)	19.64	19.58	21.09	20.10
FCMAC-BYY	10.75	9.95	15.36	11.68
FCMAC-EBYY	**10.22**	**8.95**	**13.34**	**10.60**

different time intervals across the 3 lanes. From Table 1, one can see that the FCMAC-EBYY network exhibited superior performance over all the other networks.

5   Conclusion

In this paper, a fuzzy cerebellar model articulation controller using evolutionary Bayesian Ying-Yang learning is proposed and applied to stream data analysis. After the initial cluster parameters were obtained through BYY learning, evolutionary computation is utilized for the simultaneous evolution of two species, namely fuzzy clusters and connection weights species. Dynamic rule construction allows the trained neural network update itself when new data comes. The performance of the proposed model was validated by a real-world application of traffic flow prediction.

References

[1] Han, J., Kamber, M.: Data Mining: Concepts and Techniques, 2nd edn. Morgan Kaufmann, San Francisco (2006)

[2] Guha, S., Meyerson, A., Mishra, N., Motwani, R., O'Callaghan, L.: Clustering data streams: theory and practice. IEEE Transactions on Knowledge and Data Engineeing 15(3), 515–528 (2003)

[3] Xu, L.: Advances on BYY harmony learning: information theoretic perspective, generalized projection geometry, and independent factor autodetermination. IEEE Transactions on Neural Networks 15, 885–902 (2004)

[4] Albus, J.S.: A new approach to manipulator control: The cerebellar model articulation controller (CMAC). Transaction of the ASME, Dynamic Systems Measurement and Control 97, 220–227 (1975)

[5] Nguyen, M.N., Shi, D., Quek, C.: FCMAC-BYY: Fuzzy CMAC Using Bayesian Ying-Yang Learning. IEEE Transactions on Systems, Man and Cybernetics-part B 36(5), 1180–1190 (2006)

[6]   Shi, D., Nguyen, M.N., Zhou, S., Yin, G.: Fuzzy CMAC with Incremental Bayesian Ying Yang Learning and Dynamic Rule Construction. IEEE Transactions on Systems, Man and Cybernetics, Part B 40(2), 548–552 (2010)

[7]   Shi, D., Quek, C., Tilani, R., Fu, J.: Product Demand Forecasting with a Novel Fuzzy CMAC. Neural Processing Letters 25(1), 63–78 (2007)

[8]   Nguyen, M.N., Shi, D., Fu, J.: An Online Bayesian Ying-Yang learning Applied to Fuzzy CMAC. Neurocomputing 72(1-3), 562–572 (2008)

[9]   Nguyen, M.N., Shi, D., Ng, G.S., Quek, C.: Traffic Prediction Using Ying-Yang Fuzzy Cerebellar Model Articulation Controller. In: Proceedings of the 18th International Conference on Pattern Recognition (ICPR), pp. 258–261 (2006)

[10] Lum, K.S., Nguyen, M.N., Shi, D.: GA Based FCMAC-BYY Model for Bank Solvency Analysis. In: Proceedings of IEEE Congress on Evolutionary Computation (CEC), Singapore (2007)

[11] Shi, D., Dong, C., Yeung, D.S.: Neocognitron's parameter tuning by genetic algorithms. International Journal of Neural Systems 9, 497–509 (1999)

[12] Pena-Reyes, C.A., Sipper, M.: Fuzzy CoCo: A cooperative-Coevolutionary Ap-proach to Fuzzy Modeling. IEEE Transactions on Fuzzy Systems 9(5), 727–737 (2001)

[13] García-Pedrajas, N., Hervás-Martínez, C., Ortiz-Boyer, D.: Cooperative coevolution of artificial neural network ensembles for pattern classification. IEEE Transactions on Evolutionary Computation 9, 271–302 (2005)

[14] Yao, X.: Evolving artificial networks. Proceedings of the IEEE 87(7), 1423–1447 (1999)

[15] Muhlenbein, H., Schlierkamp-Voosen, D.: Predictive Models for the Breeder Genetic Algorithm: I. Continuous Parameter Optimization. IEEE Transactions on Evolutionary Computation 1, 25–49 (1993)

[16] Tung, W.L., Quek, C., Cheng, P.: GenSo-EWS: a novel neural-fuzzy based early warning system for predicting bank failures. Neural Networks 17, 567–587 (2004)

UNIFAC Group Interaction Prediction for Ionic Liquid-Thiophene Based Systems Using Genetic Algorithm

Surya Pratap Singh, Ramalingam Anantharaj, and Tamal Banerjee

Department of Chemical Engineering, Indian Institute of Technology Guwahati
Guwahati – 781039, Assam, India
tamalb@iitg.ernet.in

Abstract. The group interaction parameter prediction of Ionic Liquids(IL's) with thiophene (C_4H_4S) and other hydrocarbons are essential to generate (Liquid Liquid Equilibria) LLE through UNIFAC model. UNIFAC model is highly non-convex and can have several local extrema. In this work, the structural group interaction parameters have been calculated for [OMIM][BF_4] + thiophene + hydrocarbons and [OMIM] [BTI] + thiophene + hydrocarbons systems through regression using GA.The obtained LLE data has been correlated with reported values and it was observed that the cumulative RMSD(root mean square deviation) of ten ternary systems used for regression were 3.01% and 3.65% for [OMIM][BF_4] and [OMIM]BTI] based system respectively. Further, the obtained interaction parameters were used to correlate the experimental LLE data for four ternary systems which were not used for regression. These systems having a total of 40 tie lines gave a very satisfactory RMSD of 1.76 to 3.99% between reported and predicted composition.

Keywords: GA, Ionic Liquid, UNIFAC, Thiophene, Liquid Liquid Extraction.

1 Introduction

By 2010, petroleum industry is bound to produce Ultra Low Sulphur Diesel (ULSD) containing maximum of 15ppm sulphur [1]. The removal of aromatic sulphur compounds from diesel oil is becoming increasingly difficult because of its resistance to hydrodesulphurization (HDS)[2]. Green solvents such as Ionic liquids provide an important alternative in removing such compounds by Liquid–Liquid Extraction (LLE) [2]. The removal of aromatic sulphur from diesel oil using ILs requires LLE data because of the numerous possible combination of ILs (to the order of 10^{18}!).Therefore, the development of effective thermodynamic model and the selection of potential ionic liquid as solvent is very crucial.

UNIFAC (**UNI**quasi **F**unctional **AC**tivity)[3] model is one such model which predicts the LLE data for ternary systems including ILs. The group interaction parameters for typical cation and anion with other hydrocarbons and aromatic sulphur compounds are unavailable. The most common way to do this is to fit the experimental data to the UNIFAC model and then use that model with fitted parameters for predicting LLE for other systems containing the same groups. But quite often the

K. Deb et al. (Eds.): SEAL 2010, LNCS 6457, pp. 195–204, 2010.

model parameter estimation process is an optimization without a unique result. It is more than likely to run into situations where the optimization problem is nonconvex with several local extrema. The focus of this work is on the application of genetic algorithm (GA) to find global extrema for the estimation of group interaction parameters. In this work, the group interaction parameters of the IL:[OMIM][BF$_4$] and [OMIM][BTI] were calculated using UNIFAC model with GA, with the help of available volume R and area Q parameter.

2 UNIFAC Model

The original UNIFAC model combines the functional group concept with a model for activity coefficients based on an extension of the quasi-chemical theory of liquid mixtures (UNIQUAC) as proposed by Fredenslund et al. [3]. This model can be applied at infinite dilution and finite concentrations and was the most widely used before several revisions and extensions were developed. The activity coefficient is expressed as a function of composition and temperature.

The data required for calculating activity coefficients at infinite dilution (nonideality) using UNIFAC are two folds, first the group specific parameters – group volume and surface area parameters, and second the group interaction parameters. Although the group surface area and volume parameters are available for a large number of groups including those of ionic liquids (ILs), it is the group interaction parameter that is largely missing. Keeping this in mind this work focuses on finding the group interaction parameters for groups present in ternary mixture of IL – thiophene-diesel. Thiophene is taken as the model sulphur component and our aim in to separate it from diesel .However in our work we have taken the hydrocarbon such as hexane, heptane etc as the diesel component.

The UNIFAC model has a combinatorial contribution ($\ln \gamma_i^C$) to the activity coefficient which is directly related to differences in size and shape of the molecules, and a residual contribution ($\ln \gamma_i^R$) to define the energetic interactions between the molecules.

$$\ln \gamma_i = \ln \gamma_i^C + \ln \gamma_i^R \tag{1}$$

The combinatorial part is given by

$$\ln \gamma_i^C = 1 - V_i + \ln V_i - 5q_i\left[1 - \frac{V_i}{F_i} + \ln\left(\frac{V_i}{F_i}\right)\right] \tag{2}$$

$$F_i = \frac{q_i}{\sum_j q_j x_j} \tag{3}$$

$$V_i = \frac{r_i}{\sum_j r_j x_j} \tag{4}$$

The pure component parameters r_i and q_i are, respectively, related to molecular van der Waals volume and molecular surface area. They are calculated as the sum of the group volume and group area parameters, R_k and Q_k.. The mole fraction of component j in the mixture is denoted as x_j. Thus

$$r_i = \sum_k v_k^{(i)} R_k \qquad (5)$$

$$q_i = \sum_k v_k^{(i)} Q_k \qquad (6)$$

Where v_k^i, always an integer is the number of groups of type k in molecule i. The group parameters R_k and Q_k are normally obtained from van der Waals group volumes and surface areas, V_k and A_k, as given below:

$$R_k = \frac{V_k}{15.17} \qquad (7)$$

$$Q_k = \frac{A_k}{2.5 \times 10^9} \qquad (8)$$

The residual part is given by:

$$\ln \gamma_i^R = \sum_k v_k^{(i)} \left[\ln \Gamma_k - \ln \Gamma_k^{(i)} \right] \qquad (9)$$

Γ_k is the group residual activity coefficient, and $\Gamma_k^{(i)}$ is the residual activity coefficient of group k in a reference solution containing only molecules of type i.

$$\ln \Gamma_k = Q_k \left[1 - \ln \left(\sum_m \theta_m \Psi_{mk} \right) - \sum_m \frac{(\theta_m \Psi_{mk})}{\sum_n \theta_n \Psi_{nk}} \right] \qquad (10)$$

$$\theta_m = \frac{Q_m X_m}{\sum_n Q_n X_n} \qquad (11)$$

$$X_m = \frac{\sum_i v_m^{(i)} x_i}{\sum_i \sum_m v_k^{(i)} x_i} \qquad (12)$$

The surface area fraction of group m in the mixture is represented by θ_m; and X_m is the mole fraction of group m in the mixture. The group interaction parameter Ψ_{nm} is defined by

$$\Psi_{nm} = \exp \left[-\left(\frac{a_{nm}}{T} \right) \right] \qquad (13)$$

The parameter a_{nm} characterizes the interaction between groups m and n at temperature T. For each group-group interaction, there are two parameters: $a_{nm} * a_{mn}$. Equations 10-12 hold true for $\ln \Gamma_k^{(i)}$, except that the group composition variable θ_k, ,is now the group fraction of group k in pure fluid i. In pure fluid, $\ln \Gamma_k = \ln \Gamma_k^{(i)}$ which means that the activity coefficient approches unity when mole fraction approaches unity. Thus γ_i^R must be close to unity because as $x_i \to 1$, $\gamma_i^C \to 1$ and $\gamma_i \to 1$. Therefore, the group parameters (R_k, Q_k, a_{nm} and a_{mn}) should be available beforehand to solve the above equations.

3 Computational Details

3.1 Liquid - Liquid Equilibrium

Many pairs of chemical species when mixed, would not satisfy the stability criterion for single phase and thus split into two phases having different composition of different components. The equilibrium criterion for LLE is uniformity of P, T and of fugacity for each component throughout both phases (I and II refers to the two phases).

$$f_i^{I} = f_i^{II};(i = 1, 2,.............n) \qquad (14)$$

In terms of activity coefficients this becomes:

$$x_i^{I}\gamma_i^{I} = x_i^{II}\gamma_i^{II};(i = 1, 2,.............n) \qquad (15)$$

In terms of K-Value(distribution ratio):

$$K_i = \frac{x_i^{I}}{x_i^{II}} = \frac{\gamma_i^{II}}{\gamma_i^{I}};(i = 1, 2,.............n) \qquad (16)$$

The above equations clearly shows that if we are able to predict activity coefficient of each component in both phases then it is easier to calculate distribution coefficient of components in both phases.

3.2 Calculation of Composition

The mole fraction of all components in both phases can be calculated using distribution coefficient K at a particular temperature. This calculation has been performed using the Rachford-Rice algorithm [4] (Fig. 1) and it is constructed by the following equations (c refers to the equilibrium condition and the number of components respectively, z being the initial molar feed fraction of mixture and $\psi = V/F$ the split between the two liquid phases as represented by V and L, such that $F = V + L$):

$$z_c = 1 - \sum_{i=1}^{c-1} z_i \qquad (17)$$

$$f(\Psi) = \sum_{i=1}^{c} \frac{z_i(1 - K_i)}{1 + \Psi(K_i - 1)} \qquad (18)$$

$$x_i^{I} = \frac{z_i}{1 + \Psi(K_i - 1)} \qquad (19)$$

$$x_i^{II} = x_i^{I} K_i \qquad (20)$$

The Rachford-Rice algorithm along with UNIFAC model predicts the distribution coefficient whenever the ionic liquid is involved in the separation of thiophene from

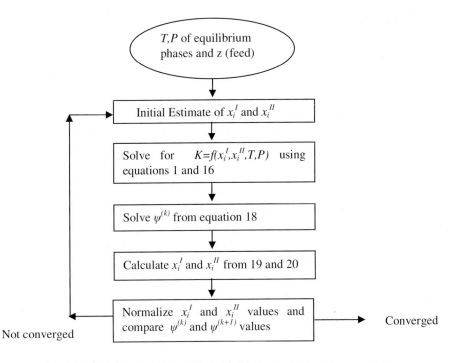

Fig. 1. Modified Rachford-Rice algorithm (k refers to iteration number)

hydrocarbon. Thereafter we will extend our work to predict group interaction parameters of ionic liquids systems containing aromatic sulphur compounds.

3.3 Group Interaction Parameters of Ionic Liquids

The group interaction parameters can be predicted by correlating large number of experimental data in combination with an objective function such as

$$\min F = \sum_{k=1}^{m} \sum_{j=1}^{2} \sum_{i=1}^{c} W_{ik}^{j} \left(x_{ik}^{j} - X_{ik}^{j} \right)^{2} \tag{21}$$

In the objective function, 'W', the weight factor has been taken as unity, c represents the number of components, m the number of tie lines ; x and X are the experimental and calculated mole fractions, respectively. The calculated mole fraction is obtained using equation 19 and 20. The global minimum of the objective function (OF) may be found until some tolerance limit is reached, which is related to accuracy of , a_{nm} and a_{mn} predicted. For this purpose we require Genetic Algorithm (GA) for regressing the experimental data. GA leads to nearly globally optimum values; it does not require any initial guess but only the upper and lower bounds of the interaction parameters. It has also been shown to perform better than inside variance estimation method (IVEM)[5] and the techniques used in ASPEN and DECHEMA.

4 Results and Discussion

We have used the Float genetic algorithm (FGA).This is better than both binary genetic algorithm (BGA) and Simulated Annealing(SA) in terms of computational efficiency and solution quality [6,7]. The methodology and the operators are described in
our earlier work[8]. Relationships for selection function and operators are also given
in our previous work[8].GA moves from generation to generation until a termination
criterion is met. The most frequently used stopping criterion is a specified maximum
number of generations (G_{max}) [6,7]. The operator values are the default values are as
used in MATLAB Toolbox[7]. For the regression GA is used with the objective function as given in equation 21. Prior to the optimization the lower and upper bounds for
the interaction parameters were given beforehand (+1000 to -1000) based on prior
literature data. Based on our earlier work on LLE [9], the number of population and
generation has been kept at 100 and 200 respectively.

Table 1. Group volume and area parameters

Parameters	CH$_2$	CH$_3$	Thiophene	[IM][BF$_4$]	[IM][BTI]
R$_k$	0.6744	0.9011	2.8569	5.6658	7.4134
Q$_k$	0.540	0.848	2.140	3.1570	6.5440

For this study, we considered ten two phase, three component systems with
[OMIM][BF$_4$] and [OMIM][BTI] based ternary systems. The groups for the three
component systems used in the prediction are listed in Table 1. The group interaction
parameter was determined from the liquid liquid equilibrium (LLE) results through
UNIFAC via regression of experimental data. The group volume R and area Q values
are taken from literature .All the group and their parameters are listed in Table 1. In
this case the ionic liquid [OMIM][BF$_4$] or [OMIM][BTI] is broken in three parts i.e.
one [IM][BF$_4$] and/or [IM][BTI] group, one octyl group and one methyl group. When
these data are used simultaneously, we need to have a grand parameter matrix. A
Grand Parameter Matrix (GPA) is a matrix of group interaction parameters which
includes all the groups (frequency matrix) present in all the compounds as shown
in Table 2.But it can be readily observed that the interaction parameter or group parameter matrix for ILs is still scarce. The accuracy of the new observed interaction
parameters (Table 3) of these values can be judged by looking at RMSDs of the five
groups used for the prediction, which are defined as

$$RMSD = 100 \times \left[\sum_{k}^{m} \sum_{j=1}^{2} \sum_{i=1}^{c} W_{ik}^{j} \frac{\left(x_{ik}^{j} - X_{ik}^{j} \right)^2}{2mc} \right]^{1/2} \qquad (22)$$

Here 'm' refers to the number of tie lines, 'c' the number of components and '2' is the
number of phases. The observed RMSD values for [OMIM][BF$_4$] and [OMIM][BTI]
are quite accurate given the cumulative RMSD of ten systems (systems 1-7 and 8-14)

considered simultaneously is 3.01% and 3.65% respectively. All the RMSD values are shown in Table 4.

A further check is provided by applying the group parameters values to system which involve compounds that are not used for the prediction. Two different ternary data sets were taken for both the ionic liquid systems to confirm the predictions. For the prediction of group interaction parameters for [OMIM][BF$_4$] based systems, tie lines corresponding to systems 1-5 of Table 4 were used for regression.

Table 2. Grand Parameter matrix & Frequency matrix

Grand parameter matrix					
Name	CH$_2$	CH$_3$	Thiophene	[IM][BF$_4$]	[IM][BTI]
CH$_2$	0	0	92.99	NA	NA
CH$_3$	0	0	92.99	NA	NA
Thiophene	-8.479	-8.479	0	NA	NA
[IM][BF$_4$].	NA	NA	NA	0	-
[IM][BTI]	NA	NA	NA	-	0
Frequency matrix					
[OMIM][BF$_4$]	7	2	0	1	-
[OMIM][BTI]	7	2	0	-	1
Thiophene	0	0	1	0	0
Cyclohexane	6	0	0	0	0
Hexane	4	2	0	0	0
Heptane	5	2	0	0	0
Dodecane	10	2	0	0	0
Hexadecane	14	2	0	0	0

NA: not available.

Table 3. Estimated group interaction parameters

[IM][BF$_4$]			
	CH	Thiophene	[IM][BF$_4$]
CH	0	92.99	1268.3
Thiophene	-8.479	0	679.7
[IM][BF$_4$]	1374.4	456.4	0
[IM][BTI]			
CH	0	92.99	313.71
Thiophene	-8.479	0	240.9
[IM][BTI]	333.81	175.06	0

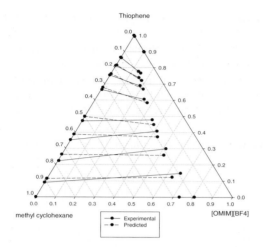

Fig. 2. Experimental and Predicted tielines for [OMIM][BF₄]-thiophene-methylcyclohexane

Table 4. RMSDs of the ternary systems used in the prediction

No.	System Name	RMSD in %	Ref	T /K
1	[OMIM][BF₄]—thiophene–cyclohexane	5.85	[12]	298.15
2	[OMIM][BF₄]—thiophene–hexane	3.33	[12]	298.15
3	[OMIM][BF₄]—thiophene–heptane	1.78	[14]	298.15
4	[OMIM][BF₄]—thiophene–dodecane	2.35	[14]	298.15
5	[OMIM][BF₄]—thiophene–hexadecane	3.60	[14]	298.15
6	[OMIM][BF₄]—thiophene–isooctane	3.99	[13]	298.15
7	[OMIM][BF₄]—thiophene–methyl-cyclohexane	2.39	[11]	298.15
8	[OMIM][BTI]—thiophene–cyclohexane	2.88	[10]	298.15
9	[OMIM][BTI]—thiophene–hexane	2.00	[15]	298.15
10	[OMIM][BTI]—thiophene–heptane	2.40	[15]	298.15
11	[OMIM][BTI]—thiophene–dodecane	2.65	[10]	298.15
12	[OMIM][BTI]—thiophene–hexadecane	4.70	[15]	298.15
13	[OMIM][BTI]—thiophene– isooctane	3.55	[13]	298.15
14	[OMIM][BTI]—thiophene–methyl-cyclohexane	1.76	[13]	298.15

Thereafter the predicted group interaction (Table 3) has been used in predicting the tie lines of systems: [OMIM][BF$_4$] + thiophene + isooctane (system 6 of Table 4) and [OMIM][BF$_4$] + thiophene + methylcyclohexane (system 7 of table 4) with RMSD of 3.99 and 2.39% respectively. For the prediction of group interaction parameters for [OMIM][BTI] based system, tie lines corresponding to systems 8-12 of Table 4 were used for regression. Thereafter the predicted group interaction (Table 3) has been used in predicting the tie lines of systems: [OMIM][BTI] + thiophene + 224trimethylpentane (system 13 of Table 4) and [OMIM][BTI] + thiophene + methyl-cyclohexane (system 14 of table 4) with RMSD of 3.55 and 1.76% respectively. The data shown in above tables can also be plotted in a ternary diagram to give a pictorial comparison of the experimental and the predicted data. Figure 3 shows the comparison between predicted and experimental tie lines for the system :[OMIM][BF$_4$]-thiophene-methylcyclohexane .It can be seen that our predicted results are in qualitative agreement with those of reported values [10-15].

5   Conclusions

In this work, UNIFAC model is used to calculate the group interaction parameter of the mixture containing IL, thiophene and hydrocarbons. New group interaction parameters are regressed using GA from 100 reported tie lines of liquid- liquid equi-librium data of three component systems. The predicted results are in good agreement with reported values. The predicted group interaction parameters and LLE results show very encouraging results in terms of RMSDs which lies between 1.5% - 5%. Thus regression using GA is a powerful tool to describe LLE especially when ex-tended to systems containing similar groups like thiophene, pyrrole or Ionic Liquid.

Acknowledgement. The authors are grateful to the Department of Science and Tech-nology (DST), Government of India for the financial support through project SR/FTP/08-08 under the Fast Track Scheme.

Nomenclature

[OMIM]	1-octyl-3-methylimidazolium
[BF$_4$]	Tetrafluoroborate
[BTI]	Bis [trifluoromethylsulfonyl] imide

References

1. Environment Protection Agency, http://www.epa.gov/otaq/tr2home.html
2. Zhang, S., Zhang, Q., Zhang, Z.: Extractive Desulfurization and Denitrogenation of Fuels Using Ionic Liquids. Ind. Eng. Chem. Res. 43, 614–622 (2004)
3. Fredenslund, Jones, R.L., Prausnitz, J.M.: Group contribution estimation of activity coeffi-cients in nonideal liquid mixtures. AIChE J. 21, 1087–1099 (1975)
4. Seader, J.D., Henley, E.J.: Separation Process Principles. John Wiley, New York (1998)
5. Vasquez, V.R., Whiting, W.B.: Regression of binary interaction parameters for thermody-namic models using an inside-variance estimation method (IVEM). Fluid Phase Equilib. 170, 235–253 (2000)

6. Houck, C.R., Joines, J.A., Kay, M.G.: Comparison of genetic algorithms, random restart, and two-opt switching for solving large location-allocation problems. Comp. Oper. Res. 23, 587–596 (1996)

7. MATLAB GA Toolbox, http://www.ise.ncsu.edu/kay/gaotv5.zip

8. Sahoo, R.K., Banerjee, T., Ahmad, S.A., Khanna, A.: Improved Binary Parameters using GA for Multi-Component Aromatic Extraction: NRTL Model without and with Closure Equations. Fluid Phase Equilib. 239, 107–119 (2006)

9. Singh, M.K., Banerjee, T., Khanna, A.: Genetic algorithm to estimate interaction parameters of multicomponent systems for liquid–liquid equilibria. Comp. Chem. Engg. 29, 1712–1719 (2005)

10. Alonso, L., Arce, A., Francisco, M., Rodriguez, O., Soto, A. (Liquid + liquid) equilibria of [C8mim][NTf2] ionic liquid with a sulfur-component and hydrocarbons. J. Chem. Thermodyn. 40, 265–270 (2008)

11. Alonso, L., Arce, A., Francisco, M., Rodriguez, O., Soto, A.: Measurement and Correlation of Liquid-Liquid Equilibria of Two Imidazolium Ionic Liquids with Thiophene and Methylcyclohexane. J. Chem. Eng. Data. 52, 2409–2412 (2007)

12. Alonso, L., Arce, A., Francisco, M., Rodriguez, O., Soto, A.: Liquid-Liquid Equilibria for Systems Composed by 1-Methyl-3-octylimidazolium Tetrafluoroborate Ionic Liquid, Thiophene, and n-Hexane or Cyclohexane. J. Chem. Eng. Data 52, 1729–1732 (2007)

13. Alonso, L., Arce, A., Francisco, M., Rodriguez, O., Soto, A.: Gasoline Desulfurization using Extraction with [C8min][BF4] Ionic Liquid. AIChE J. 53, 3108–3115 (2007)

14. Alonso, L., Arce, A., Francisco, M., Soto, A.: Solvent extraction of thiophene from n-alkanes (C7, C12 and C16) using the ionic liquid [C8mim][BF4]. J. Chem.Thermodyn. 40, 966–972 (2008)

15. Alonso, L., Arce, A., Francisco, M., Soto, A.: Phase behaviour of 1-methyl-3-octylimidazolium bis[trifluoromethylsulfonyl] imide with thiophene and aliphatic hydrocarbons: The influence of n-alkane chain length. Fluid Phase Equilib. 263, 176–181 (2008)

HIER-HEIR: An Evolutionary System with Hierarchical Representation and Operators Applied to Fashion Design

Abhinav Malhotra[1] and Varun Aggarwal[2]

[1] Netaji Subhas Institute of Technology, Delhi, India
abhinav.malhotra89@gmail.com
[2] Aspiring Minds, Gurgaon, India
varun.aggarwal@gmail.com

Abstract. There has been considerable interest in using evolutionary algorithms based techniques to design creative systems. However,these techniques suffer from either being too creative and violating design constraints of the domain or those catering to a limited search space, but operating within design constraints. We have designed a new evolutionary system 'HIER-HEIR', which is not only creative (searches a large space effectively), but creates only such designs which are *valid* with respect to the design domain. Inspired by human design methodology, the representation is a hierarchy of components and variation acts at all levels of hierarchy intelligently, facilitating effective search in the design space with explicit control over exploitation and exploration. We have explained our technique with the metaphor of automatic design of a fashion dress in this paper. The experimental results validate our hypotheses with regard to the system. With regard to previous work, our technique is new both with regard to previously published hierarchical systems and those designed for evolving fashion designs.

1 Introduction

Automatic design of creative and engineering systems using evolutionary algorithms has been of considerable interest [11,5,6,4,3,7]. For evolving designs which are useful in a real scenario, these techniques require not only to be creative, but the resulting design must follow certain design constraints to allow actual fabrication and use. The current systems for evolutionary design suffer from a few challenges. While some algorithms design very flexible and creative systems [3,4], others look at a coarse library of few designs or degenerate to simple parameter optimization for set designs [8,11,9]. The second challenge with flexible systems is that they not only lead to many *invalid* designs in the course of evolution [7], but lack of direct mapping between genotype and phenotype leads to a coarse fitness landscape [2] and possible random search.

Our aim is to design an algorithm (i) which creates designs from a large design search space, (ii) has a rich set of variation operators capable of creating every valid design in the search space, (iii) without resulting in any invalid design,

K. Deb et al. (Eds.): SEAL 2010, LNCS 6457, pp. 205–214, 2010.
© Springer-Verlag Berlin Heidelberg 2010

(iv) having a smooth fitness landscape (v) with control over exploitation and exploration. Such a system can be used to evolve designs which are not only creative, but also respect the constraints of the specific domain of design.

We have identified that encoding a design as a hierarchy of components solves the challenges discussed above. The knowledge of the hierarchy, the do's (variations) and don'ts (constraints) at each level of the hierarchy is what enables a designer to make *valid* yet creative designs. Whereas changing components at a higher level of hierarchy helps the designer take larger jumps in the search space, tweaking components at the lower level of the hierarchy helps in fine-grained search. The designer respects the component integrity at each level of hierarchy and thus never creates an invalid design.

We translate this domain knowledge of hierarchy of components into an effective evolutionary technique. Our system HIER-HEIR[1] for evolving designs is analogous to STGP (Strongly Typed Genetic Programming) as used in evolving functions (analogy discussed in Section 5). The system has the following properties: (a) Each design is one instance from a hierarchical library of designs where options and constraints, at each level are defined. (b) A hierarchical mutation operator that can create any new design in the search space from a given design without creating any invalid design. (c) A hierarchical crossover operator that can create all possible valid mixes of two designs without creating any invalid design. (d) The algorithm has a knob to control exploration vs. exploitation. The hierarchical nature of the operators provides this power. (e) The fitness landscape is fine-grained since genotype closely maps to the phenotype.

With regard to previous work, [1,9] come closest to our technique. In [1], bit-wise parameters are organized in a hierarchy to identify which bits between two individuals are of the same 'type' and can be exchanged. There isn't a notion of swapping 'blocks' at different levels of hierarchy and exchanges only happen at the leaf nodes implying only *parameter* evolution and no *component-wise* evolution. Mutation is used simply, to vary the number of blocks by adding/deleting blocks. Our operators, on the contrary, facilitate component level evolution with both mutation and crossover capable of operating at different levels of the hierarchy. For instance, in [1], the crossover cannot create every possible valid design formed by exchanging parts of two parents. In [9], the system is designed and discussed in context of analog design. The notion of operators is not well-developed and there is no discussion on how to create exploitative or explorative operators or respect constraints at different levels of the hierarchy. The authors uses simple hierarchies in his experiments, which allows them to use simple operators.

Our primary contribution is to develop a technique, HIER-HEIR, having hierarchical representation with hierarchical variation operators, which uses the domain knowledge of design, to efficiently evolve creative yet valid designs. In this paper, we explain HIER-HEIR with a metaphor of a fashion design system.

With regard to the previous attempts to fashion design using evolutionary algorithms, our system is novel showing that it is non-obvious. In [6], the system uses a flat representation of dress parts and bit-wise operations, unlike our

---

[1] 'HIER' stands for hierarchy, whereas 'HEIR' for evolution.

hierarchical design. In [11], the dress is considered as a sketch, for which coordinates can be moved to create new dresses.

The paper unfolds as follows: Our approach to implement a fashion designing system using HIER-HEIR is discussed next. It is followed by description of hierarchical representation and variation operators. Next, we elucidate the Experimental Results of our system. Last, we summarize our work.

2   System Design

A dress design could be considered as a 2-D or 3-D drawing which can be modified by re-sketching [11]. On the other hand, it may be understood as being composed of many parts which are all seamed together to make the dress. These parts can be organized in a *flat* structure or otherwise, in a hierarchical structure. Let us consider traversing such a hierarchy bottom-up: collars, sleeves, pocket design and *center* design all are joined together to make up a shirt. Similarly, belt, fly, pocket and trouser-length seam together to make up trousers. Shirt and trousers, in turn, rendered together, make up a dress. This simple illustration of the hierarchy has the shirt and trousers at a higher level, whereas the belt, fly, etc. are at a lower level and are parts of the shirt/trousers.

HIER-HIER evolves this hierarchy of components. The input to the system is a hierarchical library of parts. Each part is composed of other parts and programmable numeric or categorical parameters (such as a parameter representing color). Secondly, there is a *rule set*, which defines any constraints on putting these parts together, for instance, one may set a rule that a button design A is never put with a collar of type B.

At each level, the system requires to choose one of the alternatives for each part at the given level and define all its programmable parameters. On the basis of the alternative chosen (say A1), at the next level, the alternatives for each of the child parts of A1 (that compose A1) require to be chosen. This iterative process goes on till the smallest indivisible part is decided and the dress is constructed.

HIER-HEIR is extensible and scalable in multiple ways. The designer can add new parts at any level of the hierarchy as per taste and discretion. The definition of a part is not restrictive. By defining multiple child parts and programmable parameters for a given part, a wide variety of parts may be represented (allowing both *parameter* and *component* search together). The designer may add *recursive* parts making the design space infinitely large and more *creative*.

3   Design Representation

In line with HIER-HEIR's ideology, we preserve the hierarchy described in the previous section in all components of our evolutionary system including the representation of the genome, variation operators and in fitness evaluation. This is not as obvious as it seems. In previous works, even though the authors have had an inkling of the hierarchy they have chosen a flat representation which is very often bit-encoded [6].

The hierarchical library of parts is represented as a polymorphic class. Every possible dress or genome is an instance or object of this polymorphic class. The genome is a hierarchy of chosen parts. A 'part' is a component/block at each level of the hierarchy. Each part contains the *necessary* and *sufficient* information to construct (or equivalently draw, in this case) the part in 2-D space and to perform effective crossovers and mutations. The structure of the polymorphic class is illustrated below in C++ syntax.

C1: Code for Polymorphic Class

```
class dress { public:
    top *t;
    bottom *b; };
class top {public: //shared members
    collar *c;
    sleeve *s; };
class shirt::top //inherits from top
{public: //unshared members
    center_shirt *cs;
    button_shirt *bs; };
class collar //basic part's class
{public:
    int type;
    int to_connect[r][c];
    int shape[r][c]; };          So on and so forth...
```

There are the following two kinds of parts in the hierarchy:

Compositional Part: Compositional parts are composed of other parts (termed as *child parts* henceforth) which themselves may be compositional or basic parts. Parts at all levels of the hierarchy except the lowermost level are compositional parts. A Compositional part is represented as a class with the following data members: (a) *Shared Members*: These are (pointers to) child parts of the given part which are shared or common to all possible choices of the current part. Put another way, a shared part is common to all derived classes of the base class of the given part. As depicted in Code C1 (code for hierarchical library), collar and sleeve are shared parts, since they remain the same for a shirt, kurta or a t-shirt. From perspective of inheritance, the shared parts may be declared in the base/parent class itself. (b) *Unshared member*: There are (pointers to) child parts which are unique to the current part and not a property of any other choices for the current part. In fact, it is the unshared members which provide *individuality* to any part and distinguish it from all the other choices. For instance, buttons and *center* are unshared members for *top*, since they are different for a kurta and a shirt. (c) *type*: This field identifies whether the particular part is a basic part (represented by a numeric 0) or a compositional part (numeric 1).

Basic part: Parts of this type belong to the lowermost level(leaf nodes) of the hierarchy. Basic parts are classes with the following fields: (a) *shape*: This field is an array of (x,y) coordinates that exactly defines how the part shall be rendered/drawn in 2-D space. For each unconnected shape in the part, a set of points are defined, which if joined sequentially will render the part. The shape field can be extended to represent a 3-D format. (b) *to_connect*: This field defines what parts connect to the given part and at exactly what coordinates of the given part the connections will happen. For instance, a 'collar' connects with the 'center' and not a 'sleeve', even though all these three parts compose a shirt and are at the same level of hierarchy. (c) *parameters*: These are any numeric values for a part, which can be chosen from a range, for instance, color. (d) *type*: Same as in compositional part.

4   Initialization

Initialization is done by a recursive process traversing the polymorphic hierarchy in a top-down, breadth-first fashion. For each part, one of the alternatives is randomly chosen. Given the alternative chosen, alternatives are randomly chosen for all its child parts. The process terminates when only basic parts are left.

5   Variation Operators

5.1   Mutation

The hierarchical mutation operator simply chooses a part of the individual and *re-initializes* it. The same initialization function, which generates the whole dress in the beginning of the algorithm, is now executed just for the chosen part. The function works iteratively to *grow* the whole hierarchy of parts under chosen part. Everything in the genome remains same except the chosen part and parts that make the chosen part. All dresses created are valid dresses.

Our mutation operator can be exploitative or explorative depending on the level of hierarchy it operates at. For instance, if it re-initializes a part at the top of the hierarchy (say shirt or trousers), it is a macro-mutation (explorative) that changes a large part of the dress. Whereas, if the mutation is performed on a part at the lower levels of the hierarchy (say collar, etc.), the operator is exploitative. In our design, whenever an individual is chosen, the level at which the mutation occurs is chosen probabilistically. A probability density function (PDF) determines the probability of choosing a part at any given level of the hierarchy. The PDF is defined by the algorithm designer and can be varied as a function of generations.

5.2   Crossover

The basic idea to ensure valid crossovers is that only the parts which are of the same *type* or equivalently, inherit from the same base class can be exchanged.

The crossover simply exchanges the parts of same *type* with all its child parts, which ensures the *validity* of the new dresses created. For instance, a shirt may be exchanged with a kurta/t-shirt, since they belong to the same type, but not with trousers. Comparing this with [1], the crossover operator therein would not be able to exchange a *kurta* with a shirt.

Given two individuals, for a given part in the first individual, there may not be any part in the second individual which is of the same type. For instance, consider one dress has trousers, whereas the other dress has shorts, then no child parts of the trousers or the shorts can exchange with each other. Given any two individuals, we can calculate a **Mating Potential** as the number of parts that could be exchanged between the two individuals. If the dresses are too distinct, the Mating Potential (MP) is low and crossover could only happen at the top levels of the hierarchy. On the other hand, two similar dresses would have high MP and the crossover could happen across the levels. Do note however, that two exactly same dresses have very high MP, but the the crossover is degenerate.

Given two individuals, we use a recursive 'coloring' algorithm to color all parts that can be mated in the two individuals and to find the MP quantitatively. The algorithm works henceforth: (1) The algorithm starts from the top of the hierarchy for both individuals, which is colored by default. The part(s) under consideration in the first iteration is *top*. (2) If the corresponding parts have the same design, both their shared and unshared members are colored. For parts of same design, we know that all their children, shared and unshared, will correspond with regard to type. Whereas if the corresponding parts (that of same type) are not of the same design[2], only its shared parts are colored. for e.g., if the corresponding part is a shirt for both chosen individuals, all their child parts are exchangeable and are hence colored. Whereas if the corresponding parts are a shirt and a t-shirt, only their shared child parts are colored. (3) All the colored child parts again go through Step 2. This process continues till the part being compared is a basic part or there are no shared children for a part.

This algorithm colors all parts of both individuals which can be exchanged. More the number of exchangeable parts between two individuals, higher is their MP. Given the colored individuals, we probabilistically choose which part to exchange. Similar to mutation, a PDF controls exploration vs. exploitation depending on which level of the hierarchy the exchange is done. MP can also be used to choose the individuals for mating: crossover of two individuals with high MP will lead to a micro-mutation and vice-versa.

Our hierarchical crossover exchanges parts of the same type leading to valid dresses. Exploration vs. exploitation is controlled both by the choice of the individual that undergo crossover and the level of hierarchy at which the crossover takes place. This can be explicitly controlled using the PDF.

---

[2] Each 'basic part' class has a field called shape which defines how exactly that part will be rendered on the screen. This field represents the 'design' of that part. The 'design' of a basic part can be defined by the position of its seams when it is or will be stitched to make a dress. The 'design' of a compositional part is based on the 'design' of basic parts it contains.

5.3   Discussion on Representation and Operators

Our hierarchical representation and operators are unique. They do not lead to any invalid designs. They are not restrictive and can potentially cross two designs in all possible valid ways and mutate a design to create any new design. The variations exchange or modify *phenotypic* building blocks, i.e. parts. This implies that a micro-variation in the genotype space corresponds to a small change in the phenotype (rendered dress) space as well and vice-versa. This leads to a smooth fitness landscape, which increases the chances of success/convergence of the algorithm.

The hierarchical nature of the operators, provides a knob to control exploration and exploitation according to the discretion of the algorithm designer. We believe this extra knob lays a lot of power in the hand of the algorithm designer to control the trajectory of the algorithm more efficiently than relying on the emergent dynamics of evolution.

With regard to genetic programming paradigms, we believe that our representation and operators are analogous to STGP [10]. Whereas in STGP, the *type* of arguments a function takes, is defined, for us the *type* of parts which compose any part are defined. In both cases, this allows only valid (or those adhering to the defined grammar) designs to be created. The crossover in STGP happens only at points where the same *type* of arguments are available, whereas in our system, they happen where the same *type* of part is available.

Our representation and operators strictly adhere to the objectives detailed in Section 1.

6   Fitness Evaluation

The system takes fitness values from the end user. The *interactive* system is displayed as a GUI, which renders/shows eight different designs per page (refer Figure 1). We first describe the GUI and later, the details of how various designs are rendered.

The GUI asks the user/designer to rate the designs on two criteria: *style* and *wearability*. The former represents the aesthetic value of the design, whereas the latter is about the ease/comfort with which the dress can be worn. Both of the above criteria have a meter of judgment that ranges from 0 to 10 with a step-size of 1. The average of these two parameters is taken as the fitness value of a fashion dress.

Render and Connect Functions: Our genome is hierarchical in nature having both compositional and basic parts. A mapping function recursively flattens this hierarchical genome to a flat array of basic parts called 'dress_phenotype'.

We implement two functions 'connect' and 'render' to connect and render the fashion dress as an appropriate connection of all the basic parts in 2-D space. The render function simply draws the part in the 2-D space at a given position.

To render a complete fashion-design/dress each of these parts is combined and connected in a specific way. The connect function takes the 'dress_phenotype' as input and renders them at appropriate positions (here co-ordinates), thus

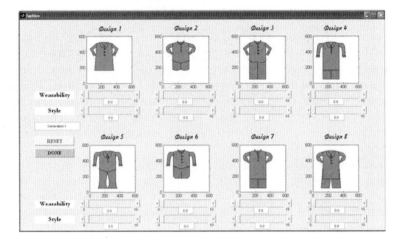

Fig. 1. Design System's GUI

connecting them. The connected assembly of parts forms the phenotype which is then assigned fitness by the user.

Algorithm for the 'connect' function: (1) We randomly pick one part and render it at the origin using the 'render' function. This part is labeled 'rendered'. (2) We pick any remaining part, such that it connects to the part labeled 'rendered' and not 'connected'. We render this part at appropriate coordinates in 2-D space so that it 'connects' to the part already rendered. We render this part next, because given the positions of the parts already rendered , we can exactly determine the position of the current part. (3) We go to Step 2 till all parts that connect to the part rendered in step 2 are rendered at appropriate positions. This part is then labeled 'connected' and all the parts that connected to it are labeled 'rendered'. (4) We go to Step 2 till all parts are labeled both 'rendered' and 'connected'.

7   Experiments

We implemented HIER-HEIR in MATLAB. Since MATLAB supports very little Object Oriented Programming and does not support polymorphism or pointers, we implemented the polymorphic classes in MATLAB ourselves.

Our hierarchical library constitutes the following parts. The dress is divided in three parts: top, belt and bottom. There are two alternatives for top: shirt and kurta. Each of these comprise of right and left collar, right and left sleeve, and a center design. There are three alternative designs each for sleeve, collar and center design. There are three alternatives for bottom: trousers, shorts and pajamas. There are three alternate designs for each of these. There are three alternatives for belt style.The total search space spanned by our library is 39,366.

We ran an instance of our algorithm to get reactions from users. The population size used is 8. We use *tournament* selection with tournament size of 2. The crossover probability is p_c (0.75), whereas the probability of mutation is p_m (0.4). We use a two-valued discrete PDF for crossover and mutation. The probability of modifying (crossover or mutation) a basic part is *pmicro*, whereas that to modify any compositional part is *pmacro*. We have set these probability values such that crossover is more exploitative, whereas mutation is more explorative. For Crossover: c_{pmacro}=0.33 and c_{pmicro}=0.66; Mutation: m_{pmacro}=0.75 and m_{pmicro}=0.25.

Experiment: We requested fourteen subjects (eight men, six women) to rate the designs created by our system on the basis of both *Style* and *Wearability*. After ten generations (80 designs), they were asked to fill up a questionnaire. Four key questions were asked and their answer endorsement rates are as follows. For each answer choice, the number of subjects that endorsed the given choice is mentioned besides it.

1. Do u think the system was able to find a design which you liked?
   Ans- Yes: 3; To a fair extent: 5; Somewhat: 5; Not Really: 1.
2. Did you feel that the system slowly understood what you like and generated more designs of your choice?
   Ans- Yes: 2; To a fair extent: 11 ;Somewhat Yes: 1; Not Really: 0.
3. Do you think the system was creative?
   Ans- Yes: 4; To a fair extent: 5; Somewhat: 5; Not Really: 0.
4. Would you like to choose such a system for designing or choosing clothes?
   Ans- Yes: 4; Yes, if designs are visually more realistic: 4; Somewhat: 4; Not Really: 2.

We group the first two choices as positive endorsement, whereas the last two as negative. Whereas, 90% of the subjects felt that the system was learning their taste of design, an encouraging 60% of the subjects thought the system found a design they liked. The former provides strong evidence to our hypothesis of a smooth fitness landscape and that the selection-variation dynamics are working as desired. The latter is a fairly strong endorsement of our system in meeting the expectations of the users.

All subjects found the system 'creative' with no one endorsing the 'Not Really' question. They enjoyed designing on such a system and were quite enthusiastic with it. Though, a consistent feedback included the need for 3-D designs and inclusion of texture and patterns.

Our current experiments show that a large proportion of subjects were *satisfied* with HIER-HEIR in its ability to create designs and they could also see the system *learning* their likes and dislikes. The current experiments along with the feedback from subjects, gives us direction and scope of future work, which will include 3-D rendering, controlled experiments and larger sample sets. In the long term, we wish to use our system as a test-bed for general hierarchical evolution. We wish to plug-in different design problems in it and test its usefulness. We wish to also try our hands at some problems where fitness is measured objectively to precisely benchmark different approaches.

8   Summary

We describe a new and powerful technique, HIER-HEIR that implements hierarchical representation and hierarchical variation operators, to create designs. The representation and operators efficiently capture the nuances in creative designing to generate creative yet *valid* designs. The paper details application of the HIER-HEIR technique to automatic design of dresses through interactive evolution. The design of hierarchical representation, crossover and mutation operators for dress evolution is discussed. Finally, experiments are conducted to validate efficient working of our technique. As future work, we aim to add more variety and aesthetics to our fashion designing system and also, use HIER-HEIR technique for automatic evolution of other creative systems.

References

1. Bentley, P.J., Wakefield, J.P.: Hierarchical crossover in genetic algorithms. In: Proceedings of the 1st On-line Workshop on Soft Computing (WSC1), pp. 37–42 (1996)
2. Grimbleby, J.B.: Hybrid genetic algorithms for analogue network synthesis. In: Proceedings of Congress on Evolutionary Computation (CEC 1999), Washington, DC. Press (1999)
3. Hemberg, M., O'Reilly, U.-M., Menges, A., Jonas, K., Gonçalves, M., Fuchs, S.: Genr8: Architects' experience with an emergent design tool. In: The Art of Artificial Evolution, pp. 167–188 (2008)
4. Hornby, G.S., Lipson, H., Pollack, J.B.: Generative representations for the automated design of modular physical robots. IEEE Transactions on Robotics and Automation 19, 703–719 (2003)
5. Kicinger, R., Arciszewski, T., De Jong, K.A.: Generative design in structural engineering. In: ASCE International Conference on Computing in Civil Engineering (2005)
6. Kim, H.-S., Cho, S.-B.: Knowledge-based encoding in interactive genetic algorithm for a fashion design aid system. In: GECCO, p. 757 (2000)
7. Koza, J.R., Bennett III, F.H., Andre, D., Martin, A., Dunlap, F.: Automated synthesis of analog electrical circuits by means of genetic programming. IEEE Transactions on Evolutionary Computation 1, 109–128 (1997)
8. Leenaerts, D., Kruiskamp, W.: Darwin: Cmos opamp synthesis by means of a genetic algorithm. In: Design Automation Conference, pp. 433–438 (1995)
9. McConaghy, T., Palmers, P., Gielen, G., Steyaert, M.: Genetic programming with reuse of known designs for industrially scalable, novel circuit design. In: GPTP V, pp. 159–184 (2008)
10. Montana, D.J.: Strongly typed genetic programming. Evol. Comput. 3(2), 199–230 (1995)
11. Ogata, Y., Onisawa, T.: Interactive clothes design support system. In: Ishikawa, M., Doya, K., Miyamoto, H., Yamakawa, T. (eds.) ICONIP 2007, Part II. LNCS, vol. 4985, pp. 657–665. Springer, Heidelberg (2008)

A Population Diversity-Oriented Gene Expression Programming for Function Finding

Ruochen Liu, Qifeng Lei, Jing Liu, and Licheng Jiao

Key Laboratory of Intelligent Perception and Image Understanding of Ministry of Education of China, Institute of Intelligent Information Processing, Xidian University, Xi'an, 710071

Abstract. Gene expression programming (GEP) is a novel evolutionary algorithm, which combines the advantages of simple genetic algorithm (SGA) and genetic programming (GP). Owing to its special structure of linear encoding and nonlinear decoding, GEP has been applied in various fields such as function finding and data classification. In this paper, we propose a modified GEP (Mod-GEP), in which, two strategies including population updating and population pruning are used to increase the diversity of population. Mod-GEP is applied into two practical function finding problems, the results show that Mod-GEP can get a more satisfactory solution than that of GP, GEP and GEP based on statistical analysis and stagnancy (AMACGEP).

Keywords: Gene expression programming, genetic programming, function finding.

1 Introduction

Among various data mining tasks, function finding is considered as a fundamental activity [1]. Gene expression programming (GEP) proposed by Ferreira [2] has gotten relatively good results in finding exact model functions for complex systems.

On the other hand, GEP [2] also has a low speed of convergence, and tends to be trapped in local optima. Taiyong Li [3] pointed out that in the process of GEP, many individuals in one population are identical. But works on these issues in GEP are very little. Based on the above facts, a modified GEP (Mod-GEP) is proposed to improve the diversity of population and then to deal with prematurity and the low convergence speed. The experimental results of two function finding problems show that Mod-GEP can find a more satisfactory function compared with GP [4] and GEP [2] and GEP based on statistical analysis and stagnancy (AMACGEP) [5].

2 An Overview of GEP Algorithm

GEP [6] uses fixed-length chromosome as its genotypes which can be expressed as phenotypes, i.e., expression trees (ETs) with different sizes and shapes. The chromosome employs the head-tail encoding method which ensures the validity of the reproduced offsprings after unconstrained modification.

K. Deb et al. (Eds.): SEAL 2010, LNCS 6457, pp. 215–219, 2010.
© Springer-Verlag Berlin Heidelberg 2010

The GEP algorithm [6] begins with an initial population of randomly generated chromosomes (individuals), which are encoded as ETs, and then evaluate each individual in the population by a predefined fitness function. According to their fitness, some individuals in the population will be selected and subjected to reproduction by using genic operations such as mutation, inversion, recombination, and so on. These reproduced individuals are subjected to the same process: fitness evaluation, selection and reproduction. The process is repeated until certain termination criteria are satisfied.

3 Mod-GEP Algorithm

3.1 The Framework of Mod-GEP Algorithm

To illustrate Mod-GEP explicitly, the whole framework of the algorithm is given as follows:

Step 1: Initialization. A population P_{old} is initialized according to a specific problem.

Step 2: Evaluation. The fitness F_{old} of each chromosome in P_{old} is evaluated.

Step 3: Selection. Select M chromosomes from P_{old} according to fitness by roulette wheel to form a new population P_{new}.

Step 4: Population Updating. Randomly generate N_{dead} chromosomes to replace the same number of chromosomes in P_{new}. After population Updating, the middle population is denoted as P_{new1}.

Step 5: Genic Operation. P_{new1} is evolved by using mutation, inversion, recombination and so on. Another middle population is denoted as P_{new2}.

Step 6: Evaluation. The fitness F_{new2} of the chromosomes in P_{new2} is evaluated.

Step 7: Population Pruning. If the condition of population pruning is not satisfied, return to step 8.

Step 8: Re-selection. The chromosomes with the highest fitness are selected from P_{old} and P_{new2} to pass to the next generation.

3.2 Population Pruning

When there is no remarkable improvement in term of the best fitness over a certain number of generations (suppose $Gen=100$), it can be believed that the population may has a low diversity. In this step, the chromosomes are sorted by descending order according to their fitness, and then we pick out the first chromosome to form the next generation, and prune the chromosomes that have a similarity with the picked one. We repeat the operation till there's no chromosome left in the current population.

What's worth noting is that what we do is just to prune the chromosomes with lower fitness than the picked one within a certain limited range (δ). There is an advantage that we can directly make full use of the fitness of the chromosomes and we don't have to calculate how many symbols are the same in the corresponding positions. If the number of the chromosomes in population is less than the predefined population size after population pruning, we randomly initialize the equal number of chromosomes to maintain the population size.

Here, another point to be put is that we do population updating after roulette wheel selection, which is different from other algorithms such as [7]. Since roulette wheel selection makes high-fitness chromosomes quickly dominate the population and thus the diversity of population is decreased.

4   Experimental Evaluations

The common way for function finding task in real life is to run several times, and accept the best solution. So, each experiment was repeated for 5 runs independently in our experiments, and the best results were compared with genetic programming (GP) [4], GEP, AMACGEP [5] in term of the mean squared error (MSE). And the results in [4] and [5] are used to make a direct comparison.

In our experiments, GEP and Mod-GEP use the same following empirical parameters: the population size is 30, the number of maximum generations is set as 10000, number of genes = 5, the head length is 8, all the modification rates are set as table 5.8 in [6], the function set $F=\{+, -, *, /, \sin, \cos, \exp, \log, \mathrm{sqrt}\}$ where 'log' and 'sqrt' deal with the absolute value of the corresponding numbers, the terminal set $T= \{t, ?\}$ or $\{I, ?\}$ where "?" represents the random numerical constant ranged from -1 to 1 in experiment 1 while "?" ranged from -2 to 2 in experiment 2. We choose the fitness function based on the absolute mean squared error (MSE) proposed by Ferreira [6]. In addition, $N_{dead} = 7$ and $\delta = 1e\text{-}3$ in Mod-GEP.

Table 1. The data of experiment 1

No.	t(min)	y(%)	Result				Relative error(%)			
---	---	---	GP	GEP	AMAC GEP	Mod-GEP	GP	GEP	AMAC GEP	Mod-GEP
1	1	4.00	3.9991	3.9820	4.0034	4.0145	0.0223	0.4502	0.0843	0.3613
2	2	6.40	6.4284	6.5253	6.4265	6.4267	0.4443	1.9583	0.4145	0.4179
3	3	8.00	7.9389	7.9221	7.9430	7.9947	0.7641	0.9734	0.7123	0.0667
4	4	8.80	8.7385	8.7527	8.8305	8.7839	0.6985	0.5376	0.3468	0.1832
5	5	9.22	9.2257	9.2199	9.2161	9.2437	0.0617	0.0011	0.0426	0.2572
6	6	9.50	9.5524	9.4871	9.4859	9.4976	0.5511	0.1353	0.1482	0.0248
7	7	9.70	9.7863	9.6730	9.7093	9.6926	0.8899	0.2781	0.0957	0.0767
8	8	9.86	9.9620	9.8499	9.8739	9.8672	1.0350	0.1029	0.1406	0.0727
9	9	10.00	10.0988	10.0334	10.0366	10.0386	0.9884	0.3340	0.3657	0.3864
10	10	10.20	10.2083	10.2001	10.2322	10.2070	0.0816	0.0012	0.3158	0.0686
11	11	10.32	10.2979	10.3257	10.3335	10.3652	0.2138	0.0557	0.1311	0.4376
12	12	10.42	10.3726	10.4077	10.4114	10.4656	0.4547	0.1182	0.0822	0.4372
13	13	10.50	10.4358	10.4646	10.4858	10.4920	0.6111	0.3368	0.1349	0.0765
14	14	10.55	10.4900	10.5216	10.5388	10.5148	0.5685	0.2704	0.1059	0.3337
15	15	10.58	10.5370	10.5887	10.5821	10.5437	0.4066	0.0825	0.0195	0.3428
16	16	10.60	10.5781	10.6552	10.6661	10.5852	0.2067	0.5208	0.6237	0.1397

4.1 Experiment 1

The data used in experiment 1 originates from one chemical experiment, the experimenter wrote down the density of a material every other minute. The data are given in Table 1. The task is to find the relationship between the density of the material $y/\%$ and the time t/min.

For the relative error, it seems a little difficult to judge which algorithm is better between AMACGEP and Mod-GEP. But we can still see that, seven out of sixteen points, the relative error obtained by Mod-GEP is smallest in four algorithms, but four out of sixteen points, the relative error obtained by AMACGEP is smallest.

4.2 Experiment 2

The data in experiment 2 originates from a physical experiment: when current I passes 2Ω resistance, the voltage on the two sides of the resistance is V. The data are given in table 3. The task of the problem is to find the relationship between V and I.

From table 4, we can vividly see that Mod-GEP has a smaller relative error than any of other compared algorithms on the whole.

Table 2. Comparison with other algorithms in term of MSE

Algorithm	GP	GEP	AMACGEP	Mod-GEP
MSE	0.0032235881	0.0019886611	0.0007948444	0.0006489875

Table 3. The data of experiment 2

No.	I(A)	V(V)	Result				Relative error (%)			
			GP	GEP	AMAC GEP	Mod-GEP	GP	GEP	AMAC GEP	Mod-GEP
1	1	1.8	1.8092	1.7981	1.7993	1.7949	0.5107	0.1080	0.0363	0.2815
2	2	3.7	3.8320	3.6977	3.6729	3.7043	3.5681	0.0623	0.7321	0.1170
3	4	8.2	7.8777	8.2019	8.2267	8.2043	3.9308	0.0226	0.3364	0.0522
4	6	12.0	11.9233	12.0346	12.0689	11.9981	0.6389	0.2882	0.5744	0.0162
5	8	15.8	15.9690	15.7954	15.8266	15.7990	1.0695	0.0294	0.1685	0.0066
6	10	20.2	20.0146	20.1492	20.0451	20.1947	0.9176	0.2514	0.7667	0.0263

Table 4. Comparison with other algorithms in term of MSE

Algorithm	GP	GEP	AMACGEP	Mod-GEP
MSE	0.0316997178	0.0006348718	0.0051572400	0.0000159727

5 Conclusions and Future Work

In this paper, we proposed a modified GEP (Mod-GEP) to perform better in function finding by increasing the diversity of population. In the future, we will focus on combining GEP with other up-to-date methods of diversity maintenance [8]

Acknowledgments

This work was supported by the National Research Foundation for the Doctoral Program of Higher Education of China (20070701022), the National Natural Science Foundation of China under Grant (No.60803098, No.60872135, No.60703108), the China Postdoctoral Science Foundation Special funded project (No. 200801426).

References

1. Witten, I.H., Frank, E.: Data mining: Concepts and Techniques. Morgan Kaufmann, San Francisco (1999)
2. Ferreira, C.: Gene Expression Programming: A New Adaptive Algorithm for Solving Problems. J. Complex System. 13, 87–129 (2001)
3. Li, T.Y., Tang, C.J., He, T., Wu, J., Qin, W.B.: Gene Expression Programming without Reduplicate Individuals. In: Fifth International Conference on Natural Computation, vol. 4, pp. 249–253. IEEE Press, New York (2009)
4. Zhou, A.M., Gao, H.Q., Kang, L.S., Huang, Y.Z.: The Automatic Modeling of Complex Functions Based on Genetic Programming. J. System Simulation. 15, 797–799 (2003)
5. Li, K.S., Pan, W.F., Zhang, W.S., Chen, Z.X.: Automatic Modeling of a Novel Gene Expression Analysis and Critical Velocity. In: 2008 IEEE Congress on Evolutionary Computation, pp. 641–647. IEEE Press, New York (2008)
6. Ferreira, C.: Gene Expression Programming, 2nd edn. Springer, Berlin (2006)
7. Gan, Z.H., Yang, Z.K., Li, G.B., Jiang, M.: Automatic Modeling of Complex Functions with Clonal Selection-based Gene Expression Programming. In: Third International Conference on Natural Computation, vol. 4, pp. 228–232. IEEE Press, New York (2007)
8. Chen, L.: An Adaptive Genetic Algorithm Based on Population Diversity Strategy. In: Third International Conference on Genetic and Evolutionary Computing, pp. 93–96. IEEE Press, New York (2009)

Evolutionary Optimization of Catalysts Assisted by Neural-Network Learning

Martin Holeňa[1], David Linke[2], and Uwe Rodemerck[2]

[1] ICS AS, Pod vodárenskou věží 2 | FIT CTU, Kolejní 2
Prague, Czech Republic
martin@cs.cas.cz
[2] Leibniz Institute for Catalysis, Albert-Einstein-Str. 29a, 18059 Rostock, Germany

Abstract. This paper presents an important real-world application of both evolutionary computation and learning, an application to the search for optimal catalytic materials. In this area, evolutionary and especially genetic algorithms are encountered most frequently. However, their application is far from any standard methodology, due to problems with mixed optimization and constraints. The paper describes how these difficulties are dealt with in the evolutionary optimization system GENACAT, recently developed for searching optimal catalysts. It also recalls that the costly evaluation of objective functions in this application area can be tackled through learning suitable regression models of those functions, called surrogate models. Ongoing integration of neural-networks-based surrogate modelling with GENACAT is illustrated on two brief examples.

Keywords: evolutionary optimization, mixed optimization, constrained optimization, neural network learning, surrogate modelling, evolutionary algorithms in catalysis.

1   Introduction

In chemical engineering, much effort is devoted to increasing the performance of industrially important reactions, i.e., to achieving a higher yield of the desired reaction products without higher material or energy costs. Over 90% of chemical processes use a catalyst to this end. Catalysts are materials that decrease the energy needed to activate a chemical reaction without being themselves consumed in it. They typically consist of several components with different purpose, which can be selected from among many substances. Chemical properties of those substances constrain the possible ratios of their proportions, but they still allow for an infinite number of catalyst compositions. Moreover, the catalyst can usually be prepared from the individual components in a number of ways, and the preparation method also influences its performance in the chemical process. Consequently, the search for new catalytic materials leading to optimal performance of a chemical reaction entails high-dimensional constrained optimization tasks. Their *objective functions* cannot be analytically described, their values must be *obtained empirically*. Commonly used smooth optimization methods are not convenient to this end. Indeed, to obtain sufficiently precise numerical

K. Deb et al. (Eds.): SEAL 2010, LNCS 6457, pp. 220–229, 2010.
© Springer-Verlag Berlin Heidelberg 2010

estimates of gradients or second order derivatives of the empirical objective function, those methods need to evaluate the function in points some of which would have a smaller distance than is the measurement error. That is why *methods not requiring derivatives* have been employed to solve those optimization tasks - both deterministic ones, in particular the *simplex method* and *holographic strategy*, and stochastic ones, such as *simulated annealing*, or *evolutionary algorithms* [1]. Evolutionary, especially *genetic algorithms* (GA) are encountered most frequently, but their application to this area is far from any standard methodology. Main obstacles on a way to such a methodology are *mixed optimization* with respect to continuous and discrete variables, and *constraints*.

This paper describes how those two obstacles are tackled in the evolutionary optimization system GENACAT, developed in recent years at the Leibniz Institute for Catalysis in Rostock, in collaboration with the Institute for Computer Science in Prague. The overall functionality of the system has been outlined, from the point of view of the application domain, in the Journal of Chemical Information and Modeling [2]. The present paper, on the other hand, explains the principles of the underlying evolutionary approach to mixed constrained optimization, which have not been published yet.

The evaluation of the empirical objective functions encountered in the optimization of catalytic materials is *costly and time-consuming*. In particular, testing a generation of materials proposed by an evolutionary algorithm typically needs several to many days of time and costs thousands of euros. Therefore, evolutionary optimization usually proceeds only for 5–10 generations in this application area. A common approach to the optimization of such objective functions is to evaluate the original objective function only sometimes, and to replace it otherwise with a suitable regression model learned from the available data and called *surrogate model* of the objective function [3,4,5,6]. Several successful applications of this approach have been already reported also in catalysis [7,8,9], inciting us recently to incorporate surrogate models based on two kinds of artificial neural networks into GENACAT. An explanation of the integration of surrogate modelling with this evolutionary system would exceed the extent of the paper. Therefore, it has been presented in a separate more comprehensible companion paper [10]. Here, we mainly document the usefulness of surrogate modelling in the evolutionary optimization of catalytic materials on examples, not included in [10].

In the next section, the optimization task entailed by the search for new catalytic materials is formalized, and the solution adopted in the GENACAT system is explained and illustrated on an example. Section 3, on the other hand, shows two brief examples of neural-network based surrogate modelling.

2   Constrained Mixed Optimization in the Search for New Catalytic Materials

In the search for new catalysts leading to the optimal performance of a chemical reaction, the individual coordinates of points in the input space of the objective function typically convey some of the following meanings:

(i) *Qualitative composition* of the catalytic material that is, of which components it consists, and what its support is.
(ii) *Quantitative composition* of the catalytic material, that is, the fractions of the various components mentioned in (i).
(iii) *Preparation* of the catalytic material, its individual steps and their quantitative characterizations, such as temperatures or durations.
(iv) *Reaction conditions* of the catalyzed reaction.

There is an intimate connection between qualitative and quantitative composition. The presence of a particular component in the material is equivalent to the fraction of that component being non-zero. In evolutionary optimization, the employed algorithm has to guarantee that this equivalence cannot become invalidated through its operations, e.g., through the crossover and mutation in GA.

Taking into account the fact that some of those coordinates are continuous and other discrete, the considered optimization task entailed by the search for new catalysts can be formulated as:

$$\text{maximize } f(x, d) \text{ subject to } c_1, \ldots, c_{n_c}, \tag{1}$$

where

- f is some empirical performance measure of the catalytic material, most frequently yield of (some of) the reaction product(s);
- $x = (x_1, \ldots, x_{n_x})$ is a vector of values of continuous variables $X_1, \ldots, X_{n_x}$, the range $\mathrm{Val}(X_i)$ of X_i is a union of intervals of non-zero lengths (typically, a single such interval);
- $d = (d_1, \ldots, d_{n_d})$ is a vector of values of discrete variables $D_1, \ldots, D_{n_d}$, thus the range $\mathrm{Val}(D_j)$ of D_j is countable (typically, finite) and not necessary a subset of real numbers;
- $c_1, \ldots, c_{n_c}$ can be equality or inequality constraints, which can include functions of any of the variables $x_1, \ldots, x_{n_x}$ and $d_1, \ldots, d_{n_d}$, as well as constraints determining the distributions of $X_1, \ldots, X_{n_x}, D_1, \ldots, D_{n_d}$.

As an example, Fig. 1 shows one of the tasks faced by the system GENACAT during the first year of its use.

2.1   Solution Using Evolutionary Optimization

The popularity of evolutionary optimization methods in the search for catalytic materials is mainly due to the fact that they tend to find global rather than local solutions, and due to the possibility to establish a straightforward correspondence between the optimization paths followed by the evolutionary algorithm and channels of the high-throughput reactor in which the materials are experimentally tested.

Due to the specific meaning conveyed by coordinates of points in the input space of the objective function, it is quite difficult to use general evolutionary optimization software, which optimizes functions with input spaces of low-level data types, such as vectors of real numbers or bit-strings. Therefore, it is not

Fitness: Y (product yield)

Continuous inputs:

X_i: proportion of the i-th component from the components pool available for the catalytic material, $i = 1,\ldots,37$, $\mathrm{Val}(X_i) = [0,0.1]$ for $i = 1,\ldots,22$, $\mathrm{Val}(X_i) = [0,0.003]$ for $i = 23,\ldots,37$;

X_{38}: overall proportion of components belonging to precious metals, $\mathrm{Val}(X_{38}) = [0,0.003]$;

X_{39}: overall proportion of components belonging to alkaline earth metals or lanthanoids, $\mathrm{Val}(X_{39}) = [0,0.05]$;

X_{40}: proportion of the lower valence element in a fixed pair of alkaline earth metals or lanthanoids, $\mathrm{Val}(X_{40}) = [0,0.01]$;

X_{41}: proportion of the higher valence element in a fixed pair of alkaline earth metals or lanthanoids, $\mathrm{Val}(X_{41}) = [0,0.05]$;

X_{42}: overall proportion of components not belonging to precious or to alkaline earth metals or lanthanoids, $\mathrm{Val}(X_{42}) = [0.003,0.05]$;

Discrete inputs:

D_1: choice of a material serving as support of the catalyst, $\mathrm{Val}(D_1) = \{material1, material2\}$;

D_2: proportion of support; $\mathrm{Val}(D_2) = \{0.95,0.99\}$;

D_3: choice of a fixed pair of alkaline earth metals or lanthanoids, $\mathrm{Val}(D_3) = \{(i\text{-th component}, i'\text{-th component}): (29 \le i \le 31 \ \& \ 32 \le i' \le 33) \vee (32 \le i \le 33 \ \& \ 34 \le i' \le 37)\}$;

D_4: number of included components belonging to precious metals, $\mathrm{Val}(D_4) = \{0,1\}$;

D_5: number of included components belonging to alkaline earth metals or lanthanoids, $\mathrm{Val}(D_5) = \{0,1,2\}$;

D_6: number of included fixed pairs of alkaline earth metals or lanthanoids, $\mathrm{Val}(D_6) = \{0,1\}$;

D_7: number of included components belonging neither to precious metals, nor to alkaline earth metals or lanthanoids, $\mathrm{Val}(D_7) = \{1,2,3,4\}$;

D_8: overall number of all included components, $\mathrm{Val}(D_8) = \{2,3,4\}$;

Constraints:

c_i: probability distribution of X_i on $[0.003,0.1]$ is uniform, $i = 1,\ldots,22$;

c_i: probability distribution of X_i on $(0,0.003]$ is uniform, $i = 23,\ldots,37$;

c_{38}: joint probability distribution of (X_{40},X_{41}) on $\{(x,x') : 0 \le x \le 0.01 \ \& \ 0 \le x' \le 0.05 \ \& \ 20x \le x' \le 50x\}$ is uniform;

c_i: $P(0 < X_{i-38} < 0.003) = 0$, $i = 39,\ldots,60$;

c_i: $P(X_1>0.03) = 3*P(X_{i-59}>0.03)$, $i = 61,\ldots,81$;

c_j: probability distribution of D_{j-60} on $\mathrm{Val}(D_{j-60})$ is uniform, $j = 82,\ldots,84$;

c_{85}: probability distribution of D_4 on $\{0,1\}$ is $(0.8,0.2)$;

c_{86}: joint probability distribution of (D_5,D_6) on $\mathrm{Val}(D_5) \times \mathrm{Val}(D_6) = \{(0,0),(0,1),(1,0),(1,1),(2,0),(2,1)\}$ is $(\frac{1}{3},\frac{1}{3},\frac{2}{9},0,\frac{1}{9},0)$;

c_{87}: probability distribution of D_8 on $\{2,3,4\}$ is $(0.45,0.45,0.1)$;

c_{88}: $|\{i : 23 \le i \le 28 \ \& \ x_i > 0\}| = d_4$;

c_{89}: $|\{i : 29 \le i \le 37 \ \& \ x_i > 0\}| = d_5$;

c_{90}: $|\{i : i = 40 \ \& \ x_i > 0\}| = d_6$;

c_{91}: $|\{i : 1 \le i \le 22 \ \& \ x_i > 0\}| = d_7$;

c_{92}: $d_2 + x_{38} + x_{39} + x_{42} = 1$;

c_{93}: $x_1 + x_2 + \cdots + x_{22} = x_{42}$;

c_{94}: $x_{23} + x_{24} + \cdots + x_{28} = x_{38}$;

c_{95}: $x_{29} + x_{30} + \cdots + x_{37} + x_{40} + x_{41} = x_{39}$;

c_{96}: $20x_{40} \le x_{41} \le 50x_{40}$;

c_{97}: $d_4 + d_5 + d_6 + d_7 = d_8$.

Fig. 1. Example optimization task encountered in the search for new catalysts

surprising that – apart from early attempts to use general evolutionary software and of a recent application of genetic programming based on general context-based crossover [11] – the application of evolutionary algorithms in this area took the route of developing them specifically for the optimization of catalytic materials [12,13,14]. However, none of those specific algorithms attempted to tackle both the obstacles mentioned in the Introduction – mixed optimization and constraints, and the experience gathered with them so far shows that they bring a difficulty of another kind: they are usable only for a narrow spectrum of particular optimization tasks and have to be reimplemented each time when different tasks emerge.

In GENACAT, that difficulty has been tackled through *automatically generating*, at run time, a specific GA precisely *tailored to the optimization task being solved*. The algorithm is generated by a program generator, based on a user specification of the task in a *catalyst description language* (CDL). The implementation of the program generator, the CDL-language, as well as the creating and processing of CDL-descriptions have been presented in [2]. Here on the other hand, we explain the method used in GENACAT to solve the mixed constrained optimization task (1). It is based on two specific features of that task pertaining to the search for new catalysts:

(i) It is sufficient to consider only linear constraints. Even if the set of feasible solutions is not constrained linearly in reality, the finite measurement precision of the involved continuous variables always allows to constrain it piecewise linearly and to indicate the relevant linear piece with an additional discrete variable. Consequently, the set of values of the continuous variables that are feasible for a particular combination of values of the discrete variables is a polyhedron, determined by some matrix A_P and vector b_P

$$P = \{x : A_P x \leq b_P\}. \tag{2}$$

(ii) If a solution polyhedron is described with (2), then its feasibility (i.e., non-emptiness) is invariant with respect to any permutation of columns of A_P, and to any permutation of rows of $(A_P b_P)$. Moreover, the relation $\approx$ defined

$$P \approx Q \text{ iff } (A_Q b_Q) \text{ can be obtained from } (A_P b_P) \text{ through}$$
$$\text{some permutation of columns of } A_P, \text{ followed by} \tag{3}$$
$$\text{some permutation of rows of the result and of } b_P$$

is an equivalence, partitioning the set of polyhedra into disjoint classes.

The property (ii) plays a crucial role in GENACAT because only one representative from each class needs to be checked for non-emptiness. For the example optimization task introduced in Fig. 1, the difference between the number of solution polyhedra and the number of their classes is shown in Fig. 2

On the set of nonempty polyhedra, discrete genetic optimization is performed, using operations selection, mutation and crossover developed specifically to this end. Each of the polyhedra forming the population obtained in this way contains a subpopulation of combinations of values of continuous variables found through continuous genetic optimization. The union of all such subpopulations combined with the combinations of values of discrete variables corresponding to the respective polyhedra, form together the final population of solutions to the optimization task. The specific genetic operations employed in the discrete optimization are defined as follows:

Selection is in the first generation uniform, in subsequent generations proportionally to the importance of the polyhedron due to points from earlier generations that it contains. As a measure of that importance, the difference between the fitness (=value of the objective function) of a point and the minimal fitness encountered in previous generations is taken, summed over points with the combination of values of the discrete variables corresponding to the polyhedron.

Mutation consists in replacing an existing polyhedron with a uniformly selected nonempty one. The values of continuous variables forming a point in that polyhedron are again obtained through continuous genetic optimization. If the mutation rate is μ, then a proportion μ of the population is selected in this way, and the proportion $1 - \mu$ is selected using the above proportional selection.

Crossover relies on the fact that a solution polyhedron P is determined on the one hand by the assignment of continuous variables to the columns of A_P, on the other hand by a particular combination of values of some or all discrete variables.

	All	Nonempty
Solution polyhedra	583 232 160	282 810
Equivalence classes	480	60

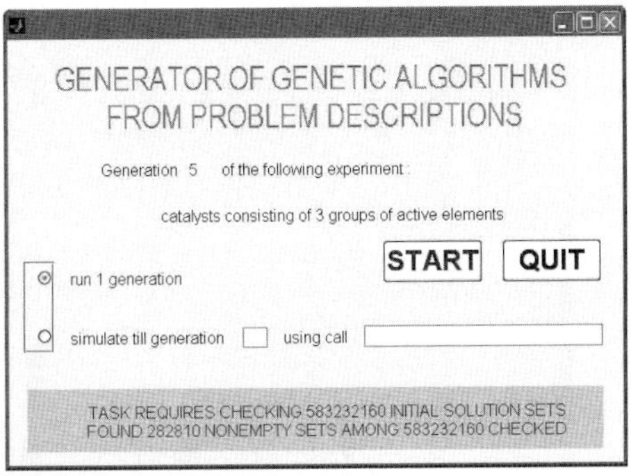

Fig. 2. Comparison of the number of all and nonempty polyhedra with the number of their equivalence classes for the example optimization task in Fig. 1 (left). End-users are not confronted with the concept of equivalence classes, that is why the graphical interface controlling the evolutionary optimization by GENACAT reports only the numbers of solution polyhedra (right).

We assume that for the optimization of catalytic materials, the assignment of variables to the columns of A_P is more important, and we suggest a crossover operation that always exchanges exactly one of the continuous variables assigned to the parent polyhedra, and attempts to include as many discrete variables corresponding to them as possible. Formally, denote m_p the number of columns of A_P, and let $\{i_{P,1}, \ldots, i_{P,m_P}\} \subset \{1, \ldots, n_x\}$ and $V_j^P \subset \mathrm{Val}(D_j)$ for $j = 1, \ldots, n_d$ be such that every solution (x, d) of (1) fulfills

$$A_P(x_{i_{P,1}}, \ldots, x_{i_{P,m_P}}) \le b_P \ \& \ d_j \in V_j \text{ for } j = 1, \ldots, n_d. \tag{4}$$

If the crossover rate is λ, then for each pair P and P' of solution polyhedra selected using the proportional selection, a set of recombination offsprings is formed with probability λ, in the following way:

(i) The set of candidate offsprings of P and P' is defined by

$$\mathcal{C}(P, P') = \{Q - \text{solution polyhedron} : Q \ne \emptyset \ \&$$
$$\& \ \{i_{Q,1}, \ldots, i_{Q,m_Q}\} \subset \{i_{P,1}, \ldots, i_{P,m_P}\} \cup \{i_{P',1}, \ldots, i_{P',m_{P'}}\} \ \&$$
$$\& \ [(m_Q = m_P \ \& \ |\{i_{Q,1}, \ldots, i_{Q,m_Q}\} \cap \{i_{P,1}, \ldots, i_{P,m_P}\}| = m_Q - 1)] \vee$$
$$\vee (m_Q = m_{P'} \ \& \ |\{i_{Q,1}, \ldots, i_{Q,m_Q}\} \cap \{i_{P',1}, \ldots, i_{P',m_{P'}}\}| = m_Q - 1)]\}.$$
$$\tag{5}$$

(ii) For each $Q \in \mathcal{C}(P, P')$, the uncertainty index of Q is computed as

$$u(Q) = |\{j : 1 \le j \le n_d \ \& \ V_j^Q \not\subset V_j^P \cup V_j^{P'}\}|. \tag{6}$$

(iii) The final set of offsprings of P and P' is defined by

$$\mathcal{O}(P, P') = \{Q \in \mathcal{C}(P, P') : u(Q) = \min_{Q' \in \mathcal{C}(P,P')} u(Q')\}. \tag{7}$$

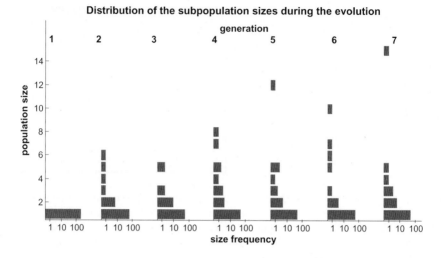

Fig. 3. The distribution of sizes of the individual polyhedra during the 7 generations of running the GA for the optimization task in Fig. 1

Due to the proportional selection, the subpopulations of combinations of continuous variables in polyhedra with higher importance tend to increase, whereas the subpopulations in polyhedra with lower importance tend to decrease or to disappear. This is illustrated in Fig.3 for the example task introduced in Fig.1. For that task, the GA was run with a population size 96, given by the number of available channels in the reactor in which the catalysts were tested. The evolution was finished after 7 generations, when the number of found materials with sufficiently high fitness (yield) was already satisfactory for the users, in view of the cost of the evaluation of another generation. Figure 3 shows the development of the distribution of subpopulation sizes.

3   Learning Neural-Networks-Based Surrogate Models

This section has been included due to its particularly high relevance to simulated evolution and learning. However, space limitations cause it to lack detail, for which the reader is referred to a more comprehensive companion paper [10], devoted specifically to surrogate modelling in evolutionary optimization of empirical objective functions.

Surrogate modelling is a general approach to the optimization of objective functions with costly or time-consuming evaluations, encountered both in traditional optimization, mainly in connection with the efficient global optimization (EGO) approach [15], and in evolutionary optimization [5,6,16]. It consists in restricting the evaluation to points considered to be most important for the progress of the employed optimization method and approximating it otherwise, using a suitable model, learned from available data. For the progress of evolutionary optimization, most important are considered points with highest values of the

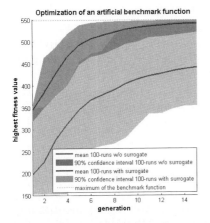

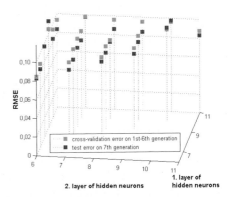

Fig. 4. Comparison of the highest values of the benchmark fitness function from [9] found by the GA from Section 2 without surrogate modelling and with an RBF-based surrogate model

Fig. 5. Comparison of the cross-validation and test RMSE of surrogate models with 21 different 2-hidden-layers MLP architectures learned from data introduced in [19]

fitness and points that most contribute to the diversity of the population. Since empirical objective functions are typically nonlinear, nonlinear regression models have been the primary choice for surrogate modelling, in particular polynomials [17], Gaussian processes [3,15], and artificial neural networks of the kinds multi-layer perceptron (MLP) or radial basis function (RBF) network [5,6,18]. Various possibilities how to combine surrogate modelling with evolutionary optimization discussed in the literature (most important among which are the individual-based strategy and the generation-based strategy) have been explained in [10].

The fact that surrogate modelling is employed in the context of costly or time-consuming objective functions effectively excludes the possibility to use those functions for tuning surrogate modelling methods, and for comparing different models and different ways of combining them with evolutionary optimization. To get round this difficulty, artificial benchmark functions can be used, computed analytically (thus in negligible time) but expected to have similar properties like the empirical objective function from the point of view of evolutionary optimization. For catalysis, two such benchmark functions were proposed in [9] and [12]. An alternative approach is to use available data from past evaluations of empirical objective functions. This allows to estimate the accuracy of predictions obtained with a surrogate model but it does not allow to test or to compare particular strategies how to combine surrogate modelling with evolutionary optimization because the values of the objective function cannot be obtained in points suggested by the strategy.

As an example of the former approach, Fig. 4 compares the highest values of the benchmark fitness function proposed in [9] found in a population of the same size by the GA outlined in Section 2 without and with surrogate modelling. A RBF-network learned with data from all previous generations is used as surrogate model, combined with the GA according to the individual-based-strategy algorithm described in [10]. An example of the latter approach is given in Fig. 5, using data from optimization of catalysts for the high-throughput synthesis of HCN [19]. It compares cross-validation RMSE of 21 architectures of MLPs with two hidden layers, learned from data obtained in the 1.-6. generation of the GA, with test RMSE of MLPS with those architectures on data from the 7. generation. Both examples clearly document the usefulness of surrogate modelling.

4   Conclusion

The paper presented several aspects of a real-world application of both evolutionary computation and learning, an application to the search for optimal catalytic materials. It explained the principles of an evolutionary algorithm underlying the recently developed optimization system GENACAT for this application area. They have not been addressed in the original, application-oriented presentation of the system in [2]. Due to the orientation of the SEAL conference to learning, it also recalled an ongoing incorporation of neural-networks-based surrogate models into GENACAT. However, it only complemented with two brief new examples the presentation of this topic in a more comprehensible companion paper [10]. Moreover, those examples illustrate only the general usefulness of the integration of GENACAT with surrogate modelling, not our own modest contribution to this area, which consists in enhancing neural-networks-based surrogate models with regression boosting [18].

Acknowledgment

The research reported in this paper was supported by the German Federal Ministry of Education and Research (BMBF), as well as by the Czech Science Foundation (GAČR) grants 201/08/0802 and ICC/08/E018.

References

1. Baerns, M., Holeňa, M.: Combinatorial Development of Solid Catalytic Materials. Design of High-Throughput Experiments, Data Analysis, Data Mining. World Scientific, Singapore (2009)
2. Holeňa, M., Cukic, T., Rodemerck, U., Linke, D.: Optimization of catalysts using specific, description based genetic algorithms. Journal of Chemical Information and Modeling 48, 274–282 (2008)
3. Büche, D., Schraudolph, N., Koumoutsakos, P.: Accelerating evolutionary algorithms with gaussian process fitness function models. IEEE Transactions on Systems, Man, and Cybernetics, Part C: Applications and Reviews 35, 183–194 (2005)
4. Jin, Y.: A comprehensive survery of fitness approximation in evolutionary computation. Soft Computing 9, 3–12 (2005)
5. Ulmer, H., Streichert, F., Zell, A.: Model assisted evolution strategies. In: Jin, Y. (ed.) Knowledge Incorporation in Evolutionary Computation, pp. 333–355. Springer, Heidelberg (2005)

6. Zhou, Z., Ong, Y., Nair, P., Keane, A., Lum, K.: Combining global and local surrogate models to accellerate evolutionary optimization. IEEE Transactions on Systems, Man and Cybernetics. Part C: Applications and Reviews 37, 66–76 (2007)
7. Baumes, L., Farrusseng, D., Lengliz, M., Mirodatos, C.: Using artificial neural networks to boost high-throughput discovery in heterogeneous catalysis. QSAR and Combinatorial Science 23, 767–778 (2004)
8. Farrusseng, D., Clerc, F., Mirodatos, C., Azam, N., Gilardoni, F., Thybaut, J., Balasubramaniam, P., Marin, G.: Development of an integrated informatics toolbox: HT kinetic and virtual screening. Combinatorial Chemistry and High Throughput Screening 10, 85–97 (2007)
9. Valero, S., Argente, E., Botti, V., Serra, J., Serna, P., Moliner, M., Corma, A.: DoE framework for catalyst development based on soft computing techniques. Computers and Chemical Engineering 33, 225–238 (2009)
10. Holeňa, M., Linke, D., Rodemerck, U., Bajer, L.: Neural networks as surrogate models for measurements in optimization algorithms. In: Al-Begain, K., Fiems, D., Knottenbelt, W. (eds.) Analytical and Stochastic Modeling Techniques and Applications. LNCS, vol. 6148, pp. 351–366. Springer, Heidelberg (2010)
11. Baumes, L., Blanché, A., Serna, P., Tchougang, A., Lachiche, N., Collet, P., Corma, A.: Using genetic programming for advanced performance assessment of industrially relevant heterogeneous catalysts. Materials and Manufacturing Processes 24, 282–292 (2009)
12. Wolf, D., Buyevskaya, O., Baerns, M.: An evolutionary approach in the combinatorial selection and optimization of catalytic materials. Applied Catalyst A: General 200, 63–77 (2000)
13. Ohrenberg, A., Törne, C., Schuppert, A., Knab, B.: Application of data mining and evolutionary optimization in catalyst discovery and high-throughput experimentation – techniques, strategies, and software. QSAR and Combinatorial Science 24, 29–37 (2005)
14. Pereira, R., Clerc, F., Farrusseng, D., Waal, J., Maschmeyer, T.: Effect of genetic algorithm parameters on the optimization of heterogeneous catalysts. QSAR and Combinatorial Science 24, 45–57 (2005)
15. Leary, S., Bhaskar, A., Keane, A.: A derivative based surrogate model for approximating and optimizing the output of an expensive computer simulation. Journal of Global Optimization 30, 39–58 (2004)
16. Ong, Y., Nair, P., Keane, A., Wong, K.: Surrogate-assisted evolutionary optimization frameworks for high-fidelity engineering design problems. In: Jin, Y. (ed.) Knowledge Incorporation in Evolutionary Computation, pp. 307–331. Springer, Berlin (2005)
17. Hosder, S., Watson, L., Grossman, B.: Polynomial response surface approximations for the multidisciplinary design optimization of a high speed civil transport. Optimization and Engineering 2, 431–452 (2001)
18. Holeňa, M., Linke, D., Steinfeldt, N.: Boosted neural networks in evolutionary computation. In: Chan, J.H. (ed.) ICONIP 2009, Part II. LNCS, vol. 5864, pp. 131–140. Springer, Heidelberg (2009)
19. Möhmel, S., Steinfeldt, N., Endgelschalt, S., Holeňa, M., Kolf, S., Dingerdissen, U., Wolf, D., Weber, R., Bewersdorf, M.: New catalytic materials for the high-temperature synthesis of hydrocyanic acid from methane and ammonia by high-throughput approach. Applied Catalysis A: General 334, 73–83 (2008)

Dominance-Based Pareto-Surrogate for Multi-Objective Optimization

Ilya Loshchilov[1,2], Marc Schoenauer[1,2], and Michèle Sebag[2,1]

[1] TAO Project-team, INRIA Saclay - Île-de-France*
[2] Laboratoire de Recherche en Informatique (UMR CNRS 8623)
Université Paris-Sud, 91128 Orsay Cedex, France
FirstName.LastName@inria.fr

Abstract. Mainstream surrogate approaches for multi-objective problems build one approximation for each objective. Mono-surrogate approaches instead aim at characterizing the Pareto front with a single model. Such an approach has been recently introduced using a mixture of regression Support Vector Machine (SVM) to clamp the current Pareto front to a single value, and one-class SVM to ensure that all dominated points will be mapped on one side of this value. A new mono-surrogate EMO approach is introduced here, relaxing the previous approach and modelling Pareto dominance within the rank-SVM framework. The resulting surrogate model is then used as a filter for offspring generation in standard Evolutionary Multi-Objective Algorithms, and is comparatively validated on a set of benchmark problems.

1 Introduction

This paper is concerned with evolutionary Multi-Objective Optimization (EMO) [2], and most specifically focuses on designing and using surrogate models in order to speed up the evolutionary search. Surrogate models, namely computationally light estimates of the objective function, have been extensively used in Evolutionary Algorithms (EAs) since the 1990's [6], as the Achilles's heel of EAs is known to be the high number of times the objective function has to be computed. This high number forbids using mainstream EAs in some application domains, e.g. Optimal Design and Numerical Engineering where the objective functions are computationally demanding. Surrogate-based EAs alleviate this limitation by iteratively estimating the objective function, fueling the EA with the estimate (aka surrogate model), acquiring new examples of the objective function and revising the surrogate model accordingly; the reader will find a comprehensive review of surrogate evolutionary optimization in [6].

Surrogate models are equally useful in evolutionary Multi-Objective Optimization, all the more so when EMO involves several computationally heavy objectives [8]. Current surrogate-based EMO approaches, with the notable exception of [13] and [10], basically extend surrogate-based standard EAs, building

---

* Work partially funded by FUI of System@tic Paris-Region ICT cluster through contract DGT 117 407 *Complex Systems Design Lab* (CSDL).

K. Deb et al. (Eds.): SEAL 2010, LNCS 6457, pp. 230–239, 2010.

one surrogate for each objective function and replacing the objective by its surrogate. The main limitation of such approaches is due to the approximation noise as the number of objectives increases. The estimation cost indeed increases linearly with the number of objectives; but the Pareto dominance test, checking whether one individual is dominated by another one, requires comparing their surrogate values over all objectives; the error thus exponentially increases in the worst case with the number of objectives.

The first mono-surrogate EMO approach was proposed by [13], aimed at characterizing the already visited region of the objective space, although this characterization hardly enables to guide evolution in the decision space. Addressing this limitation, another mono-surrogate approach defined in the decision space aimed at characterizing the current Pareto front and dominated region in the decision space in order to guide further evolution [10] (more in section 2). While this approach, referred to as Aggregated Surrogate Model (ASM), yields significant savings in terms of computational cost on benchmark problems, it relies on a complex adaptation of the Support Vector Machine framework [12], involving regression- and one-class-like constraints. The ASM limitations, related to the diversity of the Pareto front or the care to be exerted when using the surrogate model to filter the offspring, are blamed on the over-constrained formulation of the mono-surrogate model.

A new and relaxed version of Aggregated Surrogate Model for EMO is proposed in this paper, inspired from rank-based SVM [4,7]. Basically, the new surrogate model referred to as RASM (Rank-based ASM) is only required to locally approximate the Pareto dominance relation, enabling to rank neighbor points within the objective space. RASM is still used to filter the offspring, through estimating whether they improve on their parents in terms of approximated Pareto-dominance.

This paper is organized as follows. Section 2 briefly discusses Aggregated Surrogate Models in EMO, detailing the approach proposed in [10] for the sake of self-containedness. Section 3 presents and discusses the RASM approach, which is experimentally validated in Section 4. Section 5 concludes and presents some perspectives for further research.

2   Aggregated Surrogate Models

Without pretending to exhaustivity, and referring the reader to [8] for a comprehensive review of surrogate-based EMO approaches, this section focuses on mono-surrogate EMO algorithms. As mentioned in the introduction, the first ever mono-surrogate EMO algorithm proposed by Yu et al. aimed at characterizing the region of the objective space visited so far [13]. The rationale for this approach, based on One-Class SVM [11], is that the envelope of the visited region excludes the Pareto front. In the general case however, the Pareto front in the objective space does not tell much about the Pareto set in decision space (except for specific problems where the Pareto front in the objective space corresponds to a set of rectangles in the decision space) and thus hardly allows one to guide the EMO search.

The only mono-surrogate EMO approach in the decision space, to our best knowledge, was referred to as Aggregate Surrogate Model [10]; it combines several variants of the Support Vector Machine framework.

In their initial formulation [12], Support Vector Machines aim at a linear model on the instance space $X \subset \mathbb{R}^d$, solution of a quadratic optimization problem:

$$\text{Argmin}_{\{w,\xi\}} \; (\frac{1}{2}||w||^2 + C \sum_k \xi_k)$$

where the norm of the sought linear solution $w \in \mathbb{R}^d$ is minimized for the sake of good generalization guarantees, and each ξ_k stands for the violation of one learning constraint, to be minimized[1]

Typically, mapping a point $x_i \in \mathbb{R}^d$ onto a desired value y_i up to some tolerance threshold ϵ (regression problem) amounts to four constraints:

$$< w, x_i > -y_i \le \epsilon + \xi_i^{up} \qquad \qquad \xi_i^{up} \ge 0$$
$$< w, x_i > -y_i \ge -\epsilon - \xi_i^{low} \qquad \qquad \xi_i^{low} \ge 0$$

Likewise, mapping a point x_i onto a half space $[a, \infty)$ (one-class problem) involves two constraints:

$$< w, x_i > \ge a - \xi_i \qquad \xi_i \ge 0$$

The ASM approach presented in [10] hybridizes the above two types of constraints as follows. Let the training set be defined as $\mathcal{E} = \{x_1, \ldots x_\ell, x_{\ell+1}, \ldots x_m\}$ where the first ℓ points belong to the current Pareto front and the following points $x_{\ell+1}, \ldots x_m$ are dominated ones (possibly sub-sampling the current population to preserve the diversity in the objective or decision space). The ASM is obtained from the following learning constraints:

- All Pareto points $x_1 \ldots x_\ell$ are mapped on some value ρ up to tolerance ϵ (regression constraints);
- All dominated points are mapped onto $(-\infty, \rho + \epsilon[$ (one-class constraints).

The intuition behind this formulation is that the *true* Pareto front would then expectedly lie in the 'half space, $]\rho + \epsilon, +\infty)$, thus enabling to guide the exploration of the search space.

The ASM problem finally reads:

$$\text{Argmin}_{\{w,\xi\}} \; (\frac{1}{2}||w||^2 + C \sum_{i=1}^{\ell} (\xi_i^{up} + \xi_i^{low}) + C \sum_{i=\ell+1}^{m} \xi_i^{up})$$

---

[1] The extension of the SVM approach to non-linear search spaces relies on the so-called kernel trick, implicitly mapping the instance space X onto a feature space [12]. See section 3 and [12] for more details.

subject to

$$
\begin{array}{lll}
<w, x_i> \le \rho + \epsilon + \xi_i^{up} & \xi_i^{up} \ge 0 & i = 1 \ldots \ell \\
<w, x_i> \ge \rho - \epsilon - \xi_i^{low} & \xi_i^{low} \ge 0 & i = 1 \ldots \ell \\
<w, x_i> \le \rho - \epsilon + \xi_i^{up} & \xi_i^{up} \ge 0 & i = \ell + 1 \ldots m
\end{array}
$$

As mentioned in the introduction, the ASM problem is overconstrained as all Pareto points must be mapped on a narrow interval $]\rho - \epsilon, \rho + \epsilon[$. Another issue is that it should make no difference whether the dominated points are mapped onto $(-\infty, \rho - \epsilon[$ or $]\rho + \epsilon, +\infty)$. Still, the experimental validation on problems ZDT1:3-6 [14] and their rotated variants IHR1:3-6 [5] shows that the most effective variant depends on the underlying benchmark problem. Whereas one can proceed by trying both variants and retaining the most effective one, the approach is clearly unsatisfactory. Some attempts at a symmetrical formulation of the ASM problem failed to address this issue.

3   Rank-Based Aggregate Surrogate Model

This section gives an overview of the Rank-based Aggregate Surrogate Model (RASM), meant to address the ASM limitations. After stating the RASM formulation and sketching its resolution, it details its use within the EMO framework.

3.1   A Surrogate Modelling Pareto Dominance

A new learning setting aimed at preference learning, a.k.a. learning to rank, has been addressed within the SVM framework [7]. While preference learning can be cast as a classification problem on $X \times X$ (the class of (x, x') is positive iff x is to be preferred to x'), it offers better generalization guarantees to formalize preference learning as an underconstrained regression problem, where the hypothesis h mapping X onto the real-valued space $\mathbb{R}$ is only required to satisfy $h(x) > h(x')$ whenever x is preferred to x'.

Let $\mathcal{E} = \{x_1, \ldots, x_m\}$ and let $\mathcal{P}$ denote the set of pairs (i, j) such that x_i is preferred to x_j; the original formulation of rank-based SVM, involving all preference constraints, is as follows:

$$
\text{Argmin}_{\{w, \xi\}} \left(\frac{1}{2} ||w||^2 + C \sum_{(i,j) in \mathcal{P}} \xi_{i,j} \right) \tag{1}
$$

$$
\text{subject to} \quad \left. \begin{array}{c} \langle w, x_i \rangle - \langle w, x_j \rangle \ge 1 - \xi_{i,j} \\ \xi_{i,j} \ge 0 \end{array} \right\} \forall (i, j) \in \mathcal{P} \tag{2}
$$

where $\xi_{i,j}$ stands for the slack variable associated to the violation of the preference constraint associated to (x_i, x_j) along the same lines as in Section 2. However, for the sake of tractability, the resolution of Eq. (1-2) proceeds iteratively, considering a set Ω_{active} of active constraints which is initially empty. Eq. (1)-(2) then only imply Ω_{active} instead of $\mathcal{P}$. At each iteration, the most violated constraint in $\mathcal{P}$ is added to Ω_{active}, and optimization proceeds [7].

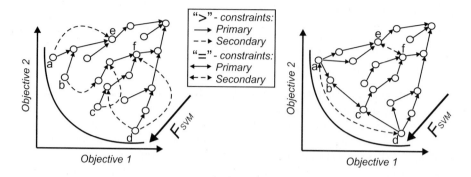

Fig. 1. Constraints involved in Rank-based Aggregated Surrogate Models. Left: The current RASM. Right: Further extensions; see section 5.

Rank-SVM is adapted to the EMO framework as follows. Let $\mathcal{E}$ define the current training set (population or archive), and consider the preference order defined by non-dominated sorting. Several requirements on the rank-based surrogate model are defined. First of all, the number of constraints should be linear or sub-linear in the population size for the sake of tractability. Secondly, the constraints should enforce an accurate model in terms of generality w.r.t. Pareto dominance. Lastly, the model should support the diversity of the population along the Pareto front. To comply with these requirements, only dominance constraints have been considered so far (Fig. 1.left)[2]. Specifically, let **primary dominance constraints** be associated to pairs (x_i, x_j) such that x_j is the nearest neighbor of x_i conditionally to the fact that x_i dominates x_j (continuous arrow, Fig. 1.left), and let **secondary dominance constraints** be associated to pairs (x_i, x_j) such that x_i belongs to the current Pareto front and x_j belongs to another front from non-dominated sorting (dotted arrow, Fig. 1.left).

Formally, the RASM is built by solving the dual problem associated to Eqs.(1-2); due to space limitations the reader is referred to [10] for more details. Let $K_{a,b}$ denote the scalar product $< \Phi(x_a), \Phi(x_b) >$ where Φ denotes the mapping from the initial to the feature space (kernel trick), and let $\alpha_{a,b}$ denote the Lagrange multiplier associated to the constraint relating x_a and x_b. Then it comes:

Dual form:

$$\text{Argmax}_{\{\alpha\}} \sum_{i,j=1}^{m} \alpha_{ij} - \frac{1}{2} \sum_{i,j,u,v=1}^{m} \alpha_{ij} \alpha_{uv} \left(K_{iu} - K_{iv} - K_{ju} + K_{jv} \right)$$
$$\text{subject to } 0 \leq \alpha_{ij} \leq C \tag{3}$$

The Lagrangian (Eq. 3) is maximized iteratively by optimizing a single α_{ij} multiplier (uniformly selected in Ω_{active}) in each iteration. Taking inspiration from [7], RASM maintains a set Ω_{active} of active constraints which is initialized to the set of primary dominance constraints. After the first $1000 \times |\Omega_{active}|$

---

[2] Other possibilities, illustrated in Fig. 1.right, will be discussed in section 5.

iterations, every $10 \times |\Omega_{active}|$ iterations the most violated secondary constraint is thereafter added to Ω_{active}, until a given number of secondary constraints have been added (typically 10% the number of primary constraints in the presented experiments, see Section 4).

3.2 RASM-Based EMO

The RASM is integrated within existing EMO algorithms along the same lines as the ASM [10]. Two different EMO algorithms have been considered, respectively NSGA-II and MO-CMA-ES. At each generation, the training set is built from the current population and an archive of points that have been visited by the algorithm (more in Section 4.1). The standard variation operators of the underlying EMO are used to generate many offspring; these offspring are thereafter filtered according to the RASM as follows. Formally, the quality of an offspring z is estimated as $RASM(z) - RASM(x_z)$ where x_z is the individual from population nearest to z in decision space.

As could have been expected, greedily selecting the offspring with maximal quality leads to premature convergence due to the approximation noise. The offspring are therefore ordered according to their quality estimate; a probabilistic selection attached to the offspring indices is achieved, using a normal distribution with variance σ^2_{sel} (Fig. 2.right). Granted that the number of offspring is sufficiently large, parameter σ^2_{sel} thus controls the selection pressure and the exploration vs exploitation trade-off. However we use at the moment a small value $\sigma^2_{sel} = 0.001$ for the normal distribution for the ranked point to be chosen (Fig. 2.right) to simulate the situation when new offspring always has the first rank, and allow comparison with results from [10].

4 Experimental Validation

4.1 Experimental Setting

For the sake of a fair comparison between the presented RASM and the ASM first presented in [10], the same experimental validation procedure is used: two state-of-the-art EMO algorithms are considered as baselines, the (100+100)-S-NSGA-II [3], NSGA-II with hypervolume indicator as second-sorting criterion, and the $100 \times (1+1)$-MO-CMA-ES, the multi-objective version of CMA-ES [5]. Both ASM and RASM are integrated within these algorithms.

Both ASM and RASM are based on the Radial Basis Functions kernel: $K(x_i, x_j) = e^{-\|x_i - x_j\|^2 / 2\sigma^2}$, where the bandwidth σ is set as the average distance between all pairs of training points. The SVM penalization constant C is set to 1000.

The training set $\mathcal{E}$ that is used at each generation to build the RASM model is an archive that contains at most $N_{archive} = 1000$ points. The current population is added to the archive at each generation. When this archive gets larger than $N_{archive}$, it is pruned by removing the worst individuals after non-dominated sorting. Furthermore, in order to improve the diversity of the training set (many

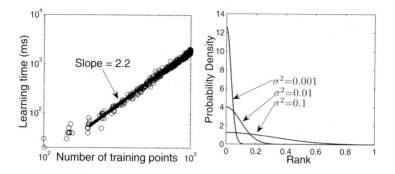

Fig. 2. Left: Learning time of the proposed dominance-based RASM on ZDT1 function. Right: Mapping the ranks of pre-children to a normal distribution.

points too close together can lead to poor surrogate model), an additional filtering procedure is applied to the archive. The 2-objective space has been divided into 100×100 boxes, and at most one point among the archived non-dominated points of each box is retained in the archive.

As detailed in section 3, RASM maintains the set Ω_{active} of active constraints, initialized to the set of primary dominance constraints. After an initial round of $1000\,|\Omega_{active}|$ iterations, Ω_{active} is incrementally enriched every $10\,|\Omega_{active}|$ iterations with the most violated constraint among the secondary dominance constraints, until a total of $0.1\,|\Omega_{active}|$ secondary constraints have been added.

Performance Measures. Many ways of measuring the performance of EMO algorithms have been proposed. After [9], this study uses Pareto-compliant quality indicators, more particularly the widely used hypervolume indicator I_H. Let P be a μ-size approximation of Pareto front and let P^* be the approximate μ-optimal distribution of optimal Pareto points [1]. The approximation error of the Pareto front is defined by $\Delta H(P^*, P) = I_H(P^*) - I_H(P)$. All reported results are averaged over 10 independent runs with at most 100,000 fitness evaluations.

4.2   Result Analysis

Some experiments are first conducted to estimate the complexity of the surrogate training on 30-dimensional ZDT1 problem. The empirical complexity with respect to the number of training points is circa 2.2 (slope on Fig. 2-Left in log scale). The fact that the complexity is super-quadratic is not surprising since the SVM procedure relies on the Gram matrix $K(x_i, x_j)$ for all points x_i and x_j in the training set. The complexity however remains bounded as the size of the training set (extracted from the archive) is less than 1,000, limiting *de facto* the computational cost of the RASM learning.

Table 1 shows the comparative results of all baseline, ASM and RASM-based EMOs; in the latter cases, both $p = 2$ and $p = 10$ pre-offspring are considered. These results first confirm that S-NSGA-II performs best on separable functions

Table 1. Comparative results of two baseline EMOAs, namely *S*-NSGA-II and *MO*-CMA-ES and their ASM and RASM variants. Median number of function evaluations (out of 10 independent runs) to reach ΔHtarget values, normalized by Best: a value of 1 indicates the best result, a value $X > 1$ indicates that the corresponding algorithm needed X times more evaluations than the best to reach the same precision.

ΔHtarget	1	0.1	0.01	1e-3	1e-4	1	0.1	0.01	1e-3	1e-4
			ZDT1					ZDT2		
Best	1100	3000	5300	7800	38800	1400	4200	6600	8500	32700
S-NSGA-II	1.6	2	2	2.3	1.1	1.8	1.7	1.8	2.3	1.2
ASM-NSGA p=2	1.2	1.5	1.4	1.5	1.5	1.2	1.2	1.2	1.4	**1**
ASM-NSGA p=10	**1**	**1**	**1**	**1**	.	**1**	**1**	**1**	**1**	.
RASM-NSGA p=2	1.2	1.4	1.4	1.6	**1**	1.3	1.2	1.2	1.5	**1**
RASM-NSGA p=10	**1**	1.1	1.1	1.5	.	1.1	**1**	**1**	1.2	.
MO-CMA-ES	16.5	14.4	12.3	11.3	.	14.7	10.7	10	10.1	.
ASM-MO-CMA p=2	6.8	8.5	8.3	8	.	5.9	8.2	7.7	7.5	.
ASM-MO-CMA p=10	6.9	10.1	10.4	12.1	.	5	.	.	.	.
RASM-MO-CMA p=2	5.1	7.7	7.6	7.4	.	5.2	.	.	.	.
RASM-MO-CMA p=10	3.6	4.3	4.9	7.2	.	3.2	.	.	.	.
			ZDT3					ZDT6		
Best	1300	3500	7100	10100	15200	2500	3600	5200	12300	.
S-NSGA-II	1.4	1.9	1.6	1.9	2.2	2.1	3.4	3.8	2.7	.
ASM-NSGA p=2	1.1	1.3	1.1	1.2	1.3	1.4	2.4	2.6	2	.
ASM-NSGA p=10	**1**	**1**	**1**	**1**	**1**	1.1	1.8	2.3	2.3	.
RASM-NSGA p=2	1.1	1.3	1.2	1.4	1.6	1.5	2.4	2.8	2.1	.
RASM-NSGA p=10	**1**	1.1	1.1	2	.	1.4	2	2.3	1.8	.
MO-CMA-ES	15.4	17.8	.	.	.	2.5	2.6	2.5	2	.
ASM-MO-CMA p=2	9	.	.	.	.	1.1	1.2	1.1	**1**	.
ASM-MO-CMA p=10	8	25.6	.	.	.	**1**	1.1	1.3	2.5	.
RASM-MO-CMA p=2	8.5	.	.	.	.	1.5	1.2	1.2	**1**	.
RASM-MO-CMA p=10	8.1	.	.	.	.	**1**	**1**	**1**	1.6	.
			IHR1					IHR2		
Best	500	2000	35300	41200	50300	1700	7000	12900	52900	.
S-NSGA-II	1.6	1.5	.	.	.	1.1	3.2	6.2	.	.
ASM-NSGA p=2	1.2	1.3	.	.	.	**1**	3.9	4.9	.	.
ASM-NSGA p=10	**1**	1.5	.	.	.	1.4	6.4	4.6	.	.
RASM-NSGA p=2	1.2	1.2	.	.	.	1.5	.	.	.	.
RASM-NSGA p=10	**1**	**1**	.	.	.	1.2	5.1	4.8	.	.
MO-CMA-ES	8.2	6.5	1.1	1.2	1.2	5.8	2.7	2.1	**1**	.
ASM-MO-CMA p=2	4.6	2.9	**1**	**1**	**1**	3.1	1.6	1.4	1.1	.
ASM-MO-CMA p=10	9.2	6.1	1.3	1.2	.	5.9	2.6	2.4	.	.
RASM-MO-CMA p=2	2.6	2.3	2.4	2.1	.	2.2	**1**	**1**	.	.
RASM-MO-CMA p=10	1.8	1.9	.	.	.	.	.	.	.	.
			IHR3					IHR6		
Best	800	.	.	.	.	16500	.	.	.	.
S-NSGA-II	1.5	.	.	.	.	5.4	.	.	.	.
ASM-NSGA p=2	1.1	.	.	.	.	3.8	.	.	.	.
ASM-NSGA p=10	**1**	.	.	.	.	.	.	.	.	.
RASM-NSGA p=2	1.3	.	.	.	.	2.2	.	.	.	.
RASM-NSGA p=10	1.1	.	.	.	.	2.6	.	.	.	.
MO-CMA-ES	9.6	.	.	.	.	2	.	.	.	.
ASM-MO-CMA p=2	7.2	.	.	.	.	2	.	.	.	.
ASM-MO-CMA p=10	12.1	.	.	.	.	.	.	.	.	.
RASM-MO-CMA p=2	3.3	.	.	.	.	**1**	.	.	.	.
RASM-MO-CMA p=10	2.6	.	.	.	.	**1**	.	.	.	.

ZDTx and *MO-CMA-ES* on non-separable functions IHRx[3]. They also show that both *RASM*-NSGA and *RASM*-MO-CMA work nearly 1.5 times faster with $p = 2$ and more than 2 times faster with $p = 10$ than the baseline versions with regards to the ΔH value and the number of function evaluations.

ASM-NSGA and *RASM*-NSGA yield comparable performances. A more thorough analysis shows that *RASM*-MO-CMA is usually faster at the beginning (up to 10000-15000 function evaluations) though it might suffer from a premature convergence thereafter: experiments on concave IHR2 (and to some extent, also on ZDT2) show that *RASM*-MO-CMA converges to the value $\Delta H = 0.1$ nearly 1.5 times faster than with ASM model, and fails to go further. This failure is blamed on the fact that the diversity of the population is hardly preserved; a small part of the optimal Pareto front is sampled. Indeed, RASM learning and the RASM-based offspring selection only aim at speeding up the convergence; further work will be required to extend the approach and approximate the μ-optimal distribution of nearly-optimal Pareto points.

5   Discussion and Perspectives

The main contribution of the present paper is to show how to train a single surrogate model to reflect Pareto dominance in an EMO framework, using a *Learning to Rank* framework. RASM, the resulting surrogate model, does not require that all Pareto points are mapped onto the same value. It is thus both more constrained in the dominated region, and less constrained on the Pareto front, than ASM, previous work of the authors along similar lines [10].

Furthermore, this approach opens new and interesting perspectives for real world multi-objective problems, enabling for instance to account for the user's preferences in a flexible way by simply adding user-defined constraints to the order-based SVM formulation. Most importantly, the rank constraint formalization enables to accommodate conflicting preferences: to the best of our knowledge, this corresponds to a significant advance on the state of the art. This property is likely to be important for simulated evolution, since the ability to predict an environment is a prerequisite for intelligent behavior.

The experimental validation of the proposed approach shows that RASM-EMO usually converges faster than ASM-EMO, with the caveat that it sometimes leads to premature convergence (e.g., on ZDT2 and IHR2 problems). This premature convergence was blamed on the selection pressure and the adjustment of parameter σ_{sel}^2. A further work will explore the adaptation of the famed 1/5-th rule to adjust σ_{sel}^2, using the hypervolume indicator ΔH as measure of success.

The number of constraints that are added to the primary constraints could also be made adaptive by considering the stability of the surrogate model. However, such potential improvement would require the computation of all ξ_{ij} and F_{svm} values for all points of the archive at each generation, and would hence be computationally costly.

---

[3] *MO-CMA-ES* penalization parameter α is 1.0 for all problems in order to prevent evolution from being biased toward exploring the boundaries of the decision space.

Another shortcoming of Aggregate Surrogate Models is how to resist the loss of diversity. It is emphasized that RASM might incorporate additional specific constraints in each generation. Some possible constraints are described in Figure 1-Right: such **non-dominance** constraints involved points on the current Pareto front, and include inequality constraints from the extremal points over their neighbors (continuous arrows), and equality constraints for all neighbor pairs on the Pareto front (continuous double arrow), as well as between extremal points (dotted double arrows). Such equality constraints can be rewritten as two symmetrical inequality constraints in order to preserve the particular form of the formulation (Eq. 2). Along the same lines, constraints could be weighted, e.g. the weight of constraints related to points with the largest hypervolume contributions can be increased online. This is however a topic for further work.

References

1. Auger, A., Bader, J., Brockhoff, D., Zitzler, E.: Theory of the hypervolume indicator: Optimal μ-distributions and the choice of the reference point. In: FOGA, pp. 87–102. ACM, New York (2009)
2. Deb, K.: Multi-Objective Optimization Using Evolutionary Algorithms. John Wiley, Chichester (2001)
3. Deb, K., Pratap, A., Agarwal, S., Meyarivan, T.: A Fast Elitist Multi-Objective Genetic Algorithm: NSGA-II. IEEE TEC 6, 182–197 (2000)
4. Herbrich, R., Graepel, T., Obermayer, K.: Large margin rank boundaries for ordinal regression. In: Smola, A., Bartlett, P., Schölkopf, B. (eds.) Advances in Large Margin Classifiers, pp. 115–132. MIT Press, Cambridge (2000)
5. Igel, C., Hansen, N., Roth, S.: Covariance Matrix Adaptation for Multi-objective Optimization. Evolutionary Computation 15(1), 1–28 (2007)
6. Jin, Y.: A Comprehensive Survey of Fitness Approximation in Evolutionary Computation. Soft Computing 9(1), 3–12 (2005)
7. Joachims, T.: A support vector method for multivariate performance measures. In: De Raedt, L., Wrobel, S. (eds.) Proc. ICML, pp. 377–384. ACM, New York (2005)
8. Knowles, J., Nakayama, H.: Meta-modeling in multiobjective optimization. In: Branke, J., Deb, K., Miettinen, K., Słowiński, R. (eds.) Multiobjective Optimization. LNCS, vol. 5252, pp. 245–284. Springer, Heidelberg (2008)
9. Knowles, J., Thiele, L., Zitzler, E.: A tutorial on the performance assessment of stochastic multiobjective optimizers. Technical report (2006)
10. Loshchilov, I., Schoenauer, M., Sebag, M.: A Mono Surrogate for Multiobjective optimization. In: Branke, J., et al. (eds.) GECCO 2010, pp. 471–478. ACM, New York (2010)
11. Schölkopf, B., Platt, J., Shawe-Taylor, J., Smola, A., Williamson, R.: Estimating the Support of a High-Dimensional Distribution. Neural Computation 13, 1443–1471 (2001)
12. Vapnik, V.: Statistical Learning Theory. Wiley, Chichester (1998)
13. Yun, Y., Nakayama, H., Arakava, M.: Generation of pareto frontiers using support vector machine. In: MCDM 2004 (2004)
14. Zitzler, E., Deb, K., Thiele, L.: Comparison of multiobjective evolutionary algorithms: Empirical results. Evolutionary Computation 8, 173–195 (2000)

Learning Cellular Automata Rules for Pattern Reconstruction Task

Anna Piwonska[1,*] and Franciszek Seredynski[2,3]

[1] Bialystok University of Technology
Computer Science Faculty
Wiejska 45A, 15-351 Bialystok, Poland
a.piwonska@pb.edu.pl
[2] Institute of Computer Science
Polish Academy of Sciences
Ordona 21, 01-237 Warsaw, Poland
[3] Polish-Japanese Institute of Information Technology
Koszykowa 86, 02-008 Warsaw, Poland
sered@ipipan.waw.pl

Abstract. This paper presents results of experiments concerning the scalability of two-dimensional cellular automata rules in pattern reconstruction task. The proposed cellular automata based algorithm runs in two phases: the learning phase and the normal operating phase. The learning phase is conducted with use of a genetic algorithm and its aim is to discover efficient cellular automata rules. A real quality of discovered rules is tested in the normal operating phase. Experiments show a very good performance of discovered rules in solving the reconstruction task.

Keywords: cellular automata, pattern reconstruction task, genetic algorithm, scalability of rules.

1   Introduction

Cellular automata (CAs) are discrete, spatially-extended dynamical systems that have been studied as models of many physical and biological processes and as computational devices [8,14]. CA consists of identical cells arranged in a regular grid, in one or more dimensions. Each cell can take one of a finite number of states and has an identical arrangement of local connections with other cells called a neighborhood. After determining initial states of cells (an initial configuration of a CA), states of cells are updated synchronously according to a local rule defined on a neighborhood. In the case of two-dimensional CAs, two types of neighborhood are commonly used: von Neumann and Moore [8]. When a grid size is finite, we must define boundary conditions.

One of the most interesting features of CAs is that in spite of their simple construction and principle of operation, cells acting together can behave in an

---

* This research was supported by S/WI/2/2008.

inextricable and an unpredictable way. Although cells have a limited knowledge about the system (only its neighbors' states), localized information is propagated at each time step, enabling more global behavior.

The main bottleneck of CAs is a difficulty of constructing CAs rules producing a desired behavior. In some applications of CAs one can design an appropriate rule by hand, based on partial differential equations describing a given phenomenon. However, it is not always possible. In the 90-ties of the last century Mitchell and colleagues proposed to use genetic algorithms (GAs) to discover CAs rules able to perform one-dimensional density classification task [7] and the synchronization task [4]. The results produced by Mitchell et al. were interesting and started development of a concept of automating rule generation using artificial evolution. Breukelaar and Back [3] applied GAs to solve the density classification problem as well as AND and XOR problem in two dimensional CAs. Sapin et al. [11] used evolutionary algorithms to find a universal cellular automaton. Swiecicka et al. [13] used GAs to find CA rules able to solve multiprocessor scheduling problem. Bandini et al. [1] proposed to use several Machine Learning techniques to automatically find CA rules able to generate patterns similar in some generic sense to those generated by a given target rule.

In literature one can find several examples of CAs applications in image processing [6,10] as well as evolving by GAs CAs rules in image processing task [12]. Some of them deal with image enhancement, detection of edges, noise reduction, image compression, etc. The authors have recently proposed to use a GA to discover CAs rules able to perform pattern reconstruction task [9].

In this paper we present results of subsequent experiments concerning evolving CAs rules to perform pattern reconstruction task. The main aim of these experiments was to analyse possibilities of discovered rules in reconstructing the same patterns but on larger grids, that is the scalability of rules. The paper is organized as follows. Section 2 presents pattern reconstruction task in context of CAs. Section 3 describes two phases of the algorithm: the learning phase and the normal operating phase. Results of computer experiments are reported in Section 4. The last section contains conclusions and some remarks about future work.

2   Cellular Automata and Pattern Reconstruction Task

We assume that a given pattern is defined on a two-dimensional array (grid of cells) of size $n \times n$. Each element of an array can take one of two possible values: 1 or 2. Let us assume that some fraction q of values of grid elements is not known. These are missing parts of a pattern. Based on such a not complete pattern, it could be difficult to predict unknown states unless one is able to see some dependencies between values of cells. Let us further assume that we have a series of such not complete patterns, created from one given pattern in a random way. These patterns will be treated as initial configurations of two-dimensional CAs.

Pattern reconstruction task is formulated as follows. We want to find a CA rule which is able to transform an initial, not complete configuration to a final complete configuration.

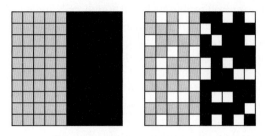

Fig. 1. The examples of a complete pattern (on the left) and an incomplete one (on the right). The incomplete pattern has 30 states unknown ($q = 0.3$).

Let us construct a two-dimensional CA of size $n \times n$, in order to describe our pattern. Our CA will be a three-state: unknown values of grid elements will be represented by state 0. That means that at each time step every cell of our CA can take a value from the set $\{0, 1, 2\}$.

In the context of CAs our task can be described as follows. Let us assume that we have a finite number of random initial configurations, each of which is an incomplete pattern. We want to find a CA rule that is able to converge to a final configuration identical with a complete pattern. That means, a rule that will be able to reconstruct a pattern. We also assume that a complete pattern is not known during searching process. The only data available during searching process is a series of incomplete patterns, randomly created from one given pattern. It is worth mentioning that related to our task is a problem from data mining field described in [5], where a heuristic CA rule was proposed.

Fig. 1 presents the example of a pattern of size 10×10 (on the left). Grid elements with value 1 are represented by grey cells and elements with value 2 are represented by black cells. On the right side of this figure one can see the example of this pattern with 30 states unknown. These are represented by white cells. Indexes of unknown elements were generated randomly. Such an incomplete pattern is interpreted as an initial configuration of a CA.

When using CAs in pattern reconstruction task, we must first define a neighborhood and boundary conndition. In our experiments we assume von Neumann neighborhood, with three possible cell states (Fig. 2). Using this neighborhood we have $3^5 = 243$ possible neighborhood states. Thus, the number of possible rules equals to 3^{243}, which means enormous search space. In our experiments we assume null boundary conditions: our grid is surrounded by dummy cells always in state 0. The interpretation of this assumption is that we do not know the state of these cells. In fact, they are not a part of our pattern.

3   The GA for Discovering CAs Rules

The proposed CA-based algorithm runs in two phases: the learning phase and the normal operating phase.

3.1   Learning Phase

The purpose of this phase is searching for suitable CAs rules with the use of the GA. The GA starts with a population of P randomly generated 243-bit CA rules. Five cells of von Neumann neighborhood are usually described by directions on the compass: North (N), West (W), Central (C), East (E), South (S). Using this convention, the bit at position 0 in the rule (the top bit in the bar in Fig. 2) denotes a state of the central cell of the neighborhood 0000000 in the next time step, the bit at position 1 in the rule denotes a state of the central cell of the neighborhood 0000001 in the next time step and so on, in lexicographic order of neighborhood.

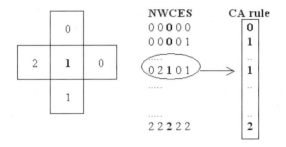

Fig. 2. The neighborhood coding (on the left) and the fragment of the rule - the chromosome of the GA (on the right, in a bar)

The next step is to evaluate individuals in the initial population for the ability to perform pattern reconstruction task. For this purpose, at each generation, starting from a complete pattern, we randomly generate an incomplete pattern with q states unknown. This process proceeds as follows. We have a complete pattern. In single step we randomly select a single cell which has not been previously chosen and this cell changes its state to 0 (unknown state). We repeat these steps until we chose $q \cdot n^2$ cells. All these cells will be in the state 0 (unknown).

Then each rule in the population is evolved on that randomly generated incomplete pattern, considered as an initial configuration of CA, for t time steps. At the final time step we compute the number of cells in the grid, with a state different from 0, that have the correct state. If a given cell is in the state 1 in an initial configuration, then the correct state for this cell in the final configuration is 1. Similarly, if a given cell is in the state 2 in an initial configuration, then the correct state for this cell in the final configuration is 2. The number of cells in a final configuration with the state 1 which are in the correct state will be denoted as n_1 and the number of cells in a final configuration with the state 2 which are in the correct state will be denoted as n_2. Since we compute the number of correct states, we deal with maximization problem. The fitness f of a rule i, denoted as f_i, is computed according to formula:

$$f_i = n\_1 + n\_2 - n\_0 , \qquad (1)$$

where n_0 denotes the number of cells in the final configuration with the state 0. Subtracting the number of cells with the state 0 is a kind of penalty factor and its task is to prevent from evolving to the final configuration with many cells in the state 0. It would be unfavorable situation from the point of view of pattern reconstruction task. The maximal fitness value equals to the number of cells in known states and equals to $n^2 - q \cdot n^2$.

After creating an initial population, the GA starts to improve it through repetitive application of selection, crossover and mutation. In our experiments we used tournament selection: individuals for the next generation are chosen through P tournaments. The size of the tournament group is denoted as t_{size}.

After selection individuals are randomly coupled and each pair is subjected to one-point crossover with the probability p_c. If crossover is performed, offspring replace their parents. On the other hand, parental rules remain unchanged.

The last step is a mutation operator. It can take place for each individual in the population with the probability p_m. When a given gene is to be mutated, we replace the current value of this gene by the value 1 or 2, with equal probability. Omitting the value 0 has the same purpose as described previously: it prevents from evolving rules with many 0s. Such rules are more likely to produce configurations containing cells with the state 0. It would be unfavorable situation.

These steps are repeated G generations and at the end of them the final population of rules is stored.

3.2   Normal Operating Phase

At the end of the learning phase we have a population of discovered rules which were trained to perform pattern reconstruction task. Let us remind that a complete pattern was not presented during this phase.

In our previous paper [9] we investigated a real quality of discovered rules by running each rule on IC random initial configurations with q states unknown. For a given final configuration produced by rule i, we counted the fraction of cells' states identical with these in a complete pattern. This value, denoted as t_i, was computed according to the formula:

$$t_i = \frac{n\_1 + n\_2}{n^2} . \tag{2}$$

An ideal rule can evolve an initial configuration to the final configuration with all cells in correct states. Thus, the maximum value t_i that such an ideal rule can obtain is 1.0. That means that the final configuration is identical with a complete pattern. Since we tested each rule on IC random initial configurations, the final value for a rule was the rule's average result over IC initial configurations. We denoted this value as $\bar{t}_i$.

In experiments described in this paper we decided to test the population of discovered rules on larger grid size, but on the same pattern. The aim of these experiments was to examine the existence of the scalability of CAs rules. Intuitively, one can suppose that testing rules on larger grid size than that on which they have been learned is much harder task. Results of these experiments are described in the next section.

4 Experimental Results

In our experiments we used two patterns with $n = 10$ presented in Fig. 3. They were denoted as pattern 1 and pattern 2. For each pattern, we tested the performance of the GA for three values of q: 0.1, 0.3 and 0.5. The maximal number of time steps t during which a CA has to converge to a desired final configuration was set to 100. Experiments showed that such a value is large enough to let good rules to converge to a desired final configuration. On the other hand, when a CA converged to a stable configuration earlier, the process of CA run was stopped.

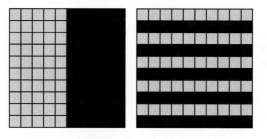

Fig. 3. Patterns used in experiments: pattern 1 (on the left) and pattern 2 (on the right)

4.1 Learning Phase

The parameters of the GA were the following: $P = 200$, $t_{size} = 3$, $p_c = 0.7$ and $p_m = 0.02$. Higher than the usual mutation rate results from rather long chromosome and an enormous search space. Experiments show that slightly greater p_m helps the GA in the searching process. The searching phase was conducted through $G = 200$ generations. Increasing the number of generations had no effect on improving results.

Fig. 4 and 5 present typical runs of the GA run for two patterns used in experiments, for the first 100 generations. On each plot one can see the fitness value of the best individual in a given generation, for three values of q.

The maximal fitness value for $q = 0.1$ equals to 90, for $q = 0.3$ equals to 70 and for $q = 0.5$ equals to 50. One can see that in case of $q = 0.1$ and $q = 0.3$, the GA is able to find a rule with the maximal fitness value.

The differences between patterns appear for $q = 0.5$. In the case of pattern 1, a fitness value of the best individual slightly oscillates around its maximal value. On the other hand, rules discovered for pattern 2 seem to be perfect. In this case, the GA quickly discovers rules with the maximal fitness value. We can conclude that in our method pattern 2 is easier in reconstruction than pattern 1.

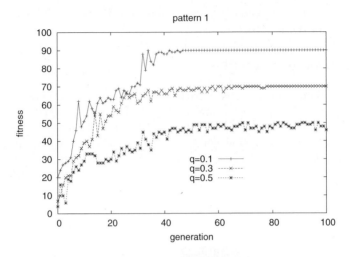

Fig. 4. The GA runs for pattern 1

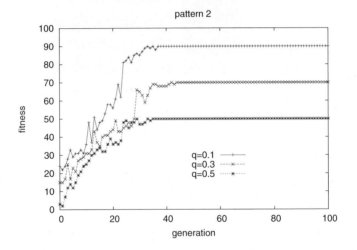

Fig. 5. The GA runs for pattern 2

4.2   Normal Operating Phase

The aim of the normal operating phase was to examine the scalability of discovered rules. For this purpose we tested populations of discovered rules on the same patterns, but on larger grid size. In our experiments we used $n = \{20, 30, 40, 50\}$. New patterns for $n = 20$ are presented in Fig. 6. Patterns for remaining n values were created in the same way.

For both patterns, we tested the final population discovered by the GA for $q = 0.1$ and $n = 10$ on $IC = 100$ random initial configurations, created on larger

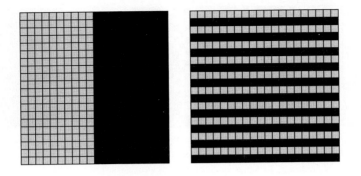

Fig. 6. Patterns for $n = 20$ used in experiments

Table 1. t_{best} values for pattern 1 and different q values

	$q = 0.1$	$q = 0.3$	$q = 0.5$
$n = 20$	0.997	0.959	0.890
$n = 30$	0.999	0.974	0.930
$n = 40$	0.999	0.980	0.945
$n = 50$	0.999	0.984	0.957

Table 2. t_{best} values for pattern 2 and different q values

	$q = 0.1$	$q = 0.3$	$q = 0.5$
$n = 20$	0.999	0.998	0.998
$n = 30$	1.0	0.998	0.994
$n = 40$	1.0	0.999	0.995
$n = 50$	1.0	0.999	0.996

grid sizes. For each rule, $\bar{t}_i$ was computed. Results of the best rules (from the whole population), denoted as t_{best} are presented in Tab. 1 and 2.

Results presented in Tab. 1 and 2 show that discovered rules have possibilities in reconstructing the same pattern on larger grids. What is interesting, this scalability does not degrade when n increases. For a given q, as n increases, t_{best} oscillates around similar values. Slight changes are caused by stochastic way of computing of t_{best}: a rule i might have different $\bar{t}_i$ values, depending on a concrete initial configuration.

As q increases, the scalability of discovered rules slightly decreases, as was to be expected. This is particularly noticable in the case of pattern 1, which proved to be more difficult in the learning phase. Higher values of q mean more unknown states in an initial configuration (and less states known). The more unknown states, the more difficult pattern reconstruction task is.

Let us look closely at the performance of one of the best rules from the normal operating phase. The leftmost picture in Fig. 7 presents the initial configuration of the CA for pattern 1, $n = 20$ and $q = 0.3$. We run the best rule found for

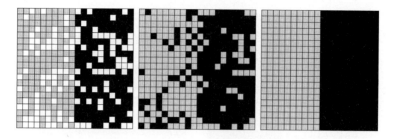

Fig. 7. Configurations of the CA in time steps: 0, 1 and 14 (from the left to the right)

$n = 10$ on this configuration, for the maximal number of time steps $t = 100$. Further configurations in time steps 1 and 14 are presented in Fig. 7.

One can see that the rule is able to quickly tranform the initial configuration to the desired final configuration. What is interesting, the number of time step sufficient to do it does not increment when n increases. Barely after one time step, the configuration of CA does not contain any cells with unknown state and has many cells in the correct state.

5   Conclusions

In this paper we have presented pattern reconstruction task in the contex of CAs. One of the aims of the paper was to present possibilities of the GA in evolving suitable CAs rules which are able to transform initial, not complete configurations to final complete configurations. Results of presented experiments show that the GA is able to discover rules appropriate to solve this task for a given instance of a problem. Found rules perform well even when the number of unknown cells is relatively high.

The most important subject addressed in this paper is the scalability discovered CA rules. Experiments show that during learning phase rules store some kind of knowledge about pattern which is reconstructed. This knowledge can be successfully reused in the process of reconstructing patterns on larger grid size.

References

1. Bandini, S., Vanneschi, L., Wuensche, A., Shehata, A.B.: A Neuro-Genetic Framework for Pattern Recognition in Complex Systems. Fundamenta Informaticae 87(2), 207–226 (2008)
2. Banham, M.R., Katsaggelos, A.K.: Digital image restoration. IEEE Signal Processing Magazine 14, 24–41 (1997)
3. Breukelaar, R., Back, T.: Evolving Transition Rules for Multi Dimensional Cellular Automata. In: Sloot, P.M.A., Chopard, B., Hoekstra, A.G. (eds.) ACRI 2004. LNCS, vol. 3305, pp. 182–191. Springer, Heidelberg (2004)
4. Das, R., Crutchfield, J.P., Mitchell, M.: Evolving globally synchronized cellular automata. In: Eshelman, L.J. (ed.) Proceedings of the Sixth International Conference on Genetic Algorithms, pp. 336–343. Morgan Kaufmann Publishers Inc., San Francisco (1995)

5. Fawcett, T.: Data mining with cellular automata. ACM SIGKDD Explorations Newsletter 10(1), 32–39 (2008)
6. Hernandez, G., Herrmann, H.: Cellular automata for elementary image enhancement. Graphical Models And Image Processing 58(1), 82–89 (1996)
7. Mitchell, M., Hraber, P.T., Crutchfield, J.P.: Revisiting the Edge of Chaos: Evolving Cellular Automata to Perform Computations. Complex Systems 7, 89–130 (1993)
8. Packard, N.H., Wolfram, S.: Two-Dimensional Cellular Automata. Journal of Statistical Physics 38, 901–946 (1985)
9. Piwonska, A., Seredynski, F.: Discovery by Genetic Algorithm of Cellular Automata Rules for Pattern Reconstruction Task. In: Proceedings of ACRI 2010. LNCS. Springer, Heidelberg (to appear, 2010)
10. Rosin, P.L.: Training Cellular Automata for Image Processing. IEEE Transactions on Image Processing 15(7), 2076–2087 (2006)
11. Sapin, E., Bailleux, O., Chabrier, J.-J., Collet, P.: A New Universal Cellular Automaton Discovered by Evolutionary Algorithms. In: Deb, K., et al. (eds.) GECCO 2004. LNCS, vol. 3102, pp. 175–187. Springer, Heidelberg (2004)
12. Slatnia, S., Batouche, M., Melkemi, K.E.: Evolutionary Cellular Automata Based-Approach for Edge Detection. In: Masulli, F., Mitra, S., Pasi, G. (eds.) WILF 2007. LNCS (LNAI), vol. 4578, pp. 404–411. Springer, Heidelberg (2007)
13. Swiecicka, A., Seredynski, F., Zomaya, A.Y.: Multiprocessor scheduling and rescheduling with use of cellular automata and artificial immune system support. IEEE Transactions on Parallel and Distributed Systems 17(3), 253–262 (2006)
14. Wolfram, S.: A New Kind of Science. Wolfram Media (2002)

Evolving Fuzzy Rules:
Evaluation of a New Approach

Adam Ghandar[1], Zbigniew Michalewicz[1,2], and Frank Neumann[3]

[1] School of Computer Science, University of Adelaide, Adelaide, SA 5005, Australia
[2] Institute of Computer Science, Polish Academy of Sciences, ul. Ordona 21, 01-237 Warsaw, Poland, and Polish-Japanese Institute of Information Technology, ul. Koszykowa 86, 02-008 Warsaw, Poland
[3] Max-Planck-Institut für Informatik, 66123 Saarbrücken, Germany

Abstract. Evolutionary algorithms have been successfully applied to optimize the rulebase of fuzzy systems. This has lead to powerful automated systems for financial applications. We experimentally evaluate the approach of learning fuzzy rules by evolutionary algorithms proposed by Kroeske et al. [10]. The results presented in this paper show that the optimization of fuzzy rules may be universally simplified regardless of the complex fitness surface for the overall optimization process. We incorporate a local search procedure that makes use of these theoretical results into an evolutionary algorithms for rule-base optimization. Our experimental results show that this improves a state of the art approach for financial applications.

1 Introduction

The combination of evolutionary algorithms and fuzzy systems has lead to many successful applications [2]. There are numerous applications for evolving fuzzy systems for applications in finance. For example, recently [1] described an application of a Takagi-Sugeno-Kang system [13] to predict stock prices. They reported impressive accuracy over 90%; the approach involved several stages to select model inputs and tune fuzzy rule parameters. These included regression analysis to select rule inputs and simulated annealing to tune rules. Also recently, [5] described a type-2 fuzzy logic system for modeling stock prices. Type-2 fuzzy systems, see [9], allow a more fine grained problem representation and may be useful. Finally, [6] describes an application of evolving fuzzy rulebases to construct adaptive asset valuation models from analysis of historic financial data and reports risk and return performance when applied for managing a portfolio of stocks. A number of analytical approaches to tune fuzzy rulebase parameters are also well known. Notably, in [11] a gradient descent method for learning fuzzy rules from input/output data is presented. Another powerful technique using the least square method is described in [12].

We describe implementation of an approach based on theoretical analysis in [10], and compare it with the method for learning fuzzy rules introduced

K. Deb et al. (Eds.): SEAL 2010, LNCS 6457, pp. 250–259, 2010.

in [6]. The new method involves division of the problem into parts to be solved using a combination of an evolutionary algorithm (EA) and local search. By analysing the partial derivatives of the evaluation function to compare solutions by the EA, the rulebase optimization problem is able to be projected onto a smooth space (with a guaranteed global optimum) for setting the importance of each rule with respect to the objective.

We show that this novel optimization technique improves the quality of solutions for prediction over a state of the art approach [6], and that the method is able to be enhanced by controlling the probability to apply local the rule weighting optimization. In testing, we focus on the distinct issues of constructing fuzzy rules for financial prediction from a large training data set.

The paper is organized as follows. Section 2 describes the problem of fuzzy rule base optimization. Section 3 describes details of the evolutionary algorithm and the local search procedure. Experimental results comparing the new approach with a standard evolving fuzzy system are reported in Section 4. Finally, we finish with some concluding remarks.

2   Problem Description

The objective of our approach is to minimize the error in predicting financial time series, for example asset price movements, using a rule base. The data comprises price series for the assets and also series of observations of factors considered in the model, for example economic or company information.

The approach involves finding asset valuation models, which are sets of *If-Then* fuzzy rules with a specific structure, that have high out-of-sample performance. Given fuzzy membership statements $A_1, \ldots, A_n$ that describe the model factors (e.g. observation is LOW) a fuzzy rulebase is a set of rules each with format

$$r = \text{If } A_1 \wedge A_2 \wedge \cdots \wedge A_n \Longrightarrow o. \tag{1}$$

The evaluation function measures error between observed percentage price movement in *training data* and prediction by rules. The rulebase output is interpreted as a prediction of percentage price movement. Note that rulebase output originally in the interval $[0.0, 1.0]$, is scaled by a factor δ appropriate for the application. The training data consists of a set $\{x_i, y_i\}_{i=1\ldots N}$, where each

$$x_i = \begin{pmatrix} x_i^1 \\ x_i^2 \\ \vdots \\ x_i^L \end{pmatrix}$$

is a vector constructed from the factors given in 1 such that each $x_i^{1\ldots L=30}$ is an observation of $f_{1\ldots L=30}$ (it is a pre-condition that $x_i \in \mathbb{R}^L$), and each y_i is the percentage price change ($y_i \in \mathbb{R}$) over a period (horizon of the prediction).

The evaluation function is a measure of squared error between predicted (ρ) and y_i (the percent price change found in the training data):

$$\epsilon = \sum_{i=1}^{N} (\rho(x_i) - y_i)^2 \tag{2}$$

$$= \sum_{i=1}^{N} \left(\frac{\sum_{j=1}^{n} a_{ij} o^j}{\sum_{j=1}^{n} o^j} - y_i \right)^2 .$$

For training data $\{x_i, y_i\}_{i=1,...,N}$, where $x_i \in \mathbb{R}^L$ and $y_i \in \mathbb{R}$ for all i and a rulebase with n rules, we define

$$a_{em} = \prod_{l=1}^{k_m} \mu_{j_l^m}^{i_l^m} (x_s^{i_l^m}),$$

where $e = 1, ..., N$ ranges over the number of training data examples N and $m = 1, ..., n$ ranges over the rules currently in the rule base. Since an equal number of membership functions are used to describe each linguistic variable (*very low*, to, *very high*), then

$$a_{em} = \prod_{l=1}^{k_m} \mu_{j_l^m} (x_s^{i_l^m}).$$

The objective is thus to minimize error, ϵ with respect to some training data. The procedure for a rule base fitness evaluation given a training window of historic data with length, *len* and a forecast horizon, s, is given in Algorithm 1. Here, $p_{a_i,t}$ refers to the price of an asset at time t (note: δ is a scaling factor to modify the range of price changes that are predicted). The number of training data points, N, is equal to the number of assets times the window length divided by the forecast horizon or step s.

Algorithm 1. Evaluation

$\epsilon \leftarrow 0$
for all (assets a_i in a market) **do**
    for $(t \leftarrow s; t < len; t \leftarrow t + s)$ **do**
        $\epsilon \leftarrow \epsilon + (\delta\rho(x_{a_i}) - p_{a_i,t}/p_{a_i,t-s})^2$
    end for
end for

---

3   Optimization Process

The optimization process involves an evolutionary algorithm and a local search procedure that makes use of the reformulated objective function. In the following, we describe the main components of our algorithm for optimizing fuzzy rules bases.

3.1   Evolutionary Algorithm

The main rule base optimization procedure is given in Algorithm 2. In the following, we describe the details of the evolutionary algorithm.

The genotype of a fuzzy rulebase is represented using three arrays I, U and O. I is an $m \times n$ matrix of integers where each $i - jth$ element corresponds to a membership function μ in $i - th$ rule for the $j - th$ linguistic variable, each variable has the same number of membership function specifications so for 5 membership functions the possible values of I are $1, 2, 3, 4$ or 5. U is an $m \times n$ matrix of boolean values, if $U_{i,j} = TRUE$ then the input in $i - th$ rule for the $j - th$ variable is switched on and used in the linguistic rule description, otherwise it is not used. O is a vector of double values with size m, one for each rule, and corresponds to the output levels, a value of 0 means the rule has zero weight and is not used.

Mutation and crossover operators are applied to vary the genotype. They are defined as follows: mutate inputs — select a random gene from either I or U with uniform probability and (with equal probability) either replace with a new random value or in/de-crement by 1 step with equal probability; mutate outputs — select and replace an output at random; crossover — uniform crossover over I and U; and rule crossover — swap rules between two different individuals, at the genotype this means the whole $i - th$ from I, U, O are swapped.

With regard to the genotype representation we can express a_{em} as follows

$$a_{em} = \prod_{l=1}^{L} U(m, l) \mu_{I(m,l)}(x_s^l),$$

where $U(m, l) = 1$ if in the m-th rule the l-th linguistic variable is switched on, otherwise it is removed from the calculation. This can be achieved in a number of other ways, for example [8] describes the use of a special *don't care* membership set that is always set to 1. Selection is with a tournament of size 2. Elitism is used.

Probabilities of the operators to create offspring are updated during the run, depending on their success (in obtaining better solutions). Further details on parameter setting methods used are able to be found in [7] and [3]. The probability of using the separate output optimization procedure oscillates in between never and always being applied $[0, 1]$ over the course of the the process. Fitness is assigned in the method applyOperators() using the specification in algorithm 1. In this way rule base antecedents (*if* parts of each rule) and consequents (*then* parts) separately for optimization. At every step it is possible that a separate optimization of rule base outputs occurs with probability $ooProb$.

3.2   Local Search Procedure

A local search procedure for the optimal weighting/individual rule output for each rule in a rulebase is integrated into the evaluation method. This is accomplished

---

Algorithm 2. EA

Require: $P_0, r_{best}, gen, operators[], opProb[], ooProb$
   while $(gen < \text{max\_gen})$ **do**
      parents[] $\leftarrow$ selectParents(P_{gen}) {t size=2}
      operator $\leftarrow selectOperator(operators, opProb)$
      offspring $\leftarrow applyOperator(parents)$
      if $(random() < ooProb)$ **then**
         $adjustOutput$(offspring)
      end if
      $P_{gen} \leftarrow$ replaceWorse$(P_{gen}, $ offspring$)$ {t size=2}
      $r_{cbest} \leftarrow best(P_{gen})$
      opProb[]$\leftarrow$update(opProb[])
      ooProb$\leftarrow$update(ooProb)
      if $(r_{cbest} > r_{best})$ **then**
         $r_{best} \leftarrow r_{cbest}$
      end if
      $gen \leftarrow gen + 1$
   end while
   return r_{best}

---

by an analysis of the properties of the evaluation function. The evaluation function is able to be rewritten as follows, the full calculation and proof may be found in [10]. The objective of the local search to set the output part of the rules (the vector O) involves the following system of equations

$$\frac{\partial \epsilon}{\partial o^q} = 0 \Leftrightarrow \left(\sum_{i=1}^{n} \sum_{j=1}^{n} (A_{iq} - A_{ij}) o^i o^j \right) = 0, \ q = 1, ..., n. \tag{3}$$

$$\text{with } A_{jk} = \sum_{i=1}^{N} (a_{ij} - y_i)(a_{ik} - y_i)$$

The constants A are calculated from the training data $(\boldsymbol{x_i}, y_i)$.

The evaluation function for the outputs of the rulebase is now to minimize this system. Furthermore, there exists a one-parameter family of solutions to the system (3) [10]. Hence the space of extremal points for ϵ is a line in $\mathbb{R}^n$ that passes through zero. We solve the problem for the optimal weights by minimizing the expression 3 using the local search procedure given in Algorithm 4.

Recall that n is the number of rules, then in the implementation, the terms $A_{i,j}$ etc, the array $A[][]$, is calculated from the training data. The objective of the local search is

The search algorithm for the output parts of the rulebase is then

---

Algorithm 3. *evaluate* (o[])

---

  result $= 0$
  for $q = 1 \ldots N$ **do**
    for $i = 1 \ldots N$ **do**
      for $j = 1 \ldots N$ **do**
        result $+= (A[i,q] - A[i,j]) * o[i] * o[j]$
      end for
    end for
  end for
  return result

Algorithm 4. Output search

---

  init $o[]$ to random
  $eval = evaluate(o[])$
  while $|(evaluate(o[]))| > 0 + \epsilon$ **do**
    $tmp[] = mutate(o[])$
    $evalTmp = evaluate(tmp)$
    if evalTmp $<$ eval **then**
      $eval = evalTmp$
      $o[] = tmp[]$
    end if
  end while

---

4   Experimental Results

This section contains experimental results obtained by application of the procedures rule learning procedures described in the previous sections. The experiments are designed to compare the performance of the novel method and a standard case. We consider statistics including best fitness, number of generations (fitness evaluations in this case of a steady state algorithm) and prediction performance.

Table 1 lists the model factors, used in these experiments. They include information derived from company balance sheets, trend and cyclical analysis of price and market trading volume volume series and accepted the, single factor, capital asset pricing model, alpha and beta. The rules can identify two types of information in these series: trend and mean reversion. The trend is information about the change in the series over the period p: $trend_A = A_t/A_{t-p}$. Distance from the mean is found using an oscillator: $oscillator_A = ma_1 - ma_2/ma1$, where ma_1, ma_2 are rolling averages of A over a short and long period (3 and 6 months). Alpha and beta, factors 1 and 2, refer to the capital asset pricing model and a standard single factor model (see [4]).

Table 2 lists the evolutionary algorithm parameters that were used. In the tests of the new method, rule weightings are optimized with an oscillating probability set by $P = cur\_gen/max\_gen$ mod 3. In addition the rulebase was repaired to

Table 1. The input factors are modeled by linguistic variables specified by triangular membership functions defined from all available observations. In addition change in each over time is considered as a separate variable. Factors 15 to 19 are "technical" indicators calculated using price data.

Variable Factor Description	
f_1	Jensen's α
f_2	β
f_3	Dividend yield.
f_4	Price to book value
f_5	Price earnings ratio
f_6	Forecast of price earnings
f_7	The market capitalization
f_8	Earnings per share over
f_9	Total debt to equity ratio
f_{10}	Long term debt to equity ratio
f_{12}	Earnings before interest and tax
f_{13}	Return on assets
f_{14}	Return on equity
f_{15}	Money flow index.
f_{16}	Price change
f_{17}	Bollinger bands.
f_{18}	Volatility (sdev)
f_{19}	Price volume oscillator

maintain < 4 inputs per rule and < 5 active rules (i.e. having output greater than zero). 5 discrete output weights were possible in the specification. Test data was sourced from Data Stream International (http://www.datastream.com/) and consisted of ASX200 listed stocks and associated data for the period $1/1/2006 - 1/1/2008$. A steady state EA with elitism was used in the normal case and altered by the addition of the local output optimization subroutine. The objective was to minimize the squared error between predicted asset price movement (from the output of fuzzy rules) and the real movement over 1 week periods.

Tables 3 contains the results after 7, 500 generations for 30 runs. Table 4 shows the error recorded when applying best solutions outside training data.

If the termination condition is to stop after 7, 500 generations the improvement is, empirically, very clear. The local output optimization (OO) test produced solutions with better fitness than the normal case (N) table 3 with a p-level less than 0.1 percent. The median and mean fitness for the OO runs were also over 35% higher. In addition, the distribution for the OO showed a lower range and and standard deviation indicating that it produces more consistent results. Specifically, the range was 192 for the OO compared with 142.

Table 4 provides compares (absolute) prediction error when applying solutions outside the training data window. The results are from applying rules for

Table 2. Parameters of the evolutionary algorithm and the testing methodology

Parameter	Value
P. OO Adj.	0.0 – 1.0 / 3 times
Population	500
Generations	7500
Elitism ?	Yes
Selection	Tournament $size = 2$
Horizon	5 days
Win len.	120 days
Initial Operator P.	0.3333 (3 Operators)
Penalties	No ranking $= 0.05$
	Direction $= 0.1$
Max Rules	5
Max Inputs	4
Output levels (d)	5

Table 3. Paired statistical comparison of the fitness values of the process (with different data) for 7500 generations

	Norm.	OO adj.
Min.	-341.1837	-242.3193
Max.	-149.2366	-100.4559
Median	-220.766	-161.1760
Mean.	-226.8101	-162.7535
Sdev	56.55486	45.69378
Mann Whit.	–	W $=205$
		$p < 0.01$ (gt.)

5000 different predictions from a pool of 201 assets considered in the investment universe over a period. For each optimization run, the best rulebase is applied to predict the price change of each stock 3 times over a subsequent 15 day period after the end of the training window; for each application the difference between the actual and the predicted change was recorded. As would be expected from the fitness results, it was the case that generalization performance of solutions produced by the OO method were better in the 7,500 generation test runs. The mean error was lower (3.8% vs. 4%) and also by parametric testing the improvement was significant.

We can see clearly that significant improvements in prediction accuracy are attained through the the implemementation of the theoretical results in [10]. This is an excellent result that demonstrates that there is promise in developing

Table 4. Paired statisical comparison of the error of prediction out of sample. The new method provided solutions that performed better with significance $P < 0.01$.

	Norm.	OO adj.
7, 500 generations		
Mean.	0.0406	0.03810
Sdev	0.0382	0.03335
Mann Whit.	$-$	W = 13256668
		$p < 0.01$ (2 sided)

methods to enable faster and more efficient restructuring of fitness surfaces and objectives to improve evolutionary algorithms in the domain of fuzzy rulebase optimization.

5 Conclusion

Evolutionary algorithms have been widely used for optimizing the rule base of fuzzy systems. This has lead to many successful applications in finance. We have examined how the state of the art system of [6] can be improved by incorporating the theoretical results obtained in [10]. The results of [10] allowed us to design an efficient approach that involves combining a local search, over a restructured objective function, with an evolutionary algorithm. Our experimental investigations show that this leads to a significant improvement in performance compared to the system given in [6].

Acknowledgments

This work was partially funded by the ARC Discovery Grant DP0985723 and by grant N516 384734 from the Polish Ministry of Science and Higher Education (MNiSW).

References

1. Chang, P., Liu, C.: A tsk type fuzzy rule based system for stock price prediction. Expert Syst. Appl. 34(1), 135–144 (2008)
2. Cordón, O., Gomide, F.A.C., Herrera, F., Hoffmann, F., Magdalena, L.: Ten years of genetic fuzzy systems: current framework and new trends. Fuzzy Sets and Systems 141(1), 5–31 (2004)
3. Eiben, G., van Hemert, J.: Saw-ing eas: adapting the fitness function for solving constrained problems, pp. 389–402 (1999)
4. Fama, E.F., French, K.R.: Multifactor explanations of asset pricing anomalies. Journal of Finance 51(1), 55–84 (1996)

5. Fazel Zarandi, M.H., Rezaee, B., Turksen, I.B., Neshat, E.: A type-2 fuzzy rule-based expert system model for stock price analysis. Expert Syst. Appl. 36(1), 139–154 (2009)
6. Ghandar, A., Michalewicz, Z., Schmidt, M., To, T.-D., Zurbruegg, R.: Computational intelligence for evolving trading rules. To appear in IEEE Transactions On Evolutionary Computation (2009)
7. Hinterding, A.E.E.R., Michalewicz, Z.: Parameter control in evolutionary algorithms. IEEE Transactions on Evolutionary Computation 3, 124–141 (1999)
8. Ishibuchi, H., Nakashima, T., Murata, T.: Performance evaluation of fuzzy classifier systems for multidimensional pattern classification problems. IEEE Transactions on Systems, Man, and Cybernetics, Part B: Cybernetics 29(5), 601–618 (1999)
9. John, R.: Type 2 fuzzy sets: an appraisal of theory and applications. International Journal of Uncertainty, Fuzziness and Knowledge-Based Systems 6(6), 563–576 (1998)
10. Kroeske, J., Ghandar, A., Michalewicz, Z., Neumann, F.: Learning fuzzy rules with evolutionary algorithms – an analytic approach. In: Rudolph, G., Jansen, T., Lucas, S., Poloni, C., Beume, N. (eds.) PPSN 2008. LNCS, vol. 5199, pp. 1051–1060. Springer, Heidelberg (2008)
11. Nomura, H., Hayashi, I., Wakami, N.: A learning method of fuzzy inference rules by descent method, pp. 203–210 (March 1992)
12. Nozaki, K., Ishibuchi, H., Tanaka, H.: A simple but powerful heuristic method for generating fuzzy rules from numerical data. Fuzzy Sets Syst. 86(3), 251–270 (1997)
13. Takagi, T., Sugeno, M.: Fuzzy identification of systems and its applications to modelling and control. IEEE Transactions on Systems, Man and Cybernetics 15(1), 116–132 (1985)

A Niched Genetic Programming Algorithm for Classification Rules Discovery in Geographic Databases

Marconi de Arruda Pereira[1,3], Clodoveu Augusto Davis Júnior[2],
and João Antônio de Vasconcelos[3]

[1] Centro Federal de Educação Tecnológica de Minas Gerais, Av. Amazonas, 7675,
Belo Horizonte, Brazil
marconi@decom.cefetmg.br
[2] Laboratório de Banco de Dados
Universidade Federal de Minas Gerais, Av. Antônio Carlos, 6627,
Belo Horizonte, Brazil
clodoveu@dcc.ufmg.br
[3] Evolutionary Computation Laboratory
Universidade Federal de Minas Gerais, Av. Antônio Carlos, 6627,
Belo Horizonte, Brazil
jvasconcelos@ufmg.br

Abstract. This paper presents a niched genetic programming tool, called DMGeo, which uses elitism and another techniques designed to efficiently perform classification rule mining in geographic databases. The main contribution of this algorithm is to present a way to work with geographical and conventional data in data mining tasks. In our approach, each individual in the genetic programming represents a classification rule using a boolean predicate. The adequacy of the individual to the problem is assessed using a fitness function, which determines its chances for selection. In each individual, the predicate combines conventional attributes (boolean, numeric) and geographic characteristics, evaluated using geometric and topological functions. Our prototype implementation of the tool was compared favorably to other classical classification ones. We show that the proposed niched genetic programming algorithm works efficiently with databases that contain geographic objects, opening up new possibilities for the use of genetic programming in geographic data mining problems.

Keywords: Classification rules, data mining, knowledge discovery in geographic databases.

1 Introduction

Classification rule mining is one of the most important tasks for knowledge discovery in databases (KDD). Recently, techniques and algorithms for classification rule mining have been intensively studied due to the large variety of practical applications for them. For instance, commercial firms want to know more about the behavior of their customers. Governments need to prioritize resource allocation and decide about

K. Deb et al. (Eds.): SEAL 2010, LNCS 6457, pp. 260–269, 2010.

public policies. Educators want to find out the factors that lead students to fail. In all of these situations, specialists are seeking unexpected patterns within data, and such patterns can emerge in the form of classification rules.

Mining classification rules usually utilizes supervised learning techniques that consist in discovering patterns in training data so that the resulting rules can be applied in the classification of other data. Goldberg [7] shows that genetic algorithms have been applied successfully in machine learning problems since the 1970s. The growth of interest in data mining has motivated the scientific community of evolutionary algorithms. Freitas [5] shows that genetic algorithms, genetic programming and, more recently, artificial immune systems, ant colony algorithms [12] and particle swarm optimization [9] have been successfully used in various data mining problems.

The main advantage of evolutionary algorithms is their robustness, that is, once the problem is correctly modeled, the algorithm is able to explore the feasible region within the space of problem solutions, looking for the best global solution. Greedy algorithms can be applied, but they usually return a *local* solution, not the *global* one [2].

The popularization of the access and use of geographical data, sponsored by big companies like Google (with Google Earth, Google Maps) and governmental departments like NASA (in USA) and INPE (in Brazil), brought up a new challenge: developing good data mining algorithms that works well with geographic tools.

There are a lot of good algorithms presented in the literature that works with geographical data mining. In section 2 the last algorithms designed to geographical data mining are shown. Nevertheless, there are not any algorithms capable to manipulate conventional data and geographical data at the same time exploring the topological relations.

In this paper, a niched genetic programming algorithm with elitism, called DMGeo, is presented. DMGeo has been designed to work with conventional and geographic data, which makes it suitable to a new range of applications.

This paper is organized as it follows. Section 2 presents related works. Section 3 introduces DMGeo and presents the proposed algorithm. Section 4 presents a case study developed to demonstrate DMGeo and shows the results. Finally, Section 5 presents a conclusion of this paper.

2 Related Works

Whigham [17] propose a first genetic programming that uses a context-free grammar to predict the density of Australians marsupials. This algorithm identifies spatial pattern behavior of these marsupials with confidence of 99% against all the non-spatial methods. However, this algorithm does not use topological relations (such as *contains*, *covers*, *crosses* and others [3]).

Bogorny et al [1] show a tool that permits integration between a classical data mining toolkit (Weka [16]) and a geographic information system. The tool, implemented as an extension of Weka, is used as it follows. The user selects a geographic database to be used and chooses a set of feature types or instances. The tool preprocesses the geographic relations between the elements of the set. These relations include topology and distance. A Weka ARFF file is thus generated, encoding a representation of the geographic relations among the selected items. Weka is then used to mine the data using conventional algorithms.

Pappa et al. [13] show a multi-objective genetic algorithm (MOGA) that is used in selecting attributes from database for data mining purposes. In particular, this genetic algorithm looks for the best set of attributes to create a decision tree to be used in a classification algorithm, namely C4.5 [11].

Just one of the mentioned papers presents a tool that is able to work with geographic and conventional data at the same time, but it cannot deal with topological relations. Moreover, the data mining specialist has to decide which kind of data is more important, that is, the tools found in literature are not capable to tell which set of attributes (combining geographical and conventional ones) is the most relevant in the target problem. The algorithm we propose and present in the next section has been designed specifically to allow the incorporation of spatial functions and operators in the genetic programming approach to rule mining, thus, avoiding complex and costly preprocessing.

3   The Proposed Algorithm - DMGeo

We propose a niched genetic programming algorithm, called DMGeo, to perform classification tasks in the presence of geographic attributes. DMGeo is capable of incorporating geographic constraints and operations in its rules, combining them with conventional operators and functions at each individual. In our approach, an individual represents a Boolean predicate, defined in the same manner as a SQL WHERE clause. The adequacy of the individual to the problem is assessed using a fitness function, which determines its chances of selection. These elements of the algorithm are presented in the next subsections.

3.1   Individual

The individual was modeled to represent a rule that should be applied to pattern selection in a database, as in the WHERE clause of a SQL query. Fig. 1 shows an example of an individual's tree, in which logical operators, attribute names and constants are combined to form a filter clause. It is important to note that the tree represents the rule to identify features of one class.

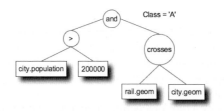

Fig. 1. Representation of an individual and its class

Tree nodes include the following information:

- **Type.** The types implemented in DMGeo are Boolean, numeric, and geographic (point, line or polygon).

- **Node body.** The body can be a constant or a geographic function call. This will be detailed next.
- **Parameters.** If the body is a function call, it must receive parameters from other nodes in the tree.

Nodes of the tree can be of the following types:

- **Function Nodes:** functions that, in this paper, have exactly two parameters, and form the set of possible inner nodes. They can be divided in two groups: conventional and geographical;
- **Terminal Nodes:** The set of terminal nodes is formed by the set of leaves in the tree. These nodes represent constants or database attributes.

The set of function nodes, as described previously, is composed by conventional functions (or operators) like =, <, >, >=, <=, AND and OR and by geographical ones. This last type of functions is implemented by geographically-extended databases. In this paper, we used the PostGIS implementations, which include the following spatial relation functions, described in [3]: *contains, covers, coveredBy, crosses, disjoint,*

Algorithm 1. Pseudocode to Generate the Individuals

Input: **Table** *target_table*	/* table with all attributes of the instances to be classified */
Output: Individual *I*	

1. Make a list of possible terminal nodes, called *possible_terminal_nodes*, with all the attributes from the list *target_table*.
2. Make a list of function nodes, called *inner_nodes*, with all possible functions (conventional and geographical);
3. Create an empty individual *I*;
4. **while** *count_nodes <= max_num_inner_nodes* **do**
5.   Select randomly a node from the *inner_nodes* list and name it as *c*.
6.   **if** (individual *I* is empty)
7.   Insert *c* as a root node and make *count_nodes := count_nodes + 1*;
8.   **else**
9.   In the tree of the individual *I*, search a function node that has an argument of the same type as node *c* and does not have this argument already set. Call this node as *cn*;
10.   **if** (*cn* ≠ null)
11.   Add node *c* as a child of node *cn*; *count_nodes := count_nodes + 1*;
12.   **end if**;
13.   **end if**;
14. **end while**;
15. **for each** function node *i* **in** the tree of individual *I* **do**
16.   *nt := retrieve_function_node(i)*; /* the function retrieve_function_node (index i) obtains the i-th function node using the tree-traversal algorithm [5] */
17.   **if** (*nt* does not have its arguments set)
18.   **if** (rand() > 0.5)
19.   Select a terminal node, called *nc*, from the *possible_terminal_nodes*;
20.   **else**
21.   Generate a terminal node, called *nc* with a random value;
22.   **end if**;
23.   Add *nc* as a argument (child) of *nt*;
24.   **end if**;
25. **end for**;
26. Evaluate the fitness of *I* using all possible classes; Choose the class that brings the greater fitness value as the class of *I*; **Return** Individual I.

equals, touches, within, and *distance.* The *distance* operator is the only one that does not generate a boolean result: it returns a numeric value that will be compared to a constant using a conventional operator.

The initial population of rules (individuals) is created randomly. Algorithm 1 shows a procedure to create an individual. The user must indicate a target table, which contains the features to be classified, with all their attributes. Training data will also have a label attribute, indicating the previously determined class.

In the Algorithm 1, an individual is generated with a max number of function nodes (loop between lines 4 and 14.) After this loop, the individual's tree is filled with terminal nodes that can have database attributes (line 19) or a constant value (line 21). Each terminal node generated has its type matching with the parameter type of the parent function node. This peculiarity is important to ensure that the individuals of DMGeo will be strongly-typed [10], that is, the DMGeo will manipulate just consistent individuals.

3.2 Rule's Performance Evaluation

The performance of a rule (individual) that classifies samples from a database is determined by evaluating a fitness function, which is totally based on SQL queries to be submitted to the database. This evaluation can be divided into two parts:

- Determination of the coefficient values of the confusion matrix;
- Calculation of the fitness function.

In the algorithm, we assume that the relation *R (Attr1, Attr2, ..., AttrN, Class)* contains all the attributes (geographic or not) that can be used in the classification, including *Class* which is the actual class of the individual.

The calculation of the confusion matrix values and the determination of the fitness are described next.

3.2.1 Confusion Matrix Coefficient Values

The number of **True Positives** for the *i-th* individual is the number of tuples selected using its rule whose class coincides with the expected class. The **True Negative** value is the number of items that are not selected using the rule prescribed by the *i-th* individual and do not belong to its class.

The number of **False Positives** is the number of items selected by the individual that do not belong to the expected class. Considering the example that was presented in Fig. 1, the False Positive number is the number of items selected by the rule "(city.population > 200.000) AND (rail.geom *crosses* city.geom)" which class is different of 'A'.

The **False Negative** value is the number of items that are not selected by the rule, which class matches the expected one.

3.2.2 Fitness Function Evaluation

A classical way to measure the effectiveness of a classifier is to obtain and compare indicators such as Accuracy, Sensitivity and Specificity. These indicators are calculated using the confusion matrix, as it follows:

$$Accuracy\ (A) = \frac{TP + TN}{TP + FP + TN + FN},$$

$$Sensitivity\ (A) = \frac{TP}{TP + FN},$$

$$Specificity\ (A) = \frac{TN}{TN + FP},$$

where A denotes the class.

Thus, after the confusion matrix coefficient evaluation, the fitness function is calculated using (1):

$$F(I, X) = Accuracy(I, X) * Sensitivity(I, X) * Specificity(I, X) \tag{1}$$

where $F(I, X)$ is the fitness function that evaluates the individual I for classifying items of class X.

3.3 Mutation

Differently of genetic algorithms, the mutation operator implemented in genetic programming is not so simple. First, it is necessary to make sure that the individual's tree remains valid after the mutation process, i.e., the mutation operation cannot replace a node (or a subtree) by a node of a different type. There are four possible outcomes after the mutation process [5]:

- **Point Mutation:** a terminal node is replaced by another terminal node.
- **Collapse Mutation:** a terminal node replaces a function node (subtree); the individual's tree decreases in size;
- **Expansion Mutation:** a function node replaces a terminal node. In this case, the individual's tree increases in size.
- **Subtree Mutation:** a function node replaces another function node. In this case, the size of the individual's tree can increase or decrease.

We also implemented a mechanism to generate changes in the class during the mutation process. It happens when another class leads to a greater value of the fitness function. Section 3.5 will show that the predicted class of an individual is used as a niche and the class change mechanism works as a niche migration.

It is important to highlight that in all four situations previously cited, the mutation works as follows:

1. Randomly select a node (terminal or function);
2. Generate another node (terminal or function) of the same type of data of the node selected in step 1;
3. Replace the selected node by the newly generated one.

3.4 Crossover

DMGeo's crossover is based on the classical crossover of genetic programming [10]. The operation is implemented as it follows: select two individuals of the same class, using roulette wheel; clone these individuals; permute a randomly selected subtree of the first individual with a randomly selected subtree from the second one.

3.5 Niches and Elitism

The number of niches should be equal to the number of classes, in such a way that each class will have, in the final process, just one rule (individual) that is expected to be the best one for selecting samples, belonging to the expected class.

To ensure that the best individual of each niche will be preserved at the next generation, the elitism technique was applied.

3.6 Storage of the Fitness Value in Cache Memory

As seen previously, the calculation of the fitness function for each individual is based on SQL queries executed in databases. As these queries demand expensive disk access, we implemented a simple cache memory to store the fitness value of some individuals. This cache consists in a hash table where the WHERE-Clause that represents the individual is the key of the table and the stored value is the performance of fitness function.

4 Tests and Results

4.1 Problems Used in Tests

The performance of the proposed algorithm DMGeo is analyzed by applying it in four classification problems: two datasets available in the UCI Machine Learning Repository[14], denoted as Heart and Wine databases, which are composed by numeric data, and two others datasets available in Geominas repository [6], City Development and Soy Aptitude, with numeric and geographic data.

The Heart database is composed of 270 instances where 120 have a heart disease (class B) and 150 are normal (class A). This problem is available in a dataset with 13 numerical attributes like age, resting blood pressure, maximum heart rate achieved, between others.

The Wine database contains data of three different groups of wine, according to its level of alcohol. The problem is composed by 178 instances where 59 are level A, 71 are level B and 48 are level C. The problem is stored in a dataset with 13 attributes, like for example, ash, alkalinity of ash and color intensity. These attributes were obtained from a chemical analysis.

The City Development database contains data of three different levels of development: high, medium and low. There are 852 cities where 264 are high (class A), 296 are medium (class B) and 292 are low (class C). This problem is stored in a dataset with 22 numerical attributes like quantity of schools (public and private), quantity of industrial electricity customers and GINI number. The dataset also contains geographical attributes like cities geometry (stored as polygons), railway and highway (stored as polygonal lines [3]).

The Soy Aptitude database contains data from two different type of soil, according to its aptitude to produce soy. There are 852 cities, where 562 have its soil appropriated to soy cultivation (class A) and 290 that have restriction in their soil for soy cultivation (class B). In this problem, we just have geographical attributes to work with:

cities geometry, rain incidence (stored as polygons), soil aptitude to beans, citrus fruits and cotton cultivation (all stored as polygons).

The datasets from UCI were used to evaluate the performance of DMGeo using just numerical attributes. The main contribution of the proposed algorithm is that it can explore well the feasible region of problems whose datasets are formed by numerical and geographical attributes. The datasets from Geominas have this property.

As a baseline, three standard classification algorithms were used in all databases: decision tree (J48), Radial Basis Function Neural Network (RBF) and Support Vector Machine (SVM). Results of these three techniques were compared to the ones obtained by DMGeo. The Soy Aptitude problem cannot be solved using the standard tools (J48, RBF and SVM) because they do not manipulate geographical attributes. Then, we did a pre-processing of the geographical attributes, using the tool presented in [1], in order to generate a dataset composed just by numeric attributes. Thus, the standard tools were applied to classify this new dataset.

It is important to notice that the Soy Aptitude problem is unbalanced, that is, one class has much more instances than the other one (66% of class A and 34% of class B). To balance this problem, copies of instances of class B were randomly generated to make the number of instances of each class similar. The experiments were made using the balanced and unbalanced datasets.

The experiments using DMGeo in the City Development dataset were conducted in the presence and in the absence of geographic data in two different runs, one using geographic data (Table 3) and one using only numeric data (Table 2).

All algorithms used the cross-validation procedure with 5 folds and the experiment was repeated 3 times with each tool using each base. DMGeo used population size = 200, generations = 200, crossover probability = 90% and mutation probability = 2%.

4.2 Results and Analysis

This section presents the main results obtained with DMGeo and with the standard techniques J48, RBF and SVM. First, we present the global index in Table 1. This index is an average of the results obtained for each class. For example, the DMGeo, in Wine Problem, classified correctly 93% as class A, 90% as B and 99% as C. In this case, the global index is the medium value, which is (93+90+99)/3 = 94%.

Table 1. Actual global index of each tool

	J48		RBF		SVM		DMGeo	
Wine	91%		98%		98%		94%	
Heart	78%		81%		84%		78%	
Soy Aptitude	Unbalanc.	Balanc.	Unbalanc.	Balanc.	Unbalanc.	Balanc.	Unbalanc.	Balanc.
	44%	68%	38%	56%	74%	73%	76%	80%
City Development	62%		58%		57%		82%	

As we can see, DMGeo performed well all these problems.

Table 2 shows the results obtained by each algorithm stratified by each class, as well as the standard deviation (σ). It is important to emphasize that a low standard deviation means that the tool was able to achieve a more homogeneous classification.

Table 2. Actual index for each class

		J48		RBF		SVM		DMGeo	
Wine	A	97%		100%		98%		93%	
	B	87%		96%		100%		90%	
	C	89%		98%		96%		99%	
σ		5.292		**2.000**		**2.000**		4.583	
Heart	A	77%		79%		83%		**85%**	
	B	78%		82%		**84%**		72%	
σ		**0.707**		2.121		**0.707**		9.192	
		Unbalan.	Balan.	Unbalan	Balan	Unbalan	Balan	Unbalan	Balan
Soy Aptitude	A	66%	68%	66%	54%	73%	74%	**93%**	**90%**
	B	22%	68%	10%	58%	**74%**	**70%**	59%	70%
σ		31.113	**0.000**	39.598	2.828	**0.707**	2.828	24.042	14.142
City Development (without geographic att.)	A	78%		78%		78%		75%	
	B	52%		47%		47%		**71%**	
	C	57%		47%		47%		**66%**	
σ		13.796		17.898		17.898		**4.509**	

Table 3. DMGeo with geographic attributes in City Development dataset

A	B	C	σ	Global index
85%	80%	82%	2.516	82%

DMGeo obtained, in this analysis, good results in the datasets that are composed by conventional data. For example, the best results of class C in the Wine dataset and class A in Heart dataset were found by DMGeo. But the biggest contribution of this algorithm is obtained when results of the classification problem can be improved by the geographical analysis. As shown in Table 3, when DMGeo used the geographical attributes the results became better. This table presents the result obtained in City Development dataset with geographic attributes. The results show that the DMGeo took advantage of the topological relations of these geographic attributes to increase its performance. When the dataset contains just geographical attributes a pre-processing tool can be applied and posteriorly a conventional classification algorithm can be used. Nevertheless, in this case, the DMGeo can present a better result.

5   Conclusion

This paper proposes a new evolutionary algorithm that can be applied in classification problems in which numeric and/or geographic data may be present. The algorithm uses niches, elitism and a cache memory to improve its performance, and represents the individual as a SQL WHERE-clause. In order to evaluate the performance of the designed algorithm, tests of classification problems were used. Classical classification algorithms, like Neural Network, Decision Tree and Support Vector Machine were also used to generate results to be compared with those obtained with DMGeo. The comparison shows that the proposed algorithm is competitive and robust, since it has presented the best results in most cases. The main contribution of the proposed algorithm is achieved when the classification problem presents regular and geographical attributes.

Acknowledgment. This work has been supported in part by the Brazilian agency CNPq – Conselho Nacional de Desenvolvimento Científico e Tecnológico.

References

[1] Bogorny, V., Palma, A.T., Engel, P.M., Alvares, L.O.: Weka-GDPM – Integrating Classical Data Mining Toolkit to Geographic Information Systems. In: SBBD Workshop on Data Mining Algorithms and Aplications (WAAMD 2006), Florianopolis, Brasil, October 16-20, pp. 9–16 (2006)

[2] Coello Coello, C.A., Lamont, G.B., Van Veldhuizen, D.A.: Evolutionary Algorithms for Solving Multi-Objective Problems, 2nd edn. Springer, Heidelberg (2007)

[3] Egenhofer, M.A.: Model for Detailed Binary Topological Relationships. Geomatica 47, 261–273 (1993)

[4] Ester, M., Kriegel, A.F.H.P., Sander, J.: Spatial Data Mining: Database Primitives, Algorithms and Efficient DBMS Support. Data Mining and Knowledge Discovery 4(3-4), 193–216 (2000)

[5] Freitas, A.A.: Data Mining and Knowledge Discovery with Evolutionary Algorithms. Natural Computing Series. Spring, Germany (2002)

[6] GeoMINAS - Programa Integrado de Uso da Tecnologia de Geoprocessamento pelos Órgãos do Estado de Minas Gerais, http://www.geominas.mg.gov.br/ (accessed in January 2010)

[7] Goldberg, D.E.: Genetic Algorithms in Search, Optimization and Machine Learning. Addison-Wesley, Reading (1989)

[8] Han, J., Koperski, K., Stefanovic, N.: GeoMiner: A System Prototype for Spatial Data Mining. In: SIGMOD Special Interest Group on Management Of Data, Arizona, EUA, pp. 553–556 (1997)

[9] Holden, N., Freitas, A.A.: Hierarchical classification of protein function with ensembles of rules and particle swarm optimization. Soft Computing Journal 13(3), 259–272 (2009)

[10] Koza, J.R.: Genetic Programming: on the programming of computers by means of natural selection. MIT Press, Cambridge (1992)

[11] Quinlan, J.R.: C4.5: Programs for Machine Learning. Morgan Kaufmann, San Francisco (1993)

[12] Parpinelli, R.S., Lopes, H.S., Freitas, A.A.: Data Mining with an Ant Colony Optimization Algorithm. IEEE Trans. on Evolutionary Computation, special issue on Ant Colony algorithms 6(4), 321–332 (2002)

[13] Pappa, G.L., Freitas, A.A.: Evolving rule induction algorithms with multi-objective grammar-based genetic programming. Knowledge and Information Systems 19(3), 283–309 (2009), http://dx.doi.org/10.1007/s10115-008-0171-1

[14] UCI Machine Learning Repository. University of California, School of Information and Computer Science, Irvine, CA, http://archive.ics.uci.edu/ml (accessed in Jun. 2010)

[15] Vasconcelos, J.A., Ramírez, J.A., Takahashi, R.H.C., Saldanha, R.R.: Improvements in Genetic Algorithms. IEEE Transactions on Magnetics 37(5), 3414–3417 (2001)

[16] Weka, http://www.cs.waikato.ac.nz/ml/weka/ (accessed in December 2010)

[17] Whigham, P.A.: Induction of a marsupial density model using genetic programming and spatial relationships. Ecological Modelling 131, 299–317 (2000)

Artificial Neural Network Modeling for Estimating the Depth of Penetration and Weld Bead Width from the Infra Red Thermal Image of the Weld Pool during A-TIG Welding

S. Chokkalingham[1], N. Chandrasekhar[2], and M. Vasudevan[2]

[1] Department of Production Engineering
PSG college of Technology, Coimbatore
[2] Advanced Welding Processes, Monitoring and Modeling Programme
Materials Technology Division
Indira Gandhi Centre for Atomic Research, Kalpakkam

Abstract. It is necessary to estimate the weld bead width and depth of penetration using suitable sensors during welding to monitor weld quality. Infra red sensing is the natural choice for monitoring welding processes as welding is inherently a thermal processing method. An attempt has been made to estimate the bead width and depth of penetration from the infra red thermal image of the weld pool using artificial neural network models. Real time infra red images were captured using IR camera during A-TIG welding. The image features such as length and width of the hot spot, peak temperature and other features are extracted using image processing techniques. These parameters along with their respective current values are used as inputs while the measured bead width and depth of penetration are used as output of the neural network models. Accurate ANN models predicting weld bead width and depth of penetration have been developed.

Keywords: Artificial neural network, Infra Red Thermal images, image processing, Depth of penetration, Weld bead width, A-TIG welding.

1 Introduction

In the recent decades, there is an extensive development in the field of welding science and technology. Robotization of welding are well investigated and many new intelligent welding robots with arc sensors and vision sensors are developed. The development of automated robotic welding systems for TIG welding is of much importance and its success lies in the development of adaptive/intelligent welding. Intelligent welding enables dynamic altering of the welding parameters to compensate for changing environment. Control of the welding process requires

- Sensing the perturbation occurring during the welding process
- Identifying the perturbation
- Providing the necessary feedback for corrective action

K. Deb et al. (Eds.): SEAL 2010, LNCS 6457, pp. 270–278, 2010.

In conventional welding, the quality of the weld is ascertained only after the welding has been completed with the help of NDT techniques (Ultrasonic or Radiography). In manual welding process, the welder ensures the quality of the weld by monitoring and suitably manipulating the process parameters according to the changing environment based on his knowledge and experience.

In the case of adaptive welding vision system, sensors play the role of welder. Vision systems use the CCD/IR/CMOS cameras for monitoring weld pool size and to control the welding parameters [1,2]. They provide process status information in real time and are an integral part of such an adaptive system. Welding, inherently being a thermal processing method, infrared sensing is a natural choice for weld process information. The basis for using infrared imaging lies in the fact that an ideal welding condition would produce surface temperature distribution that show a regular and repeatable pattern. Any variation or perturbation should result in discernible change in the thermal profiles. Welding being a thermal processing method, thermographic infrared sensing (thermography) using IR sensors are most extensively used technique for sensing, monitoring and control of weld process.

Thermocouples were used in various configurations to measure temperature distributions during the welding process [3-5]. The thermal changes during welding are rapid, the inherent slow response and low spatial response of thermocouples present significant problems in their use for process control. Non-contact measurement of the emissions in the infrared portion of the electromagnetic spectrum is used to extract information about the condition of the weld. Point, line and image analysis techniques have been used to measure the infrared emission during the welding [6]. Using image analysis techniques, seam tracking, bead width control, penetration control and cooling rate control can be successfully implemented using extracted information, which ensures satisfactory weld quality.

Infra red sensing is not without its difficulties, such as the interference of arc radiation and welding electrode emissions, these interferences are mitigated by using either mechanical or optical means to filter the unwanted thermal emissions[7]. To overcome the interference of radiation from the arc, this is done by selecting the sensors, which ignores the wavelength range. Scanning infrared sensor with the spectral response greater than 2µm ignores the radiation of the arc [8]. By employing CCD cameras with band pass filter operating at a wavelength where the arc intensity is low, the camera will be able record the image of the weld pool because of the reduced intensity of the arc light [2]. The reported control systems are based on the assumption that the surface temperature distributions are indication of the conditions below the surface of the weldment. Joint tracking, joint depth penetration, weld parameter variations and surface contaminants are the key variables that must be controlled during online welding to ensure satisfactory weld quality. To ensure reliable weld quality, joint depth penetration and bead width are the two key variables to be sensed and controlled.

The estimation of weld bead geometry such as weld bead width and depth of penetration of the weld pool from the infrared (IR) thermal video of a welding process is an intermediate step towards online control of the welding process. Among the various approaches used to calculate the weld bead geometry parameters, infrared sensing technique has been widely used [9-13]. IR sensing technique has been used for controlling the weld penetration depth [12]. Menaka et al [13] have estimated bead width and depth of penetration during welding by IR imaging. Bead width was computed based on sharp temperature gradients and change in emissivity (perceived as dip-inflection) in the profile. This point of inflection was more clearly seen from the 1st derivative plot [8,11,13]. Another approach for bead width measurement is by calculating the full width half maximum (FWHM) from the line profile across the weld [9,13]. Rider et al[14-15] used a linear array of silicon photodiodes as the sensor for measuring weld bead width.

Menaka et al[13] estimated that as the depth of penetration increases, the peak temperature attained the width and area under the temperature curve also increases. Chen and Chin [9] found out that there exists an exponential relationship between depth of penetration and area under the peak temperature profile. Malmuth et al[16] were one of the first to report on the use of a full two dimensional image of the surface temperature distribution during arc welding. They related weld penetration depth to thermal distribution of the plate. Nagarajan et al[10] and Chen et al [9] proposed and implemented a penetration control strategy based upon the size and area controlled by the surface isotherms, where an ellipse was fitted to the lower half of the selected isotherm. The length of the minor axis and the sum of all the intensity values in this region were found to be proportional to the penetration depth. Saeed and zhang [17] have calculated the depth of the weld pool surface using images captured by a calibrated charge couple device sensor. Pietrzak and Packer [18] have designed methods for controlling weld penetration using weld pool using geometry sensing. Baneerjee et al [19] and Govardhan [20] implemented bead width and penetration control systems based upon the thermal gradient across the weld pool. Vasudevan et al [21] enhanced the penetration performance of TIG welding process with the help of a specific activated flux. Sheela Rani et al [22] developed an image segmentation algorithm for characterizing the lack of penetration from the thermographs for online weld monitoring.

In recent years, artificial neural networks (ANNs) and fuzzy logic have been used for predicting weld microstructure, weld bead geometry parameters, weld recognition and weld defect detection [23]. A similar approach in which a model on artificial neural network for predicting the welding parameters (current voltage and torch speed) from the weld geometry was done by Chan et al [24]. Nandhitha et al [25] predicted the torch current using feed forward artificial neural network for monitoring lack of penetration from IR thermal images. Lee and Um [23] have used neural networks to predict welding process parameters by predicting back bead geometry. Ghanty et al [26] have used neural networks for predicting the weld bead geometry parameters after analyzing the weld control features using an online feature selection (OFS) technique. With the help of Bayesian neural network, a model predicting the ferrite content in stainless steel was developed by Vasudevan et al [27].

The present work attempts to develop accurate artificial neural network based models for estimating weld bead width and depth of penetration from the acquired infra red thermal images of weld pool during A-TIG welding. From the acquired images, various image features are extracted using image processing techniques. The extracted image features and along with the current used during welding form the input variables of the ANN models. The measured weld bead width and depth of penetration are the out put variables. The objective of the work also included validating the model predictions by carrying out experiments.

2   Experimental

A-TIG welding is carried out on 3mm thick 316LN Austenitic Stainless Steel plates of dimension 125x50mm to make square butt joints with the help of precision TIG 375 welding machine. Before welding, the plates are preprocessed. During the welding process, the surface temperature distributions of the plate being welded are measured with the help of an IR camera. It uses a high band filter to permit only a portion of the emitted energy within a wavelength range of 4.99 μm-5.1 μm thereby minimizing the interference from the arc light and hot tungsten electrode on the image quality.

2.1   Image Acquisition

The thermal radiation from the object is focused by the built-in-lens in which the image is acquired. Each scan of the camera is transferred as a video frame consisting of 320x240 discrete intensity measurements. The camera is capable of measuring temperature in the range of 473– 1973 K with an accuracy of +/-1% over the entire range. Each frame consists of the thermal images, these images has a high-resolution color image appearing on the external monitor. These images are analyzed using the Altair software. It has a number of features and tools for quantitative analysis of the images.

The plates were cleaned after welding and then polished sequentially using Automatic polishing machine. They are sectioned transversely from the beginning of the plate at 8mm, 55mm and 92mm, for measuring the bead width (BW) and depth penetration (DOP) using a machinist microscope. The experiment is conducted with various current values starting from 50 amperes to 95 amperes with increment of 5 amperes and at constant torch speed of 120 mm/min to achieve different values of depth of penetration and weld bead width.

2.2   Feature Extraction

The video frames of the surface temperature distributions corresponding to a particular cut section are determined from the torch speed and the video frame rate. For example: Torch speed = 2mm/s (120mm/min), Frame rate = 25 frames/s, Video frame corresponding to the cut section at 8mm = (8x25)/2 =100th frame, which is referred as the key frame (t). Similarly, the cut sections are selected at 55mm and 92mm, so their corresponding key frames will be 688 and 1150. At each key frame, 8 features are extracted from ten welding experiments with the help of Altair and Origin software. They have a number of features and tools for quantitative analysis of the images. For feature extraction, only the weld pool (hot spot) region is processed in order to accurately measure the length and width of the weld pool. Around 8 features are extracted from the weld pool region.

The extracted eight features are as follows,

- Length and width of the weld pool, L(t) and W(t) respectively.
- Peak temperature, T(t) based on line of scan.
- Mean μ(t) and Standard deviation σ(t) of the Gaussian temperature profile.
- Thermal area (A) under the Gaussian temperature profile.

- Full Width Half Maximum of the Gaussian temperature curve
- Bead width computed from the first derivative curve of the Gaussian temperature curve.

These features are extracted for 3 scan lines in every frame (100, 688, 1150). Along with these eight image-extracted features, their corresponding current value is also considered as another input variable. There are totally 9 input variables and one out put variable in each of the models. Weld bead width and depth of penetration are the out put variables which are measured from the cut sections of the weld joints. Totally 90 data sets are generated and are used for developing the artificial neural network models. Among the 90 data sets, 70 data sets with their corresponding known weld bead geometry are chosen randomly as the training input to train the neural network model and the remaining 20 data sets are used to test the neural network model.

3   Results and Discussion

3.1   Artificial Neural Network Model on Depth of Penetration

Number of neurons/nodes in the hidden layer is varied from 3 to 15 and the RMS error was determined for test data set. The network architecture exhibiting minimum RMS error for predicting depth of penetration was identified as 9-9-1. The RMS error value determined was 0.12934. Then the above network was used to predict the depth of penetration for the entire 90 dataset. Figure 1 compares the predicted and measured values of depth of penetration for the entire dataset. There was good agreement between the predicted and measured depth of penetration values.

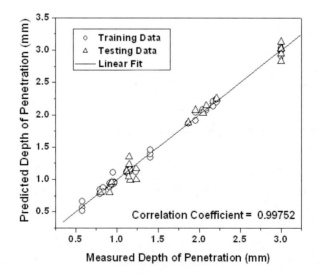

Fig. 1. Comparison between predicted and measured depth of penetration for entire data set

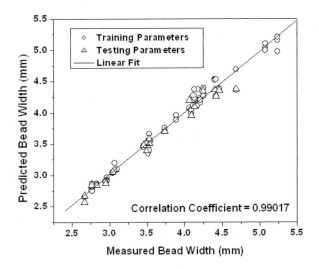

Fig. 2. Comparison between the predicted and measured weld bead width for the entire dataset

3.2 Artificial Neural Network Model on Weld Bead Width

Among all these models, the model with 11 hidden nodes exhibited minimum RMS error value of 0.10542 for the test dataset. The model with 9-11-1 architecture is selected as the best for predicting weld bead width. The optimized network was used to predict weld bead width for the entire 90 dataset. Figure 2 compares the predicted and measured weld bead width for the entire dataset. There was again good agreement between the predicted and measured values.

3.3 Validation of the Neural Network Models

During validation of the model predictions, two plates of 316LN stainless steel of same dimensions are butt welded with 60 and 85 amperes of current. Infra red images were captured online. By sectioning the weld joints at 30mm, 50mm, 70mm, 90mm, 110mm, measurements were carried out on depth of penetration and weld bead width. Respective to the sectioned regions, the frame numbers are identified and all the eight image features are extracted as explained earlier. Those image features along with their respective current values are tabulated and randomized. They are fed as input to the optimized model predicting the depth of penetration and bead width in order to validate them. The physically measured bead width and depth of penetration at the sectioned regions with their corresponding frame number are tabulated in the table 1.

The depth of penetration and weld bead width predicted by the optimized neural network models are compared with the actual depth of penetration and the weld bead width measured in figures 3 (a) & (b) respectively. There was good agreement between the values implying that the artificial neural network models are good in prediction. The slightly high error values on depth of penetration for the validation data was due to more noise in the captured images. Filtering of noise in the captured images may reduce the error further. Artificial neural network models correlating weld bead width and depth of penetration with input variables as the image features extracted from the infrared thermal images and the current values have been developed. The above models may find application during real time monitoring and control of weld quality in terms of weld bead geometry during A-TIG welding.

Table 1. Measured values of weld bead width and depth of penetration

Current (A)	Sectioned Region (mm)	Frame No.	Measured Bead Width (mm)	Measured Depth of Penetration (mm)
60	30	377	3.377	1.143
	50	630	3.397	1.118
	70	880	3.529	1.286
	90	1132	3.243	1.271
	110	1384	3.416	1.264
85	30	377	4.337	3
	50	630	4.39	3
	70	880	4.435	3
	90	1132	4.304	3
	110	1384	3.932	3

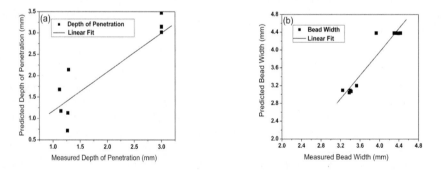

Fig. 3. Comparison between predicted and measured (a) depth of penetration (b) weld bead width for validation experiments

4   Conclusions

1. Artificial neural network models have been developed for predicting the weld bead width and the depth of penetration from the image features extracted from IR thermal images of weld pool and the current used for welding.
2. The optimized neural network model for predicting depth of penetration accurately consists of 9 input parameters, 9 neurons in the hidden layer and 1 output

(9-9-1). The RMS error for the test data set was 0.129. The correlation coefficient obtained between the predicted and measured depth of penetration was 0.99184 for the test dataset.

3. Similarly, another optimized neural network model for predicting weld bead width consists of the same 9 input parameters, 11 neurons in the hidden layer and 1 output (9-11-1). The RMS error for the test data set was 0.11. The correlation coefficient obtained between the predicted and measured weld bead width was 0.98862 for the entire test dataset.

4. Validation of the optimized models predicting depth of penetration and weld bead width were carried out and the correlation coefficient obtained between the predicted and measured depth of penetration was 0.92255. Similarly, the correlation coefficient obtained between the predicted and measured weld bead width on validation was 0.95525.

All the developed ANN models are found to be accurate and may find application during on-line monitoring and control of weld bead geometry during A-TIG welding of type 316 LN stainless steels.

References

1. Smith, J.S., Lucas, W.: Vision Based Systems for Controlling the Arc Welding operation and Inspecting the Weld Bead Profile. Welding in the World 43(suppl. issue), 10–22 (1999)
2. Houghton, M.A., Lucas, J.J., Lucas, W.: Vision Systems for Monitoring and Control of Arc Welding Operations. Soldagem Insp. Sao Paulo 12(4), 293–299 (2007)
3. Barry, J.M., Paley, Z., Adams Jr., C.M.: Heat conduction from moving Arc in Welding. Welding Journal 42(3), 97-s–104-s (1963)
4. Dorschu, K.E.: Control of Cooling rate in Steel Weld metal. Welding Journal 50(11), 49-s–62-s (1968)
5. Kannatey-Asibu Jr., E., Kikuchi, N., Jallard, A.R.: Experimental Finite Element Analysis of Temperature distribution during Arc Welding. Journal of Engineering Materials and Technology 111, 9–18 (1989)
6. Wickle, H.C., Chen, F., Nagarajan, S., Chin, B.A.: Survey of Infrared Sensing Techniques for Welding process Monitoring and Control. Journal of Chinese Institute of Engineers 21(6), 645–657 (1998)
7. Nagarajan, S., Chen, W.H., Chin, B.A.: Infrared Sensing for Adaptive Arc Welding. Welding Journal 68(11), 462-s – 466-s (1989)
8. Nagarajan, S., Banerjee, P., Chin, B.A.: Thermal imaging for Weld quality Control in Arc Welding processes. Transport Phenomena in Materials Processing 146, 171–178 (1990)
9. Chen, W., Chin, B.A.: Monitoring Joint Penetration using Infrared Sensing Techniques. Welding Journal 67, 181s–185s (1988)
10. Nagarajan, S., Banerjee, P., Chen, W.H., Chin, B.A.: Control of Welding Process using Infrared Sensors. IEEE Transactions on Robotics and Automation 8(1), 86–92 (1992)
11. Ghanty, P., Vasudevan, M., Chandrasekhar, N., Mukherjee, D., Maduraimuthu, V., Pal, N.R., Bhaduri, A.K., Bharat, P., Raj, B.: An Artificial Neural Network Approach for estimating weld bead width and depth of penetration from Infra red Thermal image of weld pool. Science and Technology of Welding and Joining 13(4), 395–401 (2008)

12. Wikle, H.C., Kottilingam, S., Zee, R.H., Chin, B.A.: Infrared Sensing Techniques for Penetration depth Control of the Submerged Arc Welding process. Journal of Materials Processing Technology 113, 228–233 (2001)
13. Menaka, M., Vasudevan, M., Venkatraman, B., Raj, B.: Estimating Bead Width and Depth of Penetration during Welding by Infrared Thermal imaging. Journal of British Institute of NDT 47(9), 564–568 (2005)
14. Rider, G.: Control of Weld pool size and position for Automatic and Robotic Sensory Welding. In: Proceedings of SPIE Third Inernational Conference on Robot Vision and Sensory Control, Cambridge, MA (November 1983)
15. Rider, G.: Measurement of Weld pool size by Self-scanned Photodiode Arrays. In: Proc. IEE International Conference on Low Light and Thermal Imaging Systems, London, pp. 3–5 (1975)
16. Malmuth, N.D., Hall, W.F., Davis, B.I., Rosen, C.D.: Transient Thermal Phenomena and Weld Geometry in GTAW. Welding Journal 53(9), 388–400 (1974)
17. Saeed, G., Zhang, Y.M.: Weld pool surface depth measurement using a Calibrated Camera and Structured light. Measurement Science and Technology 18, 2570–2578 (2007)
18. Pietrzak, K.A., Packer, S.M.: Vision based Weld Pool Width Control. ASME Journal of Engineering for Industry 116, 86–92 (1994)
19. Baneerjee, P., Liu, J.Y., Chin, B.A.: Infrared Thermography for Non-destructive Monitoring of Weld Penetration Variations. In: Liu, M. (ed.) Proceedings of the Japan U.S.A. Symposium on Flexible Automation, pp. 291–295. American Society of Mechanical Engineers (1992)
20. Govardhan, S.M., Chin, B.A.: Monitoring GTA Weld Puddle Geometry using Measured Temperature Gradients. In: David, S.A., Vitek, J.M. (eds.) Recent trends in Welding Science and Technology, pp. 383–386. American Society for Metals International Materials Park OH (1990)
21. Vasudevan, M.: Computational and Experimental Studies on Arc Welded Austenitic Stainless Steels. PhD Thesis, Indian Institute of Technology, Chennai (2007)
22. Nandhitha, N.M., Manoharan, N., Sheela Rani, B., Venkataraman, B., Vasudevan, M., Chandrasekar, Kalyana Sundaram, P., Raj, B.: Euclidean Distance Based Colour Image Segmentation Algorithm for Dimensional Characterization of Lack of Penetration from Weld Thermographs for Online Weld Monitoring. In: GTAW, 17th world Conference on Non-destructive testing WCNDT, Shanghai, China, October 26-28 (2008)
23. Lee, J.I., Um, K.W.: A Prediction of Welding process parameters by Prediction of Back bead geometry. Journal of Materials Processing and Technology 108, 106–113 (2000)
24. Chan, B., Pacey, J., Bibby, M.: Modeling Gas Metal Arc Weld Geometry using Artificial Neural Network Technology. Canadian Metallurgical Quarterly 38(1), 43–51 (1999)
25. Nandhitha, N.M., Manoharan, N., Sheela rani, B., Venkataraman, B., Vasudevan, M., Kalyana Sundaram, P., Raj, B.: Prediction of Torch Current Deviation using Feed Forward Neural Network for Monitoring Lack of Penetration from Thermal Images in GTAW of AISI Stainless Steel 316. International Journal of Intelligent Information Processing 3(2), 271–279 (2009)
26. Ghanty, P., Paul, S., Mukherjee, D.P., Vasudevan, M., Pal, N.R., Bhaduri, A.K.: Modelling Weld Bead geometry using Neural Networks for GTAW of an Austenitic Stainless Steel. Science and Technology of Welding and Joining 12(7), 649–658 (2007)
27. Vasudevan, M., Muruganath, M., Bhaduri, A.K., Raj, B., Prasad Rao, K.: Bayesian Neural Network Analysis of Ferrite number in Stainless Steel. Science and Technology of Welding and Joining 9(2), 109–120 (2004)

Swarm Reinforcement Learning Method Based on an Actor-Critic Method

Hitoshi Iima and Yasuaki Kuroe

Kyoto Institute of Technology,
Matsugasaki, Sakyo-ku, Kyoto, Japan
{iima,kuroe}@kit.ac.jp

Abstract. We recently proposed swarm reinforcement learning methods in which multiple agents are prepared and they learn not only by individual learning but also by learning through exchanging information among the agents. The methods have been applied to a problem in discrete state-action space so far, and Q-learning method has been used as the individual learning. Although many studies in reinforcement learning have been done for problems in the discrete state-action space, continuous state-action space is required for coping with most real-world tasks. This paper proposes a swarm reinforcement learning method based on an actor-critic method in order to acquire optimal policies rapidly for problems in the continuous state-action space. The proposed method is applied to an inverted pendulum control problem, and its performance is examined through numerical experiments.

1 Introduction

In ordinary reinforcement learning methods [1], a single agent learns to achieve a goal through many episodes. Since the agent essentially learns by trial and error, it takes much computation time to acquire optimal policies especially for complicated learning problems. Meanwhile, for optimization problems, population-based methods such as genetic algorithms and particle swarm optimization (PSO) [2] have been recognized that they are able to find rapidly global optimal solutions for multi-modal functions with wide solution space. It is expected that by introducing a concept of population-based methods into reinforcement learning methods, optimal policies can be found rapidly even for complicated learning problems.

We recently proposed reinforcement learning methods using the concept of the population-based methods, and call them *swarm reinforcement learning methods* [3,4]. In the methods, multiple agents are prepared and they all learn concurrently with two learning strategies: individual learning and learning through exchanging information. In the former strategy, each agent learns individually by using a usual reinforcement learning method such as Q-learning method [5]. In the latter strategy, the agents exchange information among them regularly during the individual learning and they learn based on the exchanged information.

K. Deb et al. (Eds.): SEAL 2010, LNCS 6457, pp. 279–288, 2010.

Learning methods called multi-agent reinforcement learning have been proposed [6]. Basically, the aim of the multi-agent reinforcement learning methods is to acquire optimal policies in tasks achieved by cooperation or competition among multiple agents, and each of the agents regards information of other agents as a part of environments. Therefore, the concept and objective of the swarm reinforcement learning methods are different from those of the multi-agent reinforcement learning methods. The swarm reinforcement learning methods could treat both tasks achieved by a single agent and achieved by cooperation or competition among multiple agents. In the methods, multiple agents are prepared in order to make some of them learn in shorter learning time even for complicated reinforcement learning problems.

The swarm reinforcement learning methods in [3,4] are applied to problems in discrete state-action space as the first stage of this study. Continuous state-action space should, however, be adopted for coping with most real-world tasks, and effective learning methods are required to be developed for solving complicated problems in the continuous state-action space. Thus, this paper proposes a swarm reinforcement learning algorithm which enables to acquire optimal policies rapidly for the problems in the continuous state-action space. In the proposed method, an actor-critic method is used in the individual learning because actor-critic methods are known to be easily applicable to the continuous problems [7,8]. The proposed method is applied to an inverted pendulum control problem of a single-agent task, and its performance is examined through numerical experiments.

2   Swarm Reinforcement Learning Methods

In this section, we briefly explain the swarm reinforcement learning methods [3,4]. The swarm reinforcement learning methods are motivated by population-based methods in optimization problems. Multiple agents are prepared and they all learn concurrently with two learning strategies: individual learning and learning through exchanging information. In the former strategy, each agent learns individually by using a usual reinforcement learning method. For the individual learning, any usual reinforcement learning method can be used. In [3,4], Q-learning method which is a typical reinforcement learning method is used, and each agent updates its own state-action values (Q-values) by using the update equation of Q-learning method.

In the latter strategy, the agents exchange their information among them and learn based on the exchanged information every after repeating the individual Q-learning a certain number of times. Each agent updates its own Q-values by referring to the Q-values which are evaluated to be more useful and superior to those of the other agents. For this purpose, the Q-values of each agent are evaluated after the individual Q-learning is performed. They are evaluated by an appropriate method so that the superior Q-values can be selected. Each agent updates its own Q-values by using the superior Q-values.

In ordinary reinforcement learning methods with a single agent, the agent often takes a useless action bringing a small reward, which makes learning time longer. On the other hand, in the swarm reinforcement learning methods, since the multiple agents are prepared, some of these agents could take useful actions bringing a larger reward. By the information exchange, all the agents could receive information of the agents who take the useful actions. It is therefore expected that some agents can acquire the optimal policy in a shorter learning time even for complicated reinforcement learning problems.

The flow of the swarm reinforcement learning method is as follows.

Step 1 Each of multiple agents updates its own Q-values by performing the individual Q-learning for a specified number of episodes.

Step 2 The Q-values of each agent are evaluated by an appropriate method so that superior Q-values can be selected.

Step 3 Based on the evaluation of Step 2, the superior Q-values are selected. The Q-values of each agent are updated by using the superior Q-values.

Step 4 If the termination condition of the swarm reinforcement learning method is satisfied, terminate the method. Otherwise, return to Step 1.

3   Proposed Method Based on an Actor-Critic Method

This section proposes a swarm reinforcement learning method based on an actor-critic method in order to acquire optimal policies rapidly for problems in continuous state-action space. First we outline the actor-critic method which is used in the individual learning, and then we propose the swarm reinforcement learning method using it.

3.1   Actor-Critic Method

Actor-critic methods [1] consist of two learning modules: actor and critic. In the actor module, the policy of each agent is updated based on state values in such a way that the agent could gain more rewards. In the critic module, the state values are estimated for the policy. The agent perceives the current state, and takes an action according to the policy. Consequently, it gains a reward and perceives the next state. By using the reward and the value of the next state, the value of the current state is updated in the critic module, and then the policy is updated based on values in the actor module.

Several learning methods in the actor and critic modules have been proposed so far, and any actor-critic method can be used as the individual learning of the proposed swarm reinforcement learning method. In this paper, we adopt an actor-critic method using eligibility traces [7].

In this actor-critic method, the value v_s of each state s is updated by means of TD(0) method [1]. Namely, when the agent who perceives the current state s^* gains a reward r and perceives the next state s_n by taking an action, the value v_{s^*} of the current state is updated by

$$ER^{\mathrm{TD}} = r + \gamma^{\mathrm{V}} v_{s_n} - v_{s^*} \tag{1}$$

$$v_{s^*} \leftarrow v_{s^*} + \alpha^{\mathrm{C}} ER^{\mathrm{TD}} \tag{2}$$

where ER^{TD} is the TD error, γ^{V} is the discount-rate parameter of the state values and α^{C} is the learning-rate parameter in the critic module.

In the actor module, the policy is expressed by a function of some parameters $\{w_k\}$ $(k = 1, 2, \cdots, N^{\mathrm{A}})$ whose number is denoted as N^{A}. As a typical example, the policy p is given by the normal distribution

$$p = \frac{1}{\sigma\sqrt{2\pi}} \exp\left(\frac{-(a-\mu)^2}{2\sigma^2}\right) \tag{3}$$

where μ and σ are the mean and the standard deviation, respectively, and they are usually expressed by functions of the actor parameters $\{w_k\}$. For instance, they are given by (11)(12) as shown in Sect. 4. According to a random value based on the normal distribution, the agent selects an action a. When the agent who perceives the current state s^* takes action a^*, the values of all the actor parameters $\{w_k\}$ are updated by

$$e_k = \left.\frac{\partial \ln p}{\partial w_k}\right|_{s=s^*, a=a^*, w_k=w_k^*} \tag{4}$$

$$et_k \leftarrow e_k + \gamma^{\mathrm{E}} et_k \tag{5}$$

$$w_k \leftarrow w_k + \alpha^{\mathrm{A}} ER^{\mathrm{TD}} et_k \tag{6}$$

$$(k = 1, 2, \cdots, N^{\mathrm{A}})$$

where w_k^* is the value of w_k before the update, e_k is the eligibility, et_k is the eligibility trace, γ^{E} is the discount-rate parameter of the eligibility traces and α^{A} is the learning-rate parameter in the actor module.

The flow of the actor-critic method for one episode is as follows.

Actor-critic method for one episode

Step 1   Initialize the current state s^*, and set $et_k \leftarrow 0$ for $\forall k$.

Step 2   The agent takes action a^* according to a random value based on the normal distribution (3).

Step 3   The agent gains a reward r and perceives the next state s_{n}. Update the value of critic parameter v_{s^*} for the current state by using (1)(2).

Step 4   Update all the eligibility traces $\{et_k\}$ and the values of all the actor parameters $\{w_k\}$ by using (4)–(6).

Step 5   If the terminate condition of episode is satisfied, terminate the episode. Otherwise, return to Step 2.

3.2   Basic Framework of the Proposed Method

This subsection describes the basic framework of the proposed swarm reinforcement learning method. In the proposed method, multiple agents learn by individual learning and learning through exchanging information among them. In the individual learning, each agent i updates its own values of actor-critic parameters $\{v_{is}, w_{ik}\}$ by using the above-mentioned actor-critic method. After the individual learning, the values of the actor-critic parameters of each agent are evaluated by an appropriate method so that the superior values can be selected, and each agent updates its own values by using the superior values.

The flow of the proposed swarm reinforcement learning method is as follows.

Step 1 Each of the multiple agents updates its own values of the actor-critic parameters by individually performing the actor-critic method for a specified number of episodes.

Step 2 The values of the actor-critic parameters of each agent are evaluated by an appropriate method so that the superior values can be selected.

Step 3 Based on the evaluation of Step 2, the superior values are selected. The values of the actor-critic parameters of each agent are updated by using the superior values, which is explained in the next subsection.

Step 4 If the termination condition of the proposed swarm reinforcement learning method is satisfied, terminate the method. Otherwise, return to Step 1.

3.3   Information Exchange Method among Agents

The performance of the swarm reinforcement learning method depends on a method of exchanging the information, which should be appropriately designed. Because population-based methods are often used for rapidly finding a global optimal solution in optimization, an optimal policy could be also found rapidly by applying an update procedure used in the population-based methods to the swarm reinforcement learning method. PSO [2] is a promising population-based method which originates in social behavior. Each agent updates its own candidate solution by utilizing its own personal best and the global best. In this paper, we adopt the update procedure of PSO for exchanging the information among the agents.

In the proposed method, each agent updates its own value of the actor-critic parameter by using the update equations of PSO. The personal best of agent i and the global best are selected and memorized. The personal best is defined as the value of the actor-critic parameter which is evaluated to be superior to any other value among the values found by agent i so far. The global best is defined as the value of the actor-critic parameter which is evaluated to be superior to any other value among the values found by all agents so far. The update equations are given by

$$V_{ik}^{A} \leftarrow WV_{ik}^{A} + C_1 R_1^{A}(P_{ik}^{A} - w_{ik}) + C_2 R_2^{A}(G_k^{A} - w_{ik}) \tag{7}$$
$$w_{ik} \leftarrow w_{ik} + V_{ik}^{A} \tag{8}$$
$$V_{is}^{C} \leftarrow WV_{is}^{C} + C_1 R_1^{C}(P_{is}^{C} - v_{is}) + C_2 R_2^{C}(G_s^{C} - v_{is}) \tag{9}$$
$$v_{is} \leftarrow v_{is} + V_{is}^{C} \tag{10}$$

where
V_{ik}^{A}, V_{is}^{C} : so-called velocities for actor-critic parameters,
W, C_1, C_2 : weight parameters,
R_1^{A}, R_2^{A}, R_1^{C}, R_2^{C} : uniform random numbers in the range from 0 to 1,

P_{ik}^A, P_{is}^C : personal best values of actor-critic parameters for agent i,
G_k^A, G_s^C : global best values of actor-critic parameters.

3.4   Flow of the Proposed Method

The flow of the proposed swarm reinforcement learning method is as follows.

I : number of agents.
T : total number of episodes.
Y : number of episodes for which the individual actor-critic method is performed
     between the information exchange among the agents.

Step 1 Set the initial values of actor-critic parameters w_{ik}, v_{is} for $\forall i, k, s$. In
        addition, set the initial values of velocities V_{ik}^A, V_{is}^C for $\forall i, k, s$. Set $t \leftarrow 1$,
        where t is the number of episodes.
Step 2 Each agent i updates its own values of the actor-critic parameters w_{ik}, v_{is}
        by individually performing the actor-critic method using the eligibility
        traces for Y episodes.
Step 3 The updated values are evaluated by an appropriate method.
Step 4 If the actor-critic parameters $\{w_{ik}, v_{is}\}$ of each agent i are superior to
        the personal best $\{P_{ik}^A, P_{is}^C\}$, set $P_{ik}^A \leftarrow w_{ik}$ and $P_{is}^C \leftarrow v_{is}$ for $\forall k, s$.
        In addition, if $\{w_{ik}, v_{is}\}$ are superior to the global best $\{G_k^A, G_s^C\}$, set
        $G_k^A \leftarrow w_{ik}$ and $G_s^C \leftarrow v_{is}$ for $\forall k, s$.
Step 5 All the values of the actor-critic parameters and the velocities for all the
        agents are updated by (7)–(10).
Step 6 If the termination condition $t \geq T$ is satisfied, terminate the proposed
        method. Otherwise, set $t \leftarrow t + IY$ and return to Step 2.

4   Numerical Experiments

The performance of the proposed swarm reinforcement learning method is evaluated by applying it to an inverted pendulum control problem, and by comparing its computational efficiency with that of ordinary actor-critic method using a single agent.

4.1   Inverted Pendulum Control Problem

In the inverted pendulum device shown in Fig. 1, the cart can move horizontally by impressing a voltage to the electric motor, and the pole is mounted on the cart. The pole swings according to movement of the cart. The inverted pendulum control problem is to design a controller to remain standing the pole upright after the inverted pendulum device is actuated from the initial state that the pole is slightly leaned. Let x_1, x_2, x_3, x_4 and u be the position [m] of the cart, the angle [rad] of the pole, the velocity [m/s] of the cart, the angular velocity [rad/s] of the pole, and the input voltage [V], respectively.

The initial values of (x_1, x_2, x_3, x_4) are denoted as $(x_{10}, x_{20}, x_{30}, x_{40})$. If $|x_1| > x_1^{\max}$, $|x_2| > x_2^{\max}$, $|x_3| > x_3^{\max}$ or $|x_4| > x_4^{\max}$ after the inverted pendulum device

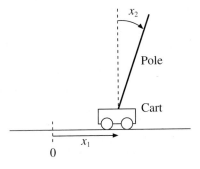

Fig. 1. Inverted pendulum

is actuated, we judge that the attempt to control the pendulum has failed. If the attempt does not fail during a specified time T_s [s], we judge that the pendulum successfully remains standing.

4.2   Experimental Set Up

For the problem parameters, the following values are used:
initial values of state variables :
$$(x_{10}, x_{20}, x_{30}, x_{40}) = (0.01[\mathrm{m}], 0.5[\mathrm{rad}], 0[\mathrm{m/s}], 0[\mathrm{rad/s}]),$$
bound in control failing condition :
$$(x_1^{\max}, x_2^{\max}, x_3^{\max}, x_4^{\max}) = (2.4[\mathrm{m}], 0.6[\mathrm{rad}], 2[\mathrm{m/s}], 1.5[\mathrm{rad/s}]),$$
time required for judging that the pendulum successfully remains standing :
$$T_s = 60[\mathrm{s}].$$

The following methods are applied to the inverted pendulum control problem, and their computational efficiency is compared:
PSO-AC : proposed swarm reinforcement learning method using the actor-critic method and PSO,
1-AC   : ordinary actor-critic method using a single agent.

In these methods, an agent is a controller and its action a corresponds to the input voltage u. The policy p is defined by (3), and the mean μ and the standard deviation σ are expressed by

$$\mu = \sum_{k=1}^{4} w_k x_k \tag{11}$$

$$\sigma = \frac{1}{1 + \exp(-w_5)} \tag{12}$$

where w_k ($k=1,2,3,4,5$) is the actor parameter [7]. From (3)(4)(11)(12), each eligibility is given by

$$e_k = \frac{(u^* - \mu)x_k}{\sigma^2} \qquad (k = 1, 2, 3, 4) \tag{13}$$

$$e_5 = \frac{\{(u^* - \mu)^2 - \sigma^2\}(1 - \sigma)}{\sigma^2} . \tag{14}$$

For the parameters of each method, the following values are used:

number of agents : $\quad\quad\quad\quad\quad\quad\quad\quad I = 10$,

total number of episodes : $\quad\quad\quad\quad T = 20000$,

number of episodes for which the individual learning is performed between the information exchange :

$$Y = 3,$$

weights in (7)(9) : $\quad\quad\quad\quad\quad W = 0.2, C_1 = C_2 = 1.8$,

learning-rate in the actor module : $\quad \alpha^A = 0.2$,

discount-rate of the eligibility traces : $\gamma^E = 0.9$,

learning-rate in the critic module : $\quad \alpha^C = 0.6$,

discount-rate of the state values : $\quad \gamma^V = 0.95$.

These values are determined through the preliminary experiments in such a way that each method works as good as possible.

If the attempt to control the pendulum fails in an episode, then the agent gains reward -1 and the episode is terminated. The episode is also terminated when the pendulum successfully remains standing for T_s seconds. If the pendulum remains standing for longer time, the values of actor-critic parameters are evaluated to be superior.

In the numerical experiments, the following model is used for the inverted pendulum device:

$$\dot{x}_1 = x_3 \tag{15}$$

$$\dot{x}_2 = x_4 \tag{16}$$

$$\dot{x}_3 = \frac{m_p \ell x_4^2 \sin x_2 - m_p g \cos x_2 \sin x_2 - c x_3}{m_c + m_p - m_p \cos^2 x_2}$$
$$+ \frac{K_t K_g}{R_a r_c (m_c + m_p - m_p \cos^2 x_2)} u \tag{17}$$

$$\dot{x}_4 = \frac{-m_p \ell x_4^2 \cos x_2 \sin x_2 + (m_c + m_p) g \sin x_2 + c x_3 \cos x_2}{\ell(m_c + m_p) - m_p \ell \cos^2 x_2}$$
$$- \frac{K_t K_g \cos x_2}{R_a r_c \{\ell(m_c + m_p) - m_p \ell \cos^2 x_2\}} u \tag{18}$$

$$c = \frac{K_t K_e K_g^2}{R_a r_c^2} \tag{19}$$

where m_c=0.522, K_t=0.00767, K_e=0.00767, R_a=2.6, r_c=0.00635, K_g=3.7, m_p= 0.212, ℓ=0.305 and g=9.8.

4.3   Results and Discussion

Figure 2 shows the variation of balancing time through the learning phase obtained by each method. The x-axis in this figure is the number of episodes. In the swarm reinforcement learning method, the x-axis is not the number of episodes

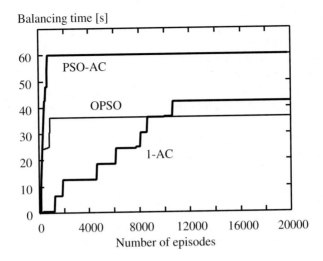

Fig. 2. Variation of balancing time through the learning phase

for each single agent, and is the sum of numbers of episodes for all the agents. The y-axis is the balancing time, namely the time the attempt to control the pendulum fails. If the pendulum successfully remains standing, the balancing time is counted as $T_s(=60)$. The balancing time in the figure is the maximum among those obtained so far. Each method is performed ten times with various random seeds, and their results are averaged.

It is confirmed from Fig. 2 that the balancing time of PSO-AC at each episode is longer than that of 1-AC. Therefore, learning with multiple agents works better. Since the balancing time of PSO-AC reaches 60 seconds at a small number of episodes, PSO-AC finds the optimal policies in shorter computation time for all the ten trials.

In PSO-AC, optimal policies can be found rapidly by using both the actor-critic method and the update equations of PSO. In order to discuss this fact, we evaluate the performance of a method in which the values of actor parameters are updated by using only the update equations of PSO. For this purpose, the original PSO (OPSO) is applied to the inverted pendulum control problem. In OPSO, a candidate solution is the actor parameters $\{w_{ik}\}$, and the objective function is the balancing time in the case where the input voltage u is given by $u = \mu$ and (11).

The experimental result of OPSO is shown in Fig. 2. Whereas both OPSO and 1-AC do not necessarily find optimal policies, PSO-AC can find them. Because the personal best and the global best in OPSO are not necessarily good policies at and shortly after the first episode, the policy of each agent is not improved by using these bests. By contrast, in PSO-AC, since the individual actor-critic method is applied before the application of the update equations of PSO, the personal best and the global best in PSO-AC are superior to those in OPSO, and the policy of each agent is improved by using these bests. Therefore, better polices can be found rapidly by PSO-AC.

5   Conclusion

This paper has proposed a swarm reinforcement learning method based on an actor-critic method for solving problems in continuous state-action space. In the proposed method, multiple agents are prepared, and the update equations of PSO are used for exchanging information among the agents. In order to evaluate the performance of the proposed method, the proposed method has been compared with the one-agent actor-critic method and the original PSO in numerical experiments. It is concluded from the experimental results that the proposed method outperforms the other methods.

Acknowledgments

This work was partly supported by the Ministry of Education, Science, Sports and Culture, Grant-in-Aid for Scientific Research (C) (22500131).

References

1. Sutton, R.S., Barto, A.G.: Reinforcement Learning. MIT Press, Cambridge (1998)
2. Kennedy, J., Eberhart, R.C.: Swarm Intelligence. Morgan Kaufmann Publishers, San Francisco (2001)
3. Iima, H., Kuroe, Y.: Reinforcement Learning through Interaction among Multiple Agents. In: SICE-ICASE International Joint Conference, pp. 2457–2462 (2006)
4. Iima, H., Kuroe, Y.: Swarm Reinforcement Learning Algorithms Based on Particle Swarm Optimization. In: IEEE International Conference on Systems, Man and Cybernetics, pp. 1110–1115 (2008)
5. Watkins, C.J.C.H., Dayan, P.: Q-Learning. Machine Learning 8, 279–292 (1992)
6. Busoniu, L., Babuska, R., Schutter, B.D.: A Comprehensive Survey of Multiagent Reinforcement Learning. IEEE Transactions on Systems, Man, and Cybernetics, Part C 38, 156–172 (2008)
7. Kimura, H., Kobayashi, S.: An Analysis of Actor/Critic Algorithms using Eligibility Traces: Reinforcement Learning with Imperfect Value Function. In: 15th International Conference on Machine Learning, pp. 278–286 (1998)
8. Doya, K.: Reinforcement Learning in Continuous Time and Space. Neural Computation 12, 219–245 (2000)

XCS Revisited: A Novel Discovery Component for the eXtended Classifier System

Nugroho Fredivianus, Holger Prothmann, and Hartmut Schmeck

Karlsruhe Institute of Technology (KIT) – Institute AIFB
76128 Karlsruhe, Germany
{nugroho.fredivianus,holger.prothmann,hartmut.schmeck}@kit.edu

Abstract. The eXtended Classifier System (XCS) is a rule-based evolutionary on-line learning system. Originally proposed by Wilson, XCS combines techniques from reinforcement learning and evolutionary optimization to learn a population of maximally general, but accurate condition-action rules. This paper focuses on the discovery component of XCS that is responsible for the creation and deletion of rules. A novel rule combining mechanism is proposed that infers maximally general rules from the existing population. Rule combining is evaluated for single- and multi-step learning problems using the well-known multiplexer, Woods, and Maze environments. Results indicate that the novel mechanism allows for faster learning rates and a reduced population size compared to the original XCS implementation.

1 Introduction

Learning Classifier Systems (LCS, [2]) are rule-based evolutionary on-line learning systems that are widely used in various application areas [1]. One of the most successful LCS is the eXtended Classifier System (XCS) that has been introduced by Wilson [9, 10].

As reinforcement learning system, XCS acquires knowledge by interacting with an unknown environment. It receives an environmental input (coded as bit string), performs an action based on its current knowledge, and receives a numerical reward in response. Its knowledge is represented by a set (or *population*) of condition-action rules that are called *classifiers*. A classifier's condition (denoted as $cl.C$ in the following) specifies to which inputs the classifier is applicable. Conditions are defined over a ternary alphabet $\{0, 1, \#\}$, where '#' represents a *don't care symbol* (or *wildcard*) that matches both '0' or '1'. Other specifications of conditions using larger alphabets or numerical values may be reasonable for special applications (see [8]), but for the topics of this paper the classical ternary alphabet is sufficient. A classifier's action $cl.A$ specifies a possible response to the matched inputs. Several additional parameters keep track of a classifier's predicted future payoff $cl.P$, its past prediction error $cl.\varepsilon$, its fitness $cl.F$ (that measures the relative accuracy of a classifier with respect to other classifiers with overlapping conditions), and other bookkeeping values.

K. Deb et al. (Eds.): SEAL 2010, LNCS 6457, pp. 289–298, 2010.

Based on the current classifier population $[P]$, XCS' *performance component* determines the action that is executed in response to a given input: Classifiers with a matching condition form a *match set* $[M]$. Since the classifiers in $[M]$ can advocate different actions, a selection has to take place. The action can be either chosen randomly to *explore* the environment or deterministically to *exploit* the previously learnt knowledge. In either case, all classifiers in $[M]$ advocating the selected action are stored in an *action set* $[A]$ before the action is executed.

Based on the action's success, XCS receives a reward that is used to update the parameters of all classifiers in $[A]$. The update of existing classifiers is the regime of the *reinforcement component*. An additional *discovery component* aims at obtaining a small population of accurate rules: Randomized covering and genetic operators help to create well-performing classifiers, a subsumption mechanism combines similar rules, and a randomized deletion mechanism removes classifiers of a low fitness from the population.

In this paper, the discovery component of XCS is revisited and a novel rule combining technique is proposed. Rule combining aims at creating maximally general classifiers that match as many inputs as possible while still being exact in their predictions. It explicitly takes into account previously learnt knowledge and infers generalized classifiers from the existing population, thereby improving the achievable learning rate and reducing the population size.

The remainder of this paper is structured as follows: Section 2 focuses on the discovery component of XCS and briefly reviews literature on generalization in XCS, before Sect. 3 presents the novel rule combining mechanism. Section 4 compares an XCS variant that uses rule combining to the original XCS. The comparison considers single- and multi-step environments using the well-known multiplexer, Woods, and Maze problems. Finally, the paper concludes with a summary and an outlook on future work in Sect. 5.

2   Generalization in XCS

XCS aims at learning a population of maximally general classifiers. Within its discovery component, several mechanisms interact to support this goal:

Covering. Covering occurs for an input s if the classifiers in $[M]$ represent less than a predefined number of different actions. During covering, additional matching classifiers are created randomly. Depending on a *don't care probability* $P_\#$, wildcards are inserted in a new classifier's condition so it matches a set of inputs including s. The classifier's action is chosen randomly among those not present in $[M]$, its prediction, prediction error, and fitness are initialized with defaults.

Genetic operators. Genetic operators are applied to classifiers in the action set if the average time since their last genetic modification exceeds a threshold. In this case, parent classifiers are selected using fitness-proportionate [9,10] or tournament [3] selection. From the selected parents, offspring classifiers are created using crossover and mutation.

Subsumption. When inserting newly created classifiers into $[P]$ or when building the action set, subsumption deletion can be applied. A classifier is subsumed by a more general classifier advising the same action if the more general classifier is sufficiently accurate and experienced. If so, the subsumed classifier is deleted and its numerosity is added to the subsumer. (The *numerosity* specifies the number of identical classifier copies that a classifier represents.)

Deletion. To keep the population within its size limit, a classifier is selected for deletion with a probability that is proportional to the average size of the action sets it was part of. In this way, classifiers in well covered environmental niches have a higher probability to be removed. Furthermore, if the classifier is experienced and its fitness is lower than the population's average fitness, its deletion probability is increased.

In combination, covering, genetic operators, subsumption, and deletion help to learn classifiers that are as general as possible without losing accuracy, as formulated in Wilson's *generalization hypothesis* [10]. Butz et al. identified and analyzed the evolutionary pressures guiding the learning process in XCS [4], thereby supporting the theoretical understanding of Wilson's hypothesis. Additionally, the analysis of the complex learning interactions in XCS provided insights on the system configuration [7] and identified room for further improvements (e. g., by modifying the deletion [6] and selection [3] schemes). In the spirit of this previous research, the following section introduces a novel rule combining mechanism for XCS that supports the generalization of classifiers.

3   A Novel Discovery Component

This section introduces several modifications to XCS that improve the learning performance and reduce the required population size. The main modification is a novel *rule combining* mechanism that supports the discovery of maximally general classifiers. Rule combining reuses the covering, subsumption, and deletion mechanisms of XCS, but removes the genetic operators. New classifiers are created by inference, instead. The resulting novel XCS variant is called XCS-RC.

3.1   Rule Combining

In XCS-RC, a rule combining mechanism infers general, but accurate classifiers. Every T_{comb} cycles, the mechanism is executed in between the regular learning cycles. Its steps are illustrated in Fig. 1: In a preparatory Step 0, classifiers advocating the same action are gathered in a combining set $[C]$. (In the figure, different actions are depicted by different shades of grey.) Afterwards, the classifiers in $[C]$ are examined pairwise to obtain a generalized candidate classifier cl_* which is then checked for potential conflicts with existing classifiers (Step 1). If no conflicts are found, existing classifiers that are less general than cl_* are marked for subsumption or deletion, while cl_*'s parameters are updated (Step 2). Finally, the generalized classifier is inserted into $[C]$ and $[P]$, while

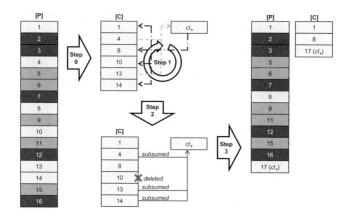

Fig. 1. Overview of the rule combining process

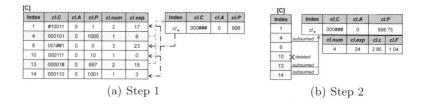

(a) Step 1

Index	cl.C	cl.A	cl.P	cl.num	cl.exp
1	#10011	0	1	2	17
4	000101	0	1000	1	6
8	001##1	0	0	3	23
10	000111	0	10	1	0
13	00001#	0	997	2	15
14	000110	0	1001	1	3

Index	cl.C	cl.A	cl.P
cl∗	000###	0	998

(b) Step 2

Fig. 2. Rule combining example

marked classifiers are deleted (Step 3). In the following, the steps are presented in detail.

Steps 0 and 1: Combining Set and Candidate Creation To prepare the generalization of existing classifiers, the *combining set* $[C]$ is created in Step 0. It contains all classifiers in $[P]$ that advocate the same action. Afterwards, Step 1 examines the classifiers in $[C]$ pairwise. If a pair cl_i and cl_j is sufficiently experienced and has similar payoff predictions, the classifiers will be combined into a new classifier $cl_∗$. To be combined, the experience of both classifiers must exceed a minimum experience threshold $minExp$. Furthermore, their prediction difference $|cl_i.P - cl_j.P|$ must not exceed a prediction tolerance $predTol$.

With these preconditions fulfilled, a generalized classifier $cl_∗$ can be created from the parent classifiers cl_i and cl_j. Its condition $cl_∗.C$ is derived according to the following rules:

- If positions p in the conditions of cl_i and cl_j are identical (i. e., if $cl_i.C[p] = cl_j.C[p]$), that position is copied to $cl_∗$ (i. e., $cl_∗.C[p] := cl_i.C[p]$).
- If positions p in the conditions of cl_i and cl_j are different (i. e., if $cl_i.C[p] \neq cl_j.C[p]$), that position becomes a wildcard in $cl_∗$ (i. e., $cl_∗.C[p] := \#$).

The generalized classifier $cl_∗$ has the same action as its parent classifiers, its prediction is determined as the numerosity-weighted average prediction of cl_i

and cl_j (see Fig. 2(a) for an example). Other values of cl_* remain unspecified at this time, as they will be determined later in the combining process (see Step 2).

Once a candidate classifier cl_* is created, all classifiers in $[C]$ are examined to prevent potential conflicts with previously learnt knowledge. A *conflict* occurs if a classifier $cl_k \in [C]$ with $cl_k.exp \geq 1$ matches one or more inputs that are also matched by cl_*, while cl_k and cl_* differ significantly in their predictions (i.e., $|cl_k.P - cl_*.P| > predTol$). In case of a conflict, the generalized classifier cl_* is abandoned. No changes are made to $[C]$ or $[P]$. The rule combining process is continued in Step 1 with a new pair cl_i and cl_j.

Steps 2 and 3: Subsumption and Population Update In case that no conflicts are found for cl_*, the classifiers in $[C]$ are checked for subsumability and deletion: A classifier $cl_k \in [C]$ is subsumable to cl_*, if cl_k's condition is less general than the condition of cl_*. In case that cl_k is experienced (i.e., $cl_k.exp \geq 1$) and the prediction difference $|cl_k.P - cl_*.P|$ of both classifiers does not exceed the prediction tolerance $predTol$, cl_k is *subsumed*. If, on the other hand, $cl_k \in [C]$ has no experience (i.e., $cl_k.exp = 0$) while its condition is less general than the condition of cl_*, it is *deleted*.

The parameters of cl_* are updated based on the subsumed classifiers. When cl_k is subsumed by cl_*, the prediction of cl_* becomes the numerosity-weighted average of $cl_k.P$ and $cl_*.P$, while cl_k's experience and numerosity are added to the corresponding values of cl_*. The prediction error and fitness of cl_* are derived according to the following equations:

$$cl_*.\varepsilon = \begin{cases} \frac{|cl_*.P - P_I|}{cl_*.exp}, & \text{if } cl_*.exp \leq 1/\beta \\ \frac{|cl_*.P - P_I|}{\lfloor 1/\beta \rfloor} \cdot (1-\beta)^{cl_*.exp - \lfloor 1/\beta \rfloor}, & \text{otherwise} \end{cases} \qquad (1)$$

$$cl_*.F = (F_I - 1) \cdot (1-\beta)^{cl_*.exp} + 1 \qquad (2)$$

Equations (1) and (2) have been derived experimentally. Their result resembles the outcome of an iterative process that starts with the default values for new classifiers (i.e., ε_I and F_I). In the iteration, $cl_*.\varepsilon$ and $cl_*.F$ are updated as if cl_* was part of an action set $[A] := \{cl_*\}$ and received a reward of $cl_*.P$ for $cl_*.exp$ times. The intention of these computations is to ensure that cl_* is considered accurate, having a low prediction error and a high fitness.

Figure 2(b) depicts an example showing the obtained parameters for cl_* after subsuming cl_4, cl_{13}, and cl_{14} (for $\beta = 0.2$, $P_I = 10.0$, and $F_I = 10.0$). The subsumed and deleted classifiers and their parameters are available in Fig. 2(a).

Once its values have been updated, cl_* is inserted into $[C]$ and $[P]$ in Step 3 of the combining process. All classifiers marked for subsumption and deletion are removed from both sets, including the parents cl_i and cl_j. Afterwards, rule combining continues with Step 1 using the updated combining set.

3.2   Further Changes

Besides the novel rule combining mechanism, several minor changes are introduced with XCS-RC:

Covering. Like in XCS, XCS-RC uses full covering, but applies a don't care probability $P_\# = 0$. As a result, classifiers created by covering match the current input, only. Generalization is the regime of the previously discussed rule combining mechanism.

Classifier deletion. XCS-RC introduces an additional deletion mechanism that is executed after updating the classifiers in the action set. A classifier cl is deleted in case it is sufficiently experienced (i. e., if $cl.exp > 2 \times minExp$), $cl.\varepsilon$ is greater than the prediction error tolerance $predErrTol$, and the update did not reduce $cl.\varepsilon$.

Explorative action selection. The last change in XCS-RC affects the performance component. In exploration mode, XCS randomly selects an action for execution in the environment. In contrast, XCS-RC favors actions advocated by classifiers with low experience. If classifiers cl with $cl.exp < minExp$ exist in $[M]$, one of their actions is chosen randomly. Otherwise, XCS-RC operates like XCS. The modified action selection in explore mode ensures that classifiers with a low experience can quickly gain the minimum experience required for taking part in the rule combining process.

All changes support the quick and accurate rule generalization in XCS-RC.

4   Experimental Evaluation

To compare XCS and XCS-RC, different single- and multi-step problems have been investigated: The boolean multiplexer that is known from Wilson's original work [9, 10] serves as an example for a single-step problem, while multi-step problems are represented by various Woods [9, 10] and Maze [7] environments. All experiments have been performed using the XCS implementation by Butz [5] that is also the basis for XCS-RC.

XCS and XCS-RC are compared based on their performance and their population size requirements. Figure 3 summarizes the experimental results by depicting average results of 20 runs for each problem. For the performance measurements, the figure contains additional error bars indicating the 25th and 75th percentiles. Invisible bars are due to small errors.

4.1   Multiplexers

Boolean multiplexer functions are defined for binary strings of length $l = k + 2^k$. The first k bits address a position in the remaining 2^k bits of the input. The addressed bit is the output of the multiplexer and is expected to be the selected action of the classifier system.

The configuration of XCS and the experimental setup for the comparison of XCS and XCS-RC are as in [10]: A random input is presented to the classifier system that has to select the correct function value as its action. A reward of 1000 is provided in case of a correct action, or otherwise the reward is zero. For every second input, the classifier systems operate in explore or exploit mode, respectively. For XCS-RC, the parameters for the discovery mechanism are set to $T_{comb} = 100$, $minExp = 1$, $predTol = 10$, and $predErrTol = 100$.

Figures 3(a) to 3(c) depict the comparison between XCS and XCS-RC for the multiplexer problems with respect to correct classification rates and population sizes. As in [9, 10], the depicted data is based on a sliding window average of the last 50 exploit trials. Therefore, a correctness rate of 50% means that 25 out of the last 50 exploit steps resulted in a correct classification.

For the 6-bit multiplexer problem, the maximum population size N has been limited to 400 classifiers. Figure 3(a) shows that XCS-RC achieves high correctness rates more quickly than XCS. XCS-RC correctly classifies all inputs after approximately 350 explore trials while XCS requires more than 2000 trials to reach a stable correctness rate of 100%. Furthermore, XCS-RC quickly reduces the population size. Less than 200 explore trials are required to obtain a population of less than 24 classifiers. In comparison, the population of XCS is still converging after 5000 trials, reaching a size of approximately 45 classifiers.

Similar results have been obtained for the 11-bit multiplexer (see Fig. 3(b)). Using $N = 800$ and the same parameters as before, XCS-RC uses about 3000 explore trials to reach a stable correctness rate of 100%, while XCS requires over 10 000 trials. As for the 6-bit multiplexer scenario, the XCS classifier population has not finished converging by the end of the experiment. After 20 000 trials, the population still contains around 112 classifiers. In contrast, XCS-RC maintains only 50 classifiers after about 1850 trials.

Figure 3(c) presents the result for 20-bit multiplexer with a parameter change of $N = 2000$ for XCS. XCS-RC, using $N = 800$, requires only 12 000 trials to achieve a correctness rate of 99.8%, while XCS needs more than 61 000 trials to reach the equal rate. With respect to the population size, XCS-RC keeps at most 200 classifiers after 31 000 trials. In contrast, XCS never reduces its population to less than 340 classifiers till the end of the simulation at 100 000 trials.

In summary, results for the single-step multiplexer problem indicate a clear benefit for classifier generalization based on rule combining.

4.2   Woods and Maze Environments

To test the performance of XCS-RC for multi-step problems, various Woods [9, 10] and Maze environments [7] have been investigated. At the beginning of every simulation, an *animat* is placed randomly at an empty position in a grid environment. The animat can move to any of the eight neighboring cells which is not occupied by a rock. Its task is to find a food source hidden in the environment while minimizing the steps taken.

The animat's movements are controlled by a classifier system that receives the status of the neighboring cells as input and decides the animat's next movement. Once the food is reached, a reward is distributed and the environment is restarted by placing the animat again at a random empty position (see [9, 10] for details).

Experiments for the Woods and Maze environments follow the setup in [10] using a maximum population size of $N = 800$. For XCS-RC, the prediction error tolerance is set to $predErrTol = 5$, other parameters are kept unchanged from

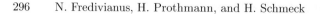

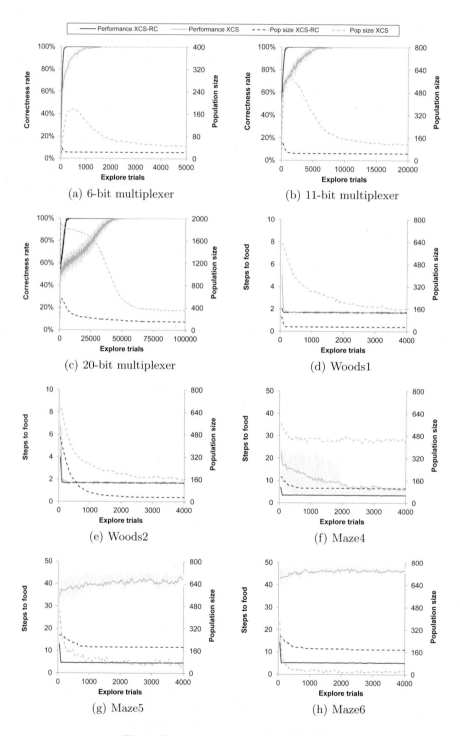

(a) 6-bit multiplexer

(b) 11-bit multiplexer

(c) 20-bit multiplexer

(d) Woods1

(e) Woods2

(f) Maze4

(g) Maze5

(h) Maze6

Fig. 3. Summary of experimental results

the multiplexer experiments. For comparability with previous results in [10], the performance of XCS and XCS-RC is evaluated in a 50/50 explore/exploit regime based on the average number of steps the animat required to reach the food in the last 50 exploit trials. Additionally, the size of the classifier population is recorded throughout the experiments. Both measures should be minimized.

Figures 3(d) and 3(e) compare XCS and XCS-RC for the Woods environments. With respect to the required steps to food, XCS and XCS-RC perform similarly for Woods1 and Woods2. After approximately 100 explore trials, the animat does not require significantly more than 1.69 steps to reach the food which is optimal for both environments. A remarkable difference, however, can be observed for the population sizes. XCS-RC successfully reduces the classifier population to less than 36 classifiers after 150 trials for Woods1 or 2200 trials for Woods2, respectively. In contrast, XCS maintains a population of approximately 160 classifiers for both environments at the end of the experiment.

Results for the Maze environments (see Fig. 3(f) to 3(h)) provide further evidence for the superiority of classifier generalization in XCS-RC. For Maze4, Fig. 3(f) shows that XCS-RC is significantly quicker in minimizing the number of steps required to reach the food. While XCS fails to generalize and maintains a population of approximately 450 classifiers by the end of the simulation (i. e., after 4000 trials), XCS-RC quickly reduces its population to a size of less than 120 classifiers in only about 450 trials.

Maze5 and Maze6 are bigger than Maze4 and the food can be reached from one or two positions, only. In both environments, XCS fails to solve the learning task (see Fig. 3(g) and 3(h)). Although XCS significantly reduces the number of classifiers over time, the required number of steps to the food is not minimized. In contrast, XCS-RC is capable to minimize the number of steps to the food, requiring less than six steps on average after 100 trials in both environments. To achieve this result, XCS-RC maintains a population that is larger than that of XCS. However, the learning result shows that the classifiers are properly generalized (and are not overly general as in the XCS population).

The observation that XCS performs badly for Maze5 and Maze6 has also been made by Lanzi [7] who suggests a modified XCS version (called XCSS) that performs better in the Maze environments. After approximately 400 trials, XCSS requires less than ten steps on average to reach the food in Maze5 and less than six steps in Maze6 (see [7] for details). In comparison, XCS-RC reaches a similar performance for both environments after approximately 100 trials (see Fig. 3(g) and 3(h)). Unfortunately, no comparison can be made with respect to the population size since no details are available in [7].

In summary, XCS-RC outperforms XCS (and XCSS) for the investigated single- and multi-step test problems. XCS-RC requires less trials to solve the learning tasks and is capable of reducing its population size quickly by using the novel rule combining mechanism in its discovery component. It performs well even in Maze environments that are not successfully tackled by XCS (like Maze5 and Maze6).

5   Conclusion

The paper presents a novel rule combining mechanism that improves the classifier discovery in XCS. Rule combining creates maximally general, but accurate
classifiers from the existing population. It does not rely on genetic operators to
create new classifiers, but infers them using an extended subsumption mechanism that combines specific, but experienced classifiers to more general rules.
The general classifiers then replace their specific counterparts unless the population contains contradicting, but experienced classifiers.

The resulting classifier system (called XCS-RC) has been compared to XCS
for various single- and multi-step learning problems. Boolean multiplexers served
as test case for single-step learning, while Woods and Maze environments of
varying sizes have been investigated as examples of multi-step problems. In all
experiments, XCS-RC outperforms XCS with respect to learning performance.
Furthermore, XCS-RC quickly reduces the size of its classifier population below the number of classifiers maintained by XCS, thereby demonstrating the
generalization capabilities of the rule combining approach. In two cases (Maze5
and Maze6) where XCS maintains a smaller population than XCS-RC, XCS
overgeneralizes the population and fails to learn.

Future work will simplify the rule combining mechanism in XCS-RC to improve its computational requirements and to support the theoretical analysis
of its generalization capabilities. Furthermore, the performance of XCS-RC in
dynamically changing environments will be investigated.

References

1. Bull, L. (ed.): Applications of Learning Classifier Systems. Springer, Heidelberg
   (2004)
2. Butz, M.V.: Rule-Based Evolutionary Online Learning Systems. Springer, Heidelberg (2005)
3. Butz, M.V., Goldberg, D.E., Tharakunnel, K.: Analysis and improvement of fitness exploitation in XCS: Bounding models, tournament selection, and bilateral
   accuracy. Evolutionary Computation 11(3), 239–277 (2003)
4. Butz, M.V., Kovacs, T., Lanzi, P.L., Wilson, S.W.: Toward a theory of generalization and learning in XCS. IEEE Transactions on Evolutionary Computation 8(1),
   28–46 (2004)
5. Butz, M.V.: XCSJava 1.0: An implementation of the XCS classifier system in Java.
   IlliGAL report 2000027, Illinois Genetic Algorithms Laboratory (2000)
6. Kovacs, T.: Deletion schemes for classifier systems. In: Genetic and Evolutionary
   Computation Conference (GECCO 1999), pp. 329–336. Morgan Kaufmann, San
   Francisco (1999)
7. Lanzi, P.L.: An analysis of generalization in the XCS classifier system. Evolutionary
   Computation 7(2), 125–149 (1999)
8. Lanzi, P.L.: Learning classifier systems: then and now. Evolutionary Intelligence 1(1), 63–82 (2008)
9. Wilson, S.W.: Classifier fitness based on accuracy. Evolutionary Computation 3(2),
   149–175 (1995)
10. Wilson, S.W.: Generalization in the XCS classifier system. In: Koza, J.R., et al. (eds.)
    Genetic Programming, pp. 665–674. Morgan Kaufmann, San Francisco (1998)

Supplanting Neural Networks with ODEs in Evolutionary Robotics

Paul Grouchy and Gabriele M.T. D'Eleuterio

University of Toronto Institute for Aerospace Studies
4925 Dufferin Street, Toronto, Ontario, Canada
{paul.grouchy,gabriele.deleuterio}@utoronto.ca

Abstract. A new approach to evolutionary robotics is presented. Neural networks are abstracted and supplanted by a system of ordinary differential equations that govern the changes in controller outputs. The equations are evolved as trees using an evolutionary algorithm based on symbolic regression in genetic programming. Initial proof-of-concept experiments are performed using a simulated two-wheeled robot that must drive a straight line while wheel response properties vary. Evolved controllers demonstrate the ability to learn and adapt to a changing environment, as well as the ability to generalize and perform well in novel situations.

Keywords: Evolutionary Robotics, Genetic Programming, Evolutionary Algorithms, Genetic Algorithms, Evolutionary Computing, Learning.

1 Introduction

In the field of evolutionary robotics, the goal is to evolve controllers artificially for robotic applications. While the most obvious metric of success is how well the final controller solves the given task, researchers and developers are also very interested in robustness, adaptability and the ability to generalize. Evaluating controller fitness on hardware is usually infeasible owing to time and hardware constraints. Thus the fitness of individuals, which is usually their ability to accomplish the specified task, is most often evaluated in simulation. This is one of the reasons that the ability for an evolved controller to generalize is so important, as it is very difficult to perfectly predict and simulate the environmental and hardware behaviors that the evolved controller will eventually see during real-world applications. Thus a good algorithm should be able to produce a controller that can take its evolved solution and apply it successfully in never-before-seen scenarios.

With these goals in mind, researchers have been looking to nature for inspiration as so far it is nature, and not human design, that has produced the most adaptable controllers. However, natural systems are extremely complex, and so initial research tried to distill the most important elements of the evolution and operation of modern lifeforms in an attempt to reproduce nature's successes.

K. Deb et al. (Eds.): SEAL 2010, LNCS 6457, pp. 299–308, 2010.

This has yielded the core idea behind evolutionary robotics: Use an evolutionary algorithm to evolve the weights of a fixed-structure artificial neural network (e.g. [8]). Further research then worked to increase the biofidelity of this approach by adding in components such as chemical signals (e.g. [1],[7]), Hebbian learning (e.g. [2]), neuromodulation (e.g. [4]) and neural structures that are non-fixed and thus able to change and grow through evolution (e.g. [5]). While these advances have increased the adaptability and evolvability of artificial controllers, the evolved systems are still a long way from natural systems in both complexity and real-world performance.

In this paper, we propose an alternative approach. Instead of looking to increase further the biofidelity (and complexity) of the algorithms, we attempt to evolve an *abstraction* of natural neural systems. Using an algorithm based on symbolic regression algorithms from genetic programming ([3]), it is hoped that complex decision-making, learning and generalization capabilities can be evolved and represented simply and concisely as a system of ordinary differential equations. Utilizing mathematical functions as a part of the decision-making process of an evolved controller has been implemented before, most notably in Compositional Pattern Producing Networks [6]; however, as far as the authors know, the artificial neural network structures have never been completely supplanted by mathematical equations. The purpose of this paper is only to present an alternative to other, neural network based methods, and to demonstrate that despite its simplicity, this new method can tackle a difficult problem and produce complex behaviors. Performance comparisons with other methods are left for future studies.

The rest of this paper is organized as follows. Section 2 presents both the controller and evolutionary algorithm of the proposed ODE-evolving evolutionary robotics algorithm. Section 3 describes the initial proof-of-concept experiments that were used for testing, while Section 4 presents and discusses the results of these experiments. Section 5 concludes with some final thoughts and directions for future research.

2   The Algorithm

This section will present the Ordinary Differential Equation Evolution (ODEvo) algorithm in detail. The structure and operation of the controller portion of the algorithm will first be described (Section 2.1), followed by the details of the evolutionary algorithm used to train these controllers (Section 2.2).

2.1   The Controller

Despite the complex behavior that an ODEvo controller can display (see Section 4), its internal structure and operation are quite simple. The controller is fully described by a system of ordinary differential equations that control how the outputs y_i change at each time step. So for a system with N outputs, the controller will have the form:

$$dy/dt = \mathbf{f}(\mathbf{x}, \mathbf{y}) \tag{1}$$

where $\mathbf{x} = [x_0, x_1, \ldots, x_M]^\mathrm{T}$ is the set of M inputs to the controller, $\mathbf{y} = [y_0, y_1, \ldots, y_N]^\mathrm{T}$ is the set of the N outputs of the controller and $\mathbf{f}(\mathbf{x}, \mathbf{y}) = [f_0(\mathbf{x}, \mathbf{y}), f_1(\mathbf{x}, \mathbf{y}), \ldots, f_N(\mathbf{x}, \mathbf{y})]^\mathrm{T}$.

The operation of the controller is straightforward and is summarized below:

1. Select a fixed step size Δt.
2. Set $y_i = 0, \forall i = 0 \ldots N$.
3. Update $\mathbf{x}$ with current sensor readings.
4. Solve for $dy_i/dt, \forall i = 0 \ldots N$ using the controller's ODEs (1).
5. Update the controller's outputs as $y_i^{t+\Delta t} = y_i^t + \Delta t\,(dy_i/dt), \forall i = 0 \ldots N$.
6. Run agent for Δt seconds using output signals $y_i^{t+\Delta t}$.
7. End of one time step, repeat operations 3-6 for subsequent time steps.

The controller runs sequentially, i.e., an action is performed, the ODEs are then solved, the outputs are updated and then the next action is performed. It is expected that with a small enough Δt the system might be able to operate in parallel, i.e. actions can be executed continuously using the latest y_i values while the ODEs update the y_i values every Δt seconds without forcing the agent to stop and wait, however this has yet to be tested.

2.2   The Evolutionary Algorithm

The evolutionary algorithm (EA) used to evolve the ODEs is based on symbolic regression algorithms from genetic programming (GP) [3]. Equations are represented in the genomes as trees (see Figure 1 for an example), with each genome containing N such equation trees, one for each output of the system. Each tree contains a collection of terminal ("leaf") and non-terminal ("internal") nodes. The set of possible terminal nodes is comprised of all potential constants and variables, while the set of possible non-terminal nodes is composed of a user-defined group of mathematical functions. The number of "child" nodes (subtrees) of a non-terminal node depends on its function. For example, arithmetic operations such as addition and multiplication require two child nodes, while operations such as sine and absolute value have just one child node.

The EA is a standard discrete-generation algorithm whose various operations will now be explained.

Initialization. The random initialization of each equation in each initial genome is done using the "ramped half-and-half" method as described in [3]. For all the experiments in this paper, the set of potential maximum depths was $\{1, 2\}$.

Crossover. If an offspring genome is to be produced through crossover, two parent genomes are required. First, a single equation tree τ_i is selected randomly from the N equation trees that each individual possesses. The offspring inherits

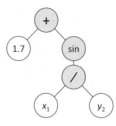

Fig. 1. Example tree representation of an equation. Assuming this is the first equation in the genome, this tree is equivalent to the ODE $dy_0/dt = 1.7 + \sin(x_1/y_2)$.

exact copies of all the equation trees $\tau_j, \forall j < i$ from the first parent and exact copies of all the equation trees $\tau_j, \forall j > i$ from the second parent.

As for τ_i, the offspring version of this equation tree is generated by crossing over the corresponding trees of the two parents using the standard tree crossover method described in [3]. For all experiments in this paper, crossover occurred with a probability of 0.7.

Mutation. The mutation operators used are similar to those suggested in [3], however unlike in GP, mutation and crossover are not mutually exclusive in ODEvo, i.e., a genome can be produced through crossover and still undergo mutation. For more details on the mutation operators described here, the reader is referred to [3].

Each of an offspring's equation trees is selected for mutation with probability μ_τ, with $\mu_\tau = 1/N$ in this paper. Once selected, a tree will undergo one of several types of mutations:

- **Point Mutation.** Occurring with probability 0.5, a point mutation performs one of several operations on a single randomly selected node in the tree:
  - *Perturbation of a constant.* This operation is performed with probability 0.5 and only if the tree in question contains one or more constants. The operation adds a random value taken from a Gaussian distribution with mean 0 and standard deviation 1.0 to a randomly selected constant.
  - *Mutation of a non-terminal function.* This operation is performed if a perturbation of a constant was not done and if a randomly selected node is a non-terminal. The new function is randomly selected from the set of functions requiring the same number of child nodes as the current function (i.e., an addition function can be mutated to multiplication but not to sine).
  - *Mutation of a terminal function.* This operation is performed if a perturbation of a constant was not done and if a randomly selected node is a terminal. One of two mutations occurs, with equal probability. The chosen terminal is either mutated to a randomly chosen variable or it is mutated to a constant randomly chosen from a uniform distribution of potential constants.

– **Subtree Mutation.** Occurring if there was no point mutation, this operation selects a random node on the original tree and replaces it with a new random subtree. A small variation to the standard subtree mutation was also added. With a probability of $\mu_s = 0.05$, the roles of the subtree and the original tree are swapped, i.e., a random node on the randomly generated subtree is replaced with the entire original tree and this becomes the new tree of the offspring.

3   Experiments

Initial proof-of-concept experiments were done in a two-dimensional simulation world using a simple robot with two independently actuated wheels, one on each side. Only step changes in wheel speeds were allowed, i.e., wheel acceleration was not modeled. The distance between the two wheels was set to 1 m. The task required of this robot was for it to drive in a straight line along the x-axis. The starting position of the robot is $(0,0)$, with an initial orientation randomly selected from $-\pi < \theta \leq \pi$. A straight line is defined here as driving in the positive x direction with a y position in the strict range of $(-0.01, 0.01)$ m, measured at the center of the robot. The fitness of a simulation run is calculated as $(d_0 - d)^2$, where d_0 is the target distance of 300 m and d is the total distance (in metres) driven in the positive x direction within the defined y boundaries. Thus, this is a minimization problem.

This may seem like a fairly straightforward task, however there is a twist. The wheels do not respond directly to the signals sent to their actuators. Their response is based on a multiplier that is an abstraction of all possible factors that could affect their real-world performance, such as motor degradation, slippage, tire pressure, etc. As there are two independently actuated wheels, a left wheel and a right wheel, there are two independent wheel multipliers, w_L and w_R. Therefore, the velocities (in m/s) produced by the left and right wheels are

$$v_L = w_L y_0$$
$$v_R = w_R y_1$$

(2)

where y_0 and y_1 are the signals sent to the left and right actuators, respectively.

For all experiments presented here, each simulation was run for $T = 300$ s and $w_L, w_R \in [0.25, 1.25]$. For a given training or test scenario, initial w_L and w_R values are chosen randomly and are changed to new random values at five randomly selected times between 1 s and $(T - 50$ s$)$. A starting θ value is also randomly selected.

These experiments were designed to be difficult and to require a solution that has the ability to adapt, learn and generalize. A simple feedforward neural network would not be able to adapt to the changing environment, however more complicated approaches could potentially solve this problem. The aim of these experiments is to demonstrate the power of ODEvo despite the relative simplicity of the algorithm itself.

3.1 Controller Parameters

The parameters of the controller used in these experiments are summarized in Table 1. The controller has three inputs $\mathbf{x} = [x_0, x_1, x_2]^{\mathrm{T}}$ and two outputs $\mathbf{y} = [y_0, y_1]^{\mathrm{T}}$, thus requiring two ODEs (one for each output).

Table 1. Controller parameters used in all experiments

Parameter	Description	Value/Range
Δt	Step Size (s)	0.1
x_0	Robot Position on x-axis (m)	$(-\infty, \infty)$
x_1	Robot Position on y-axis (m)	$(-\infty, \infty)$
x_2	Robot Orientation θ (rad)	$(-\pi, \pi]$
y_0	Signal to Left Actuator	$[-1.0, 1.0]$
y_1	Signal to Right Actuator	$[-1.0, 1.0]$

The allowed mathematical operations were **addition**; **multiplication**; **division**, where the denominator of the operation is forced to be in the range $(-\infty, -10^{-6}] \cup [10^{-6}, \infty)$ and values outside of this range are set to the nearest acceptable value; **differentiation**, calculated as $g'(t) = (g(t) - g(t - \Delta t))/\Delta t$; **absolute value**; **sine**; **cosine**; **base-e exponential**; **natural logarithm**, which will only operate on values in the range $[10^{-6}, \infty)$ and values less than 10^{-6} are set to 10^{-6}.

3.2 Evolutionary Algorithm Parameters

All experiments ran populations of 1000 for 100 generations. Tournament selection with a tournament size of 12 was used, and constants were real-valued and uniformly drawn from $[-1000, 1000]$. Elitism, where the top individual from the current generation is cloned for the next generation, was also used. A collection of simple equation-reducing operations was performed on the entire population every 10 generations to help reduce bloat.

When generating random trees, be it for the initial population or for sub-tree mutation, "ramped half-and-half" was always used, generating trees with a maximum depth of 1 or 2 with equal probability.

The training set consisted of 10 randomly generated scenarios, was fixed for all generations within an evolutionary run and was also held constant between evolutionary runs. Every evolutionary run started with a randomly generated population. When comparing experiments with different parameter settings, the same set of random initial populations was used.

4 Results

4.1 Training Results

A summary of the training performance of the evolutionary algorithm with various parameter settings can be found in Table 2. Each configuration was run 50

Table 2. Training results summary

EA Settings	Best Fitness (x10³)				Gen. Converged				Evo. Time (s)			
	μ	σ	Min.	Max.	μ	σ	Min.	Max.	μ	σ	Min.	Max.
baseline	301	105	157	593	81	20	24	100	479	247	223	1260
no crossover	426	177	172	819	78	22	25	100	315	55	221	445
$\mu_s = 0.0$	342	98	158	589	73	23	26	100	431	257	220	1282
no eqn. reduction	359	151	155	821	81	21	19	100	539	374	252	2504

times. The generation converged is the generation where the best solution for that run was first found and the evolution time was measured as the real-world execution time of an evolutionary run with multithreaded fitness evaluations on an Intel® Core™2 Quad CPU Q9550 @ 2.83GHz.

From these results, we can clearly see that crossover is an important part of the ODEvo algorithm. Furthermore, μ_s does indeed seem to provide some benefits for average fitness. Finally, simple equation reduction every 10 generations has a strong impact on running times and also seems to improve average fitness results.

A manually simplified and re-arranged version of the best ODEvo controller found during the baseline training runs is

$$dy_0/dt = 2\,(dx_2/dt) + 3\,|dx_2/dt| + 2\sin\,(dx_2/dt) + e^{348.24x_1}$$
$$- \left((x_1y_1)\,/566.17\right) + 4\,|x_2| + 65x_1 + 24x_2 - 9362.96x_1x_2 \tag{3}$$

$$dy_1/dt = y_1 + 0.82 - 232.16\,(x_1 + x_2)\ .$$

It should be noted that it contains rounding errors and thus is not the exact form of the actual controller used for testing. The paths that this controller takes in the first four training scenarios are shown in Figure 2. While a detailed analysis of these equations is outside the scope of this paper, a few interesting properties can be noted. First of all, the two equations obviously have very little in common, despite the symmetrical nature of the simulated robot. Furthermore, the equation for the left actuator seems more complex than that of the right. This is corroborated by data showing the values of the two signals y_0 and y_1 over the course of a run (e.g. Figure 3). It is clear that the right signal is usually set to the maximum, while the left signal is responsible for the majority of the maneuvers. These equations also contain the equivalent of recurrent connections in artificial neural networks, with the right side of both equations containing the current value of y_1 and with dy_0/dt also depending on the time derivative of x_2 (the orientation θ of the robot).

4.2   Test Results

While the results on the training set are strong, it is the ability to generalize that is of true concern in evolutionary robotics. All the results that follow are

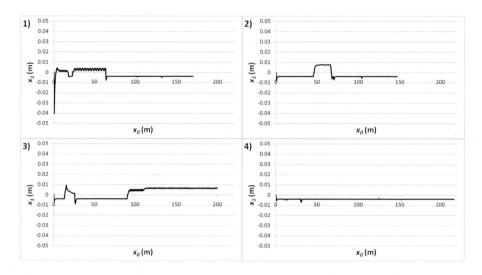

Fig. 2. Performance over the first four training scenarios of the best evolved controller. It is important to note the scales, as what at first may appear to be a zigzag path usually only has a vertical displacement measureable in millimetres, while its horizontal displacement is in metres.

from experiments using the best evolved controller from the baseline runs (see (3)). Note that this controller was only trained on 10 different scenarios.

When tested in new scenarios (generated in the same manner as the training scenarios, see Section 3 for details), the ODEvo controller performs quite well. Visual inspection of the paths of the controller in 100 new scenarios revealed that the robot successfully drove along the x-axis in all 100 cases. Table 3 shows statistics on the distance driven along the x-axis during both training and testing. In a novel scenario, the ODEvo controller can on average drive along the x-axis for 94% of the distance of an average training run. This percentage becomes even more significant when one considers average maximum possible speeds (assuming straight-line driving) in both sets of scenarios, as this value in the test set (0.58 m/s) is roughly 94% of that in the training set (0.62 m/s).

Table 3 also contains results from an identical test set, but with varying distance between wheels on the robot. As can be seen, even though the controller

Table 3. Comparison of performance of best controller on training and test sets

	Dist. on x-axis (m)	
Scenario Set	μ	σ
Training Set (10 runs)	185	52
Test Set (100 runs)	174	38
Wheel Dist. 0.75 m (100 runs)	157	51
Wheel Dist. 1.25 m (100 runs)	174	38

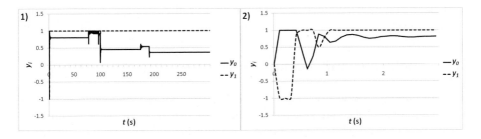

Fig. 3. 1) y_0 and y_1 values of the best evolved controller over the course of its operation in the second training scenario. 2) A close-up of the first three seconds.

was evolved on a robot with 1 m between wheels, it can operate successfully on different simulated hardware. With the wheel distance set at 0.75 m, the controller failed on one test scenario (i.e. it could not successfully drive a straight line) and in 14 others the robot was not always within the strict y boundaries, however it never strayed more than 0.10 m from the x-axis and always drove in the desired direction. The other 85 runs were completely successful. With a wheel distance of 1.25 m, the controller was successful on all 100 test scenarios and performed almost identically to when it had its original wheel distance. One final test was performed on this controller. The starting position was $x = y = \theta = 0$ and the controller was given 30 s from $t = 0$ to learn to operate with its randomly generated wheel settings. After these 30 s, the controller was artificially moved to a new y position randomly selected from $[-2, 2]$ m, rotated to a random orientation ($\theta \in (-\pi, \pi]$) and given a new set of random wheel parameters. From here, the controller was given 270 s to return to the x-axis and continue to drive along it.

In all 100 cases, the controller was successfully able to return the robot to within 0.10 m of the x-axis, and in 89 of these cases, it managed to get back to within 0.01 m of the x-axis. Two such runs are shown in Figure 4.

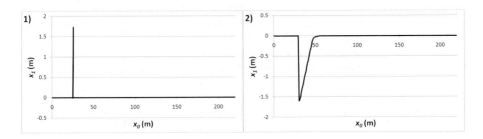

Fig. 4. 1) started with $w_L = 0.95$, $w_R = 0.88$ and at $t = 30$ s was sent 1.66 m above the x-axis, rotated to 0.93 rad and given new wheel parameters of $w_L = 0.83$, $w_R = 0.75$. 2) started with $w_L = 1.05$, $w_R = 1.03$ and at $t = 30$ s was sent 1.6 m below the x-axis, rotated slightly to 0.07 rad and given new wheel parameters of $w_L = 1.09$, $w_R = 0.93$.

This controller has demonstrated a remarkable ability to generalize, learn and adapt, especially considering that it was only evolved over 10 training scenarios.

5   Conclusion and Future Work

ODEvo presents a new approach to evolutionary robotics, completely dispensing with neural networks and showing immense promise in the areas of learning and generalizability, all while sidestepping the controller and evolutionary algorithm complexity that usually accompanies these terms.

This paper only presented initial proof-of-concept experiments, and so ODEvo still has much to prove. These experiments assumed perfect position and orientation sensors and ignored wheel acceleration. Further work is needed to evolve proper controllers for hardware testing, more complex tasks need to be attempted and comparisons need to be done with current neural network based methods.

Acknowledgments. Many thanks to Graham Holker for the introduction to GP and to the reviewers for their helpful comments. This work was in part supported by the Natural Sciences and Engineering Research Council of Canada.

References

1. Federici, D., Downing, K.: Evolution and Development of a Multicellular Organism: Scalability, Resilience, and Neutral Complexification. Artificial Life 12, 381–409 (2006)
2. Floreano, D., Mondada, F.: Evolution of Plastic Neurocontrollers for Situated Agents. In: 4th International Conference on Simulation of Adaptive Behavior. MIT Press, MA (1996)
3. Poli, R., Langdon, W.B., Mcphee, N.F.: A Field Guide to Genetic Programming (2008), Published via http://lulu.com and freely available at http://www.gp-field-guide.org.uk
4. Soltoggio, A., Durr, P., Mattiussi, C., Floreano, D.: Evolving Neuromodulatory Topologies for Reinforcement Learning-like Problems. In: IEEE Congress on Evolutionary Computation, pp. 2471–2478. IEEE Press, Los Alamitos (2007)
5. Stanley, K.O., Miikkulainen, R.: Evolving Neural Networks Through Augmenting Topologies. Evolutionary Computation 10, 99–127 (2002)
6. Stanley, K.O.: Compositional Pattern Producing Networks: A Novel Abstraction of Development. Genetic Programming and Evolvable Machines Special Issue on Developmental Systems 8, 131–162 (2007)
7. Thangavelautham, J., D'Eleuterio, G.M.T.: A Coarse-Coding Framework for a Gene-Regulatory-Based Artificial Neural Tissue. In: Capcarrère, M.S., Freitas, A.A., Bentley, P.J., Johnson, C.G., Timmis, J. (eds.) ECAL 2005. LNCS (LNAI), vol. 3630, pp. 67–77. Springer, Heidelberg (2005)
8. Wieland, A.: Evolving Neural Network Controllers for Unstable Systems. In: Proceedings of the International Joint Conference on Neural Networks, pp. 667–673. IEEE Press, Los Alamitos (1991)

Parallel Distributed Implementation of Genetics-Based Machine Learning for Fuzzy Classifier Design

Yusuke Nojima, Shingo Mihara, and Hisao Ishibuchi

Dept. of Computer Science and Intelligent Systems, Graduate School of Engineering,
Osaka Prefecture University, Sakai , Osaka, 5998531 Japan
{nojima@,mihara@ci,hisaoi@}cs.osakafu-u.ac.jp

Abstract. Evolutionary algorithms have been successfully applied to design fuzzy rule-based classifiers. They are used for attribute selection, fuzzy set selection, rule selection, membership function tuning, and so on. Genetics-based machine learning (GBML) is one of the promising evolutionary algorithms for classifier design. It can find an appropriate combination of antecedent sets for each rule in a classifier. Although GBML has high search ability, it needs long computation time especially for large data sets. In this paper, we apply a parallel distributed implementation to our fuzzy genetics-based machine learning. In our method, we divide not only a population but also a training data set into subgroups. These subgroups are assigned to CPU cores. Through computational experiments on large data sets, we show the effectiveness of the proposed parallel distributed implementation.

Keywords: fuzzy classifier design, genetics-based machine learning, parallel distributed implementation, large data sets.

1   Introduction

Evolutionary computation has been frequently and successfully used for knowledge acquisition [1]. Genetics-based machine learning (GBML) is one of the largest branches of evolutionary computation. It can search for a set of if-then rules as understandable knowledge [2]. Fuzzy rule-based systems have also been optimized in the framework of GBML which is referred to as genetic fuzzy systems (GFS) or evolutionary fuzzy system [3,4]. A hot research issue in the field of fuzzy GBML (and GBML in general) is the scalability improvement of evolutionary algorithms to large data sets. This is because even an evaluation of a single fuzzy rule-based classifier needs a long computation time in the case of large data sets. Since the execution of a fuzzy GBML algorithm involves tens of thousands of evaluations of fuzzy rule-based classifiers, its application to large data sets is very difficult.

To reduce the computation time, parallel implementation of evolutionary computation is a promising approach [5,6]. When we use island models for parallel implementation, each sub-population is assigned to a different processing node (e.g., CPU core). Let N_{CPU} be the number of processing nodes which are available for parallel

K. Deb et al. (Eds.): SEAL 2010, LNCS 6457, pp. 309–318, 2010.
© Springer-Verlag Berlin Heidelberg 2010

implementation. In an ideal case, the total computation time could be reduced to $1/N_{\text{CPU}}$ of that by non-parallel implementation. On the other hand, from the viewpoint of data mining, data reduction such as feature selection and instance selection [7-10] is a frequently-used approach. The computation time depends on the data reduction rate linearly. However, there is a negative side-effect that useful information in the training data set for designing good classifiers is accidentally removed from the original data sets by data reduction.

We have already proposed parallel distributed implementation of genetic fuzzy rule selection for pattern classification problems in [11-13]. Our genetic fuzzy rule selection is a two-stage method [14]. At the first stage, a number of candidate fuzzy if-then rules are extracted from numerical data in a heuristic manner. Then, at the second stage, a subset of the extracted candidate rules is selected by a genetic algorithm. In our parallel distributed implementation [11], the second stage is performed in parallel. We divide not only a population but also a training data set into subgroups. A pair of a sub-population and a training data subset is assigned to one CPU core. In order to avoid the overfitting of each sub-population to a specific training data subset, training data subset re-assignment is periodically performed (e.g., every 10 generations). In [11], we used a workstation with four CPU cores (a CPU core was used as a server, and the others were used as clients). The experimental results showed that we can reduce the computational time to 1/9 with no deterioration of the test data accuracy.

In this paper, we apply the parallel distributed implementation of our genetic fuzzy rule selection [11] to our fuzzy genetics-based machine learning (GBML) for pattern classification problems [15]. Our fuzzy GBML is a hybrid version of Pittsburgh approach and Michigan approach. It can optimize the combination of antecedent conditions and the number of rules. Although there is a high possibility that fuzzy GBML can obtain more accurate classifiers than genetic fuzzy rule selection thanks to the larger search space of fuzzy GBML, the computation cost of fuzzy GBML is much heavier than genetic fuzzy rule selection when we apply them to the design of fuzzy classifiers for large data sets.

This paper is organized as follows. First we briefly explain fuzzy rule-based classifier and our fuzzy GBML algorithm in Section 2. Next we explain the parallel distributed implementation of our fuzzy GBML in Section 3. Then experimental results are reported in Section 4. Finally we conclude this paper in Section 5.

2   Fuzzy Genetics-Based Machine Learning

Let us assume that we have m training (i.e., labeled) patterns $\mathbf{x}_p = (x_{p1}, ..., x_{pn})$, $p = 1, 2, ..., m$ from M classes in an n-dimensional pattern space where x_{pi} is the attribute value of the pth pattern for the ith attribute ($i = 1, 2, ..., n$). For the simplicity of explanation, we assume that all the attribute values have already been normalized into real numbers in the unit interval $[0, 1]$. For our classification problem, we use fuzzy if-then rules of the following type:

$$\text{Rule } R_q\text{: If } x_1 \text{ is } A_{q1} \text{ and } ... \text{ and } x_n \text{ is } A_{qn} \text{ then Class } C_q \text{ with } CF_q, \tag{1}$$

where R_q is the label of the qth fuzzy rule, $\mathbf{x} = (x_1, ..., x_n)$ is an n-dimensional pattern vector, A_{qi} is an antecedent fuzzy set ($i = 1, 2, ..., n$), C_q is a class label, and CF_q is a rule weight. We denote the antecedent fuzzy sets of R_q as a fuzzy vector $\mathbf{A}_q = (A_{q1}, A_{q2}, ..., A_{qn})$.

We use 14 fuzzy sets in four fuzzy partitions with different granularities in Fig. 1. In addition to those 14 fuzzy sets, we also use the domain interval [0, 1] as a special antecedent fuzzy set for representing a *don't care* condition.

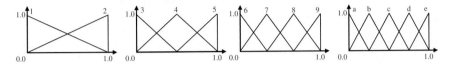

Fig. 1. Homogeneous fuzzy partitions used in this paper

The consequent class C_q and the rule weight CF_q of each fuzzy rule R_q are specified from training patterns compatible with its antecedent part $\mathbf{A}_q = (A_{q1}, A_{q2}, ..., A_{qn})$ in a heuristic manner [16]. Since the consequent class and the rule weight of each fuzzy rule can be specified by compatible training patterns, only the combinations of antecedent fuzzy sets for fuzzy rules in a classifier are optimized in our fuzzy GBML algorithm.

2.1 Genetics-Based Machine Learning

Our fuzzy GBML is a hybrid version of Pittsburgh approach and Michigan approach [15]. Its main framework is based on Pittsburgh approach in which a rule set is codified as a string. Michigan approach is used as a kind of local search.

Each fuzzy rule R_q is represented by its antecedent fuzzy sets A_{qi} ($i = 1, 2, ..., n$) as an integer substring of length n, where n is the dimensionality of the pattern space (i.e., n is the number of attributes of each pattern). We use 15 symbols (e.g., 0, 1, ..., 9, a, b, ..., e) to represent *don't care* and the 14 antecedent fuzzy sets as shown in Fig. 1.

A rule set S is handled as an individual and coded as a concatenated integer string where each substring of length n represents a single fuzzy if-then rule. It should be noted that the number of fuzzy rules in each rule set is not fixed in our fuzzy GBML. This means that we use strings of variable length as individuals.

In order to generate an initial population, first we randomly select N_{rule} training patterns. Next we generate a fuzzy if-then rule R_q from each selected training pattern $\mathbf{x}_p = (x_{p1}, ..., x_{pn})$ by probabilistically choosing an antecedent fuzzy set A_{qi} for each attribute value x_{pi} from the 14 antecedent fuzzy sets B_k ($k = 1, 2, ..., 9, a, b, ..., e$) in Fig. 1. Each antecedent fuzzy set B_k has the following selection probability for the attribute value x_{pi}:

$$P(B_k) = \frac{\mu_{B_k}(x_{pi})}{\sum\limits_{j=1}^{e} \mu_{B_j}(x_{pi})}, \quad k = 1, 2, ..., 9, a, b, ..., e. \tag{2}$$

That is, each antecedent fuzzy set B_k has a selection probability which is proportional to its compatibility grade with the attribute value x_{pi}. Then each antecedent fuzzy set of the generated fuzzy rule is replaced with a *don't care* condition using a prespecified probability $P_{don't care}$. This process makes the rule more general.

In this manner, N_{rule} initial fuzzy rules are generated. An initial rule set consists of the generated fuzzy if-then rules. By iterating this procedure, we generate N_{pop} initial rule sets (i.e., an initial population).

A newly generated rule set S is evaluated by the following three objective functions:

$f_1(S)$: The misclassified rate (%) on the training patterns by S,

$f_2(S)$: The number of fuzzy rules in S,

$f_3(S)$: The total number of antecedent conditions of fuzzy rules (i.e., total rule length) in S.

To calculate $f_1(S)$, a single winner rule R_w is identified by using the compatibility grade and the rule weight of each fuzzy rule in the rule set S. The input pattern $\mathbf{x}_p$ is classified as the consequent class C_w of the winner rule R_w. When multiple fuzzy rules with different consequent classes have the same maximum value, the classification of $\mathbf{x}_p$ is rejected. If there is no compatible fuzzy rule with $\mathbf{x}_p$, its classification is also rejected.

Whenever a new rule set is generated by not only the above procedure directly from training patterns but also Pittsburgh-type genetic operations and Michigan-type operation, each rule set is evaluated by the same three objective functions.

We use the following weighted-sum fitness function for obtaining accurate and simple fuzzy classifiers.

$$fitness(S) = w_1 f_1(S) + w_2 f_2(S) + w_3 f_3(S), \tag{3}$$

where w_1, w_2 and w_3 are non-negative weights.

Two parents (i.e., two rule sets) are selected from the current population by binary tournament selection with replacement. Let the selected rule sets be S_1 and S_2. Some fuzzy rules are randomly selected from each parent to construct an offspring rule set by crossover. The number of fuzzy rules to be inherited from each parent to the new rule set is randomly specified. We use an upper bound on the number of fuzzy rules in each rule set (e.g., 60 in our computational experiments). When the number of fuzzy rules is larger than the upper bound, we randomly remove fuzzy rules from the offspring rule set until the upper bound condition is satisfied. The above-mentioned crossover operation is applied to the selected pair of parent rule sets with a prespecified crossover probability P_C. When the crossover operation is not applied, one of the two parent rule sets is randomly chosen as their offspring rule set. Each antecedent fuzzy set of fuzzy rules in the offspring rule set is randomly replaced with a different antecedent fuzzy set in Fig. 1 by mutation. This mutation is applied to each antecedent fuzzy set with a prespecified mutation probability P_M.

After the crossover and mutation operations, a single iteration of the following Michigan-style algorithm is applied to the newly generated offspring rule set S:

Step 1: Classify all training patterns by the rule set S. The fitness value of each fuzzy rule in S is the number of correctly classified training patterns by that rule.

Step 2: Generate $N_{replace}$ fuzzy rules. Some rules are generated from the existing rules in S by genetic operations. The others are generated directly from misclassified and/or rejected training patterns.

Step 3: Replace the worst $N_{replace}$ fuzzy rules in S with the newly generated $N_{replace}$ fuzzy rules.

In Step 2, $N_{replace}$ fuzzy rules are to be generated. The number of replaced fuzzy rules (i.e., $N_{replace}$) is specified as $\lceil 0.2 \times |S| \rceil$ for each rule set S where $\lceil 0.2 \times |S| \rceil$ is the minimum integer not smaller than $0.2 \times |S|$. We generate at least a half of new fuzzy rules (i.e., at least $N_{replace}/2$ fuzzy rules) by genetic operations from the existing fuzzy rules in S. The probabilistic specification of each antecedent fuzzy set by (2) and the replacement with *don't care* using the probability $P_{don't care}$ are used to generate the other fuzzy rules.

Let N_{MR} be the sum of the number of misclassified and rejected training patterns by the rule set S. When N_{MR} is less than or equal to $N_{replace}/2$, all the N_{MR} training patterns are used to generate new fuzzy rules. In this case, N_{MR} fuzzy rules are generated from the N_{MR} training patterns. The other fuzzy rules (i.e., ($N_{replace} - N_{MR}$) fuzzy rules) are generated by genetic operations. On the other hand, when N_{MR} is larger than $N_{replace}/2$, $N_{replace}/2$ patterns are randomly chosen from the N_{MR} misclassified or rejected training patterns. Then $N_{replace}/2$ fuzzy rules are directly generated from the chosen patterns. The other $N_{replace}/2$ fuzzy rules are generated by genetic operations.

When we generate a new fuzzy rule from existing rules in S by genetic operations, first a pair of fuzzy rules is selected from S using binary tournament selection with replacement as parents. Then the standard uniform crossover operation is applied to the selected pair to generate a new fuzzy rule. Finally each antecedent fuzzy set is randomly replaced with a different one using a prespecified mutation probability. This procedure is iterated to generate a required number of new fuzzy rules (i.e., $N_{replace}$ fuzzy rules including directly generated fuzzy rules from training patterns).

As we have already explained, in our fuzzy GBML, a new rule set S is generated by selection, crossover, mutation, and a single iteration of the Michigan-style algorithm. These operations are iterated N_{pop} times to generate an offspring population of N_{pop} rule sets. The next population is constructed by choosing the best N_{pop} rule sets from the current and offspring populations with respect to the fitness function in (3). Generation update is iterated until a prespecified stopping condition is satisfied. The total number of generations was used as the stopping condition in our computational experiments.

3 Parallel Distributed Implementation

We use a workstation with multiple CPU cores for parallel distributed implementation of our fuzzy GBML. It is also possible to use a cluster system with several computers. Our implementation can be written as follows.

Step 1: Generate N_{pop} integer strings as an initial population by the heuristic rule initialization mechanism in (2) where N_{pop} is the population size.

Step 2: Randomly divide the current population P with N_{pop} integer strings into N_{CPU} sub-populations of the same size $\{P_1, P_2, ..., P_{N_{CPU}}\}$. Randomly divide the training data set D with m patterns into training data subsets of the same size $\{D_1, D_2, \ldots, D_{N_{CPU}}\}$ in Fig 2.

Step 3: Assign a sub-population and a training data subset to each of the N_{CPU} CPU cores.

Step 4: Perform the fuzzy GBML algorithm in each CPU core.

Step 5: If the prespecified termination condition is satisfied, go to Step 8.

Step 6: If the prespecified rotation interval is satisfied, rotate the N_{CPU} training data subsets as shown in Fig. 2 over the N_{CPU} CPU cores.

Step 7: Return to Step 4.

Step 8: Choose the best string S_{best} from the whole population in terms of the training data accuracy (i.e., $f_1(S)$) for the whole training data set.

It should be noted that in Step 6 the consequent class and the rule weight of each rule in the current sub-population at each CPU are updated by the newly assigned training data subset. Whenever these parameters are specified or updated, the compatibility grade of each training pattern in the training data subset with each rule is stored in order to reduce their repeated calculation.

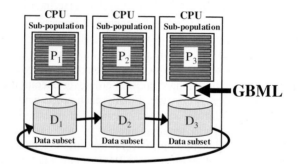

Fig. 2. Parallel Distributed Implementation of GBML

4 Computational Experiments

In our computational experiments, we used three data sets in Table 1 available from the KEEL-dataset repository [17]. The generalization ability was evaluated by iterating the ten-fold cross validation procedure 3 times (i.e., 30 runs in total).

We used a workstation with two Xeon 2.93 GHz quad processors (i.e., eight CPU cores in total). We codified the parallel distributed fuzzy GBML by Java on Windows 7 operating system. We used five CPU cores in this paper as preliminary experiments, because the operation system needs at least one CPU core for the basic operations. In this paper, we specified the population size as 210. That is, the size of each sub-population for parallel distributed implementation was 42. The remaining setting was as follows:

The number of fuzzy rules in each initial rule set: 30,
Upper limit on the number of fuzzy rules: 60,

Lower limit on the number of fuzzy rules: 1,
Probability of *don't care* ($P_{don't\ care}$): $(n\text{-}1)/n$, (n is the number of attributes),
Probability of the Michigan-style algorithm: 0.5,
Crossover probability in the main part (P_C): 0.9,
Crossover probability in the Michigan-style part: 0.9,
Mutation probability in the main part (P_M): $1 / (n\ |S|)$,
Mutation probability in the Michigan-style part: $1/n$,
Termination condition: 10,000 generations,
The weight vector in Eq. (3): $\mathbf{w} = (100, 1, 1)$.

We used a large value for the first element of the weight vector to find accurate classi-fiers. Since obtained classifiers strongly depend on the weight vector specification, we may need further experiments with different specifications as a future study.

We examined the following specifications of the rotation interval for training data subsets rotation.

Rotation interval (generation): 50, 100, 200, 500, none,

where "none" means that training data subsets were not rotated. When the rotation interval was specified as 50, training data subsets were rotated every 50 generations. After the rotation, each rule set in the current sub-population was re-evaluated for the newly assigned training data subsets.

Table 1. Data sets used in our computational experiments

Data set	Number of attributes	Number of patterns	Number of classes
Phoneme	5	5404	2
Satimage	36	6435	6
Pendig	16	10992	10

In Tables 2-4, we summarize experimental results (i.e., training data accuracy, test data accuracy, the number of rules, total rule length, and computation time). Each entry in the tables is the average over 30 runs. The first five rows of each table represent the results by the parallel distributed implementation of our fuzzy GBML algorithm with different rotation intervals of training data subsets. The last row (i.e., Standard) repre-sents the results by the standard non-parallel implementation of our fuzzy GBML algo-rithm with a single population and no data subdivision. With respect to training data accuracy and test data accuracy, improved results from the non-parallel implementa-tion are highlighted by the bold face.

Table 2 shows the results for Phoneme data set. From Table 2, we can see that the parallel distributed implementation with the rotation of training data subsets searched for classifiers with higher test data accuracy (i.e., higher generalization ability) than those by the non-parallel method. With respect to the test data accuracy, similar re-sults were obtained over a wide range of rotation intervals. The main advantage of the parallel distributed implementation is the reduction of the computation time. Whereas the parallel distributed implementation of genetic fuzzy rule selection [11] was about

Table 2. Average results over 3x10CV by the parallel distributed and non-parallel implementation of fuzzy GBML for Phoneme data set

Rotation interval	Training data accuracy [%]	Test data accuracy [%]	Number of rules	Total rule length	Computation time [min]
50	84.58	**83.31**	16.50	61.83	9.4
100	84.22	**83.12**	17.33	65.27	9.3
200	84.33	**83.27**	18.57	68.27	9.4
500	84.25	**83.44**	17.80	64.60	9.1
None	83.00	81.62	17.90	50.00	7.0
Standard	84.59	82.98	16.50	50.30	60.3

$N_{CPU}{}^2$ faster than the non-parallel one, the proposed method was only about seven times faster ($N_{CPU} < 7 < N_{CPU}{}^2 = 25$) than the non-parallel one. That is, the speeding-up rates were not so high.

Table 3 shows the results for Satimage data set. It is showed that the parallel distributed implementation with the rotation of training data subsets searched for classifiers with higher training and test data accuracy than those by the non-parallel method. The proposed method was 10-13 times faster than the non-parallel one. Table 3 also shows that the average number of rules was decreased by the frequent rotation of training data subsets.

Table 4 shows the results for Pendig data set. With respect to the training data accuracy and test data accuracy, the parallel distributed implementation did not work well for Pendig data set. There exists a large gap between the training data accuracy by the parallel distributed implementation and that by the non-parallel method. The same observation was not seen for the other data sets. Although there exists a clear deterioration in the classification accuracy, our parallel distributed implementation was 9-11 times faster than the non-parallel one. The detailed analysis is necessary to investigate the reason. We may need longer generations for Pendig data set.

Table 3. Average results over 3x10CV by the parallel distributed and non-parallel implementation of fuzzy GBML for Satimage data set

Rotation interval	Training data accuracy [%]	Test data accuracy [%]	Number of rules	Total rule length	Computation time [min]
50	85.66	**84.46**	11.47	89.67	9.8
100	**85.95**	**84.85**	11.93	112.00	10.0
200	**85.87**	**85.15**	13.73	106.90	10.6
500	85.04	83.82	14.33	94.67	10.9
None	83.65	82.88	18.97	65.33	11.9
Standard	85.72	83.94	18.10	72.93	128.4

Table 4. Average results over 3x10CV by the parallel distributed and non-parallel implementation of fuzzy GBML for Pendig data set

Rotation interval	Training data accuracy [%]	Test data accuracy [%]	Number of rules	Total rule length	Computation time [min]
50	94.41	93.85	31.10	141.20	36.5
100	94.67	93.88	28.50	147.97	35.4
200	94.65	93.95	29.37	159.97	38.0
500	94.21	93.31	30.00	162.73	38.6
None	93.09	92.36	36.67	159.13	42.4
Standard	96.42	95.22	34.50	164.2	392.9

5 Conclusion

In this paper, we applied the parallel distributed implementation proposed in [11] to our fuzzy GBML algorithm. Through the computational experiments on three large data sets with a large number of patterns, we examined the effects of the proposed parallel distributed implementation of our fuzzy GBML algorithm. For Phoneme and Satimage data sets, the proposed method worked very well. Classifiers with higher generalization ability were obtained in shorter time than the non-parallel method. For Pendig data set, we observed the deterioration in the test data accuracy by the proposed method. But the computation time was drastically reduced by the parallel distributed implementation in all the three data sets.

As a future study, we need further experiments with other data sets to quantitatively clarify the effectiveness of our parallel distributed approach. The effects of the (sub) population size on the search ability should be studied. As reported in [18] for standard data sets (i.e., not with a large number of patterns), there exists a tradeoff between accuracy and complexity of fuzzy classifiers. We should also examine the effect of different specifications of the weight vector in the fitness function on the search ability of the proposed method. The use of evolutionary multiobjective optimization will also be an important step to avoid the weight specification.

Acknowledgment. This study is partially supported by the Japanese Grant-in-Aid for Young Scientists (B): KAKENHI (22700239).

References

1. Freitas, A.A.: Data Mining and Knowledge Discovery with Evolutionary Algorithms. Springer, Heidelberg (2002)
2. Bull, L., Bernado-Mansilla, E., Holmes, J.: Learning Classifier Systems in Data Mining. Springer, Heidelberg (2008)
3. Cordon, O., Herrera, F., Hoffmann, F., Magdalena, L.: Genetic Fuzzy Systems. World Scientific, Singapore (2001)
4. Herrera, F.: Genetic fuzzy systems: taxonomy, current research trends and prospects. Evolutionary Intelligence 1(1), 27–46 (2008)

5. Alba, E., Tomassini, M.: Parallelism and evolutionary algorithms. IEEE Transactions on Evolutionary Computation 6(5), 443–462 (2002)
6. Cantu-Paz, E.: A survey of parallel genetic algorithms. IlliGAL Report no. 95003 (1997)
7. Liu, H., Motoda, H.: Instance Selection and Construction for Data Mining. Kluwer Academic Publishers, Dordrecht (1998)
8. Liu, H., Motoda, H.: Feature Selection for Knowledge Discovery and Data Mining. Kluwer Academic Publishers, Dordrecht (1998)
9. Cano, J.R., Herrera, F., Lozano, M.: Stratification for scaling up evolutionary prototype selection. Pattern Recognition Letters 26(7), 953–963 (2005)
10. Cano, J.R., Herrera, F., Lozano, M.: On the combination of evolutionary algorithms and stratified strategies for training set selection in data mining. Applied Soft Computing 6(3), 323–332 (2006)
11. Nojima, Y., Ishibuchi, H., Kuwajima, I.: Parallel distributed genetic fuzzy rule selection. Soft Computing 13(5), 511–519 (2009)
12. Nojima, Y., Ishibuchi, H.: Effects of data reduction on the generalization ability of parallel distributed genetic fuzzy rule selection. In: Proc. of 9th International Conference on Intelligent Systems Design and Applications, pp. 96–101 (2009)
13. Nojima, Y., Ishibuchi, H., Mihara, S.: Use of very small training data subsets in parallel distributed genetic fuzzy rule selection. In: Proc. of 4th International Workshop on Genetic and Evolutionary Fuzzy Systems, pp. 27–32 (2010)
14. Ishibuchi, H., Nozaki, K., Yamamoto, N., Tanaka, H.: Selecting fuzzy if-then rules for classification problems using genetic algorithms. IEEE Trans. on Fuzzy Systems 3(3), 260–270 (1995)
15. Ishibuchi, H., Yamamoto, T., Nakashima, T.: Hybridization of fuzzy GBML approaches for pattern classification problems. IEEE Trans. on Systems, Man, and Cybernetics-Part B 35(2), 359–365 (2005)
16. Ishibuchi, H., Nakashima, T., Nii, M.: Classification and Modeling with Linguistic Information Granules: Advanced Approaches to Linguistic Data Mining. Springer, Berlin (2004)
17. KEEL dataset repository, http://keel.es/
18. Ishibuchi, H., Nojima, Y.: Analysis of interpretability-accuracy tradeoff by multiobjective fuzzy genetics-based machine learning. International Journal of Approximate Reasoning 44(1), 4–31 (2007)

Multi-Objective Evolutionary Algorithms for Feature Selection: Application in Bankruptcy Prediction

António Gaspar-Cunha[1], Fernando Mendes[1], João Duarte[2], Armando Vieira[2], Bernardete Ribeiro[3], André Ribeiro[4], and João Neves[4]

[1] IPC/I3N - Institute of Polymers and Composites, University of Minho, Guimarães, Portugal
{agc,fmendes}@dep.uminho.pt
[2] Department of Physics, Instituto Superior de Engenharia do Porto, R. S. Tomé, 4200 Porto, Portugal
{jmmd,asv}@isep.ipp.pt
[3] Department of Informatics Engineering, Center of Informatics and Systems, University of Coimbra, Coimbra 3030-290, Portugal
bribeiro@dei.uc.pt
[4] ISEG School of Economics and Management, Technical University of Lisbon, Portugal
andremsr@mail.pt, jcneves@iseg.utl.pt

Abstract. A Multi-Objective Evolutionary Algorithm (MOEA) was adapted in order to deal with problems of feature selection in data-mining. The aim is to maximize the accuracy of the classifier and/or to minimize the errors produced while minimizing the number of features necessary. A Support Vector Machines (SVM) classifier was adopted. Simultaneously, the parameters required by the classifier were also optimized. The validity of the methodology proposed was tested in the problem of bankruptcy prediction using a database containing financial statements of 1200 medium sized private French companies. The results produced shown that MOEA is an efficient feature selection approach and the best results were obtained when the accuracy, the errors and the classifiers parameters are optimized.

Keywords: Multi-Objective, Evolutionary Algorithms, Feature Selection, Bankruptcy Prediction.

1 Introduction

The problem of feature selection from databases with high amount of data is of crucial importance, specially when dealing with problems such as bankruptcy prediction given its consequences for banks, insurance companies, creditors and investors. Thus, the ability to discriminate between faithful customers from potential bad ones is thus crucial for commercial banks and retailers [1].

Traditional methods used to study this problem, such as discriminant analysis [2] and Logit and Probit models [3], have important limitations. Discriminant

K. Deb et al. (Eds.): SEAL 2010, LNCS 6457, pp. 319–328, 2010.

analysis is limited due to its linearity, restrictive assumptions, for treating financial ratios as independent variables and can only be used with continuous independent variables. In non-linear models the choice of the regression function creates a bias that restricts the outcome, they are very sensitive to exceptions, and most conclusions have an implicit Gaussian distribution on data, which is inappropriate in many cases. To overcome these problems other approaches have been applied recently in the problem of bankruptcy classification, such as Artificial Neural Networks (ANN) [4,5], Evolutionary Algorithms (EA) and Support Vector Machines (SVM) [6]. Usually, complementary tools based on ANN, EA and SVM are used to classify credit risk. In some studies it is shown that ANN outperforms discriminant analysis in bankruptcy prediction [5,6,7,8,9]. Moreover these promising results, it is generally recognized that further research is needed [10].

Due to the large number of variables usually present, and due to the high correlation between these variables, it is of fundamental importance the existence of a feature selection method able to reduce the number of features considered for analysis [11]. A possible approach to deal with this problem consists on the use of Multi Objective Evolutionary Algorithms (MOEA). Bi in [12] proposed a framework for SVM based on multi-objective optimization with the aim of minimize the risk of the classifier and the model capacity (or accuracy). Igel in [13] followed an identical approach, but replaced the objective concerning the minimization of the risk by the minimization of the complexity of the model (i.e., the number of features). Oliveira et al. in [14] used an hierarchical MOEA operating at two levels: performing a feature selection to generate a set of classifiers (based on artificial neural networks) and selecting the best set of classifiers. Hamdani et al. in [15] used the NSGA-II [16] algorithm to optimize simultaneously the number of features and the global error obtained by a neural network classifier. Alfaro-Cid et al. in [17] applied a MOEA to take into account individually the errors of type I (false positive) and type II (false negative). Finally, Handl and Knowles in [18] studied the problem of unsupervised feature selection by formulating them as a multi-objective optimization problem.

This work follows the main ideas of a previous work proposed by the authors, were a methodology based on MOEA was used to accomplish simultaneously two objectives: the minimization of the number of features used and the maximization of the accuracy of the classifier used [19]. In the present case different accuracy measures, such as maximization of the F measure (F_m) and the minimization of errors (type I and type II), will be tested. Simultaneously, the parameters required by the classifier will be optimized. This is an important issue since parameter tuning is not an easy task [20].

2   Bankruptcy Prediction

The Problem and Dataset. In the bankruptcy prediction problem the aim is to infer the probability that a company will become distressed, over a specified period, given a set of financial statements. This can be done from over one, or several years. In general this task is performed by dividing the data into two groups:

healthy and bankrupted companies, and then training a binary classifier, either supervised or unsupervised, to learn the pattern that discriminate between the two cases. Often, the database needs some previous treatment, prior to training the classifiers, in order to create a well balanced and unbiased sample. Usually, a full dataset is composed by tenths of accounting features, or ratios, measuring different characteristics of a company (e.g., the profitability, liabilities, cash-flow and equity). These features are often highly correlated and confusing, being important to use just some of them. This will simplify considerably the problem. However, in order not to loose important information, special care must be taken during the process of reducing the number of features. Thus the performance of the classifier will not decrease. It is clear that these ideas can be generalized to other type of classification problems than bankruptcy prediction.

In this work a sample obtained from the DIANE database was selected. The initial database consisted of financial ratios of about 60 000 industrial French companies, for the years of 2002 to 2006, with at least 10 employees. From these companies, about 3000 were declared bankrupted in 2007 or presented a restructuring plan ("Plan de Redressement") to the court for approval by the creditors. No distinction between these two categories has been made since both categories signals companies in financial distress. The dataset includes information about 30 financial ratios, as defined by COFACE (Table 1), of companies covering a wide range of industrial sectors.

Classification and Metrics. In the methodology proposed a SVM classifier will be used while a MOEA is used to determine the best compromise between the two and/or the three conflicting objectives. Support Vector Machines (SVMs) are a set of supervised learning methods based on the use of a kernel, which can be applied to classification and regression. In the SVM a hyper-plane or set of hyper-planes is (are) constructed in a high-dimensional space. In this case, a

Table 1. Set of features considered (as defined by COFACE)

F1 Number of employees	F2 Capital Employed/Fixed Assets
F3 Financial Debt/Capital Employed	F4 Depreciation of Tangible Assets
F5 Working capital/current assets	F6 Current ratio
F7 Liquidity ratio	F8 Stock Turnover days
F9 Collection period	F10 Credit Period
F11 Turnover per Employee	F12 Interest / Turnover
F13 Debt Period days	F14 Financial Debt/Equity
F15 Financial Debt/Cashflow	F16 Cashflow/Turnover
F17 Working Capital/Turnover (days)	F18 Net Current Assets/Turnover
F19 Working Capital Needs/Turnover	F20 Export
F21 Value added per employee	F22 Total Assets/Turnover
F23 Operating Profit Margin	F24 Net Profit Margin
F25 Added Value Margin	F26 Part of Employees
F27 Return on Capital Employed	F28 Return on Total Assets
F29 EBIT Margin	F30 EBITDA Margin

good separation is achieved by the hyper-plane that has the largest distance to the nearest training data points of any class. Thus, the generalization error of the classifier is lower when this margin is larger. SVMs can be seen an extension to nonlinear models of the generalized portrait algorithm developed by Vapnik in [21]. In this work the SVM from LIBSVM was used [22]. The selection of the right kernel, as well the definition of the best kernel parameters, is of primordial importance for the SVM performance [13]. In the present study only the C-SVC method using as kernel the Radial Basis Function (RBF) was tested [22]. Thus, two different SVM parameters are to be selected carefully: the regularization parameter (C) and the kernel parameter (γ). Simultaneously, another important parameter for training the SVM is the Learning Rate (LR), which was also taking into account in this study. Another important issue concerns the performance metrics used to evaluate the learning methods [23,24,25]. The most straightforward way is to use the accuracy given by the ratio between the number instances correctly evaluated and the total number of instances, i.e.:

$$Accuracy = \frac{TP + TN}{TP + TN + FP + FN} \tag{1}$$

where, TP are the positives correctly classified, TN are the negatives correctly classified, FP are the positives incorrectly classified and FN are the negative incorrectly classified.

It is also important to know the level of the errors accomplished by the classifier, mainly on problems where the existence of errors is critical. Two different error types can be defined, type I and type II, given respectively by:

$$e_I = \frac{FP}{FP + TN} \tag{2}$$

and

$$e_{II} = \frac{FN}{FN + TP} \tag{3}$$

Another traditional way to evaluate the information is using the sensitivity or recall (R) and the precision (P) of the classifier:

$$R = \frac{TP}{TP + FN} \tag{4}$$

and

$$P = \frac{TP}{TP + FP} \tag{5}$$

In the present work F_m, which represents the harmonic mean of R and P, was adopted here to evaluate globally the classifier:

$$F_m = \frac{2PR}{P + R} \tag{6}$$

The selection of the best learning algorithm to use and the best performance metric to measure the efficiency of the classifier is nowadays the subject of many studies [23,25].

3   Multi-Objective Evolutionary Algorithms

MOEAs have been recognized in the last decade as good methods to explore and find an approximation to the Pareto-optimal front for multi-objective optimization problems. This is due to the difficulty of traditional exact methods to solve this type of problems and by their capacity to explore and combine various solutions to find the Pareto front in a single run. A MOEA must provide a homogeneous distribution of the population along the Pareto frontier, together with an improvement of the solutions along successive generations [26,27]. In this work, the Reduced Pareto Set Genetic Algorithm (RPSGA) is adopted [27,28], where a clustering technique is applied to reduce the number of solutions on the efficient frontier. Detailed information about this algorithm can be found elsewhere [27,28]. In the present study the RPSGA algorithm was adapted to deal with the features selection problem, so it can be considered as a combinatory optimization task. Concerning the definition of the decision variables, two possibilities were considered. Initially, a pure feature selection problem was analyzed. In this case the parameters of the classifier, such as type of training (in the present study only k-fold cross validation was used) and learning rate and the SVM parameters (C and γ), were initially set. In a second approach, these parameters were also included as variables to be optimized. The latter approach has the advantage of obtaining in a single run the best features and, simultaneously fine tuning the classifier parameters. This approach will be illustrated in the next section.

4   Results and Discussion

4.1   Case Studies

The MOEA methodology presented above will be used in a problem of finding the minimum number of features while maximizing F_m and minimizing e_I. Based on the data from a given year, the classifier is trained to predict whether the company will survive over the following year. Table 2 shows the different experiments tested. In all cases the C-SVC method using as kernel the Radial Basis Function (RBF) and 10-fold validation training method were used. First four experiments, using only as decision variables the features, were performed (experiments c-svc1 to c-svc04 in Table 2). In this case the Learning Rate (LR), C and γ are set to 0.01, 1 and 10, respectively. In experiments c-svc11 to c-svc14 and c-svc21, LR, C and are also considered as decision variables (i.e., they are parameters to be optimized). The range of variation allowed for these variables is shown on Table 2. The RPSGAe was applied using the following parameters: 100 generations, crossover rate of 0.8, mutation rate of 0.05, internal and external populations with 100 individuals, limits of the clustering algorithm set at 0.2 and the number of ranks (N_Ranks) at 30. Due to the stochastic nature of the initial population several runs were performed (in the present case 16 runs) for each experiment. Thus, a statistical method based on attainment functions

Table 2. Experimental setup

Exp.	LR	C	γ	Objectives
c-svc1	0.01	1	10	$NF + F_m$
c-svc2	0.01	1	10	$NF + e_I$
c-svc3	0.01	1	10	$NF + e_I I$
c-svc4	0.01	1	10	$NF + F_m + e_I$
c-svc11	[0.001,0.1]	[1,1000]	[0.005,10]	$NF + F_m$
c-svc12	[0.001,0.1]	[1,1000]	[0.005,10]	$NF + e_I$
c-svc13	[0.001,0.1]	[1,1000]	[0.005,10]	$NF + e_I I$
c-svc14	[0.001,0.1]	[1,1000]	[0.005,10]	$NF + F_m + e_I$

was applied to compare the final population for all runs [29,30]. This method attributes to each objective vector a probability that this point is attaining in one single run [29]. It is not possible to compute the true attainment function, but it can be estimated based upon approximation set samples, i.e., different approximations obtained in different runs, which is denoted as Empirical Attainment Function (EAF) [31]. The differences between two algorithms can be visualized by plotting the points in the objective space where the differences between the empirical attainment functions of the two algorithms are significant [32].

4.2   Correspondence between Optimization Objectives

First, in order to have an idea about the shape of the Pareto fronts a population of 1000 individuals generated randomly was initially evaluated using the classifier and the different metrics (equations 1 to 6). Figure 1 shows these results. As can be seen it is possible to obtain identical value for F_m with different number and combinations of features. However, when approaching the top left corner (maximization of Fmeasure and minimization of the number of features) this does not happen, as expected. In the case of the graph e_I versus F_m the best location is the bottom right corner, but, as expected, the tendency is to go to the top right corner. This means that these objectives are conflicting.

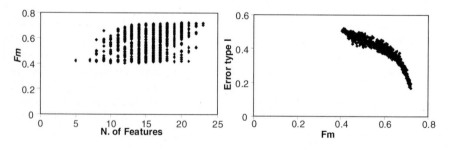

Fig. 1. Pareto plots for 1000 solutions generated randomly

Table 3. Results for Run 1 of experiment C-SVC14

N.F.	F_m	e_I	LR	γ	C	Features
2	0.885	0.0467	0.0953	9.99	985	F11, F28
3	0.962	0.0054	0.0992	9.98	983	F11, F18, F28
3	0.967	0.0173	0.0970	9.99	975	F 8, F11, F28
4	0.997	0.0000	0.0941	9.98	959	F 8, F11, F16, F28
4	0.998	0.0017	0.0960	9.99	967	F 8, F11, F13, F28
5	1.000	0.0000	0.0986	9.99	967	F 1, F 8, F11, F13, F2
5	1.000	0.0000	0.0937	9.98	971	F 3, F 8, F11, F16, F28
5	1.000	0.0000	0.0971	9.94	905	F 3, F 8, F11, F13, F28
5	1.000	0.0000	0.0969	9.97	986	F 3, F 8, F11, F28, F30
5	1.000	0.0000	0.0917	9.98	968	F 8, F11, F13, F16, F28
5	1.000	0.0000	0.0943	9.98	975	F 4, F 8, F11, F16, F28
5	1.000	0.0000	0.0908	9.99	965	F 8, F 9, F11, F16, F28
5	1.000	0.0000	0.0962	9.96	977	F 6, F 8, F11, F13, F28
5	1.000	0.0000	0.0958	9.99	969	F 8, F11, F13, F22, F28
5	1.000	0.0000	0.0984	9.95	905	F 5, F 8, F11, F13, F28
5	1.000	0.0000	0.0979	9.99	980	F 8, F11, F18, F22, F28

Table 4. Results with 3 features for Runs of experiment C-SVC14

Run	F_m	e_I	LR	γ	C	Features
1	0.967	0.0173	0.0970	9.99	975	F8, F11, F28
2	0.963	0.0151	0.0805	9.93	737	F8, F16, F22
3	0.953	0.0168	0.0443	9.95	942	F8, F9, F28
4	0.962	0.0194	0.0739	9.95	990	F6, F8, F23
5	0.956	0.0135	0.0529	9.90	961	F8, F14, F16
6	0.964	0.0226	0.0182	9.81	984	F8, F11, F29

4.3   Optimization Results

EAFs graphs were vused to compare the performance between experiments c-svc1 and c-svc11, c-svc2 and c-svc12 and c-svc3 and c-svc13, were objective 1 is the number of features and objective 2 is F_m, e_I or e_{II}, respectively. These plots were not presented here due to a lack of space. The analysis of these plots allows concluding that the best performance is always obtained when the classifier parameters are optimized simultaneously (i.e., experiments c-svc11 to c-svc13). This indicates that the optimization algorithm is able to find the best classifier parameters for the case under study. The same is true for the experiments with three objectives (c-svc4 and c-svc14), to which is more difficult to obtain the EAFs graphs and, thus, they are not present. Figure 2 plots the Pareto-fronts after 100 generations of a single run of experiments c-svc4 (left) and c-svc14 (right). It is clear from these plots that the run corresponding to experiment c-svc is more efficient. In this case fewer solutions are found since the MOEA was able to reach to solutions were the Fmeasure and eI converge to its best value (1 and 0, respectively).

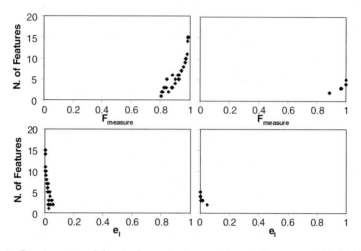

Fig. 2. Pareto optimal fronts for a single run of experiment c-svc14 (Table 2)

The results obtained for a single run of experiment c-svc14 (i.e., the solutions shown in the right graphs of Figure 2) are presented on Table 3. As can be seen there are several solutions were the number of features obtained are the same, mainly in the case of the solutions with five features. This is due to small changes produced in the classifier parameters values (LR, C and γ) also optimized. If considered that a good solution will be the one with an F_m higher than 90%, the two solutions with three features were selected: the solution with F11, F18 and F28 and the solution with F8, F11 and F28 features, respectively. The difference between these two solutions is due to features F11 and F8. When F8 is present both F_m and e_I increase.

Table 4 presents for ilustrating purposes the best results accomplished only with three features selected for some of the runs of experiment c-svc14. In this case the MOEA converged to a different set of features. Also, in this case, the features selected are able to cluster the companies. Therefore, for this set of data there is more than one solution able to attain the objectives defined.

5   Conclusion

In this work a MOEA was used for feature selection in the bankruptcy prediction problem using Support Vector Machines classifier. The methodology proposed was able not only to reduce the features necessary but is able also to provide relevant information to the decision maker. The algorithm does not only provide the best features to be used but, also, with the best parameters of the classifier. The best performance only is attained when the classifier parameters are optimized simultaneously with the features to be selected, since the classifier performance is strongly dependent on these parameters. Finally, the MOEA was able to provide more than one set of features able to optimize the objectives defined.

Acknowledgments. The financial support of the Portuguese science foundation (FCT) under grant PTDC/GES/70168/2006 is acknowledged.

References

1. Atiya, F.: Bankruptcy prediction for credit risk using neural networks: A survey and new results. IEEE Transactions on Neural Networks 12, 12–16 (2001)
2. Eisenbeis, R.A.: Pitfalls in the Application of Discriminant Analysis in Business, Finance and Economics. J. of Finance 32, 875–900 (1997)
3. Martin, D.: Early Warning of Bank Failure: A Logit Regression Approach. J. of Banking and Finance 1, 249–276 (1977)
4. Charitou, A., Neophytou, E., Charalambous, C.: Predicting corporate failure: empirical evidence for the UK. European Accounting Review 13, 465–497 (2004)
5. Neves, J.C., Vieira, A.S.: Improving Bankruptcy Prediction with Hidden Layer Learning Vector Quantization. European Accounting Review 15, 253–271 (2006)
6. Fan, A., Palaniswami, M.: Selecting bankruptcy predictors using a support vector machine approach. In: Proceedings of IJCNN 2000, pp. 354–359 (2000)
7. Coats, P.K., Fant, L.F.: Recognizing Financial Distress Patterns Using a Neural Network Tool. Financial Management 22, 142–155 (1993)
8. Yang, D.T.: Urban-biased policies and rising income inequality in China. American Economic Review Papers and Proceedings 89, 306–310 (1999)
9. Tan, C.N.W., Dihardjo, H.: A Study on Using Artificial Neural Networks to Develop an Early Warning Predictor for Credit Union Financial Distress with Comparison to the Probit Model. Managerial Finance 27, 56–77 (2001)
10. Vieira, A.S., Duarte, J., Ribeiro, B., Neves, J.C.: Accurate Prediction of Financial Distress of Companies with Machine Learning Algorithms. In: Kolehmainen, V., Toivanen, P., Beliczynski, B. (eds.) ICANNGA 2009. LNCS, vol. 5495, pp. 569–576. Springer, Heidelberg (2009)
11. Guyon, I., Gunn, S., Nikravesh, M., Zadeh, L.: Feature Extraction Foundations and Applications. Springer, Heidelberg (2006)
12. Bi, J.: Multi-Objective Programming in SVMs. In: Proceedings of the Twentieth International Conference on Machine Learning, ICML 2003, Washington, DC (2003)
13. Igel, C.: Multi-Objective Model Selection for Support Vector Machines. In: Coello Coello, C.A., Hernández Aguirre, A., Zitzler, E. (eds.) EMO 2005. LNCS, vol. 3410, pp. 534–546. Springer, Heidelberg (2005)
14. Oliveira, L.S., Morita, M., Sabourin, R.: Feature Selection for Ensembles Using the Multi-Objective Optimization Approach. SCI, pp. 49–74 (2006)
15. Hamdani, T.M., Won, J.-M., Alimi, A.M., Karray, F.: Multi-objective Feature Selection with NSGA II. In: Beliczynski, B., Dzielinski, A., Iwanowski, M., Ribeiro, B. (eds.) ICANNGA 2007. LNCS, vol. 4431, pp. 240–247. Springer, Heidelberg (2007)
16. Deb, K., Pratap, A., Agarwal, S., Meyarivan, T.: A fast and elitist multi-objective genetic algorithm: NSGA-II. IEEE Transaction on Evolutionary Computation 6, 181–197 (2002)
17. Alfaro-Cid, E., Castillo, P.A., Esparcia, A., Sharman, K., Merelo, J.J., Prieto, A., Mora, A.M., Laredo, J.L.J.: Comparing Multiobjective Evolutionary Ensembles for Minimizing Type I and II Errors for Bankruptcy Prediction. In: CEC 2008, Washington, USA, pp. 2907–2913 (2008)

18. Handl, J., Knowles, J.: Feature subset selection in unsupervised learning via multiobjective optimization. Int. J. of Computational Intelligence Research 2, 217–238 (2006)
19. Gaspar-Cunha, A., Mendes, F., Duarte, J., Vieira, A., Ribeiro, B., Ribeiro, A., Neves, J.: Feature Selection for Bankruptcy Prediction: A Multi-Objective Optimization Approach. Int. J. of Natural Computing Research 1, 71–79 (2010)
20. Kulkarni, A., Jayaraman, V.K., Kulkarni, B.D.: Support vector classification with parameter tuning assisted by agent-based technique. Computers and Chemical Engineering 28, 311–318 (2008)
21. Cortes, C., Vapnik, V.: Support-Vector Networks. Machine Learning 20, 273–297 (1995)
22. Chang, C.-C., Lin, C.-J.: LIBSVM a library for support vector machines (Tech. Rep.). Dept. of Computer Science and Information Engineering, National Taiwan University, Taipei, Taiwan (2000)
23. Caruana, R., Niculescu-Mizil, A.: Data Mining in Metric Space: An Empirical Analysis of Supervised Learning Performance Criteria. In: KDD 2004, Seattle, Washington, pp. 69–78 (2004)
24. Provost, F., Fawcet, T.: Analysis and Verification of Classifier Performance: Classification under Imprecise Class and Cost Distributions. In: KDD 1997, Menlo Park, CA, pp. 43–48 (1997)
25. Fawcet, T.: An introduction to ROC analysis. Pattern Recognition Letters 27, 861–874 (2006)
26. Deb, K.: Multi-Objective Optimization using Evolutionary Algorithms. Wiley, New York (2001)
27. Gaspar-Cunha, A., Covas, J.A.: RPSGAe - A Multiobjective Genetic Algorithm with Elitism: Application to Polymer Extrusion. In: Dorigo, M., Birattari, M., Blum, C., Gambardella, L.M., Mondada, F., Stützle, T. (eds.) ANTS 2004. LNCS, vol. 3172, pp. 221–249. Springer, Heidelberg (2004)
28. Gaspar-Cunha, A.: Modelling and Optimization of Single Screw Extrusion. Published doctoral dissertation, 2000. Lambert Academic Publishing, Köln (2009)
29. Fonseca, C., Fleming, P.J.: On the performance assessment and comparison of stochastic multiobjective optimizers. In: Ebeling, W., Rechenberg, I., Voigt, H.-M., Schwefel, H.-P. (eds.) PPSN 1996. LNCS, vol. 1141, pp. 584–593. Springer, Heidelberg (1996)
30. Knowles, J.D., Thiele, L., Zitzler, E.: A tutorial on the performance assessment of stochastive multiobjective optimizers. TIK-Report No. 214 (2006)
31. Fonseca, V.G., Fonseca, C., Hall, A.: Inferential performance assessment of stochastic optimisers and the attainment function. In: Zitzler, E., Deb, K., Thiele, L., Coello Coello, C.A., Corne, D.W. (eds.) EMO 2001. LNCS, vol. 1993, pp. 213–225. Springer, Heidelberg (2001)
32. López-Ibañez, M., Paquete, L., Stützle, T.: Hybrid population based algorithms for the bi-objective quadratic assignment problem. J. of Math. Modelling and Algorithms 5, 111–137 (2006)

Tile Pasting P System with Multiple-Edge Pasting

T. Robinson[1], Atulya K. Nagar[2], and S. Jebasingh[3]

[1] Madras Christian College, Chennai – 600 059, India
robin.mcc@gmail.com
[2] Liverpool Hope Univeristy, Liverpool, L16 9JD, UK
nagara@hope.ac.uk
[3] Saveetha Engineering College, Chennai - 602 105, India
jebasingh21@yahoo.com

Abstract. Pasting Systems and Tile Pasting P Systems are syntactic techniques using pasting operation generating tiling patterns and tessellations. A variant of the Parametric Tile Pasting P System equipped with multiple-edge pasting operations and geometric transformation operations has been presented here. This extension, Tile Pasting P System with Multiple-Edge Pasting (TPPSMEP) discussed in this paper with the enhanced operations increases the generative capacity to produce patterns with recurring primitives and rotated/ scaled tiling patterns unlike the earlier versions.

Keywords: Membrane computing, Picture languages, Syntactic methods.

1 Introduction

The Pasting System (PS) and its variants such as Extended Pasting Scheme (EPS) and Tabled Pasting System (TPS) are syntactic methods, introduced by Robinson et al. in [14,15,16], to generate tessellation and tiling patterns. The technique of gluing two tiles introduced in [16] specifies the edges of tiles that are to be glued or pasted the tiles thus pasted side to side with such pasting rules result in bigger pattern of tiles known as tiling pattern. With reference to the generative capacity of these systems, it has been proved that in [14] PS and EPS are incomparable and PS is properly contained in TPS.

Membrane Computing [8, 9], a branch of nature inspired systems is an area of computer science aiming to abstract computing models from the structure and functioning of the living cell. The basic model of membrane systems (also known as P Systems) which consist of membrane structure, are distributed parallel computing devices, processing multisets of symbol-objects, synchronously, in the compartments defined by cell-like membrane structure. The evolution rules also govern the modification of these objects in time, and transfer of objects from one membrane to another membrane. In recent years, generation of two dimensional picture languages using various syntactic techniques has been is extensively studied. P Systems using array-objects, picture–objects have also been defined. P Systems generating arrays and Indian folk designs (Kolam patterns) are considered in [11]. Tissue like P-systems with active membranes that generate local picture languages are considered in [1].

K. Deb et al. (Eds.): SEAL 2010, LNCS 6457, pp. 329–338, 2010.

Theoretical models of P Systems are introduced in [12, 14] for generating two dimensional tiling patterns that are formed by gluing square tiles, wherein the application of pasting rule to a picture pattern is by sequential i.e. one rule is applied at a time [12] or in a maximally parallel manner [14] i.e., all the rules are applied to the edges of the tiling pattern simultaneously. In [14] it has been proved that the picture languages generated by TPS are contained in the family of picture languages generated by Evolution-Communication Tile Pasting P System (ECTPPS) and Tile Pasting P System with Active Membrane (TPPS–AM). In this paper we propose an extension to the Tile Pasting P System, namely Parametric-Multiple Tile Pasting P System with Multiple-Edge Pasting (TPPSMES), allowing multiple edge pasting and geometric operations on the tiling patterns. Tessellation patterns made up of primitive tiling patterns are generated by this newly proposed system..

2 Preliminaries

In this section we recall certain notions relating to Pasting Systems [12, 14, 15]. A tile is a 2-dimensional topological region (disk) with a single simple closed bounding curve whose ends join up to form a loop without crossing or branching. A tiling in the Euclidean plan is a countable family of tiles that cover the plane without gaps or overlaps. Hence the intersection of any finite set of tiles in a plane tiling has zero area. Such an intersection will consist of a set of points (vertices) and lines (edges). Two adjacent tiles have an edge in common. Regular polygon tiles such as triangles, squares and hexagons tile the plane without crossing. A tile, its vertices and edges may be labeled distinctively. A square tile with label a for its region and edge labels x_1, x_2, x_3 and x_4 respectively for its right, top, left and bottom side edges can be expressed as a tuple $(a, x_1x_2x_3x_4)$. A decorated tile is a disk with any colour or pattern engraved in its region. A planar tiling is said to be non-periodic, if it admits no translations. A set of tiles is said to be aperiodic, if it admits only non-periodic tilings.

A pasting rule [12, 15] is a pair (x, y) of edges of two (not necessarily different) labeled tiles a and b. If x is the label of the right side edge of a square tile a and y is the label of the left side edge of another square tile (say, b), then an application of the rule (x, y) pastes side by side the two tiles i.e the tile b will be pasted to the right side of the tile a. Similarly pasting rules for top and bottom sides of square tiles are also defined. Pasting of tiles gluing or pasting edges result in formation of new tiling patterns.

2.1 Pasting System (PS)

Definition: A Pasting System (PS) [12, 15] is a 3-tuple, S = (Σ, P, t_0), where Σ is a finite non-empty set of (labeled regular polygons) tiles, P is a finite set of pasting rules and t_0 is the axiom tile or tiling.

A tiling pattern p_2 is obtained from a tiling pattern p_1 by applying the pasting rules in parallel to all the boundary edges of p_1 where pasting is possible. Note that the labels of pasted edges in a pattern are ignored once the tiles are pasted. Rotation of tiles is not allowed in the PS while pasting two tiles. The set of all patterns generated from the axiom t_0 using the pasting rules of P is called the language of tiling patterns of S, and is denoted by T(S).

Example: Consider the Pasting System $S_1 = (\Sigma, P, t_0)$ with a single tile (as well as axiom)

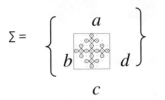

$$\Sigma = \left\{ \begin{matrix} a \\ b \quad \blacksquare \quad d \\ c \end{matrix} \right\}$$

A member generated by S_1 at second generation of pasting is shown in Figure 1.

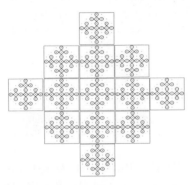

Fig. 1. A member of the family generated by S_1

2.2 Extended Pasting Scheme (EPS)

Definition: An Extended Pasting Scheme (EPS) is a 4-tuple (Σ, P, t_0, Δ), where Σ is a finite non-empty set of (labeled regular polygons) tiles, P is a finite set of pasting rules, t_0 is the axiom and Δ is a finite set of constraints on the edge labels of Σ.

Example: Consider the Extended Pasting Scheme $S_2 = (\Sigma, P, t_0, \Delta)$ where

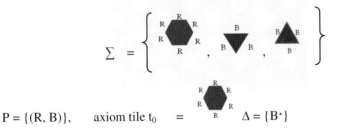

$P = \{(R, B)\}$,   axiom tile t_0 $=$ $\Delta = \{B^*\}$

The language of tiling patterns generated in parallel is given in figure 2.

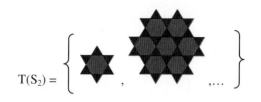

$$T(S_2) = \left\{ \quad , \quad , \dots \right\}$$

Fig. 2. The language of tiling patterns by S_2

2.3 Tabled Pasting System (TPS)

Definition: A Tabled Pasting System (TPS) [14] is a 4-tuple (Σ, P, t_0, C), where is a finite non empty set of (labeled regular polygons) tiles, P is a finite set of tables $\{T_1, T_2, \dots, T_k\}$ where each table T_i contains pasting rules, t_0 is the axiom pattern and C is a control language over P.

Example: Consider the Tabled Pasting System $S_3 = (\Sigma, P, t_0, C)$ where

$$\Sigma = \left\{ \quad , \quad , \quad \right\}$$

$P = \{T_1, T_2, T_3\}$, $T_1 = \{(R_2, G_5), (R_4, G_1), (R_6, G_3), (G_6, B_3), (G_4, B_1), (G_2, B_5)\}$,
$\qquad T_2 = \{(B_1, G_4), (B_2, R_5), (B_3, G_6), (B_4, R_1), (B_5, G_2,), (B_6, R_3)\}$,
$\qquad T_3 = \{(R_1, B_4), (R_2, G_5), (R_3, B_6), (R_4, G_1), (R_5, B_2), (R_6, G_3)\}$

$$t_0 = \left\{ \quad \right\} \quad \text{and } C = \{(T_1 T_2 T_3)^n / \ n \geq 1\}.$$

A tessellation obtained by S_3 is shown in the figure 3.

Fig. 3. A view of a tessellation generated by S_3

2.4 Tile Pasting P System (TPPS)

Definition: A Tile Pasting P System (TPPS) is a construct $\Pi = (\Sigma, \mu, F_1, \dots, F_m, R_1, \dots, R_m, i_0)$ where Σ is a finite set of tiles; μ is a membrane structure with m membranes labeled in a one-to-one way with 1,...,m; F_1,...,F_m are finite sets of picture patterns over tiles of Σ associated with the m regions of μ; R_1,...,R_m are finite sets of pasting rules (t; $(x_i; y_i)$; $1 \leq i \leq n$) associated with the m regions of μ and i_0 is the output membrane which is an elementary membrane.

Example: Consider the TPPS Π with $\Sigma = \{t_0 = (a, cbvh),\ t_1 = (a, chch),\ t_2 = (a, vbvb),\ t_3 = (a, vdvb)\}$; $\mu = [_1[_2[_3]_3]_2]_1$, indicating that the system has three regions one within another i.e. region 1, denoted by $[_1]_1$ is the 'skin' membrane which contains region 2, $[_2]_2$ which in turn contains region 3, $[_3]_3$; $i_0 = 3$ indicating that region 3 is the output region; $F_1 = \{t_0\}$, $F_2 = F_3 = \varphi$; R_1 contains a pasting rule $(t_1, (c, c), in)$ (with target indication in), R_2 contains two pasting rules $(t_2, (b, b), out)$, $(t_3, (b, b), in)$, $R_3 = \varphi$.

Starting with the initial object t_0 in region 1, the rule $(t_1, (c, c), in)$ allows tile t_1 to be pasted to t_0 growing one tile vertically but the target indication in sends the pattern formed to region 2. In this region 2 if the rule $(t_2, (b, b), out)$ is applied, then the pattern grows one tile horizontally giving rise to a shape of the letter L but the pattern is sent back to region 1 and the process repeats. On the other hand if rule $(t_3, (b, b), in)$ is applied in region 2, again the pattern grows horizontally one tile and the pattern formed in the shape of L is sent to the inner output region 3 wherein it is collected in the picture pattern language formed by the TPPS Π.

3 Tile Pasting P System with Multiple-Edge Pasting (TPPSMEP)

The Tile Pasting P Systems considered so far have generated tessellation or tiling patterns by pasting of tiles at the edges specified by the rules. At each stage individual tiles are pasted. We extend the pasting rules specified for edges to a sequence of edges. In this section we introduce a variant of the Parametric Tile Pasting P System (PMTPPS) introduced in [17] which allows non edge-edge tiling, multiple edge pasting and geometric transformations of tiles/tiling patterns such as translation, rotation, reflection, replication, scaling and edge rewriting and relabeling operations. These operations facilitate the rotation of tiles/tiling patterns and pasting of multiple tiles to tiling pattern unlike the earlier models.

The following geometrical transformation operators were included in [17]:
Translation: A tile labeled t subject to a translation of (h, k) in Cartesian plane and moved into a target region j is expressed as

$$[t] \rightarrow [\text{tra}(t, <h, k>)]_j$$

Rotation: A tile labeled t subject to a rotation of θ with respect to its centre and moved into a target region j is expressed as
$$[t] \rightarrow [\text{rot}(t, <\theta>)]_j.$$

Positive values for the rotation angle define counterclockwise rotations about the pivot point and negative values rotate tiles in the clockwise direction.

Reflection: A tile labeled t subject to a vertical reflection in Cartesian plane and moved into a target region j is expressed as

$$[t]_i \rightarrow [\mathrm{refv}(t)]_j .$$

Similarly horizontal reflection is expressed by

$$[t] \rightarrow [\mathrm{refh}(t)]_j .$$

Replication: A tile labeled t_p subject to replication where finite number (k) of copies are made and placed in a target region j is expressed as

$$[t_p] \rightarrow [t_{p1}, t_{p2,,...,} t_{pk}]_j .$$

Scaling: A tile labeled t subject to scaling by an integer k > 0 and placed in a target region j is expressed as

$$[t] \rightarrow [kt]_j.$$

For k < 1, the tile is reduced in its area.

Edge/path rewriting (linear scaling): An edge or path bounding a tile or tiling is rewritten when a sequence of vertices and edges $v_i e_i v_{i+1} e_{i+1} v_{i+2} ... v_{n-1} e_{i+k} v_n$ is replaced by a new sequence $v_i e_j v_{i+1} e_{j+1} v_{i+2} ... v_{n-1} e_{i+l} v_n$ for a finite $k, l \geq 1$. The length of the path is increased for k < l, decreased for k > l and maintained for k = l. The operation is expressed as

$$(v_i e_i v_{i+1} e_{i+1} v_{i+2} ... v_{n-1} e_{i+k} v_n) \rightarrow (v_i e_j v_{i+1} e_{j+1} v_{i+2} ... v_{n-1} e_{i+l} v_n).$$

3.1 Parametric Tile Pasting P System (PTPPS)

Definition: A Parametric Tile Pasting P system (PTPPS) Π is a construct with (Σ, μ, F_1, ..., F_m, R_1,..., R_m, i_0) where Σ is a finite set of tiles with vertices and edges labeled; μ is a membrane structure with m membranes labeled in a one-to-one way with 1,...,m; F_1,...,F_m are finite set of tiles associated with the m regions of μ; R_1,..., R_m are finite sets of evolution rules containing multiple edge pasting rules and geometric operators associated with the m regions of μ and i_0 is the output membrane which is an elementary membrane.

A computation in a PTPPS is defined in a way similar to TPPS with the successful computations being the halting ones. The computation starts with the membrane m_i, containing the axiom pattern t_0. The pasting (evolution) rules present in the set R_i are applied to the edges of the tiles in a maximally parallel manner. For instance during a computation if pasting is possible at all the edges of the given pattern p_i then the evolution rules present in the table R_i are applied to all possible edges simultaneously. If R_i contains two or more pasting rules/geometric operators, then any one of them, chosen non deterministically is applied in the membrane.

To each picture pattern, from each region of the system, for which geometric operations and pasting rules could be applied, should be applied; The picture pattern thus evolved is communicated to the region j indicated by the target associated with the rule used. $j \in \{here, out, in_m\}$. *here* refers to the same region, *out* means that the pattern exits the current membrane and *in* means that the pattern is immediately sent to one of the directly lower membranes (m), non deterministically chosen if several exist; if no internal membrane exists, then a rule with the target indication *in* cannot be used). A tiling pattern is evolved where no operator or pasting rule can be applied to the existing picture patterns. The result of a halting computation consists of the picture patterns placed in the membrane with label i_o in the output membrane. The set of all such picture patterns computed or generated by a PTPPS Π is denoted by PTPPL(Π). The family of all such languages PTPPL(Π) generated by systems Π as above, with at most m membranes, is denoted by PLm(PTPPS).

Example: The language of chair like non-periodic tiling patterns, whose first four members shown in figure 5 is generated by the following PTPPS.

$\Pi = (\Sigma, [_1[_2[_3[_4]_4]_2]_1, F_1, F_2, F_3, F_4, R_1, R_2, R_3, R_4, 4)$ where F_1 has the tile T_0 shown in figure 4. The vertices of T_0 are labelled 1.

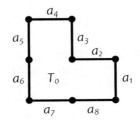

Fig. 4. Axiom of Π

$$F_2 = F_3 = F_4 = \phi,$$

$$R_1 = \{[T_0] \rightarrow [T_0, T_1, T_2, T_3]_2\} \, .$$

The edges of the tile T_1 are b_i where $b_i = \varphi_1(a_i)$, $a_i \in T_0$.
Similarly, the edges of the tile T_2 are c_i where $c_i = \varphi_2(a_i)$, $a_i \in T_0$ and the edges of the tile T_3 are d_i where $d_i = \varphi_3(a_i)$, $a_i \in T_0$.

$$R_2 = \{[T_0] \rightarrow [T_0]_3, [T_1] \rightarrow [T_1]_3, [T_2] \rightarrow [rot(T_2, <90^0>]_3,$$

$$[T_3] \rightarrow [rot(T_3, <270^0>]_3,$$

$$[(1a_7 1a_8 1 \rightarrow 1a_7 1, 1a_5 1a_6 1 \rightarrow 1a_6 1, 1b_2 1c_1 1 \rightarrow 1a_2 1, 1b_3 1d_4 1 \rightarrow 1a_3 1, 1c_5 1c_6 1 \rightarrow 1a_8 1, 1c_7 1c_8 1 \rightarrow 1a_1 1, 1d_5 1d_6 1 \rightarrow 1a_4 1, 1d_7 1d_8 1 \rightarrow 1a_5 1)]_{out},$$

$$[(1a_7 1a_8 1 \rightarrow 1a_7 1, 1a_5 1a_6 1 \rightarrow 1a_6 1, 1b_2 1c_1 1 \rightarrow 1a_2 1, 1b_3 1d_4 1 \rightarrow 1a_3 1, 1c_5 1c_6 1 \rightarrow$$
$$1a_8 1, 1c_7 1c_8 1 \rightarrow 1a_1 1, 1d_5 1d_6 1 \rightarrow 1a_4 1, 1d_7 1d_8 1 \rightarrow 1a_5 1)]_4 \}.$$

$$R_3 = [T_0, T_1, T_2, T_3, \{(d_3, d_2; b_4, b_5), (a_4, a_3, a_2; d_1, b_6, b_7), (c_4, c_3, c_2; a_1, b_8, b_1) \}]_{out}$$

$$R_4 = \phi$$

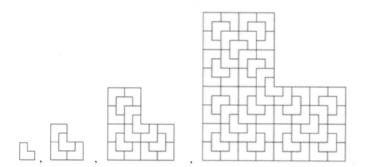

Fig. 5. First four members of the language of Π

The computation starts with the membrane m_1 having the tile T_0 and the evolution rules present in R_1 (associated with m_1) are applied to the tile T_0, thereby producing three new tiles with a different labelling as defined in R_1 and communicating all the tiles to m_2. In m_2 the geometric operator rules rotates the tiling patterns and communicates it to m_3, edge rewriting rules rewrite the edges of the tiling patterns and communicate it to m_1 and m_4. In m_3 the multiple edge pasting rules are applied to the four tiles T_0, T_1, T_2, T_3 thereby producing the chair like tiling patterns and communicated to m_2. In the successive pasting operation the primitive chair like tiling patterns are used for assembling the larger chair like tiling patterns. The tiling patterns reached in the output membrane m_4 are collected in set PMPPL(Π).

3.2   Tile Pasting P System with Multiple-Edge Pasting (TPPSMEP)

A Multiple-edge pasting rule is a pasting operation defined between two sequences of n-edges of a tiling pattern. A Multiple edge pasting rule $(a_1 a_2...a_n; b_1 b_2...b_n)$ is a pasting operation defined between two sequences of n-edges of a tiling pattern. If $a_1 a_2...a_n$ are the sequence of n-edges of a tiling pattern t_i and $b_1 b_2...b_n$ are the sequence of n-edges of a another tiling pattern t_{i+1}, then the application of the multiple edge pasting rule $(a_1 a_2...a_n; b_1 b_2...b_n)$ pastes the two tiling patterns t_i and t_{i+1} side by side. Multiple edges of tiling patterns are thus pasted with another tiling pattern having the same number of edges. Tile Pasting P System with Multiple-Edge Pasting (TPPSMEP) is the enhanced PTPPS given in 3.1. Tiles pasted to tiling patterns in a membrane are pushed to another membrane where rules for multiple edges assemble the tiling patterns to form bigger tiling patterns. For example, the pattern shown in figure 6 (b) is obtained

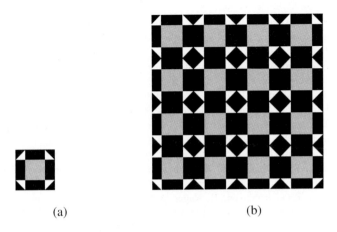

(a)   (b)

Fig. 6. A recurring pattern and a tessellation obtained

from two triangle tiles (black and white), a black and a grey square tiles. The recurring primitive pattern shown in figure 6 (a) is generated and pumped into another region where multiple copies of these primitives are pasted with edge sequence pasting rules.

4 Conclusion

The Tile Pasting P System for tiling and tessellation pattern generation has been extended with multiple-edge pasting operation and geometric transformation operations on the tiles and tiling patterns. This generalized form of PTPPS has the generative capacity of patterns where tiling patterns are pasted with other tiling patterns.

Acknowledgments. The first two authors would like to acknowledge the support rendered by the Leverhulme Trust, UK.

References

1. Ceterchi, R.,, Gramatovici, R., Jonaska, N., Subramanian, K.G.: Tissue-like P Systems with Active Membranes for Picture Generation. Fundamenta Informaticae, 311–328 (2003)
2. Hearn, D., Pauline Baker, M.: Computer Graphics. Prentice–Hall of India, New Delhi (1999)
3. Martin-Vide, C., Paun, G., Pazos, J., Rodriguez-Paton, A.: Tissue P Systems. Theoretical Computer Science 296(2), 295–326 (2003)
4. Fu, K.S.: Syntactic Pattern Recognition and Applications. Prentice-Hall Inc., Englewood Cliffs (1982)
5. Castellanos, J., Paun, G., Iguezpaton, A.R.: Computing with Membranes: P Systems with Worm-Objects, Centre for Discrete Mathematics and Theoretical Computer Science (CDMTCS)-123 (2000)
6. Krithivasan, K., Rama, R.: Introduction to Formal Languages, Automata and Computation. Pearson Education, New Delhi (2009)

7. Kalyani, T., Sasikala, K., Dare, V.R., Abisha, P.J., Robinson, T.: Triangular Pasting System. Formal Models, Languages and Application. Series in Machine Perception and Artificial Intelligence 66, 195–211 (2006)
8. Paun, G.: Computing with Membranes: An Introduction. Springer, Berlin (2002)
9. Paun, G., Perez-Jimenez, M.J., Riscos-Nunez, A.: P Systems with Tables of Rules. In: Karhumäki, J., Maurer, H., Păun, G., Rozenberg, G. (eds.) Theory Is Forever. LNCS, vol. 3113, pp. 235–249. Springer, Heidelberg (2004)
10. Bottoni, P., Labella, A., Martin-Vide, C., Paun, G.: Rewriting P Systems with Conditional Communication. In: Brauer, W., Ehrig, H., Karhumäki, J., Salomaa, A. (eds.) Formal and Natural Computing. LNCS, vol. 2300, pp. 325–353. Springer, Heidelberg (2002)
11. Subramanian, K.G., Saravanan, R., Robinson, T.: P System for Array Generation and Application to Kolam Patterns. Forma 22, 47–54 (2007)
12. Subramanian, K.G., Robinson, T., Nagar, A.K.: Tile Pasting P System model for Pattern Generation. In: Proc. Third Asia International Conference on Modelling and Simulation, Indonesia (2009)
13. Subramanian, K.G., Saravanan, R., Geethalakshmi, M., Chandra, P., Margenstern, M.: P System with array objects and array rewriting rules. Progress in Natural Science 17, 479–485 (2007)
14. Robinson, T., Jebasingh, S., Nagar, A.K., Subramanian, K.G.: Tile Pasting Systems for Tessellation and Tiling Patterns. In: Barneva, R.P., Brimkov, V.E., Hauptman, H.A., Natal Jorge, R.M., Tavares, J.M.R.S. (eds.) CompIMAGE 2010. LNCS, vol. 6026, pp. 72–84. Springer, Heidelberg (2010)
15. Robinson, T.: Extended Pasting Scheme for Kolam Pattern Generation. Forma 22, 55–64 (2007)
16. Robinson, T., Dare, V.R., Subramanian, K.G.: A Parallel Pattern Generating System. In: Proc. of the 6th International Workshop on Parallel Image Processing and Analysis (IWPIA-1999), Madras, pp. 23–32 (1999)
17. Jebasingh, S., Robinson, T., Nagar, A.K.: A variant of Tile Pasting P System for Tiling Patterns. In: Proc. of 2010 IEEE Fifth International Conference on Bio-Inspired Computing: Theories and Applications, BIC-TA 2010(UK), vol. 2, pp. 1568–1576 (2010)

Modeling and Automation of Diagnosis and Treatment of Diabetes

Abhirami Baskaran*, Dhivya Karthikeyan, and Anusha T. Swamy

PES Institute of Technology, Bangalore, India

Abstract. The present work aims at designing and implementing an automated decision making system for the treatment of diabetes. The automated medical tool has been equipped to handle the decisions regarding the care plan of the patient and also helps in diagnosis. It takes in essential parameters like glucose, cholesterol, blood pressure and devises a care plan for the patient. Fuzzy logic was used to implement the medical decision support system. A knowledge base for diabetes containing the essential concepts, treatment algorithms was created. The fuzzy logic based system used the knowledge base for constructing the collection of rules. The essential parameters from the patient database were provided as input and the decisions like the type of diabetes, diet plans, medication etc were recorded. The tool also takes the decisions and the parameters that led to the decisions to build an optimal care pathway for the patient.

Keywords: Clinical Pathways, Diabetes mellitus, Fuzzy logic.

1 Introduction

Diabetes mellitus ranks high in the list of wide spread diseases in today's society. The International Diabetes Federation has indicated that India will become the diabetes capital of the world, signalling the need for a uniform care plan in the treatment and diagnostic process throughout the world. Currently, there exists variance in treatment for patients with the same condition. Clinical pathways [1] are useful in standardizing the treatment processes. Also in rural areas and primary care centres, they act as guiding tools to nurses and care workers.

The current research focuses on an expert system developed to aid the specialist. An expert system is a computer program that simulates the judgement and behaviour of a human or an organization that has expert knowledge and experience in a specific field. For example, Chang et al. [2] have employed artificial neural network [ANN] based expert system for the diagnosis of breast cancer. Dey and Bajpai [3] have devised a diagnostic tool based on ANN for diabetes. Similar efforts have been reported by Sivakumar for diabetic retinopathy [4] and Phuong [5]. While these researchers have only focussed on the diagnostic aspects of the treatment, the current paper covers the treatment as well. The usage of

---

* Scholars, Computer Science Department.

K. Deb et al. (Eds.): SEAL 2010, LNCS 6457, pp. 339–348, 2010.

ANN does not lay emphasis on the intermediate decisions taken to arrive at the final outcome. The intermediate decisions taken are crucial for building clinical pathways. Hence a fuzzy control system has been designed to take medical decisions related to diagnosis and treatment. It works on the basis of standard treatment algorithms for diabetes.

Fuzzy logic [6] is an ideal choice in decision making as it is similar to human thinking and decision making in spirit. Since medical diagnosis and treatment of diabetes involves a certain degree of uncertainty and imprecision, fuzzy logic is an ideal choice to implement a smart system for diabetes treatment. The rules have linguistic variables replacing quantitative values. Fuzzy logic enables us to produce definitive outputs or decisions for fuzzy input sets.

This paper talks about how the action plans for the treatment are arrived at based on fuzzy logic. Firstly it gives an insight into the diabetes domain and the fuzzy logic framework. Consequently the system architecture and the patient database are discussed. This is followed by the implementation of the expert system. The expert system employs the concepts in diabetes domain to bring about intelligent decision making. It constructs a pathway for each patient based on various input parameters like laboratory tests, previous diagnosis etc. It includes the response parameters that decide the treatment course and traces the exercise, diet plans etc. of the patient.

2   Overview of Diabetes and Fuzzy Logic

2.1   Diagnosis and Treatment of Diabetes

The blood glucose level is regulated with the help of insulin, a hormone made in the pancreas. Diabetes mellitus develops when the pancreas stops producing insulin (Type 1 diabetes) or when the body does not respond properly to insulin (Type 2 diabetes). Insulin injections are necessary to treat Type 1 diabetes. Type 2 diabetes can usually be controlled in the first instance by regular exercise and diet. Tablets and eventually insulin injections may be needed as the disease progresses [7].

Diagnosis of the Diabetes type and pre-diabetes is done by fasting plasma glucose(FPG) test. An oral glucose tolerance test (OGTT) measures blood glucose after a person fasts at least 8 hours and 2 hours after the person drinks a glucose-containing beverage. HbA1c, present in our blood, is a minor component of hemoglobin to which glucose is bound. Therefore, HbA1c is a useful indicator of how well the blood glucose level has been controlled in the recent past and may be used to monitor the effects of diet, exercise, and drug therapy on blood glucose in diabetic patients, as per [8] and [9].

The less known Type 1 diabetes usually occurs in people under the age of 30.It can be managed only by replacement of insulin via injections. The premeal dose of insulin can be adjusted based on the content of the meal and the patient's blood glucose level. In our work we have taken into account only Rapid-Acting insulin, as described in [11] and [12]. The more prevalent Type 2 diabetes usually occurs in people over 30 years of age but it may occur in overweight and inactive teenagers, and children with a family history of diabetes. Most people who are newly diagnosed with type 2 diabetes will immediately begin a drug monotherapy.In case HbA1c goals are not met, a combination of drugs are given. Insulin shots might be recommended if the A1C is higher than 8.5 percent. Low-dose aspirin therapy,Angiotensin-Converting Enzyme (ACE) inhibitors and statins are prescribed for heart disease, high blood pressure ad cholesterol respectively.

2.2 Working with Fuzzy Logic

Fuzzy logic [13] is a system that is multivalued with its reasoning being approximate rather than concrete. In other words fuzzy logic is based on fuzzy sets. A fuzzy set,F in a universe of discourse can be characterized by a membership function(μ) which take values between 0 and 1 namely, $\mu : U \rightarrow [0,1]$. This value of a variable is referred to as the degree of membership. The key components are:

- Knowledge base: It is a repository of information consisting of all the data, concepts and policies required to define the control rules and build a control system.
- Fuzzification scheme: Its main role is to map the range of input variables to their appropriate universes of discourse which are represented by linguistic labels. Crisp values are converted into fuzzy sets using a fuzzifying operator. Suppose Cr is the crisp value and F is the fuzzy set then,

$$F \leftarrow fuzzifier(Cr) \qquad (1)$$

- The decision making system: It makes inferences based on the rules formulated based on the knowledge base. The rule base is a set of linguistic statements structured in IF X AND Y THEN Z form. The output given out by the decision making system is a particular range in a fuzzy set.
- Defuzzification scheme: This scheme performs the reverse of fuzzification. It converts the fuzzy output to a non fuzzy or crisp value. Many methods are used in this procedure like max-min, averaging,root mean square method etc.

$$z_0 = defuzzifier(z) \qquad (2)$$

where z_0 is the non-fuzzy control output and *defuzzifier* is the defuzzification operator.

2.3 System Architecture

The overall systems architecture is shown in figure 1. An Oracle database functions as the storage and retrieval base for the patient records. MATLAB fuzzy logic toolbox has been used to model the decision making system for diabetes. The required patient data is imported into a CSV (comma separated values) file. These records serve as inputs to the fuzzy logic toolbox. The principal steps in

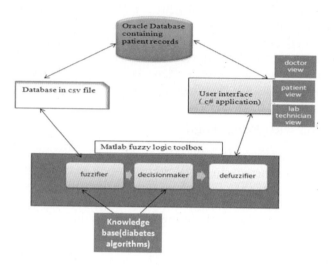

Fig. 1. System architecture

fuzzy logic like fuzzification, decision making and defuzzification are performed in MATLAB using the rule set available in knowledge base. The front end has been designed as a Visual Studio C# application.

3 Implementation

3.1 Expert System Implementation

Diagnosis of Diabetes. Diagnosis of the diabetes has the input parameters FPG (Fasting Plasma Glucose), OGTT(Oral Glucose Tolerance test) and family history. Based on the ranges of FPG and OGTT, diagnosis is made.Hence the output is a membership function value of Type1, Type2 or PreDiabetes.

```
Input Parameters fuzzy sets = {LowFPG, HighFPG, NormalFPG, LowOGTT,
                              normalOGTT, highOGTT, familyHistory }
Output Parameters fuzzy sets = {Type1, PreDiabetes, Type2, normal }
```

Figure 2 shows the method of diagnosing the disease while figure 3 depicts the membership functions and fuzzy sets for FPG. The rule base is determined on the basis of diagnostic criteria by WHO. For example,

```
If FPG is HighFPG and OGTT is HighOGTT  and  FamilyHistory  is type1
               then   diagnosis is Type1.
```

There are similar rules for type-2 and Prediabetes. Here the input values are fuzzified into the above mentioned sets. The membership functions go on to produce values for each parameter. Let the membership functions of FPG, OGTT, Familyhistory and the result be represented by F,O,FH and TD respectively. The

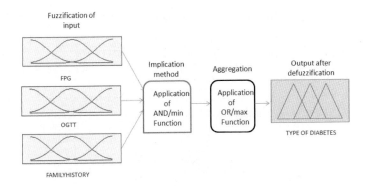

Fig. 2. Diabetes Diagnosis

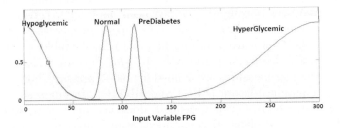

Fig. 3. Membership functions for FPG

input parameters are I_1, I_2, I_3 and output parameter can be O_1. The inference from the rule is made as follows:

$$TD(O_1) \leftarrow min\{F(I_1), O(I_2), FH(I_3)\} \tag{3}$$

$TD(O_1)$ constitutes the matching value of the type-1 rule. Similarly there exist $TD(O_2)$ for type-2 rule and $TD(O_3)$ for PreDiabetes. Aggregation is performed on these variables by the OR/Max function and output fuzzy set is obtained.

$$fuzzyoutput \leftarrow max\{TD(O_1), TD(O_2), TD(O_3)\} \tag{4}$$

Let

$$Y_1 \leftarrow TD(O_1), Y_2 \leftarrow TD(O_2), Y_3 \leftarrow TD(O_3) \tag{5}$$

Defuzzification is performed using centroid formula as follows:

$$DiagnosisType = \sum_{i=1}^{3} w(i) * Y_i / \sum_{i=1}^{3} Y_i \tag{6}$$

where Y_i refers to the output vector, w(i) is the membership co-efficient vector. The defuzzification procedure returns the probability of the type of diabetes diagnosed. The crisp output value can be again evaluated and conclusions can be made about the diagnosis.

Table 1. Grouping based on values

Input parameters	Ranges	Membership function	Label
LDL	0-100	trapmf	Normal
(Low Density Lipoprotein)	100-160	trapmf	Borderline
	160-190	trapmf	High
	190-320	smf	very high
HDL	0-40	trimf	Low
(High Density Lipoprotein)	40-130	trimf	Normal
Triglycerides	0-150	zmf	Normal
	150-200	trapmf	Borderline
	200-500	trapmf	High
	500-625	smf	very high

Type 1 and Type 2 Treatment Review. A fuzzy control system has been designed for monitoring clinical parameters relevant to diabetes. This fuzzy control system performs decisions based on the Texas diabetes algorithms [10]. Here the Texas Diabetes Association algorithms have been employed to construct the rules relevant to the system. The input parameters, ranges, membership functions and their labels for cholesterol are shown in the table 1. In the table, *trapmf* denotes trapezoidal shaped membership function and *smf* denotes S shaped membership function. Similar measures are defined for blood pressure, body mass index (BMI), fasting plasma glucose (FPG) and HbA1C. In our system, five output variables have been similarly defined along the lines of treatment. They are diet plans, exercise plans, glycemic medication, cholesterol medication and blood pressure medication. Consider for example, a fuzzy system for glycemic management. The management of glucose is dependent on three main factors namely fasting blood glucose levels, HbA1c level and the progress of the patient in the careplan. These values from the patient record fires the rules related to the parameter. The steps involved are similar to the diagnosis implementation. First the FPG and HbA1c values are mapped into a membership value by a membership function. 'AND' operation is performed to give a membership value to the output vector. The combined ouput from the fuzzy inference system give the drug therapy as the output.

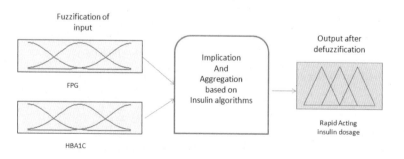

Fig. 4. Insulin dosage determination for type 1 diabetes

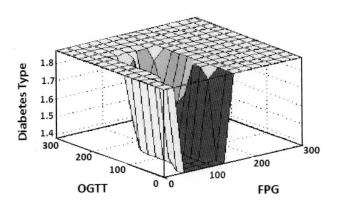

Fig. 5. Surface view for diagnosis

Drug therapy value ← (FBGvalue) AND(HBa1c)

Finally defuzzification is performed by the centroid method to get a probabilistic value for the type of drug treatment. The output indicates how strongly recommended the drug is. Based on the output a decision regarding the medication can be made.

Insulin Dose for Type 1 and Type 2. A fuzzy system was developed for calculating the increase or decrease in insulin dose for type 1 patients during pre-meals according to the blood glucose level observed at that time. The inputs to the insulin dosage calculator are FPG and HbA1c level and the output is unit of increase or decrease in the insulin dose as displayed in the figure 4. The membership function used for the input FPG is trapmf, HbA1c is trimf and the output insulin dose is trapmf. The ranges for FPG to calculate amount of insulin were used from the tables 2 and 3. Similarly another fuzzy system was developed to calculate insulin dosage for type 2 patients with the same input and output variables of same membership functions.

Table 2. Action Plan for Insulin Lispro (Rapid-Acting) Supplements

Premeal blood glucose level, mg/dL (mmol/L)	Insulin dose (when HbA1C level ≥ 7.%)
≤ 50 (2.8)	Decrease by 2 units
50 to 80 (2.8 to 4.4)	Decrease by 1 unit
80 to 130 (4.4 to 7.2)	No change
130 to 150 (7.2 to 8.3)	Increase by 1 unit
150 to 200 (8.3 to 11.1)	Increase by 2 units
200 to 250 (11.1 to 13.9)	Increase by 3 units
250 to 300 (13.9 to 16.7)	Increase by 4 units
300 to 350 (16.7 to 19.4)	Increase by 5 units

Table 3. Action Plan for Insulin Therapy with Glargine or basal insulin daily

Premeal blood glucose level, mg/dL (mmol/L)	Insulin dose (when HbA1C level ≥ 7.%)
≤80 (4.4)	Decrease by 2 units
80 to 100 (4.4 to 5.5)	No change
100 to 120 (5.6 to 6.6)	Increase by 1 unit
121 to 140 (6.7 to 7.7)	Increase by 2 units
141 to 180 (7.8 to 10)	Increase by 4 units
≥180 (10)	Increase by 6 units

4   Results

4.1   Surface Views

Surface view is a 3-D representation of the fuzzy inference system with two input parameters and 1 output parameter. The surface view of the fuzzy system for the diagnosis of the diabetes disease is shown in Figure 5. The x-axis shows the FPG, y-axis the OGTT. The output is shown along the z-axis, i.e., Diagnosis Type. The surface view of the fuzzy System for the Insulin Therapy for type 1 diabetes was obtained as shown in Figure 6. The x-axis shows the FPG, y-axis the HbA1c level which are the inputs. The output is shown along the z-axis, i.e., the Insulin Dosage. As seen, the mapping relationships conform to intuition as well as the rules.

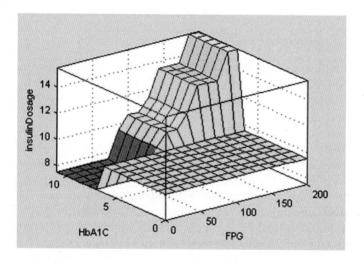

Fig. 6. Surface viewer of insulin dosage for type 1

4.2   Pathways

The clinical pathways are built in the front end based on the decisions undertaken by the fuzzy system. The pathway is a flow based diagram which shows the parameters and their values at each decision stage. A pathway generated for a type-1 diabetes patient is shown in Figure 7.

4.3   Accuracy of Prediction

The patient datasets of 100 records from a private hospital, of which 50 were diabetic and 50 non-diabetic, were used to test the Fuzzy Systems. The datasets were fed from a csv file to a Matlab file which used the fuzzy systems to get

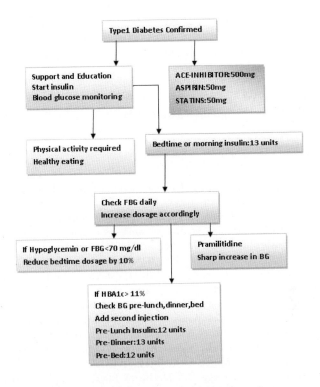

Fig. 7. Pathway for type 1 diabetic patient

the values for various output parameters. The results obtained predicted the diabetes cases with accuracy rate of 97% of which 5 were diagnosed as type-1 diabetic and the rest as type-2. Also, the type 1 and 2 treatment review and their insulin dosage fuzzy systems have produced 98% accurate results as per the physician's prescription.

5   Conclusions and Future Work

In this paper a clinical decision making system for the diagnosis and treatment of diabetes is modeled and implemented. The fuzzy logic based expert system is devised for diagnosis, patient review and insulin administration. It takes in patient records and gives an integrated care plan which includes diets, exercise regime, medication, insulin dosage etc. An optimal clinical pathway is traced for the patient. Such automated systems provide standard and quality in treatment of diabetes and can help reduce the diabetic population. They can act as support systems or guides for the specialists in health care centres, military bases etc. It provides the patient a ready system that calculates the insulin level and tracks the medication history and the progress levels. Hence an automated clinical system is the way to go to ensure a risk free and standard medical treatment domain.

The system can be enhanced further to include risk factors like heart disease etc as output. Adaptability in case of unexpected changes in a patient's condition such as sudden liver failure and emergence of new technology like improved drugs, medical procedures need to be addressed.

Acknowledgments. We like to acknowldege Prof. Srinivas, Computer Science Department, PESIT for his guidance and help.

References

1. Kitchiner, D.J., Bundred, P.E.: Clinical Pathways. Medical Journal of Australia 170, 54–55 (1999)
2. Chang, Y.H., Zheng, B., Wang, X.H., Good, W.F.: Computer Aided Diagnosis of Breast Cancer using Artificial Neural Networks: Comparison of Backpropagation and Genetic Algorothms. IEEE Xplore, 3674–3679 (1999)
3. Dey, R., Bajpai, V., Gandhi, G., Dey, B.: Application of Artificial Neural Network(ANN) technique for Diagnosing Diabetes Mellitus. In: IEEE Region 10 Colloquium and the Third ICIIS, pp. 1–4 (2008)
4. Sivakumar, R.: Neural Network Based Diabetic Retinopathy Classification using Phase Sprectal Periodicity Components. ICGST-BIME Journal 7, 23–28 (2007)
5. Phuong, N.H.: Fuzzy Set Theory and Medical Expert Systems: Survey and model. In: Bartosek, M., Staudek, J., Wiedermann, J. (eds.) SOFSEM 1995. LNCS, vol. 1012, pp. 431–436. Springer, Heidelberg (1995)
6. Warren, J., Beliakov, G., Zwaag, B.V.D.: Fuzzy Logic in Clinical Practice Decision Suppot Systems. In: 33rd Hawaii International Conference on System Sciences, pp. 1–10 (2000)
7. International Diabetes Institute, http://www.idf.org/node/1203
8. National Institute of Clinical Excellence: Diagnosis and Management of Type 1 Diabetes in Adults. NHS, pp. 7–14 (2004)
9. National Diabetes Information Clearing House, http://diabetes.niddk.nih.gov
10. Texas Diabetes Council, http://www.dshs.state.tx.us/diabetes/PDF/09toolkit/Appendix.pdf
11. Hirsch, I.B., Bergenstal, R.M., Parkin, C.G., Wright, J.E., Buse, J.B.: A Real World Approach to Insulin Therapy in Primary Care Practice. Clinical Diabetes 23, 78–86 (2005)
12. Burge, M.R., Castillo, K.R., Schade, D.S.: Meal Composition is a determinant of Insulin Lispro-induced Hypoglycemia in IDDM. C Diabetes Care 20, 152–155 (1997)
13. Lee, C.C.: Fuzzy Logic In Control Systems: Fuzzy Logic Controller -Part 1. IEEE Transactions on Systems, Man and Cybernetics 20, 404–418 (1990)

The Evolution of Fuzzy Classifier for Data Mining with Applications

Václav Snášel[1,2], Pavel Krömer[1], Jan Platoš[1], and Ajith Abraham[2]

[1] Department of Computer Science
Faculty of Electrical Engineering and Computer Science
VŠB – Technical University of Ostrava
17. listopadu 15, 708 33 Ostrava – Poruba, Czech Republic
{vaclav.snasel,pavel.kromer,jan.platos}@vsb.cz
[2] Machine Intelligence Research Labs (MIR Labs),
Washington 98071, USA
ajith.abraham@ieee.org

Abstract. Fuzzy classifiers and fuzzy rules can be informally defined as tools that use fuzzy sets or fuzzy logic for their operations. In this paper, we use genetic programming to evolve a fuzzy classifier in the form of a fuzzy search expression to predict product quality. We interpret the data mining task as a fuzzy information retrieval problem and we apply a successful information retrieval method for search query optimization to the fuzzy classifier evolution. We demonstrate the ability of the genetic programming to evolve useful fuzzy classifiers on two use cases in which we detect faulty products of a product processing plant and discover intrusions in a computer network.

Keywords: genetic programming, information retrieval, classifier evolution, fuzzy systems.

1 Introduction

Genetic programming is a powerful machine learning technique from the wide family of evolutionary algorithms. In contrast to the traditional evolutionary algorithms, it can be used to evolve complex hierarchical tree structures and symbolic expressions. It has been used to evolve Lisp S-expressions, mathematical functions, symbolic expressions, decision trees, and recently to infer search queries from relevance ranked documents in a fuzzy information retrieval system.

The last application is interesting for the general data mining as well. It can be directly applied in the data mining domain. Extended Boolean queries (i.e. fuzzy queries) can be interpreted as symbolic fuzzy classifiers that describe a fuzzy subset of some data set by means of its features. Moreover, a fuzzy classifier evolved over a training data set can subsequently be used for efficient classification of new data samples and e.g. predict quality of products or detect harmfull actions in computer network. Artificial evolution of search expressions is a promising approach to data mining because genetic programming yields very

K. Deb et al. (Eds.): SEAL 2010, LNCS 6457, pp. 349–358, 2010.
© Springer-Verlag Berlin Heidelberg 2010

good ability to find symbolic expressions in other problem domains. The general process of classifier evolution can be used to evolve custom classifiers for many data sets with different properties.

2 Fuzzy Information Retrieval

The evolution of fuzzy classifiers for data mining is implemented within the framework for search query optimization designed for efficient information retrieval. Data records are interpreted as documents and data features are mapped to index terms.

The area of information retrieval (IR) is a branch of computer science dealing with storage, maintenance, and searching in large amounts of data [1]. It defines and studies IR systems and models.

An IR model is a formal background defining the document representation, query language, and document-query matching mechanism of an IR system. The proposed classification algorithm builds on the extended Boolean IR model, which is based on the fuzzy set theory and fuzzy logic [1,2]. Documents are interpreted as fuzzy sets of indexed terms, assigning to every term contained in the document a particular weight from the range $[0, 1]$ expressing the degree of significance of the term for document representation. A formal description of a document collection in the extended Boolean IR model is shown in Eqn. (1) and Eqn. (2), where d_i represents $i-$th document and t_{ij} $j-$th term in $i-$th document. An index matrix of the entire document collection is denoted $\boldsymbol{D}$.

$$\boldsymbol{d}_i = (t_{i1}, t_{i2}, \ldots, t_{im}), \ \forall \ t_{ij} \in [0, 1] \tag{1}$$

$$\boldsymbol{D} = \begin{pmatrix} t_{11} & t_{12} & \cdots & t_{1m} \\ t_{21} & t_{22} & \cdots & t_{2m} \\ \vdots & \vdots & \ddots & \vdots \\ t_{n1} & t_{n2} & \cdots & t_{nm} \end{pmatrix} \tag{2}$$

Each IR model defines a query language to specify search requests. The query language in the extended Boolean model allows weighting search terms in order to attribute them different levels of importance. Moreover, the aggregation operators (most often AND, OR and NOT) can be weighted to parameterize their impact on query evaluation [1,2].

In this study, we adopt the threshold interpretation of the query term weights. In the threshold interpretation, an atomic query (i.e. a query with one weighted term representing single search criterion) containing term t_i with the weight a is a request to retrieve documents having index term weight t_i equal or greater than a [1].

The effectiveness of an information retrieval system can be evaluated using the measures precision P and recall R. Precision corresponds to the probability of retrieved document to be relevant and recall can be seen as the probability of retrieving relevant document.

Precision and recall in the extended Boolean IR model can be defined using the Σ−count $\|A\|$ [3]:

$$\rho(X|Y) = \begin{cases} \frac{\|X\cap Y\|}{\|Y\|} & \|Y\| \neq 0 \\ 1 & \|Y\| = 0 \end{cases} \quad (3)$$

$$P = \rho(REL|RET) \qquad R = \rho(RET|REL) \qquad (4)$$

where REL stands for the fuzzy set of all relevant documents and RET for the fuzzy set of all retrieved documents.

For an easier IR effectiveness evaluation, measures combining precision and recall into one scalar value were developed. The F-score $F = \frac{2PR}{(P+R)}$ [4] is among the most used scalar combinations of P and R.

3   Genetic Programming for the Evolutionary Query Optimization

Genetic programming (GP) [5,6], an offshoot of genetic algorithms [7], facilitates the efficient artificial evolution of symbolic expressions. In this work, we use the genetic programming originally developed for search query optimization (see e.g. [8,9]) to evolve general fuzzy classifiers for data mining. Genetic programming was chosen due its ability to evolve symbolic tree-like expressions that correspond to search queries. The symbolic nature of the algorithm output also allows good possibility of verification and feedback from the side of system users.

It was shown that the GP was able to optimize search queries so that they described a set of relevant documents. In the fuzzy information retrieval model, the relevant documents formed a fuzzy subset of the collection of all documents and the extended Boolean queries that would describe them were evolved.

An information retrieval system based on the extended Boolean IR model was implemented to validate evolutionary query optimization. The $tf \cdot idf_t$ term statistics [10] was used for document indexing and the threshold interpretation of query term weights was implemented. The query language in the IRS supported the standard Boolean operators AND, OR, and NOT.

The information retrieval system served as a test bed for the evolutionary query optimization. The GP evolved tree representations of the search queries with the Boolean operators as function nodes and terms as leaves. Both, operator nodes and term nodes, were weighted. In order to generate a random initial population for the GP, the system was able to generate random queries. The parameters of the random query generator showing the probabilities of generating a particular query node are summarized in Table 1a.

The crossover operator was implemented simply as a mutual exchange of two randomly selected branches of parent tree chromosomes. The mutation operator

Table 1. Random query generation an mutation probabilities

(a) Probabilities of generating random
query nodes

Event	Probability
Generate term	0.5
Generate operator AND	0.24
Generate operator OR	0.24
Generate operator NOT	0.02

(b) Probabilities of mutation operations

Event	Probability
Mutate node weight	0.5
Insert or delete NOT node	0.1
Replace with another node or delete NOT node	0.32
Replace with random branch	0.08

selected a node from the processed chromosome at random and performed one
of the mutation operations summarized in Table 1b.

The query mutation types that were implemented included:

- change of selected node weight
- replacement of selected node type by a compatible node type (e.g. operator
  OR replaced by an AND, term replaced by another term)
- insertion of the NOT operator before selected node
- removal of the NOT operator if selected
- replacement of selected node by a randomly generated branch

The F-Score was used as a fitness function. An experimental evaluation of such
an information retrieval system showed that the GP can find search expression
describing fuzzy sets of relevant documents [8,9].

The extended queries evolved by the algorithm can be seen as fuzzy classifiers
describing the fuzzy set of relevant documents, or, more generally, a fuzzy set of
data records. The fuzzy classifier evolved over a training data set can be easily
used to classify new data samples.

4   Applications

The algorithm for evolutionary query optimization was applied to the evolution
of a general symbolic fuzzy classifier. In this work, we have evolved a fuzzy
classifier for quality prediction in an industrial manufacturing process and for
intrusion detection in an intrusion detection system.

Table 2. Product features data set

(a) Normalized product features

Id	Feat. 1	Feat. 2	...	Feat. 839	Prod. class
1	0.846	0.951	...	0.148	1
2	0.856	0.9452	...	0.160	1
3	0.882	0.968	...	0.160	0
⋮	⋮	⋮	⋮	⋮	⋮
204	0.618	0.861	...	0.025	0

(b) Product features data set as an IRS index matrix D.

$$D = \begin{pmatrix} 0.846 & 0.951 & \cdots & 0.148 \\ 0.856 & 0.9452 & \cdots & 0.160 \\ \vdots & \vdots & \ddots & \vdots \\ 0.618 & 0.861 & \cdots & 0.025 \end{pmatrix}$$

4.1 Genetic Evolution of Fuzzy Classifier

In heavy industry, a product is created. During its processing, a number of product features are measured and recorded. The features include the chemical properties of the raw material, density, temperature at several processing stages, and many other indicators that are recorded several times during the production. At the end, the product is classified as either flawless or defective. The data and classification for a number of product samples are known and the goal of the algorithm is to find a fuzzy classifier that could be used for product quality prediction during product processing.

The problem differs from the query optimization task only semantically. We interpret products as documents and product features as terms. The product feature value then corresponds to the index weight of a term in a document (feature weight in a product). The product class corresponds to document relevance.

We have obtained a test dataset from a product processing plant. The dataset contained 204 samples with 839 features each. 200 samples described flawless products (class 0) and 4 samples described deffective products (class 1). The raw product features values were normalized to the interval $[0, 1]$. A sample of product features data after normalization is shown in Table 2a. The mapping of normalized data onto an IRS index matrix is demonstrated in Table 2b. The goal of the optimization algorithm was to find a search expression (fuzzy classifier) that would describe the set of defective products as good as possible. As the data set is very small and contains only 4 samples of deffective products, the results of presented experiment should be seen as a proof of concept rather than a rigorous evaluation of the algorithm.

We have implemented the GP for the evolution of fuzzy classifiers. The fuzzy classifier that was evolved by the algorithm corresponds to a search expression that describes the class of defective products in terms of product features. The parameters of the executed GP (found after initial tuning of the algorithm) are shown in Table 3.

During 12 independent optimization runs, the GP delivered a best classifier with a fitness of 0.9996 and a worst classifier a with fitness of 0.399872. Every fuzzy classifier reaching a fitness of 0.5 and higher was able to identify all defective products

Table 3. GP parameters used to evolve fuzzy classifier for quality prediction

Parameter	Value
Population size	100
Generations limit	1000
Fitness	F-Score
Mutation probability	0.8
Crossover probability	0.8
Independent runs	12

Table 4. Example of evolved fuzzy classifiers for quality prediction

Label	Query		Fitness
Q1 (Best)	(Feat308:0.79	and:0.95	0.9996
	(Feat295:0.36	or:0.34	
	Feat413:0.99))		
Q2	Feat641:0.998113		0.5759
Q3	(Feat641:0.97	and:0.06	0.6066
	(Feat593:0.76	and:0.81	
	Feat421:0.80))		
Q4 (Worst)	Feat426:0.999203		0.3999

without an error or without false positives (i.e. without flawless products being marked as defective). A fuzzy classifier with a fitness higher than 0.5 was evolved in 10 cases out of 12 independent runs. An example of several evolved fuzzy classifiers is shown in Table 4.

The best classifier found by the algorithm was Q1. It is indeed a perfect expression describing defective products in the available data set. It is superior in terms of its F-Score, but also in terms of precision and recall because it describes defective products only.

The symbolic nature of the GP output gives us valuable information about the features that indicate product defectiveness. From Q1, we can see that the product can already be classified as faulty or flawless after the value of feature 413 (out of 839 measured product features) was read. Therefore, a defective product can be removed from production at an earlier stage and costs can be saved. Moreover, it is also a good clue telling us what features are really worth measuring. The other sensors can be suspended and savings can be made. Last but not least, the classifier provides also an important feedback on the production process. Production specialists can focus on adjusting the technology so that the reason for the problematic values of identified key features are eliminated in the future.

4.2   Genetic Evolution of Fuzzy Classifier for Intrusion Detection

Next, we have implemented an evolution of fuzzy classifier for intrusion detection. In this experiment, larger-scale data were processed by the algorithm.

To investigate the ability of discussed algorithm to find useful classifiers, a test sytsem implementing evolution of fuzzy expressions was implemented. The 10% sample of the KDD Cup 1999 intrusion detection dataset[1] was used to evolve classifiers and test their ability to detect illegal actions. It contains 10% of the large intrusion detection data set created in 1998 by the DARPA intrusion detection evaluation program at MIT. The full data set contains 744 MB data with 4,940,000 records with 41 nominal and numerical features. For our experiments, all features were converted to numeric and normalized.

The data describes normal traffic and 4 attack classes called DoS (Denial of Services), U2R (User to Root), R2L (Remote to User), and Probe (Probing). The records for each class are divided into training (40%) and testing (60%) data set. For each class, the training data set was used to evolve the fuzzy classifier and testing data set was used to evaluate the detection capabilities of the classifier. The attack classes contained following number of records: DoS contained 195,494 training and 293,242 testing records, U2R consisted of 38,931 training and 58,399 testing records, R2L included 39,361 training and 59,043 testing records, and finally the Probe class consisted of 40,553 training and 60,832 testing records.

Intrusion detection classifier evolution. We interpret data samples in intrusion detection data set as documents and its features as terms. The normalized feature value corresponds to the index weight of a term in a document (feature weight in a data sample) while the class of the record corresponds to document relevance. In the testing data, there are only 2 crisp product classes: normal traffic (class 0) and attack (class 1). The goal of the algorithm was to find an expression (fuzzy classifier) that will describe the set of records describing an attack.

The settings for the GP are summarized in Table 5.

Table 5. GA parameters used for fuzzy classifier evolution

Parameter	Value
Population size	100
Generations limit	5000
Fitness	F-Score
Mutation probability	0.8
Crossover probability	0.8

The F-Score parameter β was set in different experiments to 1, 0.5 and 5 to see detection capabilities of evolved classifiers with different priorities of precision and recall in fitness function. We have observed overall accuracy of the classification (OA) as the percent of correctly classified records in the test collection, detection rate (DR), e.g. the percent of correctly classified attacks and

---

[1] http://kdd.ics.uci.edu/databases/kddcup99/kddcup99.html

Table 6. Calssification results for different attack classes

(a) Results for attack class DoS

	β		
	0.5	1	5
OA	93.95	99.31	95.22
DR	99.42	99.27	94.04
FP	28.07	0.53	0.05

(b) Results for attack class U2R

	β		
	0.5	1	5
OA	99.95	99.96	99.95
DR	50	34.34	50
FP	0.02	0	0.02

(c) Results for attack class R2L

	β		
	0.5	1	5
OA	93.95	98.87	99.09
DR	99.42	38.17	31.07
FP	28.07	0.43	0.12

(d) Results for attack class Probe

	β		
	0.5	1	5
OA	94.02	98.46	98.34
DR	90.34	63.2	59.11
FP	5.83	0.05	0.01

false positives (FP), e.g. the percent of regular records missclassified as attacks. Obviously, good classifier would feature high OA, high DR and low FP.

The results of experiments are summarized in Table 6.

We can see that the evolved classifier reached in all cases and for all attack types good accuracy higher than 93 percent. However, DR and FP are for some attack classes not so good. The best combination of high DR and low FP was reached for DoS and $\beta = 1$. The classifier managed to detect 99.27 percent of attacks and misclassified only acceptable 0.53 percent of harmless connections. For the U2R attack class, the best classifiers managed to detect 50 percent of the attacks. In R2L experiment, the classifier evolved with $\beta = 0.5$ reached DR 99.42 percent, but it also misclassified close to 30 percent of harmless connections. The classifiers with low FP managed to detect only 38 and 31 percent of attacks. Finally, the classifiers evolved for Probe attacks managed to detect fair 90 percent of attacks at the cost of 5.83 percent of false positives for $\beta = 0.5$ and around 60 percent of attacks with FP percent below 0.05 for $\beta = 1$ and $\beta = 5$.

The different results for different attack classes suggest that the nature of the fetaures describing the attacks varies and different settings for GP (e.g. the value of β) needs to be used. Moreover, we have seen that high overall acuuracy of classification does not imply good detection rate and low misclassification of legitimate traffic.

5   Conclusions

We have implemented a genetic programming to evolve fuzzy classifiers for data minig. In contrast to previous efforts in this area (see e.g. [11]), our approach is inspired by information retrieval. We interpret data classes as fuzzy sets and evolve fuzzy search expressions that would describe such sets rather than traditional rule-based fuzzy classifiers. The data mining problems were reformulated as information retrieval tasks and the search query optimization algorithm was used to infer symbolic fuzzy classifiers describing classes of data records.

The evolution of fuzzy classifier for data mining is an ongoing project. We have used the genetic programming originally developed for query optimization and the results are encouraging. However, a number of tasks deserves attention. The choice of the best fitness function (are IR measures really the best fitness function for classifier evolution?) or the interpretation of the fuzzy weights in the classifier (is the IR retrieval status value the optimal choice?) are among the most appealing open questions.

Acknowledgement

This work was supported by the Ministry of Industry and Trade of the Czech Republic, under the grant no. FR-TI1/420.

References

1. Crestani, F., Pasi, G.: Soft information retrieval: Applications of fuzzy set theory and neural networks. In: Kasabov, N., Kozma, R. (eds.) Neuro-Fuzzy Techniques for Intelligent Information Systems, pp. 287–315. Springer, Heidelberg (1999)
2. Kraft, D.H., Petry, F.E., Buckles, B.P., Sadasivan, T.: Genetic Algorithms for Query Optimization in Information Retrieval: Relevance Feedback. In: Sanchez, E., Shibata, T., Zadeh, L. (eds.) Genetic Algorithms and Fuzzy Logic Systems. World Scientific, Singapore (1997)
3. Larsen, H.L.: Retrieval evaluation. In: Modern Information Retrieval Course, Aalborg University Esbjerg (2004)
4. Losee, R.M.: When information retrieval measures agree about the relative quality of document rankings. Journal of the American Society of Information Science 51(9), 834–840 (2000), http://citeseer.ist.psu.edu/losee00when.html
5. Koza, J.: Genetic programming: A paradigm for genetically breeding populations of computer programs to solve problems. Dept. of Computer Science, Stanford University, Technical Report STAN-CS-90-1314 (1990)
6. Koza, J.R.: Genetic Programming: On the Programming of Computers by Means of Natural Selection. MIT Press, Cambridge (1992)
7. Mitchell, M.: An Introduction to Genetic Algorithms. MIT Press, Cambridge (1996)
8. Húsek, D., Owais, S.S.J., Snášel, V., Krömer, P.: Boolean queries optimization by genetic programming. Neural Network World, 359–409 (2005)
9. Snasel, V., Abraham, A., Owais, S., Platos, J., Kromer, P.: Emergent Web Intelligence: Advanced Information Retrieval. In: User Profiles Modeling in Information Retrieval Systems. Advanced Information and Knowledge Processing, Springer, London (2010), http://www.springerlink.com/content/p74163276h776457/, doi:10.1007/978-1-84996-074-8_7 iSBN 978-1-84996-073-1 (Print) 978-1-84996-074-8 (Online)

10. Salton, G., Buckley, C.: Term-weighting approaches in automatic text retrieval. Information Processing and Management 24(5), 513–523 (1988)
11. Carse, B., Pipe, A.G.: A framework for evolving fuzzy classifier systems using genetic programming. In: Proceedings of the Fourteenth International Florida Artificial Intelligence Research Society Conference, pp. 465–469. AAAI Press, Menlo Park (2001)

PID Step Response Using Genetic Programming

Marcus Henrique Soares Mendes[1,3], Gustavo Luís Soares[2],
and João Antônio de Vasconcelos[3]

[1] Universidade Federal de Viçosa, Campus de Florestal, Rodovia LMG818, km 6
35690-000 Florestal-MG, Brasil
[2] Pontifícia Universidade Católica de Minas Gerais, Av. Dom José Gaspar, 500
30535-901 Belo Horizonte-MG, Brasil
[3] Evolutionary Computation Laboratory
Universidade Federal de Minas Gerais, Av. Antônio Carlos, 6627
31270-010 Belo Horizonte-MG, Brasil
marcus@ufv.br, gsoares@pucminas.br, jvasconcelos@ufmg.br

Abstract. This paper describes an algorithm that generates analytic functions for
PID step response characteristics (i. e. rise time, overshoot, settling time, peak
time and integral of time weighted absolute error) in an application of a third-
order plant. The algorithm uses genetic programming for symbolic regressions
and provides formal expressions composed of variables, constants, elementary
operators and mathematical functions. Results show a good fitting between the
desired and obtained step response for DC motor positioning problem.

Keywords: PID step response, genetic programming, symbolic regression.

1 Introduction

The Proportional-Integral-Derivative (PID) controller has been widely used in the
industrial processes because of its simple structure and high-quality performance in
several operational conditions [1]. Astrom and Hagglund [2], pointed out that in a
survey over than 11000 regulatory controllers, 97% of them had the PID structure. In
few words, a PID controller attempts to minimize the error by adjusting three *gain
parameters*: the proportional gain k_p, the integral gain k_i and the derivative gain k_d [3].
According to [4] the transfer function of a PID controller in the frequency domain is
as given by Eq. (1):

$$C(s) = k_p + \frac{k_i}{s} + k_d s . \tag{1}$$

The process of finding the best set of gain parameters is called PID tuning. Even
trial and error can be used to achieve the best tuning inspite being time-consuming.
On the other hand, as presented in [4-6], the PID tuning can be also formulated as an
optimization problem for one or more *step response characteristics* such as *rise time,
overshoot, settling time, peak time, Integral of the Absolute Error* (IAE), *Integral of
the Squared Error* (ISE) and *Integral of Time weighted Absolute Error* (ITAE).

Sharing this line of thought, Herreros et al. [5] propose a genetic multiobjective algorithm, Boubertakh et al. [7] present an ant colony optimization algorithm and Visioli

K. Deb et al. (Eds.): SEAL 2010, LNCS 6457, pp. 359–368, 2010.
© Springer-Verlag Berlin Heidelberg 2010

[8] compares different fuzzy methodologies for PID tuning. Soares [6], in particular, proposes a robust multiobjective formulation and uses Interval Analysis to deal with uncertainties in the gain parameters. In this approach, it is mandatory to deduce interval functions from analytic expressions of the step response characteristics. The task for deducing such interval functions grows with the complexity of the mathematical model. Hence, we decided to use Symbolic Regression (SR) to cope with this problem.

Some methods in regression analysis have been used to find out the function structure set up by coefficients and variables in order to represent the behavior of a diverse phenomenon. The traditional ones utilize a linear, quadratic or polynomial framework. Alternatively, SR is a flexible way to find a mathematical expression, in symbolic form, which best fits a given finite sample of data from an unknown curve [9].

Genetic Programming (GP) [10] can be successfully applied in several fields, for example, sequence induction, optimal control, forecasting, SR and others. Like the genetic algorithms, GP is based on the Darwinian principle of survival of the fittest. However, the structures undergoing adaptation are hierarchical computer programs, which dynamically change size and shape. This flexibility provides a powerful symbolic expressiveness. For instance, the computer program evolved in SR is a mathematical expression that is not a priori defined, unlike other regression techniques.

Therefore, considering the previous discussions, we present an algorithm that uses GP for SR in order to obtain analytic functions to step response characteristics of PID controller.

The paper is organized as follows. Section 2 reviews the GP. In section 3 the algorithm is described. Section 4 is divided in two parts: i) the first one introduces the PID controller study-case and ii) the second one presents and discusses the results of the experiment. Finally, the final considerations are presented in Section 5.

2   Genetic Programming in Symbolic Regression

As shown in [9], the GP is a stochastic approach that searches in the space of *computer programs* (individuals) for one of them, which best solves the problem at hand. According to Fig. 1 based in [10], GP starts with an initial random population of *programs* and, generation by generation, transforms the programs in new ones, hopefully better. GP determines how well a program works by running it, and then comparing its behavior to some reference value. GP uses the standard genetic operations (crossover, mutation and reproduction) to evolve the new generation of programs.

Specifically, to handle SR problems, it is indispensable to define in GP: the *set of functions,* the *set of terminals,* the *fitness measure*, the *controlling parameters*, the *stopping criteria*, the *training* and the *validation sets* [11].

```
1.  Initialize a population of programs.
2.  Until achieve an stopping criteria, repeat
    2.1 Run the programs and determine their fitness.
    2.2 Select programs.
    2.3 Generate child programs by applying genetic operations.
3.  Return the best program.
```

Fig. 1. Basic GP Algorithm

In GP, the convergence to the correct mathematical expression consists in evolving the sets of functions and terminals taken together. The set of functions is formed by any arithmetical operator and the set of terminals is composed of algebraic terms (variables) and ephemeral constants. Both are important to provide elements to the mathematical expressions that in GP, generally, are represented by trees. Each internal tree node is a member of the set of functions and the leaves nodes belongs to the set of terminals. The constants are ephemeral because they remain to the future populations if and only if they belong to the group of selected programs.

Typically, the fitness measure is computed by sum or average of the error between the result produced by the expression and the reference value over different combinations of the inputs and outputs, called *fitness cases*. Keijzer [12] proposed an improvement in this calculation, that is denominated fitness scaling, which consists of the computation of the intercept (a) and slope (b) for the expression. Given that $y = \text{expression}(x)$ is an output of an expression produced by GP on the input x. A linear regression on the reference values (r) can be evaluated by Eqs. (2-3) [12]:

$$a = \bar{r} - b\bar{y}, \tag{2}$$

and

$$b = \frac{\sum_{i=1}^{n}[(r_i - \bar{r})(y_i - \bar{y})]}{\sum_{i=1}^{n}[(y_i - \bar{y})^2]}. \tag{3}$$

In Eqs. (2-3) the parameter n is the number of fitness cases and $\bar{r}$ and $\bar{y}$ denote the average reference and average output value, respectively. Thus, the scaled version of the most common fitness measure that is the Mean Squared Error (MSE) is given by Eq. (4):

$$\frac{1}{n}\sum_{i=1}^{n}(a + by_i - r_i)^2. \tag{4}$$

The main controlling parameters in GP are the population size, the maximum number of generations, maximum height tree for expressions created during the run, maximum height tree for initial random expressions, the generative method for initial random population, the selection method and the probabilities of crossover, mutation, among others.

The two more common criteria to terminate a GP run are: to reach the maximum number of generations or to obtain an expression with zero error for all fitness cases.

A SR involves two phases: training and validation. The first one performs GP in one subset of samples, called training set. The second one estimates the accuracy of the expressions obtained in the previous phase when evaluated in other subset, called validation set. The training and the validation subsets are complementary, in other words, their union is the full sample of data.

An example of GP training phase in a SR problem is presented hereafter. The goal is to find an expression in which output y is equal to the values of the $x^2 + 2x + 4$ with $x \in [-3, 4]$. Table 1 summarizes the GP settings.

Running the GP algorithm once generated the expression ((x)+(2))*((x)/(5)) with slope and intercept equal to 5 and 4, respectively. Note that the returned expression is equivalent to $5(x + 2)\frac{x}{5} + 4$, which is identical to the expression $x^2 + 2x + 4$.

Table 1. GP parameters for the SR example

Parameter	Value
Maximum number of generations	10
Number of programs	500
Crossover, mutation and reproduction rates	0.9, 0.1 and 0.2, respectively
Rate of crossover on internal nodes	0.8
Function set	+, -, *, /, ^, $\sqrt{}$, sin, ln and exp
Terminal set	x and ephemeral constant
Real and integer ephemeral constant ranges	]-10, 10[and {1, 2, …, 5}
Population tournament size	2
Maximum tree height during the run	7
Maximum tree height for initial generation	2
Generative method for initial population	Ramped half-and-half as in [9]
Stopping criteria	Maximum number of generations
Fitness measure	MSE with fitness scaling
Fitness cases – training set	25 random pairs (x,y), x ∈ [-3,4]

3 The Algorithm

The GP implemented in this paper follows the structure shown in **Fig. 2**. We highlighted two points in the algorithm: 1) in the 5.4 step, where each program (individual) is converted to an expression in order to perform its evaluation. But, the fitness value is calculated just for valid programs. Thus, a preprocessing stage is executed as proposed in [13], which uses Interval Analysis to make sure that the mathematical expression discovered by GP does not contain any undefined values. 2) In the 5.6 step, where two programs are considered similar if both have the same fitness value, the same quantity of terminal and internal nodes.

```
1.  Initialize a population of programs.
2.  Define the fitness cases.
3.  Run the programs and determine their fitness.
4.  Calculate the ranking to each program.
5.  Until achieve an stopping criteria, repeat
    5.1. Save the best programs from current generation (elitism).
    5.2. Select programs.
    5.3. Generate child programs by applying genetic operators.
    5.4. Run the child programs and determine their fitness.
    5.5. Append the best programs with the child programs (elitism).
    5.6. Eliminate similar programs.
    5.7. Calculate the ranking to each program.
    5.8. Select the best ranked programs to next generation.
6.  Return the best program.
```

Fig. 2. Customized GP algorithm

The algorithm in Fig. 2 is applied in the training phase of the SR. Fig. 3 shows the algorithm proposed to the validation phase. In the step 3.2 the *percent error* is given by Eq. (5):

$$\delta = \frac{|r - y|}{|r|} * 100 , \tag{5}$$

where r is the reference value and y is the value returned by the mathematical expression that was obtained in the training phase. In the step 4 the *tolerance* is defined as 51th percentile of the percent error considering all validation's samples. This method was proposed, because it was noticed in the preliminary experiments that best (smallest MSE) expression obtained in the training, frequently was not the same in the validation, considering all expressions in last generation population.

```
1.  Take an expression obtained in training.
2.  Define the validation samples.
3.  For each validation sample
    3.1. Evaluate the expression.
    3.2. Calculate the percent error.
4.  Return the tolerance.
```

Fig. 3. Validation algorithm

4   Experiment and Results

4.1   Modeling of DC Motor Positioning - Case Study

At this subsection, a real problem is presented. It consists in the DC motor, which is a common actuator in control systems. The problem specifications were based in [14]. Table 2 presents the values for its physical parameters. The main objective is set the position of the motor precisely using a PID controller. In this experiment the transfer function of the closed-loop system is given by Eq. (6) [14]:

$$\frac{\theta}{V} = \frac{K}{s\big((Js + B)(Ls + R) + K^2\big)} , \tag{6}$$

where the shaft position (θ) is the output and the voltage (V) the input of the system. Note that Eq. (6) is a third-order function.

4.2   Training and Validation

The GP parameters were set as shown in the Table 3. The choice of elements in the function set was made so as to result in expressions which contain elementary operators and continuous functions, which are desirable in composing interval functions. The sample data were obtained by using numerical methods in Matlab® simulations, as shown in [14]. We ran 1800 simulations with randomly k_p, k_i and k_d ranging from

0 until 1000. The data were equality divided between training and validation sets. The validation set was split in nine complementary groups each one with one hundred samples. The GP algorithm was executed around fifty times for each time response characteristic. It was used one thousand programs in the GP population to the rise time, two thousand to the settling time and ITAE, and five hundred to the overshoot and peak time.

Table 2. Physical parameters values for the DC motor position problem

Physical parameter (acronym)	Value
Moment of inertia of the rotor (J)	3.2284E-6 kg.m^2
Damping ratio of the mechanical system (B)	3.5077E-6 Nms
Electromotive force constant (K)	0.0274 Nm/Amp
Electric resistance (R)	4 Ω
Electric inductance (L)	2.75E-6 H

Table 3. GP parameters

Parameter	Value
Maximum number of generations	50
Number of programs	500, 1000 or 2000
Crossover, mutation and reproduction rates	0.9, 0.1 and 0.2, respectively
Rate of crossover on internal nodes	0.8
Function set	+, -, *, /, ^, $\sqrt{}$, sin, ln and exp
Terminal set	k_p, k_i, k_d and ephemeral constant
Real and integer ephemeral constant ranges	]-10, 10[and {1, 2, ..., 5}
Population tournament size	2
Maximum tree height during run	7
Maximum tree height for initial generation	2
Generative method for initial population	Ramped half-and-half as in [9]
Stopping criteria	Maximum number of generations
Fitness measure	MSE with fitness scaling
Fitness cases – training set	900 samples
Validation set	900 samples divided in 9 groups

The last generation population of the training was verified by validation algorithm for each group. Hence, it was computed for each expression the mean tolerance and the mean MSE (considering all validation groups). Finally, the expression with the smallest mean tolerance was declared the best one. Table 4 summarizes the best running for each time response characteristic. The algorithm was run on a core i3 CPU with 4 GB of RAM memory.

According to Table 4 the overshoot has the best (just 1.43% to mean tolerance) expression among time response characteristics. On the other hand, the peak time has the worst expression (mean tolerance = 6.21%). Nevertheless, it is an acceptable value, mainly if we consider the order of magnitude (10^{-6}) in the peak time values. The peak time was tested with more programs in the population, but better results were not reached. Also, for all time response characteristics, that mean validation MSE is greater than training MSE. However, the median has the same order of magnitude in both situations.

Table 4. Results considering the best running

Time Response Characteristic	Value of the best obtained expression to					Run Time (s)
	Training MSE	Validation MSE		Tolerance (%)		
		Mean	Median	Mean	σ	
Rise time	4.96E-10	2.94E-9	1.50E-10	4.48	0.42	510
Overshoot	4.06E-5	5.08E-4	3.68E-5	1.43	1.04	376
Settling time	6.66E-12	3.57E-7	8.23E-12	5.89	1.70	1740
Peak time	1.38E-9	1.55E-8	2.94E-9	6.21	0.83	240
ITAE	1.09E-6	2.14E-4	1.54E-6	5.05	1.88	1371

The equations (7), (8), (9), (10) and (11) show the best analytic expressions (with some mathematical simplifications) obtained for rise time, overshoot, settling time, peak time and ITAE, respectively.

$$\frac{\left(\frac{1}{e^{0.77669}}\right)\left(\left(\frac{1}{e^{0.75797}}\right)^{\frac{1}{kd}}\right)^{\frac{1}{kd}}+\frac{\ln(2.23)}{e^{kd}}+\frac{1}{e^{0.77669}}}{(\ln(2.23)^{0.5789992464})\left(\frac{1}{e^{0.77669}}\right)^{\ln(1.678)(e^{0.69874})kd^{\frac{5}{kd}}}} \tag{7}$$

$$-\frac{58527\sin\left(\frac{5000\sqrt{kd}}{32829}\right)\ln(kd)^2}{10000}+\frac{17513\ln(ki+5)}{2000\,kd}+\left(\sqrt{\sqrt{kd}}+2\right)\ln(kd)+kd \tag{8}$$

$$-\frac{4}{\frac{23559\sin(4)\ln(kd)}{10000}+\sqrt{kd}+e^3+\frac{1}{kd^{2.3559}}}+\frac{\ln(kd)+\sin\left(\frac{94079\,e^3}{30000\,kd}\right)+e^3}{kd} \tag{9}$$

$$\frac{e^{\frac{\sqrt{2887}}{50}}\,(\ln(kp)+2\,kd+108.6219)}{(kd+1.4475)\left(\frac{400\,kd}{979}+19.3877\right)(\ln(kp)+2\,kd+101.6191)} \tag{10}$$

$$\left(\frac{\left((2.885) - \left(\sin\left(\frac{7.634}{kd}\right)\right)\right)(\sqrt{kp})}{(\sqrt{kd})^{\ln(kd)}}\right) + \left(\frac{9.2088}{2.7183\ kd}\right) \tag{11}$$

Although, according to control system theory, there are individual effects of PID adjustable parameters on time response characteristics. In the all best obtained expressions we can note that at least one PID adjustable parameter is not present. It suggests that GP algorithm selected the most significant parameters according to the case study peculiarities. It would be possible to add a constraint in GP algorithm that obliges the inclusion of all parameters. However, it was not our intention in this work.

It is necessary to know the slope and the intercept in order to use the obtained expressions. Both are calculated during the training because the fitness scaling technique was used. Table 5 shows its values.

Using the obtained expressions with the respective slope and intercept it was possible to build the validation graphics. Fig. 4 shows, for each step response characteristic, the percent error to the four hundred fifty nine samples that have the smallest percent error values considering the union of all validation groups. For all step response characteristic, expect settling time, we can clearly see by inspection that biggest percent error value (samples number 459) is smaller or equal the mean tolerance value presented previously in Table 4. Thus, we can realize an improvement in the tolerance value of the rise time, overshoot, peak time and ITAE, when all samples are taken together.

Table 5. The slope and the intercept for the obtained expressions

Characteristic	Slope	Intercept
Eq. (**7**) – Rise time	0.00924428311695392	-0.00926197893941111
Eq. (**8**) – Overshoot	0.000187149002201421	-0.00985609977638255
Eq. (**9**) – Settling time	8.20756558513056e-5	1.09948493977817e-5
Eq. (**10**) – Peak time	0.0481524972937903	1.9290865493687e-6
Eq. (**11**) – ITAE	0.000498795690522851	1.62323177760705e-5

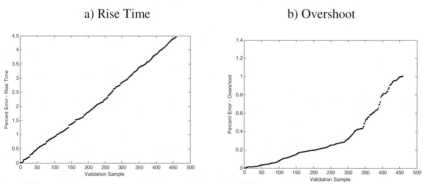

Fig. 4. The percent error (in ascending order) to the 459 samples that have the smallest percent error according to each step response characteristic: a) Rise time, b) Overshoot, c) Settling time, d) Peak time and e) ITAE

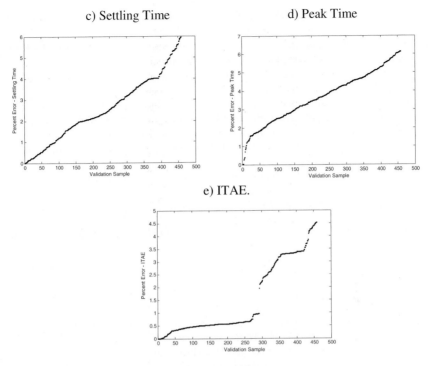

c) Settling Time   d) Peak Time

e) ITAE.

Fig. 4. (*Continued*)

5 Conclusion

The formulation by SR was appropriated to obtain the analytic functions for step response. The results show that the best selected analytic expressions were obtained without at least one gain parameter to the DC motor positioning problem. This suggests that the missing parameters were not representative for the step response characteristics taking into account of case study specifications. Besides, for all expressions, the mean tolerance in the validation is between 1.43% and 6.21% that shows a good fit, mainly if we consider the magnitude of the data.

Therefore, the described algorithm is a helpful and feasible way to allow or to become less hard, depending on the transfer function order, the use of PID tuning approaches that require analytic expressions for step response characteristics such as the interval ones.

Acknowledgment. This work has been supported in part by CNPq and CAPES.

References

1. Astrom, K.J., Hagglund, T.: The future of PID control. Control Engineering Practice 9, 1163–1175 (2001)
2. Astrom, K.J., Hagglund, T.: Revisiting the Ziegler–Nichols step response method for PID control. Journal of Process Control 14, 635–650 (2004)

3. Li, Y., Ang, K.H., Chong, G.C.Y.: PID control system analysis and design. IEEE Control Systems Magazine 26(1), 32–41 (2006)
4. Gaing, Z.: A particle swarm optimization approach for optimum design of PID controller in AVR system. IEEE Trans. Energy Conversion 19(2), 384–391 (2004)
5. Herreros, A., Baeyens, E., Peran, J.R.: Design of PID-type controllers using multiobjective genetic algorithms. ISA Transactions 41(4), 457–472 (2002)
6. Soares, G.L.: Algoritmos Intervalares e Evolucionário-Intervalar para Otimização Robusta Multi-Objetivo. Phd Thesis, Universidade Federal de Minas Gerais (2008)
7. Boubertakh, H., Tadjine, M., Glorennec, P., Labiod, S.: Tuning Fuzzy PID Controllers using Ant Colony Optimization. In: 17th Mediterranean Conference on Control & Automation, Greece, pp. 13–18 (2009)
8. Visioli, A.: Tuning of PID controllers with fuzzy logic. In: IEE Proceedings-Control Theory and Applications, vol. 148, pp. 1–8 (2001)
9. Koza, J.R.: Genetic Programming - On the Programming of Computers by Means of Natural Selection. MIT Press, London (1992)
10. Poli, R., Langdon, W.B., McPhee, N.F.: A field Guide to Genetic Programming (2008), http://www.gp-field-guide.org.uk
11. Koza, J.R.: Genetic Programming II - Automatic Discovery of Reusable Programs. MIT Press, London (1994)
12. Keijzer, M.: Scaled Symbolic Regression. Genetic Programming and Evolvable Machines 5, 259–269 (2004)
13. Keijzer, M.: Improving symbolic regression with interval arithmetic and linear scaling. In: Ryan, C., Soule, T., Keijzer, M., Tsang, E.P.K., Poli, R., Costa, E. (eds.) EuroGP 2003. LNCS, vol. 2610, pp. 275–299. Springer, Heidelberg (2003)
14. University of Michigam – Michigam Enginnering - Control Tutorials for Matlab, http://www.engin.umich.edu/group/ctm/examples/motor2/motor.html

Divide and Evolve Driven by Human Strategies

Markus Borschbach, Christian Grelle, and Sascha Hauke

Chair of Optimized Systems, Faculty of Computer Science
University of Applied Sciences
Hauptstr. 2, 51465 Bergisch Gladbach, Germany
{Markus.Borschbach,Christian.Grelle,Sascha.Hauke}@fhdw.de

Abstract. This work describes the incorporation of human strategies into a genetic algorithm. Human competence and machine intelligence are merged, creating a divide and evolve approach. The approach is applied to the restoration problem of Rubik's Cube and successfully solves this task.

Keywords: Human strategy, genetic optimizer, discrete optimization, performance evaluation.

1 Introduction

This work illustrates the incorporation of human strategies into a genetic algorithm, thereby introducing a method of divide and evolve called HuGO! – the Human strategy based Genetic Optimizer.[1][3][7]

As a use case, an application is analyzed that is simple regarding its principle functionality, yet invokes complex mathematics – Rubik's Cube. This three-dimensional puzzle is widely known and represents an easy to grasp exemplification of the problem class.

In 1994, MICHAEL HERDY and GIANNINO PATONE solved the cube using evolution strategies.[4] Therefore, they introduced a quality function for the evaluation of the cube state. The quality function to be minimized consists of three parts, Q1, Q2 and Q3, combined by addition. Q1 is increased for a wrong facelet, while Q2 and Q3 penalize wrong positioned edge- and corner cubies. 10 different mutations, as introduced by [4] are realized using swaps and turns of individual cubies. On the one hand, this enables a rapid solution search because dependencies are minimized. But on the other hand, the results will be fairly long solution sequences, because accomplishing a single swap already incorporates around 10 cube rotations. Since our approach is supposed to find short solutions, HERDY and PATONE's approach is inappropriate for this work's goals. CYRIL CASTELLA built an evolutionary approach for solving the cube, seeking short solution sequences, that would be much more convenient to adapt.[2] The program uses a genetic algorithm, which is based on a one-point crossover, omitting a selection operator and mating pool. Unfortunately, the approach suffers from missing integrity so that no performance comparison could be conducted, as of yet. More precisely, the solution output seems to fail if the cube is already in the two-generator group. However, it incorporates some useful functions that are enhanced in and applied to HuGO!.

K. Deb et al. (Eds.): SEAL 2010, LNCS 6457, pp. 369–373, 2010.
© Springer-Verlag Berlin Heidelberg 2010

2 HuGO!

The most obvious way of solving the cube by evolutionary means is to use a trivial fitness function that compares the scrambled cube to the solved one and counts the number of facelets consistent with the solution. Tests showed that this approach turns out to be inefficient. While the algorithm quickly reaches around 70% consistent facelets, further computational efforts only provide marginal improvements. This is due to the $4.3 \cdot 10^{19}$ interdependent permutations that are possible states of the cube group, which lead to an enormously jagged fitness landscape containing numerous local optima. Consequently, the algorithm repeatedly becomes trapped in local optima, considerably prolonging calculation time.

The integration of a common human strategy for solving Rubik's Cube, called two-generator method, provides a solution to this problem.[2] The two-generator method is inefficient in respect to the fast generation of solutions, but is often used by contestants of fewest moves challenges. This method restores a scrambled cube by transforming it into the two-generator subgroup first and then solving the cube in this group. The two-generator subgroup contains only about 73,483,200 different states, a number that is significantly smaller than that of the entire cube group. Thus, it is a promising approach to split up the cube-solving search algorithm into partial solutions.

In the first phase a 2x2x3 subcube, i.e. the entire cube except two adjacent layers, is solved in one of the twelve possible locations. The two remaining layers are left scrambled. Since not all cube states that only have two adjacent layers unsolved are automatically in the two-generator, the second phase transforms the cube into this group. For this purpose tests are necessary that check, whether the edge cubie orientations and corner cubie permutations stick to the states achievable in the two-generator. In the third phase, the remaining two layers are turned until the entire 3x3x3 cube is completely restored. The three phases of the algorithm realizing this human strategy are three independent algorithms that are based on the common canonical genetic algorithm and share a lot of analogies. The individual components, as well as their interaction in the composed workflow are explained in detail in the succeeding sections.

According to these suggestions, the discrete characteristics of Rubik's Cube as optimization process allow a rather direct incorporation of the search space into the solution representation. All three phases of the algorithm have in common that instead of binary strings, each gene of a solution contains a number ranging from 0 to 17 representing one of the 18 potential cube turns.[2]

In the case of Rubik's Cube, the scrambled cube represents the environment and the solution string represents the individual that is customized to fit to the environment. The fitness function assigns a fitness value to the individual judging the quality of adaptation, i.e. ability to solve the cube. To test this ability, the particular solution operates on the scrambled cube and the resulting cube undergoes evaluations that determine the individual's fitness. It becomes clear that the implementation of the cube has to fulfill two main capabilities. First it must be receptive to the individual's modifications, i.e. turns. Then the resulting cubes must be distinguishable in quality, respecting the number of turns applied.

To simplify implementation, the facelets of the cubies are labeled with integer numbers.[2] Since there are 54 facelets on the cube, the numbers from 0 to 53 are used to identify them. Allocated row by row on the faces, the entire cube is encompassed.

During evaluation, the individuals are traversed successively from gene to gene while on each step the substring's (first to current gene) fitness is determined. The best fitness and the step number, i.e. number of turns, are stored. The overall fitness of the potential solution is determined by calculating the fitness and subtracting the number of needed steps. The fitness describes the progress of restoring the cube. The counted number of steps allows to differentiate between longer and shorter turn sequences of cube states of the same fitness. The distinct determination of the best fitness is different in each algorithm phase due to different goals of the phases.[2]

In the first phase, the *solve 2x2x3 cube phase*, the fitness is determined by counting the number of facelet pairs of the same color on the 2x2x3 surface. Particularly, the pairs are corner-edge pairs and center-edge pairs. A *corner-edge pair* is the conformance of the corner cubie facelets and the adjacent edge cubie facelets of a corner cubie and an edge cubie. Thus there exist six corner-edge pairs on the 2x2x3 surface. A *center-edge pair* is the conformance of a center cubie facelet and an edge cubie facelet. There are 10 center-edge pairs on the 2x2x3 surface, adding up to a total number of 16 pairs that constitute the maximum fitness value when the entire 2x2x3 subcube is solved. The pair technique encompasses the entire 2x2x3 surface and no facelet is ignored.[2]

In the second phase, the *transformation to two-generator phase*, fitness is determined by the number of corner-edge pairs and center-edge pairs as in the first phase, as well as the fact, whether the cube is in the two-generator group. The check for the pairs is done because in this phase, the cube is allowed to temporary leave the solved 2x2x3 state. Only in this way it is possible to manipulate the two not yet solved layers beyond the turns allowed in the two-generator to enter the two-generator subgroup. So the maximum fitness value achievable is 17 and signifies that the cube is successfully transformed to the two-generator group. To determine, whether the cube is in the two-generator, the cubies that are in the edge cubicles and in the corner cubicles of the not yet solved layers are requested. The center cubies are not directly involved in this transformation process because their position is fixed a priori.

The third phase, the *solve two-generator phase*, only allows turns of the two still scrambled sides, so that the cube remains in the two-generator subgroup. This phase performs the integrity determination, which equals the fitness value, for the entire cube surface. Integrity is always at least 16 because the already solved 2x2x3 cube stays unaffected. The maximum value the integrity can achieve is 48 pairs, which means that the two-generator is solved and the cube is completely restored.

As selection operator, stochastic universal sampling is used to propagate individuals to the mating pool. The selection operator incorporates two selection points to reduce spread. The best solution is maintained as elite. Elitist strategies link the lifetime of individuals to their fitness. They are techniques to keep good solutions in the population for longer than one generation. The use turns search more exploitative rather than explorative. Elitist strategies are guessed to be necessary when genetic algorithms are used as function optimizers and the goal is to find a global optimal solution as it is the case in the cube optimization process regarding integrity restoration.[1][6] The genetic operations in the reproduction stadium are slightly adapted standard operators. The crossover operator is realized by a uniform crossover, eliminating positional bias. As explained before, the gene values are, in contrast to binary values, composed of numbers 0 to 17. A standard mutation is performed after the crossover using the adjustable mutation probability p_m. In contrast to binary string individuals, where during mutation a gene value is just swapped from 0 to 1 or from 1 to 0, here the genes have values from 0 to 17. So if a gene is chosen to become mutated, the gene value will be replaced by a random integer number between 0 and 17.

3   Test Results

Performance tests are conducted that give information about the efficiency of the algorithm. For this purpose, a measure called *complexity* is introduced with the help of the program CubeExplorer[5], which provides the length of an optimal solution to a scrambled cube, to classify it according to the difficulty of restoration. For each complexity class, the algorithm's performance is worked out. The performance tests investigate the efficiency of HuGO!. To display the efficiency in terms of solution lengths, it is essential to differentiate between different difficulties of scrambled cubes. A feasible way to gain information on the scrambled cube is to count the number of turns that led to the cube state. However, the number of turns is limited in information content. For example, a scramble sequence of 20 turns might need no turns as solution, if after 10 turns, the first 10 turns are inverted and repeated in reverse order. For this reason, a measure called *complexity* is introduced that does not consider the scramble sequence but orientates on the actual solution length. The actual solution length can be computed with a program generating optimal solutions to a given cube configuration. This is the case in CubeExplorer[5]. So, for the performance tests, the complexity of each cube configuration considered is derived from CubeExplorer beforehand. The resulting optimal solution length is used to develop classes. Since each cube, according to the general scientific assumption, should be solvable in up to 20 turns, complexity has a range of 1 to 20, whereas any cube can be classified into one of the 20 complexity classes. For each of the 20 classes, three random scrambles are solved 10 times resulting in 600 test runs that evaluate the performance of the algorithm. As generation numbers 200, 5000 and 50000 are used. The third phase is run with the highest amount of generations, because the third phase produces the longest part-solutions and therefore has the highest potential for solution improvements.

The performance tests reveal four distinct efficiency characteristics depending on the introduced complexity measure. From complexity 1 to 3, HuGO! reaches optimal solutions. From 4 to 7, a heavy variation in solution lengths occurs. From 8 to 10, HuGO! loses the capability of optimal solutions and in the last category from 11 to 20, a relatively equal amount of around 38 turns establishes itself. Consequently, the algorithm performs well for lower complexities, while it shows disadvantages in the handling of complex scrambles.

4 Conclusion

The Human strategy based Genetic Optimizer is applied to the restoration problem of Rubik's Cube and successfully solves this task. The introduction of HuGO! is not only theoretically reasoned but also empirically. Tests to solve the cube using a plain genetic algorithm show the inappropriateness of this optimization technique to the chosen discrete optimization problem. Only when a human strategy was implemented that splits up the process to distinct intermediate stages, the algorithm's solving capability became enabled. The strategy consists of three phases that allow the cube to be gradually solved. To summarize, this work shows, how a human strategy can be incorporated in a genetic algorithm. The goal was not to develop an outstanding fast or efficient algorithm, but to demonstrate the advantageous adaptation of genetic algorithms to human induced strategies using a divide and evolve approach. While an external strategy guides the algorithm, the evolutionary process produces solutions no human had thought of.

References

1. Borschbach, M., Grelle, C.: Empirical Benchmarks of a Genetic Algorithm Incorporating Human Strategies. Technical Report, no. 2009/01 (2009)
2. Castella, C.: Rubik's Cube, méthodes pour tous (2005),
   http://www.francocube.com (Call date 2009-01-20)
3. De Garis, H.: Genetic Programming: building artificial nervous systems using genetically programmed neural networks modules. In: Porter, R., Mooney, B. (eds.) Proceedings of the 7th International Conference on Machine Learning, pp. 132–139 (1990)
4. Herdy, M., Patone, G.: Evolution Strategy in Action – 10 ES-Demonstrations (1994),
   http://www.bionik.tu-berlin.de/user/giani/esdemos/evo.html
   (Call date 2009-02-23)
5. Kociemba, H.: CubeExplorer (2009), http://kociemba.org/cube.htm (Call date 2009-02-21)
6. Sarma, J., De Jong, K.: Generation gap methods. In: Bäck, T., Fogel, D., Michalewicz, Z. (eds.) The Handbook of Evolutionary Computation, Bristol, pp. C2.7:1–C2.7:5 (2002)
7. Schoenauer, M., Savéant, P., Vidal, V.: Divide-and-Evolve: a Sequential Hybridization Strategy using Evolutionary Algorithms. In: Siarry, P., Michalewicz, Z. (eds.) Advances in Metaheuristics for Hard Optimization, Berlin, pp. 279–298 (2007)

A Genetic Algorithm for Assigning Individuals to Populations Using Multi-locus Genotyping

Avnish K. Bhatia[*], Monika Sodhi, Dinesh K. Yadav, and B. Prakash

National Bureau of Animal Genetic Resources, Karnal – 132001, India
Fax: +91-184-2267654
avnish@lycos.com

Abstract. This paper reports a genetic algorithm (GA) for individual assignment using multi-locus microsatellite genotyping. Its performance has been compared with existing frequency, Bayesian and distance-based methods using simulated as well as actual data. Simulated data has been generated with SIMCOAL program. Actual data has been generated from genotypes of four cattle breeds from India. The GA showed lower accuracy while assigning individuals from simulated data. Its performance was comparable to that of existing methods using actual data.

Keywords: Genetic algorithm, Individual assignment, Breed, Microsatellite genotyping.

1 Introduction

Genetic algorithms (GAs) [1] are general purpose procedures that work on principle of natural selection. It starts from a randomly generated population of strings representing potential solutions, which are selected according to their fitness values. Selected strings reproduce and create population at the next generation. Genetic operators, mainly crossover and mutation, alter composition of strings and make them ready to be part of the population in later generations. Strings in a generation are having higher average fitness than predecessors. GA marches towards the global optimum over a number of generations.

Multi-locus genotyping refers to detection of alleles on a number of loci in individuals. Highly polymorphic microsatellites are being used for evaluating genetic variability of populations [2]. Individual assignment problem [3] consists of assigning individuals to one of the candidate populations of likely origin on the basis of multi-locus genotyping. Assignment accuracy is percentage success of classification of untested individuals to one of the baseline populations. It depends on number of loci, number of alleles per locus, baseline population size, and degree of differentiation between populations as measured by F_{ST}. Assignment of individuals to specific breed / strain / variety is useful in conservation and utilization of animal, wildlife, plant and aquatic genetic resources.

---

[*] Corresponding author.

K. Deb et al. (Eds.): SEAL 2010, LNCS 6457, pp. 374–378, 2010.

Existing methods for individual assignment are categorized into three groups. First group utilizes likelihood estimation of multi-locus genotype of individuals to be assigned to each of the two or more candidate populations [3]. It includes frequency and Bayesian methods, which rely on two assumptions - loci should be at Hardy-Weinberg equilibrium and each locus should be independent. Second group includes methods that utilize genetic distances to assign individuals to the nearest population [3]. Various inter-population distances such as Nei's distances and Cavalli-Sforza & Edwards chord distance are available. Shared-allele distance is an inter-individual distance. The last group includes supervised machine learning approaches such as k-NN classifier, artificial neural networks, decision tree, etc. [4]. These methods involve evaluation of baseline data set to predict classes of cases in test data set. There are no assumptions on data in distance-based and machine learning methods.

GAs are placed in the category of machine-learning approaches. These are applied in solving diverse range of problems either by optimizing parameters of other approaches or by formulation of candidate problem as an optimization problem. GAs have been used to solve complex problems including bioinformatics problems [5] due to their flexibility in tuning genetic parameters. Guinand et al. [6] and Topchy et al. [7] utilized GA for selecting subset of loci among available genotypes, which provided maximum accuracy of individual assignment based on likelihood classification. Authors are not aware of any method based on GA for solving the individual assignment problem.

2 Methodology

2.1 Genetic Algorithm for Individual Assignment

Genetic algorithm for individual assignment has been designed so as to evolve a position matrix in the form of a binary matrix of order $n.l$, where n is size of baseline population and l is number of loci. Objective function maximizes accuracy of assignment within baseline population. Each individual in the baseline population is compared with all the individuals including it. Accuracy is calculated on the basis of deviation of score of an individual from average score of baseline population calculated with the function:

$$\frac{1}{n}\sum_{k=1}^{n}\left[\sum_{i=1}^{n}\sum_{j=1}^{l}a_{ij}s_{i_j}^{k_j}\right],\tag{1}$$

where $a_{ij}=0/1$ is the number at a position in evolved position matrix, $s_{i_j}^{k_j}$ is the score of matching j^{th} allele of individual 'k' to the jth allele of individual 'i'. A value of '1' at a position in the position-matrix indicates that an allele at the position will be considered for scoring, and a value '0' indicates that it will not be taken.

The GA also evolves haploid matching and no-matching scores. Diploid matching score is obtained by multiplying haploid matching score by two. Scores are evolved in the range (d, 1.0), where 'd' is pre-specified level of deviation of score of an individual from average score of baseline population. An individual is assigned to baseline population if deviation of its score from average score of individuals in the baseline population is less than or equal to 'd'.

Finally, evolved values of scores, position-matrix and average score of baseline population are stored for utilization in assignment of new individuals. Score of an unknown individual is calculated for each candidate population using formula (1) and it is assigned to the population with minimum deviation of its score.

2.2 Implementation of Genetic Algorithm

GA execution involves generation of initial population and decision on genetic parameters and operators, which are generally chosen on the basis of experiences of researchers. Initial population in the GA consisted of bits. Sixteen bits were used each for evolving a real number for haploid-matching score as well as for no-matching score. Position-matrix required $n*l$ bits. Thus, the overall string size was equal to 32 plus $n*l$ bits. Single-point crossover operator was used. Mutation operator was traditional bit operator. Values of other genetic parameters were fixed as: population size = 100, crossover probability = 0.5, and total number of generations = 1000.

Mutation probability (p_m) was varied by a decaying function $p_m = 0.1 * \exp(-1.0 * (t-1)/(T-1) * (\log(L) - \log(h)))$, where t is generation number, T is total number of generations, L is string size, h = 0.7 is a mutation factor [8], and 'log' calculates natural algorithm. It varies p_m from 0.1 at generation one to (0.7/L) at last generation.

2.3 Experimental Setup

Experiments were conducted on simulated as well as on actual test data. Results obtained using the GA were compared with frequency, Bayesian, Nei's Standard distance and Chord distance methods as implemented in GeneClass2 software [9].

SIMCOAL program [10] was used for generating simulated haploid individuals, converted to diploids by randomly selecting two individuals and pairing corresponding alleles at a locus. Data was generated for eight combinations of population features: population differentiation (low F_{ST} - 0.01 and high F_{ST} - 0.1), allelic diversity (low - four alleles per locus and high - eight alleles per locus), number of loci (10 and 20). Simulated population consisted of 1100 diploids.

Actual microsatellite genotyping data was obtained from laboratory experiments at 22 loci for four Indian cattle breeds - Red Kandhari, Deoni, Hariana and Sahiwal [11, 12]. To avoid missing alleles, allelic frequencies were calculated from genotype data. Populations consisting of 1100 diploids for each breed were generated from allelic frequencies for six combinations of two breeds. Allelic diversity values of all the six breed combinations were near 8.0.

We utilized separate 'holdout' method [6] for comparing the results. Size of baseline population was taken equal to 50 and test population equal to 1000 for each population in both simulated and actual data.

3 Results

The proposed GA was executed to maximize accuracy of assignment at 5% level of deviation (d=0.05) of individual score from average-score of baseline population.

Table 1 shows the results as summary of five experiments for simulated data. Success of assignment using the GA was somewhat lower in comparison to other methods. For the population with high F_{ST}, 20 loci and eight alleles per locus, accuracy provided by the GA was 88.2% in comparison to accuracy of 93.5% with frequency method and 92.1% with chord distance method.

Table 2 shows the results for data generated from actual allelic frequencies of four cattle breeds. Success rates of GA for this data were comparable with other methods. In the case of assignment of individuals between Hariana and Sahiwal cattle breeds, GA provided accuracy value equal to 98.75%. Accuracy values with frequency, Bayesian, Nei's standard distance and chord distance methods were 100%, 100%, 99.6% and 99.7 % respectively.

Table 1. Accuracy (%) of assignment as average (s.d.) reached by genetic algorithm and other methods on simulated data

F_{ST}	No. of Loci	Allelic diversity	Genetic Algorithm	Frequency method	Bayesian	Nei's Standard	Chord distance
Low (0.01)	10	4	53.7 (2.6)	58.7 (2.9)	59.1 (2.6)	59.7 (0.8)	56.7 (2.6)
		8	56.1 (1.5)	67.0 (3.7)	66.7 (4.8)	65.8 (3.5)	65.8 (3.5)
	20	4	51.6 (2.0)	62.2 (3.4)	62.3 (3.4)	61.0 (2.8)	60.6 (2.4)
		8	51.1 (2.1)	66.1 (2.6)	66.9 (2.8)	64.5 (1.4)	65.1 (1.7)
High (0.1)	10	4	63.1 (5.0)	77.2 (1.7)	77.4 (2.0)	77.3 (0.7)	71.6 (3.6)
		8	71.8 (3.8)	87.6 (5.8)	88.5 (5.3)	86.9 (3.5)	76.0 (4.5)
	20	4	61.0 (2.8)	77.6 (1.8)	76.9 (2.5)	78.1 (1.4)	67.9 (3.3)
		8	88.2 (1.3)	93.5 (2.8)	94.2 (2.2)	91.9 (2.4)	92.1 (2.7)

Table 2. Accuracy (%) of assignment reached by genetic algorithm and other methods on the data generated from actual genotypes

Breeds	F_{ST}	Genetic Algorithm	Likelihood method	Bayesian	Nei's Standard	Chord distance
Red Kandhari, Deoni	0.11	99.90	100	100	100	100
Red Kandhari, Hariana	0.13	100	100	100	100	100
Red Kandhari, Sahiwal	0.19	100	100	100	100	99.95
Deoni, Hariana	0.08	98.65	100	100	99.90	99.95
Deoni, Sahiwal	0.13	99.05	100	100	99.85	100
Hariana, Sahiwal	0.10	98.75	100	100	99.60	99.70

4   Conclusion

Genetic algorithms are appropriate procedures for solving individual assignment problem using micosatellite genotypes due to its flexible parameter tuning capabilities. These are not dependent on any assumptions about candidate populations like other machine learning approaches. A methodology based on genetic algorithm has been proposed for individual assignment. It has been compared with existing frequency, Bayesian, Nei's standard distance and chord distance methods. Results show lower success for genetic algorithm on simulated data. Performance of genetic algorithm is comparable to other methods in case of data generated from actual microsatellite genotypes of four cattle breeds. Further experiments using systematic tuning of genetic parameters might improve performance of the genetic algorithm.

References

1. Goldberg, D.E.: Genetic algorithms in search, optimization and machine learning. Addison-Wesley, Reading (1989)
2. Neilsen, E.E., Hansen, M.M., Bach, L.A.: Looking for a needle in a haystack: discovery of indigenous Atlantic salmon(*Salmo salar L.*) in stocked populations. *Conservation Genetics 2, 219–232 (2001)*
3. Cornuet, J.-M., Piry, S., Luikart, G., Estoup, A., Solignac, M.: New methods employing multilocus genotypes to select or exclude populations as origins of individuals. Genetics 153, 1989–2000 (1999)
4. Guinand, B., Topchy, A., Page, K.S., Burnham-Curtis, M.K., Punch, W.F., Scribner, K.T.: Comparison of likelihood and machine learning methods of individual classification. Journal of Heredity 93(4), 260–269 (2002)
5. Narayanan, A., Keedwell, E.C., Olsson, B.: Artificial intelligence techniques for bioinformatics. Applied Bioinformatics 1(4), 191–222 (2003)
6. Guinand, B., Scribner, K.T., Topchy, A., Page, K.S., Punch, W., Burnham-Curtis, M.K.: Sampling issues affecting accuracy of likelihood-based classification using genetical data. Environmental Biology of Fishes 69, 245–259 (2004)
7. Topchy, A., Scribner, K., Punch, W.: Accuracy-driven loci selection and assignment of individuals. Molecular Ecology Notes 4, 798–800 (2004)
8. Bhatia, A.K., Basu, S.K.: Implicit elitism in genetic search. In: King, I., Wang, J., Chan, L.-W., Wang, D. (eds.) ICONIP 2006. LNCS, vol. 4234, pp. 781–788. Springer, Heidelberg (2006)
9. Piry, S., Alapetite, A., Cornuet, J.-M., Paetkau, D., Baudouin, L., Estoup, A.: GENECLASS2: A software for genetic assignment and first-generation migrant detection. Journal of Heredity 95(6), 536–539 (2004)
10. Excoffier, L., Novembre, J., Schneider, S.: SIMCOAL: A general coalescent program for the simulation of molecular data in interconnected populations with arbitrary demography. Journal of Heredity 91(6), 506–509 (2000)
11. Mukesh, M., Sodhi, M., Bhatia, S., Mishra, B.P.: Genetic diversity of Indian native cattle breeds as analysed with 20 microsatellite loci. Journal of Animal Breeding and Genetics 121, 416–424 (2004)
12. Sodhi, M., Mukesh, M., Mishra, B.P., Prakash, B., Ahlawat, S.P.S., Mitkari, K.R.: Evaluation of genetic differentiation in *Bos indicus* cattle breeds from Marathwada region of India using microsatellite polymorphism. Animal Biotechnogy 16, 1–11 (2005)

Extended Q-Learning Algorithm for Path-Planning of a Mobile Robot

Indrani Goswami (Chakraborty)[1], Pradipta Kumar Das[2],
Amit Konar[1], and R. Janarthanan[3]

[1] ETCE Department, Jadavpur University, Kolkata, India
[2] Dhaneswar Rath Institute of Engineering and Management Studies
Tangi, Cuttack, Orissa
[3] Jaya Engineering College, Tamilnadu
daspradipta78@gmail.com, konaramit@yahoo.co.in,
srmjana_73@yahoo.com

Abstract. In this paper, we study the path planning for Khepera II mobile robot in a grid map environment using an extended Q-learning algorithm. The extension offers an additional benefit of avoiding unnecessary computations involved to update the Q-table. A flag variable is used to keep track of the necessary updating in the entries of the Q-table. The validation of the algorithm is studied through real time execution on Khepera-II platform. An analysis reveals that there is a significant saving in time- and space- complexity of the proposed algorithm with respect to classical Q-learning.

Keywords: Q-learning, Reinforcement learning, Motion planning.

1 Introduction

Motion planning is one of the important tasks in intelligent control of a mobile robot. The problem of motion planning is often decomposed into path planning and trajectory planning. In path planning, we need to generate a collision-free path in an environment with obstacles and optimize it with respect to some criterion [8], [9]. However, the environment may be imprecise, vast, dynamical and partially non-structured [7]. In such environment, path planning depends on the sensory information of the environment, which might be associated with imprecision and uncertainty. Thus, to have a suitable path planning scheme in a cluttered environment, the controller of such kind of robots must have to be adaptive in nature. Trajectory planning, on the other hand, considers time and velocity of the robots, while planning its motion to a goal. It is important to note that path planning may ignore the information about time/velocity, and mainly considers path length as the optimality criterion. Several approaches have been proposed to address the problem of motion planning of a mobile robot. If the environment is a known static terrain and it generates a path in advance, it is said to be off-line algorithm. It is called on-line, if it is capable of producing a new path in response to environmental changes.

K. Deb et al. (Eds.): SEAL 2010, LNCS 6457, pp. 379–383, 2010.

A good path-planning algorithm requires *a priori* knowledge of the robot's world map. Usually a learning algorithm is employed to make the robot aware about its environment. Reinforcement learning [1-5] helps a robotic agent to adapt it experience throughout its life. Q-learning falls under the reinforcement class. This research is aimed at improving the performance of the Q-learning algorithm. It examines the scope of the improved algorithm in path planning application of mobile robots.

The rest of the paper is organized as follows. Classical Q-Learning is introduced in section 2. The algorithm for conditional Q-learning is presented in detail in section 3. Experimental instances are given in section 4, and conclusions are listed in section 5.

2   The Classical Q-Learning Algorithm

In classical Q-learning, all possible states of an agent and its possible action in a given state are deterministically known. In other words, for a given agent A, let $S_0, S_1, S_2,...,$ S_n, be n- possible states, where each state has m possible actions $a_0, a_1, a_2,...,a_m$. At a particular state-action pair, the specific rewards that the agent may acquire is known as immediate reward. For example r (S_i, a_j) denotes the immediate reward that the agent A acquires by executing an action a_j at state S_i. An agent selects its next state from its current states by using a policy. This policy attempts to maximize the cumulative reward that the agent could have in subsequent transition of states from its next state. For example, let the agent be in state Si and is expecting to select the next best state. Then the Q-value at state S_i due to action of a_j is given in equation 1.

$$Q(S_i,a_j)=r(S_i,a_j)+\gamma \underset{a'}{Max}\ Q(\delta(S_i,a_j),a') \tag{1}$$

where $\delta(S_i, a_j)$ denotes the next state due to selection of action a_j at state S_i. Let the next state selected be S_k. Then $Q(\delta(S_i, a_j), a')= Q(S_k, a')$. Consequently selection of a' that maximizing $Q(S_i, a_j)$ is an interesting problem. One main drawback for the above Q-learning is to know the Q-value at a state S_k for all possible action a'. As a result, each time it accesses the memory to get Q-value for all possible actions at a particular state to determine the most appropriate next state. So it consumes more time to select the next state. Since the action a' for which $Q(S_k, a')$ is maximum needs to be evaluated, we can remodel the Q-learning equation by identifying the a' that moves the agent closer to the goal.

In the extended Q-learning to be presented, we have created only one field for action of each state. In this way, we save the space required for the Q-table. Thus the Q-table storing the Q-values for the best action in each state, requires small time for retrieval of Q-values, and hence saves significant time complexity.

3   The Extended Q-Learning

Let for any state S_k, the distance between the goal state and the next feasible state of S_k are known. Let the next feasible state of S_k be {S_a, S_b, S_c, S_d}. Let G be the goal and the distance between S_a, S_b, S_c, S_d and G be $d_{aG}, d_{bG}, d_{cG}, d_{dG}$ respectively. Let the distance in order be $d_{bG} < d_{aG} < d_{cG} < d_{dG}$. Then the agent should select the next state S_b from its current state S_k, if the Q-value of the state S_b is known. We can evaluate the Q-value of state S_k by the following approach.

$$Q(S_k, a') = r(S_k, a') + \gamma \underset{a''}{Max} Q(\delta(S_k, a'), a'')$$

$$= 0 + \gamma \underset{a''}{Max} Q(\delta(S_k, a'), a'') \qquad (2)$$

Now $\delta(S_k, a') = S_a | S_b | S_c | S_d$, where $|$ denotes OR operator. Therefore,

$$\underset{a''}{Max} Q(\delta(S_k, a'), a'')$$

$$= \underset{a''}{Max} Q\{S_a | S_b | S_c | S_d, a''\}$$

$$= Q(S_b, a''). \quad (\because d_{bG} < d_{aG} < d_{cG} < d_{dG}) \qquad (3)$$

Combining (2) and (3) we have:

$$Q(S_k, a') = 0 + \gamma Q(S_b, a'').$$

Thus if the next state having the shortage distance with the goal is known, and the Q-value of this state is also known, then the Q-value of the current state is simply $\gamma \times$Q-value of this next state.

Let S_p, S_n and S_G be the present, next and the goal states respectively. Let Q_p and Q_n be the Q-value at the present and the next states S_p and S_n respectively. Let d_{xy} be the Euclidean distance between the states S_x and S_y. We use a Boolean variable Lock: L_x to indicate that the Q_x value of a state is fixed permanently. We set lock $L_n=1$ if the Q-value of the state n is fixed, and won't change further after L_n is set to 1. The Lock variable for all states except the goal will be initialized zero in our proposed Q-learning algorithm.

We now study some interesting property of Q-learning algorithm, which would provide us an insight to extend classical Q-learning algorithm for both speed up and minimum space complexity. The proof of the properties is not given here for space limitation.

Property1 : If $L_n = 1$ and $d_{pG} > d_{nG}$ then $Q_p = \gamma \times Q_n$ and set $L_p = 1$.

Property2 : If $L_n = 0$ and $d_{pG} > d_{nG}$ then $Q_p = Max(Q_p, \gamma \times Q_n)$.

Property3 : If $L_p = 1$ and $d_{nG} > d_{pG}$ then $Q_n = \gamma \times Q_p$ and set $L_n = 1$.

Property 4 : If $L_p = 0$ and $d_{nG} > d_{pG}$ then $Q_n = Max(Q_n, \gamma \times Q_p)$.

It is apparent from the property 1 that if L_n is 1, we on updating Q_p, set L_p to 1. Similarly, in property 3, we set L_n to 1 when $L_p =1$ is detected. When $L_n= 0$ in property 2 and $L_p =0$ in property 4, we just update Q_p and Q_n respectively. When L_i for all state i=1, we need not update Q-table further. These properties have been employed in the extended Q-learning algorithm.

Time-Complexity: In classical Q-learning, the updating of Q-values in a given state requires determining the largest Q-value, in that cell for all possible actions. Thus if there are m possible actions at a given state, maximization of m possible Q-values, require m −1 comparison. Consequently, if we have n states, the updating of Q values of the entire Q table by classical method requires n(m −1) comparisons. Unlike the classical case, here we do not require any such comparison to evaluate the Q values at a state S_p from the next state S_n. But we need to know whether state n is locked i.e., Q-value of S_n is permanent and stable. Thus if we have n number of states, we require n number of comparison. Consequently, we save $n(m-1) - n = nm - 2n = n(m-2)$.

Proposed Algorithm for Extended Q-Learning

1. Initialization

 For all $S_i, i = 1$ to n except $S_i = S_G$

 {set $L_i = 0; Q_i = 0$; action $= \varphi$};

 L_G (for goal S_G) $= 1; Q_G$ (for goal) $= 100$;

2. Assign γ in $(0,1)$ and initial state $= S_p$;

3. Update Q – table :

 Repeat

 {   a)Select a_i from $A = \{a_1, a_2, ..., a_m\}$;

 b)Determine d_{nG} and d_{pG};

 If $(d_{nG} < d_{pG})$

 Then if $L_n = 1$

 Then $\{Q_p = \gamma \times Q_n; L_p = 1$; save action of $S_p = "a_i"\}$

 Else if $(Q_p > \gamma \times Q_n)$ then {save $Q_p = \gamma \times Q_n$ and action of $S_p = "a_i"\}$;

 Else if $L_p = 1$ then $\{Q_n = \gamma \times Q_p$; save action of S_n = opposite action of "a_i"; $L_n = 1\}$;

 Else if $(Q_n > \gamma \times Q_p)$ then $\{Q_n = \gamma \times Q_p$ and save action of S_n = opposite action of"a_i"$\}$

 }

 Until $L_i = 1$ for all i;

Space-Complexity: In classical Q-learning, if we have n states and m action per state, then the Q-table has a dimension $(m \times n)$. In the conditional Q-learning, we only store Q-value at a state for the best action. Further, we need to store whether the Q-value is already stable or changing. Naturally, we need to store the status of lock variable for each state. Consequently, for each state we require 3 storages, such as Q-value, lock and the best action at that state. Thus for n number of states, we require $(3 \times n)$ number of storage. The saving in memory in the present context with respect to classical Q thus is given by $mn - 3n = n(m - 3)$.

4 Experiments

Experiment on a large maze of 200×200 show that the run time for planning is less and the optimal path obtained in classical Q-learning is not lost by the present scheme. Snapshots of the planning steps realized on Khepera II are given in Fig. 1 below. The extended Q-learning algorithm is used in the first phase to learn the movement steps from each grid in the map to its neighbor. After the learning phase is over, the planning algorithm is executed with the snapshots as indicated in the Fig.1 to reach the goal.

Fig. 1. Snapshots of experimental planning instances

5   Conclusions

The paper presented an extension of classical Q-learning algorithm, so as to reduce both space and time-complexity by economically updating the Q-table, only when such updating is mandatory. The economic selection of specific entries of the Q-table is performed by certain properties of the extended Q-learning algorithm. The claim of this paper is four fundamental properties, and utilization of these properties to set conditional checking on updating of Q-table entries. This is realized in a pseudo code given in the paper. The reduction in both space-and time-complexity is indicated. Verification of the algorithm has been performed on Khepera II platform. The explanation of the algorithm cannot be included for space limitation.

References

1. Dean, T., Basye, K., Shewchuk, J.: Reinforcement learning for planning and Control. In: Minton, S. (ed.) Machine Learning Methods for Planning and Scheduling. Morgan Kaufmann, San Francisco (1993)
2. Bellman, R.E.: Dynamic programming, p. 957. Princeton University Press, Princeton (1957)
3. Watkins, C., Dayan, P.: Q-learning. Machine Learning 8, 279–292 (1992)
4. Konar, A.: Computational Intelligence: Principles, Techniques and Applications. Springer, Heidelberg (2005)
5. Busoniu, L., Babushka, R., Schutter, B.D., Ernst, D.: Reinforcement Learning and Dynamic Programming Using Function Approximators. CRC Press, Taylor & Francis Group, Boca Raton, FL (2010)
6. Chakraborty, J., Konar, A., Jain, L.C., Chakraborty, U.: Cooperative Multi-Robot Path Planning Using Differential Evolution. Journal of Intelligent & Fuzzy Systems 20, 13–27 (2009)
7. Gerke, M., Hoyer, H.: Planning of Optimal paths for autonomous agents moving in in homogeneous environments. In: Proceedings of the 8th Int. Conf. on Advanced Robotics, pp. 347–352 (July 1997)
8. Xiao, J., Michalewicz, Z., Zhang, L., Trojanowski, K.: Adaptive Evolutionary Planner/Navigator for Mobile robots. IEEE Transactions on Evolutionary Computation 1(1) (April 1997)
9. Bien, Z., Lee, J.: A Minimum–Time trajectory planning Method for Two Robots. IEEE Trans. on Robotics and Automation 8(3), 443–450 (1992)

An XML Format for Sharing Evolutionary Algorithm Output and Analysis

Dharani Punithan[1], Jerome Marhic[2], Kangil Kim[1], Jakramate Bootkrajang[3],
RI (Bob) McKay[1], and Naoki Mori[4]

[1] Structural Compexity Laboratory, Seoul National University, Korea
[2] Computer Science Dept, INSA de Rennes, France
[3] School of Computer Science, University of Birmingham, UK
[4] Computer and Systems Science Dept, Osaka Prefecture University, Japan

Abstract. Analysis of artificial evolutionary systems uses post-processing to extract information from runs. Many effective methods have been developed, but format incompatibilities limit their adoption. We propose a solution combining XML and compression, which imposes modest overhead. We describe the steps to integrate our schema in existing systems and tools, demonstrating a realistic application. We measure the overhead relative to current methods, and discuss the extension of this approach into a community-wide standard representation.

1 Introduction and Background

Evolutionary algorithms are used for a vast range of problems in sciences and engineering. Each system has its own data format and analysis tools. As the field progresses, the need for fair comparisons and detailed analysis grows. We need a standard format to enable the development of easily configurable post-processing tools for smooth data exchange and integration. We propose a representation based on XML [1]. The major concern with XML as a data representation is the size of XML files – far larger than raw data. Here, we provide some technical background on XML compression and XML interfacing, our proposed XML schema for output data, an application example, and tests on comparative overheads. Full detail is available in our technical report [2].

Genetic Programming and XML: A number of authors have used XML in connection with GP. Martens [3] and Garcia-Sanchez et al [4] use GP to generate XML transformation scripting. Johnson and Kumara [5] use XML as an evolutionary representation within the evolutionary system, as do Amarteifio [6] and Tanev and Shimohara [7]. Tanev [8] uses it to describe communication between evolutionary agents. All have very different aims from this proposal.

XML Compression and Interfacing to XML: Many XML compressors have been proposed in the literature in recent years. Among them, XMill [9] compressor achieves the best compression ratio for so-called 'structural' documents of the kind that arise in GP output. Though XMill is based on gzip, it achieves about twice the compression ratio at roughly the same speed. XMill is not only XML-conscious, but also an archival

K. Deb et al. (Eds.): SEAL 2010, LNCS 6457, pp. 384–388, 2010.

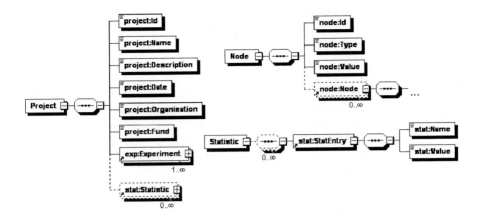

Fig. 1. XML Schema

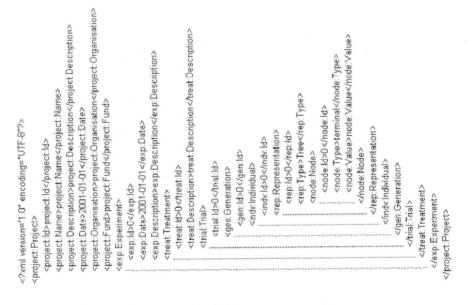

Fig. 2. Sample XML output representation

compressor. Since a good archival compressor is exactly what we need, we chose XMill to compress our XML output. For interfacing, we chose Java Simple Exchange Format API (JSefa, [10]) providing high-level streaming over an iterator interface. Our framework for interfacing XML, built on top of JSefa, is available from http://sc.snu.ac.kr (click on tab Research and then Software).

Equivalent Decision Simplification: We tested the flexibility of this format with a typical complex problem: simplification of GP expression trees to remove bloat. This requires both input and output of the standard format. We used Equivalent Decision Simplification (EDS) [11], a semantic simplifier. We also tested it on a simpler post-processing problem, computing mean best fitness.

Our XML Schema is designed to support many hierarchical sub-levels, each with a statistics element. A sample XML schema for Project, Node and Statistics levels is shown in Figure 1. Due to space constraints, most sub-levels are omitted. For details, please refer to [2]. A small example of the schema is shown in Figure 2. While the schema is currently only instantiated for standard GP, it is readily extensible to most grammar-based or linear GP representations, and to other evolutionary algorithms.

2   Workflow Tests

We took data from a typical GP experiment on a simple problem, symbolic regression to find approximations to the polynomial $x^4 + x^3 + x^2 + x$ [12]. For these trials, we performed a typical GP workflow: running 30 trials, and saving the output data, including full representations of all trees generated, followed by two analytical processes: computing mean best fitnesses and equivalent decision simplification, requiring both reading the original data, and writing the simplified output. We compared four treatments: XML output vs our original (and typical) prefix representation; using compressed vs uncompressed output. All work was performed in pipes, so that for the compressed versions, uncompressed files never needed to be stored. For each stage of processing, we recorded the total run time of the stage, the memory used and the storage required. We then aggregated over the 30 trials, and computed means and standard deviations. We used bzip2 compression for the prefix notation, XMILL for XML. Experiments were conducted on a cluster of identical 2GHz computers. We used our C++ implementation of GP. It is reasonably efficient, with no known memory leaks. The GP parameter settings are shown in Table 1.

Table 1. Evolutionary Parameter Settings

Fitness Cases	20 points in $[-1 \dots 1]$
Fitness	Sum of Errors
Genetic Operators	Tournament selection(5)
	Subtree crossover
	Subtree mutation
♯ of Runs	30
♯ of Generations	1001
♯ of Individuals	250
Crossover prob.	0.9
Mutation prob.	0.1
Maximum Depth	15

Table 2. GP Processing Stage

Com-pressed	XML	Time (secs)	Memory (miB)	File Size (MB)
Yes	Yes	123 ± 20	48 ± 2	6.2 ± 1.0
Yes	No	438 ± 188	55 ± 8	1.4 ± 0.3
No	Yes	51 ± 10	33 ± 2	240 ± 44
No	No	324 ± 130	49 ± 9	149 ± 59

Results of these treatments are shown in Tables 2, 3 and 4. Though CPU processing times are less for uncompressed XML, it is a poor strategy resulting in enormous files compared with other treatments. Compressed XML is more storage-efficient than uncompressed prefix-tree storage. Even with the overheads of compression, XML based processes used much less resources than the classic 'uncompressed prefix tree' approach, with the exception of the small 'mean best' computation stage (refer Table 4), where memory requirements and processing time were both substantially higher for XML based processes. When the two 'compressed' strategies are compared, the prefix tree compression (i.e. bzip2) achieved substantially greater compression than XMILL, but at the cost of much greater computation time, especially for compression. Thus even when issues of standardisation are ignored, compressed prefix storage is worse than compressed XML based storage. In using these methods, all processing must be done in pipes (to avoid creation of immense intermediate file structures); and close attention must be paid to compiler and operating system I/O limits (often in the 1-4GB range).

Table 3. Simplification Stage

Com-pressed	XML	Time (secs)	Memory (miB)	File Size (MB)
Yes	Yes	188 ± 47	2408 ± 26	2.4 ± 0.5
Yes	No	438 ± 532	384 ± 30	0.4 ± 0.1
No	Yes	98 ± 33	2397 ± 2	702 ± 172
No	No	498 ± 302	2372 ± 22	40 ± 23

Table 4. Mean Best Computation Stage

Com-pressed	XML	Time (secs)	Memory (miB)	File Size (MB)
Yes	Yes	47 ± 27	2412 ± 21	NA
Yes	No	14 ± 5	40 ± 2	NA
No	Yes	40 ± 20	2410 ± 19	NA
No	No	10 ± 4	36 ± 3	NA

3 Discussion and Conclusions

The testing demonstrates that XML-based data representation and storage for GP output data, combined with XMILL compression technology, is an efficient and viable strategy. The potential advantages, in sharing data and analyses, are immense. The main cost of transition to an XML representation is modification of current systems to output data in this XML format, and rewriting existing analysis tools to support it. Essentially, it requires new output drivers for GP systems, and new input drivers for processing systems. We are currently developing generic APIs that will allow these to slot in seamlessly.

We have argued the need for a standard XML output representation for GP, and presented a suitable XML schema, supporting the gradual development by the research community of a reservoir of post-processing tools based on it. The natural concern with this proposal, especially in GP, lies in the resource issues, especially file storage and processing costs. We have tested these in typical experimental settings, and confirmed that modern XML processing tools provide a remedy. We believe that there will be negligible resource costs in typical scenarios; those costs almost certainly do not outweigh the potential advantages of a standard output format.

Acknowledgment

We thank Prof. Xuan Hoai Nguyen for early access to his TAG3P system for our experiments. Seoul National University institute for Computer Science and Technology provided facilities for this research. This research was supported by Basic Science Research Program through the National Research Foundation of Korea(NRF) funded by the Ministry of Education, Science and Technology(313-2008-2-D00943).

References

1. XML: Extensible markup language(XML), http://www.w3.org/XML/
2. Punithan, D., Marhic, J., Kangil, K., Bootkrajang, J., McKay, R.B., Mori, N.: An xml format for sharing evolutionary output and resources. Technical Report TRSNUSC:2010:002, SNU Structural Complexity Laboratory, Seoul National University, Seoul, Korea (September 2010)
3. Martens, S.: Automatic creation of XML document conversion scripts by genetic programming. In: Koza, J.R. (ed.) Genetic Algorithms and Genetic Programming at Stanford 2000. Stanford Bookstore, Stanford, California, 94305-3079 USA, pp. 269–278 (June 2000)
4. García-Sánchez, P., Guervós, J.J.M., Laredo, J.L.J., Mora, A., Castillo, P.A.: Evolving xslt stylesheets for document transformation. In: Rudolph, G., Jansen, T., Lucas, S., Poloni, C., Beume, N. (eds.) PPSN 2008. LNCS, vol. 5199, pp. 1021–1030. Springer, Heidelberg (2008)
5. Johnson, J., Kumara, S.: Coadaptation of cooperative players in an iterated prisoners dilemma game using an XML based GA. In: Whitley, D. (ed.) Late Breaking Papers at the 2000 Genetic and Evolutionary Computation Conference, Las Vegas, Nevada, USA, Las Vegas, Nevada, USA, pp. 147–154 (July 8, 2000)
6. Amarteifio, S.: Interpreting a genotype-phenotype map with rich representations in XMLGE. Master of science in computer science, University of Limerick, University of Limerick, Ireland (2005)
7. Tanev, I., Shimohara, K.: XGP: XML-based genetic programming framework. In: Proceedings of the 34th Symposium of the Society of Instrument and Control Engineers (SICE) on Intelligent Systems, Japan, SICE, pp. 183–188 (2007)
8. Tanev, I.: DOM/XML-based portable genetic representation of morphology, behavior and communication abilities of evolvable agents. Artificial Life and Robotics 8(1), 52–56 (2004)
9. Liefke, H., Suciu, D.: XMILL: An efficient compressor for XML data. In: ACM SIGMOD Conf. 2000, pp. 153–164 (May 2000)
10. JSefa: Java Simple Exchange Format API, http://jsefa.sourceforge.net/
11. Mori, N., McKay, B., Hoai, N.X., Essam, D., Takeuchi, S.: A new method for simplifying algebraic expressions in genetic programming called equivalent decision simplification. Journal of Advanced Computational Intelligence and Intelligent Informatics 13(3), 237–244 (2009)
12. Koza, J.R.: Genetic programming: On the programming of computers by means of natural selection. Statistics and Computing 4(2), 87–112 (1994)

Car Setup Optimisation

Moisés Martínez, Gustavo Recio, Pablo García,
Emilio Martín, and Yago Saez

Department of Computer Science, University Carlos III de Madrid,
Avenida de la universidad 30, 28911 Leganés, Spain
{moises.martinez,emilio.martin,gustavo.recio,yago.saez}@uc3m.es,
100065306@alumnos.uc3m.es
http://www.uc3m.es

Abstract. Computational intelligence competitions have recently gained a lot of interest. These contests motivate and encourage researchers to participate on them. Computer games are interesting test beds for research in artificial intelligence that motivate researchers to apply their work areas to specific games. In this paper a structural parameter set of a car agent is optimised using particle swarm optimisation and evolution strategies. The change was for were to the TORCS competition held during the Car Setup Optimization Competition EvoStar 2010.

Keywords: Car Racing, Evolution, Optimisation, Evolutionary Strategies, Particle Swarm Optimisation, Torcs, Simulation.

1 Introduction

Designing a racing car is quite complex and expensive. The fact that the optimisation of the final design in most cases depends on environmental variables, such as the weather, makes it a difficult task. In addition, the structural configuration of a car does not behave in the same way over different circuits, this implies that parts of its structure should allow modifications to exploit the caracteristic of the circuit. The main problem is that this type of modifications are normally very expensive and require a long time. Most racing teams do not have the financial resources or the time to properly perform this type of configuration.

This paper focuses on the use of Artificial Intelligence techniques to optimise the structural setup of a car agent for The Open Racing Car Simulator (TORCS) [1], a car racing simulator that provides a physics engine, a 3D visualization and different models of car and tracks. The optimisation of the parameter set that defines the structural configuration of a car (Rear Wing, Brake System, Rear Anti-Roll Bar, Front Left-Right Wheel, etc) was done by applying two evolutionary techniques: a particle swarm Optimisation algorithm and an evolutionary strategy $\{\nu + \lambda, \alpha\}$-ES. The performance of these controllers are compared and the results obtained are presented here and analysed individually.

K. Deb et al. (Eds.): SEAL 2010, LNCS 6457, pp. 389–393, 2010.
© Springer-Verlag Berlin Heidelberg 2010

2   The Simulation

For developing the different algorithms, the organizers of the Car Setup Optimization Competition EvoStar 2010 (Luigi Cardamone, Daniele Loiacono, Markus Kemmerling and Mike Preuss) have provided a server module for TORCS that provides the communication to the remote controller called opt-Server through an implemented API allowing to send the parameters that defined the car setup and replies with the result of the simulation which is a vector of four parameters (lap time, max speed, distance, damage).

There is an important problem that needs to be taken into account. It is very important to know that the simulation results not only depend on the parameters set or simulated time, but also on the position of the car in the final instant of the previous simulation. This is a very important issue which will influence the outcome of the simulation. For example if a car is located pointing backwards the agent will have to turn around in order to continue the race with the corresponding waste of time, however, a car driving at a very high speed at the end of the previous simulation will have a positive influence over the current one.

if anyone interested then look for reference `http://cig.dei.polimi.it/?page id=103`.

3   Techniques

In this section, the tecniques that have been selected to solve the parameter optimisation of the car agent are dealt with, these are particle swarm optimisation and evolution strategies.

3.1   Particle Swarm Optimisation

The PSO implementation used in this worrk, is based on Jswarm [7], which has been adapted and changed to meet the dynamic fitness evaluation time problem mentioned above and the car optimisation competition contest requirements. This technique uses a group of 200 particles (solutions) and then searches for optima by consecutive generations. At every iteration, each particle is updated depending on the value of three parameters: global best, local best and personal best. The algorithm finishes when the maximum number of game tics allowed by the simulator (1 million game tics) is reached.

The fitness of an individual is obtained by running the simulation process once and computing the fitness through the following equation

$$f(x) = MaxSpeed - Damage \ . \tag{1}$$

which tends to select individuals that maximise the speed and minimise the damage suffered by the car.

3.2   Evolutionary Strategies

The second artificial intelligence technique used here to optimise a vector of parameters of a car agent is an evolutionary strategy, in particular a $\{\nu + \lambda, \alpha\}$-ES. with ν parents and λ offspring per generation. Again, the algorithm finishes when the maximum number of game tics allowed by the simulator is reached. The fitness of an individual is obtained by running the simulation process in random order 5 times with different number of tics, and then computing the average fitness values of each individual as

$$f(x) = (MaxSpeed + AverageSpeed) - Damage \ . \qquad (2)$$

The parameter α in the definition of the ES was introduced to account for the number of iterations that an individual remains in the population without being evaluated. It is used to avoid the effects of the noise in the evaluation and increase the number of iterations of the process.

The above equation (2) selects individuals with maximum top speed, those with more stable speed during the entire race and those with less damage. Initially a variant of this equation was used, the average damage was substracted from the sum of the average speed and the best top speed, however, it was observed that in many cases the car speed was decreased significantly and thus it was not usefull.

4   Experiments

4.1   Evaluation

To validate the two techniques presented in the paper, every controller is tested on three different tracks (E-Track 5, Dirt 4 and E-Track 1), each one of them has different characteristics to prove the efficiency of the algorithm. Each circuit presents a different structure and characteristics which allows checking the potential of the algorithm to find a good setup for the car controller.

Figure 1 shows the three different circuits selected for the experiments, the first circuit, called E-track 5, is a simple fast track, that has been selected to find the maximum speed of the optimised vehicles, the second track, called Dirt 4, is probably the most difficult track, due to the large number of bends which allows finding vehicles that deal without problems with the need to turn. The last circuit, called E-Track 1, is a difficult environment with areas where the maximum speed can be reached followed by sharp bends. All experiments shown in this paper have been executed using the version 1.3.1 of TORCS.

4.2   Results

The evaluation process is performed in three steps. First, for the PSO technique three agents are obtained, one at each circuit. Then, the process will be repeated for the ES technique. Next, each of the six technique-circuit pair is tested on

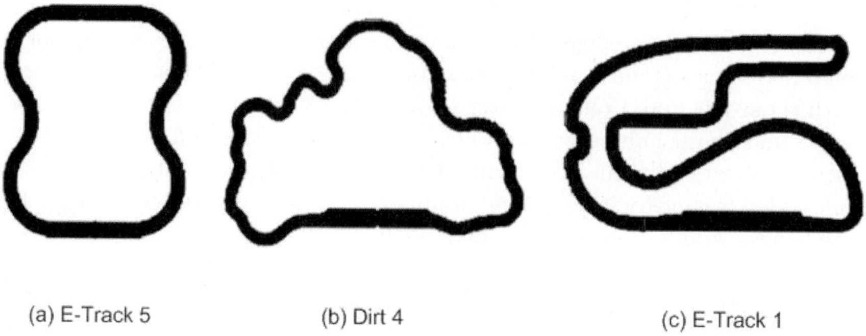

(a) E-Track 5   (b) Dirt 4   (c) E-Track 1

Fig. 1. Structure of the different circuits used to test the setup vector generated for each technique

each circuit for 50 runs of 10,000 game tics (about 3 minutes and 20 seconds) and the results are averaged. Finally, the different solution are tested on their corresponding circuit running together with other cars to observe the influence of other vehicles in the solutions.

The following tables show the results obtained after making the tests, in most cases the PSO gets better results but the difference between the two techniques is quite small.

Table 1. Results for individual tests

Track	ES		PSO	
	Average	Standard Deviation	Average	Standard Deviation
E-Track 5	9726.66	625.06	**9903.63**	494.33
E-Track 1	8071.80	468.17	**8116.94**	395.04
Dirt 4	**7984.93**	402.88	7924.65	411.38

Table 1 shows the results obtained after executing the different car agents in the different circuits.

In table 2 the results of driving with other cars on the track are presented. In this case the PSO technique dominates over the ES.

After analysing the results show inthe tables 1 and 2, it seems to be a good idea to perform multiple assessments to obtain the most reliable fitness or use a small population, although this does not guarantee a quality vehicle. Since the best results are obtained by the swarm of particles this represents the most stable solution regardless of circuit conditions and evolves with large populations and single individual assessment.

Table 2. Results for test with traffic

Track	ES		PSO	
	Average	Standard Deviation	Average	Standard Deviation
E-Track 5	9736.13	543.65	**9918.84**	483.57
E-Track 1	7883.70	1239.06	**8153.21**	417.50
Dirt 4	7388.48	394.74	**7416.74**	450.83

5   Conclusions

In this paper, two different artificial intelligence techniques, PSO and ES, have been used to successfully optimise a setup vector for a racing car agent. This setup vector represents certain structural characteristics of a vehicle.

The results suggest that both techniques are able to generate configurations that offer similar results regardless of the circuit, but in general the PSO technique obtained better results. Although the results were quite good, it has been observed that the optimisation process incurred in some problems. These were due to the way the fitness is evaluated, fitness is based on the results obtained in several sections of the circuit which leads to optimised parameters that produce very fast vehicles, but not downshifting effectively in the curves, because the optimisation process always selects the fastest vehicle with minor damage.

Acknowledgments. This work was supported in part by the Spanish projects Mstar TIN2008-06491-C04-04 and eInkPlusPlus TSI-020110-2009-137.

References

1. The open racing car simulator website, http://torcs.sourceforge.net
2. Wloch, K., Bentley, P.J.: Optimizing the performance of a formula one car using a genetic algorithm. In: Yao, X., Burke, E.K., Lozano, J.A., Smith, J., Merelo-Guervós, J.J., Bullinaria, J.A., Rowe, J.E., Tiňo, P., Kabán, A., Schwefel, H.-P. (eds.) PPSN 2004. LNCS, vol. 3242, pp. 702–711. Springer, Heidelberg (2004)
3. Cardamone, L., Loiacono, D., Lanzi, P.J.: Evolving competitive car controllers for racing games with neuroevolution. In: GECCO 2009: Proceedings of the 11th Annual Conference on Genetic and Evolutionary Computation, pp. 1179–1186. ACM, New York (2009)
4. Deb, K., Saxena, V.: Car suspension design for comfort using genetic algorithms. In: Proceedings of the Seventh International Conference on Genetic Algorithms, pp. 553–560. Thomas Bäck (1997)
5. Kazancioglu, E., Wu, G., Ko, J., Bohac, S., Filipi, Z., Hu, S.J., Assanis, D., Saitou, K.: Robust optimisation of an Automobile Valvetrain Using a Multiobjective Genetic Algorithm. In: Proceedings of ASME 2003 Design Engineering Technical Conferences, Illinois, Chicago (2003)
6. Poli, R.: An Analysis of the publications on the applications of particle swarm optimisation. Journal of Artificial Evolution and Applications, 1–10 (2008)
7. Jswarm-PSO Framework, http://jswarm-pso.sourceforge.net

Robustness of Multi-objective Optimal Solutions to Physical Deterioration through Active Control

Gideon Avigad and Erella Eisenstadt

Braude College of Engineering
{gideon,erella,miri}@braude.ac.il

Abstract. In this paper, we suggest a novel problem within the context of multi objective optimization. It concerns the control of solutions' performances in multi objective spaces. The motivation for controlling these performances comes from an inspiration to improve the robustness of solutions to physical deterioration. When deterioration occurs, the solution performances degrade. In order to prevent extended degradation and loss of robustness, an active control is implemented. Naturally, in order to enable such a control, the solution (product) should have tunable parameters that would serve as the controlled variables. Optimizing the solution for such a problem means that the tunable parameters should be found and their manipulation determined. Here the optimal solutions and the controller are designed using multi and single objective evolutionary algorithms. The paper is concluded with a discussion on the high potential of the approach for research and real life applications.

Keywords: Evolutionary multi-objective, Physical deterioration.

1 Introduction

Multi Objective search is an important research topic. It concerns the search for solutions to many real world problems, dubbed appropriately enough Multiobjective Problems (MOPs). According to Mattson and Messac [1], successful engineering design generally requires the resolution of various conflicting design objectives. In the case of contradictory objectives, there is no universally accepted definition of an 'optimum' as in a single-objective optimization, [2]. In such cases, there is no single global solution, often rendering it useful to determine a set of solutions that fits a predetermined definition for an optimal set and lets a decision maker choose among them. The predominant concept in defining such a set is Pareto optimality. By definition, Pareto solutions, which belong to the Pareto optimal set, are considered optimal because there are no other designs that are superior in all objectives, [3].

Most Multi-Objective Evolutionary Algorithms (MOEAs) use the non-dominance relation, [4], to direct the evolutionary search towards a Pareto front. For instance, they use niching to support the spreading of solutions along the front. According to [5], the second generation of Pareto-based algorithms, such as NSGA-II, [6], involves three major elements: The first element concerns the creation of a search pressure towards the Pareto set. This is commonly achieved by one of the known Pareto-based

K. Deb et al. (Eds.): SEAL 2010, LNCS 6457, pp. 394–403, 2010.
© Springer-Verlag Berlin Heidelberg 2010

fitness assignment (dominance-based) techniques. The second element is set to avoid convergence to a single solution and preserve diversity. The third element is elitism, which helps to prevent losing non-dominated solutions that are diversified. Detailed descriptions of multiobjective evolutionary techniques can be found in [7].

Robust design tries to ensure that performance requirements are met and constraints are not violated due to system's uncertainties and variations (e.g., [1]). Fundamentally, robust design is concerned with minimizing the effect of such variations without eliminating the source of the uncertainty or variation, [8]. Taguchi, (e.g., [9]), has contributed tremendously to the development of this field of interest by introducing several approaches (e.g., Loss Function, Orthogonal Arrays and Linear Graphs). There are many possible ways to treat robustness by using Evolutionary Computation (EC), with a few possible heuristics having been suggested in [10].

Dealing with robustness within MOPs has been rarely studied. Not long ago Deb and Gupta, [11], introduced a formulation for the different aspects of robustness and suggested an approach to evolve a robust front based on the mean of an effective fitness function. Teich, [12], defined a probabilistic domination relation to represent the noise objective functions. Luo et al., [13], used an EMO approach to evolve robust fronts that are a result of taking into consideration possible market changes.

According to the current survey, no research concerning multiobjective robustness to physical deterioration has been conducted. Nevertheless, modeling of such deteriorations exists. Physical deterioration is defined as the loss in the value of assets, commodities, goods, materials, etc., due to exposure to the elements and wear and tear associated with their storage and use. Examples of models for such deteriorations include mathematical models of tools' wear, [14], deterioration models for effective bridge management, [15] and statistical models for analyzing failure.

Although the current paper is related to all of the above issues, here a totally novel, previously untreated (as far as our review goes) problem is suggested. Here control is implemented for attaining robustness for physical deterioration by actively controlling the performances of the solutions within multi objective spaces.

2   The Active Control Approach

The idea behind the methodology of the current paper might be elucidated through depicting Figure 1. The left panel of the figure depicts four Pareto optimal solutions' performances in a bi-objective space, designated by capital letters, A-D. Physical deterioration of the solutions' design parameters (due to e.g., wear) might cause the performances to degrade to new performance vectors, designated by small letters, a-d. As a result, the Pareto front alters and solution B does not belong to the new, shifted, front (it is dominated by solution C) and, according to [11], it should not be considered a robust solution. Moreover, although solutions C and D do not dominate each other (also after deterioration), solution C seem to be more robust than D as its performances are less effected (the distance C-c seems closer than D-d). As a last note, for all solutions A-D, the robustness could not suffice due to the possibility that the loss of performances is higher than the designers are ready to reconcile with.

Now suppose that there is a way to actively change some (at least one) of the solutions' parameters by actively controlling its value. If this is done properly, the

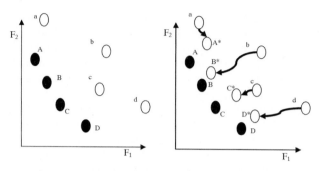

Fig. 1. Left Panel: Pareto front (bold circles) and related performances after deterioration (white circles), Right Panel: Active control shifts performances closer to the original performances

performances might be improved to new performance vectors (designated by A*-D*). Clearly the performances are more robust than before (the influence on performances due to deterioration has been reduced). Moreover, it is possible that a solution that was previously (when no control action has been taken) the worst in terms of robustness is now the best in that respect (see solution B*). In the following section, the way to achieve this improvement in robustness through active control is explained and formulated.

3 The Control Problem

In the current paper, a solution is associated with two types of decision variables. The first $x = [x_1, x_2, ..., x_n]^T \in \Omega \subset R^n$ are variables with values that, once they are chosen, may not be intentionally altered any more. These variables are subjected tochanges due to physical deterioration, such that $x = [x_1^o + f_1(x^o, y^o, t), x_2^o + f_2(x^o, y^o, t), ..., x_n^o + f_n(x^o, y^o, t)]^T$, where x_i^o is the initial value of the i-th variable (before deterioration occurs) and f_i is the function that models the physical deterioration of the i-th variable (see Section 1). The second type: $y = [y_1, y_2, ..., y_m]^T \in \Gamma \subseteq R^m$, are decision variables, whose values might be changed with time (time dependent variables) by actively tuning them. This means that y might be expressed as $y = [y_1^o + C_1(t), y_2^o + C_2(t), ..., y_m^o + C_m(t)]^T$, where y_i^o is the initial value of the i-th tunable variable (before tuning it) and C_i represents a controller acting to tune it.

A solution z, is comprised of both types of variables, $z = [x, y]^T \in \Psi, \Psi \subseteq \Omega \times \Gamma$ where Ψ is the compound design space. The mapping of a solution from the compound design space to the objective space is done by utilizing the MOP's objective functions: $F: \Psi \mapsto \Phi \subseteq R^K$ where $F(z) = [F_1(z), F_2(z), ..., F_K(z)]^T$, and K is the number of objectives of the MOP.

As pointed out before, at the initial design stage, the decision vectors x and y are set such that $x^o = [x_1^o, x_2^o, ..., x_n^o]^T$, $y^o = [y_1^o, y_2^o, ..., y_m^o]^T$. We suggest that these would be set to comprise an optimal solution. This means that $z^o = [x_1^o, x_2^o, ..., x_n^o, y_1^o, y_2^o, ..., y_m^o,]^T$ would be determined by solving the following MOP:

$$\text{Find } (z) \text{ in order to} \qquad (1)$$
$$\text{Min } (F(z))$$

The solution would therefore be a Pareto set P *and Pareto front PF*.

$$P^* := \{z^o \subseteq \Psi | \neg \exists \, z'^o \in \Psi : F(z'^o) \preceq F(z^o)\} \qquad (2)$$
$$PF^* := \{s^* \in \Phi | s^* = F(z^o) : z^o \in P^*\}$$

Once these are found, the motivation of keeping robust solutions (undesired changes would cause a small as possible change in the initially designed performances) is treated. With that motivation in mind, we suggest that once the Pareto set and front of the problem are found, the control problem, may be defined over a period of time t, as:

$$\text{Find } (z^o, C) \text{ where C is } C = [C_1, C_2, ..., C_n]^T \qquad (3)$$

In order to: $\text{Min } (e)$ where $e = \int_0^t \|F(z^o) - F(z)\|$

The control schema is shown in Figure 2. It is depicted that the set-point for the control is an optimal vector, which is found by solving Equation 1. Note that there are several control loops, each designed for one solution of the Pareto set. The schema is in fact a regulation control for keeping the error at zero.

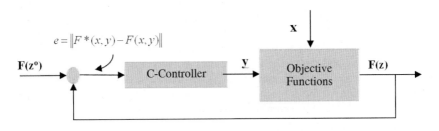

Fig. 2. The active control schema

The initial performances belong to the Pareto front. As a result of physical deterioration (change in x), the performances alter. The error signal is a normalized distance in objective space between the set-point and the new performances. The controller (that is to be designed) should change y in order to minimize the error caused by the disturbance, x. Such a minimization would increase the robustness of a solution as it diminishes its performances' changes due to the physical deteriorations.

The controller design has to be considered as an a-posteriori step to finding the optimal set. Commonly, when designing a controller, the relations between the inputs (x and y) and the outputs (the multi objective performances) should be modeled and analyzed. When considering an MOP, the model is inherently given by the objective functions. Yet, in order to design a successful controller, the influence of changes in x and separately those of y on the performances should be found. This influence might be modeled through transfer functions, state space representations artificial neural nets, fuzzy relations, etc. In the context of the current methodology, we try not to confine the solutions to any of these modeling techniques as this might harm the generic nature of the suggested problem. This also implies that we try to avoid any specific controller design approach. Yet, in order to demonstrate the applicability of the problem, the next section contains several examples in which we design controllers by utilizing an evolutionary search.

4 Examples

4.1 An Artificial Example

In the current sub-section, an artificial academic example is considered. It involves a robust optimization within a bi-objective space. The problem is to design optimal robust solutions to the problem:

Find x^o, y^o, C in order to: $Min\left(\dfrac{x}{y}, xy\right)$,   Subject to : $2 \le x \le 10, 1 \le y \le 5$ (here,

$n = m = 1$) and $f(x) = x^o + 1.2t$ (t is time).

As the first step of solving the problem, NSGA-II, [6] is implemented in order to find the Pareto set and the Pareto front (see Equations 1 and 2). The latter is depicted in the left panel of Figure 3. A population of nine individuals (such a small number has been chosen for the sake of visibility) are run for 500 generations using a real valued coding with uniform cross over (50%) and mutation of 5%. The right panel of Figure 3 depicts the Pareto front (stars) and the degraded performances (circles) after deterioration period of t=1. It is depicted that some of the Pareto solutions are more affected by the degradation (e.g. the boundary solutions). This might be interpreted such that a boundary solution is not robust as say the center one to such deterioration.

In order to improve the robustness, active control is implemented by tuning y. In the current paper, we chose a straightforward approach to tune this variable. We chose a proportional controller such that, $y = y^o + k_c \cdot e$, where k_c is the proportional controller gain and e is defined in Equation 3.

The tuning of the controller's gain is sequentially implemented for each Pareto set's solution. This is done by running a simple genetic algorithm to find the best kc within an interval: $-5 \le k_c \le 5$ that minimizes e for each solution. The different gains (one for each solution) starting from the upper left boundary of the Pareto front (which is depicted in figure 3), are: -3.23, -2.74, -2.24, -1.55, -0.56, 0.46, 0.93, 0.92, 0.71.

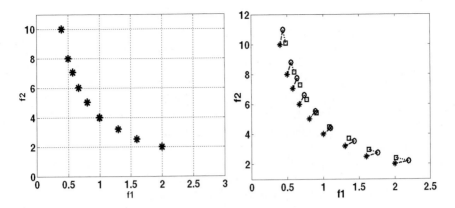

Fig. 3. Left panel: The Pareto front. Right panel: the deterioration affect on the performances (circles) and the control resulting performances (squares).

Implementing the control mends the performances to performances with smaller deviations from the original Pareto front. The performances that are the result of the control are depicted in the right panel of Figure 3, designated by squares. The improvement in performances is clear (reduced error). This improvement may be profoundly depicted for the boundary solutions. Extending the use of the same gains for further time steps is tested next. Figure 4 depicts one Pareto solution performances (designated by a star), the shifted deteriorated performances for t = [0, 4] when no control action is taken as diamonds, the deteriorated performances when control is implemented by circles, and the results of the control action by squares. It is unambiguous that with all time samples, the suggested approach enhances robustness. This may be depicted by the concentration of circles and squares in the vicinity of the star in contrast to the remoteness of the diamonds from it.

It is noted that, although the same gain computed for the first step has been used, the robustness has been enhanced for four time steps. Yet, it is expected that utilizing such a controller, which is a linear controller, within non-linear functions would perform well in the vicinity of the initial conditions (around the Pareto front) and would not perform as well if the work conditions change substantially. This inherent problematic issue might be depicted if the same controller is used for six time steps. The left panel of Figure 5 depicts the instability occurring at the 5-th time step. Until the 4-th time step the robustness is the one of Figure 4; yet starting from the 5-th time, step, the controller is unable to halt back the degradation (see the circle at [1.0, 8.1] and square at point [2.2, 3.8]).

These results might be further supported by depicting the right panel of the figure. Until the 5-th time step, higher robustness is enhanced by the control approach, which is designated in Green when compared to an uncontrolled deterioration, which is designated in blue. Yet, starting from the 5-th step, this advantage is disrupted by an unsuccessful control action.

In order to improve the performances of the control, adaptive control might be considered. As highlighted in the methodology, here the purpose is to demonstrate the

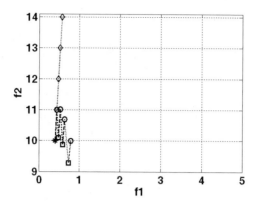

Fig. 4. Samples at t=[0,4], with the control (circles and squares) and without the control (diamonds)

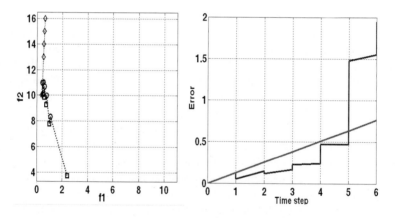

Fig. 5. Incapability of the control to halt back the increase in error (from the 5-th step)

approach rather than develop or improve controllers. Therefore, we took a simple and straightforward tuning approach. At each time step, a genetic search for the best kc has been conducted. This resembles the search for the initial kc, repeated for each solution, several times (each time for another time sample). The results are depicted in Figure 6. Comparing Figures 5 and 6, shows a clear improvement in robustness.

When all controllers are designed (one for each Pareto solution), the decision maker, may be presented with more knowledge concerning the alternative optimal solutions. Now, each solution may be accompanied with a measure of robustness (the overall error, which is computed from 3). The robustness of solutions, as defined in Equation 3, depends on the time, t. Therefore, this may result in preference of different solutions when robustness over different time durations is tested.

For the current problem, adaptive control over five time steps and over ten time steps has been tested for all nine Pareto solutions. This ended up with two different most

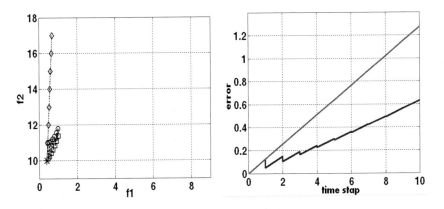

Fig. 6. An adaptive tuning of gains, showing major improvement

robust solutions: those with performances at [0.6, 6.0] and [1.0, 4.0] (see Figure 3), respectively.

4.2 An Engineering Example

The engineering problem is depicted Figure 7.

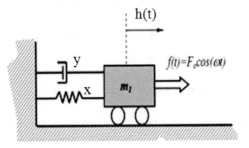

Fig. 7. A schema of the engineering problem

The dynamic model for the system is: $m\ddot{h}(t) + y\dot{h}(t) + xh(t) = F_o \cos(\omega t)$, where m=20 kg, is the mass, y is the viscous damping coefficient, x is the spring constant, F_o=1000N is the excitation amplitude and ω its frequency. It may be shown that in the steady state response the amplitude of oscillations would be:

$$h = \frac{F_o}{\sqrt{(x - m\omega^2)^2 + (y\omega)^2}},$$

It is assumed that the spring's deterioration might be modeled such that $x(t) = x^o - 0.1x^o \cdot t$, which might be a result of hardening of the spring due to extended

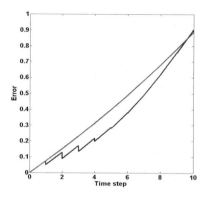

Fig. 8. Robustness of one Pareto solution with active control (blue) and without it (green)

use. The objectives of the MOP are to minimize the response overshoot and steady-state- error, that is:

$$Overshoot = 100\% \times \exp\left(-\frac{\pi y}{2\sqrt{mx}}\middle/\sqrt{1-\frac{y^2}{4mx}}\right), \quad \begin{array}{l} 1000 \le x \le 10000[N/M] \\ 20 \le y \le 100[kg/sec] \end{array},$$

The deterioration might be compensated by altering the outlet orifice (change of y). The improvement in robustness may be highlighted by comparing the suggested approach' results with the no control approach by comparing the blue and green curves of Figure 8, respectively.

5 Summary and Conclusions

In the current paper, we suggest a novel research direction. It deals with the idea of controlling the performances of solutions within multi objective spaces in order to achieve robustness to physical deterioration. The current approach assumes that some of the design parameters are subjected to such deterioration, while others may be tuned in order to improve the degraded performances. The problem is posed as a regulatory feedback control schema. The main purpose of the current paper is to highlight the possibility of such a control and to demonstrate its potential in improving the robustness of solutions. The approach is to initially evolve the Pareto set and then to tune the controlled parameters in order to maintain minimal changes to these initial performances. It has been shown that the approach may improve the robustness of solutions and prolong the serviceability of products. It is clear that the current work is just the launch of an elaborated research, which may be further elaborated by e.g., a. Designing a global controller that may cope with all Pareto solutions, b. Exploring other more applicable real life examples from various fields of interest other than engineering, c. Developing an evolutionary algorithm that simultaneously searches for optimal robust solutions, d. Suggesting servo related problems and controllers.

References

1. Mattson, C.A., Messac, A.: Pareto frontier Based concept selection under uncertainty, with visualization. Optimization and Engineering 6, 85–115 (2005)
2. Van Veldhuizen, D.A., Lamont, G.B.: Multiobjective evolutionary algorithms, analyzing the State-of-the-Art. Evolutionary Computation 8(2), 12–147 (2000)
3. Miettinen, K.M.: Non-linear Multiobjective Optimization. International Series on Operations Research and Management Science. Kluwer Academic Publishers, Dordrecht (1999)
4. Goldberg, D.E.: Genetic algorithms in search, optimization and machine learning. Addison-Wesley, Reading (1989)
5. Coello, C.A.C.: Recent Trends in Evolutionary Multiobjective Optimization. In: Ajith, A., Jain, L., Robert, G. (eds.) Evolutionary Multiobjective Optimization: Theoretical Advances and Applications, pp. 7–32. Springer, London (2005)
6. Deb, K., Pratap, A., Agarwal, S., Meyarivan, T.: A Fast and elitist multiobjective genetic algorithm: NSGA–II. IEEE Transactions on Evolutionary Computation 6(2), 182–197 (2002)
7. Deb, K.: Multi-objective optimization using evolutionary algorithms. J. Wiley & Sons, Ltd., Chichester (2001)
8. Phadke, M.S.: Quality engineering using robust design. Prentice-Hall, Englewood Cliffs (1989)
9. Taguchi, G., Taguchi, S., Tugulum, R.: Taguchi robust engineering: worlds' best practice for achieving competative advantage in the new millennium. McGraw-Hill, New York (1999)
10. Branke, J.: Reducing the sampling variance when searching for robust solutions. In: Proceedings of Genetic and Evolutionary Computation Conference (GECCO 2001), pp. 235–242. Morgan Kaufmann Publishers, San Francisco (2001)
11. Deb, K., Gupta, H.: Searching for robust Pareto-optimal solutions in multi-objective optimization. In: Coello Coello, C.A., Hernández Aguirre, A., Zitzler, E. (eds.) EMO 2005. LNCS, vol. 3410, pp. 150–164. Springer, Heidelberg (2005)
12. Teich, J.: Pareto Font exploration with uncertain objectives. In: Zitzler, E., Deb, K., Thiele, L., Coello Coello, C.A., Corne, D.W. (eds.) EMO 2001. LNCS, vol. 1993, pp. 314–328. Springer, Heidelberg (2001)
13. Luo, L., Kannan, P.K., Besharati, B., Azarm, S.: Design of Robust New Products under Variability: Marketing Meets Design. Journal of Product Innovation Management 22, 177–192 (2005)
14. Singh, R., Khamba, J.S.: Mathematical modeling of tool wear in ultrasonic machining of titanium. Int. J. Adv. Technol. 4, 573–580 (2009)
15. Lounis, Z., Madanat, S.M.: Integrating mechanistic and statistical deterioration models for effective bridge management. In: 7th ASCE International Conference on Applications of Advanced Technology in Transportation, Boston, pp. 513–520 (2002)

Non-dimensional Multi-objective Performance Optimization of Single Stage Thermoelectric Cooler

P.K.S. Nain[1], S. Sharma[2], and J.M. Giri[3]

[1] Department of Mechanical Engineering,
Amity School of Engineering & Technology
Amity University, Sector 125, Noida - 201 303, India
pksnain@amity.edu
http://www.amity.edu

[2] Department of Mechanical Engineering,
Noida Institute of Engineering & Technology
Greater Noida - 201 306, India
sudhanshu.shr@gmail.com
http://www.niet.co.in

[3] Department of Mechanical Engineering,
Skyline Institute of Engineering & Technology, Greater Noida - 201 306, India
jmgiri@engineer.com
http://www.skylineinstitute.com

Abstract. Thermoelectric devices are indeed device of future as they are green cooling devices. Tough still under research, performance of these devices is main concern to engineers for their suitability for practical use. In the present work, the two main concern i.e. Coefficient of Performance (COP) and Rate of Refrigeration (ROR) of such devices are simultaneously addressed. NSGA-II is used for finding Pareto-optimal solutions under three different settings for ambient conditions. Mathematical model is considered and effect of ambient conditions on optimal performance is also highlighted.The results of optimization are verified by theoretical governing equations for Thermo-Electric Coolers (TEC). It is concluded that Bi-Objective optimization of performance of single stage TEC is possible, relevant and have huge potential for practical use by designers of TEC.

Keywords: TEC, Practical Performance Optimization, Multi-Objective Optimization, NSGA-II.

1 Introduction

Thermoelectric generation technology, as one entirely solid state energy conversion way can directly transform thermal energy into electricity by using thermoelectric transformation materials. A thermoelectric power converter has no moving parts, and is quite, compact, highly reliable and environment friendly.

K. Deb et al. (Eds.): SEAL 2010, LNCS 6457, pp. 404–413, 2010.

Due to these merits, this generation technology is presently becoming a notice-able research direct [4].

Thermoelectric coolers (Also known as thermoelectric refrigerators or Peltier modules) have been employed in various applications of small volume devices, typical of which are to stabilize the temperature of solid state lasers, to cool infrared detectors and charge coupled devices, and to increase the operating speed and reduce unwanted noise of integrated circuits [5]. Peltier modules have also been employed in portable cool boxes for medicine/serum transport and for picnic items storage [6].

Thermoelectric refrigeration and generation devices occupy a niche market, because they are quite and reliable and friendly to our environment. However, thermoelectric refrigeration appears to has made little impact on the domes-tic refrigeration market. The main factors that determine the marketability of a thermoelectric refrigerator are price and running cost, together with reliabil-ity, quietness, flexibility and temperature stability are important considerations. Price reflects the manufacturing cost, while the running cost is mainly deter-mined by the coefficient of performance (COP) of the cooling unit. Although the COP of a Peltier module is lower than that of conventional compressor units, efforts have been made to develop the thermoelectric refrigerators to exploit the advantage associated with this solid state energy-conversion technology [7].

2   Past Studies

There has been a considerable interest during the past ten years in finding new materials and structures for use in green, highly efficient cooling and energy conversion systems [8,9]. A good thermoelectric material should possess large Seebeck coefficient (α), low thermal conductance (K) to retain the heat at the junction and maintain a large temperature gradient and low electrical resistance (R) to minimize Joule heating. These desirable properties are embodied in a so-called thermoelectric material property figure of merit Z. To describe material more useful method is dimensionless figure of merit (ZT), where T is the absolute temperature of interest. ZT provides a measure of the quality of such materials for applications and is defined by $ZT = \alpha^2 T/RK$ [8]. The increase in ZT leads directly to the improvement in the cooling efficiency of Peltier modules and in energy conversion efficiency of thermoelectric generators [10]. Much effort has been made to raise Z of thermoelectric materials using various methods, so that they have some improvements in Z (For example, $3.20 \times 10^{-3} K^{-1}$ at 300 K [11] and $3.99 \times 10^{-3} K^{-1}$ at 298 K [12] for the n type Bi-Te alloys and $3.70 \times 10^{-3} K^{-1}$ at room temperature [13] and $4.58 \times 10^{-3} K^{-1}$ at 308 K [14] for the p-type Bi-Te alloys). But their values are not sufficient to improve dramatically the cooling efficiency.

In addition to Thermoelectric (TE) material improvements, there are many areas in which research is continuing to improve the performance of thermo-electric cooler (TEC). An assembly technique which affects the temperature difference across the TE module is also being developed that may improve the device performance. There is a technique that involves the thermal isolation of each of the thermoelectric couples in the direction of fluid flow [15]. Individual thermoelectric elements which are thermally isolated can additionally heat or cool a working fluid to increase system performance compared to a standard device. Coefficient of performance (COP) approaches the theoretical maximum of a continuous system by increasing the number of TE elements. COP increases up to 120% over conventional TE devices. Experimentally, the thermal isolation technique has shown measured performances ranging from 60 to 90% of the theoretical COP and at least 50% greater than the theoretical best COP for conventional TE systems [16]. More experimentation in this area is on the way.

3   Mathematical Modeling

A simple thermoelectric refrigeration system consists of a single stage thermo-electric device, which is composed of large number of p-and n-type semiconductor elements connected electrically in series and thermally in parallel as shown in Fig.1, where Q_h and Q_c are rate of rates of heat rejection and input from the device to the high temperature sink and from low temperature cold space to the devices respectively. M indicates the pair number p- and n-type semiconductor elements. The refrigeration system is operated between the heat reservoirs at the temperatures T_h and T_c. The thermoelectric device is insulated; both thermally and electrically, from its surroundings except its junction reservoir contacts. When the electric current I flows through the thermoelectric device, the heat $M\alpha IT_c$ and $M\alpha IT_h$ are absorbed from the low temperature cooled space and released to the high temperature heat sink by the thermoelectric device, respectively, due to Peltier effect. Where, α is the Seebeck coefficient of a pair of semiconductor elements.

Heat abstracted from low temperature

$$Q_c = M\alpha IT_c - \frac{1}{2}I^2 MR - MK(T_h - T_c) , \qquad (1)$$

Heat rejected to high temperature

$$Q_h = M\alpha IT_h + \frac{1}{2}I^2 MR - MK(T_h - T_c) . \qquad (2)$$

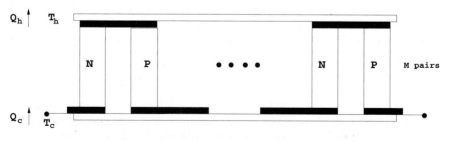

Fig. 1. General schematic diagram of thermoelectric refrigerator

Here, I^2MR is the Joule heat which goes half to high temperature sink and half to low temperature source. $MK(T_h - T_c)$ is the heat transfer due to temperature difference. Where, K and R are the thermal conductance and the electrical resistance of a pair of semiconductor elements.

Using Eqn.1 and 2 and the definition of the figure of merit $Z = \alpha^2/(RK)$ of the semiconductor elements, we obtain the Coefficient of Performance (COP) as

$$\varepsilon = \frac{Q_c}{Q_h - Q_c} = \frac{j - \frac{\theta}{2}\frac{j^2}{ZT_h} + (1 - \theta)}{(\theta - 1)j + \frac{\theta}{Z}\frac{j^2}{T_h}} . \tag{3}$$

where $j = \alpha I/K$, $\theta = T_h/T_c$, and $P = Q_h - Q_c$ is the power input. It is seen from Eq.3 that putting COP to be zero for a thermoelectric refrigerator, we get two values of j (called dimensionless current)

$$j = ZT_c[1 - \sqrt{1 - 2(\theta - 1)/ZT_c}\,] = j_{min} , \tag{4}$$

$$j = ZT_c[1 + \sqrt{1 - 2(\theta - 1)/ZT_c}\,] = j_{max} . \tag{5}$$

For a practical refrigerator, the value of ε (COP) must be real and positive. It is obvious that only if the dimensionless electric current of a single- stage thermoelectric refrigerator lies in the region between j_{min} and j_{max}, i.e.

$$j_{min} \leq j \leq j_{max} , \tag{6}$$

and the temperature of the heat sink to the cooled space satisfies the following condition:

$$\theta < \frac{1 + \sqrt{1 + 2ZT_h}}{2} . \tag{7}$$

If Eqn. 6 and 7 are not satisfied simultaneously, a single stage thermoelectric refrigerator would lose its role. Further, Rate of refrigeration for a single stage thermoelectric refrigerator

$$R = Q_c$$

Dimensionless rate of refrigeration:

$$r = \frac{Q_c}{MKT_h} = \frac{M\alpha IT_c - 0.5I^2MR - MK(T_h - T_c)}{MKT_h} = \frac{j}{\theta} - \frac{j^2}{2ZT_h} + (\frac{1}{\theta} - 1) . \tag{8}$$

There are two important performance criteria of thermoelectric refrigerator, first, rate of refrigeration (ROR) and second, coefficient of performance (COP) which can be optimized to enhance the utility of the device.

3.1   COP and ROR Characteristic Curves

Following the governing equations as mentioned in the mathematical modeling part, a characteristic plot for COP against dimensionless current j at three different values of θ namely, 1.1, 1.2 and 1.3 for single stage thermoelectric refrigerator is drawn in Fig. 2. Similarly, a plot of ROR at same values of θ is also drawn in Fig.3.

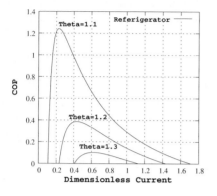

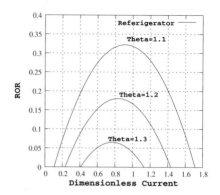

Fig. 2. COP Vs. Dimensionless Current of single-stage thermoelectric refrigerator

Fig. 3. ROR Vs. Dimensionless Current of single-stage thermoelectric refrigerator

The fact which is observed with these values given in Table 1 and Fig. 2 are:

1. The peak value of COP is shifting towards higher values of dimensionless current (from 0.22 to 0.6) with increase in value of $\theta = T_h/T_c$. It shows that as we increase the temperature ratio for obtaining peak value of COP we have to supply more current.
2. The peak value of COP is also decreasing with increase in value of θ. It is decreasing from 1.22 to 0.1. It indicates the performance of single stage TEC is deteriorating with increase in source and sink temperature ratio (θ) of TEC.
3. Now if we try to plot COP Vs. j curves for further higher values of $\theta = 1.4$ and so on, we find no curve. It shows that with increase in the temperature ratio single stage refrigerator becomes ineffective.

Table 1. COP Vs. Dimensionless current

θ	j	COP
	0.1	0
1.1	0.22	1.22
	1.7	0
	0.22	0
1.2	0.4	0.38
	1.42	0
	0.4	0
1.3	0.6	0.1
	1.15	0

Table 2. ROR Vs. Dimensionless current

θ	j	ROR
	0.1	0
1.1	0.9	0.32
	1.7	0
	0.22	0
1.2	0.81	0.18
	1.42	0
	0.4	0
1.3	0.75	0.07
	1.15	0

Observing the values given in Table 2 and Fig. 3, it is noted that:

1. The peak value of ROR is shifting towards lower values of dimensionless current (from 0.9 to 0.75) with increase in value of $\theta = T_h/T_c$. It shows that as we increase the temperature ratio θ, for obtaining peak value of ROR we have to supply less current.
2. The peak value of ROR is also decreasing with increase in value of θ. It is decreasing from 0.32 to 0.07. It indicates the performance of single stage TEC is deteriorating with increase in source and sink temperature ratio (θ) of TEC.
3. Now if we try to plot ROR Vs. j curves for further higher values of $\theta = 1.4$ and so on, we find no curve. It shows that with increase in the temperature ratio single stage refrigerator becomes ineffective.

Hence, as we observe that it shows a conflicting behavior of single stage thermo-electric refrigerator as one tries to maximize COP and ROR simultaneously for a given value of θ, because it has shifted peaks of maximum COP and maximum ROR with respect to values of dimensionless current, which makes impossible to optimize both objectives simultaneously. It implies possibility of Pareto-optimal solutions as neither of the objectives cannot be improved without making a sacrifice on other. It makes this problem of optimizing performance of single stage TEC, a perfect case for multi-objective optimization[1,2,3].

It also indicates that for higher values of θ greater than 1.3, the single stage TEC becomes ineffective as no feasible values of COP and ROR is observed. It indicates a possible use of multi-stage TEC in place of single stage TEC for cooling situations having high ambient temperatures such that T_h/T_c is greater than 1.3.

4   Simulation Results for Performance Optimization of Single Stage TEC

The problem of the multi-objective optimization of ROR and COP of thermo-electric refrigerator is formulated as the following Bi-Objectives optimization problem:

$$\text{Max ROR},$$
$$\text{Max COP},$$
$$\text{Subject to:}$$
$$j_{min} \leq j \leq j_{max}.$$

Three different values of θ, namely 1.1, 1.2 and 1.3 are taken as three separate optimization problems, the results for which are presented in this section in same order. Non-dominated Sorting Genetic Algorithm-II (NSGA-II) [1,2,3] has been used for multi-objective optimization. General parameter settings for multi-objective optimization of COP and ROR are follows. A population size of 200 for 100 generation is run with distribution index for crossover set at 10 while for mutation is taken 20. The probability of crossover is taken 0.8 while for mutation is taken 0.4. The variable bounds on dimensionless current is taken from Table 1.

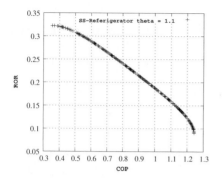
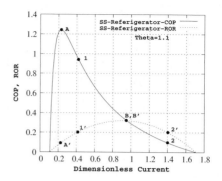

Fig. 4.   Pareto-optimal   front   for   **Fig. 5.** COP and ROR Vs. Dimension-
theta=1.1   for   single-stage   thermo-   less Current at Theta=1.1 of single-stage
electric refrigerator   thermoelectric refrigerator

   The algorithm gives result in the form of Pareto-optimal front shown in Fig.4. in order to validate simulation results, the corresponding characteristic plot of COP and ROR is given in Fig.5. It is observed from Fig.5 that if we have a target value of a parameter, say 0.2 value of ROR, this value can be achieved at two different values 0.4 and 1.4 of dimensionless current. These are shown in Fig.5 by points $1'$ and $2'$. Corresponding to point $1'$, the COP is 0.95 (Point 1) at dimensionless current value of 0.4. While, corresponding to point $2'$ the COP is 0.1 (Point 2) at 1.4 value of dimensionless current. It is better to select 0.4 value of dimensionless current because it gives target value of ROR (0.2) with higher value of COP(0.95). In fact, if we observe Fig.4, NSGA-II has picked up this point in its results.

   It can be observed from Fig.5 that there are two values of dimensionless current for a single value of COP or ROR. Between these two values of dimensionless current, there is range where COP and ROR have conflicting nature. At $\theta=1.1$, the region from the dimensionless currents 0.22 to 0.9 is the range where both the parameters have conflicting nature. This range can be seen from Fig.5 where at 0.22 value of dimensionless current COP & ROR are at points A & A' respectively and at 0.9 value of dimensionless current COP & ROR are at points B & B' respectively. It is clear that we can optimize both the parameters (COP & ROR) and NSGA-II has picked up these two extremities in its results. It is worth noticing that the dimensionless current can be used as variable with lower and upper bounds to get feasible optimized solutions. It should be considered that making dimensionless current as variable in optimization problem we can minimize the complication since dimensionless current is a value with combinations of Seebeck coefficient, thermal conductance and electric current.

   For implementation of NSGA-II at $\theta=1.2$, GA settings remain same except the lower limit and upper limit of real variable which is 0.22 and 1.42 respectively. The NSGA-II gives the Pareto-optimal front as shown in Fig.6. The corresponding characteristic plot is given in Fig.7. The extremities of NSGA-II result

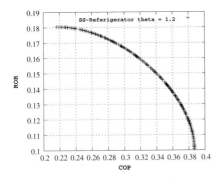

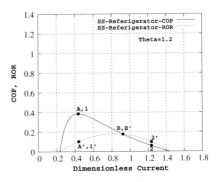

Fig. 6. Pareto-optimal front for theta=1.2 for single-stage thermoelectric refrigerator

Fig. 7. COP and ROR Vs. Dimensionless Current at Theta=1.2 of single-stage thermoelectric refrigerator

matches with extremities of that range where two parameters have conflicting nature (point (A, A') and point (B, B')). This means that the algorithm successfully converged to the Pareto-optimal front. Similar to our previous discussion, we find that this time also NSGA-II has picked up point $(1, 1')$ in preference to point $(2, 2')$ of Fig.7.

For implementation of NSGA-II at $\theta=1.3$, GA settings remain same except the lower limit and upper limit of real variable which is 0.4 and 1.1 respectively. The NSGA-II gives the Pareto-optimal front as shown in Fig.8. The corresponding characteristic plot is given in Fig.9. The extremities of NSGA-II result matches with extremities of that range where two parameters have conflicting nature (point (A, A') and point (B, B')). This means that the algorithm successfully converged to the Pareto-optimal front.

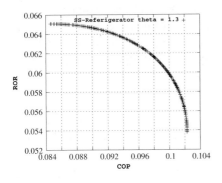

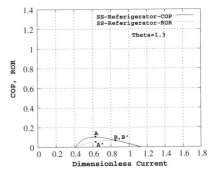

Fig. 8. Pareto-optimal front for theta=1.3 for single-stage thermoelectric refrigerator

Fig. 9. COP and ROR Vs. Dimensionless Current at Theta=1.3 of single-stage thermoelectric refrigerator

5   Conclusions and Scope of Future Work

In this work, the need for optimization of single stage TEC performance is discussed. A mathematical model for single stage TEC is discussed. Later, by studying characteristic curves for COP and ROR, a framework of multi-objective optimization problem is developed. The Pareto-optimal solutions for three different settings of $(\frac{T_h}{T_c})$ values up to 1.3 are reported.

The potential use of single stage TEC can be significantly improved by choosing optimum values of current to device. In this work, a dimensionless approach in optimization is employed, which combines several significant material properties into meaningful dimensionless groups. Such an approach simplifies optimization problem as now the optimizer (NSGA-II) has to deal with reduced problem variables. Using such an approach leads to greater flexibility to the designer of TEC as the can manipulate any of the variables present in the dimensionless group to attain Pareto-optimality. In present work, both objectives as well as problem variable are dimensionless, making the presented optimization approach very powerful, significant, meaningful to a TEC designer. Current work deals with dimensionless COP, dimensionless ROR and dimensionless current. Even the setting of θ, which is the only limitation of the present work is dimensionless as it deals with ratio of temperature. In era of microprocessors, which are finding increased applications in automobile industries for enhancing performance of vehicle by storing and controlling timing and quantity of fuel injection and ignition advances, a dream to control performance of TEC by microprocessor based current controller is not very far. In such case, a look-up table containing Pareto-optimal settings for TEC can be stored and be used to enhance the performance of TEC. As we know that, when initially a cooling device is switched on, θ value is low, as temperature of refrigerated space is almost equal to ambient temperature. As time progresses this difference and hence θ value starts increasing till it saturates. This makes the very strong case of use of Pareto-optimal solutions for simulations presented in this work in conjunction of microprocessor based current controllers. It will ensure that under each operating condition, TEC will be operating at nearest possible Pareto-optimal point and hence will be a makeup for their low COP and ROR when compared to traditional cooling devices. It is also found that beyond $(\frac{T_h}{T_c}) = 1.3$, the single stage TEC lose their effectiveness as coolers and hence multi-stage TEC suitability for such task should be explored.

References

1. Deb, K., Pratap, A., Agarwal, S., Meyarivan, T.: A Fast and Elitist Multi-Objective Genetic Algorithm: NSGA-II. IEEE Transactions on Evolutionary Computation 6(2), 182–197 (2002)
2. Deb, K.: Multi-Objective Optimization Using Evolutionary Algorithms, 1st edn. John Wiley & Sons, Chichester (2001)
3. Deb, K.: Optimization for Engineering Design: Algorithms & Examples. Prentice-hall of India Private Limited, New Delhi (1995)

4. Bell, L.E.: Cooling, heating, generating power and recovering waste heat with thermoelectric systems. Science 321, 1457–1461 (2008)
5. Goldsmid, J.H.: Electronic refrigeration, pp. 198–219. Pion, London (1986)
6. Min, G., Rowe, D.M., Zhang, J.S.: Thermoelectric technology and applications, pp. 215–247. Defense Industries Publishing, Beijing (1996)
7. Min, G., Rowe, D.M.: Experimental evaluation of prototype thermoelectric domestic refrigerators. Applied Energy 83, 133–152 (2006)
8. Wood, C.: Materials for thermoelectric energy conversion. Reports on Progress in Physics 51, 459–539 (1908)
9. Mahan, G., Sales, B., Sharp, J.: Thermoelectric materials: new approaches to an old problem. Physics Today 50, 42–47 (1997)
10. Tritt, T.M. (ed.): Semiconductors and semi-metals. Recent trends in thermoelectric materials research Part one, vol. 69. Academic Press, USA (2000)
11. Yim, W.M., Rosi, F.D.: Compound tellurides and their alloys for Peltier cooling- a review. Solid state electronics 15, 1121–1134 (1972)
12. Yamashita, O., Tomiyoshi, S.: Effect of annealing on thermoelectric properties of bismuth telluride compounds. Japanese Journal of Applied Physics 42, 492–500 (2003)
13. Ettenberg, M.H., Jesser, W.A., Rosi, F.D.: A new n-type and improved p-type pseudo-ternary $(Bi_2Te_3)(Sb_2Te_3)(Sb_2Se_3)$ alloy for Peltier cooling. In: Proceeding of the 15th International Conference on Thermoelectric, Piscataway, NJ, pp. 52–56 (1996)
14. Yamashita, O., Tomiyoshi, S., Makita, K.: Bismuth telluride compounds with high thermoelectric figures of merit. Journal of Applied Physics 93, 368–374 (2003)
15. Bell, L.E.: Use of Thermal Isolation to Improve Thermoelectric System Operating Efficiency. In: 21st International Conference on Thermoelectrics, pp. 477–487. IEEE, Long Beach (2002)
16. Diller, R.W., Chang, Y.W.: Experimental Results Confirming Improved Performance of Systems Using Thermal Isolation. In: 21st International Conference on Thermoelectrics, pp. 548–550. IEEE, Long Beach (2002)

Multi-Objective Optimization of Particle Reinforced Silicone Rubber Mould Material for Soft Tooling Process

Arup Kumar Nandi[1] and Shubhabrata Datta[2]

[1] Central Mechanical Engineering Research Institute (CSIR),
Durgapur-713209, WB, India
[2] Birla Institute of Technology, Deoghar Campus Jasidih, Deoghar 814142,
Jharkhand, India

Abstract. Multi-objective optimizations of various conflicting objectives in designing particle reinforced silicone rubber are conducted using evolutionary algorithms to reduce the processing time of soft tooling process. A well-established evolutionary algorithm based multi-objective optimization tool, NSGA-II is adopted to find the optimal values of design parameters. From the obtained Pareto-optimal fronts, suitable multi-criterion decision making techniques are used to select one or a small set of the optimal solution(s) of design parameter(s) based on the higher level information of soft tooling process for industrial applications.

Keywords: Multi-objective optimization problem; evolutionary algorithm; particle reinforced flexible mould material; soft tooling; cooling time.

1 Introduction

In soft tooling (ST) process, there are some polymeric flexible (mould) materials (namely Silicone rubber (SR), Polyurethane, etc) used for making mould. The cooling time in soft tooling process using such conventional mould materials is normally high (as those materials possess poor thermal conductivity) which is not beneficial in competitive market. This is overcome by the practitioners through inclusion of highly thermal conductive filler particles into the mould material. From experimental studies [1], it has been observed that equivalent thermal conductivity of particle reinforced flexible mould material (FMM) increases with the enhancement of filler content in the composite like other polymer composites. On the other hand, the effective modulus of elasticity of the composite material is also rising simultaneously with increasing the filler content. But, in this case high modulus of elasticity is not desirable as increase in stiffness of mould box, creates many difficulties particularly removing the pattern from the mould box, etc. Moreover, high content of fillers in the melt mould material reduces its flow-ability by enhancing its effective viscosity [2]. Therefore,determination of optimum amount of filler content is an important issue in

K. Deb et al. (Eds.): SEAL 2010, LNCS 6457, pp. 414–423, 2010.

order to design and develop a suitable particle reinforced flexible mould material for useful application in soft tooling process. In the present work, SR composite mould materials are designed using aluminium (Al) and Graphite (Gr) filler particles based on an evolutionary algorithm for improvement of soft tooling process.

In order to accomplish this, the effective thermal conductivity of mould material should be high enough along with limiting the effective modulus of elasticity and viscosity to an acceptable/minimum value for a given configuration of wax/plastic pattern. It can be achieved by keeping the controlling parameters, namely volume fraction, size and shape factor of filler particle to optimum values for a given particle reinforced flexible mould material. Since, in this optimization process there are three primary conflicting objectives (maximization of thermal conductivity, minimization of modulus of elasticity and minimization of viscosity), it can be solved by suitable multi-objective optimization (MOO) tool. In the present study, we have adopted a well-established evolutionary algorithm (EA) based MOO tool, NSGA-II (an elitist non-dominated sorting genetic algorithm) proposed by Deb et al. [3]. After obtaining the Pareto-optimal fronts by solving the multi-objective problem (MOP),a suitable (multi-criterion decision making) MCDM technique is used to select one or a small set of the optimal solution(s) of design parameter(s) based on the higher level information of the process of soft tooling for industrial applications.

2   Problem Definition

The proposed procedure of designing the particle reinforced mould material is carried out through two kinds of approaches. The formulations of MOPs corresponding to these approaches are presented in the following. In the problem I, we have considered the maximization of equivalent thermal conductivity (k_c) and minimization of modulus (E_c) of elasticity are the objectives where the equivalent viscosity (η_c) is assumed as constraint for simplicity of the problem. On the other hand, all the three properties of mould material are treated as objectives to carry out a deeper investigation by treating them with equal importance in designing the mould material.

2.1   Problem I

The optimization problem evolved in ST process in order to make it cost effective by reducing cooling time as well as keeping more ease in casting process is described below.

$$f_1 = Maximize \; k_c(k_f, k_p, P_s, S_f, T, V_f) \tag{1}$$

$$f_2 = Minimize \; E_c(k_f, k_p, P_s, S_f, T, V_f) \tag{2}$$

Subject to (for a given composite system of flexible mould material and conductive filler material)

$$0 \leq P_s \geq 500 \tag{3}$$

$$0 \leq S_f \geq 4.0 \tag{4}$$

$$0 \leq V_f \geq 100 \tag{5}$$

$$23^\circ C \leq P_s \geq 200^\circ C \tag{6}$$

$$\frac{\phi_m(P_s, S_f) - V_f}{\phi_m(P_s, S_f)} \geq 0 \tag{7}$$

$$\eta_c^{Limiting} - \eta_c(P_s, V_f, \eta_p) \geq 0 \tag{8}$$

and k_f, k_p, G_f, K_f, K_p and η_p are constant.

2.2 Problem II

$$f_1 = Maximize \; k_c(k_f, k_p, P_s, S_f, T, V_f) \tag{9}$$

$$f_2 = Minimize \; E_c(k_f, k_p, P_s, S_f, T, V_f) \tag{10}$$

$$f_3 = Minimize \; \eta_c(P_s, S_f, V_f, \eta_p) \tag{11}$$

Subject to Equ. (3) to Equ. (7) for a given composite system of flexible mould material and conductive filler material.

The function $\phi_m(P_s, S_f)$ is called the maximum loading level (maximum packing fraction) which is determined using the empirical expressions stated in [4]. The upper and lower limits of P_s and S_f are fixed based on the fact that the particles having these sizes and shape factors are most suitable for handling in practical applications.Furthermore, the size and shape factors of common non-metallic filler materials those are commercially available in particulate form fall in those ranges. In soft tooling process, the preparation of mould is generally carried out in room temperature. Therefore, in this work, the range of temperature is set from $23^\circ C$ to $30^\circ C$.

3 Various Models/Objective Functions

In the authors' previous work, a genetic-fuzzy model of equivalent thermal conductivity of particle reinforced polymer composites was developed based on fuzzy logic rules [5]. The developed fuzzy rule based model is adopted here in order to investigate the role of Al and Gr fillers and design of composite mould materials so as to reduce cooling time in ST process. On the other hand, it was found that the model proposed by Lielens [6] provides close agreements with experimental data of particle reinforced mould materials compare to other models [1]. The relative viscosity model presented by Kriger and Dougherty [7] is adopted in the present study. The values of the controllable parameter of Kriger-Dougherty model are found as 2.8697 and 10.3459 for the Al and Gr filled flexible mould material composite systems, respectively [2].

4 Evolutionary Multi-Objective Optimization Algorithm

The evolutionary algorithms are now well known optimizers influenced by the principles of natural selection and natural genetics [8]. In case of MOPs, the genetic search is performed following the concept of Pareto-opti-mality [9]. Having several conflicting objective functions, the concept of optimum changes, from the unique global optimum, as used in the single objective problems, to a set of solutions providing the best possible compromises between the objectives, known as the Pareto set. The very definition the Pareto-optimality entails that no other solution could exist in the feasible range that is at least as good as some member of the Pareto set, in terms of all the objectives, and strictly better in terms of at least one. The Pareto set thus offers a number of equivalent optimum solutions, out of which a decision maker can easily pick and choose the most suitable ones. NSGA-II (an elitist non-dominated sorting genetic algorithm) is an efficient evolutionary algorithm to find a set of Pareto-optimal solutions [3].

In NSGA-II, the offspring population, Q_t is first created using the parent population, P_t. However, instead of finding the non-dominated front of Q_t only, first the two populations (Q_t and P_t) are combined together to form R_t of size $2N$. Then, a non-dominated sorting is performed based on the measure of non-domination rank of the individuals to classify the enter population R_t. Once the non-dominated sorting is over, solution(s) of different non-dominated fronts, one at a time fill the new population. The fillings start with the best non-dominated front, and so on. Since the overall population size of R_t is $2N$, not all fronts may be accommodated in N slots available in the new population. All fronts, which could not be accommodated, are simple deleted. When the last allowed front is being considered more solutions may be present in the last front than the remaining slots in the new population. Instead of arbitrarily discarding some members from the last front, it would be wise to use a niching strategy to choose the members of the last front, which decide in the least crowded region in that front. Based on this new population another offspring population Q_t is created using genetic operators like selection, (crowded) crossover and mutation in the next generation, and this cycle is continued till a specified number of generations is reached. In the present work, we have adopted two different niching strategies to select the solutions from the last considered front, crowding distance approach which is associated with original NSGA II [3] and clustering approach [10]. In order to solve the problem I defined in Section 2.1, original NSGA II utilizing crowding distance approach is used, whereas to solve the problem II defined in Section 2.2, we have replaced the crowding distance approach by clustering approach in the NSGA-II by by realizing that crowding distance approach does not work well with higher number of objective functions.

Since the decision variables involved in optimization of equivalent thermal conductivity and modulus of elasticity are continuous type, we have used real-coded GA and the values of GA-parameters are adopted in NSGA-II for solving both the problems stated in Section 2 are Population size=500; Number of generations=200; Mutation probability=1/no of decision variable; Crossover probability=0.98; Distribution Index for real-coded crossover=20; and Distribution Index for real-coded mutation=100.

5   MOO with Considering Viscosity as Constraint

The definition of this MOP while considering viscosity of melt particle reinforced FMM as constraint is explained in Section 2.1. In this study, three levels of viscosity melt FMM composite are considered such as Low (10Pas), Medium (50Pas), and High (200Pas).

Fig. 1 describes the Pareto-optimal fronts of optimum values of k_c-E_c combinations for different composite systems of SR when the viscosity is considered as a constraint in the MOO process carried out using NSGA II. For all three viscosity levels, it has been found that the Pareto-front of the Gr-SR composite overlaps with that of the Al-SR system. Thus, given a choice between the two combinations of composite systems, one cannot give preferences on any of them.

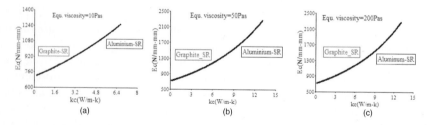

Fig. 1. Pareto-optimal fronts in optimization of k_c and E_c of FMM composites for different viscosity labels: (a) 10Pas (b) 50Pas and (c) 200Pas

From the above studies; it is found that all the decision variables (P_s, S_f and T) seem constant except V_f for both the composite systems. After analyzing the variations of V_f with k_c in different viscosity levels it is found that V_f seems to be varying linearly with k_c with the empirical expressions, Equ. (12) and Equ. (13) in Al and Gr filled SR composites, respectively:

$$V_f = 2.8965 \, K_c - 0.4385 \tag{12}$$

$$V_f = 3.2453 \, K_c - 0.574 \tag{13}$$

Relationships between V_f and k_c are valid only for the Pareto solutions, i.e. for some fixed values of other variables. It has been found that the upper limit of V_f attains lesser value for the lower viscosity level. Increase of upper limit of V_f is found more in the initial stage of increasing the viscosity level and gradually it diminishes. Later, it does not change for any further increase of viscosity level.

6   MOO with Considering Viscosity as an Objective

The Pareto-optimal fronts for maximization of k_c, minimization of E_c and minimization η_c in Al particle reinforced SR composite systems is shown in Fig. 2. From Fig. 2, it is observed that E_c and η_c are correlated to each other, because it has been found that minimization of E_c also minimizes η_c. However, in k_c-E_c combination, E_c is not correlated with k_c. Similarly in k_c-η_c combination η_c is also not correlated to k_c. Thus, to find the optimum values of decision variables any of the k_c-E_c or k_c-η_c can be considered. Form the above studies; it is found that all the decision variables (P_s, S_f and T) seem constant except V_f. and the variation of V_f with k_c is illustrated using Equ. (14) and Equ. (15).

$$V_f = 2.8737 \ K_c - 0.4265 \tag{14}$$

$$V_f = 3.1097 \ K_c - 0.4473 \tag{15}$$

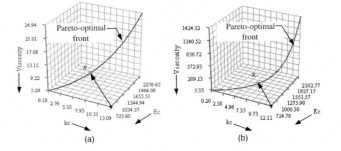

Fig. 2. Pareto-optimal fronts in optimization of k_c, E_c and η_c of silicone rubber composite filled with (a) Al (b) Gr

Note that, the variation of k_c with respect to V_f looks identical with that found in Equ. (12) and Equ. (13)., respectively which suggests the existence of correlation among E_c and η_c in Al filled SR composite system.

7   Selection of Preferred Solution(s) Based on MCDM

Although there are advantages of knowing the range of each objective for Pareto-optimality and the shape of the Pareto-optimal frontier itself in a problem for an adequate decision-making, the task of choosing a single preferred Pareto-optimal solution is an important task because the practitioners finally adopt the chosen preferred solution for implementation in industrial applications. Nowadays various MCDM techniques are available and following a classification by Veldhuizen and Lamont [11] the articulation of preferences may be done either before search (a priori) or after search (a posteriori) or during search (progressive) of the optimization process. In the present problem, we have adopted two such MCDM techniques after the optimization process: reference point based technique [12] to identify a single preferred solution, and selection of a set of (few) preferred solutions by identifying knee-region [13] on the Pareto-front based on trade-off and then DM chooses a solution from the knee zone based on higher level information. Typical other methods of selecting a preferred solution from the set of Pareto-optimal solutions are the compromise programming approach [14], the marginal rate of substitution approach [12], the pseudo-weight vector approach [9], etc.

The DM can decide a reference point based on various aspects concerning to the problem domain. A well-accepted reference point suitable to all kinds of MOP is the ideal point. An ideal point is referred to the vector of individual optimal objective values. In the present application, the ideal point is considered as a reference point and a single preferred solution is determined by finding the solution on the Pareto-front having minimum (Euclidian) distance from the reference point.

After a thorough study on the shape of Pareto-optimal front, sometimes it is found that there may be solutions where a small improvement in one objective would lead to a large deterioration in any of the other objectives. These solutions are called knee-points. The area of the Pareto-front surrounded by such solutions is termed as knee-region. Without any knowledge about the users preferences, it may be argued that the region around that knee is most likely to be interesting for the DM.

Ideal point of Al filled SR composite system is identified from the obtained Pareto-optimal front shown in Fig. 2(a), as follows: $k_c = 13.0912$; $E_c = 723.8$; $\eta_c = 5.29452$. The preferred Pareto-optimal solution obtained based on this reference point is designated by R as shown in Fig. 2(a) and the values of corresponding objective functions and decision variables are referenced in solution no. 1 in Table 1. The preferred solution, R is determined by considering each of the three properties of al filled SR composite k_c, E_c and η_c play as an objective in the decision-making procedure. In Section 6, we have seen that there exist a correlation between the objectives, E_c and η_c. Therefore, the preferred solution can also be decided by taking the projection of the Pareto-optimal front obtained in $3D$ into either k_c-E_c or k_c-η_c plane. The preferred Pareto-optimal solutions which are the solutions with minimum distance from the ideal (reference) point obtained in the k_c-η_c and k_c-E_c planes are referred to P and Q, respectively as shown in Fig. 3(a) and Fig. 3(b). The related objective function values and decision variable values corresponding to the preferred points P and Q are enlisted in Table 1. From the Figures 3(a) and 3(b), it is observed knee-zone is not clearly visible in the entire Pareto-optimal front (except at both the extreme

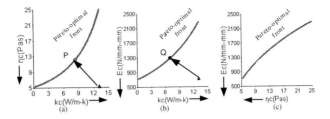

Fig. 3. Projection of Pareto-optimal front of Al filled Silicone rubber composite system as shown in Figure 2(a) in (a) k_c-η_c plane (b) E_c-k_c plane and (c) E_c-η_c plane

Table 1. List of optimal solutions based on minimum distance from ideal point: Al-SR composite

Ideal point	Optimal values of objectives			Optimal values of decision variables				Preferred solution
	k_c	E_c	η_c	P_s	S_f	V_f	T	
$k_c = 13.091$ $E_c = 723.8$ $\eta_c = 5.294$	6.733	1258.25	10.02	68.871	1.625	18.95	24.0001	Denoted as R in Fig. 2(a)
$k_c = 13.091$ $\eta_c = 5.294$	8.342	1451.93	12.17	68.871	1.625	23.57	24.0006	Denoted as 'P' in Fig. 3(a)
$k_c = 13.091$ $E_c = 723.8$	7.641	1363.72	11.16	68.872	1.624	21.56	24.0001	Denoted as 'Q' in Fig. 3(b)

ends) in both the k_c-η_c and k_c-E_c planes. The correlation between E_c and η_c is characterized by the curve as shown in Fig. 3(c).

Ideal point ($k_c = 12.1101$; $E_c = 724.781$; $\eta_c = 5.33096$) of Gr filled SR composite system considering all the three properties k_c, E_c and η_c as objectives is identified from the obtained Pareto-optimal front shown in Fig. 2(b). The preferred Pareto-optimal solution, R is obtained based on this reference point as shown in Fig. 2(b). The values of corresponding objective functions and decision variables are cited in Table 2. Like Al filled composite, existence of correlation between E_c and η_c is observed in Section 6, and the preferred solution is decided by taking the projection of the Pareto-optimal front obtained in $3D$ into either k_c-E_c or k_c-η_c plane. The preferred Pareto-optimal solutions found based on the minimum distance from the ideal point in the k_c-η_c and k_c-E_c planes are referred to P and Q, (in Fig. 4(a) and Fig. 4(b)) respectively. The related objective function and decision variable values corresponding to the preferred points P and Q are presented in Table 2. From the Fig. 4(b) no such knee-zone is found in the Pareto-optimal front in k_c-E_c plane. However, in Fig. 4(a), a pronounced knee-zone is observed (marked by a dotted ellipse) in the middle of Pareto-optimal front in the k_c-η_c. The solutions in this zone are characterized by the fact that a small improvement in either objective will cause a large deterioration in the other objective, which makes moving in either direction not very attractive. If we

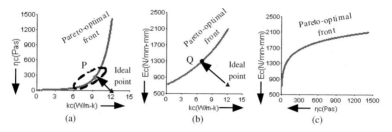

Fig. 4. Projection of Pareto-optimal front of Graphite filled Silicone rubber composite system as shown in Fig. 2(b) in (a) k_c-η_c plane (b) E_c-k_c plane and (c) E_c-η_c plane

Table 2. List of optimal solutions based on minimum distance from ideal point: Graphite-SR composite

Ideal point	Optimal values of objectives			Optimal values of decision variables				Preferred solution
	k_c	E_c	η_c	P_s	S_f	V_f	T	
$k_c = 12.110$ $E_c = 724.78$ $\eta_c = 5.330$	6.96	1294.2	74.3	68.865	1.623	21.25	24.802	Denoted as R in Fig. 2(b)
$k_c = 12.110$ $\eta_c = 5.330$	9.41	1622.5	262.1	68.867	1.623	28.87	24.0003	Denoted as'P' in Fig. 4(a)
$k_c = 12.1101$ $E_c = 724.78$	7.01	1299.1	75.8	68.871	1.623	21.38	24.802	Denoted as'Q' in Fig. 4(b)

assume linear preference functions, as (due to the lack of any other information) furthermore assume that each preference function (kc and η_c) is equally likely, the solutions at the knee-region are most likely to be the optimal choice of the practitioners. It is also noticed that the preferred point P as obtained based on the minimum distance from the ideal point is positioned on the right end of the dotted ellipse, which may suggest the extent of knee-zone on the right side is limited to the preferred point P. The correlation between E_c and η_c is found by the curve as shown in Fig. 4(c).

8   Conclusions

In order to design and develop a suitable particle reinforced silicone rubber mould material and at the same time to reduce the cycle time of soft tooling process, multi-objective optimizations of various conflicting mould material properties are conducted using evolutionary algorithms. In this regard, two different problems with or with out considering viscosity as an objective are solved using NSGA II and it is observed that modulus of elasticity and viscosity of particle reinforced mould material are correlated. Pareto-optimal fronts obtained by solving the MOPs are presented and the roles of various decision variables are analysed using innovization technique. The preferred Pareto-optimal solution(s) selected by using reference point MCDM technique as well as finding knees on the Pareto-fronts may be used for making composite silicone rubber mould in industry.

Acknowledgements

Authors acknowledge DST, New Delhi and DIT, New Delhi, India for supporting this research work as well as Prof. K. Deb for his scientific contributions.

References

1. Nandi, A.K., Datta, S.: Effective Properties of Particle Reinforced Polymeric Mould Material towards Reducing Cooling Time in Soft Tooling Process. J. of Appl. Polym. Sci. (accepted)
2. Nandi, A.K., Vesterinen, A., Cingi, C., Seppala, J., Orkas, J.: Studies on Equivalent Viscosity of Particle-Reinforced Flexible Mold Materials Used in Soft Tooling Process. J. of Reinforced Plast. Compos. 29(14), 2081–2098 (2010)
3. Deb, K., Agrawal, S., Pratap, A., Meyarivan, T.: A fast elitist non-dominated sorting genetic algorithm for multi-objective optimisation: NSGA-II. In: Deb, K., Rudolph, G., Lutton, E., Merelo, J.J., Schoenauer, M., Schwefel, H.-P., Yao, X. (eds.) PPSN 2000. LNCS, vol. 1917, pp. 849–858. Springer, Heidelberg (2000)
4. Ouchiyama, N., Tanaka, T.: Porosity Estimation for Random Packings of Spherical Particles. Industrial & Engineering Chemistry Fundamentals 23, 490–493 (1984)
5. Nandi, A.K., Datta, S., Deb, K., Orkas, J.: Studies on Effective Thermal Conductivity of Particle Reinforced Polymeric Flexible Mould Material Composites: A Genetic Fuzzy based approach. In: Proceedings of the 3rd Int. Conference on Recent Advances in Composite Materials (ICRACM 2010), France, December 13-15 (2010) (accepted)
6. Lielens, G., Pirotte, P., Couniot, A., Dupret, F., Keunings, R.: Prediction of thermo-mechanical properties for compression moulded composites. Composites Part A: Applied Science and Manufacturing 29(1-2), 63–70 (1998)
7. Krieger, I.M., Dougherty, T.J.: A mechanism for non-newtonian flow in suspensions of rigid spheres. Transactions of the Society of Rheology III, 137–152 (1959)
8. Goldberg, D.E.: Genetic Algorithms in Search, Optimization and Machine Learning. Pearson-Education, New Delhi (2002)
9. Deb, K.: Multiobjective Optimization Using Evolutionary Algorithms. John Wiley & Sons Ltd., Chichester (2001)
10. Zitzler, E., Thiele, L.: An evolutionary algorithm for multiobjective optimization: The strength Pareto approach. Technical Report 43, Zürich, Switzerland: Computer Engineering and Networks Laboratory (TIK), Swiss Federal Institute of Technology (ETH) (1998)
11. Veldhuizen, D.V., Lamont, G.B.: Multiobjective evolutionary algorithms: Analyzing the state-of-the-art. Evolutionary Computation Journal 8(2), 125–128 (2000)
12. Miettinen, K.: Nonlinear multiobjective optimization. Kluwer, Boston (1999)
13. Branke, J., Deb, K., Dierolf, H., Osswald, M.: Finding knees in multi-objective optimization. In: Yao, X., Burke, E.K., Lozano, J.A., Smith, J., Merelo-Guervós, J.J., Bullinaria, J.A., Rowe, J.E., Tiño, P., Kabán, A., Schwefel, H.-P. (eds.) PPSN 2004. LNCS, vol. 3242, pp. 722–731. Springer, Heidelberg (2004)
14. Yu, P.L.: A class of solutions for group decision problems. Management Science 19(8), 936–946 (1973)

Comparative Application of Multi-Objective Evolutionary Algorithms to the Voltage and Reactive Power Optimization Problem in Power Systems

S.B.D.V.P.S. Anauth and Robert T.F. Ah King

Dept. of Electrical and Electronic Engineering, University of Mauritius, Reduit, Mauritius
anauthsingh@gmail.com, r.ahking@uom.ac.mu

Abstract. This study investigates the applicability of two elitist multi-objective evolutionary algorithms (MOEAs), namely the Non-dominated Sorting Genetic Algorithm-II (NSGA-II) and an improved Strength Pareto Evolutionary Algorithm (SPEA2), in the voltage and reactive power optimization problem. The problem has been formulated mathematically as a nonlinear constrained multi-objective optimization problem where the real power loss, the load bus voltage deviations and the installation cost of additional reactive power (VAR) sources are to be minimized simultaneously. To assess the effectiveness of the proposed approach, different combinations of the objectives have been minimized simultaneously. The simulation results showed that the two algorithms were able to generate a whole set of well distributed Pareto-optimal solutions in a single run. Moreover, fuzzy logic theory is employed to extract the best compromise solution over the trade-off curves obtained. Furthermore, a performance analysis showed that SPEA2 found better convergence and spread of solutions than NSGA-II. However, NSGA-II found more extended trade-off curves in some cases and required less computational time than SPEA2.

Keywords: Optimal VAR dispatch; Elitist multi-objective evolutionary algorithms; Fuzzy logic theory.

1 Introduction

The classical reactive power optimization problem is to improve the voltage profile of the electric power system through optimal adjustments in voltage controllers while minimizing the transmission power losses and the allocation cost of additional VAR sources. However, these objectives can no longer be considered alone due to voltage security concerns. Voltage stability should be taken into consideration because voltage collapse may occur if ever the power system fails to supply the reactive power demand. Moreover, the use of shunt compensation for larger power transfer can bring the instability point closer to normal values. With the growing concerns of voltage security, voltage collapse has been considered as a reactive power problem which is mainly influenced by load behavior. Since, voltage collapse is characterized by a progressive and uncontrollable drop in voltage magnitudes, the load bus voltage deviations has been used as an objective during the reactive power optimization. Therefore,

K. Deb et al. (Eds.): SEAL 2010, LNCS 6457, pp. 424–434, 2010.

the aim of the voltage and reactive power optimization problem is to achieve a correct balance between economic and voltage security concerns.

The problem is formulated mathematically as a nonlinear constrained multi-objective optimization problem. The traditional approach to solve the problem by combining the objectives using the weighted-sum or parameterized method can no longer be considered because such an approach may either result in a single optimal solution or may require multiple runs to generate the set of non-dominated solutions by varying the parameters. Moreover, the use of the load bus voltage deviations as an objective to improve the stability of the system makes the problem a multi-objective problem with conflicting objective functions. The ability of MOEAs to explore the solution space and to find tradeoffs between multiple conflicting objectives in one single run have made them attractive for solving such problem.

Some researchers have carried out simultaneous optimization of multiple conflicting objectives using MOEAs. Among these methods, excellent results have been obtained by Abido [1] by using the Strength Pareto Evolutionary Algorithm (SPEA) [2] to generate a set of well distributed Pareto-optimal solutions with satisfactory diversity. However, the clustering technique used in SPEA to reduce the size of the archive, though maintaining a good diversity, does not guarantee the preservation of the boundary solutions. An improved version of SPEA, known as SPEA2 [3], has been proposed to eliminate the main weaknesses of its predecessor.

NSGA-II proposed by Deb et al. [4] and SPEA2 proposed by Zitzler et al. [3] provide excellent results as compared with other MOEAs proposed in the literature. This paper presents a comparative application of these two state-of-the-art MOEAs to the voltage and reactive power optimization problem. Simulation results considering different combinations of the three objectives are presented for a sample test system. Moreover, fuzzy logic theory [5] is applied to extract the best compromise solution over the trade-off curves obtained. Ultimately, some performance metrics are used to evaluate which one of the two implemented MOEAs performs better.

2 Problem Formulation

The voltage and reactive power optimization problem involves the simultaneous optimization of the real power loss, the load bus voltage deviations and the allocation cost of additional VAR sources which are conflicting objectives. Generally, the problem can be formulated as follows.

2.1 Objective Functions

Real Power Loss

The classical reactive power optimization problem of minimizing the real power loss in the transmission lines can be mathematically expressed as follows:

$$P_{loss} = \sum_{\substack{k \in N_{br} \\ k \in (i,j)}} g_k \left[V_i^2 + V_j^2 - 2V_i V_j cos(\delta_i - \delta_j) \right] \tag{1}$$

where N_{br} is the set of numbers of transmission lines in the system; g_k is the conductance of the kth transmission line between buses i and j; $V_i \angle \delta_i$ is the voltage at bus i of the kth transmission line.

<u>Voltage Deviation</u>

This objective is to minimize the sum of the magnitude of the load bus voltage deviations that can be expressed as follows:

$$V_d = \sum_{i \in N_{PQ}} |V_i - 1.0| \tag{2}$$

where N_{PQ} is the set of numbers of load buses.

<u>Investment Cost</u>

The allocation cost of additional VAR sources consists of a fixed installment cost and a variable purchase cost that can be stated as follows [6]:

$$I_c = \sum_{i \in N_C} \left(C_{fi} + C_{ci} Q_{ci} \right) \tag{3}$$

where N_C is the set of numbers of load buses for the installation of compensators; C_{fi} is the fixed installation cost of the compensator at bus i ($); C_{ci} is the per unit cost of the compensator at bus i ($/MVAR); Q_{ci} is the compensation at bus i (MVAR).

2.2 Constraints

The optimization problem is bounded by the following constraints:

<u>Equality Constraints</u>

These constraints represent typical load flow equations as follows:

$$P_{Gi} - P_{Di} - V_i \sum_{j \in N_B} V_j \left[G_{ij} \cos \delta_{ij} + B_{ij} \sin \delta_{ij} \right] = 0 \qquad i \in N_{B-1} \tag{4}$$

$$Q_{Gi} - Q_{Di} - V_i \sum_{j \in N_B} V_j \left[G_{ij} \sin \delta_{ij} - B_{ij} \cos \delta_{ij} \right] = 0 \qquad i \in N_{PQ} \tag{5}$$

where N_B is the set of numbers of total buses; N_{B-1} is the set of numbers of total buses excluding the slack bus; P_G and Q_G are the generator real and reactive power, respectively; P_D and Q_D are the load real and reactive power, respectively; G_{ij} and B_{ij} are the transfer conductance and susceptance between bus i and bus j, respectively.

Inequality Constraints

The control and state variables are bounded as shown in Table 1.

Table 1. Inequality Constraints

Control Variables		State Variables	
a) Generator voltage limits		a) Reactive power generation limits	
$V_{Gi}^{min} \leq V_{Gi} \leq V_{Gi}^{max}$ $i \in N_G$	(6)	$Q_{Gi}^{min} \leq Q_{Gi} \leq Q_{Gi}^{max}$ $i \in N_G$	(9)
b) Transformer tap setting limits		b) Load bus voltage limits	
$T_i^{min} \leq T_i \leq T_i^{max}$ $i \in N_T$	(7)	$V_{Li}^{min} \leq V_{Li} \leq V_{Li}^{max}$ $i \in N_{PQ}$	(10)
c) Reactive power injection limits		c) Transmission line flow limit	
$Q_{Ci}^{min} \leq Q_{Ci} \leq Q_{Ci}^{max}$ $i \in N_C$	(8)	$S_k \leq S_k^{max}$ $k \in N_{br}$	(11)

2.3 Multi-Objective Formulation

The voltage and reactive power optimization problem can be formulated mathematically as a nonlinear constrained multi-objective optimization problem by aggregating the objective functions and constraints as follows:

$$\text{Minimize } [P_{loss}, V_d, I_c] \tag{12}$$

$$\text{Subject to: Equality constraints} \qquad h(x, y) = 0 \tag{13}$$

$$\text{Inequality constraints} \qquad g(x, y) \geq 0 \tag{14}$$

where

$x^T = \left[Q_{Gi\ i \in N_G}, V_{Lj\ j \in N_{PQ}}, S_{k\ k \in N_{br}} \right]$ is the vector of dependent state variables and $y^T = \left[V_{Gi\ i \in N_G}, T_{j\ j \in N_T}, Q_{Cn\ n \in N_C} \right]$ is the vector of control variables.

3 Simulation Results and Discussion

IEEE 30-bus Test System

The two elitist MOEAs were tested on the standard IEEE 30-bus test system [7]. The system consists of 6 generator buses, 24 load buses, 41 transmission lines of which four branches are in-phase transformers with assumed tapping ranges of 10% and 2 installed shunt capacitor banks at bus 10 and bus 24. The candidate buses for reactive power compensation are 10, 12, 15, 17, 20, 21, 23, 24 and 29. The lower voltage magnitude limits at all buses are 0.95 p.u while the upper limits are 1.1 p.u for PV buses and 1.05 p.u for load buses and the slack bus.

<u>Settings of the Proposed Approach</u>

The optimization techniques used in this study were implemented using C language on an Intel Core 2 Duo 2.80 GHz processor having 1GB of RAM. In all the simulations, the population size was chosen as 100; crossover and mutation probabilities were 0.99 and 0.01 respectively. The distribution index for crossover and mutation were set at 5 and 50 respectively. The simulations were run for 500 generations. The cost settings and parameters used to calculate the annual energy loss cost (E_c) and the total cost (T_c) of the power system are given in Ref. [6].

3.1 Case 1: P_{loss} and V_d Minimization

The two MOEAs are used to minimize the real power loss and the load bus voltage deviations simultaneously. The Pareto-optimal fronts (trade-off curves) obtained are shown in Figs. 1 and 2. It can be seen that the two MOEAs were able to locate the Pareto-optimal solutions with excellent diversity. Moreover, it can be noted that voltage deviation and real power loss are two conflicting objectives such that any attempt to minimize the former would increase operation cost considerably. This is because conduction loss decreases strongly with increasing generator voltage.

Table 2 shows the best solutions obtained as compared to SPEA from [1]. The optimum values of the VAR sources were not included in this paper. The best compromise solution was extracted from each trade-off curve at the maximum value of the normalized membership function, as given in Ref. [5]. It is evident that the results obtained using NSGA-II and SPEA2 are better than that obtained using SPEA in Ref. [1]. This proves that the use of the clustering technique in SPEA for diversity preservation does not guarantee the preservation of the boundary solutions.

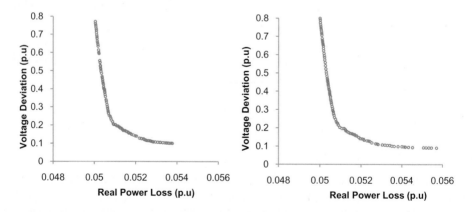

Fig. 1. Pareto-optimal front for P_{loss} and V_d using NSGA-II

Fig. 2. Pareto-optimal front for P_{loss} and V_d using SPEA2

Table 2. Best solutions for P_{loss} and V_d minimization

Control Variables & Objectives	SPEA [1]			NSGA-II			SPEA2		
	Best P_{loss}	Best V_d	Best Comp-romise	Best P_{loss}	Best V_d	Best Comp-romise	Best P_{loss}	Best V_d	Best Comp-romise
V_{G1} (p.u)	1.050	1.037	1.050	1.0500	1.0294	1.0500	1.0500	1.0173	1.0500
V_{G2} (p.u)	1.044	1.027	1.044	1.0441	1.0188	1.0416	1.0436	1.0147	1.0421
V_{G5} (p.u)	1.024	1.013	1.023	1.0240	1.0106	1.0178	1.0222	1.0164	1.0182
V_{G8} (p.u)	1.026	1.008	1.022	1.0259	1.0003	1.0195	1.0244	1.0019	1.0192
V_{G11} (p.u)	1.093	1.030	1.042	1.0016	0.9976	0.9905	1.0255	1.0366	0.9992
V_{G13} (p.u)	1.085	1.007	1.043	1.0693	0.9997	1.0210	1.0621	0.9975	1.0178
T_{11} (p.u)	1.078	1.054	1.090	1.0183	1.0123	1.0273	1.0622	1.0601	1.0322
T_{12} (p.u)	0.906	0.907	0.905	0.9538	0.9796	1.0267	0.9036	0.9442	1.0307
T_{15} (p.u)	1.007	0.928	1.020	1.0060	0.9674	1.0168	0.9932	0.9627	1.0130
T_{36} (p.u)	0.959	0.945	0.964	0.9713	0.9657	0.9889	0.9654	0.9688	0.9872
P_{loss} (MW)	5.1065	5.5161	5.1995	5.0045	5.3771	5.0912	**4.9996**	5.5707	5.0963
V_d (p.u)	0.7126	0.1477	0.2512	0.7715	0.1008	0.2075	0.8070	**0.0909**	0.2031

SPEA2 found a slightly more extended trade-off curve, as shown in Fig. 2, than that obtained by NSGA-II in Fig. 1 and, hence, it produced both the best real power loss and the best voltage deviation. Therefore, SPEA2 was capable of exploring better non-dominated solutions at the extremities of the trade-off curve than NSGA-II.

3.2 Case 2: P_{loss} and I_c Minimization

The two MOEAs are used to minimize the real power loss and the investment cost simultaneously, to assess the cost effectiveness. The trade-off curves obtained are shown in Figs. 3 and 4. Table 3 shows the best solutions obtained as compared to EP from [6]. It can be noted that the results obtained using NSGA-II and SPEA2 are better than that obtained using EP [6]. This proves that it is much better to minimize the real power loss and investment cost simultaneously using MOEAs rather than combining them linearly and using a single-objective algorithm as in Ref. [6].

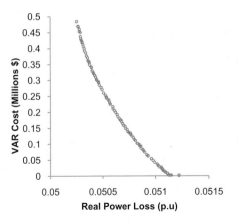

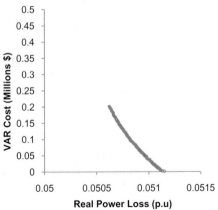

Fig. 3. Pareto-optimal front for P_{loss} and I_c using NSGA-II

Fig. 4. Pareto-optimal front for P_{loss} and I_c using SPEA2

Fig. 3 shows that NSGA-II produced a more extended trade-off curve compared to that obtained by SPEA2 in Fig. 4. This is because SPEA2 concentrated much more at maintaining a good diversity of solutions over the trade-off curve instead of refining its search to extend the latter. Hence, NSGA-II produced both the best real power loss and the best investment cost while SPEA2 produced the best total cost.

It should be noted that real power loss and VAR cost are two conflicting objectives such that any attempt to minimize the operation cost of the power system would increase investment cost considerably. This is because reactive power compensation improves the system power factor and reduces the reactive (unproductive) component of the current, thus reducing the Ohmic energy losses. Hence, Table 3 shows an increase in the expenses of the power system for the best P_{loss} solutions. Moreover, SPEA2 produced the best saving on the initial operating cost of the power system.

Table 3. Best solutions for P_{loss} and I_c minimization

Control Variables & Objectives	EP [6] Best T_c	NSGA-II Best P_{loss}	Best I_c	Best Comp-romise	SPEA2 Best P_{loss}	Best I_c	Best Comp-romise
V_{G1} (p.u)	1.050	1.0500	1.0498	1.0500	1.0500	1.0500	1.0500
V_{G2} (p.u)	1.044	1.0443	1.0427	1.0439	1.0432	1.0433	1.0436
V_{G5} (p.u)	1.023	1.0238	1.0230	1.0236	1.0220	1.0218	1.0220
V_{G8} (p.u)	1.025	1.0258	1.0234	1.0254	1.0240	1.0236	1.0240
V_{G11} (p.u)	1.050	1.0745	1.0838	1.0849	1.0811	1.0848	1.0814
V_{G13} (p.u)	1.050	1.0633	1.0646	1.0679	1.0720	1.0737	1.0737
T_{11} (p.u)	0.950	1.0310	1.0288	1.0283	0.9963	0.9945	0.9968
T_{12} (p.u)	1.100	0.9524	0.9177	0.9329	0.9831	0.9836	0.9694
T_{15} (p.u)	1.025	0.9806	0.9696	0.9797	0.9857	0.9808	0.9863
T_{36} (p.u)	1.050	0.9531	0.9463	0.9525	0.9530	0.9483	0.9489
P_{loss} (MW)	5.159	**5.0252**	5.1222	5.0698	5.0621	5.1151	5.0828
E_c ($)	2711570	**2641239**	2692218	2664692	2660618	2688520	2671522
I_c ($)	0	484110	**1058**	158577	201183	1249	95395
T_c ($)	2711570	3125349	2693276	2823269	2861801	**2689769**	2766917
% Saving	3.64	-11.06	4.29	-0.33	-1.70	**4.42**	1.67

3.3   Case 3: P_{loss}, V_d and I_c Minimization

Considering the three conflicting objective functions: real power loss, load bus voltage deviations and VAR cost simultaneously, the Pareto-optimal solutions were obtained as shown in Fig. 5 and the best solutions extracted are tabulated in Table 4.

It can be noted that SPEA2 yielded better diversity of Pareto-optimal solutions than NSGA-II. This is because SPEA2 considers density information during fitness assignment to achieve a better spread of solutions. However, Table 4 shows that NSGA-II was able to refine its search to find better extreme solutions than SPEA2. Hence, NSGA-II produced the best solutions for all three objective functions while SPEA2 found the best total cost due to the excellent diversity of its solutions.

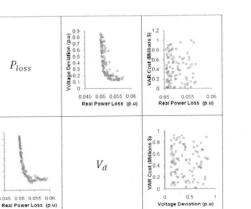

Fig. 5. Pareto-optimal solutions for P_{loss}, V_d and I_c minimization. The upper diagonal plots are for SPEA2 and lower diagonal plots are for NSGA-II.

Table 4. Best solutions for P_{loss}, V_d and I_c minimization

Control Variables & Objectives	NSGA-II				SPEA2			
	Best P_{loss}	Best V_d	Best I_c	Best Comp-romise	Best P_{loss}	Best V_d	Best I_c	Best Comp-romise
V_{G1} (p. u)	1.0500	1.0291	1.0253	1.0489	1.0500	1.0470	1.0487	1.0486
V_{G2} (p. u)	1.0449	1.0271	1.0152	1.0425	1.0435	1.0329	1.0448	1.0451
V_{G5} (p. u)	1.0240	1.0057	0.9976	1.0218	1.0212	1.0113	1.0201	1.0202
V_{G8} (p. u)	1.0247	0.9963	0.9973	1.0192	1.0244	1.0001	1.0240	1.0218
V_{G11} (p. u)	1.0326	1.0627	1.0024	1.0179	1.0453	0.9857	1.0834	1.0568
V_{G13} (p. u)	1.0682	0.9974	1.0106	1.0378	1.0688	0.9915	1.0624	1.0376
T_{11} (p. u)	1.0345	1.0906	0.9993	1.0355	1.0055	1.0049	1.0371	1.0360
T_{12} (p. u)	0.9366	0.9354	0.9012	0.9348	0.9826	0.9473	0.9321	0.9876
T_{15} (p. u)	0.9950	0.9478	0.9293	1.0039	0.9936	0.9313	0.9886	0.9896
T_{36} (p. u)	0.9646	0.9541	0.9286	0.9707	0.9584	0.9643	0.9450	0.9602
P_{loss} (MW)	**5.0003**	5.4028	5.6022	5.2023	5.0085	5.4410	5.1444	5.2125
V_d (p.u)	0.8374	**0.1149**	0.1940	0.2441	0.8559	0.1220	0.6590	0.2790
E_c ($)	**2628152**	2839699	2944541	2734348	2632457	2859800	2703916	2739712
I_c ($)	847717	505104	**9796**	96819	807450	601216	18688	23414
T_c ($)	3475869	3344803	2954337	2831167	3439907	3461016	**2722604**	2763126

4 Performance Analysis

The two main goals in a multi-objective optimization are to minimize the generation distance of the solutions to the true Pareto-optimal set and to maximize the diversity of the solutions along the Pareto-front. Twenty one independent runs were carried out using each algorithm and some performance metrics were used to test how far these goals have been attained and to evaluate which one of the two MOEAs performed better. Moreover, a reference Pareto-optimal front was obtained by selecting the best non-dominated solutions from the resulting Pareto-optimal solutions.

4.1 Generation Distance and Spread Metrics

The generation distance metric [8] evaluates the closeness of the non-dominated set obtained by an algorithm to the reference Pareto-optimal front while the spread metric [4] evaluates how evenly the non-dominated solutions are distributed in the objective space. From Table 5, it can be deduced that SPEA2 has the smallest generation distance and it gives better convergence and diversity of solutions than NSGA-II.

Table 5. Generation Distance and Spread Metrics

	Generation Distance Metric		Spread Metric	
	NSGA-II	SPEA2	NSGA-II	SPEA2
Mean	0.001103	**0.000698**	0.455250	**0.386633**
Variance	2.0600E-07	**1.2605E-07**	0.000222	**0.000699**

4.2 Statistical Analysis

A statistical analysis was performed using the Mostats5 toolbox [9] which samples the attainment surfaces of the two MOEAs throughout the fitness space to determine the percentage by which each algorithm outperforms the other. The results in Table 6 show that SPEA2 was unbeaten in the entire fitness space covered by the two MOEAs while in 52.5% of the fitness space it outperformed NSGA-II. NSGA-II did well in part of the fitness space but not enough to beat SPEA2 at the 95% confidence level.

Table 6. Statistical Analysis

	NSGA-II	SPEA2
Unbeaten (%)	47.5	**100**
Beats all (%)	0	**52.5**

Table 7. Computational Time of NSGA-II and SPEA2

	Computational Time (seconds)					
	Case 1		Case 2		Case 3	
	NSGA-II	SPEA2	NSGA-II	SPEA2	NSGA-II	SPEA2
Mean	**94.911**	195.702	**81.729**	158.317	**97.783**	345.043
Standard Deviation	1.299	11.725	2.337	14.001	1.187	11.292

4.3 Computational Complexity

Table 7 shows the computational time of the two algorithms during the different simulation cases considered. It can be noted that NSGA-II required the least computational time while SPEA2 took about twice as much time as the former for the bi-objective simulation cases and about 3.5 times as much time for the tri-objective simulation case. This is because the truncation operator used in SPEA2 is more computationally expensive than the non-dominated sort used in NSGA-II.

5 Conclusions

A comparative application of two elitist MOEAs, namely NSGA-II and SPEA2, was presented for the voltage and reactive power optimization problem in power systems. Firstly, a bi-objective optimization problem was considered on the IEEE 30-bus test system where the real power loss and the load bus voltage deviations were minimized. The results obtained were better than that obtained using SPEA in Ref. [1]. Moreover, the two MOEAs were able to locate well distributed Pareto-optimal solutions with excellent diversity. Then, to assess the cost effectiveness of the proposed approach, the real power loss and investment cost were minimized simultaneously. The results obtained during this simulation were better than that obtained using EP in Ref. [6], in which the two objectives were combined linearly. Finally, the tri-objective optimization problem considering real power loss, load bus voltage deviations and investment cost simultaneously has been considered. The results showed that SPEA2 concentrated much more at maintaining a good diversity of solutions over the trade-off curves rather than exploring better solutions at the extremities. Moreover, fuzzy logic theory was employed to extract the best compromise solution over the trade-off curves obtained during the different simulation cases considered. Ultimately, a performance analysis showed that SPEA2 found better convergence and spread of solutions than NSGA-II. However, NSGA-II found more extended trade-off curves in some cases and required less computational time than SPEA2.

References

1. Abido, M.A., Bakhashwain, J.M.: Optimal VAR Dispatch Using a Multiobjective Evolutionary Algorithm. International Journal of Electrical Power & Energy Systems 27(1), 13–20 (2005)
2. Zitzler, E., Thiele, L.: Multiobjective evolutionary algorithms: A comparative case study and the strength pareto approach. IEEE Transactions on Evolutionary Computation 3(4), 257–271 (1999)
3. Zitzler, E., Laumanns, M., Thiele, L.: SPEA2: Improving the Strength Pareto Evolutionary Algorithm for Multiobjective Optimization. In: Giannakoglou, K., et al. (eds.) Evolutionary Methods for Design, Optimisation and Control with Application to Industrial Problems (EUROGEN 2001), International Center for Numerical Methods in Engineering (CIMNE), pp. 95–100 (2002)
4. Deb, K., Pratap, A., Agrawal, S., Meyarivan, T.: A Fast and Elitist Multiobjective Genetic Algorithm: NSGA-II. IEEE Transactions on Evolutionary Computation 6(2), 182–197 (2002)
5. Dhillon, J.S., Parti, S.C., Khotari, D.P.: Stochastic Economic Load Dispatch. Electric Power Systems Research 26, 179–186 (1993)
6. Lai, L.L., Ma, J.T.: Evolutionary Programming Approach to Reactive Power Planning. IEE Proceedings-Generation Transmission Distribution 143(4), 365–370 (1996)

7. Alsac, O., Scott, B.: Optimal Load Flow with Steady-State Security. IEEE Transactions on Power Apparatus and Systems PAS-93, 745–751 (1974)
8. Veldhuizen, D.: Multiobjective Evolutionary Algorithms: Classifications, Analyses, and New Innovation. PhD thesis, Department of Electrical Engineering and Computer Engineering, Airforce Institute of Technology, Ohio (1999)
9. Knowles, D., Corne, W.: Approximating the non-dominated front using the Pareto archived evolution strategy. Evolutionary Computation Journal 8(2), 149–172 (2000)

Bayesian Reliability Analysis under Incomplete Information Using Evolutionary Algorithms

Rupesh Kumar Srivastava and Kalyanmoy Deb

Kanpur Genetic Algorithms Laboratory
Department of Mechanical Engineering
Indian Institute of Technology Kanpur
Kanpur, U.P. - 208016, India
{rupeshks,deb}@iitk.ac.in

Abstract. During engineering design, it is often difficult to quantify product reliability because of insufficient data or information for modeling the uncertainties. In such cases, one needs a reliability estimate when the functional form of the uncertainty in the design variables or parameters cannot be found. In this work, a probabilistic method to estimate the reliability in such cases is implemented using Non-Dominated Sorting Genetic Algorithm-II. The method is then coupled with an existing RBDO method to solve a problem with both epistemic and aleatory uncertainties.

Keywords: probability, epistemic uncertainty, design optimization.

1 Introduction

Deterministic optimal results are usually unreliable because there exist uncertainties in the design variables and parameters which result in output variation. These uncertainties can be classified into two major types. Aleatory uncertainties are those due to unpredictable variability in the value of a quantity, and a characterization of this variability is usually available. Epistemic uncertainties are due to lack of information about the variability, so that such a variation can not be characterized explicitly. Input variation due to aleatory uncertainty is fully accounted for in Reliability Based Design Optimization (RBDO). RBDO does not, however, consider the fact that in actual design, much of the information regarding the uncertain quantities is available in the form of a set of finite samples. These samples are usually not enough to infer probability distributions, and in many cases there is no reason to assume that they would follow any standard distribution. Also, collecting more samples is often not possible due to cost or time constraints. Techniques like the Bayesian probability based [1], Possibility based [2] and Evidence based [3] methods have been suggested to deal with low levels of information about the uncertainties.

In this work, we have implemented the Bayesian approach to reliability estimation using an evolutionary algorithm. We have then combined the Bayesian approach with an RBDO technique to yield a general Bayesian RBDO (BRBDO)

K. Deb et al. (Eds.): SEAL 2010, LNCS 6457, pp. 435–444, 2010.

algorithm. The BRBDO analysis results for a well known test problem involving variables with aleatory uncertainty, parameters with aleatory uncertainty and a variable with epistemic uncertainty are shown. Using an evolutionary algorithm, our results show how a decision making process can be facilitated for such a problem.

2   Reliability Based Design Optimization

In reliability-based optimization, uncertainties in the design are embodied as random design variables $\mathbf{X}$ and random design parameters $\mathbf{P}$, and the problem is formulated as:

$$\begin{aligned} &\underset{\mu_{\mathbf{X}}}{\text{minimize}} \quad f\left(\mu_{\mathbf{X}}, \mu_{\mathbf{P}}\right) \\ &\text{subject to:} \quad Pr\left[g_j\left(\mathbf{X}, \mathbf{P}\right) \geq 0\right] \geq R_j, \, j = 1, \ldots, J \end{aligned} \quad (1)$$

The objective of the problem is to minimize f with respect to the means of the random variables given the means of the random parameters . The problem is subject to the constraints that the probability of design feasibility is greater than or equal to some value R_j corresponding to each constraint g_j, for all $j = 1, \ldots, J$. The quantity R_j is the target reliability for the jth probabilistic constraint. Solution to a reliability-based optimization problem is called the optimal-reliable design.

3   Bayesian Approach

3.1   Bayesian Inference Method

In this section, we outline the Bayesian approach as suggested by Gunawan and Papalambros [1]. For a Bernoulli sequence [4], the probability of having r successful trials out of N trials follows a Binomial distribution, where p is the probability of success in each trial. Thus,

$$Pr\left(r \mid N, p\right) = \binom{N}{r} p^r \left(1 - p\right)^{N-r} \quad (2)$$

When the probability of success p is unknown, its distribution can be calculated using Bayes' rule by a process known as Bayesian inference. Given r successes out of N trials and a prior distribution $Pr\left(r \mid p\right)$, the posterior distribution of p given by $Pr\left(p \mid r\right)$ can be calculated as

$$Pr\left(p \mid N, r\right) = \frac{Pr\left(p\right) \times Pr\left(r \mid p\right)}{\int_0^1 Pr\left(p\right) \times Pr\left(r \mid p\right) dp} \quad (3)$$

When the trials are done for the first time, no previous information is available and $Pr\left(r \mid p\right) = U\left(0, 1\right)$, a uniform distribution can be used. If a prior distribution is known from a previous set of trials, the same can be used to calculate the posterior distribution. Thus, the distribution of p can be *updated* as more and more trials are done and more information is available. For the uniform prior distribution, the posterior distribution is the following Beta distribution where $\alpha = r + 1$ and $\beta = N - r + 1$,

$$Pr\left(p \mid N, r\right) = Beta(\alpha, \beta) = \frac{\Gamma\left(\alpha + \beta\right)}{\Gamma\left(\alpha\right)\Gamma\left(\beta\right)} p^{\alpha-1}\left(1 - p\right)^{\beta-1} \tag{4}$$

We can partition the random variables and parameters in our formulation into two vectors: $\mathbf{X} = [\mathbf{X_t}, \mathbf{X_s}]$ and $\mathbf{P} = [\mathbf{P_t}, \mathbf{P_s}]$. The vectors $\mathbf{X_t}$ and $\mathbf{P_t}$ are those random variables and parameters whose probability density functions (pdf's) are known. The vectors $\mathbf{X_s}$ and $\mathbf{P_s}$, on the other hand, are those random variables and parameters whose pdf's are not known, and instead, only some samples are available.

For a probabilistic constraint $Pr\left[g_j\left(\mathbf{X}, \mathbf{P}\right) \geq 0\right] = P_{g_j}\left(0\right)$, we can obtain a distribution estimate of $P_{g_j}\left(0\right)$ using $E(r)$, the expected number of feasible realizations of the design.

$$E(r) = \sum_{k=1}^{N} Pr\left[g_j\left(\mathbf{X_t}, \mathbf{P_t}\right) \geq 0 \mid (\mathbf{X_s}, \mathbf{P_s})_k\right] \tag{5}$$

The posterior distribution estimate of $P_{g_j}(0)$ is given by:

$$P_{g_j}\left(0\right) = Beta\left(E\left(r\right) + 1, N - E\left(r\right) + 1\right) \tag{6}$$

For any design, the confidence of that design with respect to the jth inequality constraint is defined to be the probability that it will meet or exceed the reliability target.

$$\zeta_j = Pr\left[P_{g_j}\left(0\right)\big|_{\mu_{\mathbf{X}}} \geq R_j\right]; \; j=1, \ldots, J \tag{7}$$

In Figure 1 the area of the shaded region represents the confidence for a hypothetical case when the desired R_j is 0.90 and $E(r) = 21$ for a case of 25 samples. A $\zeta_j = 0$ means that the design is certainly unreliable, while a $\zeta_j = 1$ means that the design certainly meets or exceeds the target. The confidence given in 7 can also be written as:

$$\zeta_j\left(\mu_{\mathbf{X}}\right) = 1 - \phi_{B_j}\left(R_j\right) \tag{8}$$

where $\phi_{B_j}\left(.\right)$ is the cumulative density function (cdf) of the jth Beta distribution, for all $j = 1, \ldots, J$.

The ζ_j's can be lumped into $\zeta_s\left(\mu_{\mathbf{X}}\right)$, a quantity called the overall confidence of a design. Using this measure the multi-objective problem becomes:

$$\begin{aligned} &\underset{\mu_{\mathbf{X}}}{\text{minimize}} \;\; f\left(\mu_{\mathbf{X}}, \mu_{\mathbf{P}}\right), \\ &\underset{\mu_{\mathbf{X}}}{\text{maximize}} \;\; \zeta_s\left(\mu_{\mathbf{X}}\right), \\ &\text{subject to: } 0 \leq \zeta_s\left(\mu_{\mathbf{X}}\right) \leq 1. \end{aligned} \tag{9}$$

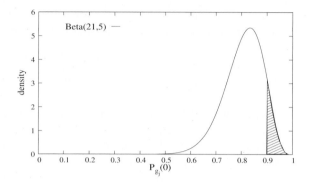

Fig. 1. Obtaining confidence (shaded region) from the distribution of $P_{g_j}(0)$

It has been shown that the maximum confidence is related to the number of samples and the reliability desired by the equation:

$$\zeta_s^{max} = 1 - R^{N+1} \tag{10}$$

This is because the maximum value of $E(r)$ for any set of samples is N, which corresponds to the rightmost distribution of $P_{g_j}(0)$ given by $Beta(N+1, 1)$. The maximum confidence for a given reliability will equal to the confidence obtained from this rightmost distribution.

3.2 Results Given by an EA for a Two Variable Problem

The following problem is solved here using the non-dominated sorting genetic algorithm (NSGA-II [5]):

$$
\begin{aligned}
\underset{\mu_{\mathbf{X}}}{\text{minimize}} \quad & f\left(\mu_{\mathbf{X}}\right) = \mu_{X_1} + \mu_{X_2}, \\
\text{subject to:} \quad & Pr\left[g_1 : 1 - \frac{X_1^2 X_2}{20} \leq 0 \right] \geq R_1, \\
& Pr\left[g_2 : 1 - \frac{(X_1 + X_2 - 5)^2}{30} - \frac{(X_1 - X_2 - 12)^2}{120} \leq 0 \right] \geq R_2, \\
& Pr\left[g_3 : 1 - \frac{80}{X_1^2 + 8X_2 - 5} \leq 0 \right] \geq R_3.
\end{aligned}
\tag{11}
$$

where $Pr_{X_1} = \dfrac{e^{-5X_1^4 + 1.5X_1^2 + 0.5X_1}}{1.614}$ was assumed to get the samples, X_2 follows $Beta(1.5, 5)$ and $R_1 = R_2 = R_3 = 0.95$ were the reliability targets.

Figure 2 shows the results obtained when 25 and 135 samples of X_1 are used. The evolutionary algorithm results in a Pareto-optimal front for each case. The front represents the trade-off between objective function value and the confidence in the design at that design point. As Equation 10 predicts, it is possible to have almost 100% confidence at some design points using 135 samples for the given reliability target.

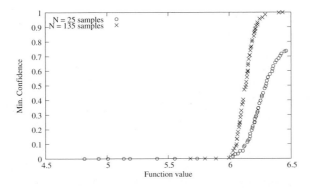

Fig. 2. Trade-off of Confidence vs. Function value for fixed reliability (95%) for all constraints (two variable problem)

4   A More Practical Approach

So far, the Bayesian approach was used to obtain designs with varying levels of confidence for a particular desired reliability. We propose an alternative approach wherein, we ask the designer to fix the confidence level he desires in the design, and perform a multi-objective optimization to yield a trade-off front between the objective function value and the maximum reliability that can be achieved with the given confidence. This will allow the designer to make a more practical decision when choosing a design. Most designers will prefer a high confidence value (depending on the number of samples, but limited by the maximum reliability they seek, according to (10)) and would prefer to see how much they will need to sacrifice on the objective function value in order to have a more reliable design.

Such a multi-objective optimization task can be readily performed using an evolutionary algorithm like NSGA-II. For this purpose, the ζ_j desired is used to obtained the R_j for the obtained P_{g_j} distribution, which is the value of R_j for which the cumulative density of P_{g_j} becomes equal to $(1 - \zeta_j)$. We then define $R_s = \min. R_j$ for $j = 1, \ldots, J$, using which the problem now becomes:

$$\begin{aligned} &\underset{\mu_{\mathbf{X}}}{\text{minimize}} \ \ f\left(\mu_{\mathbf{X}}, \mu_{\mathbf{P}}\right), \\ &\underset{\mu_{\mathbf{X}}}{\text{maximize}} \ \ R_s\left(\mu_{\mathbf{X}}\right), \\ &\text{subject to:} \ \ g_j\left(\mathbf{X}, \mathbf{P}\right) \geq 0 \, j = 1, \ldots, J \end{aligned} \quad (12)$$

Figure 3 shows a sample result of the two variable problem obtained using NSGA-II using a population size of 60 and 50 generations. The confidence in this case was fixed at 95% and 25 samples of X_1 were used.

To perform a more detailed analysis, it may be useful to compare the trade-offs between reliability and objective function value for various confidence levels. Such a comparison for the same two variable problem is shown in Figure 4. It can be clearly seen that if one needs a higher confidence for the same reliability, one must sacrifice on the function value.

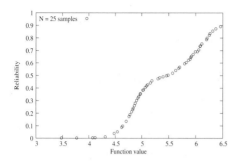
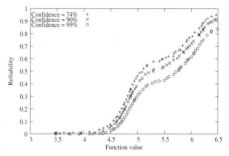

Fig. 3. Bayesian Reliability front for 95% confidence in each constraint and 25 samples of X_1 (two variable probem)

Fig. 4. Comparison of Bayesian Reliability fronts for different confidence values and 25 samples of X_1 (two variable problem)

5   Proposed Bayesian RBDO Algorithm for a General Problem

A general design optimization problem will involve several variables/parameters, some with aleatory uncertainty while others with epistemic uncertainty. Also, some constraints might involve variables with epistemic uncertainty while others may not. In such a situation, we suggest the use of Fast Reliability Index Approach (FastRIA [6]) to find the probability of feasibility of a constraint. FastRIA method is used to find a point on the constraint boundary which is closest to the solution, called the Most Probable Point (MPP [7]) of failure. Assuming linearity for the constraint in the vicinity of the MPP, a value for the reliability can be obtained which is a good approximation for non linear constraints as well. This method has earlier been used for RBDO when the variables have aleatory uncertainty characterized by a normal distribution.

5.1   Overall Algorithm

We now outline the overall algorithm for performing reliability based optimization for a problem with several constraints and variables/parameters having different types of uncertainties. Algorithm 1 shows the steps to be followed. The outlined algorithm takes into account different types of constraints when some variables have aleatory uncertainty while others have epistemic uncertainty.

5.2   Car Side-Impact Problem

The problem formulation [8] involves the minimization of the weight of the car subject to EEVC restrictions on safety performance in terms of 10 constraints. There are 11 design variables. For our study, we partition the 11-variable vector into 3 sets: variables with aleatory uncertainty $\mathbf{X_t} = \{X_1, X_2, X_3, X_4, X_6, X_7\}$, variable with epistemic uncertainty $\mathbf{X_s} = \{X_5\}$, and parameters with aleatory

---

Algorithm 1. Proposed Bayesian RBDO algorithm

---

foreach *constraint* **do**
    if *constraint involves variables with epistemic uncertainty* **then**
        foreach *sample set of* $(\mathbf{X_s}, \mathbf{P_s})_k$ **do**
        | use FastRIA to find $Pr\left[\, g_j\left(\mathbf{X_t}, \mathbf{P_t}\right) \geq 0 \,\middle|\, (\mathbf{X_s}, \mathbf{P_s})_k \,\right]$;
        end
        obtain $E(r)$ using Equation 5;
        use $E(r)$ (sum of the reliabilities obtained) to find the posterior
        distribution of $F_{g_j}(0)$ using Equation 6;
        using the desired confidence value, calculate the reliability at the point
        using Equation 8;
    end
    else
    | use FastRIA to directly find the reliability at the point;
    end
end
Find minimum reliability $R_s\left(\mu_\mathbf{X}\right)$ over all constraints;
Perform a multi-objective optimization procedure with an additional goal to
maximize $R_s\left(\mu_\mathbf{X}\right)$;

---

uncertainty $\mathbf{P_t} = \{X_8, X_9, X_{10}, X_{11}\}$. The variables/parameters with aleatory uncertainty are normally distributed about their means with standard deviations as given in Table 1, along with the description of variables and parameters and their upper and lower bounds. All quantities are in mm.

The parameters X_8, X_9, X_{10} and X_{11} are assumed to have a normal distribution with the given standard deviations and a fixed mean of 0.345, 0.192, 0 and 0 mm respectively. Thus there are seven decision variables. The Bayesian RBDO procedure is applied to the problem using NSGA-II for the optimization task. 150 generations are run for a population size of 100. A set of 25 samples of X_5 generated from a normal distribution using a standard deviation of 0.03 is used. The results obtained as a trade-off between Weight and Reliability are shown in Figure 5 for a confidence value of 90%. For 25 samples, the maximum reliability that can be achieved with 90% confidence is 0.915247 which is the limit of the Pareto-optimal front.

5.3 Analysis of Results Obtained

Figure 6 shows the trade-off fronts obtained for different confidence levels when the number of samples of X_5 is fixed at 25. The same sample set is used to obtain the fronts. The front from RBDO analysis is also shown for comparison. It can be seen that as we try to have more confidence in the design for the same number of samples, we have to compromise more and more on the weight for the same reliability. The maximum reliability obtainable also decreases as the confidence increases. The RBDO front is the one with complete information and thus is the lowest one, while Bayesian analysis has less information due to epistemic uncertainty, leading to worse fronts than the RBDO fronts.

Table 1. Description of variables and parameters for the car side-impact problem

Variable	Description	Uncertainty	Standard Deviation	Lower Bound	Upper bound
X_1	B-pillar inner	Aleatory	0.03	0.5	1.5
X_2	B-pillar reinforcement	Aleatory	0.03	0.45	1.35
X_3	Floor side inner	Aleatory	0.03	0.5	1.5
X_4	Cross member	Aleatory	0.03	0.5	1.5
X_5	Door beam	Epistemic	–	0.875	2.625
X_6	Door belt line	Aleatory	0.03	0.4	1.2
X_7	Roof rail	Aleatory	0.03	0.4	1.2
X_8	Material of B-pillar inner	Aleatory	0.006	–	–
X_9	Material of floor side inner	Aleatory	0.006	–	–
X_{10}	Barrier height	Aleatory	10	–	–
X_{11}	Barrier hitting position	Aleatory	10	–	–

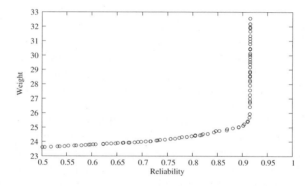

Fig. 5. Weight vs. Reliability Pareto-optimal front using 25 samples of X_5 with 90% confidence

A comparison of fronts for varying number of samples of X_5 with a fixed level of confidence (90%) is shown in Figure 7, along with the RBDO front for comparison. In this case we observe that the Bayesian RBDO front approaches the RBDO front as the number of samples increases. More samples add to the available information about the variable and the result approaches the case of complete information. It must be kept in mind that Bayesian RBDO results depend on the sample set, and will change if the samples change, even if their number remains the same. This is a consequence of having epistemic uncertainty in the design variables. The behaviour of the fronts is also interesting in that points around C seem to be *preferred* points. To the right of C, one needs to make a large sacrifice on weight in order to gain some reliability while towards the left of C, a very small improvement in weight is obtained for a large compromise on reliability. Such points are called knee points [9]. A comparison of knee

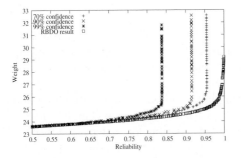

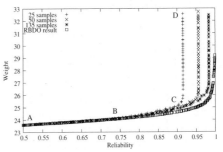

Fig. 6. Pareto-fronts obtained using 25 samples of X_5 with different confidence levels

Fig. 7. Pareto-fronts obtained using different number of samples of X_5 with 90% confidence

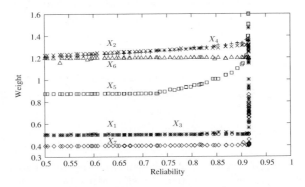

Fig. 8. Variation of design variables (for 25 samples of X_5, 90% confidence)

points in Table 2 shows that although the fronts for different number of samples change, the preferred solutions are quite close in terms of objective function value (weight).

The variation of the variable values with reliability is shown in Figure 8. We also note in Figures 7 and 8 that in the region from A to B, only the variables X_2 and X_4 are non-constant while B onwards, X_5 begins to contribute to the change in the front. The location of B is found to be a function of the number of samples of X_5 used.

Table 2. Knee Points for different confidence levels

	X_1	X_2	X_3	X_4	X_5	X_6	X_7	Weight	Reliability
25 samples	0.500	1.309	0.500	1.334	1.135	1.199	0.405	25.130	0.902
50 samples	0.500	1.333	0.500	1.388	1.178	1.196	0.405	25.724	0.944
135 samples	0.518	1.348	0.500	1.477	1.198	1.199	0.404	26.159	0.978
RBDO Result	0.509	1.341	0.500	1.406	0.875	1.199	0.400	25.197	0.967

6 Conclusion

In this paper we have demonstrated the application of an EA to a Bayesian technique of finding optimal-reliable designs with a given reliability for varying levels of confidence. Using an evolutionary approach, we then proposed finding an optimal-reliable non-dominated front, which can be used by a decision maker for a trade-off analysis between target reliability and optimal objective function value. The Bayesian analysis method and an RBDO technique (FastRIA) were then coupled to develop a Bayesian RBDO algorithm for a general problem, and the proposed algorithm was then applied to the car side-impact problem. Such analysis methods are much more versatile than traditional methods of reliability analysis which are not flexible in terms on handling different types of uncertainties, since many real world design optimization problems involve variables with different levels of information available about their uncertainty. Further research in this direction will lead to more practical methods capable of capturing various information levels in design problems.

Acknowledgements

The study is supported by Department of Science and Technology, Government of India under SERC-Engineering Sciences scheme (No. SR/S3/MERC/091/2009).

References

1. Gunawan, S., Papalambros, P.Y.: A Bayesian Approach to Reliability-Based Optimization With Incomplete Information. ASME J. Mech. Des. 128(4), 909–919 (2006)
2. Du, L., Choi, K.K., Youn, B.D., Gorsich, D.: Possibility-Based Design Optimization Method for Design Problems With Both Statistical and Fuzzy Input Data. ASME J. Mech. Des. 128(4), 925–935 (2006)
3. Mourelatos, Z.P., Zhou, J.: A Design Optimization Method Using Evidence Theory. ASME J. Mech. Des. 128(4), 901–908 (2006)
4. Leonard, T., Hsu, S.J.: Bayesian methods. Cambridge Series in Statistical and Probabilistic Mathematics. Cambridge University Press, Cambridge (1999)
5. Deb, K., Agrawal, S., Pratap, A., Meyarivan, T.: A fast and elitist multi-objective genetic algorithm: NSGA-II. IEEE Transactions on Evolutionary Computation 6(2), 182–197 (2002)
6. Deb, K., Gupta, S., Daum, D., Branke, J., Mall, A.K., Padmanabhan, D.: Reliability-Based Optimization Using Evolutionary Algorithms. IEEE Transactions on Evolutionary Computation 13(5), 1054–1074 (2009)
7. Du, X., Chen, W.: A Most Probable Point Based Method for Uncertainty Analysis. Journal of Design and Manufacturing Automation 4, 47–66 (2001)
8. Gu, L., Yang, R.J., Tho, C.H., Makowski, L., Faruque, O., Li, Y.: Optimization and Robustness for Crashworthiness of Side Impact. International Journal of Vehicle Design 26(4) (2001)
9. Branke, J., Deb, K., Dierolf, H., Osswald, M.: Finding Knees in Multi-objective Optimization. In: Yao, X., Burke, E.K., Lozano, J.A., Smith, J., Merelo-Guervós, J.J., Bullinaria, J.A., Rowe, J.E., Tiňo, P., Kabán, A., Schwefel, H.-P. (eds.) PPSN 2004. LNCS, vol. 3242, pp. 722–731. Springer, Heidelberg (2004)

Optimisation of Double Wishbone Suspension System Using Multi-Objective Genetic Algorithm

Aditya Arikere, Gurunathan Saravana Kumar, and Sandipan Bandyopadhyay

Department of Engineering Design, Indian Institute of Technology Madras
Chennai - 600036, India
a.arikere@gmail.com, {gsaravana,sandipan}@iitm.ac.in
http://www.ed.iitm.ac.in

Abstract. This paper presents an application of multi-objective optimisation for the design of an important component of automobiles, namely the suspension system. In particular, we focus on the *double wishbone* suspension, which is one of the most popular suspensions in use today and is commonly found on mid-range to high-end cars. The design of such mechanical systems is fairly complicated due to the large number of design variables involved, complicated kinematic model, and most importantly, multiplicity of design objectives, which show conflict quite often.

The above characteristic of the design problem make it ideally suited for a study in optimisation using non-classical techniques for multi-objective optimisation. In this paper, we use NSGA-II [5] for searching an optimal solution to the design problem. We focus on two important performance parameters, namely *camber* and *toe*, and propose objective functions which try to minimise the variation of these as the wheel travels in *jounce* and *rebound*. The *pareto-optimal* front between these two objectives are obtained using multiple formulations and their results are compared.

Keywords: Double Wishbone Suspension, Optimisation, GeneticAlgorithms.

1 Introduction

Computer aided design, analysis and optimisation has become an integral part the design of components and systems required to deliver good performance and satisfy multiple design requirements. The suspension system of an automobile is a good example of such a system, as it is a subsystem having multiple parts, and has to satisfy several needs such as ride comfort, good handling, cornering stability etc., some of which conflict mutually.

Standard suspension design methodology involves taking a basic design and then simulating, testing and modifying it repeatedly till the target ride characteristics are reached. The process can be expedited somewhat by performing design optimisations on a mathematical model of the same. This is the objective of the present work, with the focus on a particular type of suspension system, namely the double wishbone suspension. The double wishbone suspension is a very popular type of suspension found on mid-range to high-end cars. It is an independent suspension design using two (occasionally parallel) wishbone-shaped arms to locate the wheel as shown in Fig. 1(a). Each

K. Deb et al. (Eds.): SEAL 2010, LNCS 6457, pp. 445–454, 2010.

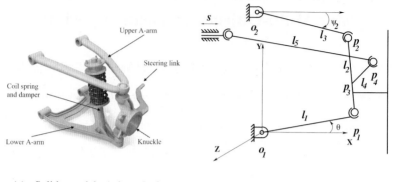

(a) Solid model (adapted from: www.carbibles.com)

(b) Kinematic model

Fig. 1. Double wishbone suspension system

wishbone or arm has two mounting points connected to the chassis and one joint at the knuckle to accept the steering input. The shock absorber and coil spring are mounted on the wishbones to control its vertical movement. Double wishbone designs allow the engineer to carefully control the motion of the wheel throughout the suspension travel, controlling parameters such as camber angle, caster angle, toe, roll centre height, scrub radius, scuff and more thereby resulting in a better tuned suspension system for good ride, handling etc. These parameters affect factors from lateral force to steering effort to anti-dive/ant-squat characteristics of the vehicle. Two of the wheel parameters that significantly effect the car handling characteristics are camber and toe, and we focus on these for the rest of the paper.

Camber is the angle the wheel plane makes with the vertical when viewed from the front of the vehicle. The camber angle of a wheel has a significant effect on the lateral force the tyre produces and hence a significant impact on the handling of the car. More importantly, the variation of camber with the wheel's vertical movement plays a big role in roll steer/bump steer. The toe is the distance between the front extremes and the back extremes of the tires when seen from the top of the vehicle. It is measured at a height from the ground that is specified by the manufacturer of the vehicle and is usually at the height of the vehicle under-body cover. Toe adds a constant slip angle to the wheels and therefore affects the lateral force generated by the tyre. It affects the agility of the vehicle and also the response of the vehicle to steering inputs. The desired performance characteristic for a suspension is to minimise the camber angle and toe in or toe out for the complete range of vertical displacement of the wheel.

A significant number of publications exist which deal with modelling and simulation of suspension systems. In [10] and [4], the authors deal with modelling and simulation of the spatial kinematics of the McPherson strut automotive suspension. Similarly, in [1], the double wishbone suspension is modelled and simulated dynamically. Neither paper deals with the design and optimisation of the suspension system. In other publications such as [6,3,7], the authors deal with the redesign or optimisation of existing suspension systems to eliminate or reduce a certain defect or fault inherent in the suspension design. These papers work on already existing and working suspension designs and refine them to obtain a better suspension system. It is important to note that these optimisation or design algorithms aim to reduce or eliminate only one defect and do not take into account other suspension characteristics. Therefore, it is possible that the optimisation in one aspect is at the cost of other performance parameters that have not been taken into consideration. In [9] and [8], the authors study the effect of the suspension kinematics on the vehicle handling characteristics and perform vehicle handling simulations. In [8], the authors also use the kinematic model of the suspension to actively adjust them for optimum ride and handling characteristics. Very few papers seem to deal with design of the suspension system from scratch, two prominent among which being [2] and [11]. While both of them do deal with suspension kinematics as a part of the design objective, their treatment is qualitative in nature. Not much is available in the public domain literature on ways to shorten the design process of a suspension system in general and double wishbone suspension in particular. Apparently, there have not been focused efforts on methods to design and optimise suspension for quantitative ride characteristics either, so far as research publications are concerned.

In this paper the authors have attempted to present an optimisation framework for the design of double wishbone suspension system for better performance, by considering two of the significant parameters, namely camber angle and toe for the complete range of vertical displacement of the wheel. The suspension kinematics problem is formed and solved in closed form using a computer algebra system, `Mathematica`. The results are used in evaluating the objective functions. This approach reduces the computational cost significantly, as symbolic simplifications are used to reduce the number of operations required. Such savings are important, as we use a Genetic Algorithm(GA)-based optimisation tool, namely `NSGA-II` [5], which require a large number of function evaluations. The choice of GA for optimisation is justified as we want a framework for a design from scratch, and as such we do not have good initial guesses that would be needed by a local search method based on the classical algorithms. Furthermore, we try to obtain the *pareto-optimal front* for the two-objective problem of camber and toe variation minimisation, representing the trade-off between the two objectives over the relevant span. This is achieved easily in the `NSGA-II` setup. The *pareto*-optimal front obtained using `NSGA-II` captures the significant region of the front used for design trade-off better as compared to that obtained using classical methods. Double wishbone suspension designs with optimal camber and toe characteristics are finally presented.

The paper is organised as follows: In section 2, the spatial kinematic model is presented. The details of the optimisation process is presented in section 3, followed by the results in section 4. Finally, the conclusions of the paper are presented in section 5.

2 Kinematic Model of the Double Wishbone Suspension

The double wishbone suspension (shown in Fig. 1(b)) can be kinematically described as a pair of coupled spatial four-bar mechanisms, resulting in three degrees of freedom. The points o_1 and o_2 describe the supports and l_1, l_2, l_3, l_4 and l_5 are the various link lengths. The first degree of freedom can be interpreted as the movement of the four-bar double wishbone loop $o_1p_1p_2o_2o_1$ which is parametrised by the orientation of the lower A-arm, θ. The second degree of freedom corresponds to the movement of the four-bar steering loop $o_1p_1p_4p_5o_1$ which is parametrised by the steering input s at point p_5. However, the present analysis is done at constant steering input. The third degree of freedom is a redundant degree of freedom that corresponds to the rotation of the link p_4p_5 about its own axis. In order to obtain the positions of all the points on the double wishbone suspension for the given values of θ and s, one needs to solve the vector loop equations for the two loops, namely, the double wishbone loop and the steering loop. To obtain the loop-closure equation for double wishbone loop, let us consider the rotation matrix to move from the frame attached to link l_1 at o_1 to the frame attached to link l_3 at o_2, given by:

$$R_1^2 = R_x(\alpha)R_y(\beta)R_z(\gamma) \tag{1}$$

where, α, β and γ are the *Euler angles* of the frame attached to link l_3 at o_2 (refer Fig. 1(b)). Here $R_x(\theta)$ stands for rotation matrix about x axis by an angle θ and so on. Let angles ψ_1 and ψ_2 be the orientation of the links l_2 and l_3. The points p_1 and p_2 on the suspension are given as:

$$p_1 = R_z(\theta)[l_1, 0, 0]^T \tag{2}$$

$$p_2 = p_1 + R_x(\alpha)R_y(\beta)R_z(\gamma)R_z(\psi_1)[l_2, 0, 0]^T \tag{3}$$

The closure equation for this loop is given by:

$$o_1 + l_1 + l_2 - l_3 - o_2 = 0 \tag{4}$$

which can to be reduced to a single scalar equation

$$l_2.l_2 - l_2^2 = 0, \quad \text{where } l_2 = p_2 - p_1. \tag{5}$$

Once the equation eq. (5) is expanded and simplified, it takes the form of eq. (6):

$$A\cos\theta + B\sin\theta + C = 0 \tag{6}$$

The coefficients of the above equation are closed-form expressions in terms of the base point coordinates, link lengths etc. For the sake of brevity, only the smallest one in size is presented below.

$$C = 2l_1o_{1x}\cos\theta - 2l_1o_{2x}\cos\theta + 2l_1o_{1y}\sin\theta - 2l_1o_{2y}\sin\theta + l_1^2 - l_2^2 + l_3^2 +$$
$$o_{1x}^2 + o_{2x}^2 - 2o_{1x}o_{2x} + o_{1y}^2 + o_{2y}^2 - 2o_{1y}o_{2y} + o_{1z}^2 + o_{2z}^2 - 2o_{1z}o_{2z} \tag{7}$$

The loop-closure equation can then be solved to obtain ψ_2 by trigonometric manipulation. Once ψ_2 is obtained, it is substituted back in l_2 to get the numerical values of l_2 and p_2. If we consider a parameter r for the knuckle link (l_2), describing the ratio of the distance of the point p_3 from p_1 to the distance of the point p_2 from p_1 then one can solve the loop closure equation for steering loop.

$$o_1 + l_1 + l_4 + l_5 - p_5 = 0$$

which can be reduced to a single scalar equation:

$$l_5.l_5 - l_5^2 = 0, \quad \text{where } l_5 = p_5 - p_4. \tag{8}$$

This loop closure equation is solved in the same way as eq. (5). Without going into details, it is noted here that the numerical values of points p_1, p_2, p_3, p_4 and p_5 can be determined as mentioned above. To solve for the performance parameters like camber, toe, etc., one needs to obtain the vector corresponding to the wheel axis. The wheel axis, is parallel to the vector w that is perpendicular to $(p_2 - p_1)$ and $(p_4 - p_3)$.

3   Optimisation of the Double Wishbone Suspension System

This section describes the framework for optimising the double wishbone suspension system for desired performance characteristics. The details of variables chosen, objective function formulation, constraints and tools for solving the same are described in the following.

3.1   Design Variables and Their Bounds

The kinematic performance of a double wishbone suspension system can be fine-tuned by choosing suitable values for several parameters which are listed in table 1. To the best knowledge of the authors, no public domain literature is available on design optimisation of double wishbone suspension system. Therefore, for the purpose of comparison, a basic design that has been reported in [1] has been taken as the reference. The values of the design parameters are listed in table 1. An initial parametric study was conducted using single objective formulation for determining the significant parameters to be considered as variables for reducing camber and toe variations separately. The combined variable set was used for multi-objective optimisation using genetic algorithm (MOGA). Variable bounds were chosen based on the initial values of the parameters from the base design given in table 1.

3.2   Objective Function

The objective function for the optimisation is formulated so as to reduce the difference between the attained variation of a given wheel parameter, say camber $g_{a,camber}(x, \theta)$, where x is the vector of design variables, to the corresponding desired variation of the same, say $g_{d,camber}(x, \theta)$, over some region of suspension operation which is parametrised by $\theta \in [\theta_{min}, \theta_{max}]$. The range of variation in θ shall correspond to the typical variation of wheel vertical displacement when encountering pot holes and bumps in a road. In the present formulation the second input from steering s is considered to be zero. The desired performance characteristic $g_d(x, \theta)$ can be kept constant with respect to θ or varied depending on the designer's intent. The objective function can be written in general as:

$$h(x, \theta) = d(g_a(x, \theta), g_d(x, \theta)), \quad \theta \in [\theta_{min}, \theta_{max}] \tag{9}$$

Table 1. Design parameters for the modelling of the suspension system and design variable and their optimised values obtained using single objective optimisation in Mathematica for camber and toe considered individually and MOGA results for toe and camber optimisation ("*" implies parameter value same as in the initial design)

Parameter, Units	Initial design from [1]	Optimised for Camber	Optimised for Toe	Solution point A on the MOGA pareto front
$[o_{1x}, o_{1y}, o_{1z}]$, m	$[0, 0, 0]$	*	*	*
$[o_{2x}, o_{2y}, o_{2z}]$, m	$[0.13, 0.16, 0.01]$	*	*	$[-0.0801, *, *]$
l_1, m	0.42	*	*	*
l_2, m	0.163	0.2	0.1807	0.1248
l_3, m	0.26	0.2929	0.2559	0.4618
$[l_{4x}, l_{4y}, l_{4z}]$, m	$[0.01, 0.02, 0]$	*	*	$[*, 0.0827, *]$
l_5, m	0.904	*	*	0.8800
r	0.125	*	*	*
$[p_{5x}, p_{5y}, p_{5z}]$, m	$[0.01, 0.02, -0.8]$	*	*	$[0.0302, 0.0372, -0.7646]$
α, radians	-0.0349	0.0334	0.0872	*
β, radians	-0.0175	0.0083	-0.0369	-0.0799
γ, radians	0.0524	0.0653	0.0543	*
Performance Characteristic				
Camber	50.6839	1.9831	-	22.0853
Toe	50.1755	-	5.7253	29.2397

where, $d(\cdot, \cdot)$ denotes any acceptable measure of distance between two functions. In the present formulation, this is done rather empirically, by sampling the domain $[\theta_{min}, \theta_{max}]$ uniformly and determining the Euclidean distance between the desired and attained values for the performance parameter under consideration.

$$f(x) = \frac{1}{n}\Sigma_{i=1}^{n}(g_a(x, \theta_i) - g_d(x, \theta_i))^2 \qquad (10)$$

The performance measure $g_a(x, \theta)$ is either camber or toe in this paper. Three multi-objective optimisation methods, namely, the weighted-sum method, the min-max method (both classical) and MOGA have been attempted for the performance optimisation of the double wishbone suspension system and results compared. In the classical methods, a multi-objective optimisation problem is converted into a series of single objective optimisation problems and the solution to the same form the *pareto*-optimal front. For the weighted sum method, the individual objectives of optimising for camber and toe have been combined using normalised weights to form a single objective as:

$$f(x) = \frac{C_{camb}}{n}\Sigma_{i=1}^{n}(g_{a,camb}(x, \theta_i) - g_{d,camb}(x, \theta_i))^2 + \frac{C_{toe}}{n}\Sigma_{i=1}^{n}(g_{a,toe}(x, \theta_i) - g_{d,toe}(x, \theta_i))^2$$

where, C_{camb}, C_{toe}, are normalised scaling weights which are systematically varied to obtain the *pareto*-optimal set. For optimisation with the min-max method, the distance between the ideal design point (in the objective space) having minimum value for both the camber and toe objectives, and the design point (again in the objective space) scaled along the different objective axes using weights, is minimised. The weights are systematically varied to obtain points on the *pareto*-optimal front. This method has been reported to provide better convergence to the desired front.

3.3 Constraints

Singularities in the mechanism, if any, will cause ψ_2 or ϕ_x (knuckle link angle) to be evaluated to complex values. This will lead to the function $g_a(x, \theta_i)$ to be evaluated to a complex value for some θ_i. This is avoided by introducing the constraint for the solution of eq. (6) for the two loops (wishbone as well as steering) as $A^2 + B^2 - C^2 \geq 0$.

4   Results and Discussion

The optimisation framework for performance enhancement of double wishbone suspension system has been accomplished. In the present work the expressions for the kinematic loop closure equations that were derived in closed form using symbolic computation in `Mathematica` aided in numerical evaluation of the same with less computational expense. The sensitivity analysis, multi-objective optimisation that has been accomplished in this work was made feasible due to the same. Initially the results of single objective optimisation of the suspension for wheel parameter camber and toe are presented. These optimisation were done considering only a limited number of variables, namely, l_2, l_3, α, β and γ. The performance parameters namely the camber and toe variations with respect to the lower A-arm angle of the suspension are plotted to visualise the performance. The desired performance for the present study has been taken as zero camber and toe for the range $\pm \pi/8$ of lower A-arm travel from the reference configuration. Single objective optimisations were performed using the tool `NMinimize` of `Mathematica` and the results are tabulated in table 1. The performance curves for camber and toe for these designs are presented in Fig. 2(a) and 2(b). As can be seen from the plots and the objective function values, the optimised suspension shows marked improvement in the wheel parameters camber and toe individually. It is important to note that while performing a single objective optimisation for one performance parameter, say, camber the other performance parameter, say, toe deteriorates (refer Fig. 2(b)) and *vise-versa* (refer Fig. 2(a)).

The design trade-off can be better addressed with multi-objective optimisation.To perform multi-objective optimisation a parametric study was done to determine the set of variables to which the two wheel parameters considered as objectives namely the camber and toe are sensitive. Camber was found to be sensitive to the parameters l_2, l_3, l_5, o_{2z}, o_{2y}, o_{2x}, p_{5z} and β. Toe was sensitive to the parameters l_3, l_{4y}, l_{4z}, l_5, o_{2x}, p_{5x}, p_{5y} and p_{5z}. These parameters were considered as variables for multi-objective optimisation. The results of multi-objective optimisation by weighted sum method, min-max method and multi-objective genetic algorithm for performance optimisation of double wishbone suspension system are presented and compared. Weighted sum method and min-max methods were solved using `NMinimize` function of `Mathematica`. As mentioned before, weights were systematically varied in both the methods to obtain a spread

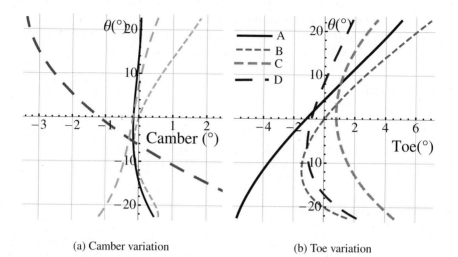

(a) Camber variation   (b) Toe variation

Fig. 2. Camber and toe variation with the lower A-arm angle (θ) for various cases of optimisation: A. Optimised for camber alone, B. Original design, C. MOGA result (point A in Fig. (3)), D. Optimised for toe alone.

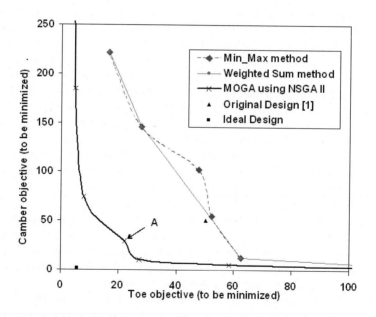

Fig. 3. *Pareto*-optimal front in the objective space for minimising camber as well as toe

of solution in the *pareto*-optimal front. In-spite of the choice of a large set of weights, several of the optimisation runs converged to the same point on the objective space. For the case of min-max method, only 5 unique points and for the case of weighted sum method 4 unique points were obtained to represent the front. These fronts are shown in the Fig. 3. The point corresponding to the original design as well as that corresponding to the ideal design where both camber and toe are minimum (refer solutions in table 1 obtained from single objective optimisation) are also depicted in the figure. The multi-objective optimisation was done using NSGA-II. A real coded NSGA-II[1] with probability of crossover as 0.9, mutation as 0.1 and 100 generations of evolution was chosen. Initially, a population size of 500 was chosen in the anticipation that at the end of optimisation there will be enough converged (i.e., rank 1) solutions representing the *pareto*-optimal front. But since the convergence was poor the same was increased to 20000 when a set of 14 solutions of rank 1 were obtained. These are also plotted in Fig. 3. The solutions obtained from NSGA-II are well spread as well as converged. The front obtained from NSGA-II is better than that obtained from weighted sum method as well as min-max methods since the same is closer to the ideal point at the knee region, which is significant for design trade-off. The trade-off solution point A on the pareto front obtained from NSGA-II shows reasonable improvements in both performance characteristics namely camber and toe (refer Fig. 2 and table 1). The convergence to the front though has been computationally expensive due to the large population size needed to capture the *pareto*-optimal front. The authors intend to do study the effect of various NSGA-II parameters on the convergence of the front and the computational efficiency.

5   Conclusions

The paper proposes a framework for optimising the design of double wishbone suspension systems. *Pareto*-optimal solutions to the mechanism synthesis problem are generated so that the designer can choose from the set of solutions under contradicting objectives of minimising camber as well as toe – the two important performance parameters leading to better vehicle handling characteristics. The results show that the method based on NSGA-II converges to solutions better than that based on classical methods. The present research can be extended to include more objectives and design variables, and the same approach may be adopted to other suspension design problems as well.

References

1. Attia, H.A.: Dynamic modelling of the double wishbone motor-vehicle suspension system. European Journal of Mechanics, 167–174 (2001)
2. Bael, S., Lee, J.M., Chu, C.N.: Axiomatic design of automotive suspension systems. CIRP Annals - Manufacturing Technology, 115–118 (2002)

---

[1] The code is developed in the Kanpur Genetic Algorithms Laboratory (KanGAL) and is freely download-able from http://www.iitk.ac.in/kangal/codes.shtml.

454   A. Arikere, G. Saravana Kumar, and S. Bandyopadhyay

3. Bian, X.L., Song, B.A., Becker, W.: The optimisation design of the McPherson strut and steering mechanism for automobiles. Forschung im Ingenieurwesen, 60–65 (2004)
4. Cronin, D.L.: MacPherson strut kinematics. Mechanism and Machine Theory, 631–644 (1981)
5. Deb, K., Pratap, A., Agarwal, S., Meyarivan, T.: A fast and elitist multi-objective genetic algorithm: NSGA-II. IEEE Transaction on Evolutionary Computation, 181–197 (2002)
6. Felzien, M.L., Cronin, D.L.: Steering error optimization of the MacPherson strut automotive front suspension. Mechanism and Machine Theory, 17–26 (1982)
7. Habibi, H., Shirazi, K.H., Shishesaz, M.: Roll steer minimization of McPherson-strut suspension system using genetic algorithm method. Mechanism and Machine Theory, 57–67 (2007)
8. Lee, S.H., Lee, U.K., Han, C.S.: Enhancement of vehicle handling characteristics by suspension kinematic control. In: Proceedings of the Institution of Mechanical Engineers, pp. 197–216 (2001)
9. Makita, M.: An application of suspension kinematics for intermediate level vehicle handling simulation. Society of Automotive Engineers of Japan (JSAE), 471–477 (1999)
10. Mántaras, D.A., Luque, P., Vera, C.: Development and validation of a three-dimensional kinematic model for the McPherson steering and suspension mechanisms. Mechanism and Machine Theory, 603–619 (2003)
11. Raghavan, M.: Number and dimensional synthesis of independent suspension mechanisms. Mechanism and Machine Theory, 1141–1153 (1996)

Self-Controlling Dominance Area of Solutions in Evolutionary Many-Objective Optimization

Hiroyuki Sato[1], Hernán E. Aguirre[2,3], and Kiyoshi Tanaka[3]

[1] Faculty of Informatics and Engineering, The University of Electro-Communications
1-5-1 Chofugaoka, Chofu, Tokyo 182-8585 Japan
[2] International Young Researcher Empowerment Center, Shinshu University
4-17-1 Wakasato, Nagano, 380-8553 Japan
[3] Faculty of Engineering, Shinshu University
4-17-1 Wakasato, Nagano, 380-8553 Japan

Abstract. Controlling dominance area of solutions (CDAS) relaxes the concepts of Pareto dominance with an user-defined parameter S. This method enhances the search performance of dominance-based MOEA in many-objective optimization problems (MaOPs). However, to bring out desirable search performance, we have to experimentally find out S that controls dominance area appropriately. Also, there is a tendency to deteriorate the diversity of solutions obtained by CDAS when we decrease S from 0.5. To solve these problems, in this work, we propose a modification of CDAS called self-controlling dominance area of solutions (S-CDAS). In S-CDAS, the algorithm self-controls dominance area for each solution without the need of an external parameter. S-CDAS considers convergence and diversity and realizes a fine grained ranking that is different from conventional CDAS. In this work, we use many-objective 0/1 knapsack problems with $m = 4 \sim 10$ objectives to verify the search performance of the proposed method. Simulation results show that S-CDAS achieves well-balanced search performance on both convergence and diversity compared to conventional NSGA-II, CDAS, IBEA$_{\epsilon+}$ and MSOPS.

1 Introduction

The research interest of the multi-objective evolutionary algorithm (MOEA) [1] community has rapidly shifted to develop effective algorithms for many-objective optimization problems (MaOPs) because more objective functions should be considered and optimized in recent complex applications. However, in general, Pareto dominance-based MOEAs such as NSGA-II [2] and SPEA2 [3] noticeably deteriorate their search performance as we increase the number of objectives to more than 4 [4,5]. This is because these MOEAs meet difficulty to rank solutions in the population, i.e., most of the solutions become non-dominated and the same rank is assigned to them, which seriously spoils proper selection pressure required in the evolution process.

To overcome this problem and induce more fine-grained ranking of solutions in MaOPs, recently some selection methods such as indicator, aggregating function

K. Deb et al. (Eds.): SEAL 2010, LNCS 6457, pp. 455–465, 2010.

and extending Pareto dominance based approaches have been proposed and verified its search performance in MaOPs. Indicator-based evolutionary algorithm (IBEA) [6] introduces fine grained ranking of solutions by calculating fitness value based on some indicators which measure the degree of superiority for each solution in the population. Multiple single objective Pareto sampling (MSOPS) aggregates fitness vector with multiple weight vectors, and reflects the ranking of solutions calculated for each weight vector in parent selection [4]. Compared with conventional Pareto dominance based NSGA-II, superiority of IBEA and MSOPS has been reported on continuous many objective optimization problems [9]. CDAS [7] relaxes the concepts of Pareto dominance by controlling dominance area of solutions using an user-defined parameter S to induce appropriate selection pressure in MOEA. CDAS shows better search performance in MaOP than NSGA-II due to convergence by the effects of fine grained ranking of solutions using $S < 0.5$ [8]. However, to bring out desirable search performance, we have to experimentally find out (select) S that controls dominance area appropriately. Also, there is a tendency to deteriorate diversity of obtained solutions by CDAS when we decrease S from 0.5.

In this work, we focus on CDAS [7] and propose a modification of CDAS, which is called self-controlling dominance area of solutions (S-CDAS), to solve the aforementioned problems and achieve well-balanced search performance between convergence and diversity toward optimal Pareto front on MaOPs. When calculating dominance relation among solutions, S-CDAS self-controls dominance area for each solution without using an external parameter, while the conventional CDAS controls using a same parameter S for all solutions. Due to this self-control of dominance area, S-CDAS realizes different fine grained ranking from conventional CDAS by considering that extreme solutions for each objective functions are never dominated by other solutions in the calculation of dominance area. In this work, we verify the search performance of the proposed method in many-objective 0/1 knapsack problems with $m = \{4, 6, 8, 10\}$ objectives by comparing with NSGA-II, IBEA$_{\epsilon+}$ [6], MSOPS [4] and CDAS [7].

2   Dominance Based MOEA and Problems in MaOPs

NSGA-II [2] is one of the well-known MOEAs that use Pareto dominance, which robust performance has been verified on a wide range of MOP especially for two or three objective optimization problems. In parent selection process, NSGA-II first classifies solutions into several layers (fronts) based on non-dominance level. Then, it selects parent solutions from higher fronts until filling up the half size of entire population. When comparing solutions that belong to a same front, NSGA-II determines superiority of solutions based on crowding distance (CD) [2] which considers solution's distribution in objective space.

When analyzing NSGA-II in MaOP, most of the solutions become non-dominated (front 1 : $\mathcal{F}_1$), and the number of solutions belonging to $\mathcal{F}_1$ exceeds the size of parent solutions $|\mathcal{P}_t|$ (half of the entire population) in early stage of the evolution. In such case, parent selection of NSGA-II becomes to rely on CD

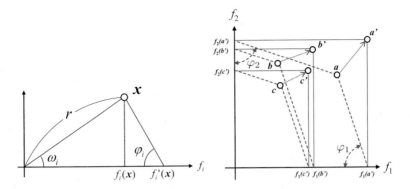

Fig. 1. Fitness modification to change the covered area of dominance

Fig. 2. Expanding dominance area by CDAS with $S < 0.5$

strongly. Consequently, the obtained POS are well-distributed in objective space, but convergence of POS towards the true POS is substantially deteriorated. To overcome this problem and enhance convergence of the obtained POS, it is necessity to improve selection pressure by discriminating non-dominated solutions using some effective manner.

3   Controlling Dominance Area of Solutions

To induce appropriate selection pressure into Pareto dominance-based MOEAs, controlling dominance area of solutions (CDAS) [7] contracts or expands the dominance area of solutions before we calculate dominance relations among solutions. For a solutions x shown in **Fig.1**, CDAS modifies the fitness value for each objective function by changing a parameter S_i in the following equation

$$f_i'(\boldsymbol{x}) = \frac{r \cdot \sin(\omega_i + S_i \cdot \pi)}{\sin(S_i \cdot \pi)} \qquad (i = 1, 2, \cdots, m), \qquad (1)$$

where $\varphi_i = S_i \cdot \pi$, r is the norm of $\boldsymbol{f}(\boldsymbol{x}) = (f_1(\boldsymbol{x}), f_2(\boldsymbol{x}), \cdots, f_m(\boldsymbol{x}))$, $f_i(\boldsymbol{x})$ is the fitness value in the i-th objective, and ω_i is the declination angle between $\boldsymbol{f}(\boldsymbol{x})$ and $f_i(\boldsymbol{x})$. In [7], a same parameter S is used for all fitness functions f_i $(i = 1, 2, \cdots, m)$. When $S < 0.5$, the i-th fitness value $f_i(\boldsymbol{x})$ is increased to $f_i'(\boldsymbol{x}) > f_i(\boldsymbol{x})$. On the other hand, when $S > 0.5$, $f_i(\boldsymbol{x})$ is decreased to $f_i'(\boldsymbol{x}) < f_i(\boldsymbol{x})$. When $S = 0.5$, $f_i'(\boldsymbol{x}) = f_i(\boldsymbol{x})$ which is equivalent to conventional dominance. **Fig.2** shows an example in the case of expanding dominance area with $S < 0.5$. In the case of conventional dominance, the solution $\boldsymbol{a}$ dominates $\boldsymbol{c}$, but $\boldsymbol{a}$ and $\boldsymbol{b}$, and $\boldsymbol{b}$ and $\boldsymbol{c}$ do not dominate each other. If we modify fitness value with $S < 0.5$ the dominance area of solutions $\boldsymbol{a}'$, $\boldsymbol{b}'$ and $\boldsymbol{c}'$ is expanded from the original one of $\boldsymbol{a}$, $\boldsymbol{b}$ and $\boldsymbol{c}$. This causes that $\boldsymbol{a}'$ dominates $\boldsymbol{b}'$ and $\boldsymbol{c}'$, and $\boldsymbol{b}'$ dominates $\boldsymbol{c}'$. That is, expansion of dominance area by smaller $S < 0.5$

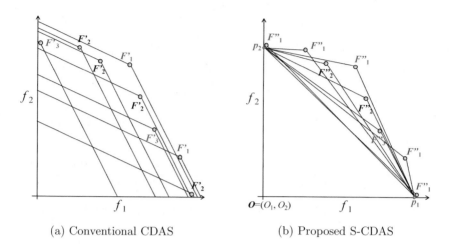

(a) Conventional CDAS   (b) Proposed S-CDAS

Fig. 3. Difference of front classification between CDAS [7] and proposed S-CDAS

works to produce a more fine grained ranking of solutions and would strengthen selection.

In case of solving MaOP, CDAS shows better search performance than conventional NSGA-II when we set $S < 0.5$. The hypervolume value for obtained non-dominated solutions is larger than NSGA-II due to convergence improvement by the effects of fine grained ranking of solutions using the optimum parameter S^* [8]. However, as a side effect, the diversity of obtained solutions deteriorates when we decrease S from 0.5. Also, to bring out desirable performance of CDAS, we have to find out appropriate parameter S experimentally.

4   Proposed Method

4.1   Motivation

To solve aforementioned problems in the conventional CDAS and achieve well-balanced search performance between convergence and diversity toward optimal Pareto front on MaOPs, the proposed S-CDAS reclassifies solutions in each front classified by NSGA-II to realize fine-grained ranking that is different from CDAS.

Fig.3 shows that difference of front classification between CDAS [7] and S-CDAS, where all solutions area included in $\mathcal{F}_1$ by the classification of NSGA-II. When we classify these solutions by the conventional CDAS, three fronts $\mathcal{F}'_1, \mathcal{F}'_2$ and $\mathcal{F}'_3$ are obtained as shown in **Fig.3** (a). In this example, since the angles of dominance area specified by all solutions are the same, only well-converged solutions distributed in a limited central region of Pareto front are included in $\mathcal{F}'_1$. Accordingly, extreme solutions having the maximum fitness values are dominated by these solutions. Thus, it becomes difficult to maintain diversity in the population. On the other hand, when we classify solutions by S-CDAS, different three fronts $\mathcal{F}''_1, \mathcal{F}''_2$ and $\mathcal{F}''_3$ are obtained as shown in **Fig.3** (b), where

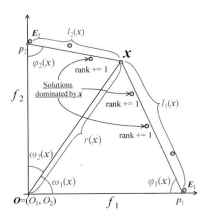

Fig. 4. Reclassification of solutions in $\mathcal{F}_1$ by the proposed algorithm

the angle of dominance area specified by each solution is different. Also, S-CDAS always guarantees the inclusion of extreme solutions in $\mathcal{F}_1''$. In other words, in the proposed method, not only highly converged solutions but also widely distributed ones are classified into higher front.

4.2   Algorithm of S-CDAS

S-CDAS reclassifies the solutions in each front $\mathcal{F}_j$ $(j = 1, 2, \cdots)$ which is classified by NSGA-II using the following procedure. **Fig.4** shows the illustration of the reclassification by the proposed algorithm.

Step 1: Move the origin to $\boldsymbol{O} = (O_1, O_2, \cdots, O_m)$ in objective space. In this work, we set $O_i = f_i^{min} - \delta$ $(i = 1, 2, \cdots, m)$, where f_i^{min} is the minimum value of the i-th objective function in $\mathcal{F}_j$ and δ is a tiny constant value.

Step 2: Create a set of landmark vectors $\mathcal{L} = \{\boldsymbol{p}_1, \boldsymbol{p}_2, \cdots, \boldsymbol{p}_m\}$, where $\boldsymbol{p}_i = (O_1, O_2, \cdots, f_i^{max} - \delta, \cdots, O_m)$, and f_i^{max} denotes the maximum value of the i-th objective function, which is derived from the extreme solution $\boldsymbol{E}_i$ in $\mathcal{F}_j$.

Step 3: Repeat the following calculation for all solutions in $\mathcal{F}_j$.

> ***Step 3-1***: For a single solution $\boldsymbol{x}$, calculate $\boldsymbol{\varphi}(\boldsymbol{x}) = (\varphi_1(\boldsymbol{x}), \varphi_2(\boldsymbol{x}), \cdots, \varphi_m(\boldsymbol{x}))$ by Eq.(2) derived from sine theorem. Here, $\varphi_i(\boldsymbol{x})$ is the angle determined by the solution $\boldsymbol{x}$ and the landmark vector $\boldsymbol{p}_i$ in the i-th objective function. $\boldsymbol{\varphi}(\boldsymbol{x})$ determines the individual dominance area of $\boldsymbol{x}$, which does never dominate extreme solutions $\boldsymbol{E}_i$ $(i = 1, 2, \cdots, m)$ having the maximum fitness value for each objective function.

$$\varphi_i(\boldsymbol{x}) = \sin^{-1}\left\{\frac{r(\boldsymbol{x}) \cdot \sin(\omega_i(\boldsymbol{x}))}{l_i(\boldsymbol{x})}\right\} \quad (i = 1, 2, \cdots, m), \qquad (2)$$

where $l_i(\boldsymbol{x})$ is Euclidean distance between the solution $\boldsymbol{x}$ and the landmark vector $\boldsymbol{p}_i$.

Step 3-2: Modify fitness values of all other solutions $\boldsymbol{y} \in \mathcal{F}_j$ by the following equation

$$f_i'(\boldsymbol{y}) = \frac{r(\boldsymbol{y}) \cdot \sin(\omega_i(\boldsymbol{y}) + \varphi_i(\boldsymbol{x}))}{\sin(\varphi_i(\boldsymbol{x}))} \quad (i = 1, 2, \cdots, m). \quad (3)$$

Step 3-3: Check dominance relations between the solution $\boldsymbol{x}$ and all other solutions $\boldsymbol{y} \in \mathcal{F}_j$. If a solution $\boldsymbol{y} \in \mathcal{F}_j$ is dominated by $\boldsymbol{x}$, the counter ($rank$) of $\boldsymbol{y}$ is incremented.

Step 4: Finally, reclassify all the solutions in $\mathcal{F}_j$ based on the accumulated $rank$ values, i.e., smaller rank corresponds to higher front, and larger rank corresponds to lower front. When multiple solutions have a same $rank$ value, they are included in the same front.

In the proposed method, reclassification procedure is performed for each non-dominated front $\mathcal{F}_j$ ($j = 1, 2, \cdots$) obtained by NSGA-II. That is, the proposed method makes front distribution fine-grained, but superiority of each solution is never overturned by S-CDAS. In other words, superiority of solutions in fronts obtained by NSGA-II is maintained even after the reclassification by S-CDAS.

After the reclassification of all fronts, the proposed algorithm selects parent solutions $\mathcal{P}_t$ from higher fronts until filling up the half size of entire population similar to NSGA-II [2]. Also, to create offspring solutions, the crowded tournament selection is applied [2].

5   Experimental Results and Discussion

5.1   Preparation

In this work, we verify the search performance of the proposed method in many-objective 0/1 knapsack problems [10] by comparing with NSGA-II [2], IBEA$_{\epsilon+}$ [6], MSOPS [4] and CDAS [7]. We generate problems with $m = \{4, 6, 8, 10\}$ objectives, $n = 100$ items, and feasibility ratio $\phi = 0.5$. For all algorithms, we adopt two-point crossover with a crossover rate $P_c = 1.0$, and apply bit-flipping mutation with a mutation rate $P_m = 1/n$. In the following experiments, we show the average performance with 30 runs, each of which spent $T = 2,000$ generations. Population size is set to $N = 200$ (size of parent and offspring population are $|P_t| = |Q_t| = 100$). In IBEA$_{\epsilon+}$, scaling parameter k is set to 0.05 similar to [6]. Also, in MSOPS, we use $W = 100$ uniformly distributed weight vectors [9], which maximizes *Hypervolume* (HV) [11] in the experiments.

In this work, to evaluate search performance of MOEA we use HV, which measures the m-dimensional volume of the region enclosed by the obtained non-dominated solutions and a dominated reference point in objective space. Here we use $\boldsymbol{r} = (0, 0, \cdots, 0)$ as the reference point. Obtained POS showing a higher value of hypervolume can be considered as a better set of solutions from both convergence and diversity viewpoints. To calculate the hypervolume, we use the improved dimension-sweep algorithm proposed by Fonseca et al. [12], which significantly reduces computational time especially for large m. To provide additional information separately on convergence and diversity of the obtained POS, in this work we also use *Norm* [13] and *Maximum Spread* (*MS*) [11], respectively. Higher value of *Norm* generally means higher convergence to true POS. On the other hand, Higher *MS* indicates better diversity in POS which can be approximated widely spread Pareto front.

5.2   Performance Comparison with Conventional CDAS

First, we observe the search performance of CDAS [7] and the proposed S-CDAS as we vary the number of objectives. **Fig.5** (a) $\sim$ (c) show results on *HV* as a combined metric of convergence and diversity, *Norm* as a measure of convergence, and *MS* as a measure of diversity. The vertical bars, overlaying the markers, represent 95% confidence intervals. For CDAS, we plot multiple results when we vary the parameter S in the range $[0.25, 0.50]$. In these figures, all the plots are normalized by the results of NSGA-II [2].

From **Fig.5** (b) and (c), the conventional CDAS increases *Norm* while decreases *MS* as we decrease S from 0.5. Obviously, there is a trade-off between convergence and diversity in the solutions obtained by CDAS. Therefore, only when we select well-balanced S between convergence and diversity, we can achieve higher *HV* values as shown in **Fig.5** (a), where the optimal parameters of CDAS to maximize HV are $S^* = \{0.50, 0.45, 0.45, 0.40\}$ for $m = \{4, 6, 8, 10\}$ objectives, respectively. On the other hand, both *Norm* and *MS* achieved by S-CDAS are higher than CDAS with $S = 0.45$. Consequently, as shown in **Fig.5** (a) the value of HV achieved by S-CDAS are higher than CDAS with any S for all $m = \{4, 6, 8, 10\}$ objectives problems.

5.3   Performance Comparison with Conventional MOEAs

Second, we compare the search performance of the proposed method (S-CDAS) with conventional MOEAs: NSGA-II, CDAS, IBEA$_{\epsilon+}$ [6] and MSOPS [4] as we vary the number of objectives. **Fig.6** (a) $\sim$ (c) show results on *HV*, *Norm* and *MS*. For CDAS, we plot only results using the optimal parameter S^* that maximizes HV. Similar to **Fig.5**, all the plots are normalized by the results of NSGA-II.

NSGA-II achieves the highest MS for $m = \{6, 8, 10\}$ objectives while *Norm* remains the minimum. That is, NSGA-II can obtain well-distributed POS, but its convergence is poor in MaOP. Consequently, the values of HV are also the lowest for $m = \{6, 8, 10\}$ objectives. IBEA$_{\epsilon+}$ achieves the highest convergence, but its diversity is the minimum. MSOPS achieves balanced search on both convergence and diversity, and consequently it achieves the highest HV among conventional MOEAs compared in **Fig.6**. On the other hand, S-CDAS achieves higher *Norm* and *MS* than conventional NSGA-II. Although *Norm* achieved by S-CDAS are lower than IBEA, MSOPS and CDAS in $m > 8$ objectives, S-CDAS achieves

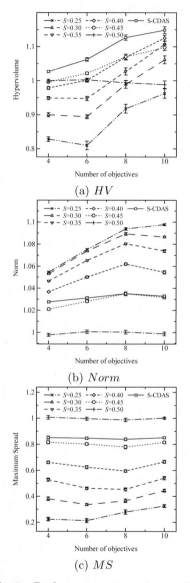

Fig. 5. Performance comparison with conventional CDAS [7]

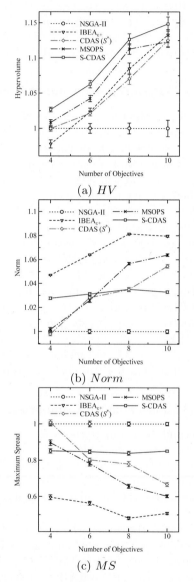

Fig. 6. Performance comparison with conventional MOEAs

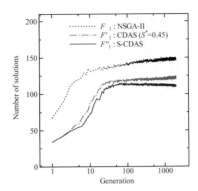

Fig. 7. Transition of front distribution over generation ($m = 8$)

high MS next to NSGA-II. Consequently, S-CDAS achieves the highest HV among all MOEAs compared in **Fig.6** for $m = \{4, 6, 8, 10\}$ objectives problems due to well-balanced search between convergence and diversity.

5.4 Algorithm's Behavior of NSGA-II, CDAS and S-CDAS

Here we observe the transition of solutions search by NSGA-II, CDAS($S^* = 0.45$) and S-CDAS in $m = 8$ objectives problem. Fist, we show the transition of the number of solutions belonging to the top front for each algorithm over generation in **Fig.7**. Among three methods, the ratio that the top front of NSGA-II ($\mathcal{F}_1$) occupied in the population is noticeably larger than CDAS and S-CDAS. This means more difficult in NSGA-II to discriminate solutions to obtain enough selection pressure. Consequently, the convergence of POS obtained by NSGA-II noticeably deteriorate especially for large m as shown in **Fig.6 (b)**. On the other hand, we can see that CDAS and S-CDAS considerably reduce the ratio of the top front ($\mathcal{F}_1'$ and $\mathcal{F}_1''$) and enhance selection pressure by using extended dominance area. Consequently, CDAS and S-CDAS improve the convergence of the obtained POS compared to NSGA-II as shown in **Fig.6 (b)**.

Next, we focus on the transition of extreme solutions $\boldsymbol{E}_i$ ($i = 1, 2, \cdots, m$), each of which has maximum value for i-th objective function. **Fig.8** shows the transition of averaged maximum objective value in the top front given by $\bar{f}^{max} = \frac{1}{m} \cdot \sum_{i=1}^{m} f_i(\boldsymbol{E}_i)$. In CDAS($S^* = 0.45$), the possibility that extreme solutions $\boldsymbol{E}_i$ ($i = 1, 2, \cdots, m$) are dominated and dismissed in parent selection is high as shown in **Fig.3 (a)**. Consequently, $\bar{f}^{max}$ obtained by CDAS becomes lower than NSGA-II over generations. On the other hand, S-CDAS achieves high $\bar{f}^{max}$ which is comparative with NSGA-II, since it has a mechanism to keep extreme solutions $\boldsymbol{E}_i$ ($i = 1, 2, \cdots, m$) in $\mathcal{F}_1''$. S-CDAS achieves better diversity for the obtained POS than conventional CDAS as shown in **Fig.6 (c)**, although S-CDAS reduces the size of the top front $\mathcal{F}_1''$ similar to conventional CDAS.

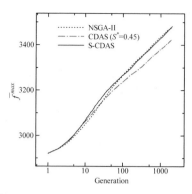 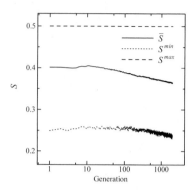

Fig. 8. Transition of average maximum objective value over generation ($m = 8$)

Fig. 9. Transition of $\bar{S}$, S^{max} and S^{min} in S-CDAS over generation ($m = 8$)

Finally, we observe the transition of dominance area which is self-controlled by S-CDAS for each solutions. For all solutions in top front $\mathcal{F}_1''$, we calculate the average angle $\bar{\varphi}$ from each solution's $\varphi_i (i = 1, 2, \cdots, m)$ which determines the dominance area, and transform $\bar{\varphi}$ to $\bar{S}$ by $\bar{\varphi}/\pi$. **Fig.9** shows the transition of $\bar{S}$ as well as the maximum S^{max} and the minimum S^{min} over generation. From this figure, $\bar{S}$ is gradually decreasing as we spend more generations. That is, S-CDAS gradually expands dominance area of solutions and strengthen selection pressure while constructing variety of dominance areas in the range $[S^{min}, S^{max}]$ for each solution. Since S^{max} is 0.5 throughout the entire evolution, S-CDAS achieves fine-grained ranking than NSGA-II similar to CDAS while keeping higher $\bar{f}^{max}$ than CDAS similar to NSGA-II.

6   Conclusions

In this work, we have proposed a modification of CDAS [7] called S-CDAS, which self-controls dominance area for each solution without the need of an external parameter. S-CDAS realizes fine-grained ranking that always guarantees the inclusion of extreme solution in top front, which is different from conventional CDAS. Through performance verification using many-objective 0/1 knapsack problems with $m = 4 \sim 10$ objectives, we have shown that S-CDAS achieves well-balanced search performance on both convergence and diversity compared to NSGA-II, IBEA$_{\epsilon+}$, MSOPS and conventional CDAS with optimal parameter S^*. Also, we have observed the transition of solutions search to see algorithm's behavior in detail.

As future works, we are planning to verify the search performance of the proposed algorithm when we vary the size of solution space and the feasibility ratio. Also, we want to apply our method to many-objective continuous optimization problems.

References

1. Deb, K.: Multi-Objective Optimization using Evolutionary Algorithms. John Wiley & Sons, Chichester (2001)
2. Deb, K., Agrawal, S., Pratap, A., Meyarivan, T.: A Fast Elitist Non-Dominated Sorting Genetic Algorithm for Multi-Objective Optimization: NSGA-II. KanGAL report 200001 (2000)
3. Zitzler, E., Laumanns, M., Thiele, L.: SPEA2: Improving the Strength Pareto Evolutionary Algorithm. TIK-Report, No.103 (2001)
4. Hughes, E.J.: Evolutionary Many-Objective Optimisation: Many Once or One Many? In: Proc. IEEE Congress on Evolutionary Computation (CEC 2005), pp. 222–227 (September 2005)
5. Aguirre, H., Tanaka, K.: Working Principles, Behavior, and Performance of MOEAs on MNK-Landscapes. European Journal of Operational Research 181(3), 1670–1690 (2007)
6. Zitzler, E., Kunzili, S.: Indicator-Based Selection in Multiobjective Search. In: Yao, X., Burke, E.K., Lozano, J.A., Smith, J., Merelo-Guervós, J.J., Bullinaria, J.A., Rowe, J.E., Tiño, P., Kabán, A., Schwefel, H.-P. (eds.) PPSN 2004. LNCS, vol. 3242, pp. 832–842. Springer, Heidelberg (2004)
7. Sato, H., Aguirre, H., Tanaka, K.: Controlling Dominance Area of Solutions and Its Impact on the Performance of MOEAs. In: Obayashi, S., Deb, K., Poloni, C., Hiroyasu, T., Murata, T. (eds.) EMO 2007. LNCS, vol. 4403, pp. 5–20. Springer, Heidelberg (2007)
8. Sato, H., Aguirre, H., Tanaka, K.: Effect of Controlling Dominance Area of Solutions in MOEAs on Convex Problems with Many Objectives. In: Proc. 7th Intl. Conf. on Optimization: Techniques and Applications (ICOTA7), in CD-ROM (2007)
9. Wagner, T., Beume, N., Naujoks, B.: Pareto-, Aggregation-, and Indicator-Based Methods in Many-Objective Optimization. In: Obayashi, S., Deb, K., Poloni, C., Hiroyasu, T., Murata, T. (eds.) EMO 2007. LNCS, vol. 4403, pp. 742–756. Springer, Heidelberg (2007)
10. Zitzler, E., Thiele, L.: Multiobjective optimization using evolutionary algorithms – a comparative case study. In: Eiben, A.E., Bäck, T., Schoenauer, M., Schwefel, H.-P. (eds.) PPSN 1998. LNCS, vol. 1498, pp. 292–304. Springer, Heidelberg (1998)
11. Zitzler, E.: Evolutionary Algorithms for Multiobjective Optimization: Methods and Applications, PhD thesis, Swiss Federal Institute of Technology, Zurich (1999)
12. Fonseca, C., Paquete, L., López-Ibáñez, M.: An Improved Dimension-sweep Algorithm for the Hypervolume Indicator. In: Proc. 2006 IEEE Congress on Evolutionary Computation, pp. 1157–1163 (2006)
13. Sato, M., Aguirre, H., Tanaka, K.: Effects of δ-Similar Elimination and Controlled Elitism in the NSGA-II Multiobjective Evolutionary Algorithm. In: Proc. IEEE Congress on Evolutionary Computation (CEC 2006), pp. 3980–3398 (2006)

Optimum Design of Balanced SAW Filters Using Multi-Objective Differential Evolution

Kiyoharu Tagawa[1], Yukinori Sasaki[2], and Hiroyuki Nakamura[2]

[1] Kinki University, 3–4–1 Kowakae, Higashi-Osaka 577–8502, Japan
tagawa@info.kindai.ac.jp
[2] Panasonic Electronic Devices Co., Ltd., 1006 Kadoma, Osaka 571-8506, Japan

Abstract. Three Multi-Objective Differential Evolutions (MODEs) that differ in their selection schemes are applied to a real-world application, i.e., the multi-objective optimum design of the balanced Surface Acoustic Wave (SAW) filter used in cellular phones. In order to verify the optimality of the Pareto-optimal solutions obtained by the best MODE, those solutions are also compared with the solutions obtained by the weighted sum method. Besides, from the Principal Component Analysis (PCA) of the Pareto-optimal solutions, an obvious relationship between the objective function space and the design parameter space is disclosed.

1   Introduction

Surface Acoustic Wave (SAW) filters have played an important role as a key device in modern mobile and wireless communication systems such as personal digital assistants (PDAs) and cellular phones[1,2]. The frequency response characteristics of SAW filters are governed primarily by their geometrical structures. Therefore, in order to decide the suitable structures of SAW filters, optimum design methods that combine optimization algorithms with computer simulations have been reported[3,4,5,6]. Evolutionary Algorithms (EAs) such as Genetic Algorithm (GA) have been also used in the optimum design methods[4,6].

In our previous paper[7], the design of the balanced SAW filter used in cellular phones was formulated as a constrained multi-objective optimization problem. Then a recent multi-objective EA based on Differential Evolution (DE)[8], i.e., Generalized DE 3 (GDE3)[9], was applied to the optimization problem. However, the optimality of the non-dominated feasible solutions obtained by GDE3 has not been confirmed. Furthermore, those solutions have not been assessed enough to get the useful knowledge about the design of the balanced SAW filter.

In this paper, in order to obtain a set of Pareto-optimal solutions of the above multi-objective optimization problem for the design of the balanced SAW filter, three variants of Multi-Objective DE (MODE) that differ in their selection schemes are employed. The three MODEs are compared in several criteria[10]. In order to verify the optimality of the Pareto-optimal solutions obtained by the best MODE, those solutions are also compared with a set of solutions obtained by using the weighted sum method. Furthermore, from the Principal Component Analysis (PCA) of the Pareto-optimal solutions, an obvious relationship between the objective function space and the design parameter space is disclosed.

K. Deb et al. (Eds.): SEAL 2010, LNCS 6457, pp. 466–475, 2010.

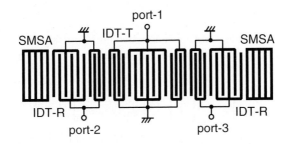

Fig. 1. Structure of balanced SAW filter

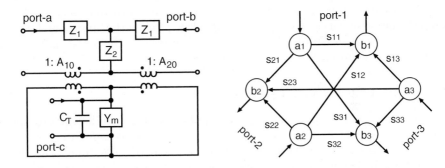

Fig. 2. Circuit model of IDT   **Fig. 3.** Network model of SAW filter

2   Balanced Surface Acoustic Wave Filter

2.1   Structure of Balanced SAW Filter

The balanced SAW filter consists of several components, namely Inter-Digital Transducers (IDTs) and Shorted Metal Strip Array (SMSA) reflectors, fabricated on a piezoelectric substrate. Fig. 1 shows a typical structure of the balanced SAW filter consisting of one transmitter IDT (IDT-T), two receiver IDTs (IDT-Rs), pitch-modulated IDTs between IDT-T and IDT-Rs, and two SMSA reflectors.

Port-1 connecting to IDT-T is an input-port of the balanced SAW filter, while a pair of port-2 and port-3 connecting to respective IDT-Rs is a balanced output-port. IDT-T converts electric input signals into acoustic signals. The acoustic signals around a resonate frequency are amplified between two SMSA reflectors. Then IDT-Rs reconvert the acoustic signals into electric output signals.

2.2   Model of Balanced SAW Filter

The behavior of each IDT can be simulated by using a circuit model in Fig. 2: port-a and port-b are acoustic signal ports, while port-c is an electric signal port[11]. The circuit model of SMSA reflector is derived from the circuit model of IDT by shorting the electric port (port-c). Therefore, an equivalent circuit model of the balanced SAW filter in Fig. 1 can be composed by linking all the components' circuit models in their acoustic signal ports. The equivalent circuit model of the balanced SAW filter is represented by a scattering matrix $\mathbf{S}$ as

$$\begin{bmatrix} b_1 \\ b_2 \\ b_3 \end{bmatrix} = \mathbf{S} \begin{bmatrix} a_1 \\ a_2 \\ a_3 \end{bmatrix} = \begin{bmatrix} s_{11} & s_{12} & s_{13} \\ s_{21} & s_{22} & s_{23} \\ s_{31} & s_{32} & s_{33} \end{bmatrix} \begin{bmatrix} a_1 \\ a_2 \\ a_3 \end{bmatrix} \tag{1}$$

From the matrix $\mathbf{S} = [s_{pq}]$ in (1), the network model of the balanced SAW filter in Fig. 1 can be described graphically as shown in Fig. 3. In the network model of the balanced SAW filter in Fig. 3, nodes a_q ($q = 1, 2, 3$) denote the electric input signals, while nodes b_p ($p = 1, 2, 3$) denote the electric output signals. Besides, scattering parameters s_{pq} labeled on edges provide the transition characteristics from the input signals a_q to the output signals b_p. The port-1, port-2 and port-3 of the network model in Fig. 3 correspond respectively to the port-1, port-2 and port-3 of the balanced SAW filter shown in Fig. 1.

2.3   Criteria of Balanced SAW Filter

It is desirable that the output signals b_2 and b_3 from a pair of balanced output-ports, namely port-2 and port-3 in Fig. 3, have the same amplitude and the opposite phases through the pass-band of the balanced SAW filter. Therefore, the amplitude balance and the phase balance of those signals are evaluated respectively by E_1 in (2) and E_2 in (3). Both criteria E_1 and E_2 should be restricted to a small value through the pass-band of the balanced SAW filter.

$$E_1 = 20 \log_{10}(|s_{21}|) - 20 \log_{10}(|s_{31}|) \tag{2}$$

$$E_2 = \varphi(s_{21}) - \varphi(s_{31}) + 180 \tag{3}$$

where, $\varphi(s_{pq})$ denotes the phase angle of the scattering parameter s_{pq}.

In order to evaluate strictly the band-pass filtering characteristics of the balanced SAW filter, differential mode signals have to be segregated from common mode signals at the balanced output-port shown in Fig. 3. Therefore, according to the balanced network theory[12], the differential mode signals (a_d and b_d) and the common mode signals (a_c and b_c) are derived respectively as follows:

$$\begin{cases} a_d = \dfrac{1}{\sqrt{2}}(a_2 - a_3) \\ b_d = \dfrac{1}{\sqrt{2}}(b_2 - b_3) \end{cases} \qquad \begin{cases} a_c = \dfrac{1}{\sqrt{2}}(a_2 + a_3) \\ b_c = \dfrac{1}{\sqrt{2}}(b_2 + b_3) \end{cases} \tag{4}$$

From (4), the matrix $\mathbf{S}$ in (1) is transformed into the matrix $\mathbf{S}_{mix}$ as

$$\mathbf{S}_{mix} = \begin{bmatrix} \sqrt{2} & 0 & 0 \\ 0 & 1 & -1 \\ 0 & 1 & 1 \end{bmatrix} \mathbf{S} \begin{bmatrix} \sqrt{2} & 0 & 0 \\ 0 & 1 & -1 \\ 0 & 1 & 1 \end{bmatrix}^{-1} = \begin{bmatrix} s_{11} & s_{1d} & s_{1c} \\ s_{d1} & s_{dd} & s_{dc} \\ s_{c1} & s_{cd} & s_{cc} \end{bmatrix} \tag{5}$$

Table 1. Design parameters of balanced SAW filter

x_j	$[\,x_j^L,\ x_j^U\,]$	e_j	design parameter
x_1	[200, 400]	–	overlap between electrodes
x_2	[10.0, 40.0]	0.5	number of fingers of IDT-R
x_3	[10.5, 40.5]	1.0	ditto of IDT-T
x_4	[1.0, 4.0]	1.0	ditto of modulated IDT
x_5	[50.0, 300.0]	10.0	number of strips of SMSA
x_6	[0.2, 0.8]	–	metallization ratio of IDT
x_7	[0.2, 0.8]	–	ditto of SMSA
x_8	[1.0, 1.1]	–	pitch ratio of SMSA
x_9	[0.9, 1.0]	–	ditto of modulated IDT
x_{10}	[1.9, 2.1]	–	finger pitch of IDT
x_{11}	[3900, 4000]	–	thickness of electrode

By using the mix-mode scattering parameters in (5), the band-pass filtering characteristics of the balanced SAW filter can be evaluated in the same way with the conventional SAW filters[5]. Therefore, the standing wave ratios of the input-port E_3 and the output-port E_4, which should be close to one through the pass-band of the balanced SAW filter, are defined respectively as follows:

$$E_3 = \frac{1 + |s_{11}|}{1 - |s_{11}|}\ , \qquad E_4 = \frac{1 + |s_{dd}|}{1 - |s_{dd}|} \tag{6}$$

The attenuation E_5 between the input- and output-ports is also defined as

$$E_5 = 20\,\log_{10}(|s_{d1}|) \tag{7}$$

3   Problem Formulation

3.1   Design Parameter

Table 1 shows $D = 11$ design parameters x_j $(j = 1, \cdots, D)$ selected for describing the structure of the balanced SAW filter in Fig. 1. Table 1 also shows the upper x_j^U and lower x_j^L bounds of the design parameter x_j. Each design parameter takes either a continuous value or a discrete value. Therefore, for discrete design parameters, the intervals $e_j \in \mathbb{R}$ of them are appeared in Table 1.

3.2   Objectives and Constraints

The values of the criteria $E_h = E_h(\boldsymbol{x}, \omega)$ $(h = 1, \cdots, 5)$ in the previous section depend on both the design parameters $\boldsymbol{x} = (x_1 \cdots, x_D)$ and the frequency ω. Therefore, a set of frequency points $\omega \in \Omega_P$ is sampled from the pass-band of the balanced SAW filter. Similarly, two sets of frequency points $\omega \in \Omega_L$ and $\omega \in \Omega_H$ are sampled respectively from the lower and higher stop-bands.

The balanced SAW filter works as a band-pass filter. Therefore, the attenuation $E_5 = E_5(\boldsymbol{x}, \omega)$ defined in (7) should be kept small through the stop-band but large through the pass-band as much as possible. In order to realize such a band-pass filtering characteristics, three objective functions are defined as

$$
\begin{cases}
f_1(\boldsymbol{x}) = \displaystyle\sum_{\omega \in \Omega_L} \frac{E_5(\boldsymbol{x},\, \omega)}{|\Omega_L|} \\[3mm]
f_2(\boldsymbol{x}) = -\left(\displaystyle\sum_{\omega \in \Omega_P} \frac{E_5(\boldsymbol{x},\, \omega)}{|\Omega_P|} \right) \\[3mm]
f_3(\boldsymbol{x}) = \displaystyle\sum_{\omega \in \Omega_H} \frac{E_5(\boldsymbol{x},\, \omega)}{|\Omega_H|}
\end{cases}
\tag{8}
$$

where, $|\Omega|$ denotes the number of frequency points $\omega \in \Omega$.

The upper $U_h(\omega)$ and lower $L_h(\omega)$ bounds are specified for the rest of criteria $E_h(\omega,\, \boldsymbol{x})$ $(h = 1, \cdots, 4)$. Then six constraints are defined with them as

$$
\begin{cases}
g_h(\boldsymbol{x}) = \displaystyle\sum_{\omega \in \Omega_P} \frac{E_h(\boldsymbol{x},\, \omega) - U_h(\omega)}{|\Omega_P|} \leq 0, \quad h = 1, \cdots, 4, \\[3mm]
g_5(\boldsymbol{x}) = \displaystyle\sum_{\omega \in \Omega_P} \frac{L_1(\omega) - E_1(\boldsymbol{x},\, \omega)}{|\Omega_P|} \leq 0 \\[3mm]
g_6(\boldsymbol{x}) = \displaystyle\sum_{\omega \in \Omega_P} \frac{L_2(\omega) - E_2(\boldsymbol{x},\, \omega)}{|\Omega_P|} \leq 0
\end{cases}
\tag{9}
$$

3.3 Optimum Design Problem

From (8), (9) and Table 1, the design of the balanced SAW filter is formulated as a constrained three-objective optimization problem as follows:

$$
\begin{bmatrix}
\text{minimize} & \{\, f_1(\boldsymbol{x}),\ f_2(\boldsymbol{x}),\ f_3(\boldsymbol{x})\,\} \\[2mm]
\text{subject to} & g_k(\boldsymbol{x}) \leq 0,\ k = 1, \cdots, 6, \\[2mm]
& x_j^L \leq x_j \leq x_j^U,\ j = 1, \cdots, D,\ (D = 11).
\end{bmatrix}
\tag{10}
$$

4 Multi-Objective Differential Evolution

4.1 Representation

MODE holds N_P individuals, or the tentative solutions of the multi-objective optimization problem defined by (10), in a population $\mathbf{P}^G$. The i-th individual $\boldsymbol{x}_i^G \in \mathbf{P}^G$ in the population of generation G is represented as follows:

$$
\boldsymbol{x}_i^G = (x_{1,i}^G, \cdots, x_{j,i}^G, \cdots, x_{D,i}^G)
\tag{11}
$$

where, $x_{j,i}^G \in \mathbb{R}$ and $0 \leq x_{j,i}^G \leq 1$ $(j = 1, \cdots, D;\ i = 1, \cdots, N_P)$.

The individual $\boldsymbol{x}_i^G$ is converted into a corresponding solution $\boldsymbol{x}$ when the values of functions $f_m(\boldsymbol{x})$ and $g_k(\boldsymbol{x})$ are evaluated. For a continuous design parameter x_j, $x_{j,i}^G$ is converted into x_j as shown in (12). On the other hand, for a discrete design parameter x_j, $x_{j,i}^G$ is converted into x_j as shown in (13).

$$x_j = (x_j^U - x_j^L) x_{j,i}^G + x_j^L \tag{12}$$

$$x_j = \mathrm{round}\left(\frac{(x_j^U - x_j^L) x_{j,i}^G}{e_j} \right) e_j + x_j^L \tag{13}$$

where, round(z) rounds a real number $z \in \mathbb{R}$ to the nearest integer.

4.2   Selection Scheme

In order to select the next population $\boldsymbol{x}_i^{G+1} \in \mathbf{P}^{G+1}$ $(i = 1, \cdots, N_P)$ from the current one $\boldsymbol{x}_i^G \in \mathbf{P}^G$, every MODE considers the dominance relation between two individuals, or their corresponding solutions $\boldsymbol{x}'$ and $\boldsymbol{x}''$. If the relation shown in (14) holds between the two solutions $\boldsymbol{x}'$ and $\boldsymbol{x}''$, $\boldsymbol{x}'$ dominates $\boldsymbol{x}''$ $(\boldsymbol{x}' \succ \boldsymbol{x}'')$ in the objective function space. Similarly, if the relation shown in (15) holds between the two solutions $\boldsymbol{x}'$ and $\boldsymbol{x}''$, $\boldsymbol{x}'$ weakly dominates $\boldsymbol{x}''$ $(\boldsymbol{x}' \succeq \boldsymbol{x}'')$.

$$\forall m \in \{1, 2, 3\}\, f_m(\boldsymbol{x}') \le f_m(\boldsymbol{x}'') \wedge \exists m' \in \{1, 2, 3\}\, f_{m'}(\boldsymbol{x}') < f_{m'}(\boldsymbol{x}'') \tag{14}$$

$$\forall m \in \{1, 2, 3\}\, f_m(\boldsymbol{x}') \le f_m(\boldsymbol{x}'') \tag{15}$$

Furthermore, if the constraint function $\hat{g}_k(\boldsymbol{x})$ $(k = 1, \cdots, 6)$ is defined as shown in (16), both the dominance and the weak dominance relations between two solutions can be also introduced into the constraint function space.

$$\hat{g}_k(\boldsymbol{x}) = \max\{\, g_k(\boldsymbol{x}),\, 0\,\} \tag{16}$$

First of all, the trial vector $\boldsymbol{u}_i^G$ $(i = 1, \cdots, N_P)$ is generated from the target vector $\boldsymbol{x}_i^G$ by using the strategy of DE[8]. Then, in the selection scheme of the first variant of MODE (MODE1), the trial vector $\boldsymbol{u}_i^G$ is compared with the target vector $\boldsymbol{x}_i^G$. As a result, if $\boldsymbol{u}_i^G$ satisfies any of the following OR-conditions, $\boldsymbol{u}_i^G$ is selected such as $\boldsymbol{x}_i^{G+1} = \boldsymbol{u}_i^G$, otherwise $\boldsymbol{x}_i^G$ is selected such as $\boldsymbol{x}_i^{G+1} = \boldsymbol{x}_i^G$.
   [OR-conditions to select the trial vector]

1) $\boldsymbol{u}_i^G$ is feasible and $\boldsymbol{x}_i^G$ is infeasible.
2) $\boldsymbol{u}_i^G$ and $\boldsymbol{x}_i^G$ are feasible and $\boldsymbol{u}_i^G \succeq \boldsymbol{x}_i^G$ in the objective function space.
3) $\boldsymbol{u}_i^G$ and $\boldsymbol{x}_i^G$ are infeasible and $\boldsymbol{u}_i^G \succeq \boldsymbol{x}_i^G$ in the constraint function space.

The second variant of MODE (MODE2) borrows the selection scheme from NSGA-II[13]. Therefore, both the target vectors $\boldsymbol{x}_i^G$ and the trial vectors $\boldsymbol{u}_i^G$ are added to the next population $\mathbf{P}^{G+1}$. As a result, the size of $\mathbf{P}^{G+1}$ is increased to $2N_P$ temporally. Then it is set back to N_P by using the selection rule based on the non-domination ranking and the crowding distance. Because feasible solutions have priority over infeasible ones, the selection scheme of NSGA-II is executed in the objective function space. However, if the number of the feasible solutions is less than N_P, it is executed in the constraint function space.

Table 2. Performance

method	N_S	H_V	τ
MODE1	70.0	76.4	61.3
MODE2	200.0	99.8	65.4
MODE3	200.0	100.0	63.5
SODE	61.8	85.5	56.2

Table 3. Dominance rate

method	MODE1	MODE2	MODE3	SODE
MODE1	85.3	78.3	75.4	98.7
MODE2	97.9	78.6	75.2	94.3
MODE3	98.3	80.0	76.7	93.9
SODE	100.0	100.0	100.0	83.3

As well as GDE3[9], the third variant of MODE (MODE3) combines two selection schemes used by DE and NSGA-II. First of all, the trial vector $\boldsymbol{u}_i^G$ is compared with the target vector $\boldsymbol{x}_i^G$. If $\boldsymbol{u}_i^G$ satisfies any of the above OR-conditions, $\boldsymbol{u}_i^G$ is selected such as $\boldsymbol{x}_i^{G+1} = \boldsymbol{u}_i^G$, otherwise $\boldsymbol{x}_i^G$ is selected. However, if $\boldsymbol{u}_i^G$ satisfies all the following AND-conditions, $\boldsymbol{u}_i^G$ is also added to $\mathbf{P}^{G+1}$ such as $\boldsymbol{x}_{i'}^{G+1} = \boldsymbol{u}_i^G$ ($i' > N_P$). As a result, the size of $\mathbf{P}^{G+1}$ comes into the range between N_P and $2\,N_P$. Then it is set back to N_P in the same way with MODE2.

[AND-conditions to keep the trial vector]

1) $\boldsymbol{u}_i^G$ has not been selected yet, namely $\boldsymbol{x}_i^{G+1} \neq \boldsymbol{u}_i^G$.
2) $\boldsymbol{u}_i^G$ is feasible and $\boldsymbol{x}_i^G \nsucc \boldsymbol{u}_i^G$ in the objective function space.

5   Experimental Results

5.1   Comparison of Three Variants of MODE

MODE1, MODE2 and MODE3 were applied to the optimization problem in (10). Besides MODEs, the weighted-sum method was also applied to the optimization problem in (10) for the comparison. For the weighted-sum method, three objective functions in (8) were integrated into one objective function as

$$f(\boldsymbol{x}) = \sum_{m=1}^{3} \alpha_m \, f_m(\boldsymbol{x}) = \sum_{m=1}^{3} f_m(\boldsymbol{x}) \qquad (17)$$

In order to solve the constrained optimization problem defined by using (9) and (17), a Single-Objective DE (SODE), i.e., GDE3[9], was employed.

For generating the trial vector $\boldsymbol{u}_i^G$, a basic strategy called "DE/rand/1/bin"[8] was used by all MODEs and SODE. The same set of control parameters' values, which was obtained experimentally, was used in every method: the scale factor $S_F = 0.9$, the crossover rate $C_R = 0.9$, the population size $N_P = 200$ and the maximum generation $G_{max} = 600$. Every method was coded by MATLAB and executed on a personal computer (CPU: Intel®Core^{TM} i7@3.33GHz).

Table 2 compares the performances of the four methods in several criteria, namely the number of non-dominated solutions N_S, the normalized hypervolume H_V, and the computational time τ [min] averaged over four runs. From Table 2, MODE3 is superior to the other methods in H_V. Table 3 shows the rate of the non-dominated solutions that are obtained by the method in the row and not dominated by any solutions obtained by the method in the column. From Table 3, MODE3 is also better than the other MODEs in the quality of solutions.

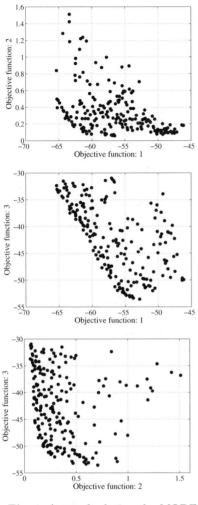

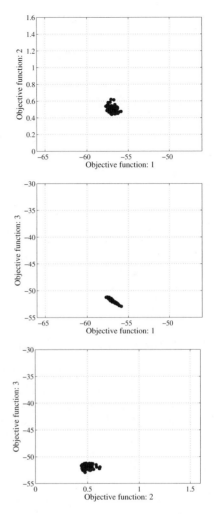

Fig. 4. A set of solutions by MODE3

Fig. 5. A set of solutions by SODE

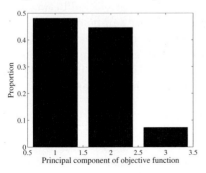

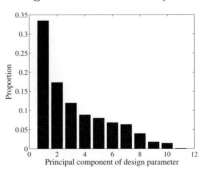

Fig. 6. PCA of objective functions

Fig. 7. PCA of design parameters

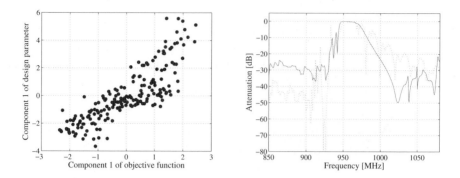

Fig. 8. Pareto-optimal solutions plotted by two principal components

Fig. 9. Attenuations achieved by two extreme Pareto-optimal solutions

Fig. 4 plots a set of non-dominated feasible solutions in the final population of MODE3 in the objective function space. Fig. 5 plots a similar set of solutions obtained by SODE in the same way. Comparing Fig. 4 with Fig. 5, the solutions by MODE3 are excellent not only in the diversity but also in the optimality.

5.2 Principal Component Analysis of Solutions

A set of the solutions obtained by MODE3 has been assessed by PCA in both the objective function space and the design parameter space. Fig. 6 shows the proportions of the principal components evaluated in the three-dimensional objective function space. Fig. 7 shows those values evaluated in the eleven-dimensional design parameter space. From Fig. 6, a difference of the Pareto-optimal solutions in the objective function space can be described by the first and the second principal components. Similarly, from Fig. 7, a difference of them in the design parameter space can be described mainly by the first principal component.

Fig. 8 plots the Pareto-optimal solutions shown in Fig. 4 by using two axes, namely the first principal component of the objective function space and that of the design parameter space. From Fig. 8, a positive correlation between the objective function space and the design parameter space can be observed.

Fig. 9 shows the attenuations $E_5(x^\star, \omega)$ achieved by two extreme Pareto-optimal solutions $x^\star$ that have the largest (solid line) and the smallest (broken line) values for the first principal component of the objective function space. From Fig. 9, the former solution is apparently better than the latter one.

6   Conclusions

Three variants of MODE that differed in their selection schemes were applied to a constrained three-objective optimization problem for the design of the balanced SAW filter. As a result, the performance of MODE3 that combined two selection schemes used by DE and NSGA-II was better than the others. Incidentally, the similar result has been reported on two-objective benchmark problems[14].

Future work will focus on the further investigation of the Pareto-optimal solutions for the optimum design problem of the balanced SAW filter that has more than three objectives. For solving the hard problem, we may need to extend the framework of MODE3[15]. Thereby we would like to clarify the relationship between the structure and the frequency response of the balanced SAW filter.

References

1. Hashimoto, K.: Surface Acoustic Wave Devices in Telecommunications – Modeling and Simulation. Springer, Heidelberg (2000)
2. Meier, M., Baier, T., Riha, G.: Miniaturization and advanced functionalities of SAW devices. IEEE Trans. on Microwave Theory and Techniques 49(2), 743–748 (2001)
3. Goto, S., Kawakatsu, T.: Optimization of the SAW filter design by immune algorithm. In: Proc. of IEEE International Ultrasonics, Ferroelectrics, and Frequency Control Joint 50th Anniversary Conference, pp. 600–603 (2004)
4. Meltaus, J., Hamalainen, P., Salomaa, M.M., Plessky, V.P.: Genetic optimization algorithms in the design of coupled SAW filters. In: Proc. of IEEE International Ultrasonics, Ferroelectrics, and Frequency Control Joint 50th Anniversary Conference, pp. 1901–1904 (2004)
5. Tagawa, K., Matsuoka, M.: Optimum design of surface acoustic wave filters based on the Taguchi's quality engineering with a memetic algorithm. In: Runarsson, T.P., Beyer, H.-G., Burke, E.K., Merelo-Guervós, J.J., Whitley, L.D., Yao, X. (eds.) PPSN 2006. LNCS, vol. 4193, pp. 292–301. Springer, Heidelberg (2006)
6. Tagawa, K., Kojima, N.: Multi-objective optimum design of DMS filters using robust engineering and genetic algorithm. In: Proc. of IEEE CEC, pp. 2208–2214 (2006)
7. Tagawa, K.: Multi-objective optimum design of balanced SAW filters using generalized differential evolution. WSEAS Trans. on System 8(8), 923–932 (2009)
8. Price, K.V., Storn, R.M., Lampinen, J.A.: Differential Evolution – A Practical Approach to Global Optimization. Springer, Heidelberg (2005)
9. Kukkonen, S., Lampinen, J.: GDE3: The third evolution step of generalized differential evolution. In: Proc. of IEEE CEC, pp. 443–450 (2005)
10. Zitzler, E., et al.: Performance assessment of multiobjective optimizations: an analysis and review. IEEE Trans. on Evolutionary Computation 7(2), 117–132 (2003)
11. Kojima, T., Suzuki, T.: Fundamental equations of electro-acoustic conversion for an interdigital surface-acoustic-wave transducer by using force factors. Japanese Journal of Applied Physics Supplement 31, 194–197 (1992)
12. Bockelman, D.E., Eisenstadt, W.R.: Combined differential and common-mode scattering parameters: theory and simulation. IEEE Trans. on Microwave Theory and Techniques 43(7), 1530–1539 (1995)
13. Deb, K., et al.: A fast and elitist multiobjective genetic algorithm: NSGA-II. IEEE Trans. on Evolutionary Computation 6(2), 182–197 (2002)
14. Zielinski, K., Laur, R.: Variants of differential evolution for multi-objective optimization. In: Proc. of IEEE Symposium on Computational Intelligence in Multi-criteria Decision Making, pp. 91–98 (2007)
15. Pal, S., Das, S., Basak, A.: Design of time-modulated linear arrays with a multi-objective optimization approach. Progress In Electromagnetics Research B 23, 83–107 (2010)

Optimizing the Risk of Occupational Health Hazard in a Multiobjective Decision Environment Using NSGA-II

Yogesh K. Anand[1], Sanjay Srivastava[1,*], and Kamal Srivastava[2]

[1] Department of Mechanical Engineering, Dayalbagh Educational Institute,
Agra 282110, India
[2] Department of Mathematics, Dayalbagh Educational Institute Agra 282110, India
ssrivastava.engg@gmail.com

Abstract. We present a novel system to lessen the risk of occupational health hazards (OHH) of workers in the labor intensive industrieswith a job-combination approach. The work is carried out in a brick manufacturing (BM) unit at Hathras, India. The risk of OHH is assessed in terms of perceived discomfort level (PDL) of workers. PDL is computed with factor rating (FR) method. It is observed based on an initial survey in the BM unit that the workers, in general, aim to maximize their earnings by subjecting themselves to extreme work conditions due to economic reasons, and hence are exposed to greater risk of OHH resulting in higher values of PDL. We employ NSGA-II, an evolutionary multiobjective optimization (EMO) technique, to search for optimal PDL-earning tradeoff (PET) profile with two conflicting objectives, viz. minimization of PDL, and maximization of earnings.

Keywords: Multiobjective Optimization, Occupational Health, NSGA-II, Factor Rating.

1 Introduction

In brick manufacturing (BM), firing workers are exposed to high risk of heat stress which adversely affects their health and work performance. Heat stress is a recognized occupational health hazard (OHH) in many industries including BM units. Current guidelines define working environment that cause an increase above 38°C (heat stress) as potentially hazardous [1]. However, the effectiveness of these guidelines is limited by the individual variation among employee and variation in work practices [2]. Hot conditions give rise to physiological heat strain [3], and cognitive decrements [4], [5]. In the present work, which is carried out in a BM unit at Hathras, India, we assess the risk of OHH in terms of perceived discomfort level (PDL) of workers for a given job-combination. We make the following observations based on an initial survey using interview method: (1) the workers, in general, are found to maximize their earnings by subjecting themselves to extreme work conditions due to economic reasons, and hence are exposed to greater risk of OHH; (2)there are three factors

---

* Corresponding author.

K. Deb et al. (Eds.): SEAL 2010, LNCS 6457, pp. 476–484, 2010.

identified to be influencing significantly the PDL of a given job, viz. number of working hours (WH), time of a rest break in minutes (RB), and number of rest breaks (NRB). BM comprises firing work, molding, and three types of lifting. Firing work is found to be the most severe job in BM as it involves undue exposure of workers to excessive heat. To reduce the risk of heat stress we propose a job-combination approach wherein the firing workers are required to do another job (brick molding in the present study) along with firing work in the same BM unit within their prescribed working hours thereby reducing their exposure to high temperature zone while maintaining their earning to a satisfactory level. Similarly molding workers go for firing work to reduce the risk of musculoskeletal disorder, a recognized occupational hazard in molding work. Therefore, workers of a BM unit at Hathras, India are trained to perform firing job along with molding job with predefined WH distribution – resulting in a firing-molding job-combination.Each job has its specific earning. High earning jobs are tedious to perform yet workers prefer such jobs due to the reasons already mentioned, which in turn create considerable health problems to them. Combining jobs is found to be a way of reducing OHH and yet maintaining the good earnings.

PDL-earning tradeoff (PET) belongs to a class of multiobjective optimization (MOO) problem wherein there is no single optimum solution rather there exists a number of solutions, which are all optimal – Pareto-optimal solutions – optimal PET profile in occupational health literature. The tradeoff between PDL and earnings gives workers wide opportunities to work out the best schedule to reduce the risk of OHH while maintaining their required earnings, and therefore PET analysis is of considerable importance from the view points of both – workers' health and owners' administration. There are six factors in a given job-combination which assume discrete values in real-life situations; therefore, the problem being tackled in this work, i.e., searching for optimal PET profile, is a combinatorial multiobjective optimization problem, which is otherwise difficult to solve using classical search and optimization techniques, and hence the use of NSGA-II, an EMO technique, is justified to solve this problem. MOO is a field reasonably explored by researchers in recent years since 1990 – as a result diverse techniques have been developed over the years [6]. Most of these techniques elude the complexities involved in MOO and usually transform multiobjective problem into a single objective problem by employing some user defined function. Since MOO involves determining Pareto-optimal solutions, therefore, it is hard to compare the results of various solution techniques of MOO, as it is the decision-maker who decides the 'best solution' out of all optimal solutions pertaining to a specific scenario [7]. Evolutionary algorithms (EAs) are meta heuristics that are able to search large regions of the solution's space without being trapped in local optima [8]. Some well-known meta heuristics are genetic algorithm, simulated annealing, and tabu search. Genetic Algorithms are search algorithms [9], which are based on the mechanics of natural selection and genetics to search through decision space for optimal solutions [10]. In genetic algorithm, a string represents a set of decisions (chromosome combination), a potential solution to a problem. Each string is evaluated on its performance with respect to the fitness function (objective function). The ones with better performance (fitness value) are more likely to survive than the ones with worse performance. Then the genetic information is exchanged between strings crossover and perturbed by mutation. The result is a new generation with (usually) better survival abilities. This process is repeated until the strings in the new generation are identical, or certain termination conditions are met. A genetic algorithm uses a population of solutions in each iteration of its search procedure, instead of a single solution. Since a population of solutions is processed in each iteration, the outcome of a GA is also a population of solutions. This unique feature of GA makes it a true multiobjective optimization technique and that is how GAs transcend classical search and optimization techniques [11]. Different versions of multiobjective GAs have been successfully employed to solve many MOO problems in science and engineering.

2 Methodology

The risk of OHH for a job-combination is evaluated based on the perceived discomfort level (PDL) of workers using interview method. We consider a job-combination consisting of two jobs – firing job and molding job in the present work. We identify three factors which influence PDL of a given job, viz. number of working hours (WH), time of a rest break in minutes (RB), and number of rest breaks (NRB) as mentioned earlier. Factor Rating (FR) is an approach that is used for evaluating PDL for a given job-combination. The importance of FR is that it provides a rational basis for evaluating PDL and facilitates comparison among PDL of different sets of a given job-combination by establishing a composite value that summarizes all related factors. FR also helps in incorporating the qualitative as well as quantitative information in evaluating PDL. The following procedure is used to develop FR for computing PDL:

1. We first identify relevant factors that influence PDL. As mentioned these are WH, RB and NRB for each job. We get six factors for a job-combination consisting of two jobs, which are WH_1, RB_1, and NRB_1 for the first job (firing work), and WH_2, RB_2, and NRB_2 for the second job (molding work).

2. Based on the perception and opinion of concerned workers and their supervisors we assign an average weight to each factor that indicates its relative importance compared with all other factors for evaluating PDL as shown below. Typically weights sum to 1.0.

Factors	WH_1	RB_1	NRB_1	WH_2	RB_2	NRB_2
Weight	0.35	0.10	0.15	0.25	0.06	0.09

3. The value of each factor is normalized in computing PDL by dividing it by its maximum value, which yields ratios ranging from 0 to 1.0.

4. We then multiply the normalized value of each factor with its corresponding weight, and sum the result, called a composite score, which is nothing but the required PDL value. The resulting expression for PDL is shown below.

$$PDL = [((WH_1/WH_{max}) * 0.35 - (RB_1/RB_{max}) * 0.10 - (NRB_1/NRB_{max}) * 0.15) + ((WH_2/WH_{max}) * 0.25 - (RB_2/RB_{max}) * 0.06 - (NRB_2/NRB_{max}) * 0.09)]$$

It is obvious that WH_1 and WH_2 contribute positively to PDL, whereas higher values of RB_1, RB_2, NRB_1, and NRB_2 would cause a decrease in PDL. Higher weights to job

1 in comparison to job 2 is attributed to the fact that job 1 (i.e. firing work) is more severe than job 2 (i.e. molding work) as mentioned earlier.

The wages of workers are based on the job they are doing. In the firing work a worker earns on an average Rs. 25 per hour while for molding work the earning is Rs. 16.66 per hour. Since the proposed system is flexible enough to allow a good amount of rest break (upto 60 minutes) and sufficient number of rest breaks (upto 8), therefore we find it lucid to deduct an amount equivalent to his total rest break time from his earnings/day for the ease of implementability of the proposed system by the owner of the BM unit. The resulting expression for earnings/day (ER/day) is illustrated below.

$$ER/day = (WH_1 - (RB_1 * NRB_1)/60) * 25) + (WH_2 - (RB_2 * NRB_2)/60) * 16.66)$$

The formal definition of PET problem is illustrated below.

> Min PDL
> Max ER/day

Subject to

$$WH_1 \in \{2, 3, 4, 5, 6, 7, 8, 9, 10\}$$
$$WH_2 = 12 - WH_1$$
$$RB_1 \in \{5, 10, 15, 20, 25, 30, 35, 40, 45, 50, 55, 60\}$$
$$RB_2 \in \{5, 10, 15, 20, 25, 30, 35, 40, 45, 50, 55, 60\}$$
$$NRB_1 = \{1, 2, 3, 4, 5, 6, 7, 8\}, \quad 0 \le NRB_1 \le WH_1$$
$$NRB_2 = \{1, 2, 3, 4, 5, 6, 7, 8\}, \quad 0 \le NRB_2 \le WH_2$$

3 NSGA-II for PET

We employ of non-dominated sorting genetic algorithm-II (NSGA-II) in solving PET, a multiobjective optimization problem. The problem tackled is a real life problem of a labor intensive BM. NSGA-II has proved its effectiveness in solving many real life MOO problems in terms of convergence to Pareto-optimal front, and in maintaining diversity within the population. The NSGA-II algorithm and its detailed implementation procedure can be found in [11] and [12]. A brief description of NSGA-II is as follows. NSGA-II uses non-dominated sorting for fitness assignments. All individuals not dominated by any other individuals, are assigned front number 1. All individuals only dominated by individuals in front number 1 are assigned front number 2, and so on. Selection is made, using tournament between two individuals. The individual with the lowest front number is selected if the two individuals are from different fronts. The individual with a higher crowding distance is selected if they are from the same front. i.e., a higher fitness is assigned to individuals located on a sparsely populated part of the front. Each individual is made to participate in exactly two tournaments, thereby making at most two copies of itself in the selected population. There are N parents in any iteration and crossover is used to generate N new individuals (offspring). This is followed by mutation which is applied on a few randomly selected individuals.

In the context of PET problem, a solution in NSGA-II is an array (g_i) where $i = 1, 2$ ..., 5 which represents an instance of a job-combination (Fig. 1). Here g_1, g_2, g_3, g_4 and g_5 represent WH_1, RB_1, NRB_1, RB_2, and NRB_2 respectively.

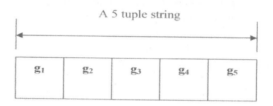

Fig. 1. An instance of job-combination schedule

The initial population consists of N solutions, where N strings are selected randomly from the feasible search space. These solutions are referred to as parents. For the crossover, two strings (say, S_1 and S_2) from the population are selected randomly. The offspring O_1 and O_2 are produced as follows: First the working hours of S_1 and S_2 are stored in the respective positions of O_2 and O_1 respectively. The remaining entries of S_1 and S_2 are copied to O_1 and O_2 respectively. To ensure the feasibility of offspring, the number of rest breaks is checked and if it exceeds the corresponding working hours, its value is reassigned randomly so as to make it feasible. Mutation is performed on randomly selected $r_m * N$ individuals from the offspring population wherein the working hours in the selected individual are reassigned randomly, where r_m is the mutation rate.

4 Simulation Results

NSGA-II implementation details are as follows. The procedure is coded in MATLAB 7.0 and run on Pentium (R)-based HP Intel (R) computer with 1.73 GHz Processor and 512 MB of RAM. The crossover rate and the mutation rate (rm) are kept as 1.0 and 0.05 respectively. The population size is chosen as 50. These parameters are decided after performing computational experiments based on faster convergence criteria. The search is set to terminate when nondominated PET profile remains unchanged for three consecutive iterations – a number is suitably decided based on extensive experiment. It takes on an average ten iterations for NSGA-II to search for the best possible PET profile. Results of an example run of NSGA-II follow to demonstrate its performance to solve PET problems. It can be seen that the initial population is well distributed over the solution space (Fig. 2). Fig. 3 illustrates the intermediate improvements in the PET profile along with different fronts of the population. In succeeding iterations NSGA-II searches for optimal PET profile. Fig. 4 depicts the nondominated PET solution points of the final generation population, which are concluded to be the best points obtained.

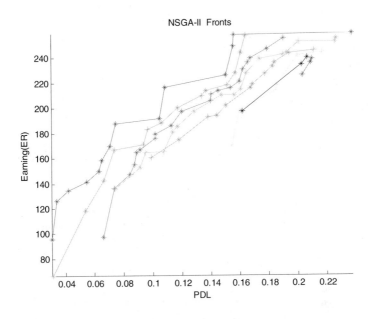

Fig. 2. An example of the Initial population with nondominated PET profile

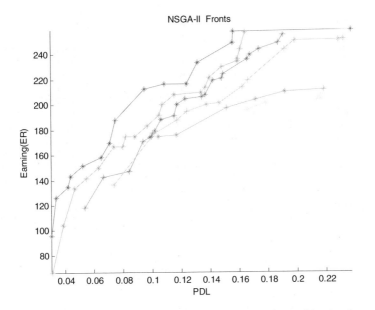

Fig. 3. Intermediate improvements in the PET profile along with other fronts

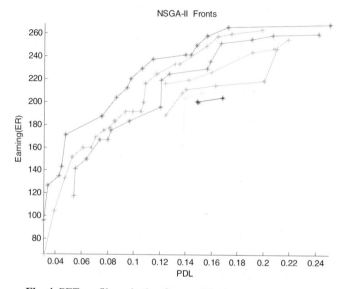

Fig. 4. PET profile and other fronts of final generation population

Table 1. Nondominated solution points of final generation

WH1	RB1	NBB1	WH2	RB2	NRB2	ER/day	PDL
4	60	6	8	50	2	55.5133	-0.0017
2	60	5	10	50	1	77.7167	0.0117
6	60	5	6	50	3	83.3100	0.0225
8	60	6	4	50	2	88.8733	0.0317
7	60	6	5	50	1	94.4167	0.0346
2	20	5	10	55	4	113.8467	0.0396
4	60	3	8	50	3	116.6300	0.0433
5	60	3	7	50	3	124.9700	0.0517
3	60	2	9	50	3	133.2900	0.0537
4	60	2	8	50	3	141.6300	0.0621
2	60	1	10	50	3	149.9500	0.0642
3	60	1	9	20	6	166.6200	0.0687
4	60	1	8	20	6	174.9600	0.0771
9	25	8	3	45	1	179.1517	0.0771
10	25	8	2	45	1	187.4917	0.0854
5	60	1	7	50	2	188.8533	0.1004
7	15	6	5	25	4	193.0333	0.1008
4	60	1	8	50	1	194.3967	0.1033
6	60	1	6	50	2	197.1933	0.1087
5	60	1	7	50	1	202.7367	0.1117
7	5	7	5	45	1	231.2217	0.1125
8	5	7	4	45	1	239.5617	0.1208
9	5	7	3	45	1	247.9017	0.1292
10	5	7	2	45	1	256.2417	0.1375
10	5	5	2	45	1	260.4083	0.1750
9	5	4	3	15	1	262.4817	0.2154
10	5	4	2	15	1	270.8217	0.2237
10	5	3	2	5	1	275.6817	0.2525

We present these points in the tabular form also (Table 1). It is interesting to note that certain PDL assume negative values which indicate that the workers are in comfortable zone for corresponding job-combination schedules.

A comparison of NSGA-II with enumeration technique follows. It takes a total of 995328 searches to compute the nondominated solution points for the problem under consideration whereas NSGA-II takes a maximum of 800 searches for the same. Further, the mean elapse time of a single run of NSGA-II is found to be 3.012 sec. – a very fast convergence. It is observed that the near optimal nondominated curve is attained in 4th or 5th iteration and in the remaining iterations more and more solutions points belonging to this front are explored thereby increasing the size of the nondominated front.

5 Conclusion

Brick kiln owners are faced with the problems of putting together and managing large number of workers while considering their health hazards and absenteeism, limited time schedules, and environment uncertainty. An analysis is done to reduce the risk of OHH by combining firing work with molding work in a BM unit. A job-combination approach is proposed so that workers' earnings are not compromised to a greater extent. Firing work is identified to be the most severe job. NSGA-II is implemented to search for the optimal PDL-Earning profile.

It has the ability to trace out Pareto-optimal front and it does not place any restriction on the form of inputs (WH, RB, and NRB of two jobs) to evaluate PDL and earnings for a given job-combination. Factor rating approach is developed to evaluate PDL. The unifying approach amalgamating job-combination concept, weighted factor rating scheme with NSGA-II in a unique way turns out to be a powerful and efficient method without losing its simplicity. For complex optimization scenario, it can effectively search for the optimal values of WH, RB, and NRB for minimum PDL and maximum earnings.

The top management faces the problem of monopoly of workers of firing work in BM units, as it is a high skill job. The system presented in this work will alleviate this problem as job-combination approach will make other workers getting trained for firing work. In fact the system will help in 'work generalization' to take over 'work specialization'. Therefore, the feasibility of implementing this system is high as it is beneficial to both the parties – workers as well as owners. In view of these facts, the work presented here forms an important basis to effectively address the issues in health management of workers. The proposed system act as an advisor to a worker to choose a job-combination and the corresponding values of WH, RB, & NRB to decide his/her occupational risks and earnings suitably.

Acknowledgement

This work is supported in part by UGC, New Delhi under Grant F. No. 36-65/2008 (SR), dated 24/03/2009.

References

1. ACGIH 2004: Threshold limit values for chemical substances and physical agent and biological exposure indices, American Conference of Governmental Industrial Hygienists, Cincinnati, OH (2004)
2. Gun, R.T., Budd, G.M.: Effects of thermal, personal and behavioral factors on the physiological strain, thermal comforts and productivity of Australian shearers in hot weather. Ergonomics 38, 1368–1384 (1995)
3. Candi, D.A., Christina, L.L., Skai, S.S., Maeen, Z.I., Thomas, E.B.: Heat strain at the critical WBGT and the effects of gender, clothing and metabolic rate. Int. J. Ind. Ergonom. 38, 640–644 (2008)
4. Hancock, P.A., Vasmatzidis, I.: Human occupational and performance limits under stress: the thermal environment as a prototypical example. Ergonomics 41, 1169–1191 (1998)
5. Enander, A.E., Hygge, S.: Thermal stress and human performance. Scand. J. Work Env. Hea. 16, 44–50 (1990)
6. Ehrgott, M., Gandibleux, X.: A survey and annotated bibliography of multiobjective combinatorial optimization. OR Spektrum 22, 425–460 (2000)
7. Coello, C.A.C.: An updated survey of GA-based multiobjective optimization techniques. ACM Comput. Surv. 32(2), 109–142 (2000)
8. Dimopoulos, C., Zalzala, M.S.: Recent developments in evolutionary computation for manufacturing optimization: problems, solutions and comparisons. IEEE T. Evolut. Comput. 4, 93–113 (2000)
9. Holland, J.H.: Adaptation in natural selection and artificial systems. Univ. of Michigan Press, Ann Arbor (1975)
10. Goldberg, D.E.: Genetic algorithms in search, optimization & machine learning. Addison-Wesley, Reading (1998)
11. Deb, K.: Multiobjective optimization using evolutionary algorithms. Wiley, Chichester (2001)
12. Deb, K., Amrit, P., Agarwal, S., Meyarivan, T.: A fast and elitist multiobjective genetic algorithm: NSGA-II. IEEE Transactions on Evolutionary Computation 6(2), 182–197 (2002)

Tuning Process Parameters of Electrochemical Machining Using a Multi-objective Genetic Algorithm: A Preliminary Study

Dilip Datta and Akan Kumar Das

Department of Mechanical Engineering, National Institute of Technology
Silchar - 788 010, Assam, India
datta_dilip@rediffmail.com, akankumardas@yahoo.co.in

Abstract. Because of their numerous and diverse ranges, the tuning of process parameters of a machining process depends heavily upon operators' technologies and experiences. Still, proper tuning cannot be expected from such a manual process, which encourages the use of an optimization tool for effective utilization of a process. In this paper, a multi-objective genetic algorithm (GA) is applied to electrochemical machining for tuning its various process parameters so that the optimum output can be achieved. An experimental dataset is used for modeling the problem through regression analysis, and then the GA is applied to a linear model and an exponential model for maximizing material removal rate and minimizing surface roughness.

Keywords: Electrochemical machining, regression analysis, multi-objective optimization, genetic algorithm.

1 Introduction

Electrochemical machining (ECM) is one of the most widely used nontraditional machining processes, which is used to process extremely hard materials that are unable to be machined by traditional machining processes. The ECM process is based on the principle of material removal by electrochemical dissolution of an anode. In this process, a direct current with high amperage and low voltage is passed between the workpiece (anode) and the tool (cathode). At the anodic workpiece surface, metal is dissolved in electrolyte as metallic ions by the electrochemical reactions, and thus the tool shape is copied on the workpiece. The tool material of the ECM process needs not to have high strength and hardness, and hence complicated shapes can be machined easily. It is a complex process, particularly in setting various process parameters for effective utilization of the process. This has necessitated the use of an optimization tool for optimizing the output by proper tuning of various process parameters, which is nearly impossible even from the experience of a skilled operator. In the case of an optimization tool, classical methods, particularly the widely used gradient-based methods [2], are not so suitable for solving such highly nonlinear and complex problems. Therefore,

K. Deb et al. (Eds.): SEAL 2010, LNCS 6457, pp. 485–493, 2010.

the *nondominated sorting genetic algorithm*-II (NSGA-II) [6], a very popular and widely applied nontraditional multi-objective optimization technique, is chosen here for optimizing the output of the ECM process. An experimental dataset, produced by the ECM process, is considered for the study. A linear model and an exponential model for the problem are first generated by multiple regression analysis of the dataset. Then, NSGA-II is applied to the models for maximizing material removal rate and minimizing surface roughness by proper tuning of the input current, voltage, flow rate of electrolyte, and inter-electrode gap size. The obtained results show very similar performance of NSGA-II in both the models. However, the results could not be compared with others due to the non-availability of any literature.

The rest of the paper is organized as follows: the specialized literature is reviewed in Sect. 2. The regression analysis of the experimental dataset is performed in Sect. 3, followed by Sect. 4 where the regression models are formulated as optimization problems. The obtained computational results are presented in Sec. 5. Finally, the paper is concluded in Sect. 6 with the present findings and the possible future scope of the theme of the present work.

2   Literature Review

The response surface methodology, hybridized with low frequency vibration of the tool, is used by Ebeid et al. [7] for deriving an ECM model from experimental data. The modeling and monitoring of the inter-electrode gap is presented by Rajurkar et al. [9]. Analytical modeling of ECM process with low frequency vibration of the tool is demonstrated by Hewidy et al. [8]. Rajurkar et al. [10] perform an experimental analysis for minimizing the machining allowance due to the sludge and the memory error. The controlled ECM through response surface methodology is studied by Bhattacharya and Sorkhel [4], in which the process is optimized by the Gauss-Jordon algorithm. Multi-objective optimization of ECM process is performed by Acharya et al. [1] using the goal partitioning algorithm. It is a complex method as a preference should be given every time to particular variables. A multi-objective model is proposed by Rao et al. [11] for optimizing process parameters of the ECM process using a particle swarm optimization algorithm. However, they solve the problem as a single-objective optimization problem by combining multiple objectives into a single function. Another multi-objective optimization model is proposed by Asokan et al. [3] for maximizing material removal rate and minimizing surface roughness, which is also solved as a single-objective optimization problem, using an artificial neural network algorithm, by mapping the two objectives into a single grade value.

3   Regression Analysis

Regression analysis is used to correlate unplanned experimental input data with output data, in which a mathematical model is generated in such a way that the model deviates minimally from the experimental data. Multiple linear regression models and exponential models are used here for generating relations between machining parameters and their responses. The experimental dataset is taken from Asokan et al. [3], in which similar regression analysis is performed and the obtained models are compared with an ANN (artificial neural network) based model. The dataset is obtained by processing a cylindrical blank, made by hardened steel, in an electrolyte composition of 10% NaCl and 0.2% H_2O_2. The dataset contains 32 observations, which are given in Table 1, where "F/rate", "MRR" and "SR" indicate "flow rate", "material removal rate" and "surface roughness", respectively.

Table 1. Experimental data used in the present study (taken from Asokan et al. [3])

Current (A)	Voltage (V)	F/rate (l/m)	Gap (mm)	MRR (mg/m)	SR (µm)	Current (A)	Voltage (V)	F/rate (l/m)	Gap (mm)	MRR (mg/m)	SR (µm)
220	24	6	0.2	2.57	2.0	260	24	6	0.2	2.85	2.5
220	32	6	0.2	2.67	2.1	260	32	6	0.2	2.96	2.5
220	24	8	0.2	2.57	2.1	260	24	8	0.2	2.70	2.1
220	32	8	0.2	2.65	2.3	260	32	8	0.2	2.91	2.3
220	24	6	0.4	2.65	2.3	260	24	6	0.4	2.70	2.1
220	32	6	0.4	2.85	2.1	260	32	6	0.4	3.10	2.5
220	24	8	0.4	2.57	2.1	260	24	8	0.4	2.62	2.0
220	32	8	0.4	2.52	2.1	260	32	8	0.4	2.71	2.1
200	28	7	0.3	2.30	2.0	280	28	7	0.3	3.05	2.5
240	20	7	0.3	2.15	2.0	240	34	5	0.3	3.01	2.5
240	28	9	0.3	2.75	2.1	240	28	7	0.3	2.91	2.5
240	28	7	0.1	2.70	2.1	240	28	7	0.3	2.82	2.5
240	28	7	0.1	2.61	2.0	240	28	7	0.5	2.64	2.1
240	28	7	0.3	2.65	2.1	240	28	7	0.3	2.59	2.2
240	28	7	0.3	2.60	2.1	240	28	7	0.3	2.59	2.0
240	28	7	0.3	2.57	2.1	240	28	7	0.3	2.55	2.0

The linear regression model, and the exponential model in the logarithmic form, can be given by (1) and (2), respectively.

$$Y_i = a_0 + \sum_{j=1}^{n} a_j x_j + \epsilon_l ; \quad i = 1, 2, \cdots, m ; \tag{1}$$

$$\ln Y_i = b_0 + \sum_{j=1}^{n} b_j \ln x_j + \epsilon_e ; \quad i = 1, 2, \cdots, m ; \tag{2}$$

where, x_j's are independent variables and Y_i's are their dependent functions, and n and m are numbers of independent variables and dependent functions, respectively. $a_0, a_1, \ldots, b_0, b_1, \ldots$ are constant terms which are to be evaluated by minimizing the error terms ϵ_l and ϵ_e. In the present study, the process parameters, $\boldsymbol{x} = \{c,v,\dot{m},h\}$, are independent variables, where c, v, $\dot{m}$ and h stand for current, voltage, flow rate of electrolyte, and inter-electrode gap, respectively. On the other hand, the material removal rate $z(\boldsymbol{x})$ and surface roughness $f(\boldsymbol{x})$ are dependent functions, which are responses to the process parameters $\boldsymbol{x}$. Using the data of Table 1 in the linear model of (1), the maximum errors of 0.3103 and 0.3256 are obtained in $z(\boldsymbol{x})$ and $f(\boldsymbol{x})$, respectively, along with their average errors of 3.646% and 5.52%, respectively. In the case of the exponential model of (2), the maximum errors obtained in $z(\boldsymbol{x})$ and $f(\boldsymbol{x})$ are 0.1229 and 0.1369, respectively, along with their average errors of 3.577% and 5.42%, respectively. Although the maximum errors are quite high, the models are acceptable because the reasonably smaller average errors. The derived linear regression models and the exponential models of $z(\boldsymbol{x})$ and $f(\boldsymbol{x})$ are given by (3) and (4), respectively.

$$\left.\begin{aligned} z(\boldsymbol{x}) &= 0.0062539c + 0.028467v - 0.053475\dot{m} - 0.046035h + 0.77816 \\ f(\boldsymbol{x}) &= 0.0046298c + 0.016201v - 0.066900\dot{m} - 0.082067h + 1.11330 \end{aligned}\right\} \quad (3)$$

$$\left.\begin{aligned} z(\boldsymbol{x}) &= 0.064635 \frac{c^{0.54941}v^{0.29730}}{\dot{m}^{0.14369}h^{0.0043271}} \\ f(\boldsymbol{x}) &= 0.127020 \frac{c^{0.47667}v^{0.19298}h^{0.0041196}}{\dot{m}^{0.20925}} \end{aligned}\right\} \quad (4)$$

It is observed in the linear models of (3) that both $z(\boldsymbol{x})$ and $f(\boldsymbol{x})$ increase with c and v, and decrease with $\dot{m}$ and h. A slightly different situation is observed in the exponential models of (4), in which $f(\boldsymbol{x})$ increases with c, v and h, and decreases with $\dot{m}$ only.

4   Formulation of Optimization Models

In order to directly use the regression models of (3) and (4) in the optimization models, the process parameters, current, voltage, flow rate, and gap size, are considered as decision variables for maximizing material removal rate and minimizing surface roughness. In addition to minimizing the surface roughness, its upper limit is also restricted to a given constant value. Accordingly, the multi-objective optimization model is given below by (5).

$$\left.\begin{aligned} &\text{Determine} && \boldsymbol{x} = (c, v, \dot{m}, h) \\ &\text{to maximize} && \text{MRR} \equiv z(\boldsymbol{x}) \\ &\text{minimize} && \text{SR} \equiv f(\boldsymbol{x}) \\ &\text{subject to:} && f(\boldsymbol{x}) - f_{\max} \leq 0 \\ & && c_{\min} \leq c \leq c_{\max} \\ & && v_{\min} \leq v \leq v_{\max} \\ & && \dot{m}_{\min} \leq \dot{m} \leq \dot{m}_{\max} \\ & && h_{\min} \leq h \leq h_{\max} \ ; \end{aligned}\right\} \quad (5)$$

where, $f_{\max}$ is the allowed maximum value of $f(\boldsymbol{x})$, and $(c_{\min}, c_{\max})$, $(v_{\min}, v_{\max})$, $(\dot{m}_{\min}, \dot{m}_{\max})$ and $(h_{\min}, h_{\max})$ are bounds of c, v, $\dot{m}$ and h, respectively. The expressions of $z(\boldsymbol{x})$ and $f(\boldsymbol{x})$ in (5) depend upon the type of models used, i.e., linear or exponential, the formulations of which are given by (3) and (4), respectively. The reason of considering $z(\boldsymbol{x})$ and $f(\boldsymbol{x})$ as objectives in (5) is that, according to the regression models of (3) and (4), they are guessed to be conflicting with each other.

It is observed in Sect. 2 that the ECM process is formulated as a multi-objective optimization problem by many researchers [1,3,11]. However, it is solved as a single-objective optimization problem by combining multiple objectives into a single function with some user-defined "weights" to different objectives. It is well established that the solution quality of such a mapped problem depends upon the user-defined "weights" to objectives. The problem, formulated in (5), is solved here as a multi-objective problem by giving equal importance to all objectives. Such a consideration will give a set of nondominated solutions, known as a *Pareto front*, in a single run, instead of a single solution per run as a single-objective formulation does. Then, a decision maker will have multiple alternatives (solutions) to choose a particular one according to his/her choice. The *nondominated sorting genetic algorithm*-II (NSGA-II) [6], a well known and widely applied non-traditional multi-objective optimization technique, is used here for solving the problem, the obtained results of which are presented in the following section.

5   Results and Discussions

Both the linear and exponential regression models of the ECM process, given by (3) and (4) respectively, are solved by NSGA-II [6] as multi-objective optimization problems, formulations of which are given in (5). The user-defined values for the problem-specific parameters of (5) are fixed as per the dataset of Table 1, which is used in Sect. 3 for formulating the regression models. These values are given in Table 2.

Table 2. User-defined values for the problem-specific parameters used in (5)

Parameter	Value
Upper limit of the surface roughness (f_{max})	2.5 (micro-meter)
Bounds of the applied current (c_{min}, c_{max})	[200, 280] (amperes)
Bounds of the applied voltage (v_{min}, v_{max})	[20, 34] (volts)
Bounds of the electrolyte flow rate ($\dot{m}_{min}, \dot{m}_{max}$)	[5, 9] (liter/minute)
Bounds of the inter-electrode gap (h_{min}, h_{max})	[0.1, 0.5] (millimeter)

Since it is well established that the performance of a stochastic optimizer, like a genetic algorithm, is usually dependent upon its various user-defined parameters [5], each model of the ECM process is solved 36 times with different initial solutions as well as other NSGA-II related parameters. Such parameter values are given in Table 3.

Table 3. NSGA-II related parameter values used for solving the problem of (5)

Parameter	Value
population size	Randomly chosen in the range of [60, 100]
Number of generations to be performed	200
Crossover probability	90%
Mutation probability	10%
Distribution index for simulated binary crossover operator of NSGA-II	Randomly chosen in the range of [2, 10]
Distribution index for polynomial mutation operator of NSGA-II	Randomly chosen in the range of [10,50]

The Pareto fronts obtained from 36 runs of each of the linear and exponential models of the ECM process are shown in Figs. 1 and 2, respectively, where NSGA-II is found very consistent in giving similar performances over all the 36 runs of both the models. As guessed in Sect. 4, the conflicting nature of the considered objectives, maximizing material removal rate (z) and minimizing surface roughness (f), is very clear in the figures as z increases with increasing f, or f decreases with decreasing z, i.e., one objective is improved at the expense of another.

As mentioned in Sect. 2 and 3, Asokan et al. [3] perform a similar study with the same dataset, in which an ANN (artificial neural network) based model

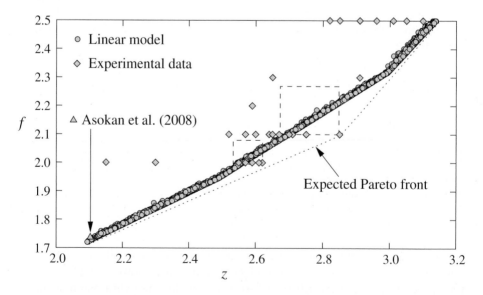

Fig. 1. Pareto fronts obtained for the linear model of (3), and their comparison with the experimental data of Table 1 and existing result of Asokan et al. [3]

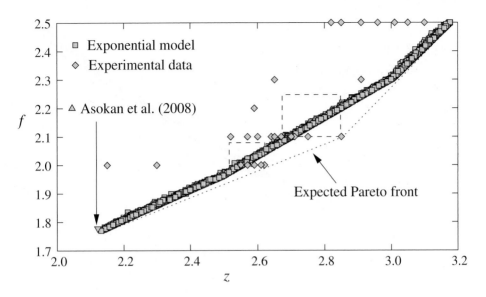

Fig. 2. Pareto fronts obtained for the exponential model of (4), and their comparison with the experimental data of Table 1 and existing result of Asokan et al. [3]

is found to be better than both linear and exponential regression models. In their models, the two objectives are mapped into a single grade value for solving the problem as a single-objective optimization problem, and the reported ANN-based optimum solution is $(c, v, \dot{m}, h) = (200\,\text{A}, 20\,\text{V}, 9\,\text{l/m}, 0.3\,\text{mm})$. According to (3) and (4), the corresponding objective values for the linear model is $(z, f) = (2.103194, 1.736560)$ and for the exponential model is $(z, f) = (2.121389, 1.778065)$. These two solutions are also plotted in Figs. 1 and 2, respectively, where it is observed that both the solutions are biased towards a single objective, i.e., minimizing the surface roughness (f) only.

The plots of the experimental data of Table 1, which are used to formulate the regression models of (3) and (4), are also shown in Figs. 1 and 2. It is observed that the computational results have missed to cover some potential experimental observations in both the models, making some portions of the Pareto fronts to be inferior to those observations. Those inferior portions of the Pareto fronts are shown inside dashed-boxes in Figs. 1 and 2. Such inferiority is certainly not caused from optimization (NSGA-II), but from the accepted errors in formulating the regression models as reported in Sect. 3. If an accurate regression model were used, what would be the expected Pareto fronts are also shown in Figs.1 and 2 by dotted curves, i.e., expected (ideal) Pareto fronts would be better than any experimental observation.

Finally, in order to study the differences between the linear regression model of (3) and the exponential regression model of (4), the Pareto fronts obtained for

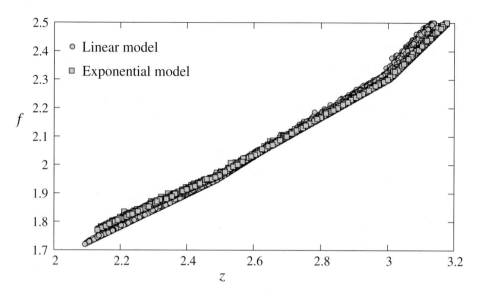

Fig. 3. Comparison of the Pareto fronts obtained for both the linear and exponential models of (3) and (4), respectively

both the models are shown in Fig. 3, where it is observed that the performances of both the models are almost similar (in the objective space), having only some negligible differences in the objective values. Such outcomes are quite appreciable as the designs for the same problem are not desirable to be different with different techniques.

6   Conclusions

The electrochemical machining process is formulated as both linear and exponential regression models using an experimental dataset, in which current, voltage, electrolyte flow rate and inter-electrode gap are considered as input process parameters, and material removal rate and surface roughness as the output responses of those process parameters. Then the models are solved by NSGA-II [6] as multi-objective optimization problems for maximizing material removal rate and minimizing surface roughness. Comparing the outcomes in the objective space, both the models are found comparable with each other. However, the models fail to cover some potential experimental observations, which has happened due to the inaccuracies accepted in the regression models. The future aim of the present work is to model the problem with higher accuracies, and then to test on different benchmark instances. Attempts will also be made for experimental validation of the computational results.

References

1. Acharya, B.G., Jain, V.K., Batra, J.L.: Multi-objective optimization of the ECM process, vol. 8(2), pp. 88–96. Butterworth Co. Ltd., Buttreworth (1986)
2. Arora, J.S.: Introduction to Optimum Design. Elsevier, San Diego (2004)
3. Asokan, P., Kumar, R.R., Jeyapaul, R., Santhi, M.: Development of multi-objective optimization models for electrochemical machining process. International Journal of Advance Manufacturing Technology 39, 55–63 (2008)
4. Bhattacharyya, B., Sorkhel, S.K.: Investigation for controlled electrochemical machining through response surface methodology-based approach. Journal of Materials and Processing Technology 86, 200–207 (1999)
5. Datta, D., Fonseca, C.M., Deb, K.: A multi-objective evolutionary algorithm to exploit the similarities of resource allocation problems. Journal of Scheduling 11(6), 405–419 (2008)
6. Deb, K., Agarwal, S., Pratap, A., Meyarivan, T.: A fast and elitist multi-objective genetic algorithm: NSGA-II. IEEE Transactions on Evolutionary Computation 6(2), 182–197 (2002)
7. Ebeid, S.J., Hewidy, M.S., El-Taweelb, T.A., Youssef, A.H.: Towards higher accuracy for ECM hybridized with low-frequency vibrations using the response surface methodology. Journal of Materials Processing Technology 149, 432–438 (2004)
8. Hewidy, M.S., Ebeidb, S.J., El-Taweela, T.A., Youssef, A.H.: Modeling the performance of ECM assisted by low frequency vibrations. Journal of Materials Processing Technology 189, 466–472 (2007)
9. Rajurkar, K.P., Wei, B., Kozak, J., McGeough, J.A.: Modeling and monitoring interelectrode gap in pulse electrochemical machining. Annals of the CIRP 44(1), 139–142 (1998)
10. Rajurkar, K.P., Zhu, D., Wei, B.: Minimization of machining allowance of electrochemical machining. Annals of the CIRP 47(1), 165–168 (1998)
11. Rao, R.V., Pawar, P.J., Shankar, R.: Multi-objective optimization of electrochemical machining process parameters using a particle swarm optimization algorithm. Journal of Engineering Manufacture 222(8), 949–958 (2008)

The Optimization versus Survival Problem and Its Solution by an Evolutionary Multi Objective Algorithm

Gideon Avigad, Erella Eisenstadt, and Miri Weiss

Braude College of Engineering
{gideon,erella,miri}@braude.ac.il

Abstract. Altruism may be found in sets (groups of solutions). In such cases, it may occur that individual/individuals degrade their chances of survival (with sacrifice in the extreme) to ensure survival of fitter individuals. The idea of altruism within group evolution is posed here as a multi objective problem. The aspiration of a group to survive (find an optimal solution) is posed versus the individual's aspiration to survive. In the paper, the problem is a trajectory planning problem with the dilemma producing a Pareto set for a decision maker to choose from. It is shown that if the decision maker is ready to forfeit some of the group members, optimality may be gained. Evolutionary multi objective algorithm is implemented in order to search for this optimal set.

Keywords: Multi objective, altruism, evolution.

1 Introduction

1.1 Biological Inspiration

In evolutionary biology, an organism is said to behave altruistically when its behavior benefits other organisms at a cost to itself, [1]. The costs and benefits are measured in terms of *reproductive fitness*, or expected number of offspring. Altruistic behavior is common throughout the animal kingdom, particularly in species with complex social structures. For example, vampire bats regularly regurgitate blood and donate it to other members of their group who have failed to feed that night, ensuring they do not starve. From a Darwinian viewpoint [2], the existence of altruism in nature is at first sight puzzling. By virtue of its relative fitness advantage within the group, the selfish mutant will out-reproduce the altruists; hence selfishness will eventually swamp altruism. Nevertheless, altruism exists in nature.

Kin selection [1] might explain the evolution of altruism. A process of group selection may thus allow the altruistic behavior to evolve. Within each group, altruists will be at a selective disadvantage relative to their selfish colleagues, but the fitness of the group as a whole will be enhanced by the presence of altruists ([3]).

Altruism is a well understood topic in evolutionary biology; the theoretical ideas explained above have been extensively analyzed, empirically confirmed, and are widely accepted. Nevertheless, this biological notion of altruism is not identical to the everyday concept. In everyday parlance, an action would only be called 'altruistic' if it was done with the conscious intention of helping another. However, in the biological

K. Deb et al. (Eds.): SEAL 2010, LNCS 6457, pp. 494–503, 2010.
© Springer-Verlag Berlin Heidelberg 2010

sense, there is no such requirement. Nevertheless 'new' group selection turns out to mathematically equivalent to kin selection, as a number of authors have emphasized (e.g., [3], [4]); Where human behavior is concerned, the distinction between biological altruism, defined in terms of fitness consequences, and 'real' altruism, defined in terms of the agent's conscious intentions to help others, does make sense. (Sometimes the label 'psychological altruism' is used instead of 'real' altruism.) What is the relationship between these two concepts? They appear to be independent in both directions, as argued in [5]. An action performed with the conscious intention of helping another human being may not affect their biological fitness at all and so would not count as altruistic in the biological sense. Conversely, an action undertaken for purely self-interested reasons, i.e., without the conscious intention of helping another, may boost their biological fitness tremendously. Therefore, evolution may well lead 'real' or psychological altruism to evolve. Contrary to what is often thought, an evolutionary approach to human behavior does not imply that humans are likely to be motivated by self-interest alone.

In the study, which is reported here, there is altruistic behavior within a group. Because we do not introduce any genes that impose an altruistic behavior, yet altruistic behavior may evolve, it appears that the altruism of the current study is related to psychological altruism and not to biological altruism.

1.2 EMO and Its Utilization for Trajectory Planning

Evolutionary Computation (EC) methods belong to a class of non-gradient methods that has grown in popularity. Searching a multi-objective design space for optimal solutions by EC approaches is commonly referred to as Evolutionary Multi-objective Optimization (EMO). Multi-Objective Evolutionary Algorithm (MOEA) is an EMO algorithm that searches for a solution in a multi-criteria space using some inspiration from evolutionary theories. According to [6], the later generation of Pareto-based algorithms, such as SPEA2, [7], and NSGA-II, [8], involves three major elements. The first element concerns the creation of a search pressure towards the Pareto set. This is commonly achieved by one of the known Pareto-based fitness assignment (dominance-based) techniques. The second element is set to avoid convergence to a single solution in order to preserve diversity. The third element is elitism, which helps to prevent losing non-dominated solutions that are diversified. Detailed descriptions of multi-objective evolutionary techniques can be found in [9].

The use of EC and EMO approaches for path and trajectory planning has been suggested in many works (e.g., [10]). In those studies, different path descriptions have been coded within the solution related chromosome code. In [11], Bezier functions were used while [12] used B-splines. [13] proposed methods for generating path planning by using EC algorithms to define composite η^3-Splines. The method generates a sequence of vertices, tangents, and curvature parameters to formulate and define basic polynomial coefficients serving the η^3-Spline curve.

1.3 Robotics and Altruism

Several works highlight the need for altruism in order to enhance better performances in multi Unmanned Vehicles (UMVs) communities. Individual robots (UMVs) must offer varying levels of "help" to their robots. In [14] and [15], a description of robot behaviors in terms of a "satisfaction   index" and transmission/reception of signals from other robots is studied. A robot's progress in a given task can be measured by its "satisfaction" in the task, which corresponds to the fitness or performance index. Thus, a robot needing help with a task may emit an attraction ("please help me") signal. Robots receiving this signal may stop their current task and aid their community members. In [16], controllers are searched for the community robots. In that work, the performance of individual robots is assessed through a fitness function. It is the difference between the rewards gained from tasks being completed and the costs involved in completing these tasks. The fitness of a community is a measure of the success measured across all community's tasks. Thus robots helping others although decreasing their self fitness may contribute to the overall fitness of the community. Recently, in [17], the evolution of different behavioral strategies has been optimized by both an EMO and by a single objective search, demonstrating the superiority of the former in producing better community performances. In [18], the authors have examined evolutionary methods that may lead to the emergence of altruistic cooperation in robot collectives. They have considered evolutionary algorithms that derive from biological theories on the evolution of altruism in nature and compare them systematically in two experimental scenarios where altruistic cooperation can lead to a performance increment. We discuss the relative merits and drawbacks of the four methods and provide recommendations for the choice of the most suitable method for evolving altruistic robots.

In the current study, we aim at posing an auxiliary MOP problem that poses the dilemma of optimality versus survival. As will be shown in cases, sacrifice of resources is needed in order to attain improved performances with the decision whether to sacrifice these resources being left for the decision maker. That means that a decision maker has to examine the Pareto front of the posed problem and decide, as common in multi criteria decision making, on a preferred solution, which might involve sacrifice (less survival). As far as the review goes, no such posing has been suggested before. Such posing is needed when there is contradiction within that MOP setting, which may only be resolved by a decision maker. Here, we suggest utilizing an MOEA to search for the optimal set for the posed MOP.

2   Methodology

2.1 Problem Definition and Solution

Altruism is associated with a cost for the community, i.e., the fitness degradation of some of its members. At the extreme, this altruism may involve total loss of these members. It is noted that, although the biological inspiration exists, the current paper deals solely with engineering products. Sacrifice should not be related to the sacrifice of human lives, but rather with a cost of losing resources. In the context of the current paper, losing resources means that some of the UMVs fall out of order or are held back for such a long period that they are useless. The cost of losing resources as assessed here in relation to the attainment of better group performances. Attaining better performances may be viewed as an aspiration for optimality. The performances of the group or its competency are a measure of the group's success in performing its task. A group of individuals is a set X of n UMVs' trajectories, $X = [x_1...x_i,.......x_n]^T, X \subseteq \Omega \subseteq R^n$ where $x_i = p(y,t)$ is the trajectory of the i-th UMV, designating the location y of that UMV in time t, $t \in [t_{initial}, t_{final}]$. The competency of the set $\Psi(X) \subseteq \Gamma \subseteq R^m$, $\Psi(X) = [\Psi_1(X), \Psi_2(X),......\Psi_m(X)]^T$, is therefore a vector of performances in an MOP with m objectives. This competency is problem specific as it is determined by the task's objective/s. The cost of losing one of the agents is $C(X)$ and is computed here directly by totaling the number of lost UMVs. Losing a UMV is a result of sacrificing it (e.g., deliberately bumping into an obstacle). The optimization versus survival (not losing resources) is posed here as an MOP of maximizing the competency of the set and minimizing sacrifice (maximizing survival). If the competency is measured based on minimization of each $\Psi(X)$, then the problem may be defined as:

$$min(\Phi(X)), \ \Phi(X) = [\Psi(X), C(X)]^T \tag{1}$$

$$\text{Subject to } X \subseteq \Omega \text{ and } \Psi(X) \subseteq \Omega$$

This MOP is associated with $m+1$ objectives. The problem does not include an inherent contradiction between the performances and the survival (cost). It is possible that optimality is gained with no need for sacrifice. Yet, in the general case, sacrifice may be needed. The solution to the problem at hand is a Pareto set C* and a Pareto front FC*:

$$C^* := \{ X^* \in C^* | \neg \exists \ X \in C : \Phi(X) \preceq \Phi(X^*) \}$$

$$FC^* := \{ z^* \in Z | z^* = \Phi(X^*) : X^* \in C^* \} \tag{2}$$

Each vector of optimal performances is represented as a point on the Pareto front.

2.2 Decision Making

By depicting FC*, which is the Pareto front of the posed MOP (see Equation 2), a decision maker has to choose the winning set. Such a decision is attained through the execution of the set's related UMVs' trajectories. While making decisions, the DM has to consider the pros and cons of improving performances as related to maintaining the available resources. It is clear that either extreme, no sacrifice whatsoever or sacrificing all but one UMV (the job has to be done), might be decided upon. Yet, a tradeoff between these extremes should also be considered. The decision should take into account the relative importance of optimality and the importance of keeping the resources for future tasks. This means that availability of substitutes should be taken into account.

3 Solution Approach and an Example

Here, each trajectory is divided into sections. While the length of the section may vary, the time of travel between the start and end point of each section is predefined and constant. This means that the speeds of the travel may vary (longer sections mean higher travel speeds). The lengths of the trajectories' sections are constrained such that the UMVs' speeds may not exceed an attainable travel speed. Figure 1 depicts the paths of two sets of UMVs. Each of the sets is associated with three UMVs (the black and white circles). The sets are designated by their related UMVs' trajectories, which are shown as bold lines for one set and dashed lines for the other. Numbers designate the times at which the UMVs arrive at those locations. In the figure, five sections are planned. It is assumed that if a UMV collides with an obstacle, it removes it, such that another UMV passing in the location where the obstacle stood before does not collide with it. The mean distance traveled by a set's UMVs (not accounting for sacrificed UMVs) is computed. This distance is utilized together with the number of lost UMVs as the set's vector of performances. The optimization is aimed at minimizing both the mean distance and the number of lost UMVs. In fact, this means that the size of $\psi(X)$ is one and together with the cost the posed MOP is a bi-objective problem. Depicting Figure 1, it is seen that for the black circles set the left most and middle UMVs has bumped into two obstacles, in the first and third sections respectively, while the third UMV bypassed the obstacles. For this set, the survival will be degraded to just one with the competency being equal to the length traveled by the right most UMV (the only survivor). Clearly the sacrifice of two UMVs was unjustified as it did not improve the performances of the set. In contrast, the white related set undertook another strategy. The left most UMV bumped into the obstacle between times 2 and 3, leaving room for the middle UMV to pass through between times 3 and 4. This means that this set will have two survivors with a distance equal to the mean distance traveled by these survivors.

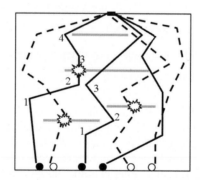

Fig. 1. Trajectories of two sets of three UMVs each. The different sets are distinguished one from the other by different circles' colors (black and white) and by different path lines (bold and dashed). The left most bold line UMV clears the way for its middle path set's member.

Each set is coded as an individual. Each individual codes all paths of the set's related UMVs trajectories. In that code, each chromosome of an individual decodes the path of one UMV. Each two adjacent genes of the chromosome decode a point along the path of that UMV. A population of such individuals is evolved using NSGA-II (Deb et al 2002).

The actual test involves a set of six UMVs instead of just three, each starting at a predefined location. It is noted that the tests shown here are easy to follow because their solution is despicable (otherwise proving optimality might be impossible). The UMVs are to reach an end point while avoiding/colliding with the four obstacles. A population of 100 individuals has been evolved for 500 generations using real coded individuals with 50% single point crossover and 5% Gaussian mutation. The evolved front is depicted in figure 2.

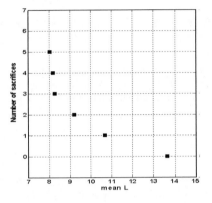

Fig. 2. The Pareto front for the posed problem. As the number of sacrifices increases, the mean distance traveled by the set decreases. Clearly sacrificing more than three UMVs does not possess merit. Decision on the strategy is left to the decision maker.

It is depicted that insuring high survivability degrades optimality. It is also clear that sacrificing more than three UMVs possesses no merit. Moreover, in fact computational inaccuracies are entirely responsible for the difference in the mean values for the distance between 3, 4 and 5 UMVs' sacrifices. A decision maker viewing the evolved front would easily detect this waste and concentrate on just the four right located points in Figure 2. In the following Figures (Figures 3 to 6), different Pareto set related cases are demonstrated. All solutions shown are evolved simultaneously, using NSGA-II.

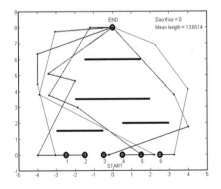

Fig. 3. Maximal survivability: The evolved set of UMVs with maximal survivability

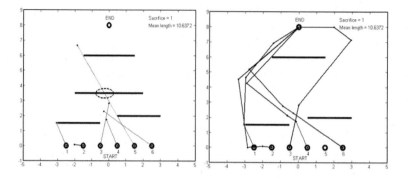

Fig. 4. One sacrifice: In the first movement (between start and time 1), UMV No. 5 bumps into one of the obstacles (left panel). Losing that UMV as an available resource allows UMVs No. 4 and No. 6 to shorten their travel, as depicted in the right panel of the figure.

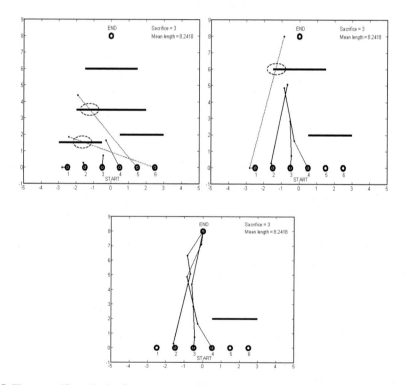

Fig. 5. Three sacrifices: In the first movement (between start and time 1), UMVs No. 5 and No. 6 dismantle two of the obstacles (upper left panel). In the second step (between time 1 and time 2), UMV No. 1 dismantles yet another obstacle (upper right panel).

Because an evolutionary approach, which is a stochastic approach, has been used we have tested the repeatability of the results. Figure 7, Depicts a box plot of these tests, where we report on the mean and average length attained for each number of

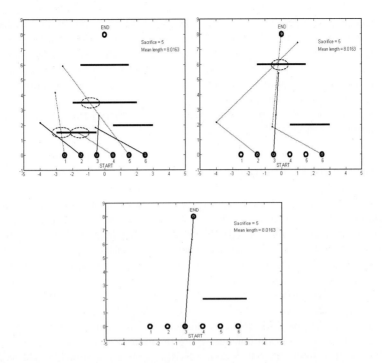

Fig. 6. Five sacrifices: In the first movement (between start and time 1), UMVs No. 1 and No. 4 dismantle one of the obstacles while No. 5 dismantles another one (upper left panel). In the second step (between time 1 and time 2), UMVs No. 2 and No. 6 dismantles yet another obstacle (upper right panel).

sacrifices. It is noted that although the actual paths may differ from one run to the other, the statistical data implies on high reliability when the multi objective problem (i.e., optimality versus sacrifice is considered).

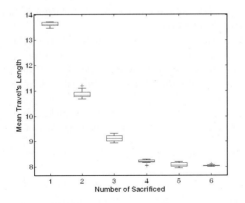

Fig. 7. A box plot describing the statistical data for running the algorithm 30 times

We have also repeated the solution for the above problem, but by dividing the trajectories to more than four sections. The only reportable difference between these runs and the one reported here is that more generations are needed when more secretions are used in order to achieve the same standard of deviations as gained here. This is understandable based on the fact that the search space increases in a direct relation to the number of added sections.

The consistency of the results shows that the approach may find different multi objective optimal solutions with high reliability.

4 Summary and Conclusions

In this paper, we suggested the posing of the optimality versus survival as an MOP. Incorporating survival within a MOP is motivated by relating such an objective to atomism. Such posing, which was never posed as such before, leads to a solution that, in the cases at hand, is a Pareto set and front. In such cases, a decision maker is presented with the Pareto front and has to decide on a preferred strategy. A preferred strategy might be one that is ready to sacrifice some of the available resources (sacrificing UMV/s) to gain better performances. In the paper, we utilized an MOEA in order to simultaneously evolve the Pareto front for trajectory planning problem. In the current problem, the posed MOP is a result of one performance objective and the cost objective. Future work should consider more than one performance objective. Moreover, the use of a simple trajectory code (division to linear sections) might be altered to a more sophisticated coding, such as B-splines. Yet, more complicated and less straight forward problems should be solved. For example, using many more obstacles with more constraints (e.g., friction affect) should be tested. As a last implication towards future work, it would be interesting to examine dynamic obstacles and may be to incorporate co-evolutionary approach where opponents interact.

References

1. Hamilton, W.D.: The Genetical Evolution of Social Behavior I and II. Journal of Theoretical Biology 7, 1–16, 17–32 (1964)
2. Darwin, C.: The Descent of Man and Selection in Relation to Sex. Appleton, New York (1871)
3. West, S.A., Griffin, A.S., Gardner, A.: Social Semantics: Altruism, Cooperation, Mutualism, Strong Reciprocity and Group Selection. Journal of Evolutionary Biology 20, 415–432 (2007)
4. Lehmann, L., Keller, L., West, S., Roze, D.: Group Selection and Kin Selections: Two Concepts but One Process'. Proceedings of the National Academy of the Sciences 104(16), 6736–6739 (2007)
5. Sober, E.: Did Evolution Make us Psychological Egoists? In: His From A Biological Point of View. Cambridge University Press, Cambridge (1994)
6. Coello, C.A.C.: Recent Trends in Evolutionary Multiobjective Optimization. In: Ajith, A., Jain, L., Robert, G. (eds.) Evolutionary Multiobjective Optimization: Theoretical Advances and Applications, pp. 7–32. Springer, London (2005)

7. Zitzler, E., Laumanns, M., Thiele, L.: SPEA2: Improving the strength Pareto evolutionary algorithm. Technical report 103, Computer Engineering and Networks Laberatory (TIK), Swiss Federal Institute of Technology (ETH) Zurich, Gloriastrasse 35, CH-8092 Zurich, Switzerland (May 2001)
8. Deb, K., Pratap, A., Agarwal, S., Meyarivan, T.A.: Fast and elitist multiobjective genetic algorithm. NSGA–II. IEEE Transactions on Evolutionary Computation 6(2), 182–197 (2002)
9. Deb, K.: Multi-objective optimization using evolutionary algorithms. J. Wiley & Sons, Ltd., Chicester (2001)
10. Saravanan, R., Ramabalan, S., Balamurugan, C.: Evolutionary multi-criteria trajectory modeling of industrial robots in the Presence of obstacles. Engineering Applications of Artificial Intelligence 22, 329–342 (2009)
11. Mittal, S., Deb, K.: Three-dimensional path planning for UAVs using multi-objective evolutionary algorithms. In: Proceedings of the Congress on Evolutionary Computation (CEC 2007), pp. 25–28 (September 2007)
12. Watanabe, K., Kiguchi, K., Izumi, K., Kunitake, Y.: Path planning for an omnidirectional mobile manipulator byevolutionary computation. In: Third International Conference Knowledge-Based Intelligent Information Engineering Systems, pp. 135–140 (1999)
13. Wei, J., Liu, J.: Collision free composite n3- splines generation for non-holonomic mobile robots by parallel variable length genetic algorithm. In: Int. Conference on Computational intelligence for Modeling control and Automation, CIMCA 2008, Vienna, Austria, December 10-12 (2008)
14. Lucidarme, P., Simonin, O., Liegeois, A.: Implementation and evaluation of a satisfaction/altruism based architecture for multi-robot systems. In: Proc. IEEE Int. Conf. on Robotics and Automation, pp. 1007–1012 (2002)
15. Simonin, O., Ferber, J.: Modeling self-satisfaction and altruism to handle action and reactive cooperation. In: SAB 2000 Proceedings, pp. 314–323 (2000)
16. Morton, R.D., Bekey, G.A., Clark, C.M.: Altruistic Task Allocation despite Unbalanced Relationships within Multi-Robot Communities. In: IEEE/RSJ International Conference on Intelligent Robots and Systems, St. Louis, USA, October 11-15 (2009)
17. Schrum, J., Miikkulainen, R.: Proceedings of the Fourth Artificial Intelligence and Interactive Digital Entertainment Conference, Stanford, California, October 22–24. AAAI Press, California (2008)
18. Floreano, D., Mitri, S., Perez-Uribe, A., Keller, L.: Evolution of Altruistic Robots. In: Zurada, J.M., Yen, G.G., Wang, J. (eds.) Computational Intelligence: Research Frontiers. LNCS, vol. 5050, pp. 232–248. Springer, Heidelberg (2008)

Synthesis of Difference Patterns for Monopulse Antenna Arrays – An Evolutionary Multi-objective Optimization Approach

Siddharth Pal[1], Aniruddha Basak[1], Swagatam Das[1], and P.N. Suganthan[2]

[1] Dept. of Electronics and Telecommunication Engg,
Jadavpur University
Kolkata 700 032, India
[2] School of Electrical and Electronic Engineering
Nanyang Technological University,
Singapore 639798, Singapore
sidd_pal2002@yahoo.com, aniruddha_ju_etce@yahoo.com,
swagatamdas19@yahoo.co.in, epnsugan@ntu.edu.sg

Abstract. Monopulse antennas form an important methodology of realizing tracking radar and they are based on the simultaneous comparison of sum and difference signals to compute the angle-error and to steer the antenna patterns in the direction of the target (i.e., the boresight direction). In this study, we consider the synthesis problem of difference patterns in monopulse antennas from the perspective of Multi-objective Optimization (MO). The synthesis problem is recast as a multi-objective optimization problem (for the first time, to the best of our knowledge), where the Maximum Side-Lobe Level (MSLL) and Beam Width (BW) of principal lobe are taken as the two objectives. The Optimal Pareto Fronts (OPF) are obtained for different number of elements and subarrays using one of the best-known evolutionary MO algorithms till date, called the Non-dominated Sorting Genetic Algorithm (NSGA-II). The quality of solutions obtained is compared with the help of Pareto fronts on the basis of the two objectives to investigate the dependence of the number of elements and the number of sub-arrays on the final solution. Then we find the best compromise solutions for 20 element array and compare the results with standard single objective algorithms such as the Differential Evolution (DE) that has been reported in literature so far for the synthesis problem.

Keywords: Antenna array, monopulse antennas, evolutionary algorithm, multiobjective optimization, non-dominated sort, genetic algorithms.

1 Introduction

The conventional way of enhancing angular accuracy amounts to taking several measurements while the antenna rotates through an area of interest, and then to compare the results. However, this method has its drawbacks even if the antenna is properly calibrated. As the measurements are taken one after the other, the target has moved

K. Deb et al. (Eds.): SEAL 2010, LNCS 6457, pp. 504–513, 2010.
© Springer-Verlag Berlin Heidelberg 2010

to another place in-between and the aspect angle has changed too. The monopulse technique was invented to eliminate this source of measurement error. A monopulse antenna [1 – 4] also takes several measurements with beams pointing into different directions, but as the name implies, these measurements are taken simultaneously, with a single pulse. The word "monopulse" implies that with a single pulse, the antenna can gather angle information, as opposed to spewing out multiple narrow-beam pulses in different directions and looking for the maximum return Therefore this technique can determine angle very precisely. Monopulse antennas are in widespread use in military applications like target-tracking radars and missile-seeker heads. Civilian applications include automotive radars, secondary radars for air traffic control and control systems, which need to know the precise whereabouts of a TV-, GPS- or other type of satellite [1, 5].

A key issue in the design of monopulse antennas is that the sum pattern and the difference pattern have to be synthesized by the same array configuration. In this context Lopez et al. [6] proposed an interesting method that is based on a subarray configuration and uses a standard binary Genetic Algorithm (GA) to determine the weights of the subarrays. Caorsi et al. [7] took a similar approach where the same synthesis problem has been faced by applying the Differential Evolution (DE) method [8], in which hybrid chromosomes (constituted by real and integer genes) are used to avoid the need for coding and decoding the real variables (weights of the subarrays). In [9] the approach of [7] is extended to the optimization of the directivity of the difference pattern by means of a hybrid real/integer DE algorithm.

As can be perceived from a literature survey, the design of monopulse antenna arrays can be formulated in several possible ways and with emphasis on various aspects of the final output expected. Under such circumstances there may not exist a single optimal solution but rather a whole set of possible solutions of equivalent quality [8]. A natural choice for handling this kind of design problems is to use Multi-objective Optimization (MO) algorithms [9, 10] that deal with such simultaneous optimization of multiple, possibly conflicting, objective functions.

Here we employ a most widely used evolutionary MO algorithm called Non-dominated Sorting Genetic Algorithm (NSGA-II) proposed by Deb *et al.* [11] for two purposes: firstly to design monopulse arrays that could simultaneously minimize the Maximum Side-Lobe Level (MSLL) and principal lobe Beam Width (BW), and secondly to study the effects of number of elements and number of subarrays on the performance of the antenna array by observing the shape of the Pareto fronts generated with NSGA-II for various combinations of these two numbers. For the multi-objective design of monopulse array, a fuzzy membership function based approach described in [12] is taken to select the best compromise solution from the Pareto front. Comparison with the single objective design results with DE and another real parameter optimizer of current interest, called Particle Swarm Optimization (PSO) [13] reflects the superiority of the multi-objective approach in terms of final accuracy of design results. Since multi-objective approach is superior to single objective cases where more than one design objectives are combined through weighted sum, the Pareto fronts generated by a reliable MO algorithm, like NSGA – II, can provide a means of identifying the optimal number of design variables (through number of elements and number of subarrays). To the best of our knowledge, such study is undertaken here for the first time in the related area.

2 Formulation of the Design Problem

An antenna array is a configuration of individual radiating elements that are arranged in space and can be used to produce a directional radiation pattern. For a linear antenna array with $2N$ isotropic radiators the array factor can be expressed as below

$$AF(\theta) = \sum_{n=-N}^{-1} a_n e^{j\left(n+\frac{1}{2}\right)kd\cos\theta} + \sum_{n=1}^{N} a_n e^{j\left(n-\frac{1}{2}\right)kd\cos\theta}, \qquad (1)$$

where a_n is the excitation of the nth radiating elements, k is the wave number of the medium in which the antenna is located, d is the distance between the elements, and θ defines the angle at which $AF(\theta)$ is calculated with respect to a direction orthogonal to the array.

The required sum pattern is obtained by the excitation a_n^s, $n = -N, \ldots, -1,1,\ldots, N$, which are assumed to be symmetric about the array centre and fixed. Thus we will have $a_n^s = a_{-n}^s$.The excitations are obtained by the Dolph-Chebyshev method [15]. Using the symmetry property the array factor reduces to the expression (2).

$$AF_s(\theta) = \sum_{n=1}^{N} a_n^s.\cos\left[\frac{1}{2}(2n-1)kd\cos\theta\right]. \qquad (2)$$

Excitations for the difference pattern are obtained by:

$$a_n^d = a_n^s \sum_{p=1}^{P} \delta_{c_n,p} g_p \qquad ,n = 1,2,3,...N. \qquad (3)$$

$\delta_{c_n,p}$ represents the Kronecker delta function [16] i.e $\delta_{c_n,p} = 1$ if $c_n = p$, otherwise $\delta_{c_n,p} = 0$. In order to obtain the difference pattern, the excitations must be anti-symmetric i.e. $a_n^s = -a_{-n}^s$. Thus the array factor for the difference pattern reduces to expression (4).

$$AF_d(\theta) = \sum_{n=1}^{N} a_n^d.\sin\left[\frac{1}{2}(2n-1)kd\cos\theta\right] \qquad (4)$$

$AF_d(\theta)$ is a function of θ which is symmetric about $0°$. Let θ_{max} be the angle at which $AF_d(\theta)$ attains global maxima. We calculate $AF_d(\theta)$ for discrete values of θ. Let those discrete values be represented by set $\psi = [0, \pi/2]$. Let the discrete steps in which $AF_d(\theta)$ is calculated be $\Delta\theta$.

For obtaining multi-objective formulation of the present problem we need to find the maximum sidelobe level and the width of the principal lobe. To calculate the maximum sidelobe we calculate where the array factor reaches local maxima, and the maximum value of all the local maxima gives us the SLL value. Let,

$$\zeta = [\theta \in \psi \,|\{ AF_d(\theta) > AF_d(\theta - \Delta\theta)\} \wedge \{ AF_d(\theta) > AF_d(\theta + \Delta\theta)\} \wedge \{\theta \neq \theta_{max}\}]$$

be the set of angles where local maxima of $AF_d(\theta)$ occur.

One null of the principal lobe is located at $0°$ because of the anti-symmetric property of monopulse antenna.

Let: $\Phi = \{\theta \in \psi \mid AF_d(\theta) < AF_d(\theta - \Delta\theta) \wedge AF_d(\theta) < AF_d(\theta + \Delta\theta) \wedge \theta \neq 0^o\}$ be   the set of angles where local minima of $AF(\theta)$ is reached. Let the local minimum closest $0°$ to be α.

Therefore $\alpha = \min(\Phi)$. Now we are at a position to define the two objective functions:

$$f_1 = \max\left(\frac{AF_d(\theta_{max})}{AF_d(\zeta)}\right). \qquad (5a) \qquad\qquad f_2 = \min(\Phi). \qquad (5b)$$

3 The NSGA-II Algorithm – An Outline

Due to the multiple criteria nature of most real-world problems, Multi-objective Optimization (MO) problems are ubiquitous, particularly throughout engineering applications. The concepts of dominance and Pareto-optimality may be presented more formally in the following way:

Definition 1: Consider without loss of generality the following multi-objective optimization problem with D decision variables X (parameters) and n objectives y:

$$\text{Minimize } \vec{Y} = f(\vec{X}) = (f_1(x_1,....,x_D),...., f_n(x_1,....,x_D)), \qquad (6)$$

where $\vec{X} = [x_1,......,x_D]^T \in P$ and $\vec{Y} = [y_1,....,y_n]^T \in O$ and $\vec{X}$ is called decision (parameter) vector, P is the parameter space, $\vec{Y}$ is the objective vector, and O is the objective space. A decision vector $\vec{A} \in P$ is said to dominate another decision vector $\vec{B} \in P$ (also written as $\vec{A} \prec \vec{B}$ for minimization) if and only if:

$$\forall i \in \{1,....,n\}: \ f_i(\vec{A}) \leq f_i(\vec{B}) \wedge \exists j \in \{1,......,n\}: f_j(\vec{A}) < f_j(\vec{B}) \qquad (7)$$

Based on this convention, we can define non-dominated, Pareto-optimal solutions as follows:

Definition 2: Let $\vec{A} \in P$ be an arbitrary decision vector.

(a) The decision vector $\vec{A}$ is said to be non-dominated regarding the set $P' \subseteq P$ if and only if there is no   vector in P' which can dominate $\vec{A}$.

(b) The decision (parameter) vector $\vec{A}$ is called Pareto-optimal if and only if $\vec{A}$ is non-dominated regarding the whole parameter space P.

During 1993-2005, plethora of different evolutionary algorithms was suggested to solve multi-objective optimization problems and an interested reader may consult [10, 11, 17, and 18]. Central to this article, we select an MO algorithm called the Non-dominated Sorting Genetic Algorithm (NSGA-II) proposed by Deb et al. [12], owing to its wide popularity among various research communities for solving practical MO problems from diverse domains [10, 17]. The NSGA-II algorithm has been illustrated through a flow-chart in Figure 1.

For a monopulse antenna array of *2N* elements and *P* subarrays we will have *N* binary coded variables and *P* real coded variables. The *N* binary coded variables signify the subarray grouping information c_n and the *P* real coded variables indicate the subarray weights g_p.

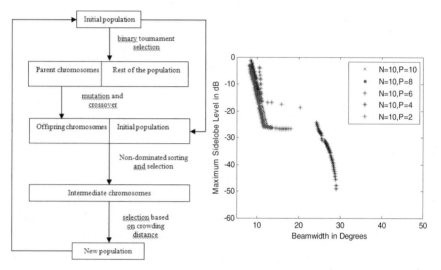

Fig. 1. Basic flow-chart of NSGA-II

Fig. 2. Trade-off Curves for 20 element array (*N*=10)

4 Study of Optimal Pareto Fronts for Designing Monopulse Antenna

This section is primarily meant to study how the parameters such as number of elements and number of subarrays affect design of monopulse antennas in terms of two important figures of merit: the BW and MSLL. For a fixed number of elements we can use an MO algorithm to decide the number of subarrays that produces a good trade-off between the two design objectives yielding an Optimal Pareto Front (OPF) that has its knee points closer to the origin. Then we fix the number of elements and number of subarrays to find the best solution from the OPF considering the same two design objectives, but this is taken up in the next section.

A. Case 1: 20 Element Array

Fixing the number of elements to 20, we run NSGA – II varying the number of subarrays *P* from 2 to 10 in steps of 2. The corresponding Pareto fronts have been shown in Figure 2.

A close inspection of Figure 2 shows that the best trade-off can be achieved for 6 subarrays as points near the knee of the OPF are closest to the origin corresponding to least values of MSLL and BW in comparison to the Pareto fronts obtained with other numbers of subarrays.

B.   Case 2: 40 Element Array

Figure 3(a) shows the trade-off curve obtained for 40 element array with number of subarrays = 2, 4, 8, 16. In this case we can observe that the best trade-off can be obtained with $P = 8$. The OPF is very close to that obtained for $P = 4$. However increasing the number of subarrays does not improve the quality of solutions. Rather the OPF obtained for $P = 16$ is far worse than the others. This is probably because for $P = 16$ the complexity of problem hyperspace is increased.

C.   Case 3: NUMBER OF SUBARRAYS CONSTANT

Here we have fixed the number of subarrays to $P = 8$. In this case we can observe that keeping P constant increasing N does not improve the solution beyond a certain limit, as is evident from the trade-off curves provided in Figure 3(b).

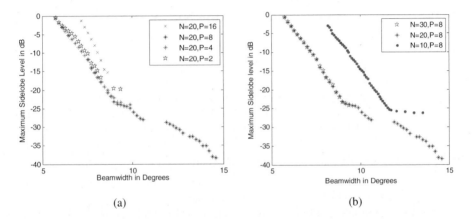

(a)   (b)

Fig. 3. (a) Trade-off Curve for 40 element array (N=20) , (b) Trade-off Curve for fixed number of subarrays (P=8)

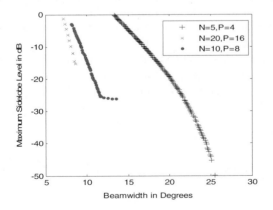

Fig. 4. Trade-off Curve for N/P=constant

D.   N/P CONSTANT

In this case we show how the quality of design improves when the number of elements and subarray are increased in the same proportion. As can be perceived from Figure 4, for $N = 10$, $P = 8$ ($N/P = 1.25$) the OPF is exhibiting a considerably good trade-off between the two design objectives, with knee points at which values of both MSLL and BW reduced more than those obtained with other values of N and P giving same ratio N/P.

The aim of this section was to investigate how the optimal combination of two vital parameters related to the design problem *viz.* number of elements and number of subarrays can be estimated using an MO algorithm. The next section is devoted to the actual design of the monopulse array using NSGA - II with the subarray weights and the element grouping kept as decision variables.

5   Design of Monopulse Antenna Using NSGA-II

In case 1 of Section 3, we discussed that for 20 element array design , 6 subarrays gives a considerably good OPF with knee points showing better compromise between the two objectives. In this section we systematically find the best compromise solution for 20 element array ($P = 2, 4, 6, 8, 10$) and compare them with the results from two standard single objective optimization algorithms namely DE [7] and PSO [9]. For DE the parametric setup is also taken from [9]. For PSO, we used swarm size = 50, acceleration coefficients $C_1 = C_2 = 2.00$, inertia weight ω linearly decreasing from 0.9 to 0.4 and for d-th component of maximum velocity $v_{d,max} = 0.9*r_d$ where r_d is the difference between the maximum and minimum values of the d-th decision variable. Both PSO and DE were run for 1500 iterations.

The best compromise solution was chosen from the OPF using the method described in [13]. Due to limitation of space we are suppressing the description of this method.

Below we provide the design results for two (among five) different cases (others are not mentioned due to space limitation) corresponding to two different numbers of subarrays. The sum pattern corresponds to a Dolph-Chebyshev array with distance between elements $d = \lambda / 2$ and SLL=-25dB. The fitness function for single objective algorithms was taken as $f_1 + f_2$ where f_1 and f_2 are given by (5a) and (5b). Note that NSGA – II was run with 100 chromosomes and run for 1500 iterations for all problems. As results we provide the best solutions found in 25 independent trials of each algorithm.

Case A: 20 Elements , 6 Subarrays

Table 1. Subarray Configuration (Case A)

Algorithms	c_1	c_2	c_3	c_4	c_5	c_6	c_7	c_8	c_9	c_{10}
NSGA - II	6	2	5	4	1	3	3	3	6	2
DE	4	3	4	2	5	1	0	1	5	6
PSO	1	4	5	6	3	3	2	5	6	2

Table 2. Subarray Weights (Case A)

Algorithms	g_1	g_2	g_3	g_4	g_5	g_6
NSGA-II	0.9976	0.4166	0.1865	0.6465	0.9064	0.9977
DE	0.9978	0.4303	0.1720	0.9976	0.7473	0.9977
PSO	0.8620	0.1842	0.4416	0.7020	0.9359	0.8621

Table 3. Design Objectives Achieved (Case A)

Algorithms	NSGA - II	DE	PSO
BW(degrees)	**12.01**	13.05	12.05
SLL	**-24.27**	-23.31	-19.14

Case B: 20 Elements , 10 Subarrays

Table 4. Subarray Configuration (Case B)

Algorithms	c_1	c_2	c_3	c_4	c_5	c_6	c_7	c_8	c_9	c_{10}
NSGA-II	6	1	7	8	10	2	5	3	9	4
DE	6	7	5	2	1	3	4	9	10	8
PSO	4	6	8	1	10	7	2	5	9	3

Table 5. Subarray Weights (Case B)

Algorithms	g_1	g_2	g_3	g_4	g_5
NSGA-II	0.3790	.9990	0.9987	0.2171	0.9988
DE	0.9022	.9023	0	0.9992	0.9023
PSO	0.9512	.9513	0.1505	0.1709	0.9513

Algorithms	g_6	g_7	g_8	g_9	g_{10}
NSGA-II	0.1538	0.6552	0.7912	0.9439	0.9988
DE	0.1627	0.5896	0.1626	0.9993	0.9023
PSO	0.0128	0.9514	0.8196	0.7864	0.9513

Table 6. Design Objectives Achieved (Case B)

Algorithms	NSGA - II	DE	PSO
BW(degrees)	**11.58**	13.34	13.74
SLL	**-23.48**	-16.93	-20.66

A keen observation of Tables 3, 6 and also Figures 5(a), 5(b) shows that in all test cases, NSGA – II achieves much better design objectives as well as array factors with lower SLL in comparison with both the single-objective algorithms - DE and PSO. Again, among all the test cases, as predicted in Section III, best results were achieved by NSGA – II for 20 elements and 6 subarrays (Table 3).

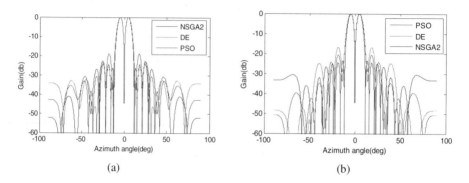

(a)   (b)

Fig. 5. (a) Normalized patterns for 20 element array (Case A), (b) Normalized patterns for 20 elements array (Case B)

6 Conclusion

This article has presented a new approach to the synthesis problem of the difference patterns of monopulse antenna arrays in a multi-objective optimization framework. One of the best known multi-objective optimization approach called NSGA – II has been over different instances of the design problem, keeping Maximum Sidelobe Level (MSLL) and Principal lobe Beam Width (BW) as two objectives to be simultaneously achieved.

The subarray grouping information and weights are obtained from the best compromise solution of the OPF corresponding to N and P as determined before. The best compromise solution for $N = 10$ and $P = 2, 4, 6, 8, 10$ are obtained from their OPF and the figure of merit of solution (i. e. MSLL and BW) are shown to beat those obtained with two well-known single objective algorithms DE and PSO. We have also demonstrated that the optimal 20 element array design should be with 6 subarrays. Increasing the number of subarrays increases the complexity of design without improving the quality of solution appreciably. In conclusion we can say that MO algorithms have a dual role in designing a monopulse antenna. They can be used both for fixing N and P and also for determining the subarray configuration and subarray weights. Moreover, other powerful MO algorithms like MOEA/D [19] can also be tested on the problem.

References

1. Skolnik, I.M.: Radar Handbook. McGraw-Hill, New York (1990)
2. Sherman, S.M.: Monopulse Principles and Techniques. Artech House, Norwood (1984)

3. Bayliss, E.T.: Design of monopulse antenna difference patterns with low sidelobes. Bell Syst. Tech. J. 47, 623–650 (1968)
4. McNamara, D.A.: Synthesis of sum and difference patterns for two section monopulse arrays. Proc. Inst. Elect. Eng., pt. H 135(6), 371–374 (1988)
5. Elliott, R.S.: Antenna theory and design. Prentice Hall, Englewood Cliffs (1981)
6. López, P., Rodríguez, J.A., Ares, F., Moreno, E.: Subarray weighting for the difference patterns of monopulse antennas: Joint optimization of subarray configurations and weights. IEEE Trans. Antennas Propag. 49(11), 1606–1608 (2001)
7. Caorsi, S., Massa, A., Pastorino, M., Randazzo, A.: Optimization of the difference patterns for monopulse antennas by a hybrid real/integer coded differential evolution method. IEEE Trans. Antennas Propag. 53(1), 372–376 (2005)
8. Price, K., Storn, R., Lampinen, J.: Differential evolution – A Practical Approach to Global Optimization. Springer, Berlin (2005)
9. Massa, A., Pastorino, M., Randazzo, A.: Optimization of the directivity of a monopulse antenna with a subarray weighting by a hybrid differential evolution method. IEEE Antennas Wireless Propag. Lett. 5, 155–158 (2006)
10. Deb, K.: Multi-Objective Optimization using Evolutionary Algorithms. John Wiley & Sons, Chichester (2001)
11. Coello Coello, C.A., Lamont, G.B., Van Veldhuizen, D.A.: Evolutionary Algorithms for Solving Multi-Objective Problems. Springer, Heidelberg (2007)
12. Deb, K., Pratap, A., Agarwal, S., Meyarivan, T.: A fast and elitist multiobjective genetic algorithm: NSGA-II. IEEE Transactions on Evolutionary Computation 6(2) (2002)
13. Abido, M.A.: A novel multiobjective evolutionary algorithm for environmental/economic power dispatch. Electric Power Systems Research 65, 71–81 (2003)
14. Kennedy, J., Eberhart, R.C., Shi, Y.: Swarm Intelligence. Morgan Kaufmann, San Francisco (2001)
15. Dolph, C.L.: A current distribution for broadside arrays. Proc. IRE 34, 335–348 (1946)
16. Abramovitz, M., Stegun, I.A.: Handbook of Mathematical Functions. Dover Publications, New York (1965)
17. Abraham, A., Jain, L.C., Goldberg, R. (eds.): Evolutionary Multiobjective Optimization: Theoretical Advances and Applications. Springer, London (2005)
18. Knowles, J.D., Corne, D.W.: Approxmating the nondominated front using the pareto archived evolution strategy. Evolutionary Computation 8(2), 149–172 (2000)
19. Li, H., Zhang, Q.: Multiobjective optimization problems with complicated Pareto sets, MOEA/D and NSGA-II. IEEE Transactions on Evolutionary Computation 13(2), 284–302 (2009)
20. Qu, B.Y., Suganthan, P.N.: Multi-Objective Evolutionary Algorithms based on the Summation of Normalized Objectives and Diversified Selection. Information Sciences 180(17), 3170–3181 (2010)
21. Zhao, S.Z., Suganthan, P.N.: Two-lbests Based Multi-objective Particle Swarm Optimizer. Engineering Optimization, doi: 10.1080/03052151003686716

Performance of Lognormal Probability Distribution in Crossover Operator of NSGA-II Algorithm

K.V.R.B. Prasad and Pravin M. Singru

BITS-Pilani, Goa Campus, Zuarinagar, Goa, India
brahmaprasad@bits-goa.ac.in,
pravinsingru@gmail.com

Abstract. This paper presents an improvement in performance of elitist non-dominated sorting genetic algorithm (NSGA-II) by modifying the probability distribution of crossover operator. The probability distribution of simulated binary crossover (SBX-A) operator, used in NSGA-II algorithm, is modified with lognormal distribution (SBX-LN). This algorithm is used to test twenty multi-objective functions. This NSGA-II (SBX-LN) algorithm performed well for different functions. This algorithm also performed well in optimizing a turbo-alternator design. It found more optimum solutions with better diversity in turbo-alternator design optimization.

Keywords: convergence, diversity, genetic algorithm, optimization.

1 Introduction

In genetic algorithm (GA), reproduction operator makes duplicates of good solutions while crossover and mutation operators create new solutions by recombination [1, 2]. The crossover operator is the main search operator in GA. The search power of a crossover operator is defined as a measure of how flexible the operator is to create an arbitrary point in the search space. The role of mutation is to restore lost or unexpected genetic material into population to prevent the premature convergence of GA to suboptimal solutions.

The performance of elitist non-dominated sorting genetic algorithm (NSGA-II) is improved by modifying the probability distribution of simulated binary crossover (SBX-A) operator [2, 3].

In this paper, the probability distribution of SBX-A operator is modified with lognormal distribution (SBX-LN). The NSGA-II algorithm with SBX-A and SBX-LN operators is used to test twenty multi-objective functions. The results are compared to find the performance of NSGA-II algorithm.

The NSGA-II algorithm with SBX-A and SBX-LN operators is also used to obtain the optimum design of a turbo-alternator. In this paper, a real-life turbo-alternator is considered for optimization. The results obtained by these two crossover operators are also compared to find the performance of NSGA-II algorithm.

K. Deb et al. (Eds.): SEAL 2010, LNCS 6457, pp. 514–522, 2010.

2 NSGA-II

The NSGA-II algorithm uses an elite preservation strategy along with an explicit diversity preserving mechanism [4]. This allows a global non-dominated check among the offspring and parent solutions. The diversity among non-dominated solutions is introduced by using the crowding comparison procedure which is used in the tournament selection and during the population reduction phase. Since, solutions compete with their crowding distance, no extra niching parameter is required here. Although the crowding distance is calculated in the objective function space, it can also be implemented in the parameter space, if so desired [5]. However, in all simulations performed in this study, the objective function space niching is used.

2.1 SBX-A Operator

The SBX-A operator works with two parent solutions and creates two offspring. This simulates the working principle of the single point crossover operator on binary strings. This operator respects the interval schemata processing, in the sense that common interval schemata between parents are preserved in children [1, 2].

The probability distribution used to create a child solution is

$$P(\beta) = 0.5(\eta_c + 1)\beta^{\eta_c} \qquad if \beta \le 1 \ . \tag{1}$$

$$P(\beta) = 0.5(\eta_c + 1)\frac{1}{\beta^{\eta_c + 2}} \qquad otherwise \ . \tag{2}$$

In equations (1) and (2), β is the spread factor and η_c is the crossover index. The value of η_c gives a probability for creating near parent solutions. A large value of η_c gives a higher probability for creating near parent solutions.

2.2 SBX-LN Operator

The SBX-A operator creates children solutions proportional to the difference in parent solutions. In this operator, near parent solutions are more likely to be chosen as children solutions than solutions away from parents. Raghuwanshi et al. [2] presented the SBX-LN operator, a new recombination operator, for contracting and expanding the crossover of real-coded GAs. In the SBX-LN operator, the probability of creating offspring away from the parents is influenced by the crossover index (η_c). This possibility decreases with the decrease in η_c and hence the SBX-LN becomes more parent centric operator. In this operator, both parents are given equal probability of creating offspring in its neighbourhood. It uses lognormal distribution for spread factor.

The lognormal distribution, defined with the probability density function, is

$$P(\beta) = \left(\frac{1}{\beta\eta_c\sqrt{2\pi}}\right)\exp\left(\frac{-1}{2}\left(\left(\frac{\log\beta - \mu}{\eta_c}\right)^2\right)\right) \qquad 0 < \beta < \infty \ . \tag{3}$$

In equation (3), β is the spread factor and μ is the mean of the variable's natural logarithm. The probability of contracting crossover is more desirable than the expanding crossover. This increases the probability of creating offspring between the parents. This is more parent centric for small value of η_c. The use of mutation may destroy already found good information. It is suggested that GAs may work well with large crossover probability (p_c) and with a small mutation probability (p_m). Hence, the mutation index (η_m) is chosen with a small value. The value of p_m, for twenty multi-objective functions, is chosen as 0.01 instead of $(1/N)$ where N is number of variables.

3 NSGA-II Algorithm to Optimize the Test Functions

The NSGA-II algorithm with SBX-A and SBX-LN crossover probability distributions is used to test five unconstrained multi-objective functions [2]. In this paper, the NSGA-II algorithm with SBX-A and SBX-LN crossover probability distributions is used to test twenty different multi-objective functions [1]. The variance generational distance (GD) is the performance metric used to measure the performance of SBX-A and SBX-LN operators. The GD of convergence metric finds an average distance of the solutions of optimum solution set obtained by the algorithm (Q) from Pareto-optimal set of solutions (P^*) [1]. The GD is obtained from the following relation.

$$GD = \frac{\left(\Sigma_{i=1}^{|Q|} d_i^p\right)^{1/p}}{|Q|} .$$

(4)

In the equation (4), the d_i is the Euclidean distance and p and i are the constants. The value of p is equal to two. The d_i (in the objective space) is the distance between solutions $i \in Q$ and the nearest member of P^*. The d_i is obtained from the following relation.

$$d_i = \min_{k=1}^{|P^*|} \sqrt{\sum_{m=1}^{M} \left(f_m^{(i)} - f_m^{*(k)}\right)^2} .$$

(5)

In the equation (5), the $f_m^{*(k)}$ is the m-th objective function value of the k-th member of P^* and $f_m^{(i)}$ is the m-th objective function value of the i-th member of Q.

The GD of diversity metric is obtained from the following relation [1].

$$\Delta = \frac{\sum_{m=1}^{M} d_m^e + \sum_{i=1}^{|Q|} \left|d_i - \overline{d}\right|}{\sum_{m=1}^{M} d_m^e + |Q|\overline{d}} .$$

(6)

In the equation (6), the d_i is the distance measure between neighboring solutions, $\overline{d}$ is the mean value of these distances and d_m^e is the distance between the extreme solutions of P^* and Q corresponding to m-th objective function.

The variance GD of convergence and diversity are computed, in each case, to find the performance of algorithm with different crossover probability distributions.

As a first step, five runs are made, for each function, with different random seeds. For all functions, the population size = 100, number of generations = 250, crossover probability (p_c) = 0.8, mutation probability (p_m) = 0.01, crossover index (η_c) = 0.05 and mutation index (η_m) = 0.5 [2]. The results are as per Table 1.

Table 1. Variance GD of convergence and diversity

S.No.	Function	NSGA-II (SBX-A)		NSGA-II (SBX-LN)		Remarks
		Convergence	Diversity	Convergence	Diversity	
1	ZDT1	5.99E-05	1.72E-02	4.55E-06	4.64E-02	SBX-LN /SBX
2	ZDT2	6.84E-05	7.90E-03	1.16E-05	4.13E-02	SBX-LN /SBX
3	ZDT3	9.50E-05	1.65E-02	1.87E-05	4.97E-02	SBX-LN /SBX
4	ZDT4	2.97E-04	1.72E-02	1.48E-04	4.64E-02	SBX-LN /SBX
5	ZDT6	2.83E-04	3.43E-02	1.81E-06	3.00E-02	SBX-LN
6	SCH1	2.18E-06	5.12E-02	1.88E-05	5.30E-02	SBX
7	SCH2	1.60E-05	1.88E-02	1.90E-04	1.25E-02	SBX/ SBX-LN
8	POL	3.72E-03	2.28E-02	9.44E-03	3.55E-02	SBX
9	FON	2.62E-05	4.09E-02	8.14E-05	3.90E-02	SBX/ SBX-LN
10	KUR	2.86E-04	2.88E-02	1.16E-03	3.30E-02	SBX
11	VNT	4.75E-05	5.41E-02	1.81E-04	8.65E-02	SBX
12	BNH	3.09E-03	2.46E-02	2.54E-03	2.71E-02	SBX-LN /SBX
13	SRN	1.42E-02	2.73E-02	1.47E-02	2.80E-02	SBX
14	CTP1	3.12E-05	2.88E-02	3.55E-05	2.73E-02	SBX/ SBX-LN
15	CTP2	5.45E-05	7.00E-03	4.04E-04	2.64E-02	SBX
16	CTP5	9.59E-05	3.61E-02	7.40E-04	2.79E-02	SBX/ SBX-LN
17	CTP6	8.95E-05	1.51E-02	1.73E-04	3.48E-02	SBX
18	CTP7	6.46E-05	3.69E-02	1.15E-04	5.56E-03	SBX
19	CTP8	2.28E-04	4.66E-03	2.85E-03	7.58E-02	SBX
20	TNK	1.24E-03	4.29E-02	2.69E-04	2.12E-01	SBX-LN /SBX

The NSGA-II algorithm with SBX-LN crossover probability distribution found better optimum solutions for different types of functions. By comparing the variance GD of convergence and diversity, it is observed that the NSGA-II algorithm with SBX-LN is having good convergence for seven functions and better diversity for five functions. Classification of these functions is shown in Table 2.

It is observed that the NSGA-II algorithm with SBX-A is having good convergence and better diversity for some functions. This is because the number of generations, for all the functions, is taken as 250. This is not acceptable for all the functions. Hence, a suitable number of generations, with sufficient number of function evaluations, are to be selected for each function to converge to the Pareto-optimal front.

From this it is observed that the performance of NSGA-II algorithm is improved by choosing a better parent centric crossover probability distribution. The OSY function with six constraints is not considered because the turbo-alternator design with six constraints is optimized using NSGA-II algorithm. In the next section, the NSGA-II algorithm is used to optimize a real-life turbo-alternator design having six constraints, with SBX-A and SBX-LN crossover probability distributions.

Table 2. Classification of functions outperformed by NSGA-II (SBX-LN)

S.No.	Parameter	NSGA-II (SBX-LN)	
		Convergence	Diversity
1	Continuous solutions	5 (ZDT1, ZDT2, ZDT4, ZDT6 and BNH)	3 (ZDT6, FON and CTP1)
2	Discontinuous solutions	2 (ZDT3 and TNK)	2 (SCH2 and CTP5)
3	Unconstrained functions	5 (ZDT1, ZDT2, ZDT3, ZDT4 and ZDT6)	3 (ZDT6, SCH2 and FON)
4	Constrained functions	2 (BNH and TNK)	2 (CTP1 and CTP5)
5	Unimodal functions	5 (ZDT1, ZDT2, ZDT3, ZDT4 and ZDT6)	3 (ZDT6, SCH2 and FON)
6	Multimodal functions	2 (ZDT3 and TNK)	2 (SCH2 and CTP5)
7	More epitasis	2 (ZDT4 and ZDT6)	1 (ZDT6)
8	One variable functions	0	1 (SCH2)
9	Two variable functions	2 (BNH and TNK)	2 (CTP1 and CTP5)
10	Five variable functions	0	1 (FON)
11	Ten variable functions	2 (ZDT4 and ZDT6)	1 (ZDT6)
12	Thirty variable functions	3 (ZDT1, ZDT2 and ZDT3)	0

4   NSGA-II Algorithm to Optimize the Turbo-Alternator Design

The NSGA-II algorithm with SBX-A and SBX-LN crossover probability distributions is used to optimize the turbo-alternator design. In this paper, the results obtained for a real-life turbo-alternator are presented. The data of turbo-alternator is as follows.

7999.35 kW, 3300 V, three-phase, 1646 A, 50 Hz, star-connected, 3000 r.p.m. machine operates at 85 percent power factor. The stator has 36 slots with 36 coils. The rated field excitation current is 252 A and the field excitation voltage is 125 V.

4.1   Turbo-Alternator Design

The efficiency and cost of turbo-alternator are the two objective functions, obtained from the given data [6-8]. A computer aided design (CAD) program is developed to obtain these parameters. The relations for efficiency and cost are as follows [9].

The efficiency of turbo-alternator, in terms of rated output power (P_0) and total losses (P_{TLoss}), is

$$\eta = \left[\frac{(P_0)}{(P_0 + P_{TLoss})} \right] . \tag{7}$$

The P_{TLoss} is divided into five major types such as stator copper loss (P_{L1}), stray load loss (P_{L2}), friction and windage loss (P_{L3}), stator iron loss (P_{L4}) and excitation loss (P_{L5}).

The cost of turbo-alternator, in terms of cost of copper (C_1), cost of iron (C_2), total weight of copper (W_1) and total weight of iron (W_2), is

$$C_T = \left[(C_1 * W_1) + (C_2 * W_2) \right] . \tag{8}$$

The actual design parameters of turbo-alternator, for a given data, are found from the CAD program.

4.2 Optimization Problem

The CAD program, developed to obtain the turbo-alternator design, is reformulated as a multi-objective optimization problem (MOOP) [9]. In this MOOP, five design parameters such as P_0, stator line voltage (V_L), stator line current (I_L), stator power factor (pf) and stator frequency (f) are considered as variables which will affect the design. The other design parameters such as stator number of phases (N_{sp}), rotor speed (N_r) and stator winding connection (W_{cs}) are kept constant in the design. Six constraints are formulated to obtain the required design of turbo-alternator. They are formed to maintain the stator slot pitch (λ) to be within maximum stator slot pitch (SP_{sm}), the temperature difference between stator copper and iron (TD_{sci}) to be within the maximum temperature difference between stator copper and iron (TD_{scim}), the rotor critical speed (N_{rc}) to be within rotor maximum critical speed (N_{mrc}), the rotor exciting current (I_{rme}) to be within its lower bound of rotor exciting current (I_{rmel}) and upper bound of rotor exciting current (I_{rmeu}) and rotor shaft deflection (DFL_{rs}) to be within its maximum rotor shaft deflection (DFL_{rsm}). The extreme values of these parameters are the $SP_{sm} = 0.07$ m, $TD_{scim} = 19^0C$, $N_{mrc} = 2400$ r.p.m., $I_{rmel} = 201.6$ A, $I_{rmeu} = 252.0$ A and $DFL_{rsm} = 5\%$ of air-gap thickness (δ).

The optimization problem of turbo-alternator design is as follows.

$$\text{Maximize } \eta = \left[\frac{P_0}{P_0 + P_{TLoss}} \right] . \tag{9}$$

$$\text{Minimize } C_T = \left[\left(C_1.W_1 \right) + \left(C_2.W_2 \right) \right] \text{ units} . \tag{10}$$

subjected to the constraints,

$$\left[1.0 - \left(\frac{\lambda}{SP_{sm}} \right) \right] \geq 0.0 . \tag{11}$$

$$\left[1.0 - \left(\frac{TD_{sci}}{TD_{scim}} \right) \right] \geq 0.0 . \tag{12}$$

$$\left[1.0 - \left(\frac{N_{rc}}{N_{mrc}} \right) \right] \geq 0.0 . \tag{13}$$

$$\left[\left(\frac{I_{rme}}{I_{rmel}} \right) - 1.0 \right] \geq 0.0 . \tag{14}$$

$$\left[1.0 - \left(\frac{I_{rme}}{I_{rmeu}}\right)\right] \geq 0.0 \; . \tag{15}$$

$$\left[1.0 - \left(\frac{DFL_{rs}}{DFL_{rsm}}\right)\right] \geq 0.0 \; . \tag{16}$$

The five design variables are allowed to vary within certain allowable range from their rated values, to obtain the optimum design of turbo-alternator. The P_0 and I_L are allowed to vary from zero to their rated values. The V_L is allowed to vary by ±5% from its rated value [10]. The pf is allowed to vary from its rated value to unity. The f is allowed to vary from 97% to 100% of the rated value.

Five best runs are chosen, for each crossover operator, with different random seeds. The population size is 1000 and number of generations is 10000. For SXB-A operator, p_c is 0.9, p_m is 0.2, η_c is 5 and η_m is 15. For SXB-LN operator, p_c is 0.8, p_m

Table 3. Parameters of turbo-alternator design

S.No.	Variable/ Constraint /Objective function	NSGA-II (SBX-A)				NSGA-II (SBX-LN)				Actual values (CAD Prog.)	Remarks
		1	2	3	4	1	2	3	4		
1	P_0 (MW)	7.99935	2.63	1.76	1.1E-03	7.99935	2.63	1.76	1.1E-03	7.99935	Variable (0.0 to 7.99935)
2	V_L (kV)	3.14	3.14	3.14	3.14	3.14	3.14	3.14	3.14	3.3	Variable (3.135 to 3.465)
3	I_L (kA)	0.67	0.58	0.38	0.16	0.67	0.58	0.38	0.16	1.646	Variable (0.0 to 1.646)
4	pf	1.0	1.0	1.0	1.0	1.0	1.0	1.0	1.0	0.85	Variable (0.85 to 1.0)
5	f (Hz)	49.88	50.0	50.0	49.85	49.88	50.0	50.0	49.85	50.0	Variable (48.5 to 50.0)
6	λ (m)	6.9E-02	6.9E-02	6.8E-02	6.6E-02	6.9E-02	6.9E-02	6.8E-02	6.6E-02	6.9E-02	Constraint (≤ 0.07)
7	TD_{sci} (^{0}C)	16.55	15.52	18.79	14.22	16.55	15.52	18.79	14.22	10.99	Constraint (≤ 19)
8	N_{rc} (r.p.m.)	1730	1710	1605	2393	1730	1710	1605	2393	1575	Constraint (≤ 2400)
9	I_{rme} (A)	229.32	230.82	207.89	201.72	229.32	230.82	207.89	201.72	248.12	Constraint ($201.6 \leq I_f \leq 252.0$)
10	DFL_{rs} (m)	2.9E-04	2.9E-04	3.3E-04	1.2E-04	2.9E-04	2.9E-04	3.3E-04	1.2E-04	3.5E-04	Constraint ($\leq 0.05\delta$)
11	δ (m)	3.2E-02	2.9E-02	2.8E-02	1.5E-02	3.2E-02	2.9E-02	2.8E-02	1.5E-02	3.3E-02	Parameter (δ)
12	η	0.98	0.965	0.961	0.02	0.98	0.965	0.961	0.02	0.91	Objective function
13	C_T (normalized by 1000000 units)	0.94	0.79	0.60	0.29	0.94	0.79	0.60	0.29	2.35	Objective function

is 0.2, η_c is 0.05 and η_m is 0.5 [2]. The optimum solutions are discontinuous in the objective space. The optimum design parameters of turbo-alternator at two points, for each crossover operator, are shown in Table 3.

The results are shown in Figure 1. The cost is normalized by 1000000 units. The results obtained by these two crossover operators are also compared to find the performance of NSGA-II algorithm [9].

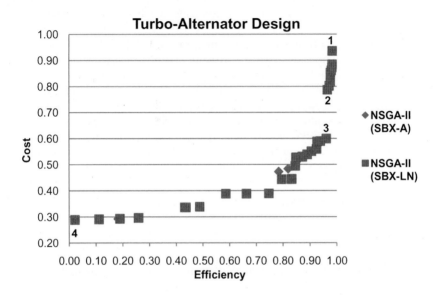

Fig. 1. Results of a real-life turbo-alternator design

4.3   Discussion of Results

The results obtained by the NSGA-II algorithm with SBX-A and SBX-LN crossover probability distributions are compared. The major observations are as follows.

The NSGA-II algorithm with SBX-LN found more optimum solutions with better diversity.

Most of the optimum solutions obtained by NSGA-II algorithm with SBX-A and SBX-LN crossover probability distributions are matching with each other.

The optimum design of turbo-alternator found by NSGA-II algorithm, with each crossover probability distribution, is having better efficiency and low cost when compared with the CAD program values.

The I_{rme} is having very large effect on the turbo-alternator design. All other constraints are not affecting the turbo-alternator design.

5   Conclusion

The following conclusions are drawn from this study.

1. The SBX-LN crossover probability distribution improved the performance of NSGA-II algorithm.
2. The NSGA-II (SBX-LN) algorithm found better optimum solutions for various functions having continuous and discontinuous solutions, unconstrained and constrained functions, unimodal and multimodal functions and different number of variables.
3. Two functions (ZDT4 and ZDT6) have more epitasis. NSGA-II (SBX-LN) has better convergence for both problems and has better diversity for one problem (ZDT6).
4. In turbo-alternator design optimization, the NSGA-II (SBX-LN) found more optimum solutions with better diversity when compared with the NSGA-II (SBX-A). These solutions are taken as Pareto-optimal solutions.
5. The I_{rme} is having very large effect on the turbo-alternator design. Hence it is identified as a hard constraint. All other constraints are not affecting the design of turbo-alternator. Hence, these are identified as soft constraints.
6. In the real-life turbo-alternator design, the number of stator slots and stator coils are equal. To satisfy this condition, the number of conductor layers in stator slot (depth) is 2. For multi-layer winding, the size of alternator is reduced. This is in turn reducing the cost considerably while reducing the efficiency marginally.

References

1. Deb, K.: Multi-Objective Optimization using Evolutionary Algorithms. John Wiley & Sons limited, Chichester (2002)
2. Raghuwanshi, M.M., Singru, P.M., Kale, U., Kakde, O.G.: Simulated Binary Crossover with Lognormal Distribution. Complexity International 12, 1–10 (2008)
3. Price, K.V., Storn, R.M., Lampinen, J.A.: Differential Evolution: A Practical Approach to Global Optimization. Springer, Heidelberg (2005)
4. Deb, K., Agarwal, S., Pratap, A., Meyariven, T.: A Fast and Elitist Multi-Objective Genetic Algorithm: NSGA-II. IEEE Transactions on Evolutionary Computation 6, 182–197 (2002)
5. Deb, K.: Multi-objective Genetic Algorithms: Problems Difficulties and Construction of Test Functions. Evolutionary Computation 7, 205–230 (1999)
6. Gray, A.: Electrical Machine Design: The Design and Specification of Direct and Alternating Current Machinery. McGraw-Hill Book Company, Inc., New York (1913)
7. Sawhney, A.K.: Electrical Machine Design. Dhanpat Rai & Sons, New Delhi (1991)
8. Shanmugasundaram, A., Gangadharan, G., Palani, R.: Electrical Machine Design Data Book. New Age International Publishers, New Delhi (2005)
9. Prasad, K.V.R.B., Singru, P.M.: Identifying the Optimum Design of Turbo-Alternator using Different Multi-Objective Optimization Algorithms. In: International Conference on Recent Trends in Information, Telecommunication and Computing 2010, IEEE CSDL, pp. 150–158. ACEEE, Kochi (2010)
10. Indian Electricity Rules,
    http://www.powermin.nic.in/acts_notification/pdf/ier1956.pdf

Metamodels for Fast Multi-objective Optimization: Trading Off Global Exploration and Local Exploitation

Enrico Rigoni and Alessandro Turco

Esteco srl,
Padriciano 99, 34149 Trieste, Italy
enrico.rigoni@esteco.com, alessandro.turco@esteco.com
http://www.esteco.com

Abstract. Metamodels can speed up the optimization process. Previously evaluated designs can be used as a training set for building surrogate models. Subsequently an inexpensive *virtual* optimization can be performed. Candidate solutions found in this way need to be *validated* (evaluated by means of the real solver).

This process can be iterated in an automatic way: this is the reason of the *fast* optimization algorithms. At each iteration the newly evaluated designs enrich the training database, permitting more and more accurate metamodels to be build in an adaptive way.

In this paper a novel scheme for fast optimizers is introduced: the virtual optimization - representing an *exploitation* process - is accompanied by a virtual run of a suited space-filler algorithm - for *exploration* purposes - increasing the robustness of the fast optimizer.

Keywords: Fast Optimization, Metamodels, Exploitation and Exploration.

1   Introduction

In real case applications, Computer Aided Engineering (CAE) solvers are usually computationally expensive: every single simulation can take hours or even days. Therefore multi-objective optimization algorithms are required to face a demanding issue: finding a satisfactory set of optimal solutions within a reduced number of evaluations. Response Surface Models (RSM) [12,8] - also known as Metamodels - can help in tackling this situation, speeding up the optimization process (e.g., see [16]).

The so-called *fast* optimization algorithms meet this demand, implementing an adaptive scheme for RSM training [9]. In this paper a novel scheme for fast optimizers is proposed: the usual virtual optimization stage (i.e. optimization performed over RSM) is completed with a virtual exploration phase (described in sec. 2.5), in order to improve both robustness and fast convergence. A detailed description of the algorithm workings is exposed in section 2. Sec. 3 briefly describes the standard optimizers incorporated within this framework, while

K. Deb et al. (Eds.): SEAL 2010, LNCS 6457, pp. 523–532, 2010.

sec. 4 exposes some details of the RSM algorithm implemented in this work. The performance of fast optimizers on some test problems is reported in sec. 5; we draw the conclusions in sec. 6.

2   Description

Fast optimizers use metamodels to speed up the optimization process. Different RSM are implemented within the algorithm: the best ones, in terms of performance on some validation points, are chosen and used for a virtual optimization. The so obtained *virtual* Pareto front is then validated (i.e., evaluated by means of the real solver), obtaining some designs that iteration by iteration progress toward the real Pareto front. At each iteration the newly evaluated designs enrich the training database, permitting a more and more accurate RSM to be build in an adaptive and iterative way.

The algorithm works on the analogy of a population of independent designs evolving through successive iterations (generations), like an ordinary genetic algorithm. So the total number of generated designs will be equal to the population size multiplied by the number of iterations.

The virtual optimization is accompanied by a virtual run of the Incremental Space Filler (ISF) algorithm, for exploration purposes: this fact increases the robustness of the fast optimizer. The relative incidence of virtual optimization vs. exploration is defined with the setting of a relevant exploration fraction parameter. Therefore the validation phase involves points generated both with virtual optimization and with virtual exploration.

For each objective and each constraint a different set of RSM is trained: considerations about fitness quality of the different models are carried out independently for each quantity. In this way the best RSM can be guaranteed in each single situation.

2.1   Iterative Loop

Fast optimizers progress toward better solutions in an iterative way. Each iteration in the main loop of the algorithm involves the following steps:

Metamodels training: RSM are trained over the existing database of evaluated designs.

Virtual exploration: ISF algorithm is employed for enriching the space exploration around the current Pareto front points.

Virtual optimization: optimization algorithm is run using the best available metamodels.

Validation process: selected points coming from the previous two steps are evaluated by means of the real solver. These new designs enrich the training database to be used in the next iteration.

Metamodels evaluation: the performance of the different RSM is evaluated over the validation points. The best RSM are selected to be used in the next iteration.

2.2   Preliminary Initialization

If a preexistent database of evaluated designs is available before the fast optimizer run (e.g., a previous optimization or exploration has already been performed), it is convenient to import it: in this way the available designs are used by RSM as training set at the first iteration. So the metamodels training occurs at this preliminary stage, and furthermore the metamodels performance assessment is performed just after the evaluation of the initial population.

This fact enhances the promptness of the adaptive scheme, since the iterative loop is custom tailored to the problem at hand just from the beginning.

Otherwise the default initialization takes place: this simply means that the iterative scheme will undergo a "cold start", reaching its running condition only from the second iteration on.

2.3   Initial Population

In this work two different optimization algorithms are implemented within fast optimizers: MOGA-II and SIMPLEX (refer to sec. 3). This leads to two different fast optimizers, i.e., FMOGA-II and FSIMPLEX, respectively.

For FMOGA-II the provided initial population determines the population size m, while for FSIMPLEX the population size has to be equal to $n + 1$, where n is the number of input variables ($n + 1$ is the number of vertices of the simplex).

Initial design are evaluated (by means of the real, expensive, solver), and they are used as training set for the RSM.

If the optional preliminary initialization occurs, RSM are trained over the preexistent database, and not over the initial population: these initial designs are used instead as the validation set for evaluating the RSM performance. In this way the assessment stage (sec. 2.8) - that usually take place after the validation process - can be anticipated: best RSM can be selected from the very beginning.

2.4   Metamodels Training

Metamodels are trained over the existing database of all designs evaluated so far. Otherwise the database size can by limited by specifying a relevant maximum size parameter: an according number of designs is randomly picked out from the full database. This fact can reduce effectively the training time of metamodels, limiting the total computational demand of the algorithm.

Four different RSM algorithms are implemented within fast optimizers: Polynomial SVD, Radial Basis Functions, Kriging, Neural Networks (refer to sec. 4). This different range of choices guarantees that for each problem - and for each single objective or constraint too - the more appropriate model can be applied.

Not always the complete list of all RSM algorithm is employed: under some particular conditions (refer to sec. 2.8 for details) some RSM algorithms - clearly the less performant ones - are automatically switched off, involving a desirable reduction of computational effort. These considerations are carried out independently for each each objective and each constraint.

At the first iteration, if preliminary initialization did not take place, default RSM are employed in the virtual exploration (sec. 2.5) and in the virtual optimization (sec. 2.6): the default choice is Radial Basis Functions (or Kriging in case of large scale problems, see sec. 2.10).

From the second iteration on, best RSM are selected by means of the procedure defined in the metamodels evaluation (sec. 2.8), according to the outcome obtained in the previous iteration.

So for exploitation and exploration purposes only the best metamodels out of all the RSM trained at this step are employed. On the contrary, all the RSM here trained are used in the metamodels evaluation step (sec. 2.8).

2.5   Virtual Exploration

The Incremental Space Filler (ISF) Design of Experiments (DOE) algorithm is used for enriching the database with designs lying in the region of interest, i.e., around the current Pareto front: this exploration stage increases the robustness of the fast optimizer.

ISF is useful for generating a uniform distribution of points in the input space. ISF is an augmenting algorithm: it considers the existing points in the database (previously generated designs), and it adds new points in order to fill the space in a uniform way. The maximin criterion is implemented: new points are added sequentially where the minimum distance from the existing points is maximized.

Here the zone filling option is enabled: new points are chosen in balls centered at the points belonging to the current Pareto front, with a specified radius. This zone radius is suitably related to a characteristic distance computed over the current Pareto designs.

There are two variants of the ISF algorithm:

- Genetic Algorithm Optimization: the internal optimization of the maximin criterion is performed by means of a Genetic Algorithm (approximated but fast and robust method).
- Voronoi-Delaunay Tessellation: the exact solution of the maximin criterion is found (precise method, but time-consuming in high dimensional case).

Here Voronoi-Delaunay is implemented when $n \leq 5$, otherwise Genetic Algorithm is used.

ISF is run in order to generate a sufficiently large number of possible candidate points. Then these designs are evaluated by means of the best available metamodels. These so obtained virtual designs are ordered according to Pareto ranking and crowding distances (refer to [4]): the best fraction of m points is extracted and selected as candidate designs for the validation process (sec. 2.7).

2.6   Virtual Optimization

The exploitation stage is performed by means of an optimization algorithm run over the best available metamodels. Clearly this virtual optimization is carried out by means of MOGA-II (multi-objective optimizer) in case of FMOGA-II, and by SIMPLEX (single-objective optimizer) in case of FSIMPLEX.

The parameters of MOGA-II and SIMPLEX, respectively, are automatically set internally the fast optimizer. The DOE (of size m) to be used as the initial population is randomly built up for 50% of designs taken from the current Pareto front (in order to improve the convergence of the optimizer) and for 50% of designs generated by Random DOE (in order to increase the robustness of the optimizer).

The best fraction - in terms of Pareto ranking and crowding distances - of $2m$ designs are extracted from the full database of virtual designs evaluated during the virtual optimization. These possible candidate points are compared with the archive of already evaluated (real) designs, in order to remove possible duplicated designs. In this way duplicates are avoided, saving computational resources. The first m designs are selected as candidate designs for the validation process (sec. 2.7). In case that - after the duplicates removal - there are less than m designs, the algorithm exits because possible convergence has been reached (in fact if the virtual optimization tends to find almost the same points, it means that possibly there is no room for further improvement).

2.7   Validation Process

In this step m candidate designs are randomly selected out of a total of $2m$ designs coming from the virtual exploration (sec. 2.5) and from the virtual optimization (sec. 2.6), according to the relative incidence defined with the setting of a relevant exploration fraction parameter.

These designs represent the new population of the current iteration: consequently they are *validated*, i.e., evaluated by means of the real solver.

2.8   Metamodels Evaluation

The validation designs obtained in the validation process (sec. 2.7) are used for evaluating the performance of all the metamodels trained in the current iteration (sec. 2.4).

Performances are expressed in terms of mean normalized error: in fact the variables (objectives and constraints) are normalized according to their range of variation, in order to avoid scale effects. The error is computed as the difference between the value predicted by the metamodel and the experimental value. By mean it should be intended the root mean square. Best RSM algorithm are selected accordingly to their performance metric.

Optionally a performance threshold for RSM competition can be specified: this value is used as a target threshold below which the competition between different metamodels is stopped. If fact if one of the competitive RSM reaches this value - in terms of mean normalized error - it is considered the definitive winner of the competition, given its excellent performance. If this is the case, in the next iterations the other RSM will not be trained, saving computational time. However, in case the performance worsens, the competition is automatically reopened.

2.9 Parallel Training

Fast optimizers are computationally demanding, since they involve time consuming operations, such as training RSM and running ISF (this is especially true for high dimensional problems and large number of designs). Paradoxically *fast* algorithm are *slow* in terms of net computational time, when compared to standard optimization algorithms. Usually this is not an issue, since the bottleneck of the optimization resides in the single design evaluation, carried out by means of an expensive CAE solver: this is indeed the domain of application of fast optimizers. However, in order to save the algorithmic computational effort, metamodels can be trained in parallel on the local machine, taking advantage of multi-core systems. In fact the training of different internal RSM - being independent tasks - can be carried out in different threads (according to the number of available processors/cores), reducing the total execution time of the algorithm.

2.10 Large Scale Problems

If the number of input variable is $n > 10$ the problem is considered to be a large scale problem: the algorithm automatically switches to a safe-mode. In this modality, the most computational intensive components of the algorithm are switched off, in order to supply an acceptable execution time. The virtual exploration is disabled, and only Kriging and Polynomial SVD (the fastest algorithms) are retained as RSM to be used in the virtual optimization.

3 Optimization Algorithms Description

The scheme outlined in this work could be applied in general to any optimization algorithm, generating its relevant fast version. In the current implementation two different optimization algorithms have been used: MOGA-II and SIMPLEX. They are briefly described in this section.

MOGA-II is a multi-objective optimization algorithm: it is an improved version of MOGA (Multi-Objective Genetic Algorithm) by Poloni et al. [16]. This genetic algorithm [5] uses a smart multisearch elitism for robustness and directional crossover for fast convergence.

SIMPLEX is a single-objective optimization algorithm [15] based on the well known Nelder and Mead downhill method [13] commonly used for multidimensional minimization problems. The SIMPLEX method does not require derivatives of the function, hence it is more robust than algorithms based on local gradients.

4 Metamodels Description

Several RSM algorithms are implemented within Fast Optimizers: this section is devoted to their general description.

Polynomial Singular Value Decomposition (SVD) produces best fitting polynomial approximations of responses minimizing the squares of the error predictions on the training dataset. The simple formulation of the model and the effectiveness of the SVD algorithm allow the training process to be very fast, compared to other metamodelling techniques.

Radial Basis Functions (RBF) are a powerful tool for multivariate scattered data interpolation [2]. Since RBF are *interpolant* response surfaces they pass exactly through training points. In our RBF implementation [18] five different radial functions are available: Gaussians (G), Duchon's Polyharmonic Splines (PS), Hardy's MultiQuadrics (MQ), Inverse MultiQuadrics (IMQ), and Wendland's Compactly Supported C^2 (W2). MQ kernel is the default choice, since it shows a fair performance on a wide set of benchmark problems. In general the leave-one-out methodology (useful for checking the goodness of an interpolant response surface) has a huge computational demand, but this is not the case in RBF framework: in fact there is a convenient way for computing the root-mean-square (rms) leave-one-out error (see [20]).

Kriging is a very popular regression methodology based on Gaussian Processes [17]. Kriging is a Bayesian methodology named after professor Daniel Krige, which is the main tool for making previsions employed in geostatistics. The formalization and dissemination of this methodology is due to Matheron [11]. The Kriging behavior (smoothness of the model) is controlled by a covariance function, called the variogram function, which rules how varies the correlation between the response values in function of the distance between different points. There are several variogram types that can be employed: Gaussian, Exponential, Matèrn, Rational Quadratic. In our implementation [10] Gaussian is the first (default) choice: the generated metamodel is infinitely many times differentiable.

Feedforward Neural Networks are a very efficient and powerful tool for function approximation purposes. Neural Networks (NN) with one single non-linear hidden layer and a linear output layer are sufficient for representing any arbitrary (but sufficiently regular) function [7]. The best known training algorithm of NN is backpropagation: a very effective approach consists in the Levenberg-Marquardt algorithm, as outlined in [6]. The Nguyen-Widrow initialization technique greatly reduces the training time (see [14]). In our implementation automatic network sizing has been adopted, using the value proposed in [19].

5   Results

We test the performances of fast optimizers, FMOGA-II and FSIMPLEX, with respect to their direct precursors MOGA-II and SIMPLEX, respectively. The comparison is focused on the ratio between quality of the best-so-far solution (point or set) and number of evaluated designs. In other words, fast optimizers are run up to a fixed maximum number of evaluations, but also some intermediate steps are analyzed, in order to study the progressive trend.

We chose two single-objective and two multi-objective problems. The first couple is taken from CUTEr [1] suite. The "linear function–full rank" problem involves nine input variables, while "Brown and Dennis" has four. Both have a unique optimal point and they are unconstrained. Their original formulation includes also a starting point for an optimization run. However, since SIMPLEX and FSIMPLEX require a set of initial points, we skip to a Random DOE. The stochasticity of this operation (and of some routines of the fast algorithm) is smoothed out by repeating 10 times the test and averaging the results. We did not tune any parameter of the algorithms, we keep standard values.

The "CTP7" [3] and "TNK" [21] problems involves two input variables and two objectives. The Pareto front of TNK presents concave and convex regions created by the two constraints added to the problem. The difficulty of CTP7 instead is the presence of local optima which are created once more by a constraint. We fix the size of the initial population to 50 points and we pick them by using a Uniform Latin Hypercube DOE algorithm which distributes samples uniformly in the configuration space. Ten independent runs of each algorithm were considered, varying only the seed for the random generator.

The performance of the multi-objective optimizers can be computed with an evaluation metric. We use the inverted generation distance (IGD) metric [22] which can measure accuracy and uniformity of a computed front with respect to a given reference one. If P is a sample of the true Pareto Set and A is an approximated set, we define:

$$\text{IGD}(A, P) = \frac{\sum_{p \in P} d(p, A)}{|P|},$$

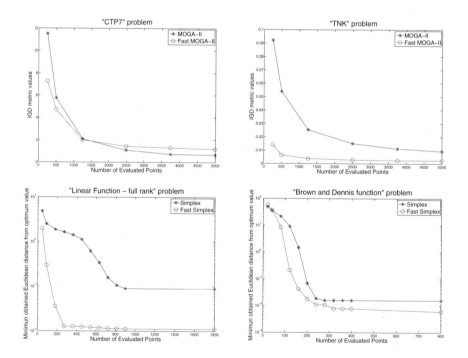

Fig. 1. Quality of the computed solution points (for single-objective problems) and sets (for multi-objective tests) with respect to the number of evaluated designs

where $d(p, A)$ is the minimum Euclidean distance between p and all the points in A. Low values of IGD are desirable, since this implies that the set A contains points near to any point in P.

A snapshot of the results can be found in Fig.1. In the majority of the cases the fast algorithms outperform the standard ones. This is much more evident for "TNK" and "linear function–full rank": in these problems the RSM managed to capture efficiently the structure of the functions involved and they speed up convergence significantly. In "Brown and Dennis" the gap is smaller although still present. The behavior observed over "CTP7" shows a difficulty for both FMOGA-II and MOGA-II in handling local optima.

It is not easy to judge whether the exploration-exploitation ratio should be moved in one direction or in the opposite one in order to reach the true Pareto front. Future studies will focus on this problem and on how the algorithm could detect a similar situation.

6 Conclusions

An automatic procedure for improving the accuracy of metamodels in an adaptive and iterative way can be devised. The fast optimizers obtain designs that iteration by iteration progress toward the real Pareto front.

Balancing the relative incidence of exploration vs. exploitation is a key issue for achieving both robustness and fast convergence. The strength of the two stages can be controlled by means of a relevant exploration fraction parameter.

Metamodels training and assessment are carried out independently for each objective and each constraint. In this way the best algorithm can be guaranteed in each single situation.

On large scale problems the virtual exploration and the metamodels training stages become computationally demanding: this issue requires proper consideration, as discussed in sec. 2.9 and 2.10. The algorithmic computational effort increases exponentially with the number of input variables and with the number of database designs: this fact represents a possible drawback of the fast optimizers, limiting their use in this high end limit.

On the contrary, the proposed tests show the possible advantages of fast optimizers. Whenever a usual algorithm can solve the problem, FMOGA-II and FSIMPLEX reduce the number of design evaluations needed to reach the solution. Moreover the exploration fraction parameter represents a possible important tool to handle more difficult problems.

References

1. Bongartz, I., Conn, A., Gould, N., Toint, P.: CUTE: constrained and unconstrained testing environment. ACM Trans. Math, Soft. 21, 123–160 (1995)
2. Buhmann, M.D.: Radial Basis Functions: Theory and Implementations. Cambridge University Press, Cambridge (2003)

3. Deb, K., Mathur, A., Meyarivan, T.: Constrained Test Problems for Multi-Objective Evolutionary Optimization. In: Zitzler, E., Deb, K., Thiele, L., Coello Coello, C.A., Corne, D.W. (eds.) EMO 2001. LNCS, vol. 1993, p. 284. Springer, Heidelberg (2001)

4. Deb, K., Pratap, A., Agarwal, S., Meyarivan, T.: A fast and elitist multi-objective genetic algorithm - NSGA-II. Report Number 2000001, KanGAL (2000)

5. Goldberg, D.E.: Genetic Algorithms in Search, Optimization and Machine Learning. Addison-Wesley Longman Publishing Co., Inc., Boston (1989)

6. Hagan, M.T., Menhaj, M.B.: Training feedforward networks with the marquardt algorithm. IEEE Trans. on Neural Networks 5(6) (1994)

7. Irie, B., Miyake, S.: Capabilities of three-layered perceptrons. In: Proceedings of the IEEE International Conference on Neural Networks, pp. I-641 (1998)

8. Jin, R., Chen, W., Simpson, T.W.: Comparative studies of metamodeling techniques under multiple modeling criteria. Structural and Multidisciplinary Optimization 23(1), 1–13 (2001)

9. Knowles, J., Nakayama, H.: Meta-modeling in multiobjective optimization. In: Branke, J., Deb, K., Miettinen, K., Słowiński, R. (eds.) Multiobjective Optimization. LNCS, vol. 5252, pp. 245–284. Springer, Heidelberg (2008)

10. Lovison, A.: Kriging. Technical Report 2007-003, Esteco (2007)

11. Matheron, G.: Les variables régionalisées et leur estimation: une application de la théorie des fonctions aléatoires aux sciences de la nature. Masson, Paris (1965)

12. Myers, R.H., Montgomery, D.C., Anderson-Cook, C.M.: Response surface methodology: process and product optimization using designed experiments. John Wiley & Sons, Inc., Chichester (2009)

13. Nelder, J.A., Mead, R.: A simplex method for function minimization. Computer Journal 7(4), 308–313 (1965)

14. Nguyen, D., Widrow, B.: Improving the learning speed of 2-layer neural networks by choosing initial values of the adaptive weights. In: Proceedings of IJCNN, vol. 3, pp. 21–96 (1990)

15. Poles, S.: The SIMPLEX method. Technical Report 2003-005, Esteco (2003)

16. Poloni, C., Pediroda, V.: GA coupled with computationally expensive simulations: tools to improve efficiency. In: Genetic Algorithms and Evolution Strategies in Engineering and Computer Science, pp. 267–288. John Wiley and Sons, Chichester (1997)

17. Rasmussen, C.E., Williams, C.K.I.: Gaussian Processes for Machine Learning. MIT Press, Cambridge (2006)

18. Rigoni, E.: Radial basis functions response surfaces. Technical Report 2007-001, Esteco (2007)

19. Rigoni, E., Lovison, A.: Automatic sizing of neural networks for function approximation. In: Proceedings of SMC 2007, pp. 2005–2010 (2007)

20. Rippa, S.: An algorithm for selecting a good value for the parameter c in radial basis function interpolation. Adv. Adv. Comp. Math. 11, 193–210 (1999)

21. Tanaka, M.: GA-based decision support system for multi-criteria optimization. In: Proc. of the Int. Conference on Systems, Man and Cybernetics - 2

22. Zitzler, E., Thiele, L., Laumanns, M., Fonseca, C., da Fonseca, V.: Performance assessment of multiobjective optimizers: an analysis and review. IEEE Transactions on Evolutionary Computation 7(2), 117–132 (2003)

Integrated Location-Inventory Retail Supply Chain Design: A Multi-objective Evolutionary Approach

Shu-Hsien Liao and Chia-Lin Hsieh

Department of Management Sciences and Decision Making, Tamkang University
No. 151, Yingjuan Road, Danshuei Jen, Taipei 251, Taiwan, ROC
michael@mail.tku.edu.tw, hsiehcl@email.au.edu.tw

Abstract. A supply chain network system is to provide an optimal platform for efficient and effective supply chain management. There's increasingly competitive, multi-channel retail world calls for a radically new strategy for evaluating supply chain network design. Retailers must abandon past practices which look to optimize the number and placement of facilities within traditional networks. A multi-objective optimization procedure which permits a trade-off evaluation for an integrated model is initially presented. This model includes elements of total cost, customer service and flexibility as its objectives and integrates facility location and inventory control decisions. Inventory control issues include economic order quantity, safety stock and inventory replenishment decisions and consider the risk pooling phenomenon to be realized from collaborative initiatives such as vendor-managed inventory. The possibility of a multi-objective evolutionary approach is developed to determine the optimal facility location portfolio and is implemented on a real large retail supply chain in Taiwan to investigate the model performance. Some preliminary results are described.

Keywords: multiobjective evolutionary algorithm, retail supply chain, facility location problem, inventory control, integrated supply chain model.

1 Introduction

Today's increasingly competitive, multi-channel retail world calls for a radically new strategy for evaluating supply chain network design. Retailers must abandon past practices which look to optimize the number and placement of facilities within traditional networks where domestic distribution centers (DCs) touch all merchandise moving to stores. In recent years, two generic strategies for supply chain design emerged: efficient and responsive supply chains. Efficient supply chains aim to reduce operational costs; responsive supply chains, on the other hand, are designed to react quickly to satisfy customer demands and thus save costs. Therefore, it has become a challenge for firms to evaluate tradeoffs among the total costs (for efficiency) and customer service (for responsiveness).

Research on integrated location-inventory distribution supply chain network systems is flourishing. Nozick & Turnquist [1] proposed a joint location-inventory model to consider both cost and service responsiveness trade-offs based on the uncapacitated facility location problem. Miranda & Garrido [2] studied an MINLP model

K. Deb et al. (Eds.): SEAL 2010, LNCS 6457, pp. 533–542, 2010.
© Springer-Verlag Berlin Heidelberg 2010

to incorporate inventory decisions into typical facility location models to solve the distribution network problem by incorporating a stochastic demand and the risk pooling phenomenon. Similarly, Gaur & Ravindran [3] studied a bi-criteria optimization model to represent the inventory aggregation problem under risk pooling. Daskin *et al.* [4] and Shen *et al.* [5] developed a single-commodity joint location-inventory model with risk pooling (LMRP) that incorporates inventory and safety stock costs at the facilities into the location problem. Liao & Hsieh [6] proposed a variation of the LMRP model: capacitated DCs and multi-objective performance metrics including customer service components. However, the single supplier assignment is usually not the practical case. When the number of suppliers increases, decisions that "where" and "which" suppliers should be identified were made. In this paper, we present an integrated location-inventory model with multiple suppliers. The basic premise of this paper is to consider inventory strategy together with facility location costs and distribution costs in determining the optimal location of suppliers and DCs, and the assignment of retailers to DCs. We consider a two-echelon location-inventory retail supply chain network design problem.

The paper is organized as follows. Section 2 describes our research problem and details the model formulation. Section 3 proposes an evolutionary algorithm for the model. Section 4 illustrates computational results of a real case problem. Finally, in section 5, we make the research conclusion.

2   Mathematical Formulation

2.1   Problem Description

Consider the problem of configuring a location-inventory distribution system, where a set of suppliers and distribution centers (DCs) are to be established to distribute various products to a set of retailers. It considers jointly both the strategic and tactical decisions in the supply chain system. The strategic decision involves the location problem, which determines the number and the locations of DCs and assigns retailers to DCs, whereas the tactical decision deals with the inventory problem which determines the levels of safety stock inventory at DCs to provide certain service levels to retailers. Fig. 1 shows our supply chain. It includes multiple suppliers (usually manufacturers but sometimes resellers or distributors) that support all retailers through DCs

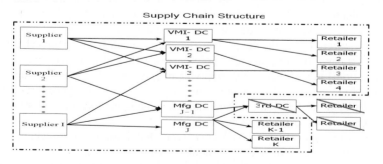

Fig. 1. Two-echelon retail supply chain network problem

of several types: supplier-owned (Mfg-DC), retailer-owned (VMI-DC) or third-party-owned (3rd-DC). Each retail-owned DC faces daily demand from the retailer's stores. The suppliers' DC receives daily orders from daily orders from retailers and places daily replenishment orders to specific suppliers. The dash-boxed area in the figure indicates our scope of interest. We tracked demand and inventory at the suppliers' DC and at the retailer's DC. We omitted third-party DC from the analysis since we were concentrating on VMI at the first-tier DC level.

2.2 Mathematical Model

Basic assumptions are used when modeling our problem. It is assumed that all the products are produced by a single supplier and one specific product for a retailer should be shipped from a single DC. Reverse flows, in-transit inventory, and pipeline inventory are not considered. All the retailers' demands are uncertain and the storage capacities of the supplier are unlimited but are capacitated at the open DCs. More assumptions will be stated when we illustrate the mathematical model. Here, the mathematical notation and formulation are as follows.

Indices. i is an index set for suppliers ($i \in$ I). j is an index set of potential DCs ($j \in$ J). k is an index set for retailers ($k \in$ K).

Decision Variables. x_{ij} is the number of units of products shipped from supplier i to DC j. y_{jk} is a binary variable to decide if DC j serves retailer k. w_i is a binary variable to see if supplier i is chosen or not. s_j is a binary variable if DC j is opened or not. Q_{jk} is the economic order quantity for retailer k at DC j.

Model Parameters. V_j is the capacity of *DC j*. P_i is the production capacity of supplier i. d_k is the mean demand rate (daily) for retailer k. σ_k is the standard deviation of daily demands for retailer k. L_{jk} is the average lead time (daily) to be shipped from DC j to retailer k. c_{ij} is the Unit cost of producing and shipping products from the supplier i to DC j and t_{jk} is the unit transportation cost of shipping product from DC j to retailer k. f_i is the fixed annual operating cost for supplier i and g_j is the facility operating cost of locating at DC j. h_j is the unit inventory holding cost at DC j. o_{jk} is the ordering cost at DC j for retailer k per order. $dis(i, j)$ and $dis(j, k)$ are the distances between supplier i and DC j and between DC j and retailer k, respectively. D_{max} is the maximal covering distance, *i.e.* retailers within this distance to an open DC are considered well satisfied. $\tau_k \overset{\Delta}{=} \{j \in \text{J} \mid dis(j, k) \le D_{max}\}$ is the set of DCs that could attend retailer k within D_{max}.

We assume that the daily demand for product k at each retailer i is independent and normally distributed, *i.e.* $N(d_k, \sigma_k)$. Furthermore, at any site of DC j, we assume a continuous review inventory policy (Q_j, r_j) to meet a stochastic demand pattern. Also, we consider that the supplier takes an average lead time L_{jk} (in days) for shipping product k from the supplier to DC j so as to fulfill an order. Considering centralized inventory system, if the demands at each retailer are uncorrelated, then the aggregate demands during lead time at the DC j is normally distributed and the total amount of safety stock pooled at any DC j is $z_{1-\alpha}\sqrt{L_{jk}\sum_k \sigma_k^2 y_{jk}}$ where 1-α is referred to the level of service and $z_{1-\alpha}$ is the standard normal value with P(z $\le z_{1-\alpha}$)= 1-α.

In our proposed model, the total cost of the system can be decomposed into the following items: (i) supplier's *operating cost*, which is the total cost incurred from the suppliers, (ii) DC's *operating cost*, which is the total cost of running DCs, (iii) *ordering cost*, which is the total annual expenses incurred in placing and order via VMI, which is the cost incurred from the suppliers, (iv) *cycle stock cost*, which is the cost of maintaining working inventory at the DCs, (v) *safety stock cost*, which is the cost of holding sufficient inventory at DCs in order to provide specific service level to their retailers, (vi) *inbound transportation cost*, which is the total cost of shipping products from suppliers to DCs, and (vii) *outbound transportation cost*, which is the total cost of shipping products from DCs to retailers. Hence, it can be represented as total cost function Z_1 as follows.

$$
\begin{aligned}
Z_1 = &\sum_{j \in J} f_i \cdot w_i + \sum_{j \in J} g_j \cdot s_j + \sum_{k \in K} \sum_{j \in J} o_{jk} \left(\frac{d_k \cdot y_{jk}}{Q_{jk}} \right) + \sum_{j \in J} \sum_{k \in K} \frac{h_j \cdot Q_{jk} \cdot y_{jk}}{2} \\
&+ \sum_{j \in J} \sum_{k \in K} h_j \cdot \left(z_{1-\alpha} \sqrt{L_{jk} \sum_{k \in K} \sigma_k^2 \cdot y_{jk}} \right) + \sum_{i \in I} \sum_{j \in J} c_{ij} \cdot dis(i,j) \cdot x_{ij} \\
&+ \sum_{j \in J} \sum_{k \in K} t_{ij} \cdot dis(j,k) \cdot d_k \cdot y_{jk}
\end{aligned}
\tag{1}
$$

Based on Z_1, the optimal order quantity Q^*_{jk} for retailer k at DC j can be obtained through differentiating Z_1 in terms of Q_{jk}, for $\forall j$ and k, and equaling to zero to minimize the total cost. We obtain $Q^*_{jk}= \sqrt{2 \cdot o_{jk} \cdot d_k \cdot y_{jk}/h_j}$ for every open DC j and every retailer k. Replacing Q^*_{jk} in the third and fourth terms of Z_1, we can obtain a non-linear cost function of Z_1. In the following, we propose our model.

$$
\begin{aligned}
\textit{Min} \quad Z_1 = &\sum_{j \in J} f_i \cdot w_i + \sum_{j \in J} g_j \cdot s_j + \sum_{k \in K} \sum_{j \in J} o_{jk} \left(\frac{d_k \cdot y_{jk}}{Q_{jk}} \right) \\
&+ \sum_{j \in J} \sum_{k \in K} \frac{h_j \cdot Q_{jk} \cdot y_{jk}}{2} + \sum_{j \in J} \sum_{k \in K} h_j \cdot \left(z_{1-\alpha} \sqrt{L_{jk} \sum_{k \in K} \sigma_k^2 \cdot y_{jk}} \right) \\
&+ \sum_{i \in I} \sum_{j \in J} c_{ij} \cdot dis(i,j) \cdot x_{ij} + \sum_{j \in J} \sum_{k \in K} t_{ij} \cdot dis(j,k) \cdot d_k \cdot y_{jk}
\end{aligned}
\tag{2}
$$

$$
\textit{Max} \quad Z_2 = \sum_{j \in J} \sum_{k \in K} d_k \cdot y_{jk} \Big/ \sum_{k \in K} d_k
\tag{3}
$$

$$
\textit{Max} \quad Z_3 = \sum_{j \in \tau_k} \sum_{k \in K} d_k \cdot y_{jk} \Big/ \sum_{j \in J} \sum_{k \in K} d_k \cdot y_{jk}
\tag{4}
$$

$$
\textit{s.t.} \quad \sum_{j \in J} x_{ij} \le p_i \cdot w_i; \quad \forall i \in I;
\tag{5}
$$

$$
\sum_{k \in K} d_k \cdot y_{jk} + z_{1-\alpha} \sqrt{L_{jk} \sum_{k \in K} \sigma_k^2 \cdot y_{jk}} \le V_j \cdot s_j; \quad \forall j \in J
\tag{6}
$$

$$
\sum_{i \in I} x_{ij} = \sum_{k \in K} d_k \cdot y_{jk} + z_{1-\alpha} \sqrt{L_{jk} \sum_{k \in K} \sigma_k^2 \cdot y_{jk}}; \quad \forall j \in J
\tag{7}
$$

$$\sum_{k \in K} y_{jk} = 1 \; ; \quad \forall k \in K \tag{8}$$

$$y_{jk}, w_i, s_j \in \{0,1\} \; ; \quad \forall j \in J; \quad \forall k \in K \tag{9}$$

$$x_{ij}, Q_{jk} \geq 0 \text{ integers}; \quad \forall i \in I; \quad \forall j \in J; \quad \forall k \in K \tag{10}$$

The objectives are referred to (2)-(4). Z_1 in (2) is to minimize the *total cost* (TC), Z_2 in (3) and Z_3 in (4) give the objectives referred to maximizing customer service by two measurements. Z_2 in (3) is referred to *volume fill rate* (VFR) which is defined as the fraction of total demand that can be satisfied from inventory without shortage. Z_3 in (4) is called *responsiveness level* (RL) which measures the percentage of fulfilled demand volume within a distance coverage for DCs. Equations in (5) and (6) are capacity restrictions on the suppliers and DCs, respectively, and permit the use of opened facilities only. Equations in (7) are product flow conservation equations at DCs, ensuring for every product that flows through the DC, a part of it is held in safety stock and the rest is used to satisfy demand at the retailers. Equations in (8) restrict a retailer's demand to be served by a single DC. Equations in (9) are binary constraints. Equations in (10) are integrality and non-negativity requirements.

3   Problem Solving Methodology

3.1   NSGAII-Based Evolutionary Algorithm

Multiobjective optimization problems give rise to a set of Pareto-optimal solutions, none of which can be said to be better than other in all objectives. Unlike most traditional optimization approaches, multiobjective evolutionary algorithms (MOEAs) work with a population of solutions and thus are likely candidates for finding multiple Pareto-optimal solutions simultaneously. Non-dominating Sorting GA (NSGA-II) [7] is one of the best techniques in MOEAs. For each solution, one has to determine how many solutions dominate it and the set of solutions to which it dominates. Thus, it ranks all solutions to form non-dominated fronts according to a *non-dominated sorting* process to classify the chromosomes into several fronts of nondominated solutions. The non-domination sorting updates a tentative set of Pareto optimal solutions by ranking a population according to non-domination. To maintain diversity in the population, NSGA-II also estimates the solution density surrounding a particular solution in the population by computing a crowding distance operator. During selection, individuals of equal non-domination rank are sorted according to their crowding distance. The selection operator selects the best individuals according to this ranking as the parents of the next generation, whereas crossover and mutation operators remain as usual. A NSGA-II-based evolutionary algorithm is proposed, as shown in Table 1. As we can see in Table I, chromosome fitness depends on the evaluation of the decoded solution in the objective functions and its comparison with other chromosomes in the selection process of the next generation. The *non-domination sorting* updates a tentative set of Pareto optimal solutions by ranking a population according to non-domination. After that, each individual p in the population is given two attributes: (1) non-domination rank in the optimization objectives; (2) local crowding distance in the objectives space directions. If both chromosomes are the same rank, the one with fewer chromosomes around in the front is preferred. Therefore, a partial order ($\geq_n$) can be defined as follows. Let p, $q \in P(t)$ be two individuals in population $P(t)$. We say that p is better fitted than q ($p \geq_n q$), either if (p.rank < q.rank) or ((p.rank = q.rank) and (p.distance > q.distance)).

Table 1. NSGA-II-based evolutionary algorithm

1: Random generating parent population P(0) of size L
2: *Non-domination* sorting P(0)
3: For each nondominated solution, assign a fitness (rank) equal to its nondomination.
4: Create a child population C(1) of size L, apply binary tournament selection, crossover, and mutation.
5: Evaluate C(1)
6: **while t ≤ T do**
7: Create the mating pool R(t) =P(t) ∪ C(t) of size 2L by combining the parent population P(t) and the child population C(t).
8: Sort R(t) using *non-domination sorting* $\geq_n$
9: Select P(t+1) from the first L chromosome of R(t)
10: Generate C(t+1) from P(t+1), apply binary tournament selection, crossover, and mutation
11: Mutate and Evaluate C(t+1)
12: t ← t + 1
13: **end while**

3.2 Hybrid Evolutionary Algorithm

Here, a hybrid evolutionary algorithm is proposed. Cycles of fitness evaluation, selection, crossover, and mutation repeat until some stopping criteria are met. However, our algorithm first focuses on fitness evaluation according to a partial order ($\geq_n$) which is used to decide which chromosomes are fitter. Suppose that $Z_k(p)$ and $Z_k(q)$ and be the k-th objective function evaluated at two decoded chromosomes p and q, respectively. Here in our model, $Z_1(.)$ indicates *cost*, $Z_2(.)$ indicates *volume fill rate* and $Z_3(.)$ indicates *responsiveness level*. We say that $p \geq_n q$ if $Z_1(p) \leq Z_1(q)$, $Z_2(p) \geq Z_2(q)$ and $Z_3(p) \geq Z_3(q)$; and $Z_1(p) < Z_1(q)$ or $Z_2(p) > Z_2(q)$ or $Z_3(p) > Z_3(q)$. The chromosome representation is represented in two parts as shown in Fig. 2. Each part has the same length m=|J| (where |J| is the number of DCs) with total length of chromosome 2J. The solution in the first part of chromosome is encoded in binary

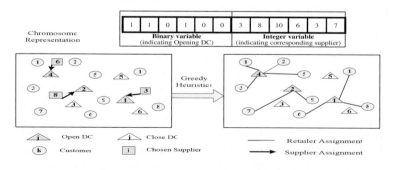

Fig. 2. Chromosome representation

variables (s_j) where the j-th position indicates if DC j is open (value of 1) or closed (value of 0). However, the second part of chromosome is encoded in integer variables where the value in it stands for the corresponding supplier that is assigned to it.

A solution also involves the assignments of retailers to open DCs (binary variables y_{jk}). This assignment is performed by a *greedy* heuristics used to obtain the single retailer-DC assignments. The retailers are firstly sorted in the descending order of their demand flows and then assign them in the sorted order to the DC according to the following rules:

Rule 1. If retailer i is covered (i.e., there are DCs within a coverage distance), it is assigned to a DC with sufficient capacity (if exists) which can serve it with the minimal difference between the remaining capacity of an open DC j and the demand flow of the retailer i through DC j. That is, the DC assignment attempts to be filled.

Rule 2. If the retailer i cannot be covered or there is no successful assignment from the coverage set τ_i, it is then assigned to the candidate DC (with sufficient capacity) that increases the total cost by the least amount, regardless of its distance to the DC.

4 Model Applications

4.1 An Illustrative Retail Supply Chain Example

C company, one of the world's largest retailer, is increased rapidly through the 1990s, and by 2010, the firm is currently 64 retail stores in Taiwan, most of them being hypermarkets. The company has set up several distribution centers in Taiwan, which distributes commodities through its national-wise retail chain. It also consigned vendor-managed inventories, consisting primarily of seasonal merchandise and such direct-to-store products as ice cream and soft drinks are held until needed. By asking them help manage inventory, the suppliers are asked to make new products available and to deliver products to stores ready to sell. It has illustrated clearly the benefits of such real-time stock management.

For this case study, 7 suppliers could be potentially chosen to make our procurement plan. According to the realistic data, there are 8 potential depots for its retail network. We also aggregate retailer's depots located in the same city or town. After aggregation, we ended up with 44 retailer stores. The maximal covering distance was set in D_{max}=150 km. Other key input parameters of the model are given in Table 2. The demand for retailer k (d_k) is set equal 50 demands per day per million people in the population. For simplicity, Euclidean distance is used for measuring distribution distances. The company intended to determine both the number of opening DCs and its corresponding suppliers for retailer's order assignments. Such assignment will be affected by DC's capacity limitation and suppliers' production capacities. Therefore, the decisions are to evaluate tradeoffs among three criteria.

4.2 Computational Results

To obtain the Pareto front, we attempted to solve the specified problem using the evolutionary approach. The Pareto front is then evaluated to find out the 'optimal' solution. We define a reference point which is a vector formed by the single-objective optimal solutions. It is the best possible solution among the Pareto front that a multi-objective problem may have. Given a reference point, the problem can be solved by locating the alternative which has the minimum distance to the reference point. The reference point can simply be found by optimizing one of the original objectives at a time subjective to all constraints. Due to incommensurability among objectives, we measure this distance by using normalized Euclidean distance between two points in k-dimensional vector space to obtain the *score* function in *Eq.* (11).

$$score = \left\{ \sum_{t=1}^{k} w_t [(f_t^* - f_t) / f_t^*]^2 \right\}^{1/2} \tag{11}$$

where f is an alternative solution in Pareto front, f^* is the reference point and w_t is the relative weight (given by prior) for the t-th objective. Then, all alternatives are ranked based on the value of d in descending orders. The highest ranked alternative (with the minimal value of *score*) is the 'optimal' solution. Input parameters are: cloning=20%; generation number=100; population size=50; crossover rate = 80%; mutation rate varies from 5% to 10%. The decision maker requires determining weights w_t by prior knowledge of objectives. The hybrid GA is evaluated with the illustrative example.

Table 2. Model parameters

Parameters	Value	Parameters	Value
Demands per unit population per day(d_k)	$50*10^{-6}$	Capacity of DC j (V_j)	$U(1.2*10^6, 1.5*10^6$
Lead time (days) (L_{jk})	5	Prod. capacity of supplier i (P_i)	$U(1.8*106, 2*10^6)$
Unit ship. cost from supplier i to DC j (c_{ij})	$0.2	Fixed ann. oper. cost for supplier i (f_i)	$U(50*10^6, 80*10^6)$
Unit ship. cost from DC j to retailer k (t_{jk})	$1	Fixed ann. oper. cost for DC j (g_j)	$U(35*10^6, 65*10^6)$
Unit ordering cost at DC j (o_{jk})	U (0.5,1)	Unit inv. holding cost at DC j (h_j)	$1.2

Fig. 3 illustrates a good evolution approach for generating the Pareto front after 200 generations in our problem. It is revealed that the population curve converges shortly after 50 generations; they are nearly overlapped among themselves. After-wards, no significant improvement is incurred. In order to illustrate the performance effects on the proposed solution procedure, we also consider four diverse scenarios by changing w_t parameters at a time as follows: (1) equal-weight scenario (S_1) with $w_1=w_2=w_3=1/3$; (2) cost-concerned scenario (S_2) with $w_1=0.8$, $w_2=w_3=0.1$; (3) respon-sive-level scenario (S_3) with $w_2=0.8$, $w_1=w_3=0.1$; (4) volume-fill-rate scenario (S_4) with $w_3=0.8$, $w_1=w_2=0.1$. Table 3 summarizes computational results of all scenarios.

In Fig. 4, we display graphically the geographical locations of three components: DCs($\triangle$) and their corresponding suppliers ($\square$) and retailers (O). All corresponding retailer assignments and supplier selections of a specific DC are represented in the same color. Fig. 4(a) illustrates the optimal assignment of alternative 27 for scenario S1 with minimal TC $1,078,800,000, maximal VFR 73.71% and maximal RL 65.14%, respectively, where 5 out of 8 potential DCs are aggregated. Most of these DCs look close to their assigned retailers. However, there are about 26.29% unas-signed retailers (especially the retailers located in southern Taiwan), indicating sales loss percentage. There are also 34.86% aggregated retailers assigned to DCs farther than the coverage distance. Fig. 4(b) represents the cost-concerned scenario. Fig. 4(c) shows the optimal assignment of alternative 6 for scenario S_3 and S_4, where 7 DCs are aggregated. The results shows that it is possible to increase VFR 13.12% and RL 17.23%, if only the percentage over TC increases 17.23% where the number of open DCs increased from 5 to 7. That is, it is necessary to spend extra costs to open DCs up to 7 to enhance customer's VFR and also to increase RL at the same time.

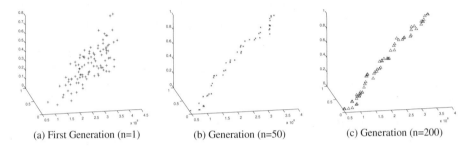

(a) First Generation (n=1)   (b) Generation (n=50)   (c) Generation (n=200)

Fig. 3. Evolutionary approach for generating Pareto front

Table 3. Summary of computational results

Scenarios	Objectives			Optimal solution			
	TC (million)	VFR	RL	Alternative	# of open DC	DC (vs. supplier)	Retailer (vs. DC)
S_1	1,078.8	73.71%	65.14%	27	5	2(4) 3(7) 4(3) 5(5) 7(2)	3(5) 4(2) 5(5) 6(3) 8(3) 9(5) 10(2) 11(5) 12(3) 15(3) 16(2) 18(5) 19(3) 20(5) 22(4) 23(3) 28(7) 30(7) 32(7) 34(4) 38(4) 39(4) 40(7) 43(4) 44(7)
S_2	388.46	28.51%	28.51%	19	2	2(2) 3(5)	3(3) 5(3) 6(3) 8(2) 13(3) 19(2) 20(3)
S_3 & S_4	1,549	90.94%	78.26%	6	7	1(6) 3(3) 4(7) 5(5) 6(1) 7(2) 8(5)	1(5) 2(5) 3(5) 4(5) 5(3) 6(3) 7(3) 8(5) 9(3) 10(1) 11(3) 12(5) 13(6) 15(5) 16(1) 18(3) 19(1) 20(1) 22(4) 23(7) 24(7) 25(7) 27(7) 28(7) 30(7) 32(1) 33(3) 34(4) 35(6) 38(4) 39(4) 40(1) 41(6) 42(8) 43(4) 44(6)

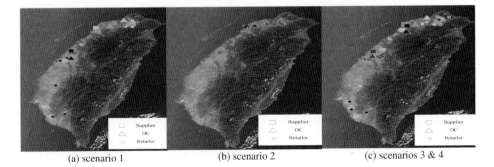

(a) scenario 1   (b) scenario 2   (c) scenarios 3 & 4

Fig. 4. Graphical display of the 'optimal' solution under scenarios

5 Conclusion

This research presented an integrated location-inventory retail supply chain network design problem which examines the effects of facility location, distribution, and inventory issues. The goal of this research is to realize the application of multi-objective evolutionary approaches to real problems. The possibility of a hybrid MOEA is developed to efficiently determine the optimal facility location portfolio and is successfully implemented on a real large retail supply chain in Taiwan to investigate the model performance. The proposed model is helpful in adjusting the distribution network to these changes.

References

1. Nozick, L.K., Turnquist, M.A.: A two-echelon allocation and distribution center location analysis. Trans. Res. Part E. 37(6), 425–441 (2001)
2. Miranda, P.A., Garrido, R.A.: Incorporating inventory control decisions into a strategic distribution network model with stochastic demand. Trans. Res. Part E. 40(3), 183–207 (2004)
3. Gaur, S., Ravindran, A.R.: A bi-criteria model for the inventory aggregation problem under risk pooling. Computers and Industrial Engineering 51(3), 482–501 (2006)
4. Daskin, M., Coullard, C., Shen, Z.: An inventory-location model: formulation, solution algorithm and computational results. Annals of Oper. Res. 110, 83–106 (2002)
5. Shen, Z.J., Coullard, C.R., Daskin, M.S.: A joint location-inventory model. Trans. Sci. 37(1), 40–55 (2003)
6. Liao, S.H., Hsieh, C.L.: A Capacitated Inventory-Location Model: Formulation, Solution Approach and Preliminary Computational Results. In: Chien, B.-C., Hong, T.-P., Chen, S.-M., Ali, M. (eds.) IEA/AIE 2009. LNCS, vol. 5579, pp. 323–332. Springer, Heidelberg (2009)
7. Deb, K., Pratap, A., Agarwal, S., Meyarivan, T.: A fast and elitist multi-objective genetic algorithm: NSGA-II. IEEE Trans. on Evolutionary Comp. 6(2), 181–197 (2002)

Multi-Objective Job Shop Scheduling Based on Multiagent Evolutionary Algorithm[*]

Xinrui Duan, Jing Liu, Li Zhang, and Licheng Jiao

Key Laboratory of Intelligent Perception and Image Understanding of Ministry of Education
of China, Institute of Intelligent Information Processing, Xidian University, Xi'an, China
xinruiduan@163.com

Abstract. With the properties of multi-objective job shop problem (MOJSP) in mind, we integrate the multiagent systems and evolutionary algorithms to form a new algorithm, multiagent evolutionary algorithm for MOJSP (MAEA-MOJSP). In MAEA-MOJSP, an agent represents a candidate solution to MOJSP, and all agents live in a latticelike environment. Making use of three designed behaviors, the agents sense and interact with their neighbors. In the experiments, eight benchmark problems are used to test the performance of the algorithm proposed. The experimental results show that MAEA-MOJSP is effective.

Keywords: Multi-objective job shop problem, multiagent system, evolutionary algorithm.

1 Introduction

Job shop problem (JSP) is one of the combinatorial optimization problems. During the last two decades, some researchers have attempted to study the multi-objective job shop problem (MOJSP). As far as we know, in [1], a crowding-measure-based multi-objective evolutionary algorithm was presented, which utilized the crowding measure to adjust the external population and assign individual's fitness. In [2], a multistage genetic algorithm (GA) and a multi-recombinated cooperative population GA were implemented to solve two-objective and three-objective JSP respectively. In [3], three approaches, GA, constraint logic programming, multi-criteria decision-making, were proposed to solve MOJSP. In [4], a memetic algorithm based on differential evolution is proposed for multi-objective job shop scheduling problems.

In this paper, we combine the multiagent systems and evolutional algorithms to form a multiagent evolutionary algorithm for MOJSP (MAEA-MOJSP). In the next section, multi-objective job shop problem is introduced and then we described MAEA- MOJSP. In section 3, experimental results are presented. Finally, section 4 makes a brief conclusion for this article.

---

[*] This work was supported by the National Natural Science Foundation of China under Grant 60872135, 60803098, and 60970067, and the National Research Foundation for the Doctoral Program of Higher Education of China under Grant 20070701022.

K. Deb et al. (Eds.): SEAL 2010, LNCS 6457, pp. 543–552, 2010.

2 MAEA-MOJSP

2.1 Multi-Objective Job Shop Problem

Generally, a multi-objective optimization problem (MOP) with N objectives can be described as follows:
Minimize:

$$f_1(x), f_2(x), ..., f_N(x) \tag{1}$$

where $f_1(x), f_2(x), ..., f_N(x)$ are N objectives to be minimized, and $x \in solution\ set$.

Consider two solutions, a and b, the solution a is said to dominated b ($a \succ b$)
If

$$\forall i \in \{1, 2, ..., N\}: f_i(a) \le f_i(b)\ and\ \exists j : \{1, 2, ..., N\}: f_i(a) < f_i(b) \tag{2}$$

If a solution is not dominated by any other solutions of the MOP, that solution is called a nondominated solution. All the nondominated solutions of the problem form optimal Pareto front.

For JSP, the following objective functions are frequently used [5]:
1)   makespan or maximum completion time:

$$makespan = \max(C_i) \tag{3}$$

where C_i is the completion time of job i.
2)   total tardiness(due date based):

$$total\_tardiness = \sum_{i=1}^{n} \max(0, L_i) \tag{4}$$

where L_i is the lateness of job i.
3)   total earliness(due date based):

$$total\_earliness = \sum_{i=1}^{n} \max(0, E_i) \tag{5}$$

where E_i is the earliness of job i.
4)   total idle time of machine:

$$total\_idletime = \sum_{k=1}^{m} I_k \tag{6}$$

where I_k is the idle-time of machine k.
5)   maximum flow time:

$$F_{max} = \max(C_i - S_i) \tag{7}$$

where C_i and S_i are completion time and start time of job i, respectively.

6)   number of tardy jobs:

$$N_T = \sum_{i=1}^{n} U_i \qquad (8)$$

where U_i is an indicator function denotes whether job i is tardy ($U_i = 1$) or not ($U_i = 0$).

For MOJSP, some objectives should be considered simultaneously. But some of them are conflicting, such as makespan and total idle time of machine, makespan and total tardiness etc.

2.2 The Agent Used in Job Shop Problem

Usually, a JSP considers n jobs to be proposed on m machines. The process of a job on a specific machine is defined as an operation. Processing time of each operation is fixed and known in advance. Each job can only have one operation processed at a time and no preemption can take place.

One encoding method for JSP is permutation with repetition, where a scheduling is described as a sequence of all $n \times m$ operation, and each operation in the sequence is described by the job-number. Thus, the search space S of a JSP consists of the elements satisfy the following conditions:

$$\forall P \in S, \ P = \langle P_1, P_2, ..., P_{n \times m} \rangle \text{ and } (P(1) = m) \text{ and } (P(2) = m) \text{ and ... and } (P(n) = m) \qquad (9)$$

where $P_i \in \{1, 2, ..., m\}$, $i = 1, 2, ..., n \times m$, and $P(j)$, $j = 1, 2, ..., n$, stand for the number of j in P. When transforming P into a schedule, P_i stands for $o_{j,k}$ if $P_i = j$ and the number of j among P_1, P_2,...,P_i is equal to k. The schedule is obtained by considering the operations in the order they occur in P and assigning the earliest allowable time to that operation. Such encoding method has the advantage that no infeasible solutions can be represented, and each element in S corresponds to a feasible schedule.

An agent for MOJSP represents an element in the search space. It is represented by the following structure:

```
Agent = Record
```
    P: $P \in$ S;
    E: The energy of the Agent, $E=Energy(P)$;
    SL: The flag for the self-learning behavior, which will
        be defined later, if SL is *True*, the self-learning
        behavior can be performed on the Agent, otherwise,
        cannot;

```
End.
```

2.3 Environment of Agents

In multiagent evolutionary algorithm (MAEA) [6], all agents live in a latticelike environment, L, which is called an agent lattice. The size of lattice is $L_{size} \times L_{size}$, where L_{size}

is an integer. Each agent is fixed on a lattice-point and can only interact with the neighbors. It can assume that the agent located at (i, j) is represented as $L_{i,j}, i, j = 1, 2, ..., L_{size}$, then the neighbors of $L_{i,j}$, $Neighbors_{i,j}$, are defined as follows:

$$Neighbors_{i,j} = \left\{ L_{i',j}, L_{i,j'}, L_{i'',j}, L_{i,j''} \right\} \tag{10}$$

where $i' = \begin{cases} i-1 & i \neq 1 \\ L_{size} & i = 1 \end{cases}$, $j' = \begin{cases} j-1 & j \neq 1 \\ L_{size} & j = 1 \end{cases}$, $i'' = \begin{cases} i+1 & i \neq L_{size} \\ 1 & i = L_{size} \end{cases}$, $j'' = \begin{cases} j+1 & j \neq L_{size} \\ 1 & j = L_{size} \end{cases}$.

The agent lattice can be seen in Fig.1. Each circle represents an agent; the number in it represents the position in the lattice. And two agents can interact with each other if only there is a line connecting them.

In MAEA-MOJSP, we use the fitness assignment of SPEA [7] as energy definition:

$$\forall a \in S, energy(a) = -(1 + \sum_{b \in P_t, b \succ a} S(b)) \tag{11}$$

where P_t is the Pareto solution set, and $S(b) = \dfrac{\left| \{ a | a \in P_t \wedge b \succ a \} \right|}{L_{size} \times L_{size} + 1}$.

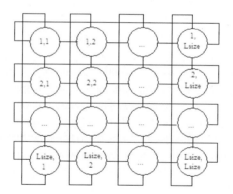

Fig. 1. Model of the agent lattice

2.4 Behaviors of Agents

Since, resources are limited and the behaviors of the agents are drove by their purpose, an agent will compete with others to get more resources. Therefore, three behaviors, competitive behavior, self-learning behavior and mutation behavior are designed for agents to realize their purpose.

(1) **Competitive behavior:** In this behavior, the energy of an agent is compared with those of the neighbors. The agent can survive if the energy is maximum, otherwise, the agent must die. Then the child of the agent with maximum energy among the neighbors will substitute the former one. The goal of the competitive behavior is to eliminate the agents with low energy, and give more chances to the potential agents.

Algorithm 1. Competitive behavior

Input: $Max_{i,j}$: $Max_{i,j}(P) = \langle m_1, m_2, ..., m_{n \times m} \rangle$;

Output: $Child_{i,j}$: $Child_{i,j}(P) = \langle c_1, c_2, ..., c_{n \times m} \rangle$;

$Random(n,i)$ is a random integer in $[0, n]$ (not equal to i)

p_c: A predefined parameter in the range of 0~1;

$U(0,1)$: A function that generates a random decimal in the range of 0~1;

---

begin

  $Child_{i,j}(P) = Max_{i,j}(P)$; $i := 1$;

  repeat

  if $(U(0,1) < p_c)$ **then**

  begin

    $l := Random(n,i)$;

    $swap(c_i, c_j)$;

  end;

  $i := i+1$;

  until $(i > n)$;

  $Child_{i,j}(SL) := True$;

end.

---

(2) **Self-learning behavior:** It is known that the performance of EAs can be boosted through the local searches. So, the self-learning behavior is designed for agents by making use of local search method. The aim of self-learning behavior is to find an exchange for the components in a permutation, so that the energy of the corresponding agent is increased after the arrangement.

Since a number in P occurs many times, the algorithm must be prevented from swapping two identical values. Let $P = \langle p_1, p_2, ..., p_{n \times m} \rangle$, o_{pi} denotes the operation corresponding to P_i, and M_{pi} is the machine on which o_{pi} is to be processed. Suppose that P_i and P_j ($P_i \neq P_j$ and $i < j$) is swapped and P' is obtained. Based on the method transforming P to a schedule, we can obtain that the two schedules corresponding to P and P' are identical if P_i and P_j satisfy (12). It is assumed that this behavior is performed on $L_{i,j}$. The details are shown in Algorithm 2.

$$\left\{ \begin{array}{l} \forall P_k, i < j < k, (M_{P_k} \neq M_{P_i}) \text{ and } (P_k \neq P_i) \text{ and} \\ (M_{P_k} \neq M_{P_i}) \text{ and } (P_k \neq P_j) \text{ and } (M_{P_k} \neq M_{P_j}) \end{array} \right\} \tag{12}$$

(3) **Mutation behavior:** In the objectives space, a solution set with high quality should be distributed uniformly. Especially duplicate individuals should not exist in the set. If two or more individuals are duplicated, individuals' places will be wasted in the solution set. This could reduce effective representation of the solution set. Moreover, the distance of two duplicate individuals is zero. This has a considerable influence on evaluating the distribution of the solution set. The

duplicate individuals in Pareto set often can be seen in some classical multi-objective evolutionary algorithm (MOEA), such as SPEA [7], NSGA-II [8] etc.

Herein, the mutation behavior is designed to prevent the emergence of the duplicate individuals. The details are shown in Algorithm 3.

Algorithm 2. Self_learning behavior

Input: $L_{i,j}: L_{i,j}(P) = \langle P_1, P_2, ..., P_{n\times m} \rangle$;

Output: $L_{i,j}$;

---

begin
 repeat
    $Repeat := False$; $k := 1$; $Iteration := 1$;
    While ($k \leq n \times m$) **do**
    begin
     $Energy_{old} := L_{i,j}(E)$;
     $l := Random(n \times m, k)$, $P_k \neq P_l$, and P_k, P_l do not satisfy (12);
     $swap(P_k, P_l)$; $Energy_{new} := L_{i,j}(E)$;
     if ($Energy_{new} < Energy_{old}$) **then** $swap(P_k, P_l)$
     else begin
        if ($Energy_{new} > Energy_{old}$) **then** $Repeat := True$;
        $k := k+1$;
        end;
     if ($Iteration < n \times m - 1$) **then** $Iteration := Iteration + 1$
     else begin $Iteration := 1$; $k := k+1$; **end**;
    end;
 until ($Repeat = True$) ;
 $L_{i,j}(SL) := False$;
end.

Algorithm 3. Mutation behavior

Input: $L_{i,j}: L_{i,j}(P) = \langle P_1, P_2, ..., P_{n\times m} \rangle$;

Output: $L_{i,j}$;

$invert(P_i, P_j)$: perform a invert mutation on $L_{i,j}(P)$ from P_i to P_j;

begin
   $start = Random(n \times m)$;
   $end = Random(n \times m)$;
   if ($end \neq start$)
     if ($end < start$) **then** $swap(start, end)$;
     else $invert(P_{start}, P_{end})$;
end.

2.5 Implementation of MAEA-MOJSP

Like most MOEA, there is an external archive beside the agent lattice in MAEA-MOJSP to maintain the solutions of high quality. But the difference here is no fixed size of the archive in MAEA-MOJSP.

Algorithm 4. MAEA-MOJSP

Step 1.   Initialize agent lattice L, and copy the nondominated solution of L into archive P; $gen:=1$;

Step 2. $Competitive(L)$;

Step 3.   In order to reduce computational cost, select the best agent ($agent_{best}$), $self\_learning(agent_{best})$;

Step 4.   For each agent $L_{i,j}$ of L, compare it with every agent p in P:

        if $L_{i,j}$ dominate p, delete p from P, and $P:=P \cup L_{i,j}$;

        else if $L_{i,j}$ and p are duplicate, $mutation(L_{i,j})$;

Step 5.   Copy the agents of P into L, randomly;

Step 6. $gen:=gen+1$;

Step 7.   **if** $gen>maxgen$, **end**;
        else go to Step 2.

Consider a circumstance that $L_{i,j}$ and p are duplicated, if $L_{i,j}$ is a agent with high quality, after the mutation on $L_{i,j}$, it may become worse and take part in the next evolution. Therefore, a replacement strategy (Step 5) is designed to prevent this. After maintenance of the archive P, suppose the size of P is $pareto\_size$, select $pareto\_size$ lattice-points randomly in L, then copy the $pareto\_size$ agents of P into the lattice-points in L. In this way, the good agents can participate in the next evolution, so that the agents of high quality can be reserved.

3   Experimental Studies

Eight benchmarks with different scales from [9] are tested to investigate the performance of MAEA-MOJSP. Makespan and the total tardiness are the two objectives to be optimized simultaneously. For that the due date is not provided in the original benchmarks, so they are set as same as in [1]. The parameter setting is given in Table 1. For each problem, the algorithm runs 30 times stochastically. Some of the nondominated solutions obtained by MAEA-MOJSP are shown in Table 2. Due to the limitation of space, here we only give the comparison with SPEA (fitness assignment of SPEA is adopted in our algorithm) of instances FT10, ORB1 and LA26 in Fig.2, Fig.3 and Fig.4, respectively.

From the results, we can see that the MAEA-MOJSP can obtain a good Pareto front, and the Pareto solutions distribute on it has an appropriate diversity. That because there is no global control exists in MAEA-MOJSP, so each agent is independent with each

Table 1. Parameter setting

MAEA-MOJSP		SPEA			
L_{size}	10	N	80	$\overline{N}$	20
p_c	0.8	p_c	0.85	p_m	0.1
maxgen	150	maxgen		150	

Table 2. Some nondominated solutions obtained by MAEA-MOJSP

FT06 6×6	FT10 10×10	ORB1 10×10	ABZ5 10×10
60, 1	1059, 830	1200, 599	1414, 0
65, 0	1174, 676	1193, 966	1362, 5
55, 28	1207, 658	1336, 391	1340, 263
59, 8	1229, 496	1423, 336	1299, 398
57, 19	1283, 284	1197, 605	1308, 371
ABZ6 10×10	ABZ7 20×15	LA26 20×10	LA27 20×10
1055, 30	798, 361	1380, 3912	1350, 3201
1022, 42	796, 383	1365, 4493	1452, 2613
979, 348	789, 426	1364, 4547	1368, 3117
981, 212	786, 430	1354, 4697	1451, 2969
979, 348	790, 397	1375, 4085	1508, 2372

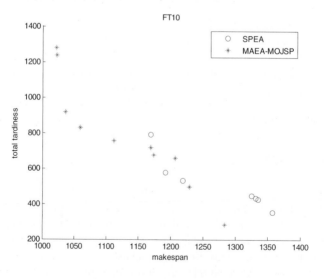

Fig. 2. Pareto front of FT10

other to some extent, which is propitious to maintaining diversity. Moreover, thanks to the mutation behavior which prevents the emergence of the duplicate individuals, the algorithm also has a great uniformity and coverage of the Pareto front. As can be seen, the latticelike which agents live is closer to the real evolutionary model than the population in traditional evolutionary algorithms.

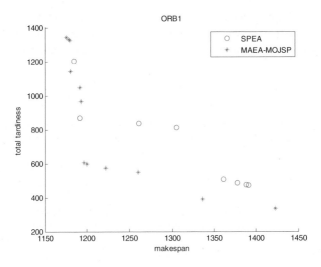

Fig. 3. Pareto front of ORB1

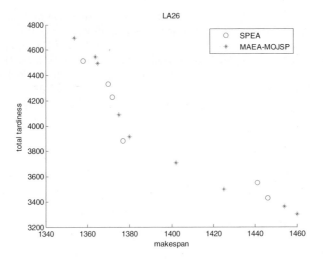

Fig. 4. Pareto front of LA26

4 Conclusion

In this paper, we integrate multiagent systems and evolutionary algorithms to form an algorithm to solve multi-objective job shop problem. Based on the properties and encoding method of job shop, we designed agent, agent environment and three behaviors for the algorithm. In the experiments, MAEA-MOJSP obtains a good performance on eight benchmark datasets. Comparing with the classical multi-objective evolutionary algorithm SPEA, experimental results demonstrate that the MAEA-MOJSP can obtain a wider Pareto front in solving multi-objective job shop problems, and the solutions which distribute on the Pareto front are of great diversity and uniformity.

References

1. Lei, D., Wu, Z.: Crowding-measure-based Multiobjective Evolutionary Algorithm for Job Shop Scheduling. The International Journal of Advanced Manufacturing Technology 30(1-2), 112–117 (2006)
2. Esuqivel, S., Ferrero, S., Gallard, R., Salto, C., Alfonso, H., Schotz, M.: Enhanced Evolutionary Algorithms for Single and Multi-objective Optimization in the Job Shop Scheduling Problem. Knowl.-Based Syst. 15, 13–25 (2002)
3. Qian, B., Wang, L., Huang, D., Wang, X.: Scheduling Multi-objective Job Shop Using a Memtic Algorithm Based on Differential Evolution. Int. J. Adv. Manuf. Technol. 35, 1014–1027 (2008)
4. Dell, A.M., Trubia, M.: Applying Tabu Search to the Job Shop Scheduling Problem. Ann. Oper. Res. 40, 231–252 (1993)
5. Pinedo, M.: Scheduling Theory, Algorithms, and Systems. Prentice-Hall, Englewood Cliffs (2002)
6. Liu, J., Zhong, W., Jiao, L.: A Multiagent Evolutionary Algorithm for Constraint Satisfaction Problems. IEEE Trans. Syst., Man, and Cybern. B 36(1), 54–73 (2006)
7. Zitzler, E., Thiele, L.: Multiobjective Evolutionary Algorithms: A Comparative Case Study and the Strength Pareto Approach. IEEE Trans. Evol. Comput. 3(4), 257–271 (1999)
8. Deb, K., Pratap, A., Agarwal, S., Meyarivan, T.: A Fast and Elitist Multiobjective Genetic Algorithm: NSGA-II. IEEE Trans. Evol. Comput. 6(2), 182–197 (2002)
9. http://people.brunel.ac.uk/~mastjjb/jeb/orlib/jobshopinfo.html

A Preference Oriented Two-Layered Multiagent Evolutionary Algorithm for Multi-Objective Job Shop Problems*

Xinrui Duan, Jing Liu, Ruochen Liu, and Licheng Jiao

Key Laboratory of Intelligent Perception and Image Understanding of Ministry of Education of China, Institute of Intelligent Information Processing, Xidian University, Xi'an, China
xinruiduan@163.com

Abstract. From the viewpoint of decision making process, it brings inconveniences for decision makers to select one (few) proper solution(s). Thus we propose preference oriented two-layered multiagent evolutionary algorithm (TL-MAEA) to meet customers' needs. The algorithm has a structure of two layers: in the top layer, preference relations among multiple objectives are calculated through interactions with the decision maker; while in the bottom layer, MAEA is employed to obtain the optimal solution corresponding to the preference relations. In the experimental, 12 benchmark problems are used to test the algorithm. The results show that the proposed algorithm is effective.

Keywords: Multi-objective job shop problems, preference, multiagent evolutionary algorithm.

1 Introduction

Job shop problem (JSP) is an important kind of actual production system, which has been demonstrated to be NP-hard [1]. During the last two decades, there were many researches on multi-objective JSPs (MO-JSPs). They are all for the purpose of finding a set of Pareto solutions. However, it is not easy for the decision maker to select one (few) solution(s) from a large number of candidate Pareto solutions.

In this paper, we designed a two-layered multiagent evolutionary algorithm (TL-MAEA) to solve the multi-objective JSPs. This algorithm aims to obtain a solution which represents the decision maker's preference rather than seek to find all Pareto solutions. We designed a hierarchical structure of three agents, director, managers and staffs. By using such a structure, it is easy for decision makers to involve into the problem solving process, and obtain proper solutions. In addition, staff agents can optimize the subproblem with given preference relation in parallel. In the experiments, we select 12 benchmark datasets to test the performance of TL-MAEA. The experimental results illustrate that TL-MAEA is effective for MO-JSPs.

---

* This work was supported by the National Natural Science Foundation of China under Grants 60872135, 60803098, and 60970067, and the National Research Foundation for the Doctoral Program of Higher Education of China under Grant 20070701022.

K. Deb et al. (Eds.): SEAL 2010, LNCS 6457, pp. 553–557, 2010.
© Springer-Verlag Berlin Heidelberg 2010

2 Preference Oriented Two-Layered Multiagent Evolutionary Algorithm

2.1 Modeling Preference by Fuzzy Logic Technique

Fuzzy preference relation [2] is used frequently in many important decision models. Generally, for N objectives problem, $N(N-1)/2$ preference information must be given by the decision maker. Cvetkovic *et al.* [3] demonstrates how to transform the fuzzy preference to weight vector.

2.2 Multiagent Evolutionary Algorithm (MAEA)

Jing Liu *et al.* [4] proposed a multiagent evolutionary algorithm (MAEA) by integrating the multiagent system and evolutionary algorithms to solve constraint satisfaction problems. In MAEA, all agents live in a latticelike environment, they sense and act on the environment by using the designed behaviors. During the process of interacting with the environment and the other agents, each agent's final goal is to increase its energy. Therefore, the MAEA can find the optimal solution.

The size of lattice is $L_{size} \times L_{size}$, where L_{size} is an integer. Each agent is fixed on a lattice-point and can only interact with the neighbors. It can assume that the agent located at (i, j) is represented as $L_{i,j}$, $i, j = 1, 2, ..., L_{size}$, then the neighbors of $L_{i,j}$, $Neighbors_{i,j}$, are defined as follows:

$$Neighbors_{i,j} = \left\{ L_{i',j}, L_{i,j'}, L_{i'',j}, L_{i,j''} \right\} \tag{1}$$

where $i' = \begin{cases} i-1 & i \neq 1 \\ L_{size} & i=1 \end{cases}$, $j' = \begin{cases} j-1 & j \neq 1 \\ L_{size} & j=1 \end{cases}$, $i'' = \begin{cases} i+1 & i \neq L_{size} \\ 1 & i = L_{size} \end{cases}$, $j'' = \begin{cases} j+1 & j \neq L_{size} \\ 1 & j = L_{size} \end{cases}$.

For JSPs, an agent, a, denotes an element in the search space S, with the energy equal to:

$$\forall a \in S, energy(a) = -\sum_{i=1}^{N} w_i f_i \tag{2}$$

where N is the number of objective function, w_i is the weight of each objective, , and f_i is value of the i-th objective function. What should be noted is that, for the maximization criterion, f_i is the value of the i-th objective function; for the minimization, f_i is the negative value of the i-th objective function.

The two designed behaviors are described as follows:

Competitive behavior: In this behavior, the energy of an agent is compared with those of the neighbors. The agent can survive if the energy is maximum, otherwise, the agent must die. Then the child of the agent with maximum energy among the neighbors will substitute the former one. The algorithm of how to generate a child of an agent can be seen in [4].

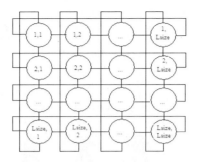

Fig. 1. Model of the agent lattice

Self-learning behavior: The self-learning behavior is designed for agents by making use of local search method. The aim of the behavior is to find an exchange for the components in a permutation, so that the energy of the corresponding agent is increased after the arrangement. More details of MAEA are given in [4].

2.3 TL-MAEA

Integrating preference information into MAEA, we propose a new algorithm TL-MAEA. In this algorithm, we adopt an interactive framework which is a procedure that transforms the preference information into quantitative relationship being considered as a priori knowledge and transmitted to evolutionary population.

There are two layers in this algorithm. The top one determines the quantitative relationship of the preferences by interacting with the decision maker. In the bottom one, we exploit this relationship to guide the optimization direction. In the TL-MAEA, four concepts are defined as follows:

(1) **Initial weight:** according the preference information, $\bar{w} = \{w_1, w_2, ..., w_N\}$ is obtained by the method in [3].

(2) **Desired value:** the best value that each objective can achieve in N_E generation evolution.

(3) **Adjust rule:** a regulation which director amends the weight of next evolution according to. Specifically, the rule is defined as a vector $\bar{c}$ which is the normalization of $\bar{w}$.

(4) **Elasticity:** the degree of the deviation from each objective's desired value that director can accept. Specifically, it is a vector $\bar{e}$ which is the normalization of $\{1 - w_1, 1 - w_2, ..., 1 - w_N\}$.

Staff agents are distributed in the bottom layer, while the other two agents are distributed in the top layer. Each staff agent can only interact with those in the same group, and only be supervised by the manager which it belongs to. Each manager is on behalf of benefit of its group, so each of them is selfishness. Therefore, in the first evolution, their energy criterion is the objective function which is assigned by the director respectively. The model of the interactive system is seen in Fig.2.

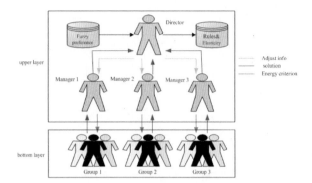

Fig. 2. Model of the agent lattice

Suppose that there is a three-objective problem. At first, three desired values are obtained. Transform the preference into initial weight, and determine adjust rule and elasticity at the same time.

Once each group gets a solution by MAEA, the manager should transmit it to the director. The director analyzes whether each solution is in the acceptable deviation (elasticity). If "yes", stops the algorithm. Otherwise, according the adjust rule, the director must give the adjust information to each manger. Then, each manager changes their energy criterion and makes its group search along the direction of the preference in *intervalgen* generation evolutioan. Finally, after *maxgen* times, the director should select a best solution with the minimum deviation from the N solutions.

3 Experiments

In this study, we apply this algorithm on MO-JSPs. 12 benchmarks from [5] with different scales are selected to test the proposed algorithm. Two regular performance measures, makespan and total tardiness which are minimum problems, are used as criteria. For that the due date is not provided in the benchmark, so they are set as same as in [6]. The parameters are set as follows: L_{size}=10, N_E=50, *maxgen*=5, *intervalgen*=20. For each problem, the algorithm runs 30 times stochastically.

For two-objective JSPs, three relationships, $H_{1,2}$ (the 1st objective is important than the 2nd one), $E_{1,2}$(the two objectives are equally important), and $L_{1,2}$(the 2nd objective is important than the 1st one) between the two objectives are tested as the preference. The results can be seen in Table 1.

Take FT06 problem as an example: when the 1_{st} objective (makespan) is important than the 2_{nd} one (total tardiness), the solution, (58, 4), indicates that minimum degree of the 1_{st} objective is greater than this of the 2_{nd} one. When the two objectives are equally important, the answer is (60, 1). And the solution, (62, 0), is obtained, when the total tardiness is important than the makespan.

From the results, we can see that the solutions vary with the preferences. In general, different solution is obtained with different preference relation, and the solution quality of this algorithm is also appropriate.

Table 1. The results of two-objective JSPs

Problem	Preference Relation		
	$H_{1,2}$	$E_{1,2}$	$L_{1,2}$
FT06 6×6	58, 4	60, 1	62, 0
FT10 10×10	1015, 205	1088, 81	1087, 60
FT20 20×5	1078, 6011	1211, 6056	1210, 5788
ORB1 10×10	1081, 607	1156, 410	1189, 362
ORB2 10×10	891,51	948,12	938,0
ORB3 10×10	1162, 857	1149, 477	1286, 429
ABZ5 10×10	1251,366	1277, 422	1297, 139
ABZ6 10×10	967, 180	974, 149	993, 61
ABZ7 20×15	702, 176	707, 195	736, 69
ABZ8 20×15	733, 217	748, 166	736, 118
LA26 20×10	1218, 2914	1230, 2872	1241, 2696
LA27 20×10	1273, 2381	1281, 2296	1285, 2273

4 Conclusion

In this paper, a preference oriented two-layered multiagent evolutionary algorithm (TL-MAEA) is proposed to solve multi-objective job shop scheduling problems. TL-MAEA uses a fuzzy method to model the preference of the decision maker on the objectives. Then use a hierarchical structure to involve the decision maker into the problem solving process. By such two-layered structure, the solutions meeting the requirements of the decision maker are obtained. The experimental results illustrates that TL-MAEA achieves a good performance, and it can find a solution which corresponds to the preference of the decision maker.

References

1. Wang, L.: Job Shop Scheduling with Genetic Algorithms. Tsinghua University Press, Beijing (2003)
2. Chiclana, F., Herrera, F., Herrrera-Viedma, E.: Integrating Three Representation Models in Fuzzy Multipurpose Decision Making Based on Fuzzy Preference Relations. Fuzzy Sets and Systems 97(1), 33–48 (1998)
3. Cvetkovic, D., Parmee, I.: Designer's Preferences and Multi-objective Preliminary Design Processes. In: Evolutionary Design and Manufacture: In ACDM 2000, Plymouth, UK (2000)
4. Liu, J., Zhong, W., Jiao, L.: A Multiagent Evolutionary Algorithm for Constraint Satisfaction Problems. IEEE Trans. Syst., Man, and Cybern. B 36(1), 54–73 (2006)
5. http://people.brunel.ac.uk/~mastjjb/jeb/orlib/jobshopinfo.html
6. Lei, D., Wu, Z.: Crowding-measure-based Multiobjective Evolutionary Algorithm for Job Shop Scheduling. The International Journal of Advanced Manufacturing Technology 30(1-2), 112–117 (2006)

Using an Adaptation of a Binary Search Tree to Improve the NSGA-II Nondominated Sorting Procedure

João Batista Mendes[1,2] and João Antônio de Vasconcelos[2]

[1] Universidade Estadual de Montes Claros, Campus Universitário Professor Darcy Ribeiro
CEP: 39401-089 - Montes Claros-MG
[2] Evolutionary Computation Laboratory
Universidade Federal de Minas Gerais, Avenida Antônio Carlos, 6627
Belo Horizonte/Minas Gerais – Brasil
jbm@cpdee.ufmg.br, jvasconcelos@ufmg.br

Abstract. In this paper, we propose an adaptation to Nondominated Sorting Genetic Algorithm (NSGA-II), introducing a data structure, called *NonDominated Tree (NDT)*. The NDT is an adaptation of a Binary Search Tree and is used to identify the nondominated fronts in only one run. This structure may be used to improve even more the performance of NSGA-II and other Evolutionary Algorithms (EAs) that use nondominated sorting procedures. It reduces the number of comparisons performed by the NSGA-II nondominated sorting routine. Some tests demonstrated that the proposed structure improves the search of fronts of nondominated solutions in an efficient way.

Keywords: NSGA-II, Search Binary Tree, Multiobjective Optimization.

1 Introduction

The use of evolutionary algorithms (EAs) to solve multiobjective optimization problems (MOPs) has been motivated mainly because the EAs allow the generation of several elements of the Pareto optimal set in a single run [1]. Several *Multiobjective Evolutionary Algorithms* (*MOEAs*) search for a subset of some solution set (Pareto Optimal Set). The *Nondominated Sorting Genetic Algorithm II (NSGA-II)* is a popular MOEA, proposed by Deb et *al.* [6], used in several MOEAs comparisons and as a foundation for other algorithm designs.

We propose in this work an adaptation in the *fast nondominated sorting* procedure of NSGA-II. We defined a data structure, called *Non Dominated Tree (NDT)*, that does nondominated sorting in an efficient manner.

This paper is organized as follows. Section 2 presents the MOEA NSGA-II. Section 3 details the NDT. In section 4 we present some results obtained in tests realized in laboratory and make some comments in the last section.

2 Nondominated Sorting Genetic Algorithm (NSGA-II)

The NSGA-II is a MOEA that uses elitism and a crowded distance operator that keeps diversity without specifying any additional parameters. In short the NSGA-II operates as: (1) Firstly, $t=0$, a random population P_t is created with n individuals; (2) P_t is

K. Deb et al. (Eds.): SEAL 2010, LNCS 6457, pp. 558–562, 2010.

sorted (by *fast nondominated sorting* procedure) into different nondomination levels (fronts) $F_1, F_2, ..., F_k$ ($k \geq 1, P_t = F_1 \cup F_2 \cup \cdots \cup F_k$). While there is no dominance between individuals in the same front F_i ($1 \leq i \leq k$), each individual in level F_{i+1} is dominated by at least one individual in level F_i: $F_1 \prec F_2 \prec \cdots \prec F_k$. (3) A population P' (size n) is generated and the population $Q^* = P_t \cup P'$(size $2n$) is sorted into nondomination fronts $F_1^\star, F_2^\star, ..., F_y^\star$ ($y \geq 1$); (4) In a last step, $t = t+1$ and a population P_t is created with the best n individuals of Q^*. Return to step 2 and repeat the operation until a stop condition is found.

The NSGA-II *fast nondominated sorting* procedure sorts a population P_t (size n) into non-dominated fronts $F_1, F_2, ..., F_k$. For a MOP with M objectives, its time complexity is $O(Mn^2)$ [3][6][7].

Due to NSGA-II popularity, some researches in enhancing NSGA-II were done. An adaptive parameter control was proposed by Carvalho and Aluizio [2] to update the static mutation operator for improving the performance of NSGA-II. Jensen [7] describes two algorithms for nondominated sorting: one for problems with only two ($M = 2$) objectives and other for problems with three or more ($M \geq 3$) objectives. Fieldsend et al. [4] proposed dominated and non-dominated trees for the storage and maintenance of archival solution sets. Berry and Vamplew [5] implemented a specialized algorithm (called Mak_Tree) for application in bi-objective problems. The Mak_Tree only accept non-dominated solutions and uses a dynamic and self-balancing binary tree structure, ordered by performance on the first objective. Fang et al. [3] proposed a data structure called dominance tree to save the dominance information among solutions. They developed a non-dominated sorting algorithm that implements the divide-and-conquer mechanism to generate the dominance tree.

In this paper we propose and describe a sorting algorithm that uses an Adapted Binary Search Tree to do nondominated sorting.

3   Non-Dominated Tree (NDT)

A *NDT* is an adapted Binary Search Tree (BST) that is used to *describe* the dominance relation between the members of a population P. Also, it *generates*, automatically, all non-dominance levels.

Each node r of the NDT is associated with a set of individuals ($r.F$) from a population P. The individuals located in the same NDT node have no dominance relation one another. The individuals in the *right subtree* of node r *dominate* the individuals of $r.F$. On the other hand, individuals of $r.F$ dominate those in left subtree of node r. In resume: (1) $r.F$: It is a list to store individuals with no dominance relation at node r; (2) *Left, Right* subtrees: They are NDT subtrees where $r.Right \prec r.F \prec r.Left$.

The main algorithms implemented by NDT are described in the following.

BuildingNDT Algorithm

The *buildingNDT* algorithm, described by Algorithm 1, gets each individual (p) of a population P and inserts it in the NDT structure. The first individual of P is inserted in the root node and the remaining ones ($2, \cdots, n$) are inserted in the NDT through the *insertIndiv* algorithm.

Algorithm 1. Algorithm to Construct the NDT Structure

<u>Input</u>: Population **P**;   // Population with *n* individuals to be ordered by dominance
<u>Output</u>: **NDT** Structure.

```
PROCEDURE buildingNDT (P);
    ndt.F ←{P(1)};
    ndt.Left ← φ;ndt.Right ← φ;
    FOR i = 2 TO |P| DO
        ndt←insertindiv (P(i), ndt); // P(i) is inserted in ndt
    END-FOR
    buildingNDT ← ndt;
END-PROCEDURE
```

The *insertIndiv* algorithm, described by Algorithm 2, works as following: (1) while solution p is dominated by none in list *ndt.F* and *ndt.Right* is not empty, it goes to *ndt.Right*. (2) But, if *any* solution in *ndt.F* dominates p, then it continues its search in *ndt.Left* (if *ndt.Left* is empty, we are in front F_k and p belongs to front F_{k+1}); (3) Otherwise, when p dominates none or dominates only k solutions in *ndt.F*, the following operations are done: (a) p is inserted in list *ndt.F*; (b)The k individuals are removed from *ndt.F* and are inserted in next front.

Algorithm 2. Algorithm to Insert an Individual in NDT structure

---

<u>Input</u>: NDT structure and an Individual **p**; <u>Output</u>: **NDT** Structure.

```
PROCEDURE insertIndiv(p, ndt);
    IF ndt = φ THEN
        ndt.F ← {p}; ndt.Left ← φ;ndt.Right ← φ;
    ELSE
        WHILE (ndt.Right <> φ) ^ (ndt.F ⊀ p) DO
            ndt ← insertIndiv(p, ndt.Right);
        END-WHILE
        IF ndt.F ≺ p THEN // p is dominated by any individual in ndt.F
            IF ndt.Left <> φ THEN
                ndt ← insertIndiv(p, ndt.Left);
            ELSE
                "We are in front F_k and p belongs to front F_{k+1}"
            END-IF
        ELSIF p ≺ ndt.F THEN //p dominates all individuals in ndt.F
            ndt ← insertIndiv(p, ndt.Right);
        ELSE // p is a nondominated solution or just dominates l_k
             // individuals in ndt.F.
            ndt.F ← ndt.F U {p}; // p is inserted in the list ndt.F
            l_k ← removeDominated(ndt.F);
            "We are in front F_k and l_k belongs to front F_{k+1}: if l_k>0"
        END-IF
        insertIndiv ← ndt;
END-PROCEDURE
```

Nondominated Fronts

The *nondominated-fronts* procedure, illustrated by Algorithm 3, shows all nondominated levels defined in the NDT ordered by non-dominance.

Algorithm 3. Algorithm to Print All non-dominanted Fronts in NDT

```
Input: NDT structure; Output: Fronts F₁, F₂, ⋯ Fₖ.
PROCEDURE nondominated-fronts (ndt);
    WHILE ndt <> φ DO
        IF ndt.Right <> φ THEN
            nondominated-fronts (ndt.Right);
        END-IF
        display (ndt.F);
        IF ndt.Left <> φ THEN
            nondominated-fronts (ndt.Left);
        END-IF
    END-WHILE
END-PROCEDURE
```

Joining Two Populations

During the NSGA-II run, a new population $Q^* = P_t \cup P'$ is created and ordered using non-dominance and crowding distance principles. As the original population (P_t)) is already ordered by non-dominance through NDT, the *combinePopulation* procedure, described by Algorithm 4, gets each individual of P' and insert it in the NDT structure. At its end, the new population Q^* will be ordered by non-dominance.

Algorithm 4. Algorithm to Join Two Populations Using NDT Structure

```
Input: NDT Structure and OffSpring Population P'; Output: Population Q* = Pt ∪ P'
PROCEDURE combinePopulation (P', ndt);
    k ← 1;
    WHILE P'(k) <> φ DO
        ndt ← insertIndiv (P'(k), ndt);
        k ← k +1;
    END-WHILE
    combinePopulation ← ndt;
; END-PROCEDURE
```

4 Results

The methodology to shown the results obtained with the proposed procedure of sorting nondominated fronts was divided in four steps: i) generation of different populations randomly, where each individual represents one point in the search space with five variables; ii) each individual was evaluated considering the problem as stated in (1); iii) both the buildingNDT and fast nondominated sorting procedures were executed to classify all nondominated fronts and iv) the results obtained with both procedures are compared, as shown in Table 1.

$$
\begin{aligned}
&\textit{minimize} \quad && F(x) = \{F_1(x), F_2(x)\} \\
&\textit{subjet to:} \quad && g_1(x) = a * \left|\sin\left(b * \pi * \left(\sin(\theta) * (F_2(x) - e) + \cos(\theta) * F_1(x)\right)^c\right)\right|^d \\
& && g_2(x) = \cos(\theta) * (F_2(x) - e) - \sin(\theta) * F_1(x) \le 0 \\
& && g_2 \le g_1 \\
& && 0 \le x_1 \le 1, -5 \le x_i \le 5, \forall i = 2,3,4,5 \\
& && \theta = -0.05 * \pi;\ a = 40;\ b = 5;\ c = 1;\ d = 6;\ e = 0.
\end{aligned}
$$
$$(1)$$
$$\textit{where } F_1(x) = x_1,\ F_2(x) = fx * \left(1 - \frac{x_1}{fx}\right) \text{ and } fx = 41 + \sum_{i=2}^{5}\left(x_i^2 - 10 * \cos(2 * \pi * x_i)\right).$$

In Table 1, the total numbers of nondominated fronts found by NDT structure, the NDT height, the number of comparisons performed by both the *buildingNDT* and *fast nondominated sorting procedures* to find all nondominated fronts are shown.

Table 1. Results obtained using a NDT to order populations of varying sizes

Population Size	Fronts	Height	NDT Comparisons	Fast-nondominated-sorting comparisons
10	6	4	136 (66.8%)	199 (100.0%)
50	11	6	2212 (88.7%)	2494 (100.0%)
100	18	14	9404 (80.8%)	11636 (100.0%)
200	23	17	41638 (82.7%)	50324 (100.0%)
500	37	25	249785 (87.8%)	284347 (100.0%)
1000	58	44	847724 (82.3%)	1030131 (100.0%)
1500	76	48	1758509 (77.6%)	2264559 (100.0%)
2000	80	70	3243565 (85.7%)	3783381 (100.0%)

5 Conclusions

The results showed that the NDT is very promising to improve algorithms that use the nondominated sorting procedure. Some important comments about this work are:

- No attempt was done to produce a balanced NDT. The individuals were inserted one by one starting from the first one. Also the strategy "delayed insertion" [3], was implemented to reduce the number of comparisons;
- The NDT structure may be applied in problems with more than 2 objective functions and adapted to MOEAs that implements nondominated sorting;
- As future works, it is interesting to evaluate the NDT performance with the structures proposed by Jensen [7] and Fang et al. [3].

Acknowledgment. This work has been supported in part by Brazilian Agencies CNPq and FAPEMIG.

References

1. Coello, C.A.C., Lamont, G.B., Van Veldhuizen, D.A.: Evolutionary Algorithms for Solving Multi-Objective Problems. Springer Science+Business Media, New York (2007)
2. Carvalho, A.G., Araujo, A.F.R.: Improving NSGA-II with an Adaptive Mutation Operator. In: GECCO 2009 (2009)
3. Fang, H., Wang, Q., Tu, Y.-C., Horstermeyer, M.F.: An Efficient Non-dominated Sorting Method. Evolutionary Computation 16(3) (2008)
4. Fieldsend, J.E., Everson, R.M., Singh, S.: Using Unconstrained Elite Archives for Multiobjective Optimization. IEEE Transactions On Evolutionary Computation 7(3) (2003)
5. Berry, A., Vamplew, P.: An efficient approach to unbounded Bi-objective archives - Introducing the Mak_Tree algorithm. In: s.l.: Genetic and Evolutionary Computation Conference, vol. 1, pp. 619–626 (2006)
6. Deb, K., Pratap, A., Agarwal, S., Meyarivan, T.: A Fast and Elitist Multiobjective Genetic Algorithm: NSGA-II. IEEE Transaction On Evolutionary Computation, 182–197 (2002)
7. Jensen, M.T.: Reducing the Run-Time Complexity of Multiobjective EAs: The NSGA-II and Other Algorithms. IEEE Transactions On Evolutionary Computation 7(5) (2003)

A Hybrid Method for Multi-Objective Shape Optimization

G.N. Sashi Kumar, A.K. Mahendra, A. Sanyal, and G. Gouthaman

Machine Dynamics Division, Bhabha Atomic Research Centre,
Trombay, Mumbai - 400 085, India
{gnsk,mahendra}@barc.gov.in

Abstract. This paper introduces a hybrid shape optimization method
(M-HYBRID) for multiple objectives (MO) using Genetic Algorithm
(GA) and Ant Colony Optimization (ACO) in combination with a mesh-
less computational fluid dynamics solver. It uses the reference point based
approach to reach the required optimum. This method was found to converge faster than MO optimizer based on GA alone. The constraint on
the handling large number of parameters with MO optimiser based on
ACO is overcome in M-HYBRID. This hybrid optimizer is good contender when a global optimum is the target.

Keywords: Multi objective optimization, shape optimization, reference
point, goal vector, M-HYBRID, GA, ACO.

1 Introduction

The problem of multi objective optimization (MOO) in context with optimal
shape design is a non-trivial task. In last few decades large variants of Genetic Algorithm (GA) [1,2,3] have been used for MOO problems extensively.
Ant Colony Optimization (ACO) is a global optimization technique based on
meta-heuristics; was introduced by Dorigo [4] and extended by Abbaspour *et
al.* [5] to parametric optimization using the route of inverse modeling (ACO-IM).

In a shape optimization problem more than 95% of the total time gets consumed by the Computational Fluid dynamics (CFD) solver in the evaluation
of the objective functions. Each set of objective functions evaluation requires a
CFD call. Thus, an optimizer that can converge in minimum number of function
evaluations, without compromising on the accuracy of the result would reduce
the time of optimization.

The authors have earlier demonstrated shape optimization with single objective using a hybrid method coupled with CFD solver [6]. In this paper the authors
have extended the hybrid method to handle multi objectives (M-HYBRID). Next
section discusses the concept of M-HYBRID method, followed by implementation details for various standard test cases. The optimal shape design problem
is discussed in section 4 followed by conclusions.

K. Deb et al. (Eds.): SEAL 2010, LNCS 6457, pp. 563–567, 2010.

2   M-HYBRID

The M-HYBRID consists of three stages. 1) MGA-multi objective genetic algorithm solver (coarse optimizer), 2) Reorganising stage from MGA to MACO and 3)MACO[7]-multi objective ACO-IM solver (fine optimizer). MGA and MACO are coupled such that the advantages of both are retained with the aim to reduce the number of function calls.

The MOO problem can be written as: Minimize $f_i(\boldsymbol{x}), i = 1, 2...m$ where parameter $\boldsymbol{x} \in X \subset R^n$ can be converted into single objective optimization problem using scalarization (weighted L_p problem).

$$\underset{x \in R^n}{Minimize} L_p(f, w, u^\star)_x = (\varSigma w_i |f_i(x) - u_i^\star|^p)^{1/p} \tag{1}$$

where $1 \leq p \leq \infty$, $w_i \geq 0 \; \forall \; i \in 1, 2, \cdots, m$. We choose $p = 2$ and $w_i = 1/m$. The reference point method proposed by Wierzbicki [8] was applied in GA by Deb *et al.* [1] and Mahendra *et al.* [2]. The method of determining the rank is (based on distance from the reference point) based on ϵ-dominance (shown in Fig. 1a) [1,2] defined by the following equation.

$$\boldsymbol{f^\star} \prec \boldsymbol{f} :\Leftrightarrow \underset{i \in 1,2,...m}{\forall}(f_i^\star \leq f_i) \wedge \underset{i \in 1,2,\cdots,m}{\exists}(f_i^\star < f_i - \epsilon_i) \qquad \epsilon_i \in (0,1) \tag{2}$$

Stage -1 : *MGA*: The GA-CFD method which was successfully validated [9] was the starting point for this work. Scalarization and ranking elaborated above have been implemented so that multiple objectives can be handled. The decision maker (DM) supplies only the reference point. *Stage* -2 : *Reorganising*: The reorganization of the results obtained from stage-1 is performed. The subdomain obtained with less number of parameters is set as the initial guess for MACO. *Stage* -3 : *MACO*: The implementation of MACO is described in detail in Sashi *et al.* [7]. Figure 1b gives the pseudo code for M-HYBRID optimizer.

Advantages of M-HYBRID method: The combination of the operators along with microGA makes MGA approach global optima very fast at the

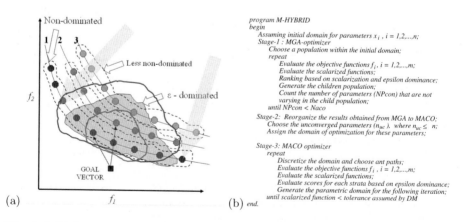

(a)   (b)

Fig. 1. (a)Reference point based approach and ranking(b)Pseudo code of M-HYBRID

begining. Once the members of the population are near to the global optima, the efficiency of MGA decreases, requiring more function evaluations to reach global optima if used alone. In M-HYBRID one switches to MACO after considerable convergence. MACO has better convergence efficiency near to the global optimum than MGA. Thus M-HYBRID method converges in less number of function calls. MACO when used alone, is very costly for problems having more than 6 (let us call it *Naco*) parameters. In M-HYBRID solver MACO is envoked once the number of unconverged parameters reduces to $< Naco$.

3   Standard Test Cases

Two test cases ZDT2 [3]and DTLZ2(with 3 objective functions) are illustrated to demonstrate the advantages of M-HYBRID method (Table 1). For ZDT2 the design variables $x_i \in (0, 1), i = 1, 2, \ldots, 30$ were the initial domain for stage-1 i.e. MGA. Stage-1 is complete when 25 design variables out of 30 variables converge (within a pre-assigned tolerance). The MGA solver is run for 138 generation with 100 members in the population. Five variables remain unconverged after 138th generation. Stage-2 prepares the input for MACO. The variables retained after stage-1 were x_1,x_2, x_{10}, x_{19} and x_{25}. Stage-3 uses these as input parameters. Each of the parameters were discretized into 5 levels and 625 ant paths were chosen randomly. The scalarized objective function with respect to reference point was used. Figure 2a shows the converged solution for various reference points using M-HYBRID. The M-HYBRID solver requires less number of function evaluations to reach the final optimum compared to MGA solver if used alone (Table 1). MACO stand alone solver cannot handle such large number of variables hence comparison with M-HYBRID cannot be made. The second test case DTLZ2 with 3 objective functions is analyzed. Figure 2b show the converged values for two

Table 1. Optimization results for all cases considered in this paper

Case 1: ZDT2	Ref pt. 1	Ref pt. 2 and 3
Goal vector	$(0.1, 0.8)^T$	$(0.5, 0.6)^T$; $(0.8, 0.2)^T$
Stage-1 function (fn) calls	14400	13800 ; 14600
Stage-3 fn calls	6250	5000 ; 7500
Best OF's	$(.2072, .9603)^T$	$(.5633, .6851)^T$; $(.849, .284)^T$
Total fn calls (M-HYBRID)	20650	18800 ; 22100
Fn calls if MGA alone used	42400	44200 ; 68900
Case 2: DTLZ2 - Goal vector	$(0.2, 0.2, 0.6)^T$,	$(0.5, 0.2, 0.2)^T$
Best OF's	$(.3752, .3752, .8751)^T$	$(.7232, .4899, .5547)^T$
Total fn calls (M-HYBRID)	5995	7255
Fn calls with MGA alone	11400	12800
Case 3: Airfoil shape optim.		
Goal vector-AOA chosen	$(0.6, 0.02)^T$ - 1 degree	$(0.25, 0.001)^T$ - 0 degree
Best OF's	$(.612, .022)^T, (0.608, 0.021)^T$	$(.245, .0012)^T, (0.24, 0.0012)^T$
Total fn calls (M-HYBRID)	1245	795

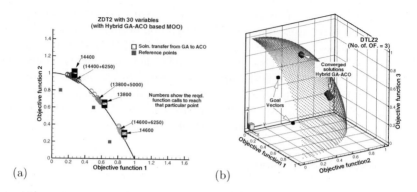

Fig. 2. Solution of (A)ZDT2 and (b) DTLZ2 problems using M-HYBRID method

reference points $(0.2, 0.2, 0.6)^T$ and $(0.5, 0.2, 0.2)^T$. In this test case as number of variables were 5, no variable was dropped while transferring the solution from stage-1 (MGA) to stage-3 (MACO). M-HYBRID is found to require less number of function evaluations compared to when MGA was alone used (Table 1).

4   Test Case 3: Aerodynamic Shape Optimization

We optimize the shape of the airfoil, such that the multiple objective functions are minimized subject to the constraints. The CFD solver is used only to evaluate the objective functions by satisfying the governing Navier-Stokes equations. Thus, the implementation of multi objective hybrid method is independent of the scheme of CFD solver used. The CFD (SLKNS) solver was run for 3 decade fall in residue. More details of the solver can be had from [7]. The airfoil is parameterized using 10 control nodes $x_i, i = 1, ..5 \in$ upper section and $x_i, i = 6, ..10 \in$ lower section of the airfoil. Two cubic splines are used to represent the top and bottom sections of the airfoil. The allowable band for the control nodes and a typical airfoil is shown in Fig. 3a. Here we deal with viscous subsonic flow around an airfoil in elliptic cloud of points at Mach = 0.6 and $7x10^5$ Reynolds number. The angle of attack (AOA) is fixed at 0 or 1 deg. It uses two objective functions. $f_1 = C_l$ and $f_2 = C_d$. 20 airfoil shapes constitute a population for a particular MGA generation. We run the MGA optimizer till the generation where 6 control variables have converged. For the case of goal vector $(C_{l,m}, C_{d,m}) = (0.6, 0.02)$ MGA is run (i.e. stage-1) for 30 generations. The four of the variables x_1, x_6, x_7, x_8 were supplied to MACO for further optimization. Each of the four parameter's domains were discretized into 5 levels and 125 ant paths were chosen randomly. Table 1 shows the details of the optimization with two different goal vectors. Figure 3b shows the change in best airfoil shape as the optimization proceeds for the case $(C_{l,m}, C_{d,m}) = (0.25, 0.001)$. The optimizer was able to converge to value near to the goal vector.

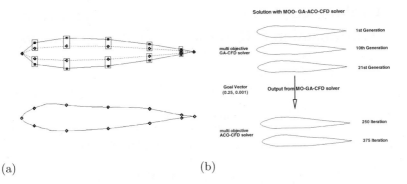

(a)   (b)

Fig. 3. (a)Allowable band and a typical airfoil shape (b)The change in shape of airfoil as optimization proceeds

5   Conclusions

The M-HYBRID method was able to solve the reference based multi objective optimization problems. We have demonstrated that M-HYBRID method can capture the member of the Pareto front, which is closest to the reference point. The M-HYBRID method based shape optimizer is a good contender in multi-objective search space.

References

1. Deb, K., Sundar, J., Bhaskara, U., Chaudhuri, S.: ISSN Intl. J. of Compl. Intel. Res. 2(3), 273–286 (2006)
2. Mahendra, A.K., Sanyal, A., Gouthaman, G., Bera, T.K.: In: Proc. of the 10th Intl. Workshop on Sepn. Phenomena in Liquids and Gases, Angra Dos Reis, Brazil, August 11-14 (2008)
3. Zitzler, E., Deb, K., Thiele, L.: Evol. Compn. 8, 173–195 (2000)
4. Dorigo, M.: IEEE trans. on systems, man and Cyber. - part B, Cyber. 261 (1996)
5. Abbaspour, K.C., Schulin, R., van Genuchten, M.T.: Advances in water resources 24, 827–841 (2001)
6. Sashi Kumar, G.N., Mahendra, A.K., Raghurama Rao, S.V.: Proc. of 18th AIAA-CFD Conference, held at Miami, FL, June 25-28 (2007)
7. Sashi Kumar, G.N., Mahendra, A.K., Gouthaman, G.: Proc. of ICFD 2010, held at Reading, U.K., April 12-15 (2010)
8. Wierzbicki, A.P.: Multiple Criteria Decision Making Theory and Applications, held at Angra dos Reis, Brazil, August 11-14 (2008)
9. Sashi Kumar, G.N., Mahendra, A.K., Deshpande, S.M.: Intl. soc. of CFD. CFD Journal 16(4), 425–433 (2008)

Evolutionary Multi-Objective Bacterial Swarm Optimization (MOBSO): A Hybrid Approach

Indranil Banerjee[1] and Prasun Das[2]

[1] Software Engineer
Videonetics Technology Pvt. Ltd.
Sector-V, Salt Lake Block EP, Kolkata 700 091, India
prime41.indra@gmail.com
[2] SQC & OR Division
Indian Statistical Institute
203, B.T. Road, Kolkata 700 108, India
dasprasun@rediffmail.com

Abstract. The field of evolutionary multi-objective optimization (MOO) has witnessed an ever-growing number of studies to use artificial swarm behavior. In this paper authors have made an endeavor to minimize the computational burden, associated with global ranking methods and local selection modules used in many multi-objective particle swarm optimizers. Two different swarm strategies were employed for global and local search respectively using particle swarms and bacterial chemotaxis. In this paper the authors have shown comparative improvements of the proposed method namely MOBSO, with a benchmark evolutionary MOO method, NSGA-II. The paper also highlights the reduction in computational complexity for large populations, due to the proposed method.

Keywords: Swarm intelligence, bacterial chemotaxis, computational complexity, particle swarm, Pareto optimality, crowding distance, multi-objective optimization, bio-inspired systems.

1 Introduction

Evolutionary multi-objective optimization (MOO) have started becoming popular with advent of algorithms like SPEA, VEGA, MOEA etc. [1], followed by NSGA (Non-dominated sorting genetic algorithm) and its subsequent improvement, namely NSGA II [2]. In recent time, there is a growing trend in using biologically inspired evolutionary techniques such as swarm based heuristics, artificial ant and bee systems, bird-flocking, bacterial foraging (BFOA) [3] etc. Current researches in this field are concerned in applying the basic concepts and ideas of Particle swarm optimization (PSO) in MOO problems [4, 5, 6 and 7]. Some of the main drawbacks of conventional PSO techniques as highlighted by the authors [5, 6], are higher computational complexity, large number of fitness function evaluation and premature convergence. Recently, there have been several studies [8, 9] on the hybridization of PSO and

K. Deb et al. (Eds.): SEAL 2010, LNCS 6457, pp. 568–572, 2010.
© Springer-Verlag Berlin Heidelberg 2010

BFOA, especially on single objective domain. Promising results from these studies encourages possible extension of these techniques to multi-objective domain.

In this paper, a fast hybrid intelligent swarm based optimizer is developed. A general purpose MOO employing some sort of global ranking scheme will have a complexity of O (MN^2) per iteration, which makes the computational time large enough so that online optimization becomes impractical. The proposed method (MOBSO) has a reduced complexity of O (kMN^2) per iteration. In MOBSO the search space was explored both globally and locally with two mutually-independent swarm based strategies.

The rest of this paper is organized as follows. Section 2 describes the details of the algorithm proposed in this work. In section 3, comparative studies with that of NSGA-II are elaborated with results obtained for the selected benchmark test problems. The paper is concluded in Section 4.

2 The Proposed Algorithm: MOBSO

MOBSO is a hybridization of PSO and BFOA in MOO domain, where the global search is carried out using social part of the standard PSO model. In standard PSO, the particles traverse in multi-dimensional search space (parameter space), having two decision making components; one of which is its' cognitive part (for local search) and another is the social part (for global search). A particle updates its velocity and position (per iteration) in the parameter space in the following way,

$$\mathbf{V}^{new}(i, d) = w * \mathbf{V}^{old}(i, d) + C_1 * R_1 * (\mathbf{X}^{\in global\ best} - \mathbf{X}^{old}(i, d)) + C_2 * R_2 * (\mathbf{X}^{\in local\ best} - \mathbf{X}^{old}(i, d)) . \tag{1}$$

$$\mathbf{X}^{new}(i, d) = \mathbf{X}^{old}(i, d) + \mathbf{V}^{new}(i, d) . \tag{2}$$

Where, V (i, d) and X (i, d) are velocity and position vectors of ith particle respectively in a d dimensional parameter space. X $^{\in global\ best}$ is a particle from global best solution set, which, in case of single objective optimization, is the global best solution. The global best set in the new approach is chosen using crowding rank, which is explained later. The term X $^{\in local\ best}$ denotes the position of a particle chosen from the neighbourhood of the particle or it can be the previous best instance of the particle itself. In MOO domain, definition of this term is often arbitrary. In the proposed solution, this term is avoided and local search is performed using bacterial chemotaxis. Here C_1 and C_2 are acceleration constants, R_1 and R_2 are random numbers between (0, 1) and w is the inertia constant.

In MOBSO cognitive local search is applied using bacterial chemotaxis and the third term of the velocity update equation is not used hence, $C_2 = 0$. The global best set is chosen in the following way. First, each non-dominated solution in the population is given a rank based on calculation of its' crowding distance [1]. This mechanism is used to maintain the diversity of the solution. At the beginning of each iteration, a fraction κ of the best (in terms of crowding distance) non-dominated solutions is chosen as the set of global best solution. The fraction κ is determined at the beginning of each iteration as $\kappa = \min (N_E / (N-N_E), 1)$, where, N_E is the number of non-dominated solutions in the population of N individuals. In the proposed method, only the dominated solutions in the population update their position and velocity using Eq. 1.

The evolution process results in creating more non-dominated solutions in the population, which search in their local space using bacterial chemotaxis. In the standard BFO model (developed from the study of e-coli bacterium [3]), each bacterium can move mainly by two mechanisms, one being *stumble*, in which the bacterium takes a short (defined by stumble length) 'jump' on a random direction in the search space. Another mechanism is called *run*, in which a bacterium that has stumbled to a better solution, continues to take short stumble in that direction until maximum run step is exhausted or no better solution is obtained, whichever comes earlier. The following equations summarize this mechanism.

$$S_t (i) = R_d / \sqrt{(R_d^T * R_d)} . \tag{3}$$

$$X^{new} (i, d) = X^{old} (i, d) + c (i, d) * S_t (i, d) . \tag{4}$$

Where, $R_d \in [0, 1]$ is a d-dimensional random vector and $(.)^T$ stands for its transpose. $S_t (i)$ is a random unit vector giving the direction of stumble of ith bacterium and c (i, d) is the stumble step size of ith bacterium in d^{th} dimension. In MOBSO independent stumble by a bacterium is considered only if it leads to a run, otherwise the stumble is discarded.

The computational complexity of the proposed approach is O (kMN^2), and k $\propto$ N_E/N, thus average k is kept below 1. Since the number of non-dominated solutions needed is more or less a pre-requisite to the user, increasing the number of solutions N does not, therefore, demand an increase in N_E in the final non-dominated set. So, for large N, $N_E/N \ll 1$ and taking k $\approx N_E/N$, the complexity becomes O $(MN^{2 - \varepsilon})$, where $0 < \varepsilon < 1$.

3   Experimental Results

To study the performance of MOBSO, authors have taken several benchmark functions from [1, 10 and 11]. These are unconstraint MOO problems, bi-objective in particular, with varying size of search space. In Table 1 the parameters chosen for simulation are given. To maintain consistency the parameters were kept constant throughout simulation. In this study two performance metrics have been used namely General Distance metric (γ) and Inverse Generational Distance metric (IGD) [1]. The comparative performance of MOBSO and NSGA II, with respect to γ, is given in Table 2. Table 3 indicates IGD metric values for selected problems from [11]. In all cases first row gives the mean value of the measure over 30 runs and second row gives the standard deviation (SD).

Table 1. Simulation parameters for MOBSO

Algorithm parameters used in simulation
Population size = 100
Maximum Chemotactic run count, $N_c = 4$
acceleration constant, $C_1 = 0.5$
inertia weight $w = 0.9$
Maximum number of iterations (generation) = 250
Maximum fitness function evaluations = 10,000
Scaling factor (μ) = 0.01
Stopping fraction of non-dominated solution (N_E/N) = 0.8
Minimum fitness function evaluations = 1,000
Number of runs to evaluate each benchmark problem was taken to be 30

Table 2. Comparative result from the γ and Δ metric

Problem	SCH	FON	KUR	ZDT1	ZDT2	ZDT3	ZDT4	ZDT5
γ metric								
NSGA II (Real Coded)								
Mean	0.003	0.002	0.0289	0.0334	0.0723	0.1145	0.5131	0.2965
SD	0.0	0.0	0.0	0.0048	0.0317	0.0079	0.1184	0.0131
MOBSO								
Mean	0.001	0.000	0.0006	0.0025	0.0015	0.0014	0.4208	0.0014
SD	0.0	0.0	0.0	0.0	0.0	0.0	0.0008	0.0

Table 3. Result of IGD metric on selected problems from CEC 2009 (special session) [11]

Problem	UF1	UF2	UF3	UF4	UF5	UF6	UF7
IGD metric							
MOBSO							
Mean	0.0265	0.0271	0.2316	0.0445	0.3732	0.6700	0.0784
SD	0.0035	0.0009	0.0822	0.0062	0.1202	0.1552	0.0020

4   Concluding Remarks

This study introduces a hybrid swarm strategy to simultaneously search for local and global solutions in the problem space. MOBSO method gives a diverse pareto-optimal front with reduced computational burden in limited number of fitness function evaluation. With problems having large search space, MOBSO can be implemented in O (kMN^2) complexity, with $k < 1$. It has been found to be competitive with benchmark algorithms in this domain. The result encourages further examination of the parameters of MOBSO along with its scalability and convergence, in order to make them problem space independent in future.

References

1. Deb, K.: Multi-Objective Optimization Using Evolutionary Algorithms. John Wiley & Sons, Chichester (2001)
2. Deb, K., Pratap, A., Agarwal, S., Meyarivan, T.: A fast and elitist multiobjective genetic algorithm: NSGA-II. IEEE Tansactions on Evolutionary Computation 6(2), 182–197 (2002)
3. Passino, K.M.: Biomimicry of Bacterial Foraging for Distributed Optimization and Control. IEEE Control Systems Magazine 22(3), 52–67 (2002)
4. Eberchart, R., Kennedy, J.: A new optimizer using particle swarm theory. In: Proceedings of International Symposium on Sym., pp. 39–43. Micro Machine and Human Science, Nagoya (1995)
5. Nebro, A.J., Durillo, J.J., Garcia-Nieto, J., Coello Coello, C.A., Luna, F., Alba., E.: SMPSO: A New PSO-based Meta-heuristic for Multi-objective Optimization. In: IEEE symposium on Computational intelligence in miulti-criteria decision-making, pp. 66–73 (2009)
6. Zhang, X.-h., Meng, H.-y., Jiao, L.-c.: Intelligent Particle Swarm Optimization in Multiobjective Optimization. In: The 2005 IEEE Congress on Evolutionary Computation, vol. 1, pp. 714–719 (2005)
7. Guzmań, M.A., et al.: A novel multi-objective optimization algorithm based on bacterial chemotaxis. Engineering Applications of Artificial Intelligence (2009), doi:10.1016/j.engappai.2009.09.010
8. Biswas, A., Dasgupta, S., Das, S., Abraham, A.: Synergy of PSO and bacterial foraging optimization: a comparative study on numerical benchmarks. In: Corchado, E., et al. (eds.) Second International Symposium on Hybrid Artificial Intelligent Systems (HAIS 2007), Innovations in Hybrid Intelligent Systems. Advances in Soft computing Series, vol. ASC 44, pp. 255–263. Springer, Germany (2007)
9. Dasgupta, S., Das, S., Abraham, A., Biswas, A.: Adaptive computational chemotaxis in bacterial foraging optimization: an analysis. Trans. Evol. Comp. 13(4), 919–941 (2009)
10. Zitzler, E., Deb, K., Thiele, L.: Comparison of Multiobjective Evolutionary Algorithms: Empirical Results. Evolutionary Computation 8(2), 173–195 (2000)
11. Li, C., Yang, S., Nguyen, T.T., Yu, E.L., Yao, X., Jin, Y., Beyer, H.G., Suganthan, P.: Benchmark Generator for CEC 2009 Competition on Dynamic Optimization. University of Leicester, University of Birmingham, Nanyang Technological University, Tech. Rep. (2008)

Multi-Objective Optimisation of Web Business Processes

Ashutosh Tiwari, Christopher Turner, Peter Ball, and Kostas Vergidis

Decision Engineering Centre, Cranfield University, Cranfield, Bedfordshire, UK, MK43 0AL
{a.tiwari,c.j.turner,p.d.ball}@cranfield.ac.uk

Abstract. This paper proposes an approach for the optimisation of web business processes using multi-objective evolutionary computing. Business process optimisation is considered as the problem of constructing feasible business process designs with optimum attribute values such as duration and cost. This optimisation framework involves the application of a series of Evolutionary Multi-objective Optimisation Algorithms (EMOAs) in an attempt to generate a series of diverse optimised business process designs for given requirements. The optimisation framework is tested to validate the framework's capability in capturing, composing and optimising business process designs constituted of web services. The results from the web business process optimisation scenario, featured in this paper, demonstrate that the framework can identify business process designs with optimised attribute values.

Keywords: Multi-objective optimisation, Business Process, EMOA, Web services.

1 Introduction

The design and management of business processes is a key factor for companies to effectively compete in today's volatile business environment. By focusing on the optimisation and continuous improvement of business processes, organisations can establish a solid competitive advantage by reducing cost, improving quality and efficiency, and enabling adaptation to changing requirements. This paper discusses how a business process optimisation framework (bpo^F), based on Evolutionary Multi-objective Optimisation Algorithms (EMOAs), can be applied to the real-world optimisation of actual web business processes. The details of bpo^F are presented in [1]. The proposed optimisation framework is tested with a web business process scenario to validate the framework's capability in capturing, composing and optimising process designs constituted of web services.

According to Davenport and Short [2] a business process is a set of logically related tasks performed to achieve a defined business outcome. A *business process* is perceived as a collective set of tasks that when properly connected perform a business operation. A web business process is therefore a process composed of web services, where the function each process task performs is enacted by an individual web service. A web service can be thought of as 'a discretely defined set of contiguous and autonomous business or technical functionality implemented over a network' [1]. A further discussion of the literature for business process optimisation can be found in

K. Deb et al. (Eds.): SEAL 2010, LNCS 6457, pp. 573–577, 2010.

[3]. There are limited approaches for optimising business processes. Such approaches can only deal with simple sequential business processes within a single objective optimisation framework [4]. The work of Ko et al. [5] highlights a number of examples of soft computing use in the business process arena.

2   Quantitative Representation of Business Process Designs

The business process optimisation framework (bpo^F), applied here, is described in detail in [1]. The main elements involved are the *tasks* and *resources* of the business process. The *attributes* of the tasks and the process are also taken into consideration in order to provide the capability of evaluating a business process design. Finally, the *patterns* that interconnect the tasks are also included. As task attributes, we consider measurable (quantitative) characteristics of the tasks. Examples involve task cost and task duration. The task attributes can be mapped to the corresponding process attributes (e.g. process cost) using a suitable aggregate function. We consider the inputs and outputs of a task as task resources. The task resources connect the various tasks based on their inputs and outputs. Also, the resources provide the requirements for a process design in the form of required *process input* and *process output*.

The chromosome of each individual is made up of task numbers which are then organised into a specific sequence by the Process Composition Algorithm (PCA). It is the PCA that creates the process graphs out of the tasks provided in an individual. Both crossover and mutation operators are used in the manipulation of individuals. The task library can be used to help repair an individual by the addition of extra tasks as that individual is built into a process by the PCA. The general parameter settings for the EMOAs used are shown in Table 1. The main steps of the proposed business process optimisation framework are detailed in [6].

3   Web Business Process Optimisation

The scenario featured in this paper describes an automated sales forecasting process (scenario provided by [7]). This process is considered as *semi-automated* as it involves the interaction of some applications but it is not streamlined and still requires human involvement in the act of generating and visualising the requested forecasts. In this process each of the tasks (marked 1- 5 in Fig. 1) is a web service carrying out a unique function, such as generating graphs etc. The aim of this scenario is to show the optimisation potential of the framework. Fig. 1 shows the generic business process design for the scenario. It involves two input resources: (a) company name and (b) market update request. The first resource is necessary for the web service to extract relevant data for the specified company. The second resource is a request for a market update that needs to be considered for the sales forecast.

3.1 Library of Tasks (Web Services)

Having sketched the initial business process design, we can compile the library of alternative web services based on the main steps of the generic process design.

Table 1. General parameters

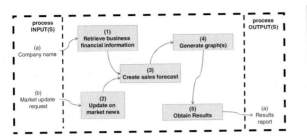

Population	250
Generations	25,000
Crossover probability	0.8
Mutation probability	0.2
Objectives	2

Fig. 1. Initial business process design for sales forecasting

Relevant research on the selected on-line libraries of web services resulted in a selection of 20 web services from different providers that can potentially contribute to implementing the scenario (an example table of web services can be found in [1] p. 213). Each task in the scenario has input and output resources (an example of the table of resources can be found in [1] p. 212). The proposed optimisation framework is tested for two objectives; Service Delivery Price (SDP) (specifies the amount of money the service customer has to pay for the consumption of distinct service volumes, i.e. the cost to use the service) and Service Fulfilment Target (SFT) (specifies the service provider's promise of effective and seamless delivery of the defined benefits to any authorised service consumer). SFT is expressed as the promised maximum number of successful individual service deliveries with respect to the total counts of individual service deliveries.

Having gathered all the necessary information (library of web services and input/output resources for each web service), the problem parameters can be defined based on the business process problem formulation (an example of parameter values can be found in [6]). This results in a complete scenario ready to be tested within the proposed optimisation framework.

3.2 Generate the Scenario's Search Space

The first step towards visualising the results is to generate the scenario's search space by producing 1000 random feasible process designs. The initial business process design in Fig. 1 involves 5 main steps. A design with less than 5 tasks shows that there is a web service that consolidates two or more tasks. A design with more tasks shows that one step requires two or more web services to be implemented. The search space for this scenario is shown in Fig. 2. The search space consists of five different regions, each corresponding to a group of designs with same number of tasks (4, 5, 6, 7 or 8).

3.3 Test the Scenario with BPO^F

The challenge for the EMOAs in the framework is to identify non-dominated (optimised) solutions in each of the regions in the search space. Fig. 2 shows the results obtained by combining the outputs of four EMOA algorithms (the four algorithms used were NSGA2 (Non-dominated Sorting Genetic Algorithm 2), PAES (Pareto Archived Evolutionary Strategy), PESA2 (Pareto Envelope-based Selection Algorithm 2) and SPEA2 (Strength Pareto Evolutionary Algorithm 2)). All the algorithms identify process designs near the Pareto front of each of the five regions of the search space. This is a strong indicator of the performance of the algorithms and the confidence in the generated designs being near optimal. Fig. 2 demonstrates two optimised business process designs (a) and (b), one with 4 tasks and one with 6 tasks (taken from the Pareto fronts indicated in Fig. 2). Each of these designs belongs to a different

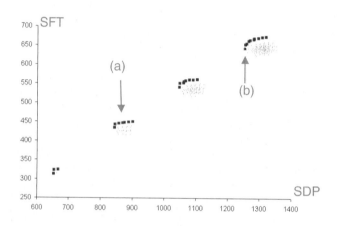

Fig. 2. Search space and EMOA results for the scenario

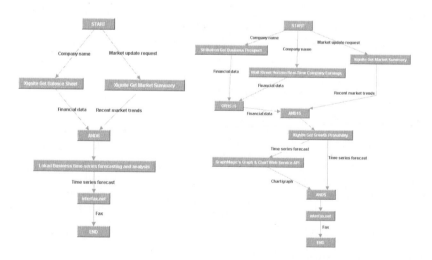

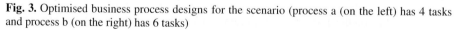

Fig. 3. Optimised business process designs for the scenario (process a (on the left) has 4 tasks and process b (on the right) has 6 tasks)

island based on its size. The arrows in Fig. 2 indicate the island from where each design in Fig. 3 originates. Fig.3. (a) shows a business process design with one of the generic steps missing. The forecasting results are not plotted into a graph but they are just faxed back to the requestor. The framework reduces cost in this instance. Therefore, in a semi-automated process the framework can take 'initiative' and alter the generic design provided that the process input and output requirements are still satisfied. Fig. 3(b) is composed of 6 services and involves two tasks for obtaining the company's financial data either from selecting one or both (OR is not exclusive choice). This provides better confidence in terms of accuracy of the data obtained and improved reliability of the process execution itself.

4   Conclusions

The scenario featured in this paper demonstrated how an optimised business process can be automatically created by the optimisation framework using web services. The framework identified the optimal designs for all the available process sizes. The generated designs select and incorporate different web services arranged with the appropriate process patterns so that (i) the process input and output requirements are satisfied and (ii) the attribute values are optimised. Results from the real-life scenario, featured in this paper, demonstrate that the optimisation framework can identify business process designs with optimised attribute values.

References

[1] Vergidis, K.: Business Process Optimisation Using an Evolutionary Multi-objective Framework, PhD Thesis, School of Applied Sciences, Cranfield University, Cranfield, Bedfordshire, UK (2008)
[2] Davenport, T.H., Short, J.E.: The New Industrial Engineering: Information Technology and Business Process Redesign. Sloan Management Review, Summer 1990, 11–27 (1990)
[3] Vergidis, K., Tiwari, A., Majeed, B.: Business Process Analysis and Optimization: Beyond Reengineering. IEEE Transactions on Applications and Reviews on Systems, Man, and Cybernetics, Part C 38(1), 69–82 (2008)
[4] Hofacker, I., Vetschera, R.: Algorithmic Approaches to Business Process Design. Computers & Operations Research 28, 1253–1275 (2001)
[5] Ko, M., Tiwari, A., Mehnen, J.: A Review of Soft Computing Applications in Supply Chain Management. Applied Soft Computing 10(3), 661–674 (2010)
[6] Tiwari, A., Vergidis, K., Turner, C.J.: Evolutionary Multi-Objective Optimisation of Business Processes. In: Gao, X.Z., Gaspar-Cunha, A., Köppen, M., Schaefer, G., Wang, J. (eds.) Advances in Intelligent and Soft Computing: Soft Computing in Industrial Applications. Springer, Heidelberg (In press 2010)
[7] Grigori, D., Casati, F., Castellanos, M., Dayal, U., Sayal, M., Shan, M.-C.: Business Process Intelligence. Computers in Industry 53, 321–343 (2004)

Multi Objective Optimization of Planetary Gear Train

Vipin K. Tripathi and Hiten M. Chauhan

Department of Mechaninical Engineering, College of Engineering, Pune – 411005
vkt.mech@coep.ac.in, hitenchauhan21@gmail.com

Abstract. In present work, multi-objective optimization of multi-stage plane-tary gear train is done. Optimization of multi-stage speed reducer is difficult due to involvement of integer variables. Minimizing surface fatigue life factor and minimization of volume of gear box are two conflicting objective functions under consideration. Two methods, one classical (SQP) and other non-traditional (NSGA-II) have been used for analysis to satisfy strength and other geometric criteria. Previous work is concentrated on optimization of spur or helical gears. This work is an extension to earlier work in a sense that planetary gear train with reduced speed involves more geometric constraints.

Keywords: Multi-objective optimization, planetary gear train, Genetic Algorithm, SQP.

1 Introduction

Optimization of multi-stage speed reducer is a complex problem as it involves design variables which are integer (number of teeth), discrete (normal module) and real (gear width).

A few studies done by Deb and Jain [1] and Shukla et. al. [2] demonstrated the use of a multi-objective evolutionary algorithm for multi-speed gear box design problem involving mixed discrete and real-valued parameters and more than one objective.

In the present work, minimization of surface fatigue life factor of gears and mini-mization of volume of gear box is done simultaneously using a traditional Sequential Quadratic Programming (SQP) optimization technique and a non traditional technique Non-dominated Sorting Genetic Algorithm - II (NSGA II).

2 Problem Formulation

In the study a generic mathematical model for a two stage planetary gear train for different gear train applications is developed [3]. Input parameters and other gear design parameters describe in current section are for grinder mixer application.

2.1 Input Parameters

These are T_o output torque 50 N-m, G over all reduction 49 and maximum volume V is $90 \times 90 \times 90$ mm^3.

K. Deb et al. (Eds.): SEAL 2010, LNCS 6457, pp. 578–582, 2010.
© Springer-Verlag Berlin Heidelberg 2010

2.2 Gear Design Parameters

Design parameters and constants necessary for the formulation of constraint equations and objective functions are provided in the following Table 1. K_s, C_H, Z_N, Y_r, K_B, S_F, $Z_{N'}$ and S_H are size factor, hardness ratio factor, stress cycle factor for contact, reliability factor, rim thickness factor, bending factor of safety,   stress cycle factor for bending and bending factor of safety respectively and all are taken as 1.

Table 1. Gear design parameters

Parameter	Description	Values	Unit
Z_E	Elastic coefficient	232	$\sqrt{N/mm^2}$
$S_{b'}$	Allowable bending stress number	1172	N/mm^2
$S_{W'}$	Allowable contact stress number	2600	N/mm^2
μ	Poisson's ratio	0.28	None
K_o	Overload factor	1.25	None
K_V	Dynamic factor	1.81	None
K_H	Load distribution factor	1.2765	None
Y_J	Geometry factor	0.2	None
E	Young's modulus	155000	N/mm^2

2.3 Gear Design Variables

Upper and a lower bound on the design variables are as follows,

$$x = [m \quad t_{s1} \quad t_{p1} \quad n_1 \quad b_1 \quad \alpha \quad t_{s2} \quad t_{p2} \quad n_2 \quad b_2]$$

$$x^{(L)} = [0.1 \quad 6 \quad 24 \quad 1 \quad 5 \quad 20 \quad 6 \quad 24 \quad 1 \quad 5]$$

$$x^{(U)} = [2 \quad 23 \quad 63 \quad 3 \quad 30 \quad 20 \quad 23 \quad 63 \quad 3 \quad 30]$$

where m is module, α is pressure angle, t_{si} and t_{pi} are number of teeth on sun and planet gear respectively, b_i face width, n_i are and number of planets, i is index denotes first and second stages.

2.4 Constraints

$$g_1(x) = (t_{p1} + t_{s1})\sin\alpha - \sqrt{(t_{p1} + 2)^2 - (t_{p1}\cos\alpha)^2} \tag{2.1}$$

$$g_2(x) = (t_{p2} + t_{s2})\sin\alpha - \sqrt{(t_{p2} + 2)^2 - (t_{p2}\cos\alpha)^2} \tag{2.2}$$

$$g_3(x) = b_1 - 5m \tag{2.3}$$

$$g_4(x) = b_2 - 5m \tag{2.4}$$

$$g_5(x) = 20m - b_1 \tag{2.5}$$

$$g_6(x) = 20m - b_2 \tag{2.6}$$

$$g_7(x) = 4\left(1 + \frac{t_{p1}}{t_{s1}}\right)\left(1 + \frac{t_{p2}}{t_{s2}}\right) - 49 \tag{2.7}$$

$$g_8(x) = 90 - m\left(t_{s1} + 2t_{p1} + 4.25\right) \tag{2.8}$$

$$h_1(x) = t_{p2} + t_{s2} - 2t_{p1} - t_{s1} \tag{2.9}$$

$$h_2(x) = \frac{2\left(t_{p1} + t_{s1}\right)}{n_1} - Integer\left(\frac{2\left(t_{p1} + t_{s1}\right)}{n_1}\right) \tag{2.10}$$

$$h_3(x) = \frac{2\left(t_{p2} + t_{s2}\right)}{n_2} - Integer\left(\frac{2\left(t_{p2} + t_{s2}\right)}{n_2}\right) \tag{2.11}$$

$$g_9(x) = (t_{s1} + t_{p1})\sin\left(\frac{180}{n_1}\right) - (t_{p1} + 2) \tag{2.12}$$

$$g_{10}(x) = (t_{s2} + t_{p2})\sin\left(\frac{180}{n_2}\right) - (t_{p2} + 2) \tag{2.13}$$

$$g_{11}(x) = 1 - 1232.12\left[\frac{t_{s2}}{4m^2 b_1 n_1 \left(t_{s1} + t_{p1}\right)\left(t_{s2} + t_{p2}\right)}\right] \tag{2.14}$$

$$g_{12}(x) = 1 - 1232.12\left[\frac{1}{2m^2 b_2 n_2 \left(t_{s2} + t_{p2}\right)}\right] \tag{2.15}$$

$$g_{13}(x) = 1 - \frac{67.82}{m}\sqrt{\left[\frac{t_{s2}}{4\left(t_{s2} + t_{p2}\right)b_1 n_1 t_{s1} t_{p1} \sin\alpha\cos\alpha}\right]} \tag{2.16}$$

$$g_{14}(x) = 1 - \frac{67.82}{m}\sqrt{\left[\frac{t_{s2}}{4\left(t_{s2} + t_{p2}\right)\left(t_{s1} + 2t_{p1}\right)b_1 n_1 t_{p1} \sin\alpha\cos\alpha}\right]} \tag{2.17}$$

$$g_{15}(x) = 1 - \frac{67.82}{m}\sqrt{\left[\frac{1}{2b_2 n_2 t_{s2} t_{p2} \sin\alpha\cos\alpha}\right]} \tag{2.18}$$

$$g_{16}(x) = 1 - \frac{67.82}{m}\sqrt{\left[\frac{1}{2\left(t_{s2} + 2t_{p2}\right)b_2 n_2 t_{p2} \sin\alpha\cos\alpha}\right]} \tag{2.19}$$

Constraints in Equations (2.1) – (2.2) are   for interference, Equations (2.3) – (2.6) are for face width, Equation (2.7) for overall gear ratio and (2.8) for maximum diameter of ring gear, Equations (2.9) – (2.13) for planetary conditions, Equation (2.14) – (2.15) for gear tooth bending and Equations (2.16) – (2.19) for contact stresses between sun-planet and planet-ring pair of both stages.

2.5 Objective Function Formulation

Volume of sun and planets gear is calculated from pitch circle diameter:

$$F_o(x) = \frac{\pi}{4} m^2 [(t_{s1}^2 + n_1 t_{p1}^2 + 8.5(t_{s1} + 2t_{p1}) + 18.0625)b_1 +$$
$$(t_{s2}^2 + n_2 t_{p2}^2 + 8.5(t_{s2} + 2t_{p2}) + 18.0625)b_2] \qquad (2.20)$$

The study seeks to minimize   volume $f_1(x)$, and surface fatigue life factor $f_2(x)$ in range from 0.8 to 1.2 in step of 0.05 for specific number of cycles.

This problem can be formulated as follows

Minimize $f_1(x) = F_o(x)$, $f_2(x) = (1 - g_{15}(x) \pm 0.2)$

Subjected to: $g_i(x) \geq 0$ $i = 1, 2..., 16$

$h_j(x) = 0$ $j = 1, 2..., 3$

$x_k^{(L)} \leq x_k \leq x_k^{(u)}$ $k = 1, 2,......, 10$

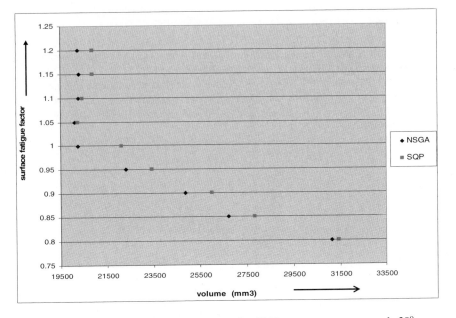

Fig. 1. Surface fatigue life versus volume for 50 N-m torque, pressure angle 20°

3 Results and Discussion

The NSGA-II and SQP methods have been used to compute the results. The NSGA-II [4] code is obtained from the KanGAL, whereas SQP codes available with Hyper study module of Altair's HyperWorks 9.0 is used.

Material Designation - FN-0208-155HT. Parameters are given in Table 1. The volume of the present gear box used by Whirlpool Corporation USA having helical and worm gears is 35092 mm^3, whereas planetary arrangement gives volume of 21660 mm^3 for surface fatigue factor 1, as shown in Fig. 1.

The NSGA-II evaluates 5000 functions, whereas SQP evaluates 31 functions to give optimum front. Modified SQP used here takes 72 minutes, while NSGA –II takes 80 minutes to give optimum solutions. But NSGA-II gives better optimum values than SQP.

4 Conclusion

The proposed work uses multi-objective optimization method for gear train design. The study makes uses of a traditional and a non-traditional algorithm to solve multi-objective optimization problem. NSGA-II gives better results than SQP and it is suitable for mixed integer type of optimization problem. Comparison of Performance of NSGA-II and SQP shows that even though time taken by both algorithms is close, but NSGA-II after more function evaluations gives better results. It is clear from the results that replacing helical and worm wheel arrangement with planetary not only makes system compact but also increases efficiency.

References

1. Deb, K., Jain, S.: Multi-speed gearbox design using multi objective evolutionary algorithms. Journal of Mechanical Design 125, 609–619 (2003)
2. Shukla, A., Thompson, D.F., Gupta, S.: Tradeoff analysis in minimum volume design of multi-stage spur gear reduction units. Mechanism and Machine Theory 35, 609–627 (2000)
3. Shigley, J.E., Mischke, C.R.: Mechanical Engineering Design, 8th edn. McGraw-Hill, New York (2008)
4. Deb, K.: Multi objective optimization using evolutionary algorithms. John Wiley and Sons, Chichester (2001)

Multi-Objective Control Systems Design with Criteria Reduction

Piotr Woźniak

Technical University of Łódź, Institute of Automatic Control, 18/22 Stefanowskiego St.,
90-924 Łódź, Poland
pwozniak@p.lodz.pl

Abstract. Control systems design may be based on many criteria. These optimization problems are nonconvex, therefore evolutionary multi-objective optimization algorithms (EMOA) are methods of choice. In engineering design problems it is desirable to find the one solution only as in single criterion optimisation. We describe a new method based on reduction of objectives while keeping relevant Pareto sets changes bounded. In the illustrative control design six objectives from optimal control, mixed norm robust optimization and standard control methods are reduced to three, which enables visualisation of the Pareto front.

Keywords: Multi-objective optimization, Pareto set analysis, genetic algorithms, criteria reduction, NSGA II, computer-aided control system design.

1 Introduction

Evolutionary Multi-Objective Algorithms (EMOA) have been successfully applied to solve control problems, when a number of design objectives are conflicting. One of the first results on EMOA application to control system design were presented by Fonseca [1]. Herreros et al. [2] presented an approach for adjusting the parameters of a PID controller. Takahashi et al. [3] showed that using EMOA one could find better solutions comparing to one obtained using the Linear Matrix Inequalities approach which is standard method for the H_2/H_∞ multi-objective robust control.

Many research projects show that EMOA are not computationally efficient in finding sets of solutions on problems having a large number of objectives. Also lack of visualization of a high dimensional front deteriorates decision-making process. These reasons, outlined in numerous works (e.g. [4],[5],[6]), explain why in high-dimensional multi-objective optimization problems, a dimension reduction of the objective space can be beneficial both for search and solution selection. This is especially convincing in engineering problems [5].

Practical engineering problems often involve many objective functions. Search for solution methods for difficult problems may lead to acceptance of some trade-offs between precision deterioration and optimization efficiency increase. Simplifications may be accepted after the initial analysis of the complete, high-dimensional, optimization problem if the formulation of the problem to be solved becomes easier.

K. Deb et al. (Eds.): SEAL 2010, LNCS 6457, pp. 583–587, 2010.
© Springer-Verlag Berlin Heidelberg 2010

The overall idea in dimension reduction in EMOA is to find justifiable ways to eliminate some of the objectives from consideration. One has to ensure that has to ensure that omission will have only minor effects on the search dynamics and the final solutions. Currently, there are mainly three approaches to deal with objectives reduction. Deb and Saxena [5] proposed method based on principal component analysis (PCA). The approach was further extended by Authors to non-linear case [6]. Jaimes et.al. [7] proposed method to reduce the objectives using feature selection technique.

One may consider additional analysis of clusters approximation of the Pareto set what may lead to reduction of the decision space. Hereafter we propose a method to identify the most non-conflicting objective in case the correlation matrix in analysis of the whole Pareto front does not suffice. The central part of the strategy is based on results of combined analysis of the Pareto front and Pareto set properties.

The exposition is organized as follows. In Section 2, the control system design as a multi-objective optimization problem is defined in high-dimensional objective space. In Section 3 we describe the objective reduction procedure and its validation is presented in Section 4. Section 5 concludes the paper.

2 Control Design as a Multi-Objective Optimization Problem

For many real-world control systems it is desirable to meet all specified goals using the controllers with simple structures like a proportional-integral (PI). In practice, these controllers are commonly tuned based on classical single norm-based performance criteria, but weight selection is a very hard task. In modern control engineering, a design problem is usually formulated as a multi-objective optimization problem with the standard notation

$$\min_{x \in X}\{f_1(x), \cdots, f_k(x)\}. \tag{1}$$

where $x \in R^n$ denotes a vector of decision variables (arguments), $X \subseteq R^n$ is the set of feasible arguments, and each $f_i : X \to R, i = 1, \cdots, k$, is a real-valued function

The most common performance measures are norms of some characteristic closed-loop properties of the control system. Therefore, different and potentially conflicting specifications, such as: good tracking for various input signals, low energy of the control signals, simultaneous attenuation of several types of disturbances, robust stability. Thanks to the flexibility of the EMOA methods all three groups may be treated as elements of one objective vector.

3 Objective Reduction by Clustering in the Decision Space

Unsupervised learning and clustering lead to extraction of information from unlabeled samples. Cluster analysis divides data into groups (clusters) such that similar data objects belong to the same cluster and dissimilar data objects to different clusters.

In this paper the *k-means* clustering algorithm is used for partitioning (or clustering) N data points approximating Pareto set into k disjoint subsets S_j containing N_j data points so as to minimize the sum-of-squares criterion

$$J = \sum_{m=1}^{M} \sum_{n \in S_j} |x_n - \mu_m|^2 , \qquad (2)$$

where x_n is a vector representing the n-th data point and m is the geometric centroid of the data points in S_j . The algorithm iteratively minimizes the criterion by updating the mean vector until there is no change in an update.

After discovering separate clusters one may decide to confine the search to the neighbourhood of the cluster. Introducing this step may facilitate final selection of the unique solution of the multi-objective problem.

4   An Illustrating Example – Dynamic Control System Design

The proportional-integral (PI) is still the most common controller [8]. The objectives of the feedback control system under consideration belong to three groups of criteria :

— *command tracking* : rise time t_n, settling time t_s, overshoot M [8],
— *functional* : Integral of Time-weighted Absolute Error (ITAE),
— *disturbance rejection* : noise attenuation , disturbance attenuation.

The first group is representative of the standard control system design approach, the second to the optimal control problem, while disturbance rejection is typical of the mixed-norm optimization problem.

The considered feedback control system is shown in Fig.1.

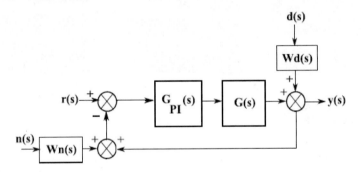

Fig. 1. Considered control system

Let $G(s) = (1+s)^{-1}$ be the plant transfer function, $W_n(s) = (s+1)^{-1}$ and $W_d(s) = 1$ are simple filters. The decision variables $x=[K_P, T_i]$ for this problem are the parameters of the PI controller, which has the following transfer function

$$G_{PI}(s) = K_P(1 + (T_i s)^{-1}). \qquad (3)$$

For the regulation problem ($r(s) \equiv 0$) the closed-loop output is described as

$$y(s) = -G_{yn}(s)W_n(s)n(s) + G_{yd}(s)W_d(s)d(s) , \qquad (4)$$

with $G_{yn}(s) = \dfrac{G_{PI}(s)G(s)}{1+G_{PI}(s)G(s)}$, $G_{yd}(s) = \dfrac{1}{1+G_{PI}(s)G(s)}$.

The elements of the six-criteria objectives (5) are based on the step response of the system, with inputs: the reference $\mathbf{r(s)}$, the disturbance $\mathbf{d(s)}$, and the noise $\mathbf{n(s)}$

$$\min_{x \in X} \left\{ \left\| G_{yn}(s) \right\|_2, \left\| G_{yd}(s) \right\|_2, t_s, t_n, M, \text{ITAE} \right\}. \tag{5}$$

The stability of the system was guaranteed by search over the set of feasible decision vectors and the search space starting from $x_1 \in [10, 100]$, $x_2 \in [10, 100]$. Optimization of all problems was performed using the conventional NSGA-II algorithm.

4.1 Cluster Analysis in the Decision Space

The multi-objective optimization of the control system design may be analyzed in the decision space. It is done using the k-means clustering which aims to partition n observations into k clusters in which each observation belongs to the cluster with the nearest mean. The results may be visualised in the form of silhouette plots [9], which displays a measure of how close each point in one cluster is to points in the neighbouring clusters. This measure ranges from +1 for points that are distant from neighbouring clusters, through 0, indicating points that are not distinctly in one cluster or another, to -1, indicating points that are probably assigned to the wrong cluster. The results for six-, five- and four-criteria optimization are shown in Fig.2.

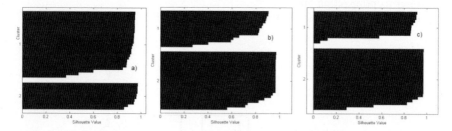

Fig. 2. Silhouette plots of Pareto fronts' clusters of a) six-,b) five-, and c) four-criteria problems

One may observe presence of two well separated clusters in Fig.2a. This enabled selection of only part from the decision space (denoted as cluster 1). This part of decision space was considered in the five criteria problem. The resultant Pareto set again formed two well separated clusters (Fig.2b). The reduction procedure was repeated for reduction to four-criteria problem (in the neighbourhood of the cluster 2 from Fig.2b) and three- criteria problem.

The final set of objectives consists of $\left\| G_{yn}(s) \right\|_2$, $\left\| G_{dn}(s) \right\|_2$, and *ITAE*. The Pareto optimal solutions are searched locally $x_1 \in [10, 20]$, $x_2 \in [10, 35]$ i.e. in the neighbourhood of the cluster 2 from Fig.2c.

The Pareto sets are shown in Fig.3a, while the Pareto front resulting from the three-criteria optimisation problem is shown in Fig.3b. The final selection of a unique solution of the three-objective problem remains open. At this stage the selection of the final solution is not discussed, but the presented approach reduced sixfold the decision subspaces to $\left\| G_{yn}(s) \right\|_2 \in [0.162, 0.213]$ and $\| G_{dn}(s) \|_2 \in [2.133, 2.938]$.

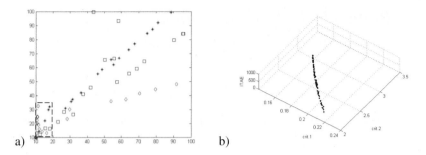

Fig. 3. a) Pareto sets for: six-(*), five-(□) and four-(◊) criteria problems with the reduced cluster (dashed rectangle) b) the Pareto front for the reduced three-criteria problem

The potential further research includes addressing the problem of robustness of the proposed algorithm.

5 Conclusions

The paper presents the application of the procedure to identify the most redundant objectives of EMOA so that one can obtain a lower-dimensional problem. To facilitate dimension reduction we proposed use of the cluster analysis of the Pareto set in the decision space. The illustrative control system design problem proves the effectiveness of this approach.

References

1. Fonseca, C.: Multiobjective Genetic Algorithms with Application to Control Engineering Problems. Ph.D. thesis, University of Sheffield (1995)
2. Herreros, A., Baeyens, E., Peran, J.: MRCD: a genetic algorithm for multiobjective robust control design. Engineering Applications of Artificial Intelligence 15, 285–301 (2002)
3. Takahashi, R., Palhares, R., Dutra, D., Goncalves, L.: Estimation of Pareto sets in the mixed H_2/H_∞ control problem. Int. Journal of Systems Science 35, 55–67 (2004)
4. Woźniak, P.: Dimensionality Reduction in Evolutionary Multiobjective Design: Case Study. In: Conference on Genetic and Evolutionary Computation, p. 913. ACM Press, New York (2007)
5. Deb, K., Saxena, D.: Searching for pareto-optimal solutions through dimensionality reduction for certain large-dimensional multi-objective optimization problems. In: 2006 IEEE Congress on Computational Intelligence, pp. 3352–3360. IEEE Press, Los Alamitos (2006)
6. Saxena, D., Deb, K.: Non-linear dimensionality reduction procedures for certain large-dimensional multi-objective optimization problems: employing correntropy and a novel maximum variance unfolding. In: Obayashi, S., Deb, K., Poloni, C., Hiroyasu, T., Murata, T. (eds.) EMO 2007. LNCS, vol. 4403, pp. 772–787. Springer, Heidelberg (2007)
7. Jaimes, A.L., Coello, C.A.C., Chakraborty, D.: Objective Reduction Using a feature Selection Technique. In: Conference on Genetic and Evolutionary Computation, pp. 673–680. ACM Press, New York (2008)
8. Astrom, K.J., Murray, R.M.: Feedback Systems: An Introduction for Scientists and Engineers. Princeton University Press, Princeton (2008)
9. MATLAB, The MathWorks, Inc., Natick, MA, http://www.mathworks.com

Probabilistic Based Evolutionary Optimizers in Bi-objective Travelling Salesman Problem

Vui Ann Shim, Kay Chen Tan, and Jun Yong Chia

Department of Electrical and Computer Engineering, National University of Singapore,
4 Engineering Drive 3, 117576, Singapore
{g0800438,eletankc,g0900313}@nus.edu.sg

Abstract. This paper studies the probabilistic based evolutionary algorithms in dealing with bi-objective travelling salesman problem. Multi-objective restricted Boltzmann machine and univariate marginal distribution algorithm in binary representation are modified into permutation based representation. Each city is represented by an integer number and the probability distributions of the cities are constructed by running the modeling approach. A refinement operator and a local exploitation operator are proposed in this work. The probabilistic based evolutionary optimizers are subsequently combined with genetic based evolutionary optimizer to complement the limitations of both algorithms.

Keywords: Estimation of distribution algorithm, evolutionary multi-objective optimization, restricted Boltzmann machine, travelling salesman problem.

1 Introduction

Travelling salesman problem (TSP) is one of the famous permutation based combinatorial optimization problems [1]. The problem aims to minimize the total distance travelled, in which each city is visited exactly once and the salesman must return to the starting depot. The adaptation of TSP into multi-objective framework (MOTSP) is another promising area which can be explored [2-3]. In the multi-objective formulation, the aim is to simultaneously optimize several conflicting objectives, such as shortest travelling distance, minimum time, minimum cost and lowest risk [4].

Probabilistic based evolutionary algorithms (EAs), commonly known as Estimation of distribution algorithms (EDAs) [1], mimic the biological evolutionary principle to guide the search. The primary difference between EDAs and genetic based EA (specifically genetic algorithm) is that no genetic operators (crossover and mutation) are implemented in EDAs. The reproduction is based on building of probabilistic model from the selected solutions and sampling from the constructed model.

Several researches have been carried out to study the single objective permutation based problems (specifically TSP) by using EDAs [1]. However, there is no research which studies multi-objective permutation based problems (specifically MOTSP) by using EDAs. In this paper, binary representation of multi-objective univariate marginal probability algorithm (MOUMDA) and multi-objective restricted Boltzmann machine (MORBM) [5] are adapted into a permutation based representation to solve Bi-TSP. The two objectives being considered are travelling distance and travelling cost. Permutation

K. Deb et al. (Eds.): SEAL 2010, LNCS 6457, pp. 588–592, 2010.

refinement operator is proposed to refine the cities in a chromosome to guarantee that no city is repeated. A local exploitation operator is also presented to enhance the search capability of the algorithms. Probabilistic based EAs are subsequently combined with genetic based EA to increase the spread of the trade-off solutions.

2   Algorithms' Framework

2.1   Modeling and Reproduction

Two modeling approaches are considered in this paper. UMDA [6] learns the distributions of the cities without considering their linkage dependencies with other cities. In the modeling, a nxn probability matrix which models the distribution of the cities is constructed, according to the following equation.

$$\text{Prob}_g(C_{i,j}) = \frac{\sum_{k=1}^{pop} \delta_k(C_{i,j}=c_i)+{}^1/_n}{pop+{}^{pop}/_n} \quad \text{where} \quad \delta_k(C_{i,j}=c_i) = \begin{cases} 1 & \text{if } C_{i,j} = c_i \\ 0 & \text{otherwise} \end{cases}.$$

$\text{Prob}_g(C_{i,j})$ is the marginal probability of city i at the jth place of the chromosome at generation g, c_i is the city i, pop is the population size, and n is the number of cities.

RBM is energy based neural network [5] which learns the distribution of the input stimuli through unsupervised learning. The probabilistic model is constructed as

$$\text{Prob}_g(C_{i,j}) = \frac{\sum_{k=1}^{pop} P_k(C_{i,j}=c_i)+{}^{Z_j}/_{(pop*numc)}}{Z_j+{}^{Z_j}/_{pop}} \quad \text{where}$$

$$P_k(C_{i,j}=c_i) = \begin{cases} \sum_{h=1}^{H} e^{-E(v=c_i,h)} & \text{if } C_{i,j}=c_i \\ 0 & \text{otherwise} \end{cases}, \quad Z_j = \sum_{x,y} e^{-E(x,y)}$$

$$E(v,h) = \sum_i \sum_j v_i h_j w_{ij} - \sum_i v_i b_i - \sum_j h_j b_j$$

where v is the input state and h is the hidden state of the network, w and b is the synaptic weights and biases, Z is the normalizing constant, and E is the energy value of the network. The simple probabilistic sampling mechanism [5] is applied to generate offspring based on the built probabilistic model.

For evolutionary optimizer [7], the variation operators are based on crossover and mutation. Single point crossover is used to create the offspring. This operator randomly selects the position to cut the chromosomes for crossing over between two parents. This single point crossover is equivalent to route inter-crossing. After which, mutation is carried out by swapping between two randomly selected alleles within the chromosome. This genetic perturbation provides exploitation capability to the optimizer to search within fitter region.

2.2   Refinement Operator and Local Exploitation Operator

After reproduction some cities may not be visited, while others are visited more than once. To overcome this problem, a refinement operator is proposed. Firstly, the repeated and unvisited cities in a chromosome are detected. An insertion is carried out by inserting the unvisited cities to the position of the repeated cities. The average distance and cost (normalized) between the adjacent cities in the permutation are calculated and served as the main criteria for insertion. In order to enhance the search, a local exploitation operator is also proposed. The process flow of this operator is as follow. Firstly, a set of k number of cities to be relocated is randomly selected. Distances and costs among all the selected cities are calculated. The permutation of the cities is re-determined according to the distances and costs information. Due to optimization of two objective functions in this paper, three types of relocation criteria are considered. The first criterion determines the permutation based on the shortest distance in first objective function. Second considers the lowest cost in second objective function, and last criterion computes the normalized average distance and cost from both objective functions.

2.3   Overall Framework

The algorithmic process flow of the proposed algorithm is shown in Fig. 1. Firstly, initialization is performed by randomly generating permutation in integer number. Then, evaluation is carried out. Based on the objective domain, a new fitness is assigned to each solution based on Pareto ranking and crowding distance [7]. Binary tournament selection is applied to choose the promising solutions. The selected solutions will undergo modeling based on univariate or RBM approach. Based on the constructed model, probabilistic sampling is carried out to produce n offspring, where n is the population size. Then, the refinement operator is performed. To further improve the routing, a local exploitation is incorporated. The local exploitation will only be performed if the generated random value is smaller than a predefined local exploitation rate. After this, an archive is created to store the promising solutions found. The same procedure is iterated until the stopping criterion is met. The same process flow is implemented in genetic based EA. As for combination of genetic and probabilistic based EAs, the algorithms starts with probabilistic based optimizers and alternated with genetic based optimizer every 500 generation.

3   Simulation Results and Discussions

The experimental settings are shown in Table 1. TSP with two objectives is studied. The information of the distance and cost among cities is randomly generated in the range of [0, 1000] as done in [8]. Two performance metrics namely inverter generational distance (IGD) and non-domination ratio (NR) are utilized. IGD measures the proximity as well as the spread of the optimal solutions to the evolved solutions. NR measures the non-dominated ratio of solutions in one algorithm compared to other algorithms. The approximate optimal solutions set is formed from all the non-dominated solutions found in all algorithms [8]. EA refer to genetic based EA. UMEA and RBMEA is the combination of UMDA and RBM with EA.

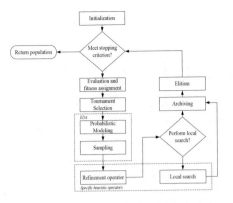

Fig. 1. Process flow of MOEDAs

Table 1. Parameter setting for the algorithms

Parameter	Value
Population size	Number of cities
Number of cities	100, 200, 500
Stopping criterion	2000 generations
Local search rate	0.5
Crossover rate	0.8
	0.05
Mutation rate	
Independent runs	10

Results for 100 cities are plotted in Fig. 2. EDAs give better performance than EA in term of IGD and NR. This is because EDAs consider the overall distribution of the population to guide the search, which is different from EA which use individual chromosome to generate offspring. From convergence trace, it is observed that RBM has the fastest convergence rate at early evolution, while EA has the lowest convergence rate. Furthermore, the Pareto front curve shows that UMDA has good proximity but poor spread. The corporation between EDAs with EA improves the spread of the algorithms with the cost of sacrificing proximity.

Results for 200 cities are presented in Fig. 3. Most of the final non-dominated solutions are generated from RBM. RBM and RBMEA take advantages from the dependencies of the cities. Thus, outperform other algorithms. Furthermore, the incorporation of EDAs with EA seems to improve the performance of sole algorithms.

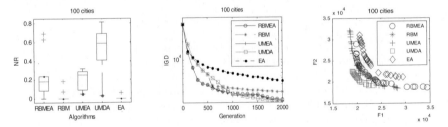

Fig. 2. NR, Evolution trace and, Evolved Pareto front with 100 cities

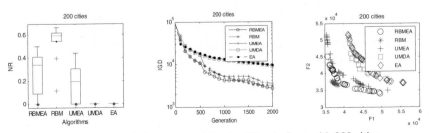

Fig. 3. NR, Evolution trace and, Evolved Pareto front with 200 cities

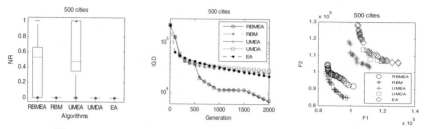

Fig. 4. NR, Evolution trace and, Evolved Pareto front with 500 cities

The simulation results for problem with 500 cities are presented in Fig. 4. Overall, the EDAs and EAs alone are unable to evolve a set of good trade-off solutions. The corporation between EDAs and EAs improve the overall performance.

4   Conclusions

This paper has studied the probabilistic based EAs in solving Bi-TSP. It is among the first attempts to employ EDAs in the study of permutation based multi-objective problems. Two probabilistic modeling techniques have been adapted. They include univariate modeling and RBM approach. A refinement operator has been proposed to make sure the hard constraints of the problem are not violated. In addition, a local search operator is defined to enhance the exploitation of the algorithms. As the limitation of EDAs in evolving a set of good spread solutions, genetic based evolutionary optimizer is incorporated with EDAs to complement their weaknesses. The empirical results show that EDAs have better proximity while EA has better spread in their final evolved solutions. The incorporation between EDAs and EA mutually complements each other's limitation; thus yielding better performance.

References

1. Larrañaga, P., Lozano, J.A.: Estimation of Distribution Algorithms: A New Tool for Evolutionary Computation. Kluwer, Norwell (2001)
2. Jahne, M., Li, X., Branke, J.: Evolutionary Algorithms and Multi-objectivization for the TSP. In: Genetic and Evolutionary Computation Conference, pp. 595–602 (2009)
3. Herrera, F., Garcia-Martinez, C., Cordon, O.: A taxonomy and an Empirical Analysis of Multiple Objective Ant Colony Optimization Algorithms for the Bi-criteria TSP. European Journal of Operational Research 180(1), 116–148 (2007)
4. Yang, M., Kang, L, Guan, J.: An Evolutionary Algorithm for Dynamic Multi-objective TSP. In: Conference on Advances in Computation and Intelligence, pp. 62–71 (2007)
5. Tang, H.J., Shim, V.A., Tan, K.C., Chia, J.Y.: Restricted Boltzmann Machine based Algorithm for Multi-objective Optimization. In: Congress on Evolutionary Computation (2010)
6. Muhlenbein, H., Paass, G.: From Recombination of Genes to the Estimation of Distributions I. Binary Parameters. In: Conference on Parallel Problem Solving from Nature, pp. 178–187 (1996)
7. Deb, K., Pratap, A., et al.: A Fast and Elitist Multiobjective Genetic Algorithm: NSGA-II. Transactions on Evolutionary Computation 6(2), 182–197 (2002)
8. Peng, W., Zhang, Q., Li, H.: Comparison between MOEA/D and NSGAII on the Multi-objective Travelling Salesman Problem. In: Multi-objective Memetic Algorithms, vol. 171, Springer, Heidelberg (2009)

Automatic Shape Independent Shell Clustering Using an Ant Based Approach

Siddharth Pal, Aniruddha Basak, and Swagatam Das

Department of Electronics and Telecommunications,
Jadavpur University, India
sidd_pal2002@yahoo.com, aniruddha_ju_etce@yahoo.com,
swagatamdas19@yahoo.co.in

Abstract. This paper presents a novel technique to detect irregular shell clusters using an algorithm that is inspired by Ant Colony Optimization (ACO). Till now major work on shell clustering has been based on regular shells using a fuzzy-based technique. However the proposed algorithm can separate irregular shell clusters from the solid clusters very efficiently. The algorithm is tested on seven test images and it is seen to give very good results.

Keywords: Shell Clustering, Ant Colony Optimization, metaheuristics, pattern, recognition.

1 Introduction

Traditional fuzzy clustering algorithms like Fuzzy C –Means (FCM) [1] and Possibilistic C – Means (PCM) [2] cannot detect clusters that lie in nonlinear subspaces of the feature space because they use points (i.e. cluster centroids) as prototypes. To find clusters in nonlinear subspaces that resemble *shells* or patches of hyper-surfaces with no interior points, prototypes like curves/hyper-surfaces have been proposed. The shell clustering techniques provide an effective means for solving the problem of fitting multiple curves/hyper-surfaces to unlabeled, sparse, and scattered data. Algorithms dedicated to detect shell type clusters have also found applications in boundary detection, surface approximation and similar computer-vision tasks [3 - 5]. A few representative fuzzy shell clustering algorithms are Adaptive Fuzzy C Shells (AFCS) algorithm [6], Fuzzy C Quadric Shells (FCQS) and its variants [7, 8], Fuzzy C Plano-Quadric Shells [9] etc. and they attempt to minimize the weighted squared sum of distances of a feature point to a prototype by updating the fuzzy membership and parameters in an alternating fashion. Yager and Filev proposed the Mountain Clustering Method (MCM) [10], to estimate the cluster prototypes in a simple way. Pal and Chakraborty [11] extended MCM for detecting circular shell-shaped clusters and proposed the Mountain Circular Shell (MCS) method. Most of the shell-clustering algorithms available in the literature are computationally quite expensive since either they need to perform matrix inversions or they solve some nonlinear equations iteratively. Usually for better results a series of algorithms need to be applied on the data. Moreover to the best of the author's knowledge no work has been done with irregular shell clusters.

K. Deb et al. (Eds.): SEAL 2010, LNCS 6457, pp. 593–602, 2010.

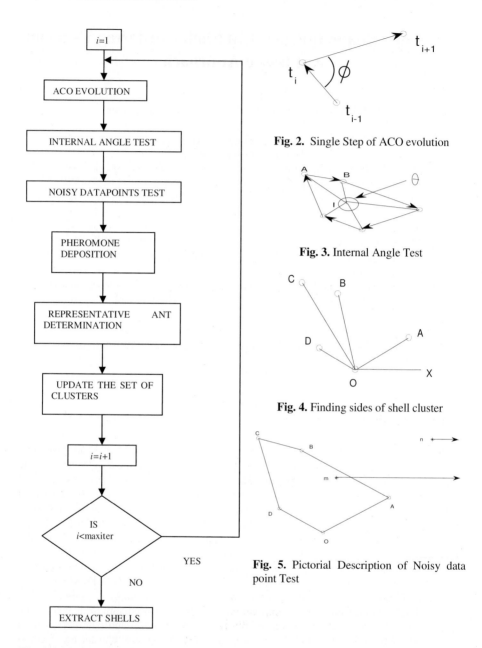

Fig. 2. Single Step of ACO evolution

Fig. 3. Internal Angle Test

Fig. 4. Finding sides of shell cluster

Fig. 5. Pictorial Description of Noisy data point Test

Fig. 1. Basic Flowchart of the proposed algorithm

In this paper we have use a hybrid intelligent technique which is inspired by Ant Colony Optimization. Proposed by Dorigo and Stutzle[12] , Ant Colony Optimization has found applications in solving Traveling Salesman Problem [17] and image processing and machine vision problems [13-16].

We have divided the paper into 4 sections. The different stages of the hybrid algorithm are explained in Section 2 followed by a description of the working of the algorithm in Section 3. In Section 4 we show the experimental results and conclude the paper in Section 5.

2 Description of Algorithm

The basic principal of the algorithm is represented with the help of a flowchart in Figure 1. The algorithm has been divided into various steps. The aim of the algorithm is to extract irregular shell clusters automatically from datasets. A modified ACO algorithm is used. After extraction of the shells normal clusters are extracted using conventional automatic clustering algorithms.

The basic steps of the algorithm were illustrated by the means of a flowchart. They are dealt in further detail now.

1. INITIALIZATION PHASE:
The starting location of the ants is fixed randomly. The various attributes of the ants are also initialized.

2. ACO EVOLUTION:
In this step all the ants complete one journey each. In the figure 2, a single step of ACO evolution is shown. t_i be the present data point of the ant at the ith iteration. t_{i-1} be the data point of the ant at the previous iteration. Let the set Ξ represent all the data points sufficiently close to data point t_i. Let $t(\in \Xi)$ be the data point which the ant considers moving to from t_i

We form another set Y . $t \in Y$ if $t \in \Xi$ and $\phi > 90^o$.
An ant goes to its next datapoint t using the following selection rule:

$$t = \underset{t \in Y}{\operatorname{argmax}} \left\{ \left[\tau_{i,t} \right]^\alpha \left[d_{i,t} \right]^\beta \right\} \quad \text{if } r \leq r_0 \left. \begin{array}{c} \\ \\ \\ \end{array} \right\} \quad (1)$$

$$= J \qquad\qquad\qquad \text{otherwise}$$

$$P_{t_i t} = \frac{\left[\tau_{i,t} \right]^\alpha \left[d_{i,t} \right]^\beta}{\sum\limits_{t \in Y} \left[\tau_{i,t} \right]^\alpha \left[d_{i,t} \right]^\beta} \qquad \text{if } t \in Y \qquad (2)$$

where,

$d_{i,t}$ =Euclidean distance between two data points

$\tau_{i,t}$ = Pheromone trail between data points t_i & t.

r denotes a random variable (0,1).

J denotes a random variable generated by the probability distribution given in equation (2).

The idea behind equation (1) is that the ant will be biased to travel through a route having high pheromone deposition. The ant should also choose to travel to a closer data point than a data point which is at a greater Euclidean distance. Hence α should be positive and β negative.

3. SHELL VALIDITY TEST 1(INTERNAL ANGLE TEST):

The contour followed by each ant is checked if it is a shell or not. If yes then the angle subtended at an internal point would be found to be very close to 360°.

While calculating the total angle subtended by the cluster at an internal point we sum the angles subtended at the internal point by all the sides forming the shell. Suppose we consider a side AB and we name the internal point as I. ABI can be considered as 3 vertices of a triangle. We need to find $\angle AIB$. We first calculate the Euclidean distances $|AB|, |BI|, |AI|$. Let $a = |IB|, b = |AI|, c = |AB|$. The required internal angle subtended by side AB $\angle AIB = \cos^{-1}\left(\dfrac{a^2 + b^2 - c^2}{2ab}\right)$. Similarly we find the angle subtended by all the sides forming the shell cluster. This will be the total internal angle.

If θ is the angle subtended, then to pass the test the following condition should be satisfied $|\theta - 360^\circ| < \varepsilon$. We chose $\varepsilon = 0.0001$. If the ant travel in a closed contour as shown in Figure 3 then θ is the sum of all the angles subtended by all the sides.

4. SHELL VALIDITY TEST 2(NOISY DATAPOINTS TEST):

After the shell has passed the internal angle test, it is subjected to the noisy data point test.

First we need to find the sides forming the shell cluster.

We first find the data point which has the minimum y-objective value. Let it be denoted by O. We know all the data points belonging to that shell. Next we calculate the angles associated with the vertices, the two arms being the line joining the vertex under consideration and O and the positive x- axis. From Figure 4 we see that $\angle AOX < \angle BOX < \angle COX < \angle DOX$. Hence the vertices are sorted as follows O,A,B,C,D. A data point is termed as a "noisy data point" if it is present inside the detected shell.

The set of all data points be Ξ. Suppose the shell to be tested, be a set of data points Z.

Each datapoint is characterized by 2 objectives *(x,y)*.

Only those datapoints can be tested for this test which belong in set $\Xi - Z$.

Suppose datapoints $m, n \in \Xi - Z$.

We consider a positive ray from m and n as shown in Figure 5. If the ray intersects the shell odd number of times then the datapoint is inside the shell and if it intersects even number of times then it figures outside the shell.

Let us describe the procedure for the datapoint m.To reduce the complexity of the process we first scan for those sides whose one vertex has greater y-objective value and the other has lesser y-objective value. The sides obtained would be AB and CD. Then we solve for the point of intersection of a horizontal line passing through m and the sides under consideration. If the point of intersection has x-objective value greater than m then it will intersect the ray in the positive x direction. Only AB will intersect the positive ray through m.

Each closed contour is tested if the no. of noisy points is less than a threshold. If yes then it is a valid shell.

5. PHEROMONE DEPOSITION:
Pheromone is deposited by successful ants (i.e. ants which detected valid shells). The Pheromone Update equation is described here.

Suppose a successful ant A_s travels from data point i to data point j then the pheromone level between the two data points τ_{ij} and τ_{ji} is updated as given in equation 3.

$$\tau_{ij} = \tau_{ji} = \tau_{ij} + \frac{1}{\left[N_p + 1\right]^\delta}, \qquad (3)$$

where N_p is the number of noisy data points detected. Now likewise the contribution of all successful ants is considered.

After the contribution of all successful ants has been considered the pheromone trail is evaporated by equation 4.

$$\tau_{ij} = \left(1 - \mu\right)\tau_{ij} + \mu\tau_0, \qquad (4)$$

where μ is the pheromone evaporation rate, $0 \le \mu \le 1$.

τ_0 is the initial pheromone level

6."REPRESENTATIVE ANT" DETERMINATION:
Let the set of clusters detected in the present iteration be represented by Ψ. Let cluster $w \in \Psi$. It is possible that more than one ant travel on cluster w. That ant is named the "representative ant" of cluster w which has maximum no. of points and minimum noisy datapoints..

7. UPDATE OF SET OF CLUSTERS:
Let the set of cluster detected till present iteration be called ζ.

Say cluster $w \in \Psi$. It may happen that many points of w are already present in cluster $w'(\in \zeta)$. Let the number of points in cluster w be p. The two shell clusters are termed as "similar" if more than p/λ points are common with cluster w'. We generally choose $\lambda = 4$. If similar shell be found, w replaces w' in set ζ if it has more no. of points and less number of noisy data points. If no such w' be found in ζ then cluster w is added to ζ.

8. EXTRACT SHELLS
As can be seen from the flowchart this step is outside the loop. After the shell clusters are returned it sometimes happens that all the data points haven't been covered. Those data points that are sufficiently close to the shells are added to the shells as long as the number of noisy data points does not increase for the new shell cluster as compared to the previous one. Now that the shell clusters have been extracted the remaining solid clusters can be taken care of by any standard clustering algorithm.

3   Basic Functioning of Algorithm

Till now the basic steps of the algorithm has been listed. In this section we will explain how the different steps of the algorithm contribute to the detection of shells. A very basic question that arises is that how does the algorithm distinguish a shell cluster from a solid cluster. This problem is solved by the Internal Angle Test. If the ant travels on a solid cluster then it will subtend more than $360°$ at an internal point. Any ant that has followed a closed path will at least subtend $360°$ at an internal point. If it is a shell cluster then exactly $360°$ is subtended at an internal point. However for a solid cluster due to formation of multiple loops an angle greater than $360°$ is subtended at an internal point.

Another question that arises is whether this algorithm can efficiently extract an irregular shell from a solid cluster which is positioned nearby.

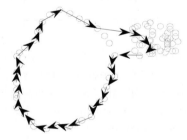

Fig. 6. Path travelled by ant with shell and solid cluster close together (case 1)

Fig. 7. Path travelled by ant with shell and solid cluster close together (case 2)

Here in figure 6, a shell cluster is shown close to a solid cluster. An ant can travel from the shell cluster to solid cluster as shown by the arrows. When the ant enters the solid cluster it moves inside the solid cluster as shown making multiple loops in the process. These solutions get immediately discarded as they don't pass the Internal Angle test.

It might also happen that the ant travel through the solid cluster as shown in figure 7.This ant passes the internal angle test. However there are a number of data points trapped inside the shell which are termed as "noisy data points". Hence this solution is either discarded if the number of datapoints exceeds maximum limit or deposits less pheromone compared to true solution. A number of ants might travel on the same shell. So a representative ant is determined which represents the shell cluster in the best possible way. Now the newly detected clusters are added to or update the earlier set of clusters.

4 Experimental Results

The algorithm was tested with 7 datasets with varying degree of complexity. The datasets that were considered are 2 dimensional and the final result has been shown in the form of an image which clearly depicts the clustered shells and the solid clusters left unrecognized. The aim of this algorithm was to detect irregular shell clusters. Extensive work has been carried out in detecting solid clusters. This algorithm is dedicated to detecting irregular shells. Hence after the extraction of shell clusters, further clustering can be performed by any standard clustering algorithm like k-means, Fuzzy C-Means, etc.

The various parameters that performed well on all the datasets are listed below.

- The parameters related to ACO Evolution were chosen as $\alpha = 1, \beta = -3, r_0 = .8$,
- Pheromone evaporation rate $\mu = .1$,
- Initial pheromone level $\tau_0 = .001$,
- Noise Coefficient of pheromone deposition $\delta = 3$,
- Upper limit on the number of steps of an ant $S_{max} = 100$,
- Upper limit on maximum number of noisy data points $= S_{max}/10$,
- Number of iterations $T=20$,
- No. of ants $n_a =60$.

To study how the test images perform 25 runs were carried out for each dataset. Table 1 shows the simulation results. Datasets 1 to 4 contain only regular clusters.

A trial is said to be feasible if all the shell clusters were correctly identified. A trial is said to be successful if all the data points were correctly clustered.

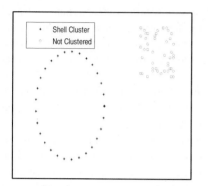

Fig. 8. Clustered Result(Dataset 1)

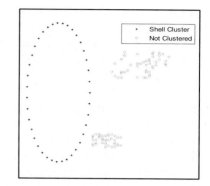

Fig. 9. Clustered Result(Dataset 2)

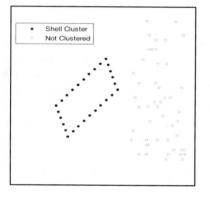

Fig. 10. Clustered Result(Dataset 3)

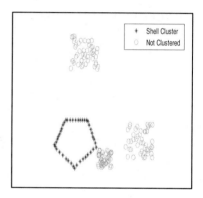

Fig. 11. Clustered Result(Dataset 4)

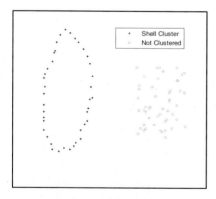

Fig. 12. Clustered Result(Dataset 5)

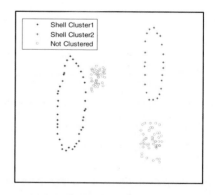

Fig. 13. Clustered Result(Dataset 6)

Table 1. Statistical Results

Dataset No.	Number of feasible runs	Number of successful runs	Clustering Efficiency
1	25	25	100%
2	25	23	92%
3	25	24	96%
4	25	23	92%
5	25	25	100%
6	25	23	92%
7	25	23	92%

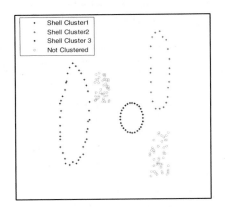

Fig. 14. Clustered Result (Dataset 7)

Clustering efficiency,

$$\eta = \frac{Trials_{successful}}{Trials}$$

The clustering efficiency is independent of the type of clusters (regular or irregular). However it is noted that if a solid cluster is present extremely close to shell clusters then the efficiency of clustering decreases. This is mostly because sometimes one or two data points belonging to a neighbouring solid cluster get wrongly classified as data points of a shell cluster. However for all the images every trial was feasible as in every trial all the shell cluster were correctly detected.

5   Conclusion

We believe that the main contribution of this paper is to propose an algorithm that can detect shell clusters independent of their geometric shapes. Most of the existing shell clustering algorithms uses a prototype to match with the data points to extract the clusters. Thus they are capable of only extracting shell clusters which can be defined by means of equations. Our algorithm is radically different from that viewpoint. We have also tested our algorithm on seven datasets. The first four datasets contain regular shell clusters and solid clusters. All the regular shell clusters were correctly identified with great efficiency. The rest of the datasets contain irregular shell clusters which were also identified correctly. Thus we can conclude that this algorithm can be used for shell clustering purposes and is also very promising for image processing applications.

References

1. Bezdek, J.C.: Pattern Recognition with Fuzzy Objective Function Algorithms. Plenum, New York (1981)
2. Krishnapuram, R., Keller, J.: The possibilistic c-means algorithm: insights and recommendations. IEEE Trans. on Fuzzy Systems 4, 385–393 (1996)

3. Dave, R.N.: Generalized Fuzzy c-Shell clustering and detection of circular and elliptical boundaries. Pattern Recognition 25(7), 713–721 (1992)
4. Balakumaran, T., Vennila, I.A., Gowri Shankar, C.: Detection of microcalcification in mammograms using wavelet transform and fuzzy shell clustering. International Journal of Computer Science and Information Security 7(1) (2010)
5. Barni, M., Mecocci, A., Perugini, G.: Application of possibilistic shell-clustering to the detection of craters in real-world imagery. In: Proceedings of the IEEE for Geoscience and Remote Sensing Symposium, vol. 1, pp. 168–170 (2000)
6. Dave, R.N., Bhaswan, K.: Adaptive fuzzy C shells clustering and detection of ellipses. IEEE Trans. Neural Networks 3, 643–662 (1992)
7. Krishnapuram, R., Frigui, H., Nasraoui, O.: New fuzzy shell clustering algorithms for boundary detection and pattern recognition. In: Proc. SPIE Conf. Intell. Robots Comput. Vision X: Algorithms Techniq., Boston, pp. 458–465 (November 1991)
8. Krishnapuram, R., Frigui, H., Nasraoui, O.: Quadratic shell clustering algorithms and their applications. Pattern Recog. Lett. 14(7), 545–552 (1993)
9. Krishnapuram, R., Frigui, H., Nasraoui, O.: Fuzzy and possibilistic shell clustering algorithms and their application to boundary detection and surface approximation: Parts I and II. IEEE Trans. Fuzzy Syst. 3, 44–60 (1995)
10. Yager, R.R., Filev, D.P.: Approximate clustering via the mountain method. IEEE Trans. Systems, Man, Cybernet. 24(8), 1279–1284 (1994)
11. Pal, N.R., Chakraborty, D.: Mountain and subtractive clusteing method: improvements and generalization. Internat. J. Intell. Systems 15, 329–341 (2000)
12. Dorigo, M., Stutzle, T.: Ant Colony Optimization. MIT Press, Cambridge (2004)
13. Vallone, U., Merigot, A.: Imitating human visual attention and reproduction optical allusion by ant scan. Internat. J. Comput. Intell., Appl. 9, 157–166 (2003)
14. Zheng, H., Wong, A., Nahavandi, S.: Hybrid ant colony algorithm for texture classification. In: IEEE Congress on Evolutionary Computation Proc., pp. 2648–2653 (2003)
15. Zhuang, X., Mastorakis, N.E.: Image processing with the artificial swarm intelligence. WSEAS Trans. Comput. 4, 333–341 (2005)
16. Lu, D.S., Chen, C.C.: Edge Detection improvement by ant colony optimization. Pattern Recognition Letters 29(4), 416–425 (2008)
17. Dorigo, M., Gambardella, L.M.: Ant Colonies for the travelling Salesman problem. Biosystems 43, 73–81 (1997)

Hybrid Search for Faster Production and Safer Process Conditions in Friction Stir Welding

Cem Celal Tutum[1], Kalyanmoy Deb[2], and Jesper Hattel[1]

[1] Technical University of Denmark (DTU), Department of Mechanical Engineering,
Produktionstorvet, 2800, Kgs. Lyngby, Denmark
[2] Kanpur Genetic Algorithms Laboratory (KanGAL),
Indian Institute of Technology Kanpur, PIN 208016, Kanpur, India
cctu@mek.dtu.dk, deb@iitk.ac.in, jhat@mek.dtu.dk

Abstract. The objective of this paper is to investigate optimum process
parameters and tool geometries in Friction Stir Welding (FSW) to minimize temperature difference between the leading edge of the tool probe
and the work piece material in front of the tool shoulder, and simultaneously maximize traverse welding speed, which conflicts with the former
objective. An evolutionary multi-objective optimization algorithm (i.e.
NSGA-II), is applied to find multiple trade-off solutions followed by a
gradient-based local search (i.e. SQP) to improve the convergence of the
obtained Pareto-optimal front. In order to reduce the number of function
evaluations in the local search procedure, the obtained non-dominated
solutions are clustered in the objective space and consequently, a post-
optimality study is manually performed to find out some common design
principles among those solutions. Finally, two reasonable design choices
have been offered based on several process specific performance and cost
related criteria.

Keywords: Evolutionary multi-objective optimization, gradient-based
local search technique, ϵ-constraint method, hybrid search, friction stir
welding, thermal simulation, material flow.

1 Introduction

The FSW process is an efficient solid-state, i.e. without melting, joining technique that is invented by Wayne Thomas and a team of his colleagues at the
Welding Institute (TWI), UK, in December 1991; especially for aluminium alloys which are difficult to weld with traditional welding techniques [1]. Figure 1a
shows a standard welding tool having a cylindrical shoulder and probe which are
in general designed in different size and shapes with/out thread features or manufactured with different materials based on work piece and process specific needs
or limitations. The process, which is schematically shown in Figure 1b, consists
of several subsequent procedures denoted as plunging, dwelling, actual welding
and pulling the tool out of the work piece. First, the tool is submerged vertically
into the joint line with high rotational speed in the plunge period and then dwell

K. Deb et al. (Eds.): SEAL 2010, LNCS 6457, pp. 603–612, 2010.
© Springer-Verlag Berlin Heidelberg 2010

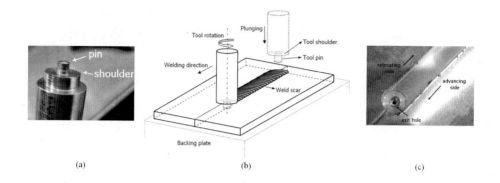

Fig. 1. (a) A standard tool having a cylindirical shoulder and probe (pin) design. (b) Schematic view of the FSW process. (c) A welded structure having a key hole at the end.

period takes place, where the the tool is held steady relative to the work piece while keeping rotation and heating the surrounding work piece material locally. Following dwelling, the tool is moved forward while stirring two work piece material to be joint (welding period) and is pulled out of the work piece leaving a key hole behind as seen in Figure 1c. These sequences have also been represented schematically in Figure 2 emphasizing different computational modelling approaches with respect to different reference frames, i.e. the Lagrangian (also known as "global approach" [9], where transient effects are captured) and the Eulerian ("local approach" [7], in general used for the steady-state conditions).

In FSW, heat is generated by friction (mainly at the interface between tool shoulder and upper surface of the work piece) and plastic deformation (by tool probe or pin in plunging stage and during welding period via stirring two work piece material along the joining line). The heat flows into the work piece as well as the tool. The amount of heat conducted into the work piece influences the quality of the weld, distortion and residual stress in the work piece [7,8,9]. Insufficient heat generation from the tool shoulder and the probe could lead to failure of the tool pin since the work piece material is not soft enough. Therefore, understanding the heat aspect of the FSW process, which is the main driving force for all consequent coupled simulations, e.g. microstructure and solid mechanics models, is extremely important, not only for understanding physical phenomena, but also for improving the process efficiency, e.g. welding faster and safer.

In the present paper, optimum process parameters and tool geometries in FSW are investigated to minimize the temperature difference between the leading edge of the tool probe and the work piece material in front of the tool shoulder, i.e. to soften the material enough to move the tool probe forward without failure, and simultaneously to maximize traverse welding speed, therefore production rate, subjected to hot and cold weld conditions. More specifically, the choices of the tool rotational speed and the traverse welding speed together with the radii of the tool shoulder and the probe have been investigated in order to achieve the

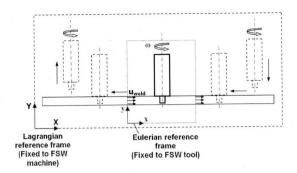

Fig. 2. Lagrangian versus Eulerian approach for simulation of the FSW process

goals mentioned above which are in essence conflicting. A Steady-state Eulerian thermal finite element model with temperature dependent thermo-physical (i.e. heat treated aluminum alloy, AA2024-T3) material data has been implemented using a commercial multi-physics simulation software COMSOL for the function evaluations. An evolutionary multi-objective optimization (EMO) algorithm, i.e. non-dominated sorting genetic algorithm (NSGA-II) is initially performed to find the Pareto-optimal front. The non-dominated solutions found so far have been clustered based on their Euclidean distances (in the objective space) in a pre-fix grid structure to reduce the number of the solutions, which will in turn be served as initial starting points for the gradient-based local search technique, i.e. sequential quadratic programming (SQP). The ϵ-constraint method is applied by fixing the second objective (i.e. welding speed) as a constraint for each clustered non-dominated solutions independently to obtain the modified optimized front. Further improvement in accuracy and confidence in the convergence of the Pareto-optimal front is achieved, and following this, a brief post-optimality study is performed to unveil some common design principles among members of the clustered Pareto-optimal set. Finally, two reasonable design solutions among those multiple trade-off solutions have been selected based on different characteristics of the temperature distribution under the tool shoulder induced by the material flow, tool selection and production rate preferences.

2   Thermal Model

Due to relatively high heat generation contribution from the surface of the tool shoulder, an assumption based only on modelling the tool shoulder is taken into account. The radius or in other words location of the tool probe is hypothetically included as a design variable to compute the first objective function (i.e. temperature difference) and temperature variation under the tool, between the tool shoulder and the probe, to be used in the decision making step. Modelling the whole welding process, i.e. plunge, dwell and pull out periods, holds some notable complexities. In order to reduce the computational cost regarding moving heat source, meanwhile preserving the applicability, only the welding period is taken into account and a moving coordinate system (i.e. Eulerian reference frame) which is located on the heat source is applied. The shear layer formed below the tool shoulder due to high tool rotational speed is also included; hence an asymmetric temperature field along the joint line is obtained in the present numerical model. Equation 1, on the right side, describes the steady-state heat transfer in the plate (transient term on the left side disappears),

$$\rho c_p \frac{\partial T}{\partial t} = \nabla(k\nabla T) + q_{vol} - \rho c_p u \nabla T \Rightarrow \rho c_p u \nabla T - \nabla(k\nabla T) = q_{vol} \quad (1)$$

where T denotes the temperature field to be solved, ρ, c_p and k are the density, the specific heat capacity and the heat conductivity of the AA2024-T3 (the work-piece material), respectively, besides u prescribing the flow vector and q_{vol} is the volumetric stationary heat source representing the tool as a circle in Figure 3a.

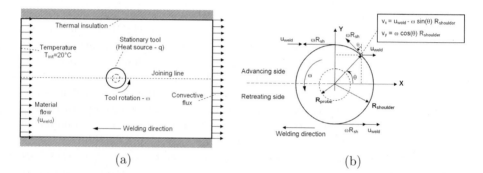

Fig. 3. (a) 2-D Eulerian Steady state thermal model. (b) Mathematical modelling of the flow field under the tool shoulder in detail.

In the present model, heat generation is a function of the tool radius and the temperature dependent yield stress of the work piece material ($\sigma_y(T)$ of AA2024-T3) and assumed to be uniform through the thickness ($t=3$ mm) of the plates to be welded, as given below in Equation 2.

$$q_{vol}(r(x,y),T) = \frac{\omega r(x,y) \frac{\sigma_y(T)}{\sqrt{3}}}{t} = \left(\frac{n_{rev}2\pi}{60}\right)\sqrt{x^2+y^2}\frac{\sigma_y(T)}{t\sqrt{3}} \quad (2)$$

The details of this temperature and position dependent heat source model entitled as Thermal-Pseudo-Mechanical (TPM) model are given in detail elsewhere [2]. The traverse motion of the tool and the relatively complex flow field under it are modelled by prescribing a material flow through the rectangular plate region, as shown in Figure 3a. Due to this flow prescription, Equation 1 includes a convective term (u) in addition to the conductive term. The derivation of the mathematical prescription of the material flow is also schematically represented in Figure 3b and components of the flow vector in the welding and the transverse directions are formulated for an arbitrary point on the periphery of tool shoulder as a function of θ (i.e. $\theta = \arccos(\frac{x}{\sqrt{x^2+y^2}})$ in Cartesian reference frame). Equation 3 generalizes the flow field description $(u(\theta) = u(x,y) = (v_x, v_y))$ for the whole domain as follows,

$$u = \begin{cases} (u_{weld} - sin(\theta)wR_{shoulder}, cos(\theta)wR_{shoulder}) & \text{if } r(x,y) \leq R_{shoulder} \\ (u_{weld}, 0) & \text{if } r(x,y) > R_{shoulder} \end{cases} \quad (3)$$

where $r(x,y)$ is the radius or the position vector, $cos(\theta) = \frac{x}{\sqrt{x^2+y^2}}$ and $sin(\theta) = \frac{y}{\sqrt{x^2+y^2}}$. As a boundary condition, the room temperature $(20°C)$ is defined at the left edge of the rectangular region where the tool is assumed to be moving towards. The heat flux on the right edge of the plate region, where the material leaves the computational domain, is dominated by convection. On the upper and lower edges of the plate boundaries, thermal insulation is enforced.

3   Hybrid EMO

Although evolutionary computation has become an important problem solving methodology with its population-based collective learning process, self-adaptation, and robustness, the performance of the algorithms still depend on proper selection of various parameters (e.g. probabilities, selection and mutation schemes, etc.), namely the proper relationship between the exploration and the exploitation capabilities avoiding premature convergence, as mentioned above. Moreover, the computational speed is relatively slow as compared to the classical (deterministic) algorithms. Therefore, as expected, the need for hybrid algorithms, which combine an evolutionary algorithm with e.g. a local search method, emerges aiming at both robust and accurate solutions with (if possible at all) less computational cost. Local search methods may be incorporated within the population members (parents) or among the offsprings. The architectures of hybrid evolutionary algorithms have been summarized by [5] as follows: hybridization between two different EAs (a GP asisted GA), an EA with a neural network, a particle swarm optimization (PSO) or an ant colony optimization (ACO) as well as hybridization between EA and other heuristics (such as local search, tabu search, simulated annealing, hill climbing, etc.). However, as the *No Free Lunch Theorem* proposes, on average, all black-box algorithms have identical behavior, thus there is not a definite answer for which local search procedure to use for any sort of problems.

4   Hybrid EMO for the Thermal FSW Model

In this section, the multi-objective optimization problem (MOP), briefly described in section 1 that is related to the thermal aspects of the FSW process, is formulated. Optimum process parameters, i.e. the tool rotational and traverse welding speeds (n_{rev} and u_{weld}), and geometrical tool parameters, i.e. tool shoulder and probe radii ($R_{shoulder}$ and R_{probe}), are investigated to minimize the temperature difference (ΔT) between the leading edge of the tool probe and the work piece material in front of the tool shoulder, and simultaneously to maximize traverse welding speed. The second objective, based on the duality principle, is reformulated as the minimization of -u_{weld} due to the way of implementation of the EMO algorithm, i.e. MATLAB implementation of the original NSGA-II [3] algorithm by the first author [6]. This MOP problem is constrained with hot and cold weld conditions, geometrical constraints (the tool shoulder radius is desired to be 5 mm larger than the tool probe radius), besides lower and upper limits of the design variables. In order to evaluate hot and cold weld conditions, average temperature (T_{avg}) is computed under the tool shoulder, in other words, the temperature values on each element inside the circular region (i.e. $Area = \pi R_{shoulder}^2$) are integrated and divided by the number of elements. The constrained multi-objective optimization problem is given below,

$$
\begin{aligned}
Minimize &: f_1(x) = \Delta T = T_{probe} - T_{ahead}, \\
Minimize &: f_2(x) = -u_{weld}, \\
subject\ to &: 450°C \leq T_{avg} \leq 500°C, \\
& R_{probe} + 5\,mm \leq R_{shoulder}, \\
& 8\,mm \leq R_{shoulder} \leq 17\,mm, \\
& 3\,mm \leq R_{probe} \leq 12\,mm, \\
& 100\,rpm \leq n_{rev} \leq 1250\,rpm, \\
& 0.5\,mm/s \leq u_{weld} \leq 15\,mm/s.
\end{aligned}
\tag{4}
$$

As mentioned above, NSGA-II, which is an EMO algorithm [3] enabling finding well-spread multiple Pareto-optimal solutions for an MOP by incorporating three substantial features, i.e. elitism, non-dominated sorting, and diversity preserving mechanism (crowding distance), is used for the proposed constrained problem. Population size is 100 and the number of generations is fixed to 10 due to relatively high computational cost of the function evaluations, i.e. simulation time for each set of designs is approximately 10 minutes on a PC having Core 2 CPU, 2.33 GHz, 2 GB of RAM. Real variable-coding is used for the design variables. Therefore the simulated binary crossover (SBX) and the polynomial mutation [4], with a distribution indices of 5 and 10, are used as a crossover and mutation operators, respectively. Figure 4a shows all NSGA-II (non-dominated) solutions composing a non-convex Pareto-optimal front, having $-u_{weld}$ on the horizontal axis and ΔT on the vertical axis. As expected, the higher welding speeds results in higher temperature difference indicating steeper gradients in front of the tool, which is not desirable in case of limitations due to improper tool or machine designs. More detailed analysis of these trade-off designs is performed after the local search procedure which aims for further improvement in the convergence of the obtained trade-off frontier.

Prior to the local search step, the non-dominated solutions found so far are clustered simply based on their Euclidean distances (i.e. minimum d_i) with respect to their mean, which is computed in each cell and in each axes, in a prefix grid structure to reduce the number of the solutions (for the sake of computational cost), as

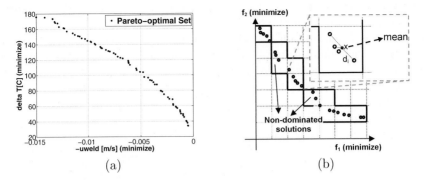

Fig. 4. (a) The non-convex Pareto-optimal front obtained with NSGA-II. (b) Clustering scheme.

represented in Figure 4b on a hypothetically distributed points in the objective space. Figure 5a shows 17 clustered solutions, indicated by cross markers, out of 72 non-dominated solutions for the FSW problem in a 10-by-10 grid.

In the next step, before investigating some common manufacturing process or design principles in the FSW process among the members of the Pareto-optimal set, NSGA-II solutions are sought to be further improved or at least validated to be true Pareto-optimum solutions. ϵ-constraint method [10] is a very suitable approach to alleviate the difficulties faced in non-convex objective spaces (a weighted objective or a Tchebyscheff metric or any other metric which will convert multiple objectives into a single objective can also be used). In this approach, the MOP is reformulated by just keeping one of the multiple objectives and restricting the rest of the objectives within user-specified values (i.e. u_{weld} is transformed into a constraint by considering $\epsilon = 10^{-6}$ in the present case). More details are given in [4]. In order to solve the scalarized optimization problem by means of searching the minimum of the aggregated objective function, a gradient-based local search technique, i.e. sequential quadratic programming (SQP; `fmincon` function in MATLAB Optimization Toolbox), is used. The modified front is shown with the blue curve in Figure 5b. The change in the convergence is not exaggerated, but on the other hand, it enhances the confidence in the trade-off front obtained by NSGA-II.

After completing the multi-objective optimization task, a set of optimal solutions specifying the design variables and their trade-offs is obtained. If these optimal solutions are sorted according to the worse order of the first objective (min. ΔT), they would also get lined up in the second objective (min. $-u_{weld}$) in an ascending order. Having such a wide variety of solutions make the decision-making process much easier compared to having only one optimal solution. This enables engineers or designers to judge or plan the performance of a product or a process in a larger perspective in terms of sacrifices and gains with respect to multiple criteria [4]. Moreover, a basic post-optimality study can unveil interesting design knowledge that is common to all of these trade-off solutions or a partial set of them [11]. This design methodology, which was originally formulated as "innovization" (innovation through optimization) [12,13], has also been applied manually to the current FSW problem. A set of design variables corresponding to some of the solutions on the Pareto-curve have been listed in Table 1.

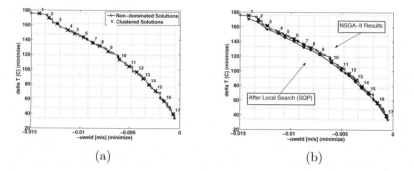

Fig. 5. (a) Clustered non-dominated solutions indicated by crosses and the corresponding numbers on top of them. (b) Pareto-optimal front modified after the local search on each clustered non-dominated solutions.

Table 1. Set of designs corresponding to some of the members on the modified Pareto-optimal front in Figure 5b

$R_{shoulder}$ [mm]	R_{probe} [mm]	u_{weld} [mm/s]	n_{rev} [rpm]
10.932	5.155	13.224	1037.8
10.819	5.377	12.359	1014.9
10.896	5.831	11.261	1119.6
14.959	9.887	9.694	974.33
13.793	8.263	7.171	856.07
13.579	8.005	4.304	718.17
10.998	5.74	3.672	1057.7
11.113	5.527	1.539	880.33
10.694	4.581	0.643	968.65

Three intervals can be distinguished looking at Table 1, i.e. $0.5mm/s \leq u_{weld} < 4mm/s, 4mm/s \leq u_{weld} < 10mm/s$ and $10mm/s \leq u_{weld} < 14mm/s$. In the first and the third intervals, tool shoulder and probe radii are approximately same (10 mm and 5 mm, respectively). In the middle interval, there is a significant increase in the tool dimensions, but there is also a common tendency to have a 5 mm difference in the radius dimensions (second constraint is active) along the Pareto-front. Moreover, in the middle interval, since the tool shoulder is getting larger, therefore heat generation is increasing due to the increase in the frictional surface area (consequently, the hot weld condition becomes active),

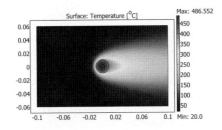

Fig. 6. Thermal field for a parameter set: u_{weld}=11.3 mm/s, $R_{shoulder}$=10.5 mm, $R_{shoulder}$=5.5 mm, n_{rev}=1100 rpm

tool rotational speed (n_{rev}) shows a decrease compared to other two intervals on the Pareto-curve. In most of the designs, distribution of the temperature field under the tool is almost uniform (i.e. cold weld condition is not active), thus standard deviation is close to the mean value, which is a desired process condition. Main criterion for the manufacturer to select one or two designs out of these possibilities would be welding speed which is related to investment and operating cost of different kinds of tool-machine combinations. In case of limited financial resources, manufacturer or engineer would like to weld slower (need to sacrifice in production rate) in order to improve the lifetime of the tools. In this case, such a design set: u_{weld}=2-3 mm/s, n_{rev}=700-800 rpm, $R_{shoulder}$=9-10 mm and R_{probe}=4-5 mm would be preferable. In an opposite case, where financial limitations are negligible, the production rate would be a dominant criterion (e.g. u_{weld} >11 mm/s), but similar tool geometries with higher rotational speeds would be sufficient (e.g. Figure 6).

5   Conclusions and Discussions

Optimum FSW process parameters and tool geometries are investigated to minimize temperature difference between the leading edge of the tool probe and the work piece material in front of the tool shoulder, and simultaneously to maximize traverse welding speed, which are in essence conflicting. An evolutionary multi-objective optimization algorithm (i.e. NSGA-II), is applied to find multiple trade-off solutions followed by a gradient-based local search (i.e. SQP) on clustered non-dominated solutions to enhance the confidence in the convergence of the obtained Pareto-optimal front. A brief post-optimality study is performed to find out some common design principles among those trade-off solutions. Three intervals in the design space have been distinguished looking at these optimal solutions. Finally, production rate is found to be a dominant criterion, and consequently, two reasonable choices have been made to solve the manufacturer's dilemma (faster production versus safer process conditions which can also be related to cheaper solutions).

References

1. The Welding Institute (TWI) homepage, `http://www.twi.co.uk`
2. Schmidt, H., Hattel, J.: Thermal modelling of friction stir welding. Scripta Materialia 58, 332–337 (2008)
3. Deb, K., Agarwal, S., Pratap, A., Meyarivan, T.: A fast and elitist non-dominated sorting genetic algorithm for multi-objective optimization: NSGA-II. IEEE Transactions on Evolutionary Computation 6, 182–197 (2002)
4. Deb, K.: Multi-Objective Optimization using Evolutionary Algorithms. John Wiley and Sons, Ltd., Chichester (2001)
5. Grosan, C., Abraham, A.: Hybrid Evolutionary Algorithms: Methodologies, Architectures, and Reviews. SCI, vol. 75, pp. 1–17 (2007)
6. Tutum, C.C.: Optimization of thermomechanical conditions in friction stir welding. PhD thesis, Technical University of Denmark, Kgs. Lyngby, Denmark (2009) ISBN:978-87-89502-89-2
7. Tutum, C.C., Schmidt, H., Hattel, J., Bendsøe, M.P.: Estimation of the Welding Speed and Heat Input in Friction Stir Welding using Thermal Models and Optimization. In: 7th World Congress on Structural and Multidisciplinary Optimization (COEX), Seoul, Korea, pp. 2639–2646 (2007)
8. Tutum, C.C., Hattel, J.: Multi-objective Optimization of Process Parameters in Friction Stir Welding. Poster in Genetic and Evolutionary Computation Conference (GECCO 2010), Portland, Oregon, pp. 1323–1324 (2010)
9. Tutum, C.C., Hattel, J.: A Multi-objective Optimization Application in Friction Stir Welding: Considering Thermo-mechanical Aspects. In: WCCI 2010 IEEE World Congress on Comp. Intell., CEC 2010, Barcelona, Spain, pp. 427–434 (2010)
10. Haimes, Y.Y., Lasdon, L.S., Wismer, D.A.: On a bi-criterion formulation of the problems of integrated system identification and system optimization. IEEE Transactions on Systems, Man, and Cybernetics 1(3), 296–297 (1971)
11. Deb, K.: Unveiling innovative design principles by means of multiple conflicting objectives. Engineering Optimization 35(5), 445–470 (2003)
12. Deb, K., Srinivasan, A.: Innovization: Innovating design principles through optimization. In: Proceedings of the 8th annual conference on Genetic and Evolutionary Computation (GECCO 2006), pp. 1629–1636. ACM, New York (2006)
13. Bandaru, S., Deb, K.: Automated Discovery of Vital Knowledge from Pareto-optimal Solutions: First Results from Engineering Design. In: WCCI 2010 IEEE World Congress on Comp. Intell., CEC 2010, Barcelona, Spain, pp. 1224–1231 (2010)

Hybrid Optimization Scheme for Radial Basis Function Neural Network

Vidyut Dey, Dilip Kumar Pratihar*, and Gauranga Lal Datta

Department of Mechanical Engineering
Indian Institute of Technology,
Kharagpur- 721302, India
Tel.: +91 3222 282992; Fax: +91 3222 282278
dkpra@mech.iitkgp.ernet.in
http://www.iitkgp.ac.in

Abstract. Radial Basis Function Neural Network (RBFNN) is a curve fitting tool in a higher dimensional space. The nature of this surface depends mainly on the number of neurons in the hidden layer. The number of hidden neurons is decided by the number of clusters into which the data-set gets divided. It has been shown that the accuracy in prediction depends upon the quality of the clusters. To obtain good quality clusters, in this study, a hybrid optimization scheme of running a genetic algorithm in the outer loop, while simultaneously running a back-propagation algorithm in the inner loop, has been adopted. The number of hidden neurons is kept the same with that of clusters formed by an algorithm proposed here, apart from the popular fuzzy-c-means and entropy-based clustering algorithms. RBFNN developed using the proposed clustering algorithm is found to perform better than that obtained utilizing the other two clustering algorithms. The method has been successfully implemented in both forward and reverse mappings of electron beam welding process.

Keywords: Electron beam welding; Radial Basis Function Neural Networks; Forward Mapping; Reverse Mapping; Clustering.

1 Introduction

Radial Basis Function Neural Network (RBFNN) was proposed by Broomhead and Lowe [1]. It is a special type of Artificial Neural Network (ANN), where an error surface can be fitted into a multidimensional space. The basic architecture of the network comprises of input nodes, a single hidden layer containing some neurons and an output layer. As the number of inputs and that of outputs remain fixed, the accuracy in prediction depends upon the number of hidden neurons, constants of the transfer functions used and connecting weights in-between the layers. There are various methods of determining the number of neurons in the hidden layer [2]. One of the methods states that the number of hidden neurons

---

* Corresponding author.

K. Deb et al. (Eds.): SEAL 2010, LNCS 6457, pp. 613–622, 2010.

can be made equal to the number of clusters formed by the data-set. The number of clusters formed depends not only on the type of clustering algorithm but also on the optimal values of parameters of the chosen algorithm. Out of various available optimization techniques, Genetic Algorithm (GA) is one of the best choices, as it searches the entire space for the best values, whereas the connecting weights between different layers and the constants of transfer functions are better minimized, if they are allowed to converge along the steepest descent direction. The combined effect of the above global and local searches, gives rise to a hybrid optimization technique for the RBFNNs. The hybrid optimization technique has been successfully utilized in both forward and reverse mappings of electron beam welding process carried out on stainless steel plates.

2   Literature Review

In RBFNN, transfer functions of the hidden neurons are assumed to be radial basis functions. It maps a multi-variable input into a multidimensional space after ensuring the best fit to all the training patterns. The performance of this network depends upon number of hidden neurons, their attributes and the type of basis function used. The training program of RBFNN basically involves the determination of center and spread of the basis functions, that is, mean and standard deviation in case of Gaussian Radial Basis Function and secondly, the weights between the hidden and output layers.

Billings and Zheng [3] provided with an extensive documentation on various algorithms used in training RBFNN before propounding the use of a GA to train the same. Since the GA can escape from local minimum, it has a lower probability of getting trapped into it. There had been similar approaches, where GA-optimized RBFNN was tuned with a back-propagation algorithm [8]. It was also shown that the training algorithm for GA-optimized RBFNN performed better, when all the input-output variables were of the same order of magnitude.

Zhang and Bai [4] proposed a training scheme of two-stages. In the first stage, a GA was employed to search for a cluster distance parameter crucial in determining the structure of the RBFNNs. Next, a self-growing algorithm was used to progressively increase the number of Radial Basis Functions (RBFs) and adjust their positions. It had been seen, mostly in pattern recognition problems that some of the inputs might not be relevant for a particular output. But, the presence of these inputs during network training might un-necessarily complicate the network. To overcome such problems, Zheng and Billings [5] developed a RBFNN, where they used suboptimal set of input variables for the desired output. The neurons in the hidden layer were determined by orthogonal least squared algorithm (OLS). There is also a study, where Self-Organizing Feature Map (SOFM) [6] was used as an unsupervised training scheme for determining the number of hidden neurons and the variables of Gaussian radial basis function. In general, the RBFs of the hidden layer used unique standard deviation (radius) for all the neurons. But, in multi-scale RBFNN, specific standard deviations were used for each neuron [7]. This algorithm gave more flexibility with more generalization capabilities.

To optimize the problems involving a large number of variables, the GA may face the well-known permutation problem [2]. In order to overcome this problem, an extensive study was performed by Amarnath and Pratihar [9], where some variables were optimized using the GA and the remaining ones were tuned by a back-propagation algorithm.

In order to automate a process, its input-output relationships are to be known both in forward and reverse directions. Not much work has been carried out in this direction. In the present study, both the problems of forward and reverse mappings have been tackled using a Gaussian RBFNN, whose architecture has been decided through clustering of the data by utilizing some fuzzy clustering algorithms. In this study, three different approaches have been illustrated. In the first two approaches, the number of neurons in the hidden layer are determined using fuzzy c-means (FCM) [12] and entropy-based fuzzy clustering (EFC) [13] algorithms, respectively. In the third approach, the newly proposed clustering algorithm has been utilized to determine the number of neurons of the hidden layer of RBFNN. All the parameters of the clustering algorithm along with the parameters and constants of back-propagation algorithm have been optimized using a GA.

3   Data Collection

An electron beam welding setup at BARC, Mumbai, India, is utilized to perform the bead-on-plate experiments on stainless steel plate (ASS-304) [10]. The experiment is conducted as per a central composite design matrix [11]. The design of experiments-proposed welds are carried out on seventeen sets of input process parameters. All the weld runs are repeated three times to take care of the experimental variations and the output data are collected. A schematic view of weld-bead profile is shown in Fig. 1.

There are three inputs for the forward mapping problem: **accelerating voltage**, **beam current** and **welding speed**. The corresponding outputs are eight in number, such as: **Bead penetration**, **Bead width**, **Bead height**, coordinates of two arbitrarily chosen points on the bead profile, namely: a_1, b_1, a_2, b_2 and Vickers micro-hardness **VH**.

4   Soft Computing Tools Used

A **RBFNN** has been utilized to model both in forward and reverse directions. In the present study, the number of neurons in the hidden layer has been determined by the number of clusters that are formed from the data set. For this purpose, three approaches have been utilized as discussed below.

In fuzzy clustering, a data may partially belong to more than one clusters at the same time. In case of FCM, the number of clusters into which the data have to be divided is first decided. All the center points along with the membership values of the data points that belong to a particular cluster are updated iteratively. The clusters obtained by this algorithm are compact but not distinct. In EFC,

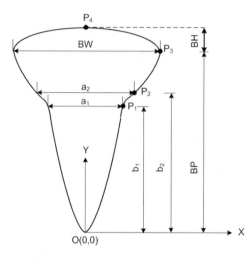

Fig. 1. A schematic view of weld-bead geometry

the number of clusters is determined by the algorithm itself. A data point is then assigned to a cluster, if its similarity value is found to be greater than a predefined threshold value. The resulting clusters are distinct but not compact. In the newly proposed algorithm, the number of clusters are determined as it is done in EFC algorithm, and then with help of iterations, the membership values are updated, as it is done in the FCM algorithm. Unlike the earlier two algorithms, clusters developed by the newly proposed algorithm are seen to be both compact as well as distinct in nature.

Approach 1:   GA-based optimization of the parameters of FCM and BP algorithms

In this approach, the data are clustered using the FCM algorithm. The parameters of clustering algorithm like number of cluster centers C (that is, the number of hidden neurons), level of fuzziness g (that is, power of the fuzzy set), constant of log-sigmoid transfer function λ along with other parameters of BPNN (learning rate η, momentum term α), have been optimized using a binary-coded GA. A uniform crossover and bit-wise mutation have been used in the GA. Moreover, a tournament selection scheme has been utilized in it. Each parameter is represented using 10 bits and the overall string is found to be 50-bits long. A typical GA-string for approach 1 is shown below.

$$\underbrace{010010\ldots\ldots}_{C}\underbrace{010011\ldots\ldots}_{g}\underbrace{100110\ldots\ldots}_{\eta}\underbrace{101101\ldots\ldots}_{\alpha}\underbrace{100111\ldots\ldots}_{\lambda}.$$

The fitness function value of a GA-string (that is, f) has been determined using the following expression (that is, mean squared deviation in prediction):

$$f = \frac{1}{Nn} \sum_{i=1}^{N} \sum_{j=1}^{n} (T_{ij} - C_{ij})^2, \tag{1}$$

where N and n represent the number of training scenarios and outputs, respectively. T_{ij} and C_{ij} denote the target and calculated values of jth output corresponding to i^{th} training scenario.

Approach 2:   GA-based optimization of the parameters of EFC and BP algorithms

In this approach, EFC has been used as the clustering algorithm. The number of cluster centers is determined by the algorithm itself. However, other parameters like the constant of similarity function a (which decides the relationship between similarity and distance), threshold value of similarity β (which decides whether a point will belong to a cluster based on its similarity with the cluster center), learning rate of the BP algorithm η, momentum term of the BP algorithm α and constant of log-sigmoid transfer function λ are coded in the GA-string. The overall string length is 50-bits, where 10-bits are assigned to denote each parameter. A GA-string used to optimize the above mentioned parameters looks as follows:

$$\underbrace{111010}_{a}\ldots\ldots\underbrace{010111}_{\beta}\ldots\ldots\underbrace{101110}_{\eta}\ldots\ldots\underbrace{111101}_{\alpha}\ldots\ldots\underbrace{001110}_{\lambda}\ldots\ldots,$$

Approach 3:   GA-based optimization of the parameters of proposed fuzzy clustering and BP algorithms

In this approach, clustering of the data is carried out using a new algorithm proposed in the present study by merging the properties of the FCM and EFC algorithms, so that the obtained clusters become distinct and compact, too. Hence, the number of cluster centers is determined by the EFC algorithm and then, the positions of the cluster centers are updated iteratively using the FCM algorithm. Ten bits have been assigned to represent each parameter. The parameters considered in this case are a, β, level of fuzziness g, learning rate η, momentum constant α and log-sigmoid transfer function λ. Thus, the GA-string is found to be 60-bits long. A typical string will look as follows:

$$\ldots\underbrace{111010}_{a}\ldots\underbrace{111010}_{\beta}\ldots\ldots\underbrace{010111}_{g}\ldots\ldots\underbrace{101110}_{\eta}\ldots\ldots\underbrace{111101}_{\alpha}\ldots\ldots\underbrace{001110}_{\lambda}\ldots\ldots,$$

5   Results and Discussion

This section presents the results of both forward and reverse mappings carried out on electron beam welded data of austenitic stainless steel plates.

5.1   Results of Forward Mapping

There are three input parameters in forward mapping, namely accelerating voltage, beam current and welding speed; and eight outputs, out of which seven related to the weld-bead geometry, such as bead penetration (BP), bead width (BW), bead height (BH), parameters : a_1, b_1, a_2, b_2 (refer to Fig. 1) and Vickers hardness (VH), have been considered.

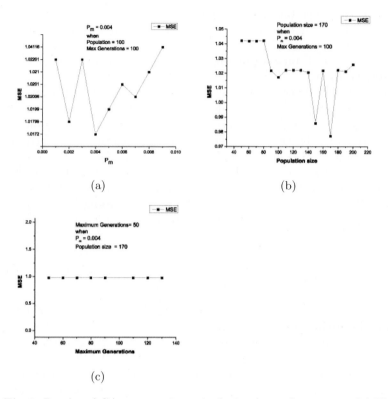

Fig. 2. Results of GA-parametric study during forward mapping of ASS-data in Approach 1

In **approach 1**, the set of optimal GA-parameters has been obtained through a parametric study, in which one parameter has been changed at a time, keeping the others fixed. Fig. 2 shows the results of this study. The following GA-parameters are found to yield the best results: Probability of mutation $P_m = 0.004$, population size $N = 170$, maximum number of generations $G = 50$. Moreover, a uniform crossover of probability equal to 0.5 has been used. Table 1 displays the ranges of variables optimized by the GA.

The optimized RBFNN parameters, such as c, g, η, α and λ are seen to be equal to 23, 1.1789, 0.1985, 0.85302 and 4.9780, respectively. In **approach 2**, the optimized GA-parameters, namely probability of mutation, population size and maximum number of generations are found to be equal to 0.08, 70 and 50, respectively. A uniform crossover as mentioned above has been used in this approach. During optimization, the ranges of constant of similarity function (a) and threshold value of similarity (β) have been kept fixed to $(0.4, 1.0)$ and $(0.2, 0.45)$, respectively. Other variables have been varied in the same ranges mentioned in Table 1. The parameters of RBFNN like a, β, η, α and λ are obtained as 0.9554, 0.3427, 0.9613, 0.8844 and 2.3167, respectively. The number of hidden neurons in this case is found to be equal to 14. In **approach 3**, the

Table 1. Ranges of the RBFNN parameters used in forward mapping

Number of Neurons (c)	2 to 25
Degree of Fuzziness (g)	1.0 to 2.0
Learning rate (η)	0.1 to 1.0
Momentum factor (α)	0.1 to 1.0
Const. of Log sigmoid function (λ)	0.5 to 5.0

Table 2. Results of test cases during forward mapping of ASS-data in terms of percent deviation in predictions

	Test Cases	Approach 1	Approach 2	Approach 3
	1	-45.15	32.02	-15.1
	2	-64.51	21.91	-33.48
BH	3	-86.22	8.61	-63.05
	4	-34.03	36.19	-8.2
	5	-3.94	50.12	16.1
	6	-39.38	32.57	-14.49
	1	-7.98	-14.27	-2.44
	2	-9.05	-14.55	-6.28
BP	3	-20.06	-10.04	-30.15
	4	5.74	10.19	-0.83
	5	3.50	4.92	1.46
	6	4.80	3.12	5.98
	1	9.66	30.32	13.72
	2	-4.53	17.33	1.97
BW	3	-10.93	9.61	-4.80
	4	5.76	26.38	6.51
	5	17.86	33.81	11.36
	6	1.48	16.28	3.63
	1	5.86	5.20	1.47
	2	4.17	6.09	6.57
VH	3	2.27	3.61	6.37
	4	6.07	5.16	1.10
	5	6.11	6.12	3.60
	6	6.66	6.07	7.18

best fitness value is obtained at 80 generations for the population size of 100 and mutation probability of 0.005. A uniform crossover with a probability equal to 0.5 has been utilized. The number of hidden neurons is found to be equal to 13. Here, the GA-string is seen to be 60-bits long, as there are six parameters to be optimized. During optimization, the ranges of these six parameters have been kept the same as mentioned above.

The values of the RBFNN parameters, such as g, η, α, λ, a and β are seen to be equal to 1.1818, 0.7501, 0.7211, 3.3020, 0.8332 and 0.3797, respectively. The performances of the above optimized networks have been tested on six different cases.

Table 3. Ranges of GA and RBFNN parameters used in reverse mapping

Parameters	Ranges	Approach 1	Approach 2	Approach 3
N	50-200	120	140	80
G	50-200	50	50	50
p_m	0.001-0.01	0.007	0.009	0.004
c	2-25	4	8	3
g	1-2	1.52	–	1.52
η	0.1-1	0.99	0.98	0.85
α	0.1-1	0.89	0.92	0.96
λ	0.5-5	4.53	4.45	4.78
a	0.7-0.85	–	0.83	0.78
β	0.3-0.4	–	0.3	0.3

Table 4. Results of test cases during reverse mapping of ASS-data in terms of percent deviation in predictions

	Test Cases	Approach 1	Approach 2	Approach 3
Acc. Voltage	1	-18.98	2.84	-14.97
	2	-17.83	6.64	-14.08
	3	-19.57	7.23	-13.43
	4	-0.30	16.65	0.45
	5	-5.20	7.83	0.04
	6	0.30	17.01	0.50
Beam Current	1	-18.46	-15.96	-14.17
	2	-18.69	-10.17	-13.86
	3	-21.06	-7.28	-13.60
	4	-14.34	-15.74	-14.14
	5	-4.19	-6.10	0.01
	6	0.53	-0.94	0.15
Weld speed	1	-29.23	-66.67	-33.33
	2	4.81	-25.00	0.00
	3	26.35	0.00	20.00
	4	-34.90	-66.67	-33.33
	5	-31.72	-66.67	-33.33
	6	-1.71	-25.00	0.00

Table 2 displays these results in terms of percent deviation in predictions of various outputs for the test cases, as obtained by the above three approaches.

Approaches 1, 2 and **3** have yielded the values of average absolute percent deviation in predictions of the outputs as 16.8178, 16.7860 and 12.7254, respectively. Moreover, the values of regression coefficients for the **approaches 1, 2** and **3** are found to be equal to 0.90, 0.90 and 0.93, respectively. Thus, **approach 3** has outperformed the other two approaches. It has happened so, due to the reason that a modified clustering technique has been used in **approach 3**. The proposed clustering technique utilized in **approach 3** combines the merits of the FCM and EFC algorithms to yield compact and distinct clusters.

5.2   Results of Reverse Mapping

In reverse mapping, eight input parameters, such as BP, BW, BH, a_1, b_1, a_2, b_2 and VH have been considered. The outputs are accelerating voltage, beam current and welding speed. The optimized values of the GA as well as RBFNN parameters along with their ranges have been displayed in Table 3.

Table 4 shows the results of reverse mapping on stainless steel welding data, as obtained by the above three approaches. **Approaches 1**, **2** and **3** have yielded the values of average absolute percent deviation in predictions of the process parameters as 14.8993, 20.2439 and 12.1889, respectively. The values of regression coefficients for the **approaches 1**, **2** and **3** are seen to be equal to 0.90, 0.86 and 0.90, respectively. Once again, **approach 3** is seen to outperform the other two approaches due to the same reasons mentioned earlier.

6   Summary

Input-output relationships of EBW process have been modeled in both forward as well as reverse directions using radial basis function neural networks. In order to decide the number of hidden neurons, input-output data set has been clustered based on the similarity among the data points using two well-known fuzzy clustering algorithms and one newly developed algorithm. Three approaches are developed for the forward and reverse mappings, each by utilizing three different clustering techniques. The performances of the developed approaches are tested on two different data sets. All these three approaches are found to carry out the forward and reverse mappings successfully within a reasonable accuracy limit. It has happened so, due to the fact that a hybrid scheme of optimization has been adopted in the said three approaches, where a global optimizer like GA has been used along with a local optimizer, namely BP algorithm. The difference in the performances of these approaches has come due to the application of different clustering algorithms. Approach 3 is found to be the best out of all the approaches. It may be due to the supremacy of the developed clustering technique over the other two well-known clustering techniques. To automate any process, its input-output relationships are to be known in both the forward and reverse directions, on-line. Thus, the present study may be considered as a significant step towards automating a process.

7   Scope of Future Work

In future, an attempt will be made to combine GA with a modern learning algorithm like Levenberg-Marquardt method to solve the same problem.

Acknowledgments. The authors acknowledge the support of EBW project (sponsored by BRNS-DAE), a joint venture of IIT Kharagpur, India and BARC Mumbai, India, for carrying out the experiments.

References

1. Broomhead, D.S., Lowe, L.: Multi-variable functional interpolation and adaptive networks. Complex Systems 11, 321–355 (1988)
2. Pratihar, D.K.: Soft Computing. Narosa Publishing House, New Delhi (2008)
3. Billings, S.A., Zheng, G.L.: Radial basis function network configuration using genetic algorithms. Neural Networks 8(6), 887–890 (1995)
4. Zhang, L., Bai, Y.F.: Genetic algorithm trained radial basis function neural networks for modelling photovoltaic panels. Engineering Applications of Artificial Intelligence 18, 833–844 (2005)
5. Zheng, G.L., Billings, S.: Radial basis function network configuration using mutual information and the orthogonal least squares algorithm. Neural Networks 9(9), 1619–1637 (1996)
6. Moradkhani, H., Hsu, K., Gupta, H.V., Sorooshian, S.: Improved stream flow forecasting using self organizing radial basis function artificial neural networks. Journal of Hydrology 295, 246–262 (2004)
7. Billings, S.A., Wei, H.L., Balikhin, M.A.: Generalized multiscale radial basis function networks. Neural Networks 20, 1081–1094 (2007)
8. Mollah, A.A., Pratihar, D.K.: Modeling of TIG welding and abrasive flow machining processes using radial basis function networks. International Journal of Advanced Manufacturing Technology 37, 937–952 (2008)
9. Amarnath, M.V.V., Pratihar, D.K.: Forward and reverse mappings of TIG welding process using radial basis function neural networks (RBFNNs). IMechE, Part B, Journal of Engineering Manufacturing 223, 1575–1590 (2009)
10. Dey, V., Pratihar, D.K., Datta, G.L., Jha, M.N., Saha, T.K., Bapat, A.V.: Optimization of bead geometry in electron beam welding using a Genetic Algorithm. Journal of Materials Processing Technology 209, 1151–1157 (2009)
11. Montgomery, D.C.: Design and analysis of experiments. John Wiley & Sons, Chichester (1997)
12. Bezdek, J.C.: Fuzzy mathematics in pattern classification, Ph. D. thesis, Applied Mathematics Center, Cornell University, Ithaca (1973)
13. Yao, J., Dash, M., Tan, S.T., Liu, H.: Entropy-based fuzzy clustering and fuzzy modeling. Fuzzy Sets and Systems 113, 381–388 (2000)

Modified Levenberg Marquardt Algorithm for Inverse Problems

Muthu Naveen[1], Shankar Jayaraman[2],
Vinay Ramanath[2], and Shamik Chaudhuri[2]

[1] PSG College of Technology, Coimbatore, India
muthu.naveen.89@gmail.com, shankar.jayaraman@ge.com,
vinay.ramanath@ge.com, shamik.chaudhuri@ge.com,
[2] GE Aviation, John F. Welch Technology Centre, Bangalore, India

Abstract. The Levenberg Marquardt (LM) algorithm is a popular non-linear least squares optimization technique for solving data matching problems. In this method, the damping parameter plays a vital role in determining the convergence of the system. This damping parameter is calculated arbitrarily in the classical LM, causing it to converge prematurely when used for solving real world engineering problems. This paper focuses on changes made to the classical LM algorithm to enhance its performance. This is achieved by adaptive damping, wherein the damping parameter is varied depending on the convergence of the objective function. To eliminate the need for a good initial guess, the idea of using an evolutionary algorithm in conjunction with the LM algorithm is also explored.

Keywords: Data Matching, Nonlinear Least squares, Levenberg Marquardt, criterion for termination of genetic algorithm.

1 Introduction

A mathematical model is often used to match actual experimental data for predicting the behavior of natural processes. Earlier, prototypes were used for prediction of behavior of engineering problems. But with the advancements in computing, computer simulations replaced this. Often, unknown parameters in the model are adjusted for obtaining a good fit. The best estimates of these are the ones that give the best fit between the test data and the model data. In order to find out the best values of these parameters, the problem is solved as a non-linear least squares problem, where the root sum squares of the error (difference between the test data and model data) is minimized iteratively. Such a problem is commonly known as an *inverse problem* [2] and is frequently encountered in the engineering field.

For example, in fluid dynamics, the test data is a set of flow values, which is experimentally determined. The scalars are the flow coefficients and the simulation model is used to match the physical flow. The scalars are optimized such that the output of the simulation model matches the experimental results. Figure 1 shows

K. Deb et al. (Eds.): SEAL 2010, LNCS 6457, pp. 623–632, 2010.
© Springer-Verlag Berlin Heidelberg 2010

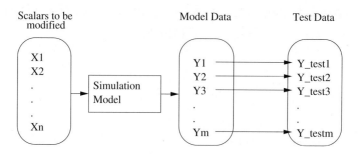

Fig. 1. Schematic of Inverse Problem Approach

the inverse problem schematically. Objective of the data-matching problem is to minimize the difference between test data and model data. The ideal value of the objective function (OF) of the data matching is zero, which implies that the experimental data and model data are matching perfectly. Data-matching problem can be solved in two different ways.

1. Least Squares minimization based method
2. Methods based on point to point match

Method one involves minimization of the OF, which is the root sum square of test data and model data. Information about the individual error of all data points are not provided to the optimizer.

$$\text{Objective Function, } OF = \sqrt{\sum (Y_{test} - Y_{model})^2}$$

In method two, the optimizer makes use of information about error in each data point and the slope to minimize the error between test data and model data. This method has been explained in detail by Akkaram et al. [2] There are two popular algorithms to solve this genre of optimization problems.

1. Gauss Newton Algorithm (GN)   2. Levenberg Marquardt Algorithm (LM)

Genetic algorithms (GA) also have been used in a few instances to solve data matching problems [4]. But they involve numerous function evaluations, which prove expensive. As a result, tackling real-life engineering problems becomes difficult. The focus of this paper is two fold:

1. Improvements on the present LM algorithm to force better convergence.
2. Usage of GA in conjunction with the LM algorithm to eliminate the dependence on a user-specified initial guess.

2   Overview of Gauss Newton Algorithm

The GN algorithm is a method used to solve non-linear least squares problems. It can be seen as a modification of Newton's method for finding a minimum of a function. The GN algorithm has the advantage that second derivatives, which can be challenging to compute, are not required. The Hessian is approximated by $J^T J$ where J is the Jacobian matrix [10]. Given m functions $r_1, \ldots, r_m$ of n variables $X = (x_1, \ldots, x_n)$, with $m \geq n$, the GN algorithm finds the minimum of the sum squares of errors. Starting with an initial guess X_i for the minimum, the method proceeds by the iterations as shown below, where the increment Δ is the solution to the normal equations.

$$\sum_{i=1}^{m} (Y_{test} - Y_{model})^2 \qquad Xi_n = Xi_{n-1} + \Delta_i \qquad (J^T J)\Delta = -J^T r$$

The GN search direction is a descent direction and is thus a suitable direction for a line search [10]. This method is very efficient when the initial guess is very close to the actual solution.

3   Overview of Levenberg Marquardt Algorithm

The efficiency of the GN algorithm is dependent on the accuracy of the initial guess. Moreover there may occur instances where $J^T J$ becomes singular and hence its inverse cannot be obtained when solving for Δ.

The LM algorithm is a damped least squares optimization technique, which is a weighted combination of the GN and the steepest descent technique [9]. Levenberg first proposed the damped GN method to avoid the weakness of GN when J becomes rank deficient [6]. Marquardt extended this idea by providing options to control the damping LM algorithm [8]. The damping parameter is added to the diagonal of the Jacobian as $(J^T J + \lambda.diag(J^T J))\Delta = -J^T r$ where λ is the damping parameter. The value of the new Xs are obtained as $Xi_n = Xi_{n-1} + \Delta_i$. The damping parameter prevents the occurrence of a singular matrix. This method is more robust than the GN, i.e. it can work well even if the initial guess is slightly off from the globally known solution.

3.1   Inefficient Damping Parameter Estimation

The damping parameter in the classical LM algorithm is updated as follows: A new set of Xs is calculated and the OF is evaluated. If this OF is less than the OF of the previous iteration, the damping parameter is updated as

New damping parameter = Previous damping parameter / Reduction constant.
If the OF is more than that of the previous iteration,
New damping parameter = Previous damping parameter * Reduction constant

Previous efforts have been made by a few authors ([3], [5]) to calculate the value of the damping parameter based on the previous and the present value of Jacobian, but most of these are not suitable for large scale engineering problems. Improper selection of the damping parameter may lead to pre-convergence. Therefore selection of an appropriate damping parameter is critical to the functioning of the algorithm. Hence a methodology has to be devised where the damping parameter is varied dynamically.

4   Modified LM: Improved Damping Parameter Updation

The behaviour of the damping parameter λ in the classical LM was extensively studied and graphically depicted in Figure 2. It was found that the convergence was rapid when λ was approximately in the range of 0.1 to 100. The improvement in the OF was very low when the value of λ was either very high or very low. A very high value of the damping parameter causes the algorithm to terminate. Based on these observations, the value of the damping parameter in the proposed method is *forced* to a specific value depending on the relative change in OF value. The relative change in the OF is computed as follows.

$$\Delta OF_1 = (OF)_{n-1} - (OF)_n \qquad \Delta OF_2 = (OF)_n - (OF)_{n+1}$$
$$\text{Relative change (RC)} = \Delta OF_2/\Delta OF_1$$

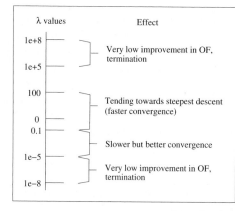

Fig. 2. Impact of λ on classical LM

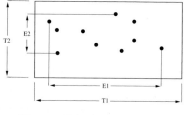

Fig. 3. Calculation of Exploration Metric

A low RC value (less than 1%) implies a very small improvement in the OF value. As a result, the damping parameter is assigned a high value forcing it to terminate. When RC is greater than 25%, the damping parameter is assigned a value between 0.1 and 100 for faster convergence. When the improvement in OF tends to slow down, i.e. RC is less than 25%, it is a signal that the algorithm is about to prematurely terminate and hence a smaller value of damping parameter is assigned to prevent it from terminating. Thus, this dynamic updation of the damping parameter effectively controls the functioning of the LM algorithm.

5 Employing GA in Conjunction with LM Algorithm

Solving a data-matching problem has always been associated with the usage of gradient-based optimization techniques. The usage of GA to solve a computationally-intensive problem is not preferred because of the large number of function evaluations required, which is true for most engineering problems. However, the main advantage of GA is that it does not require an initial guess and converges to the global optimum.

Gradient-based techniques such as the LM and the GN have a high dependence on the initial guess, which is generally not known to the user. Thus the user has to undertake many trials with various initial guesses to arrive at the global optimum. Hence, an effective solution to this problem would be to use GA to explore the whole design space in order to provide a good initial guess to the LM algorithm.

Since GA requires a large number of function evaluations to converge to the global optimum, it may be advisable to terminate it when the improvement in the OF is small and the design space has been explored significantly. A detailed investigation on this front has led to the development of two innovative metrics: *1. Exploration Metric* and *2. Mean Metric.*

5.1 Exploration Metric

Genetic Algorithms use a set of solutions to search the design space exhaustively to find the global optimum. During the search process GA try to explore the whole design space and finally concentrate on the location where global optimum resides. Since exploration is computationally costly, in this paper we propose an *Exploration Metric*, which measures the extent of exploration of the design space. The exploration metric is defined as the ratio of the explored hyper-volume to the total hyper-volume of the design space.

Exploration Metric = Explored Hyper-volume/Total Hyper-volume
Total Hyper-volume = $(x_1^{UB} - x_1^{LB})(x_2^{UB} - x_2^{LB}) \ldots (x_n^{UB} - x_n^{LB})$
Explored Hyper-volume = $(x_1^{max} - x_1^{min})(x_2^{max} - X_2^{min}) \ldots (x_n^{max} - X_n^{min})$

x_n^{LB} and x_n^{UB} are the lower and upper bound of the variable x_n and X_n^{min} and x_n^{max} are the minimum and maximum achieved values of x_n till the current generation. An example with two variables has been shown in the Figure 3.

Total Hyper-volume = $T1 * T2$
Explored Hyper-volume = $E1 * E2$
Exploration Metric = $(E1 * E2)/(T1 * T2)$

5.2 Mean Metric

This metric is used to estimate the convergence of the GA, based on the change in the mean OF value over a given number of generations.

Mean Metric = Mean OF of Nth gen − Mean OF of $(N − k)$th gen,
where $5 \leq k \leq 10$ and $N > k$.

If the Mean Metric value is below a user-specified $MeanTol\%$ of the cumulative OF mean (mean OF across all generations), the GA is assumed to have converged reasonably well. Figure 4 below shows the mean value of the OF of each generation for a typical GA run. It can be seen that during the first few generations, the reduction in mean OF is very rapid whereas after the 17th generation, the reduction is slow. The mean metric condition will indicate that suitable convergence has been achieved at this stage.

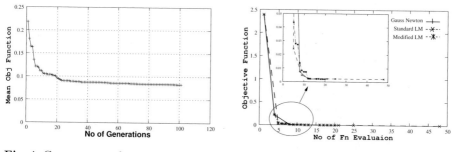

Fig. 4. Convergence of mean OF in Genetic Algorithm

Fig. 5. Covergence of Trial 1 for Test Problem 1

The termination of the GA is enforced when the desired percentage of exploration is reached and the condition for the mean metric is satisfied.

6   Results and Inference

The proposed algorithms, modified LM and GA-LM combined algorithm have been tried out on a transient algebraic function and on a transient engine model. The performance of these algorithms has been compared with GN algorithm and classical LM algorithm.Comparison has been based on the final OF value and the number of function evaluations required to achieve that. Moreover to understand the search power of the algorithms, each of them has been tried out on three different test cases defined as follows.

 − The initial guess is very close to the known solution.
 − The initial guess is moderately close to the known solution.
 − The initial guess is far off from the known solution.

During performance evaluation, the GN algorithm suggested by Heath [9] has been used along with a line-search algorithm. For classical LM algorithm, MINPACK package has been used for evaluation. The modified version of the LM algorithm has been developed from the levmar [7] implementation.

6.1   Test Problem 1: Algebraic Transient Function

Function : $f(t,x) = x_1 e^{x_2 t}$ Test Data :

t	0	1	2	3
y	2.0	0.7	0.3	0.1

Actual Solution: $x1 = 1.995, x2 = -1.01$

The inverse problem was solved with four different test cases based on the initial guess selection. The initial guess is very close to the optimum solution in case 1 and case 2, moderately close in case 3, and far off from the optimum in case 4. Results obtained from the different algorithms are presented in Table 1.

Table 1. Performance Comparison of Proposed Algorithms

Sl. No	Initial Guess	Gauss Newton		Classical LM		Modified LM	
		No of Fn Evals	Final OF value	No of Fn Evals	Final OF value	No of Fn Evals	Final OF value
1	(1, 0)	16	0.001996	48	0.00223	20	0.001998
2	(2, 2)	34	0.001997	69	0.00223	36	0.002005
3	(-3, 3)	Singular		81	0.002233	37	0.002003
4	(20,20)	Singular		50	1.07	121	4.58

GA in conjunction with LM : Objective Function Value : 0.001996
Total no. of function evaluations : GA 70 + LM 36 = 106

Observations:

1. It can be seen from Figure 5 (Case 1) and Figure 6 (Case 2) that with a good initial guess, the GN works very efficiently since it makes use of a line search routine. It is found that the classical LM consumes almost twice the number of function evaluations as that of the proposed modified LM. The modified LM effectively converges due to the effect of the dynamic damping parameter as discussed in Section 4.

2. In Case 3 and Case 4, the Jacobian in the GN algorithm becomes singular and hence the algorithm fails. This shows that GN is not suitable for an arbitrary initial guess. It is again seen in Case 3 that the modified LM converges in almost half the iterations as compared to the classical LM. Figure 7 shows the convergence of these two algorithms.

3. If the initial guess is far away from the actual solution (Case 4), it is seen that the classical LM performs better than the modified LM. Further studies will need to be done to determine the exact cause of this behaviour.

4. The use of a Genetic Algorithm in conjunction with modified LM ensured convergence in 106 (70 from GA, 36 from LM) function evaluations without

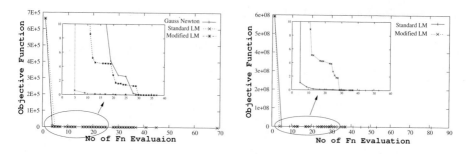

Fig. 6. Covergence Plot for Case 2   **Fig. 7.** Convergence Plot for Case 3

the need for an initial guess. This method is suitable when the user has very little idea of a good initial guess.

5. It is observed that all algorithms converge to approximately the same OF value.

6.2 Test Problem 2: Transient Engine Model

The second test problem used for comparison is a single spool thermodynamic performance model called AIR4548 [1]. There are 6 input scalars (Xs), and 9 transient outputs (Ys) with 52 time points for each Y. Since the values of the transient outputs are in different scales, they are normalized. Again, like in Test Problem 1, four test cases based on different initial guesses have been considered for studying the behaviour of the algorithms. Simulation results are presented in Table 2. Another test case (test case 5) has been considered, where LM algorithm is run in conjunction with GA.

Observations:

1. It is observed that the classical LM converges prematurely in all four cases (as the OF values are relatively high). This is evident in Figures 8 - 11.
2. In all four cases, GN is able to converge to a better solution than the classical LM, although the number of function evaluations is higher. When the initial guess is very good (Case 1), the number of function evaluations required by GN is comparable to that of the classical LM.
3. The modified LM is found to be converge much better (lowest OF value) than the other two algorithms, irrespective of the initial guess.
4. Even when the initial guess is away from the optimum (Case 3 and Case 4), the modified LM converges very rapidly, as seen in Figure 10 and Figure 11 respectively.
5. When GA is used along with the modified LM, a far superior convergence is achieved. Though this method requires more function evaluations as compared to pure gradient based techniques, it is more robust and eliminates the need for trials with different initial guesses.

Table 2. Comparison Chart for AIR4548 Problem

Sl. No	Initial Guess	Gauss Newton		Classical LM		Modified LM	
		No of Fn Evals	Final OF value	No of Fn Evals	Final OF value	No of Fn Evals	Final OF value
1	*Initial guess very close to the optimum*						
	(0.945, 0.945, 0.089, 1.01, 1.05, 1.01)	54	0.00261476	31	0.0033626	118	0.000162
2	*Initial guess close to the optimum*						
	(0.93, 0.97, 0.085, 0.98, 1.02, 1.02)	234	0.00359118	33	0.014319	100	0.000406
3	*Initial guess moderately close to the optimum*						
	(0.9, 0.9, 0.095, 1.05, 1.05, 1.1)	333	0.000633724	46	0.157508	106	0.000102
4	*Initial guess far off from the optimum*						
	(1.1, 1.1, 0.08, 0.9, 0.9, 1.1)	288	0.613374	57	0.726287	111	0.000322
5	*GA in conjunction with LM*						
	Total no of function evaluations : 346			Final OF Value : 3.452e-005			

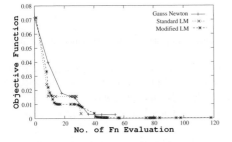

Fig. 8. Covergence Plot of Case 1

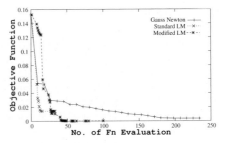

Fig. 9. Covergence Plot of Casel 2

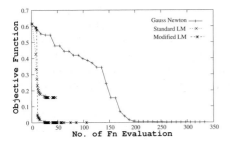

Fig. 10. Covergence Plot of Case 3

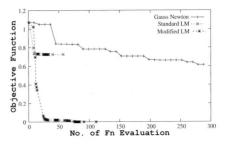

Fig. 11. Covergence Plot of Case 4

7   Conclusions and Scope for Further Work

The paper provides a comparison between the GN and the LM algorithms, which are commonly used for inverse optimization problems. The advantages and disadvantages of both the algorithms are discussed and the need to refine the classical LM algorithm is identified. In the revised LM algorithm, the damping parameter calculation is modified. Though this change makes the LM more efficient, it is still dependent upon a user defined initial guess. The idea of using GA in conjunction with LM is explored. The dynamic termination of the GA is brought about by introducing two new metrics. This method of using both GA combined with LM is more robust and can solve inverse problems without any need for an initial guess. Even though this method requires many function evaluations, it can converge to the global solution in a single trial as compared to the LM or the GN algorithms which may need multiple trials with different initial guesses.

The present work calculates only a single damping parameter which is applied for all the diagonal entries. Future work relating to arriving at a mathematical formulation to calculate the damping parameters for each diagonal entry is desired. Further work can be carried to out to enable the algorithm to handle large data sets. Research related to bringing down the computational expense of GA integrated with LM would be very beneficial to the engineering community.

References

1. Real-time modeling methods for gas turbine engine performance, document number air4548 (2007)
2. Akkaram, S., Beeson, D., Agarwal, H., Wiggs, G.: Inverse modeling techniques for parameter estimation in engineering simulation models. In: Proceedings of ASME Turbo Expo 2006 Power for Land, Sea, and Air, pp. 63–72 (2006)
3. Evelyn, A.: A variation of the levenberg marquardt method. an attempt to improve effeciency (2004)
4. Jiang, M., Xia, L., Shou, G.: The use of genetic algorithm for solving the inverse problem of electrocardiography. In: Conf. Proc. IEEE Eng. Med. Biol. Soc., pp. 3907–3910 (2006)
5. Lampton, M.: Damping undamping stratagies for the levenberg marquardt non linear least squares method. Computers in Physics 11(1), 110–115 (1997)
6. Levenberg, K.: A method for the solution of certain non-linear problems in least squares. Quarterly Journal of Applied Mathmatics II(2), 164–168 (1944)
7. Lourakis, M.: levmar: Levenberg-marquardt nonlinear least squares algorithms in C/C++ (July 2004), http://www.ics.forth.gr/~lourakis/levmar/ (Accessed on 31 Jan. 2005)
8. Marquardt, D.W.: An algorithm for least squares estimation of nonlinear parameters. Journal of the Institute of Mathematics and its Applications 11(2), 431–441 (1963)
9. Michael, H.T.: Scientific Computing An Introductory Survey. The McGraw-Hill Companies Inc., New York (2002)
10. Shidong, S.: A levenberg marquardt method for large scale bound constrained non linear least squares (2008)

Constrained Engineering Design Optimization Using a Hybrid Bi-objective Evolutionary-Classical Methodology

Rituparna Datta

Kanpur Genetic Algorithms Laboratory (KanGAL),
Department of Mechanical Engineering,
Indian Institute of Technology Kanpur, Pin 208016, India
rdatta@iitk.ac.in

Abstract. Constrained engineering design optimization problems are usually computationally expensive due to non-linearity and non convexity of the constraint functions. Penalty function methods are found to be quite popular due to their simplicity and ease of implementation, but they require an appropriate value of the penalty parameter. Bi-objective approach is one of the methods to handle constraints, in which the minimization of the constraint violation is included as an additional objective. In this paper, constrained engineering design optimization problems are solved by combining the penalty function approach with a bi-objective evolutionary approach which play complementary roles to help each other. The penalty parameter is approximated using bi-objective approach and a classical method is used for the solution of unconstrained penalized function. In this methodology, we have also eliminated the local search parameter which was needed in our previous study.

Keywords: Constrained Optimization, Engineering Design, Penalty Function and Bi-objective evolutionary algorithms.

1 Introduction

Most real-life scientific and engineering design problems involve a large number of non-linear, non-convex, and discontinuous constraints and objective function. Evolutionary population based algorithms are widely used to solve constrained optimization problems. Many researchers have already proposed methodologies using different evolutionary algorithms to solve constrained optimization problems in engineering design [1], [2] and [3].

Penalty function method has been found to be the most popular constrained handling technique due to its simple principle and easy implementation. A penalty-parameter-less strategy was proposed by Deb in 2000 [4], in which an infeasible solution is always treated worse than a feasible solution and the approach avoided the need of any penalty parameter. Coello [5] proposed a self-adaptive penalty approach by using a co-evolutionary model to adapt the penalty factors.

K. Deb et al. (Eds.): SEAL 2010, LNCS 6457, pp. 633–637, 2010.

Some studies proposed the conversion of original problem into a bi-objective optimization problem in which a measure of an overall constraint violation is used as an additional objective [6]. Recently we proposed [7] an extention of the bi-objective idea that used the reference point approach to focus its search near the constrained minimum solution.

Although many other ideas are suggested, researchers realized that the task of finding the constrained optimum by an EA can be made more efficient and accurate, if it is hybridized with a classical local search procedure [8]. Recently, we have suggested a bi-objective evolutionary optimization strategy [9] to estimate the penalty parameter R for a problem from the obtained two-objective non-dominated front. Thereafter, an appropriate penalized function is constructed and solved using a classical local search method.

In the remainder of this paper, the proposed algorithm is described, which is an extension of our bi-objective optimization procedure [9]. Two standard engineering design optimization problems considered from the existing literature and showed the results obtained from our proposed algorithm. Results are compared with some of the best-known results from the literature. Finally, the paper ends with the conclusions of this study.

2   Proposed Algorithm

This section describes the algorithm based on the principles of bi-objective handling of a constrained optimization problem and the use of a penalty function approach. First, the generation counter is set at $t = 0$.

Step 1: An Evolutionary Multi-Objective Optimization (EMO) algorithm (NSGA-II [10]) applied to the bi-objective optimization problem to find the non-dominated Pareto-optimal front. Here CV(x) is defined as follows:

$$
\begin{aligned}
&\text{minimize } f(x),\\
&\text{minimize } \text{CV}(x),\\
&\text{subject to } \text{CV}(x) \le c,\\
&\qquad x^{(L)} \le x \le x^{(U)}.
\end{aligned}
\tag{1}
$$

Since CV(x) is the normalized constraint violation, we have used $c = 0.2J$, where J is number of constraints in all our studies and $x^{(L)}$, $x^{(U)}$ are the lower and upper variable bounds.

Step 2: If $t > 0$ and $((t \bmod \tau) = 0)$, compute R from the current non-dominated front as follows. A cubic curve is fitted for the non-dominated points $(f = a + b(CV) + c(CV)^2 + d(CV)^3)$ and the penalty parameter is estimated by finding the slope at CV=0, that is $R = -b$. Since this is a lower bound on R, we use twice this value as R or $R \leftarrow -2b$.

Step 3: Thereafter, following local search problem is solved with R computed from above and starting with the current minimum-CV solution:

$$
\begin{aligned}
&\text{minimize } P(x) = f(x) + R\sum_{j=1}^{J}\langle \hat{g}_j(x)\rangle,\\
&\qquad x^{(L)} \le x \le x^{(U)}.
\end{aligned}
\tag{2}
$$

Say, the solution is $\bar{x}$.

Step 4: If $\bar{x}$ is feasible and the difference between $f(\bar{x})$ and the objective value of the previous local searched solution is smaller than a small number δ_f (10^{-4} is used here), the algorithm is terminated and $\bar{x}$ is declared as the optimized solution. Else, we increment t by one and proceed to Step 1.

It is interesting to note that, the penalty parameter R is no more a user-tunable parameter and gets adjusted from the obtained non-dominated front. We use Matlab's `fmincon()` procedure to solve the penalized function with reasonable parameter settings. We have eliminated the frequency of local search as a parameter and use local search at every generation.

3   Results on Engineering Design Optimization Problems

Our proposed methodology was applied to a number of engineering design optimization problems, taken from [1], [2], [3]. Following parameters are used: population size $= 20n$, SBX probability$= 0.9$, SBX index $= 10$, polynomial mutation probability $= 1/n$, and mutation index $= 100$. The termination criterion is described in Section 2. In each case, we run our algorithm 50 times from different initial populations.

3.1   Three-Bar Truss Design

The volume of the truss structure is to be minimized subject to the stress constraints.

Figure 1 shows the variation in the population-best objective value for a particular run. All solutions are feasible right from the initial population. The adaptation of R is also shown in the figure. At generation 2, the first local search is performed and optimal solution is found. However, the algorithm continues for another generation and local search to satisfy our termination criteria. The value of R at the end of generation 3 is found to be 2.306753 for this simulation run. All

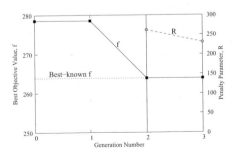

Fig. 1. Function value reduces with generation for three-bar truss design problem

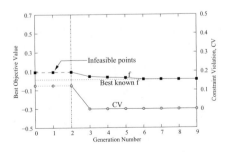

Fig. 2. Function value reduces with generation for spring design problem

Table 1. Comparison of objective value and function evaluations for three-bar truss design problem

Algorithms	Best	Median	Mean	Worst	FE
ADE[3]	263.895843	263.895843	263.895843	263.895843	45,000
DSS-MDE[1]	263.895843	263.895843	263.895843	263.895849	15,000
Ray & Liew[2]	263.895846	263.8989	263.9033	263.96975	17,610
proposed method	263.895843	263.895843	263.895843	263.895844	
FE	246	327		493	
EMO+LOCAL	120 + 126	120 + 207		160 + 333	

50 runs find a feasible solution close (with in 10^{-4}) to the best-known optimum. The best performance of our algorithm requires only 246 evaluations, whereas the best-reported existing EA methodology takes at least 70 times more function evaluations to achieve similar accurate solutions. Table 1 shows the comparison of results taken from literature with our approach. In addition to total function evaluations (FE), the sub-division of FE in EMO (NSGA-II) and local search are shown in the next line.

3.2   Tension/Compression Spring Design

The objective is to minimize the weight of a tension/compression spring.

Figure 2 shows the best objective value of the population and the corresponding constraint violation value with the generation for a typical simulation out of 50 runs. The figure shows that initially no feasible solution is found up to the second generation. At generation 3, the first feasible solutions appear. The corresponding CV value is zero, indicating that the obtained solution is feasible.

Table 2 is the comparison of our approach with that of the best, median, mean and worst function values taken from literature.

Table 2. Comparison of objective value and function evaluations for spring design problem

Algorithms	Best	Median	Mean	Worst	FE
ADE[3]	0.012665	0.012665	0.012933	0.020643	60,000
DSS-MDE [1]	0.012665	0.012665	0.012669	0.012738	24,000
Ray & Liew[2]	0.012669	0.012922	0.012922	0.016717	25,167
Coello[5]	0.012704	0.012755	0.012769	0.012822	9,00,000
Proposed method	0.012665	0.012665	0.012666	0.012668	
FE	1,826	3,886		8,774	
EMO+LOCAL	840+986	1,860+2,026		5,520+3,254	

4   Conclusions

This paper is an integration between a bi-objective evolutionary approach with the penalty function based classical optimization, which alleviates the drawback of each other. The main limitation in accurate convergence, to find the optimum by an EMO procedure is alleviated by the use of a local search involving a classical optimization procedure and the difficulty of the commonly-used penalty function based approach is alleviated by estimating the penalty parameter adaptively by the EMO procedure. The promising results on two engineering design test problems are compared with some state-of-the-art algorithms and results indicate that the proposed procedure is robust, faster and accurate than the existing ones.

Acknowledgments: The study is funded by Department of Science and Technology, Government of India under SERC-Engineering Sciences scheme (No. $SR/S3/MERC/091/2009$).

The author is highly grateful to Prof. Kalyanmoy Deb, Director, Kanpur Genetic algorithms Laboratory (KanGAL), IIT Kanpur for discussion and advice on these ideas and also the computational facilities at KanGAL.

References

1. Zhang, M., Luo, W., Wang, X.: Differential evolution with dynamic stochastic selection for constrained optimization. J. Information Sciences 178(15), 3043–3074 (2008)
2. Ray, T., Liew, K.M.: Society and civilization: An optimization algorithm based on the simulation of social behavior. J. IEEE Transactions on Evolutionary Computation 7(4), 386–396 (2003)
3. Youyun, A.O., Hongqin, C.H.I.: An Adaptive Differential Evolution Algorithm to Solve Constrained Optimization Problems in Engineering Design. J. Engineering 2, 65–77 (2010)
4. Deb, K.: An efficient constraint handling method for genetic algorithms. J. Computer methods in applied mechanics and engineering 186(2-4), 311–338 (2000)
5. Coello Coello, C.A.: Use of a self-adaptive penalty approach for engineering optimization problems. J. Computers in Industry 41(2), 113–127 (2000)
6. Surry, P.D., Radcliffe, N.J., Boyd, I.D.: A multi-objective approach to constrained optimisation of gas supply networks: The COMOGA method. In: Fogarty, T.C. (ed.) AISB-WS 1995. LNCS, vol. 993, pp. 166–180. Springer, Heidelberg (1995)
7. Deb, K., Lele, S., Datta, R.: A hybrid evolutionary multi-objective and SQP based procedure for constrained optimization. In: Kang, L., Liu, Y., Zeng, S. (eds.) ISICA 2007. LNCS, vol. 4683, pp. 36–45. Springer, Heidelberg (2007)
8. Myung, H., Kim, J.H.: Hybrid interior-lagrangian penalty based evolutionary optimization. In: Porto, V.W., Waagen, D. (eds.) EP 1998. LNCS, vol. 1447, pp. 85–94. Springer, Heidelberg (1998)
9. Deb, K., Datta, R.: A Fast and Accurate Solution of Constrained Optimization Problems Using a Hybrid Bi-Objective and Penalty Function Approach. In: Congress on Evolutionary Computation - CEC (2010)
10. Deb, K.: Multi-objective optimization using evolutionary algorithms. Wiley, Chichester (2001)

A Many-Objective Optimisation Decision-Making Process Applied to Automotive Diesel Engine Calibration

Robert J. Lygoe[1], Mark Cary[1], and Peter J. Fleming[2]

[1] Powertrain Calibration & Development, Ford Motor Company, 15/2A-F13-C, Dunton Technical Centre, Laindon, Basildon, Essex. SS15 6EE. U.K.
{blygoe,mcary}@ford.com
[2] Automatic Control & Systems Engineering, The University of Sheffield, Mappin Street, Sheffield. S1 3JD. U.K.
p.fleming@sheffield.ac.uk

Abstract. A novel process has been developed for reducing complexity in real-world, high-dimensional, multi-objective optimisation problems. This approach relies on being able to identify and exploit local harmony between objectives to reduce dimensionality. To achieve this, a systematic and modular process has been designed to cluster the Pareto-optimal front and apply a rule-based Principal Component Analysis including preference articulation for potential objective reduction. This many-objective optimisation decision-making process is demonstrated on a real-world, automotive diesel engine calibration optimisation problem comprising six objectives. The complexity reduction process resulted in three- and four-objective sub-problems. In the former, a significant improvement was achieved in one of the retained objectives at very little cost to the others.

Keywords: optimisation, many-objective, dimension reduction, engine calibration.

1 Introduction

During the process of carrying out automotive engine calibration, it is common to come across trade-off problems, that is, optimisation problems comprising two or more competing objectives. In the automotive market there is ever-increasing customer demand for more fuel efficient, higher performance, increased refinement and reliability at low cost. These requirements combined with evermore stringent exhaust emissions legislation and fierce competition amongst automotive manufacturers has led to more complex engine technologies. This has, in turn, driven the development of correspondingly complex control systems, with more actuator variables and more engine responses or objectives to be optimised and traded off.

Historically, optimisations were formulated as single objective problems, which were solved using methods available at the time, *e.g.* gradient-based or direct search algorithms. These approaches have a number of weaknesses including a tendency to get stuck in local optima, the fact that they are often designed to be problem-specific and they require multiple runs to generate a family of solutions as required for a

K. Deb et al. (Eds.): SEAL 2010, LNCS 6457, pp. 638–646, 2010.

multi-objective optimisation by definition [1]. By contrast, Evolutionary Algorithms (EAs) evolve a population of solutions to search for the optimal trade-off or Pareto-optimal front. Such methods are able to produce a diverse set of solutions in one run of the optimiser and are well suited to multi-objective problems.

EAs have been mostly applied to two- or three-objective optimisations, the results of which are straightforward to visualise in low dimensions. However, real-world and in particular, modern, complex engine calibration optimisation problems can involve significantly more than two or three objectives, termed *many-objective* optimisations. For such high-dimensional problems, multi-objective EAs (MOEAs) have issues with lack of effective search, potentially large population size required (may be computationally expensive) and visualisation of the solutions, which may be sparse [2]. Reviews of some recent and relevant potential counter-measures is provided in [3,4,5].

Research to address these issues has primarily concentrated on algorithmic developments to improve the search effectiveness. Nevertheless, even if a Pareto-optimal population has been generated, the subsequent decision-making process to select a preferred solution has received little attention, particularly in the case of many-objective problems.

2 Background

There exists an opportunity to reduce the dimensionality in many-objective problems if, for the Pareto-optimal solutions in the Decision-Maker's (DM's) region of interest, objectives are sufficiently positively correlated, *i.e.* in harmony [6,7]. In this case, improvement in one objective would automatically improve another positively correlated objective.

In order to discover such local objective dependency, if it exists, it is necessary to partition or cluster the Pareto-optimal front into groups of like-solutions. This will allow any local objective harmony to be exploited for local objective reduction. In addition, other studies [8,9] suggest that sub-dividing the Pareto-optimal front is useful for visualizing high dimensional Pareto-optimal fronts and grouping similar solutions.

In order to be effective as part of a many-objective optimisation decision-making (MOODM) process, a clustering algorithm must fulfill the following requirements: i) be efficient to run, as sufficient data density is required to determine the number and location of clusters and it may need to be run a number of times to generate reliable results, ii) generate the correct number of clusters in high-dimensional, real-world problems given that the number of clusters is not known *a priori* and iii) produce a *valid* clustering structure consistent with the downstream objective reduction process, *e.g.* Principal Components Analysis (PCA) assumes the clusters are hyper-ellipsoidal. Assessed against these requirements, the k*-Means algorithm [10] appears to be a good choice and has two main steps. Firstly, a pre-processing procedure is carried out, which assigns at least one cluster centre to each of the initial clusters. The second step involves applying a learning rule to adaptively adjust each centre to a cluster while penalising rival centres. It generates elliptical clusters and determines the number of output clusters. Simulation testing on known data was carried out in [3] on multi-variate normal and non-normal data. This provided evidence that the k*-Means algorithm did meet the requirements for clustering in a MOODM process and was a suitable choice for partitioning a Pareto-optimal front. Furthermore, estimation of cluster mean and covariance matrices robust to outliers was provided by the FAST-MCD algorithm [11].

Having partitioned the Pareto-optimal front into clusters, an appropriate method is required to in order to identify any objective dependency and hence potential dimension reduction. Dimension reduction approaches are described and reviewed in [12,13]. With regard to linear dimension reduction approaches, PCA is a widely-used method and comprises a linear transformation of the objectives to a new set of uncorrelated variables (or Principal Components) that account for the majority of the variation in the original objectives (full details are provided in [14]). Linear PCA is cited as not being suitable for non-linear data such as that typical of Pareto-optimal fronts. However, if the Pareto-optimal front is partitioned into groups of like solutions, then PCA may be useful in identifying local harmony for objective reduction. Previously, Deb and Saxena [23] have used PCA as part of an objective reduction process, but this was applied to the whole Pareto-optimal front and the authors stated that this approach showed some vulnerability in that not all conflicting objectives could be correctly identified. This approach was applied in [3], but resulted in a somewhat drastic objective reduction, one of the high priority objectives being discarded and many of the solutions violating a constraint.

With non-linear methods, the DM may need to specify additional information such as the non-linear transform required or distributional assumption. In addition, there are known problems with Multi-Dimensional Scaling not being able to project onto lower dimensions [12,16]; with Self-Organising Maps [17], which have issues with subjectivity involved in hierarchical clustering, convergence and interpretation; and with Vector Quantisation, where the DM must specify target dimension *a priori* and no consideration is given to objective harmony and conflict [18].

3 Proposed Process

The clustering and PCA elements previously discussed can be combined with an efficient search method to produce a systematic dimension reduction process. The goal is to aid the DM in discovering opportunities for progressively simplifying the optimisation by reducing the number of objectives at each step. The proposed MOODM process is displayed in Figure 1 and shows the elements of optimisation, robust clustering and objective reduction via PCA-based heuristic rules combined in a proposed MOODM process.

The first step is to efficiently generate a Pareto-Optimal Population (POP). There are many MOEAs that can be used to achieve this, but for many-objective problems, careful consideration must be given to overcome the known issues of lack of search efficiency and very large population sizes required. On completion of the optimisation, the resulting POP is partitioned into groups of like-solutions using a clustering algorithm. It is important that the DM is satisfied that there is evidence of the number and location of the clusters and so, cluster verification rules have been developed. Subsequently, for each POP cluster, PCA together with some heuristic rules are applied to

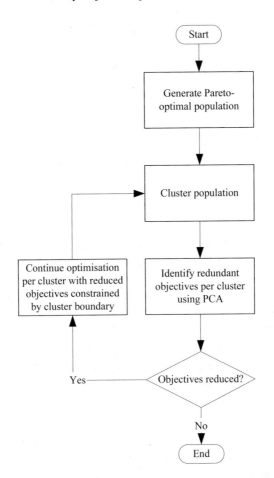

Fig. 1. Proposed Many-Objective Optimisation Decision-Making (MOODM) Process

reduce objective dimensionality. If no objective reduction is achieved, then the process is terminated. If objective reduction is possible within a cluster, then the process repeats with firstly, optimisation with the reduced objectives subject to the constraint of remaining within the cluster hyper-ellipsoidal boundary in an attempt to preserve objective correlations. The process ends when no further objective reduction can be achieved within each cluster.

As the similarity metric for clustering, the k*-Means algorithm uses Mahalanobis distance, which follows a Chi-squared distribution. Large Mahalanobis distances can be used to detect outliers by comparing them to a quantile of the Chi-squared distribution [22] and in so doing can be used as a cluster boundary.

Preference articulation can be used to direct the optimiser towards the DMs region of interest and so improve search efficiency [2]. Furthermore, a progressive approach allows the DM to specify these preferences interactively as information emerges from successive optimisations. The Progressive Preference Articulation method of [19]

(PPA$_{FF}$) provides an intuitive and efficient specification of objective goals and priorities [20]. The multi-objective optimisation algorithm, NSGAII [24] is widely used [21] and it was decided to modify this to incorporate the PPA$_{FF}$ approach.

As is the case with MOEAs, the stochastic nature of the k*-Means clustering algorithm justifies running it a number of times to gain confidence that the results are reliable. Consequently, clustering verification rules are defined [3] to ensure consistent and correct results. With many objectives and possible large population sizes, the computational demands may be significant and deserve consideration in deciding how many clustering runs are to be run.

4 Diesel Engine Calibration Optimisation

Diesel base engine calibration involves adjustment of the control actuator settings to achieve optimal trade-offs between competing objectives. Such objectives include fuel consumption as well as legislated emissions and combustion noise measures.

A six-objective diesel problem was formulated comprising the minimisation of Specific Fuel Consumption (SFC), Oxides of Nitrogen (NOx), Particulates (Parts), Hydrocarbons (HC), Carbon Monoxide (CO) and combustion noise (Noise). These objective functions were formed from empirical engine models based on engine test data, the boundary of which formed a constraint on the six continuous decision variables to ensure there was no extrapolation. A version of NSGAII modified with the PPA$_{FF}$ method was used for the optimisation with objective goals and priorities specified and initially in Stage 1, a population of 4000 run for 5000 generations to provide a reasonable computational effort, i.e. this took approximately 3 hours. In addition, the relatively large population supported the downstream clustering and sub-sampling process.

Clustering was carried out using the k*-Means algorithm and the clustering verification rules. This resulted in two clusters and a sub-sampled POP size of 1000, which gave good agreement with the parent POP in terms of cluster means and correlation matrices. Both clusters were retained as they both had significant membership and each was better than the other in different objectives. Subsequent application of the PCA-based objective reduction rules in each cluster resulted in three and four objectives being retained in Cluster 1 and 2, respectively. The objectives retained were those in each Principal Component whose eigenvector coefficients had the highest magnitude and objective priority. Detailed explanation is provided in [3].

Further optimisation in Stage 2 with the reduced objectives was carried out within each cluster resuming from the sub-sampled POPs and run for 2000 generations. A hyper-ellipsoidal cluster constraint [22] was used keep the search within each cluster and to preserve objective correlations. The resulting POPs were then plotted in parallel coordinates and scatter plot formats and these are shown for Cluster 1 in Figures 2 and 3, respectively.

Figure 2 shows for the retained objectives, both NOx and HC have improved, but partially at the expense of Parts (due to the fact that all three objectives conflict), while the discarded Objectives, SFC, CO and noise broadly show no deterioration.

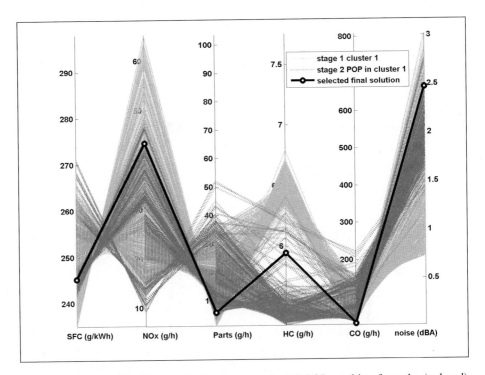

Fig. 2. Parallel coordinates plot of Cluster 1, the Stage 2 POP resulting from the (reduced) three-objective optimisation of NOx, Parts and HC and the selected final solution

The plot indicates that the objective correlations resulting from the initial optimisation have been maintained after further optimisation. The two-dimensional scatter plots in Figure 3 show this more clearly and suggest that a trade-off solution for NOx and Parts can be chosen with simultaneously improved HC. The Data Cursor feature in the Matlab® Figure environment was used to select such a solution and hence determine from the final population the corresponding values for the discarded objectives.

In Cluster 2, for the retained objectives, NOx improved along with noise at the expense of Parts and HC, whilst the discarded Objectives (SFC and CO) have both maintained their objective correlations. In summary, no overall improvement was achieved with this objective reduction.

Table 1 summarises the results from the objective reduction process applied in each of the two clusters. Of the two highest priority Objectives (NOx and Parts), final solution 1 in Cluster 2 has achieved the lowest NOx, but this was at the expense of Parts and deteriorated SFC, HC and especially CO. If Parts is prioritised higher (*i.e.* is more important) than NOx, final solution 2 in Cluster 2 is improved for SFC, Parts, HC and CO and only marginally deteriorated for noise. By comparison, final solution 1 in Cluster 1 is further improved in Parts (again at the expense of NOx), SFC, HC and CO and although noise has deteriorated, it is relatively low.

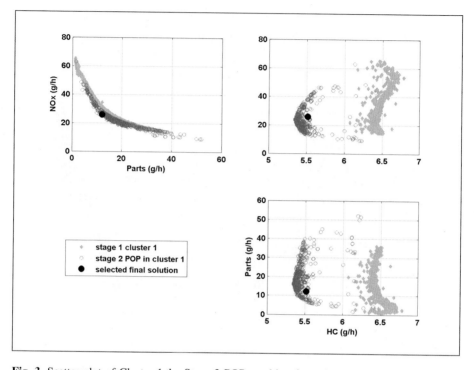

Fig. 3. Scatter plot of Cluster 1,the Stage 2 POP resulting from the (reduced) three-objective optimisation of NOx, Parts and HC and the selected final solution

Table 1. Final solutions selected

Stage	Cluster	Final Solution	SFC (g/kWh)	NOx (g/h)	Parts (g/h)	HC (g/h)	CO (g/h)	noise (dBA)
2	1	1	249.1	25.8	12.1	5.5	44.7	2.1
2	2	1	294.0	7.3	73.3	7.5	434.0	0.5
1	2	2	261.1	16.2	31.6	6.4	122.2	0.8

5   Conclusions and Future Work

In summary, the original six-objective optimisation has been split into two smaller optimisation problems via the complexity reduction process summarised above. Of these two sub-problems and with further preference articulation, the three-objective sub-problem was able to provide an improved solution in comparison to the four-objective sub-problem.

A number of observations from this case study have relevance to higher dimensional optimisation problems. Firstly, while this six-objective problem involved only one stage of objective reduction, it is possible that for problems with a larger number of objectives, the number of stages increases also. In such a scenario, the application of the clustering verification and objective reduction rules will become lengthy. A more compact form for these rules, which lends itself to being automated, would be useful. Secondly, higher dimensional problems may require larger populations to provide effective search. Larger populations in more objectives may generate more clusters. Both place significant demands on computational efficiency. Parallel computing is one approach to address this requirement. Finally, as the number of objectives increases so does the number of PCs. A PCA on a larger number of objectives may reveal a finer gradation in the percentage of variation represented by the PCs. In other words, it may be possible that the threshold for selecting PCs could be varied slightly to retain a different number of PCs and potentially, a different degree of objective reduction.

Acknowledgments. The first two authors are grateful to Ford Motor Company Limited for their resources and support to research, develop and apply this MOODM process. In addition, the authors appreciate the helpful and useful comments of the anonymous reviewers.

References

1. Deb, K.: Multi-Objective Optimization Using Evolutionary Algorithms. John Wiley & Sons, Chichester (2001)
2. Fleming, P.J., Purshouse, R.C., Lygoe, R.J.: Many-Objective Optimization: An Engineering Design Perspective. In: Coello, C.A.C. (ed.) EMO 2005. LNCS, vol. 3410, pp. 14–32. Springer, Heidelberg (2005)
3. Lygoe, R.J.: Complexity Reduction in High-Dimensional Multi-Objective Optimisation. Ph.D. thesis, University of Sheffield, Sheffield, U.K (2010)
4. Ishibuchi, H., Tsukamoto, N., Nojima, Y.: Behavior of evolutionary many-objective optimization. In: UKSIM 2008, pp. 266–271 (2008)
5. Zou, X., Chen, Y., Liu, M., Kang, L.: A new evolutionary algorithm for solving many-objective optimization problems. IEEE Transactions on Systems, Man, and Cybernetics, Part B 38(5), 1402–1412 (2008)
6. Deb, K., Saxena, D.K.: Searching for Pareto-optimal solutions through dimensionality reduction for certain large-dimensional multi-objective optimization problems. In: CEC 2006 (2006)
7. Purshouse, R.C., Fleming, P.J.: Conflict, harmony, and independence: Relationships in evolutionary multi-criterion optimisation. In: Fonseca, C.M., Fleming, P.J., Zitzler, E., Deb, K., Thiele, L. (eds.) EMO 2003. LNCS, vol. 2632, pp. 16–30. Springer, Heidelberg (2003)
8. Yoshikawa, T., Yamashiro, D., Furuhashi, T.: A Proposal of Visualization of Multi-Objective Pareto Solutions - Development of Mining Technique for Solutions. In: IEEE Symposium on Computational Intelligence in Multicriteria Decision Making (MCDM 2007), pp. 172–177. IEEE Press, Honolulu (2007)
9. Müller, H., Biermann, D., Kersting, P., Michelitsch, T., Begau, C., Heuel, C., Joliet, R., Kolanski, J., Krller, M., Moritz, C., Niggemann, D., Stber, M., Stnner, T., Varwig, J., Zhai, D.: Intuitive Visualization and Interactive Analysis of Pareto Sets Applied on Production Engineering System. In: Yang, A., Shan, Y., Bui, L.T. (eds.) Success in Evolutionary Computation. SCI, vol. 92, pp. 189–214. Springer, Heidelberg (2008)
10. Cheung, Y.M.: k*-Means: A New Generalized k-Means Clustering Algorithm. Pattern Recognition Letters 24(15), 2883–2893 (2003)

11. Rousseeuw, P., Van Driessen, K.: A Fast Algorithm for the Minimum Covariance Determinant Estimator. Technometrics 41, 212–223 (1999)
12. Carreira-Perpinan, M.A.: A Review of Dimension Reduction Techniques, Technical Report CS-96-09, Dept. of Computer Science, University of Sheffield (1997)
13. Fodor, I.K.: A survey of dimension reduction techniques, Technical report, Center for Applied Scientific Computing, Lawrence Livermore National Laboratory (2002)
14. Jolliffe, I.T.: Principal Component Analysis, 2nd edn. Springer, New York (2002)
15. Hyvärinen, A.: Survey on Independent Component Analysis. Neural Computing Surveys 2, 94–128 (1999)
16. Morrison, A., Ross, G., Chalmers, M.: Fast Multidimensional Scaling through Sampling, Springs and Interpolation. Information Visualization 2(1), 68–77 (2003)
17. Kohonen, T.: Self-Organizing Maps. Springer, Berlin (1995)
18. Kambhatla, N., Leen, T.K.: Dimension reduction by local principal component analysis. Neural Computation 9(7), 1493–1516 (1997)
19. Fonseca, C.M., Fleming, P.J.: Multiobjective Optimization and Multiple Constraint Handling with Evolutionary Algorithms — Part I: A Unified Formulation. IEEE Transactions on Systems, Man, and Cybernetics, Part A: Systems and Humans 28(1), 26–37 (1998)
20. Adra, S., Griffin, I., Fleming, P.J.: A Comparative Study of Progressive Preference Articulation Techniques for Multiobjective Optimisation. In: Obayashi, S., Deb, K., Poloni, C., Hiroyasu, T., Murata, T. (eds.) EMO 2007. LNCS, vol. 4403, pp. 908–921. Springer, Heidelberg (2007)
21. Adra, S., Dodd, T.J., Griffin, I.A., Fleming, P.J.: A Convergence Acceleration Operator for Multiobjective Optimisation. IEEE Transactions on Evolutionary Computation 13(4), 825–847 (2009)
22. Filzmoser, P.: A multivariate outlier detection method. In: Seventh International Conference on Computer Data Analysis and Modeling, Minsk, Belarus, vol. 1, pp. 18–22 (2004)
23. Deb, K., Saxena, D.K.: On Finding Pareto-Optimal Solutions Through Dimensionality Reduction for Certain Large-Dimensional Multi-Objective Optimization Problems, Technical Report 2005011, Kanpur Genetic Algorithms Laboratory (KanGAL), Indian Institute of Technology, Kanpur, India (2005)
24. Deb, K., Pratap, A., Agarwal, S., Meyarivan, T.: A Fast and Elitist Multiobjective Genetic Algorithm: NSGA-II. IEEE Transactions on Evolutionary Computation 6(2), 182–197 (2002)

A Modular Decision-Tree Architecture for Better Problem Understanding

Vineet R. Khare* and Halasya Siva Subramania

Diagnosis and Prognosis Group
India Science Lab, General Motors Global Research and Development
GM Technical Centre India Pvt Ltd, Creator Building, International Tech Park Ltd.
Whitefield Road, Bangalore - 560 066, India
{vineet.khare,halasyasiva.subramania}@gm.com

Abstract. In this paper, we propose a sequential decomposition method for multi-class pattern classification problems based on domain knowledge. A novel modular decision tree architecture is used to divide a K-class classification problem into a series of L smaller (binary or multi-class) sub-problems. The set of all K classes $c = \{c_1, c_2, \ldots c_K\}$ is divided into smaller subsets $(c = \{s_1, s_2, \ldots s_L\})$ each of which contains classes that are related to each other. A modular approach is then used to solve (1) the binary sub-problems $(p_i = \{s_i, \bar{s}_i\})$ and (2) the smaller multi-class problem $s_i = \{c_{i1}, c_{i2}, \ldots c_{in}\}$. Problem decomposition helps in a better understanding of the problem without compromising on the classification accuracy. This is demonstrated using the rules generated by the $C4.5$ classifier using a monolithic system and the modular system.

Keywords: Problem Decomposition, Decision Trees, Confusion matrix, Multi-class Classification.

1 Introduction

With the growth in complexity of the electrical architecture in automobiles, fault diagnosis is gaining more and more importance in the automotive domain. Generic diagnostic data includes standard (OBD-II government regulations [2]) and manufacturer specific diagnostic trouble code (DTC) definitions and diagnostic Parameter Identifier (PID) lists. DTCs are binary logic check points in a vehicle's electrical architecture to indicate out-of-threshold range behavior in circuits, their connections and circuit components. The various physical parameters or sensors that monitor the state of the vehicle at any given instance of time are called PIDs. The PIDs are the parametric values representing the state of the vehicle at the instance when a DTC is set. The PID to DTC mapping logic for every DTC is embedded in the vehicle's Electronic Control Units (ECUs) and is also partly reflected in the service manual. However, with increasing complexity of the on-board electrical system and common PIDs involved in various DTCs,

---

* Corresponding author.

K. Deb et al. (Eds.): SEAL 2010, LNCS 6457, pp. 647–656, 2010.

the generic PID to DTC mapping is non-trivial. The knowledge of this generic mapping can help the design experts validate the system design and discover anomalies.

In this paper we analyze the PID-DTC mapping for a specific vehicle sub-system, as a case study, to demonstrate the sequential decomposition method for multi-class pattern classification problems. Although, the overall 20-class classification task can be performed with reasonable accuracy using a $C4.5$ decision tree [11], the resulting rules are too complex for human experts to comprehend. In a physical sense, however, these rules should be much simpler. We argue that the spatial interference between the PIDs are causing the rules to be complex. We propose a modular system architecture (Sect. 3) that reduces the interference between the PIDs and produces simpler rules without compromising on the classification accuracy (Sect. 4). A brief background on the relevant literature is presented in Sect. 2.

2   Background

Problem decomposition can be viewed as the process of discovering any in-built structure in the given problem. It involves three steps [7]: (1) Decomposition (decomposing a complex problem into smaller and simpler problems), (2) Subsolution (designing modules to solve these simpler problems) and (3) Combination (combining these individual modules into a solution to the original problem). The optimal decomposition for a problem depends on the associated objective. For instance, for a classification problem one might try to decompose the problem such that the classification accuracy on the overall task or the robustness towards noisy data can be improved. There are different types of problem decompositions [7] available in machine learning literature, including parallel and sequential problem decompositions. The best decompositions corresponding to these objectives might not always match with the in-built structure of the problem. There are various benefits of problem decomposition discussed, especially, in machine learning literature. Examples of these include, learning speed, minimizing spatial interference between features [5], minimizing forgetting (temporal interference), efficient learning of related tasks [6] and better generalization. However, it is also argued [3] that by using efficient learning algorithms and sophisticated cost functions many of these advantages can be accounted for even with a non-modular architecture.

In this work we present a modular approach to solve a multi-class classification problems and explore the benefits of problem-structure-based decomposition for a better understanding of the problem. For interpretability we are constrained to use decision trees, rules from which can be understood by a human. Most of the modular approaches to the multi-class classification problems primarily involve artificial neural network (ANN) learning [1,4,9]. The multi-class classification problems are divided based on the inherent class relations among the training data. In [1,4] a K-class classification problem is divided into K two-class classification problems by using the class relations. In [9], it is divided into a series of

$\binom{K}{2}$ binary classification problems. These binary classification problems are to discriminate class c_i from class c_j for $i = 1, 2, \ldots, K$ and $j = i+1$ with using only the data specific to the two classes. Each of the binary classification problem is learnt in parallel using ANNs. The output of each of the ANN is then combined using a min-max modular network (refer to the paper [9] for details). Although these approaches highlight the benefits of problem decomposition, they have limited applications in domains where interpretability is of crucial importance. For this reason, we prefer decision trees over ANNs for the classification tasks. In addition to interpretability, decision trees (in particular $C4.5$ [11] used in this study) can also handle nominal data and provide rapid classification [12].

3   Sequential Problem Decomposition

Using problem decomposition we aim to reduce spatial interference among the attributes and obtain classification rules which are easier to understand compared to the rules obtained from the monolithic system. Following are the steps involved in the problem decomposition.

3.1   Related Classes Based on Domain Knowledge

First and the most crucial step in problem decomposition is to divide the K-class classification problem into a series of L smaller (binary or multi-class) sub-problems. The set of all K classes $c = \{c_1, c_2, \ldots c_K\}$ is divided into smaller subsets $(c = \{s_1, s_2, \ldots s_L\})$ each of which contains classes that are related to each other. This can be done in consultation with domain experts or using other sources of domain knowledge (e.g. service manuals used in Sect. 4.2). Separating related classes into different groups helps in reducing interference between attributes. Each sub-problem is solved in a sequence using a separate module.

3.2   Ordering of the Modules

The very first module receives all the data and classifies the first set of classes s_1 against the rest $\bar{s}_1$. Let the set of instances classified, by module 1, as s_1 be S_1 and $\bar{s}_1$ be $\bar{S}_1$. The second module receives only $\bar{S}_1$ as input. Out of these $\|\bar{S}_1\|$ instances, module 2 classifies s_2 against the rest $\bar{s}_2$. In general, module i receives $\|\bar{S}_{i-1}\|$ as inputs and classifies s_i against the rest $\bar{s}_i$ (Fig. 1). The ordering of sub-problems is a design issue and a greedy heuristic approach is used in this work. While deciding the ordering, two objectives can be considered – (1) classification accuracy and (2) module complexity. In the current implementation, at every step, one module is chosen that has the highest classification accuracy (s_i against $\bar{s}_i$). For further enhancements a weighted sum of classification accuracy and module complexity (measured in terms of module structure – e.g. size of C4.5 tree in Sect. 4.3) needs to be considered as the objective function for the module orderings. Initially the first module is supplied with the complete dataset. For subsequent modules (i), the input data is derived from the previous module ($\|\bar{S}_{i-1}\|$ instances from module $i - 1$).

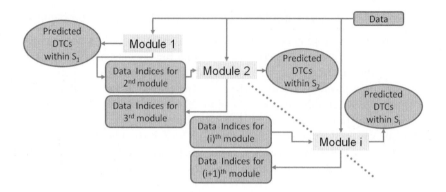

Fig. 1. Proposed modular architecture representing sequential problem decomposition

3.3   Module Design

Each module has to perform two classification tasks. The first task (binary) is to classify all instances belonging to s_i against $\bar{s}_i$ from all the input instances received from the previous module. The second task (binary or multi-class) is to identify individual classes c_{ik} from the subset of classes $s_i = \{c_{i1}, c_{i2}, \ldots c_{in}\}$, where $k \in \{1, 2, \ldots n\}$. The design of module i is illustrated in Fig. 2 and the two classification tasks are discussed below.

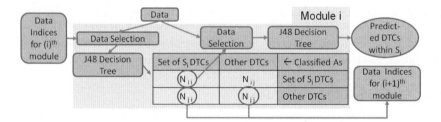

Fig. 2. Schematic illustrating the design of Module i in the proposed modular architecture (Fig. 1)

Inter-set Classification Task – The classifier receives $\|\bar{S}_{i-1}\|$ instances from module $i-1$, performs the binary classification task and produces a confusion matrix [8] with the following:

- True positives (N_{ii} in Fig. 2) – Instances correctly classified as s_i and are passed on to the intra-set classification task.
- False positives (N_{ji} in Fig. 2) – Instances incorrectly classified as s_i and and are passed on to the next module[1] (module $i + 1$).

---

[1] While testing, outputs of test instances are not known. Hence all instances classified as s_i ($N_{ii} + N_{ji}$) are passed on to the intra-set classification task.

- False negatives (N_{ij} in Fig. 2) – Instances incorrectly classified as $\bar{s}_i$ and are ignored[2].
- True negatives (N_{jj} in Fig. 2) – Instances correctly classified as $\bar{s}_i$ and are passed on to the next module (module $i + 1$).

Intra-set Classification Task – The classifier receives N_{ii} instances from the intra-set classification task and predicts individual classes c_{ik} from the subset of classes $s_i = \{c_{i1}, c_{i2}, \ldots c_{in}\}$.

4   Results and Discussion

In this section we compare a monolithic solution (Sect. 4.1) with the modular solution (Sect. 4.2) for a PID-DTC classification problem (described here). The modular solution is obtained using the problem decomposition presented in Sect. 3 and the monolithic solution is obtained using the standard $C4.5$ decision tree algorithm. For the vehicle sub-system chosen for this study, there are 387 PIDs which represent the state of the vehicle subsystem that can set one or more of 20 DTCs. Out of the 387 PIDs, only 221 have non-constant values in the dataset chosen for this study. Hence the rest of the PIDs are ignored. Out of the 221 PIDs used in the study, 141 are nominal and 80 are numeric attributes. There are $7,340$ instances in the training dataset and $3,343$ instances in the testing dataset. The DTC class distribution of the training and the testing datasets are shown in Fig. 3.

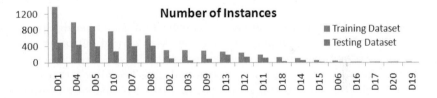

Fig. 3. Class distribution for the PID-DTC Classification Problem. Symbol D** represents DTC**

4.1   The Monolithic Solution

For the given problem and the nature of the data, decision trees are an obvious choice for variety of reasons: a mixture of numeric and nominal data, interpretability and rapid classification [12]. Hence we choose $C4.5$ [11] classifier as our monolithic solution. The implementation of $C4.5$ in WEKA [13], a data mining software, is used in this work. For C4.5, multi-way splits are used for nominal data. Pruning is performed, including the sub-tree raising option. The confidence factor c for pruning is decided based on a set of experiments with

---

[2] For testing all instances classified as $\bar{s}_i$ ($N_{ij} + N_{jj}$) are passed on to the next module.

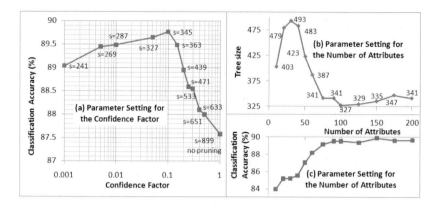

Fig. 4. (a) Parameter Setting for the confidence factor c related to the decision tree pruning. 10-fold cross-validation and testing dataset's classification accuracy for different c are plotted. s indicates the size of corresponding pruned tree. (b) and (c) Discovering the minimum number of attributes required for the PID-DTC classification task. The two plots show various tree sizes and classification accuracies corresponding to different number of attributes. Significant attributes are chosen based on the chi-squared statistic with respect to the class.

various c values. Lower values of c result in higher pruning and smaller decision trees. Figure 4(a) shows the 10-fold cross-validation classification accuracy using all 221 attributes. From the figure it is evident that $c = 0.1$ is a good choice for the confidence factor which provides a good classification accuracy and a reasonable size tree.

The attribute/feature selection for the classification is another design choice. Experiments with various number of attributes were carried out using top n_{att} significant attributes based on the chi-squared statistic with respect to the class on the training dataset. Various tree sizes and classification accuracies corresponding to different choices of n_{att} are shown in Fig. 4(b) and 4(c), respectively. Using more than 150 attributes does not provide any improvement in classification accuracy. In addition, the number of nodes in the resulting trees for 150 and more attributes also stabilizes at 150. For this reason we choose only the top 150 attributes for the classification task. The corresponding confusion matrix and the classification accuracies are presented in Tables 1 and 3, respectively.

4.2   The Modular Solution

The first step in the modular solution to the PID-DTC classification problem is to divide the whole set of 20 DTCs into subsets, each containing related classes. This is done using domain knowledge obtained using the service manuals associated with each DTC. These manuals indicate which PID values should influence the DTC in question. In addition, the DTC descriptions (e.g. DTC P0496: Evaporative Emission (EVAP) System Flow During Non-Purge and DTC P0446: Evaporative Emission (EVAP) Vent System Performance [10]) also help in identifying

Table 1. Confusion matrix for the monolithic model (test dataset). Symbol D** represents DTC**

	Classified As																				
	D01	D02	D03	D04	D05	D06	D07	D08	D09	D10	D11	D12	D13	D14	D15	D16	D17	D18	D19	D20	Class
	483	0	0	0	0	0	0	0	0	0	0	0	0	16	0	0	3	0	0	0	D01
	3	72	0	0	0	0	1	0	8	11	11	0	0	0	0	0	0	0	0	0	D02
	0	0	53	0	1	0	0	0	0	0	0	0	0	0	0	0	0	0	0	0	D03
	1	0	0	439	2	0	0	0	0	3	0	0	5	3	0	0	0	0	0	0	D04
	0	0	0	3	405	1	0	1	0	1	0	0	1	0	0	0	0	0	0	0	D05
	0	0	0	0	1	12	0	0	0	0	0	0	1	0	0	0	0	0	0	0	D06
	0	0	0	1	3	0	406	3	0	0	0	0	0	0	0	0	0	0	0	0	D07
	0	0	0	0	1	0	232	184	0	0	0	0	1	0	0	0	0	0	0	0	D08
	1	1	0	0	0	0	0	0	76	9	6	0	1	0	0	0	0	0	0	0	D09
	0	14	0	4	1	0	0	4	8	234	13	0	5	3	0	0	0	0	0	0	D10
	0	4	0	0	0	0	0	2	10	103	0	0	1	0	0	0	0	0	0	0	D11
	0	0	0	1	0	0	0	1	0	0	0	139	6	0	0	0	0	0	0	0	D12
	0	8	0	20	0	5	0	0	1	0	1	4	153	8	0	0	0	0	0	0	D13
	0	0	0	3	0	0	0	1	0	4	0	1	17	35	0	0	0	0	0	0	D14
	9	0	0	0	0	0	0	0	0	0	0	0	0	0	7	0	0	1	0	0	D15
	0	3	0	0	0	0	0	0	0	1	2	0	0	0	0	0	0	0	0	0	D16
	0	1	0	0	0	0	0	0	3	1	2	0	0	0	0	0	2	0	0	0	D17
	2	0	0	0	0	0	0	0	0	0	0	0	0	0	2	0	0	26	0	0	D18
	0	0	0	0	0	0	0	0	0	0	0	0	0	0	0	0	0	0	0	0	D19
	0	0	0	0	0	0	0	1	0	0	0	0	0	0	0	0	0	0	0	0	D20

related DTCs. Descriptions of P0446 and P0496 suggest that both are related to the EVAP system and should be related. Based on this domain knowledge, the set of 20 DTCs are divided into eight subsets (s_i where $i \in \{1, 2, \ldots, 8\}$; refer to column 2 in Table 2 for details). These subsets represent the sub-problems in the decompositions. Eight modules are then used in a sequence (obtained using the ordering presented in Sect. 3.2) to solve each of these sub-problems. As discussed in Sect. 3.3, each module i performs two classification tasks – inter-set classification and intra-set classification. Inter-set classification identifies instances specific to all s_i DTCs. Whereas intra-set classification identifies DTCs within s_i.

The confusion matrices for both inter and intra-set classification tasks are listed in Table 2 for the testing dataset. Confidence factor of $c = 0.1$ was used in all the modules. Correctly classified instances from all intra-set classification tasks are collected to obtain the classification accuracy for the overall classification task. The incorrectly classified instances (both inter and intra-set classification tasks) contribute to the overall error. Using the modular solution, we have observed improvement in classification accuracy (Table 3). This, however, was not our primary objective. In Sect. 4.3 we will compare the classification rules generated for each DTC from the monolithic and modular solutions to explain how the modular solution helps in a better understanding of the PID-DTC mapping.

Table 2. Confusion matrices for the Modular model. Symbol D** represents DTC**

Module Index (i)	Module DTC Set (s_i)	Inter-set Classification Classified As		Class	Intra-set Classification Classified As				Class
		s_1	$\bar{s}_1$		D06	D07	D08		
1	D06	840	5	s_1	11	0	0		D06
	D07	6	2492	$\bar{s}_1$	0	409	4		D07
	D08				0	238	184		D08
		s_2	$\bar{s}_2$		D03	D05			
2	D03	464	2	s_2	53	1			D03
	D05	4	2027	$\bar{s}_2$	1	413			D05
		s_3	$\bar{s}_3$		D01	D04			
3	D01	933	18	s_3	504	22			D01
	D04	31	1047	$\bar{s}_3$	0	438			D04
		s_4	$\bar{s}_4$						
4	D12	143	4	s_4					
		10	908	$\bar{s}_4$					
		s_5	$\bar{s}_5$		D09	D15	D18	D20	
5	D09	119	13	s_5	91	4	3	0	D09
	D15	17	763	$\bar{s}_5$	0	10	2	0	D15
	D18				1	3	22	0	D18
	D20				0	0	0	0	D20
		s_6	$\bar{s}_6$						
6	D02	85	18	s_6					
		24	649	$\bar{s}_6$					
		s_7	$\bar{s}_7$		D11	D13	D16		
7	D11	281	10	s_7	100	2	5		D11
	D13	29	347	$\bar{s}_7$	10	189	1		D13
	D16				2	1	0		D16
		s_8	$\bar{s}_8$		D10	D14	D17	D19	
8	D10	308	0	s_8	257	4	0	0	D10
	D14	49	0	$\bar{s}_8$	45	45	0	0	D14
	D17				0	0	6	0	D17
	D19				0	0	0	0	D19

Table 3. Classification accuracies on training and testing datasets for the monolithic and modular models

	Dataset Size	Classification Accuracy Monolithic Model	Modular Model
Training Set	7, 430	89.81%	92.69%
Test Set	3, 343	84.62%	88.54%

4.3   Benefits of Problem Decomposition

There are three benefits of the problem decomposition observed in this work – (1) Better classification accuracy; (2) Better problem understanding and (3) More flexible system design.

Table 4. Number of PIDs required for various DTC classification rules. Symbol D** represents DTC**

Classes	Number of PIDs Modular	Monolithic	Classes	Number of PIDs Modular	Monolithic	Classes	Number of PIDs Modular	Monolithic
D01	11	19	D08	6	6	D15	8	16
D02	22	44	D09	6	29	D16	7	16
D03	6	11	D10	13	43	D17	10	13
D04	12	21	D11	8	37	D18	8	10
D05	7	16	D12	5	10	D19	10	9
D06	6	19	D13	7	31	D20	6	7
D07	6	14	D14	13	27			

1. The classification accuracy improved from 84.62% (monolithic solution) to 88.54% (modular solution) on the testing dataset. This however is not the major benefit of decomposition. Of greater significance here is that the benefit in terms of better problem understanding (see below) is realized without compromising on the classification accuracy.

2. Using the trees generated from the monolithic and modular solutions, rules are generated in the form of

$$OR \left\{ \begin{array}{l} AND(p_{11}, p_{12}, \ldots, p_{1i}, \ldots), \\ AND(p_{21}, p_{22}, \ldots, p_{2i}, \ldots), \\ \qquad \ldots \\ AND(p_{j1}, p_{j2}, \ldots, p_{ji}, \ldots), \\ \qquad \ldots \end{array} \right\} \rightarrow DTC_k,$$

where p_{ji} is a boundary condition on PID p_i. Table 4 lists the number of distinct PIDs required for various DTC classification rules. Lower numbers imply simpler rules. For 18 out of 20 DTCs, the modular solution provides smaller simpler rules, thereby helping domain experts understand the PID-DTC mapping better. The other two DTCs – DTC08 has the same number of PIDs, whereas DTC19 has one additional PID in the classification rule.

3. Decomposing the problem also provides more flexibility in terms of system design. Separate modules can be designed as per the requirements of the sub-problems. This is not possible with the monolithic system. Example of this flexibility is the number of PIDs used for various sub-problems. Although, in all instances top 150 attributes are used, however, this number can be chosen based on the sub-problem being solved.

5   Conclusion

A modular approach to multi-class classification problems is presented. Problem decomposition is achieved using the domain knowledge about the problem. It is argued that using the modular approach the spatial interference between attributes is reduced. This results in a much better understanding of the original problem and the sub-problems. The PID-DTC classification problem is used to illustrate the benefits of problem decomposition. These include, better classification accuracy, better problem understanding (much simpler classification rules compared to rules obtained without the decomposition) and a much more flexible system design.

References

1. Anand, R., Mehrotra, K., Mohan, C., Ranka, S.: Efficient classification for multiclass problems using modular neural networks. IEEE Transactions on Neural Networks 6(1), 117–124 (1995)
2. Baltusis, P.: On board vehicle diagnostics (2004), SAE 2004-21-0009
3. Bullinaria, J.A.: To modularize or not to modularize? In: Bullinaria, J. (ed.) Proceedings of the 2002 U.K. Workshop on Computational Intelligence (UKCI 2002)., Birmingham, pp. 3–10 (2002), citeseer.ist.psu.edu/535117.html
4. Chen, C.H., You, G.H.: Class-sensitive neural network. Neural Parallel Scie. Comput. 1(1), 93–96 (1993)
5. Jacobs, R.A., Jordan, M.I., Barto, A.G.: Task Decomposition Through Competition in a Modular Connectionist Architecture: The What and Where Vision Tasks. Cognitive Science 15, 219–250 (1991)
6. Khare, V.R., Sendhoff, B., Yao, X.: Environments Conducive to Evolution of Modularity. In: Runarsson, T.P., Beyer, H.G., Burke, E., Merelo-Guervós, J.J., Whitley, L.D., Yao, X. (eds.) PPSN 2006. LNCS, vol. 4193, pp. 603–612. Springer, Heidelberg (2006)
7. Khare, V.R., Yao, X., Sendhoff, B.: Multi-network evolutionary systems and automatic decomposition of complex problems. International Journal of General Systems 35(3), 259–274 (2006)
8. Kohavi, R., Provost, F.: Glossary of terms. Machine Learning 30(2/3), 271–274 (1998)
9. Lu, B.L., Ito, M.: Task decomposition and module combination based on class relations: A modular neural network for pattern classification. IEEE Transactions on Neural Networks 10, 1244–1256 (1999)
10. OBD-II: Trouble Codes (November 18, 2009), www.obd-codes.com
11. Quinlan, R.J.: C4.5: Programs for Machine Learning (Morgan Kaufmann Series in Machine Learning). Morgan Kaufmann, San Francisco (January 1993)
12. Richard, O., Duda, P.E., Hart, D.G.S.: Pattern Classification, 2nd edn. John Wiley & Sons, Chichester (2001)
13. Witten, I.H., Frank, E.: Data Mining: Practical Machine Learning Tools and Techniques, 2nd edn. Morgan Kaufmann, San Francisco (2005)

Virtual Manufacturing Cell Design Using a PSO Approach with Alternative Neighbourhood Topologies

Rahul Caprihan[1], Jannes Slomp[2], Gursaran Srivastava[3], and Khushboo Agarwal[4]

[1] Faculty of Engineering, Dayalbagh Educational Institute, Dayalbagh, Agra, India
[2] Faculty of Business & Economics, University of Groningen, Groningen, The Netherlands
[3] Faculty of Science, Dayalbagh Educational Institute, Dayalbagh, Agra, India
[4] Senior Partner, Dynamic Pistons Ltd., Agra, India
RCaprihan@gmail.com, J.Slomp@rug.nl, Gursaran.db@gmail.com

Abstract. In this paper an application of conventional Particle Swarm Optimization (PSO) approach with alternative neighborhood topologies is proposed for the design of virtual manufacturing cells within which machines and jobs are assigned to the cells with a view to maximize productive output, whilst simultaneously minimizing the inter-cell movements due to the limited availability of machines. The PSO results are then compared with the following approaches: Binary PSO (BPSO) and Preemptive / Lexico Goal Programming. It is observed that the PSO topological variants perform well for the assumed VCM design problem.

Keywords: Virtual Cellular Manufacturing Design, Particle Swarm Optimization, Neighborhood Topologies.

1 Introduction

The present day manufacturing environment is characterized by profusion in product variety, exacting standards of quality, and a relentless pursuit to reduce costs to ward of competition. Cellular manufacturing (CM) has often been touted as the panacea by operations managers in the light of the above manufacturing dilemma. As a well established manufacturing paradigm, CM is known to offer notable advantages, including the reduction in setup times on account of the similarity between the part types produced within the constituent manufacturing cells, together with a reduction in lead times and work-in-process inventories [1].

However, although CM offers important advantages, there are other reasons why several firms prefer the (conventional) functional layout. In contrast to the cellular layout, the functional layout (FL) is more robust to changes in product mix, and also offers a degree of routing flexibility which may improve shop performance significantly [2][3][4].

Clearly then, both functional and cellular layouts have their respective advantages and disadvantages. With a view to harness the potential benefits of both CM and FL systems, the recent years have seen the introduction of virtual cellular manufacturing systems [2].

K. Deb et al. (Eds.): SEAL 2010, LNCS 6457, pp. 657–666, 2010.
© Springer-Verlag Berlin Heidelberg 2010

The virtual cells are a logical grouping of workstations which is only present in the planning and control system and in the minds of workers. Virtual cells are created periodically in response to changes in demand, and this imparts VCM systems their characteristic dynamic nature. Nomden et al. [2] provide a detailed account of the motivation for pursuing a VCM philosophy as a viable manufacturing option in the present day turbulent environment.

Slomp et al. [3] propose a general mathematical model for the design of virtual manufacturing cells, with a view to optimize productive throughput and minimize inter-cell dependencies. They indicate the various objectives and constraints that play a role in the design of virtual cells. In order to serve the solvability of the problem, they have partitioned the general model into two Integer Programming (IP) problems to be solved in successive stages. They propose Preemptive / Lexico Goal Programming to solve both problems and illustrate the applicability of this approach by means of a small illustrative example. Although the example problem can be solved, the required computing time also indicates the NP-hardness of both sub-problems. Most IP problems are NP-hard and heuristic approaches are needed to solve reasonably sized problems. In this paper, we suggest an optimization methodology based on the PSO approach, which is a metaheuristic search procedure to solve this VCM design problem. We illustrate the performance of our solution method by solving the first sub-problem presented by Slomp et al. [3]. We compare results obtained by the conventional PSO, with the Binary PSO (BPSO) and the Preemptive / Lexico Goal Programming proposed by [3]. Performance criteria concern the model objectives and, for the PSO algorithms, the number of iterations needed to gain the proposed solution.

The paper is organized as follows. Section 2 details the mathematical formulation for VCM design model considered. Section 3 provides an overview of swarm optimization techniques which comprise the conventional Particle Swarm Optimization (PSO) and its neighborhood topological variants (Star topology, Ring topology, and the Von Neumann topology,) and the Binary PSO (BPSO) method. Section 4 presents the PSO algorithm for the VCM design problem and provides details of the assumed experimental setup. In Section 5 we present results and conclude the paper.

2   Problem Definition

In this paper, we apply the problem definition presented in [3] for the assignment of jobs, machines, and a number of workers to the Virtual Cells. Consider a set of jobs, denoted by the index set $\{I\}$, which has to be produced in the upcoming planning period. Each job requires processing on various machine types. The required processing time for job i on machine type m is given by T_{im}. Each job belongs to a family of part types. The set of jobs ($i \in I$) belonging to family f is given by J_f. The setup time needed for family J_f on machine m is given by S_{fm}, and only one machine setup is needed if two or more jobs of the same family are manufactured sequentially on the same machine. The available number of machines of type m in the shop is denoted by θ_m, and the number of workers present in the shop at each moment in time is L. Workers are needed to operate the machines and to handle the setups. Table I provides details of the assumed symbols used in the model.

Table 1. VCM design model inputs & outputs

Inputs	
i	Job index: i = 1,2,..I (set of all jobs)
f	Family index: f = 1,2,..F (set of all families)
m	Machine type index: m = 1,2,..M (set of all machine types)
c	Cell index: c=1,2,..C (set of all cells)
θ_m	Number of type m machines available in the system
S_{fm}	Major setup time required for family f on machine type m
T_{im}	Processing time for job i on machine type m
J_f	Set of jobs belonging to family f
R	Length of planning period
L	Number of available workers
α	Setup factor indicating the ineffectiveness to reduce the need for setups
MAXW	Maximum size of a virtual cell, as the number of full-time-equivalents (FTEs)
MINW	Minimum size of a virtual cell, as the number of full-time-equivalents (FTEs)
Important outputs (decision variables)	
x_{ic}	Allocation of job i to cell c
y_{icm}	Allocation of job i to cell c on machine m
z_{fcm}	Allocation of family f to cell c on machine m
nn_{mc}	Number of type m machines needed for cell c
T_{cm}	Time needed to complete the jobs assigned in cell c on machine m (including setup times)
w_c	Number of full-time-equivalent workers (FTEs) needed to perform the operations in cell c
v_m^+	Number of machines of type m needed in more than one cell
v_m^-	Number of machines of type m not needed in any cell

It is the objective to create efficient Virtual Manufacturing Cells which are as independent as possible during the planning period R. Cells are independent if they are able to perform all operations needed for a set of jobs without the need to use the same machine stations. Jobs belonging to the same part family are ideally assigned to the same VMC in order to enable dedication and setup time savings. This can be realized by focusing on maximizing the total processing time assigned to the VMCs, given the available capacity of machines and workers.

Following the problem description, Slomp et al. [3] developed the following goal programming model formulation:

$$\text{Maximize } Z = \Pi_1 \sum_i \sum_c \sum_m (T_{im} x_{ic}) - \sum_m \Pi_{2m} v_m^+ + \sum_m \Pi'_{2m} v_m^- \tag{1}$$

Subject to:

$$\sum_c x_{ic} \leq 1, \forall i \tag{2}$$

$$T_{im} x_{ic} \leq \Omega y_{icm}, \forall i, c, m \tag{3}$$

$$\sum_{i \in Jf} y_{icm} \leq \Omega z_{fcm}, \forall f, c, m \tag{4}$$

$$LB_{fcm} = z_{fcm}, \forall f, c, m \tag{5}$$

$$UB_{fcm} = \sum_{i \in Sf} y_{icm}, \forall f, c, m \tag{6}$$

$$T_{cm} = [\sum_i T_{im} y_{icm} + \sum_f (LB_{fcm} + \alpha(UB_{fcm} - LB_{fcm})) S_{fm}], \forall c, m \tag{7}$$

$$T_{cm} \leq n_{mc} R, \forall c, m \tag{8}$$

$$\Sigma_c n_{mc} \leq \theta_m, \ \forall \ m \tag{9}$$

$$n_{mc} \leq nn_{mc}, \ \forall \ m, c \tag{10}$$

$$\Sigma_c \ nn_{mc} \leq \theta_m + v_m^+ - v_m^+, \forall \ m \tag{11}$$

$$\Sigma_m \ T_{cm} \leq w_c R, \forall \ c \tag{12}$$

$$\Sigma_c \ w_c \leq L \tag{13}$$

$$w_c \leq MAXW, \ \forall \ c \tag{14}$$

$$w_c \geq MINW, \ \forall \ c \tag{15}$$

x_{ic}, y_{icm} and z_{fcm} are 0/1 variables $\qquad \forall$ i, c, f, m

nn_{mc}, w_c, v_m^+, v_m^+, LB_{fcm}, and UB_{fcm} are integer variables $\forall$ c, m, f

n_{mc} is a real variable $\forall$ m,c

The objective function consists of three terms, one for each of the following objectives:

1. Maximize productive output, in terms of machining hours (as opposed to non-value-added time caused by machine setups and imbalances) processed in virtual cells in release period R. The parameter Π_1 indicates the importance of this objective.
2. Minimize total number of additional machines of type m needed for creating independent virtual cells. This objective implicitly minimizes the extent of intercell movements. The parameter Π_{2m} reflects the importance this objective.
3. Maximize the number of machines that are not needed in any cell. The more machines in the virtual cells, the more difficult it will be to allocate workers to only one cell and to gain sufficient machine coverage of all machines. The parameter Π_{3m} reflects the importance this objective.

Constraint (2) in the formulation ensures that each job will maximally be assigned to one virtual manufacturing cell. Constraint (3) ensures that a machine m is assigned to cell c if needed for job i (y_{icm}). Constraint (4) determines whether or not a machine of type m is needed in cell c for the processing of jobs belonging to family f (z_{fcm}). The values of y_{icm} and z_{fcm} are needed to determine the lower bound (LB_{fcm}) and the upper bound (UB_{fcm}) of the number of setups to be performed on a machine of type m at cell c for family f. This is done in constraints (5) and (6). The lower bound (LB_{fcm}) equals 1 if one or more jobs of family f require an operation on a machine of type m in cell c. The upper bound (UB_{fcm}) equals the total number of jobs of family f that has to be produced on a machine of type m in cell c. The required number of setups will lie between LB_{fcm} and UB_{fcm}. Here the parameter α is introduced in the model which indicates the ability to schedule jobs of the same family sequentially on a machine within a cell ($0 \leq \alpha \leq 1$). If $\alpha = 0$ then the workers are not able to realize any sequence of jobs belonging to the same family. If $\alpha = 1$ then the workers are able to realize maximal setup time savings within the VMCs. The value of α need to be fixed before solving the model. It is conceivable that α is larger in case of smaller VMCs. Constraint (7) calculates the total time t_{cm} needed at each VCM. This total time includes processing times (T_{im}) and setup times (S_{fm}). Constraints (8) to (10) concern the division of machine capacity among the VMCs. Constraint (8) takes care of sufficient capacity of type m machines for cell c. Constraint (9) concerns the limitation of the available capacity of machine type m.

Constraint (10) ensures that the integer number of type m machines needed in cell k exceeds the real (fractional) number of required type m machines in cell k. This integer number (nn_{mc}) is needed to calculate the excess of machines and/or the need to share machines between VMCs. This is done in constraint (11). Constraint (12) determines the minimal number of workers needed in cell c. Constraint (13) takes care of the number of available workers in the whole system. Constraints (14) and (15) limit the cell size by means of the number of FTEs.

The above model is solved in this paper by using the PSO algorithmic procedure and its topological variants detailed in the Section 3. The PSO results are then compared with those obtained using the Binary PSO approach. We also compare the results with those obtained in [3].

3 The PSO Approach and Its Variants

This section presents a brief description of the conventional PSO and its topological variants, as well as the Binary PSO approach.

3.1 Particle Swarm Optimization

Particle Swarm Optimization (PSO) was initially proposed to find optimal solutions for continuous space problems by Kennedy and Eberhart [5]. In the PSO approach, search starts with a randomly generated population of solutions called the *swarm* of particles in a d-dimensional solution space. Particle i is represented as $X_i = (x_{i1}, x_{i2}, \ldots, x_{id})$ which is called the position of the particle i in the d-dimensional space. With every particle i is associated a velocity vector $V_i = (v_{i1}, v_{i2}, \ldots, v_{id})$ that plays an important role in deciding the next position of the particle and is updated at each iteration. For updating the velocity of each particle, the particle's best position, $P_{ibest} = (p_{i1}, p_{i2}, \ldots, p_{id})$ which is the best position of particle i achieved so far, together with the global best position, $P_{gbest} = (p_{g1}, p_{g2}, \ldots, p_{gd})$ which is the best position of the entire swarm achieved so far by any particle of the swarm, are used. The following equations are used to update the velocity and position of particle i in iteration $t+1$ using values obtained through iteration t.

$$V_i(t+1) = wV_i(t) + c_1\varphi_1(p_{ibest} - X_i(t)) + c_2\varphi_2(p_{gbest} - X_i(t)) \tag{16}$$
$$X_i(t+1) = X_i(t) + V_i(t+1) \tag{17}$$

In equation (16), w is the inertia weight which controls the impact of the previous history of velocity on the global and local search abilities of the particles, while c_1 and c_2 are the positive learning constants which determine the rate by which the particle moves towards an individual's best position and the global best position respectively. Usually, c_1 and c_2 are chosen in a way so that their sum doesn't exceed the value '4'. If it does exceed that value at any instant, then both the velocities and positions, will explode toward infinity. φ_1 and φ_2 are random numbers drawn from a uniform probability distribution between $(0,1)$. Using the above update equations, the positions and velocities of the constituent (swarm) particles evolve at each subsequent iteration until the optimal solution is obtained or some termination criterion is achieved.

3.2 Neighborhood Topologies

In equation (16), the term '$c_2\varphi_2(p_{gbest}-X_i(t))$' is the social component (or *social influence*) and represents the force emerging from the attraction of the best position found so far in the entire swarm. This model of the PSO in which a particle is attracted towards the best position found in the entire swarm is also called the *gbest* PSO. This version of the PSO, however, is susceptible to premature and/or false convergence over multi-modal fitness landscapes [7, 8]. In order to overcome the problems faced with this *gbest* version of the PSO, a *neighborhood* is identified for *each particle*. The PSO is then modified so that the social influence is dictated by the best position found in the neighborhood of each individual. The relationship of influence is thus defined by a social network, which is called *population topology* [7]. Results in [7, 9] show that the performance of the PSO can be improved using different neighborhood topologies and different topologies perform differently on a given problem.

The most common topologies used with PSO are Star Topology (ST), Ring Topology (RT), and the Von Neumann Topology (VNT) [7, 8]. Fig. 1 shows these topology structures. The ST is simply the gbest model in which the entire population comprises the neighborhood. In other topologies such as the RT and VNT the neighborhood is some smaller portion of the swarm. In RT each particle is connected with two other particles while in the VNT each particle is connected with four other particles in a cubic type lattice.

Fig. 1. The standard neighbourhood topologies (Star, Ring, and Von Neumann)

3.3 Binary Particle Swarm Optimization

Kennedy and Eberhart [6] introduced the *binary particle swarm optimization* (BPSO) algorithm in 1997. In this version of the PSO, every particle is represented by a bit string, and each bit in turn is associated with a velocity which is the probability of changing the bit to 1. Particles are updated bit by bit and, in contrast to PSO, in the BPSO the velocity must necessarily be restricted within the range [0,1]. If P is the probability of changing a bit from 0 to 1, then this probability can be represented as the following function:

$$P(x_{id}(t)=1) = f(x_{id}(t), v_{id}(t-1), p_{id}, p_{gd})$$

where, $P(x_{id}=1)$ is the probability that an individual particle i will choose 1 for the bit at the dth site in the bit string, $x_{id}(t)$ is the current state of the particle i at bit d, $v_{id}(t-1)$

is a measure of the string's current probability to choose a *1*, p_{id} is the best state found so far for bit *d* of individual *i*, (i.e., a *1* or a *0*), p_{gd} is *1* or *0* depending on what the value of bit *d* in the global best particle is.

The most commonly used measure for *f* is the sigmoid function which is defined as follows:

$$f(v_{id}(t)) = \frac{1}{1+e^{-v_{id}(t)}}$$

where,

$$v_{id}(t) = w v_{id}(t-1) + (\varphi_1)(p_{id} - x_{id}(t-1)) + (\varphi_2)(p_{gd} - x_{id}(t-1)) \tag{18}$$

Equation (18) gives the update rule for the velocity of each bit, where φ_1 and φ_2 are random numbers drawn from a uniform distribution such that their sum is four. The v_{id} value is sometimes limited so that *f* does not approach *0.0* or *1.0* too closely. In this case, constant parameters *[Vmin, Vmax]* are used. When v_{id} is greater than *Vmax*, it is set to *Vmax* and if v_{id} is smaller than *Vmin*, then V_{id} is set to *Vmin*. This simply limits the ultimate probability that bit x_{id} will take on a zero or one value. A higher value of *Vmax* makes new vectors less likely. Thus *Vmax* in the discrete particle swarm plays the role of limiting exploration after the population has converged [6], i.e., it can be said that *Vmax* controls the ultimate mutation rate or temperature of the bit vector. Smaller *Vmax* leads to a higher mutation rate [6].

4 PSO Implementation Details for VCM Design

The applicability of the PSO (and BPSO) algorithm(s) for the design of virtual manufacturing cells is now described.

4.1 PSO VCM Design Algorithm

The PSO (and BPSO) algorithmic procedure(s) for assigning jobs and machines to the virtual cells is described in the following steps.

1. Choose an appropriate particle string (chromosome) representation for jobs assignments.
2. Define the fitness function (numerically measures the closeness of the solution to the optimal solution) on the basis of the objective function and constraints (per equations (1) through (15)).
3. Select suitable search parameters.
4. Execute the PSO algorithm:
   a. Generate the candidate solutions.
   b. Check constraint violation; if the constraint is violated, repair candidate solution.
   c. Compute fitness.
   d. Update 'self best' and 'global best' particles.
   e. Update the state of the particle.

4.2   Experimental Setup

To test the application of the suggested PSO procedure with its topological variants, we use the example problem detailed in [3]. The experiments are carried out for different combinations of number of virtual cells (1 & 3) and for α values (0 & 1.0).

4.3   Implementation Details for BPSO and PSO Topological Variants

In this model, jobs are assigned to the virtual cells using the PSO algorithm and the BPSO approach described above. Given that there is a pool of 12 jobs to be processed (see [3] for details), each cell can process at most 12 jobs in all. The chromosome length in case of the BPSO (which is a binary string) required for representation is then computed as follows: Binary string length = (Number of Cells × Total number of jobs that a cell can process × Bits required to represent a job). Accordingly, the BPSO binary string length for the 1-cell, 2-cell, and 3-cell cases are respectively, 48, 96 and 144. The representation can be illustrated through Fig. 2. For the case of the PSO topological variants, the chromosome length is computed thus: (Number of Cells × Total number of jobs that a cell can process) because in this case, the chromosome is a chunk of real numbers instead of bits 0 or 1. Accordingly, the chromosome length for the 1-cell, 2-cell, and 3-cell cases are respectively, 12, 24 and 36.

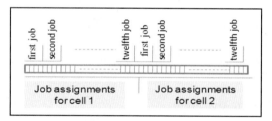

Fig. 2. Binary string representation for job assignment in the 2-cell case

The selection of the PSO algorithm's parameters need to be carefully done as this may have a major impact on the convergence towards the optimal solution. Table 2

Table 2. PSO and BPSO Parameters

Population size	50
PSO Parameters	
Vmax, Vmin	6, -6
Xmax, Xmin	12, 0
c1, c2	2, 2
BPSO Parameters	
Vmax, Vmin	1, 0

details the assumed parameters used for the PSO as well as the BPSO algorithm for the VCM design problem in this paper.

Following the assignments of jobs into cells, the machines are next assigned into the cells using constraint (3) and then nn_{mc} is computed. Next, using constraints (5), (6) and (7), the LB_{fcm}, UB_{fcm} and T_{cm} values are computed. It may be noted that the n_{mc} values are then assigned in such a way that the constraints (8), (9) and (10) are not violated. To effect this, first the ratio (T_{cm}/R) is compared with nn_{mc}. If (T_{cm}/R) becomes larger than nn_{mc} then some additional machines of type m are assigned into cell c so as to increase the value of nn_{mc}. As soon as the requirement that $(T_{cm}/R \leq nn_{mc})$ for each of the formed virtual cells is ensured, then for each cell the n_{mc} value is set equal to (T_{cm}/R). In this way the constraints (8) and (10) are satisfied.

Now only w_c values remain to be assigned in order to satisfy constraints (12), (13), (14) and (15). Constraints (12) and (14) require that $(\Sigma_m T_{cm}/R \leq MAXW)$ for all the cells and machines. If this does not happen then again a repair procedure ie effected by removing some jobs from cell c. Once this is done, then the w_c values are assigned by taking the minimum of MINW and $\Sigma_m T_{cm}/R$. In this way the constraints (12), (14) and (15) get satisfied. Also, if constraint (13) gets violated for some cells then again some jobs are removed from the cell c in order to decrease the value of $\Sigma_m T_{cm}/R$ and w_c. Proceeding sequentially in this way, for each candidate solution, all the constraints are satisfied.

After application of the repair procedure, the fitness value for each candidate solution is computed in order to estimate the solution quality obtained. For the VCM design problem at hand, the fitness value is computed by estimating the objective function value Z.

5   Results and Conclusion

The results obtained from the application of the conventional PSO algorithm with its alternative neighborhood topologies, as well as for the BPSO approach for the assumed VMC design problem are detailed in Table 3. The table also contains the results reported in [3] (shown highlighted) obtained using the Preemptive / Lexico Goal Programming approach. It is seen that the PSO results compare well with the reported results in [3]. It was also observed that the PSO results were obtained very quickly (often in a few seconds) whereas those from the Goal Programming procedure sometimes took several hours when solved using the LINGO software package on a Intel Core 2 Duo CPU (2.4 GHz) PC. It is also interesting to note that for the 3-cell VCM design problem, the Goal Programming solution is not the general best. This may be explained by the fact that the solver used for implementing the Goal Programming approach applies a branch and bound methodology and that some of the presented solutions in [3] are intermediate results obtained after several hours of running on a PC.

In conclusion, the PSO algorithm with its topological variants are efficient approaches for the VCM design problem that can easily be implemented within real-world VCM-type manufacturing environments.

Table 3. Results for the PSO VCM design procedure

No. of Cells	a	Cell Size (Range)	Applied Search Algorithm	Required Iterations	z	Jobs Assigned To			Machines In			Vm	
						Cell 1	Cell 2	Cell 3	Cell 1	Cell 2	Cell 3	+	−
1	0	8	GP	--	227	1,2,3,4,5,6,8,9,10,11,12	—	—	1,2,2,3,3,4,4,5,5	—	—	0	1,3,4
			PSO (ST)	111	227	1,2,3,4,5,6,8,9,10,11,12	—	—	1,2,2,3,3,4,4,5,5	—	—	0	1,3,4
			PSO (RT)	50	227	1,2,3,4,5,6,8,9,10,11,12	—	—	1,2,2,3,3,4,4,5,5	—	—	0	1,3,4
			PSO (VNT)	73	227	1,2,3,4,5,6,8,9,10,11,12	—	—	1,2,2,3,3,4,4,5,5	—	—	0	1,3,4
			BPSO	8000	219	2,4,5,6,7,8,9,10 11 12	—	—	1,1,2,2,3,3,4,4,5,5	—	—	0	3,4
	1		GP	--	165	1,2,3,5,6,7,8,10	—	—	1,2,2,3,3,3,4,4,5,5	—	—	0	1,4
			PSO (ST)	10	167	2, 3, 5, 6, 7, 8, 10, 11	—	—	1,2,2,3,3,3,4,4,5,5	—	—	0	1,4
			PSO (RT)	7	167	2, 3, 5, 6, 7, 8, 10, 11	—	—	1,2,2,3,3,3,4,4,5,5	—	—	0	1,4
			PSO (VNT)	12	167	2, 3, 5, 6, 7, 8, 10, 11	—	—	1,2,2,3,3,3,4,4,5,5	—	—	0	1,4
			BPSO	7	167	2, 3, 5, 6, 7, 8, 10, 11	—	—	1,2,2,3,3,3,4,4,5,5	—	—	0	1,4
3	0	2-3	GP	--	210	9,10,12	7,8,11	1,2,3,4	2,3,4,5	1,3,4,5	1,2,3,4,5	5	0
			PSO (ST)	2333	210	3,5,6,8	9,10,1	2,4,7	2,3,4,5	2,3,4,5	1,2,3,4,5	2,5	1
			PSO (RT)	330	210	3,9,10,12	5,6,8	2,4,7	2,3,3,4,5	1,2,3,4,5	2,3,4,5	2,3,5	1
			PSO (VNT)	698	210	1,2,3,4	9,10,1	7,8,11	1,2,3,4,5	2,3,4,5	1,3,4,5	5	0
			BPSO	8000	207	3,9,11,12	5,6,8	1,2,10	1,2,3,4	1,2,3,4,5	1,2,3,5	1,2	4
	1		GP	--	149	3,7,8	2,10	1,12	1,3,4,5	2,3,5	1,2,3,4	0	4
			PSO (ST)	5418	147	5,9	3,7,12	2,10	1,2,3,4,4,5	1,2,3,4	2,3,5	2	0
			PSO (RT)	1263	151	3,8,12	2,10	7,11	2,3,4,5	2,3,5	1,3,4	0	1,4
			PSO (VNT)	5065	151	2,3,8	10,12	7,11	2,3,4,5	2,3,4,5	1,3,4	0	1
			BPSO	620	148	3,7,11	9,12	2,10	1,3,4	2,3,4	2,3,5	0	1,4,5

References

1. Burbidge, J.L.: Production flow analysis. Prod. Engg. 42, 742–752 (1963)
2. Nomden, G., Slomp, J., Suresh, N.C.: Virtual manufacturing cells: a taxonomy of past research and identification of future research issues. Int. J. of Flex Manuf. Syst. 17, 71–92 (2006)
3. Slomp, J., Chowdary, B.V., Suresh, N.C.: Design of virtual manufacturing cells: a mathematical programming approach. Robotics and Comp. Integ. Manuf. 21, 273–288 (2005)
4. Suresh, N.C., Meredith, J.R.: Coping with the loss of pooling synergy in cellular manufacturing systems. Manage Sci. 40, 466–483 (1994)
5. Kennedy, J., Eberhart, R.: Particle swarm optimization. In: Proceedings of the IEEE International Conference on Neural Networks, Piscataway, NJ, pp. 1942–1948 (1995)
6. Kennedy, J., Eberhart, R.: A discrete binary version of the particle swarm algorithm. In: Proceedings of the IEEE International Conference on Systems, Man, and Cybernetics, pp. 4104–4108 (1997)
7. Mendes, R.: Population topologies and their influence in particle swarm performance. Ph.D. Dissertation, University of Minho (April 2004)
8. Kennedy, J., Mendes, R.: Neighborhood topologies in fully informed and best of neighborhood particle swarms. In: Proceedings of the 2003 IEEE International Workshop Soft Computing in Industrial Applications, pp. 45–50 (June 2003)
9. Kennedy, J.: Small world and mega minds: effects of neighborhood topology on particle swarm performance. In: Proceeding of the 1999 Conference on Evolutionary Computation, pp. 1931–1938 (1999)

Energy Saving System for Office Lighting by Using PSO and ZigBee Network

Wa Si[1], Harutoshi Ogai[1], Tansheng Li[1], Masatoshi Ogawa[1],
Katsumi Hirai[2], and Hidehiro Takahashi[2]

[1] Graduate School of Information, Production and Systems, Waseda University
2-7 Hibikino, Wakamatsu-ku, Kitakyushu, Fukuoka, Japan, 808-0135
ariel@fuji.waseda.jp, Ogai@waseda.jp,
anvaya@ruri.waseda.jp, stream@aoni.waseda.jp
[2] Hakutsu Technology Corporation
2-6-5 Kuriki, Asou-ku, Kawasaki city ,Kanagawa, Japan, 215-0033
{Khirai,hidehirox}@hakutsu-tech.co.jp

Abstract. In order to reduce the amount of wasted energy in office lighting and provide a major contribution to lowering overall energy consumption, we are developing a new energy saving system for office lighting by using adjustable lamp, ZigBee Wireless Sensor Network (WSN) and Particle Swarm Optimization (PSO).In this paper we make a prototype system which consists of one ZigBee control module, four fluorescent lamps with dimming capacity and three illumination sensors.The illumination sensors collect and send the data to the control module. After the PSO (Particle Swarm Optimization) process, the module finally sets the power of the lamps according to the PSO result. After real experiments in both sunny day and cloudy dayin a small-size office, it was proved that the system can successfully control the lights, and save considerable energy.

Keywords: Office Lighting, Energy Saving System, PSO, ZigBee, Wireless Sensor Network.

1 Introduction

Nowadays, due to the shortage of energy resources and greenhouse effect, energy saving becomes an emergence subject all over the world. However, the demand of energy is still growing rapidly each year. So improving the energy efficiency becomes a core strategy for slowing down the growth of energy consumption.

The amount of energy used from lighting varies from industry to industry but, typically, lighting accounts for approximately 15% of the electrical load in industry[1].

As a major economic power which has the world's second-largest economy by nominal GDP and the third largest in purchasing power parity, Japan consumes more than 900 billion kWh per year. And buildings are responsible for up to 40% of the total energy[2]. It is reported that lighting accounts for around 40% of the primary energy use in Japanese office buildings[3]. So energy saving for office lighting may be one of the most efficient ways to save the energy among all the electrical systems.

K. Deb et al. (Eds.): SEAL 2010, LNCS 6457, pp. 667–676, 2010.
© Springer-Verlag Berlin Heidelberg 2010

In our daily life, light plays a very important role. However, it is excessively used, not only at night but also in the daytime in most office buildings for the reasons like unused natural illumination and un-full occupancy.

Great potential of energy savings can be achieved by methods like daylight harvesting, load shedding, scheduling, etc[4].However, it is very expensive for legacy buildings to use traditional wired technology[5]. Thus, wireless sensor network (WSN) becomes a promising economic technology without complicated rewiring. Wireless sensor network (WSN) also have the advantages of convenient and economic which is especially good for small-size office usage.

2   Lighting Control Algorithm

The core of the proposed system is an intelligent lighting optimization algorithm which can get the optimal power settings for each luminaire according to the illumination contribution from environment and luminaires in target positions.

2.1   Framework

Fig. 1 shows the framework of the lighting system. The illumination sensors collect the data of illumination contribution from both environment (mostly sunlight) and luminaires (Fluorescent lamps) in their own positions and send the data to the network control module wirelessly. After processing the Particle Swarm Optimization (PSO), the network control module sends the wireless actuation commands to the luminaires and finally sets the illumination in the office.

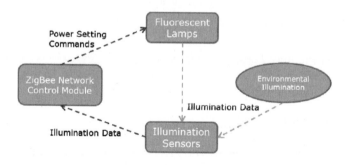

Fig. 1. Framework of the lighting system

2.2   Control Algorithm

Fig. 2 shows the framework of the control algorithm.

$N_i$represents the natural illumination contribution to each sensor whilea_{ij} represents the illumination contribution from lamp j to sensor i. F_j is the power proportion of each lamp. Sosuppose there are n lamps and m sensors in the system, we can get the formula to calculate the total illumination in each sensor:

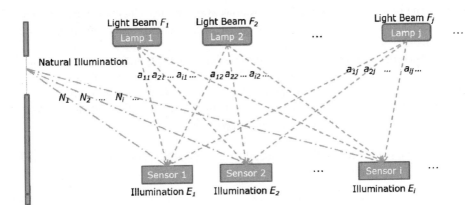

Fig. 2. Framework of the control algorithm

$$\begin{bmatrix} E_1 \\ E_2 \\ \vdots \\ E_m \end{bmatrix} = \begin{bmatrix} a_{11} & a_{12} & \cdots & a_{1n} \\ a_{21} & a_{22} & \cdots & a_{2n} \\ \vdots & \vdots & \ddots & \vdots \\ a_{m1} & a_{m2} & \cdots & a_{mn} \end{bmatrix} \begin{bmatrix} F_1 \\ F_2 \\ \vdots \\ F_n \end{bmatrix} + \begin{bmatrix} N_1 \\ N_2 \\ \vdots \\ N_m \end{bmatrix}. \qquad (1)$$

The final power consumption P is:

$$P = f(F_1) + f(F_2) + \cdots + f(F_n). \qquad (2)$$

The system is aim at making the illumination in each sensor E_i higher than the target illuminationTand minimizing the total power consumption that is to avoid useless illumination.

So there is a cost function to evaluate the lamp settings (F_1, F_2,···, F_n):

$$\text{Cost} = \sum_{i=1}^{m} |E_i - T| \times \text{penalty},$$

$$\text{penalty} = \begin{cases} 1, & \text{if } E_i - T \geq 0 \\ 10000, & \text{if } E_i - T < 0 \end{cases}. \qquad (3)$$

2.3 Particle Swarm Optimization (PSO)

The convergence speed of evolutionary algorithms to the global optimal results is better than that of traditional techniques [6]. Among which particle swarm optimization (PSO) is anevolutionary computation technique which can be applied to solve most optimization problems and problems that can be converted to optimization problems [7]. It is efficient and easy conducted.

Fig. 3 shows the flow chart of Particle Swarm Optimization (PSO). There are mainly 5 steps:

Step.1. The program first initiates the positions of n particles which are n sets of lamp power ratio settings (F_1, F_2,...,F_n) and the velocity (the changing speed of particle value every time).

Step.2. Then the program begins to evaluate each particle according to the cost function (the less cost, the better particle).

Step.3. Compare and update every particle's own current best position (p_{best}).

Step.4. Compare and update the best position in the group, global best position (g_{best})

Step.5. Update the positions and velocities of particles.

After repeating k times of this process, the final global best position will be the final settings of the luminaires.

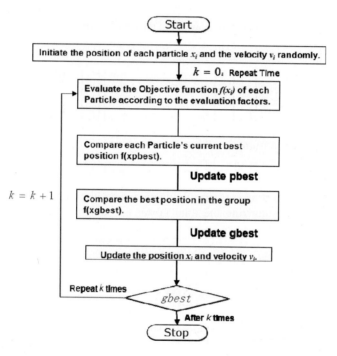

Fig. 3. Flow chart of Particle Swarm Optimization (PSO)

3   Prototype of the System

The prototype of the wireless networked lighting system consists of one ZigBee control module, four fluorescent lamps with dimming capacity and three illumination sensors and is applicable to small-sizeoffice using. Fig.4 shows the actual look of ZigBee control module and illumination sensor.

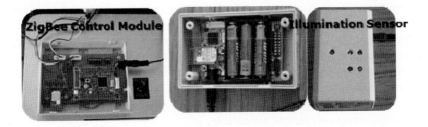

Fig. 4. Photos of ZigBee control module and illumination sensor

3.1 ZigBee Control Module

The module is implemented with the control program which is the core of the system where the optimization algorithm resides. It is responsible of receiving data from the illumination sensors, processing the optimization and issuing the commands to the lamps.

3.2 Illumination Sensor

When calculating the illumination coefficients, the lamps are turned off one by one, and during each time, the sensor gets the illumination contribution a_{ij} from each lamp and thus calculates the environmental illumination N_i according to the total illumination E_i and illumination from luminaires a_{ij}:

$$N_i = E_i - \sum_{j=1}^{n} a_{ij} \ . \tag{4}$$

The sensor can be powered by both wired source and battery which means that its position can be changed anytime thus makes the system more adaptable.

3.3 Fluorescent Lamp with Dimming Capacity

The fluorescent lamp is connected with wireless equipment which enables the lamp to receive the power setting commands wirelessly. There is also a dimming circuit which makes the lamp capable of changing luminosity continuously. The luminosity of the lamp is linear with the input power and cannot light below certain input power.

4 Experimental Verification

The experiments were taken on both sunny day and cloudy day. Electricity meter and luminosity meter are used to check the power consumption and the actual illumination in the positions of sensors. And a data collecting device implemented in computer was used to record the sensor data.In our experiments we set the target illumination 800 lux to ensure the enough luminosity.

The minimum dimming ratio for the luminaire to illuminate is 20% of the maximum power.The initial dimming ratio of each lamp in the experiments was 80% of the maximum power. Thus the theoretical final dimming ratio setting for each lamp is 0.25 ~ 1.25(0.2/0.8=0.25 ~ 1/0.8=1.25).

In our experiments, we set the maximum final setting of the lamp 1.2. And in the PSO program, we set all the positions 0 if they were less than 0.25.

The illumination contribution from one luminaire to one sensor will theoretically be the same as long as the position has not been changed. It was proved in our experiments and fig.5 shows the changing of the illumination coefficients in one day. We can see that the illumination coefficients were stable.

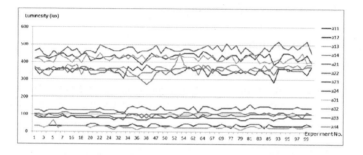

Fig. 5. Changing of the Illumination Coefficients in One Day

According to the data in fig.5, we calculate the average illumination coefficients showed in fig.6.

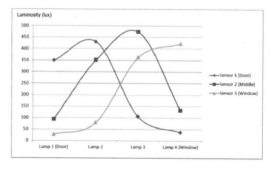

Fig. 6. Average Illumination Coefficient

In the PSO program, the number of particles was limited to 6 due to the limitation of the memory shortage of the network control module. The optimization repeats 1000 times.

The process took only about 3 seconds in each test.

In order to see if the PSO program works well, we do some pre-test which repeat only 100 times and record the particle change history. Fig.7 shows the particle position changing history of the second particle in one pre-test.

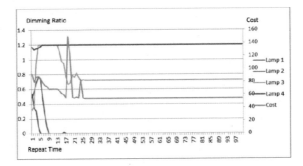

Fig. 7. Particle Position Changing History of the Second Particle in one Pre-test

We can see from the graph that the optimization reached the best solution after re-peating 26 times and the final dimming ratio settings for lamp 1, lamp 2, lamp 3 and lamp 4 were: 0 which meant turned off, 0, 0.72, and 1.2 which was the maximum value.

Fig.8 shows one example of the experiment. The natural illumination of sensor 1, sensor 2 and sensor 3 were 772 lux, 479 lux and 251 lux. After the optimization, lamp 1 and lamp 2 were turned off, while lamp 3 was set 1.2, and lamp 4 was set 0.35 which is about 1/3 of the original power. The controlled illumination of sensor 1, sensor 2 and sensor 3 were 896 lux, 854 lux and 888 lux. In this experiment, the power consumption of the system was 148.8 w, and more than half of the energy (55.4%) was saved.

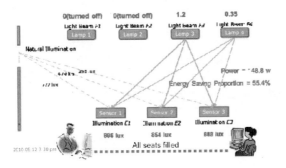

Fig. 8. Example of the experiment when all seats filled

Fig.9 shows the nature illumination and energy saving proportion on sunny day and cloudy day. The experiment results prove that the energy saving proportion is linear with the nature illumination (when there are higher natural illuminations, the system can save more energy). The system can save around 50% energy during daytime when the weather is fine. The system can save nearly 30% of energy during cloudy days when the natural illumination is low. When there is no nature illumination, the system can still save 10% of the energy by avoiding useless illumination.

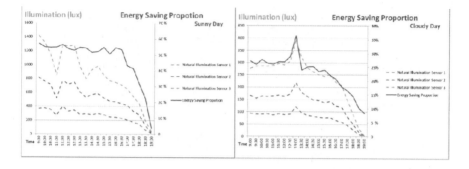

Fig. 9. Nature illumination and energy saving proportion on sunny day and cloudy day

Fig.10 shows the comparison of natural illumination, full power illumination and controlled illumination on sunny day and cloudy day. When all lamps on, the total illumination is over 1000 lux and sometimes over 2000 lux, which means that part of the energy is wasted.

In the experiments, after the controlling process, the illumination inside the office is well controlled above 800 lux which was the target illumination.

The nature illumination differs greatly from different positions. But the system makes the illumination uniform in different places and thus creates a good working environment inside the office.

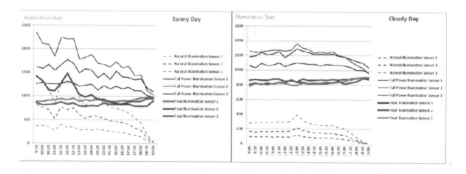

Fig. 10. Comparison of natural illumination, full power illumination and controlled illumination on sunny day and cloudy day

The experiments also considered the condition in which empty seats exist (the illumination in some sensors' position do not need to meet the target illumination).

Fig.11 shows one example of the experiment when one of the three seats was empty. This time, the natural illumination of sensor 1, sensor 2 and sensor 3 were 296 lux, 168 lux and 93 lux. After the optimization, lamp 1 was turned off, while lamp 2, lamp 3 and lamp 4 were set 0.42, 0.96 ad 0.31. The controlled illumination of sensor 1 and sensor 2 were 842 lux and 825 lux while the illumination of sensor 3 where empty seat exist was

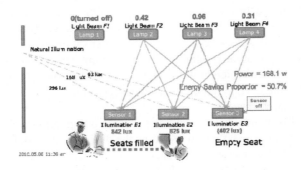

Fig. 11. Example of the experiment when one seat was empty

402 lux. In this experiment, the power consumption of the system was 168.1 w, and half of the energy (50.7%) was saved.

Fig.12 shows the comparison of energy saving proportion when all 3 sensors, 2 sensors, and only one sensor turned on. It is shown that the system can save even more energy when there are empty seats in the office: 35-50% of the energy can be saved when one of the three sensors is off and 45-60% can be saved when two of the three sensors are off.

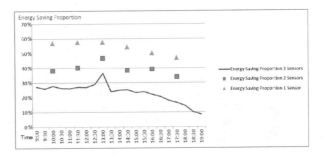

Fig. 12. Comparison of energy saving proportion when all 3 sensors, 2 sensors, and only one sensor turned on

5 Conclusion and Future Research

This paper presents a prototype of energy saving system for office lighting which can efficiently avoid useless illumination and thus save considerable proportion of the total energy. It can also recognize the occupancy condition and react properly to save energy in shared-space office.

The experiments prove the feasibility and usefulness of the energy saving system and the opportunities of future research.

By using PSO, the control time was limited within 3 seconds, which allow the system to react fast when conditions change.

In order to improve the accuracy and efficiency of the system further, the program of illumination sensors should be modified to get more accurate data. Besides, the PSO program can be improved so the system can get better solutions.

In the proposed prototype system, sensors in different positions share the same target illumination. However, different individuals may prefer different illumination conditions. So in the improved system, changeable target illumination may be set for each sensor to meet various needs.

Large-size offices may have more complicated configuration with much more luminaires and sensors (seats) which increases the difficulty of optimization especially the calculation of illumination coefficients.

In this prototype system, before each optimization, every lamp needs to be turned off one by one to let the illumination sensors to calculate the illumination coefficients. So every time the natural illumination change or occupancy condition change, this need to be done again. However, as proved in the system, the illumination coefficients stay the same as long as the position of the sensor is not changed, so the calculation of illumination coefficients will be meaningless. The sensors only need to measure the total illumination, and calculate the latest natural illumination according to the old data of illumination coefficients.So a database may be introduced to the system to store the coefficient and the history of sensor changing.

Acknowledgments. This research receives support from Kitakyushu city as an EnvironmentalFuture Technology Development project.

References

1. Ali, N.A.M., Fadzil, S.F.S., Mallya, B.L.: Improved illumination levels and energy savings by uplamping technology for office buildings. In: 2009 International Association of Computer Science and Information Technology - Spring Conference (2009)
2. EIA. Annual energy outlook 2009 (aeo2009). Technical report, DOE/EIA-0383, US Department of Energy, Washington, US (2009)
3. Energy Consumption in Japanese Office Building, The Energy Conservation Center, Japan (2009), http://www.eccj.or.jp/
4. Mills, E.: Global Lighting Energy Savings Potential. Light and Engineering 10, 5–10 (2002)
5. Best Practice Design, Technology and Management, State Government of Victoria (2002)
6. Yildiz, A.R.: A novel particle swarm optimization approach for product design and manufacturing. International Journal of Advanced Manufacturing Technology 40(5-6), 617–628 (2009)
7. Shi, Y., Eberhart, R.C.: A modified particle swarm optimizer. In: Proceedings of IEEE International Conference on Evolutionary Computation, pp. 69–73 (1998)

EcoSupply: A Machine Learning Framework for Analyzing the Impact of Ecosystem on Global Supply Chain Dynamics

Vikas K. Garg[1] and N. Viswanadham[2]

[1] IBM Research - India
vikaskga@in.ibm.com
[2] Global Logistics and Manufacturing Strategies (GLAMS)
Indian School of Business (ISB), Hyderabad, India
N_Viswanadham@isb.edu

Abstract. A global supply chain spans several regions and countries across the globe. A tremendous spurt in the extent of globalization has necessitated the need for modeling global supply chains in place of the conventional supply chains. In this paper, we propose a framework, *Eco-Supply*, to analyze the supply chain ecosystem in a probabilistic setting unlike the existing methodologies, which presume a deterministic context. EcoSupply keeps track of the previous observations in order to facilitate improved prediction about the influence of uncertainties in the ecosystem, and provides a coherent mathematical exposition to construe the new associations, among the different supply chain stakeholders, in place of the existing links. To the best of our knowledge, EcoSupply is the first machine learning based paradigm to incorporate the dynamics of global supply chains.

Keywords: Supply Chains, Global Sourcing, Machine Learning.

1 Introduction

Global outsourcing has acquired a central role in the contemporary manufacturing and service industries. Multinational companies invest at different locations across the globe to gain competitive advantage by exploring new markets, availing cutting edge technology, and harnessing skills at sustainable costs. Therefore, the need for making an effective decision regarding the selection of locations and global business partners from a plausible set of candidates can not be overemphasized.

The literature abounds in techniques for modeling the supply chain formation. Walsh et. al [1] proposed a combinatorial protocol, consisting of a one-shot auction and a strategic bidding policy, to study the negotiations on production relationships among multiple levels of production in a distributed setting. Prior to this work, auction mechanisms were proposed, such as in [2], to address the complementarities or the mutual dependencies among values of obtaining inputs and producing outputs. Typically, a global supply chain is the result of trade by

K. Deb et al. (Eds.): SEAL 2010, LNCS 6457, pp. 677–686, 2010.

a firm across national borders by means of either foreign direct investment (FDI) or outsourcing, though other levels of operational strategies such as licensing, joint venture, and acquisition etc. also exist. Consequently, a lot of effort has gone into investigating the decision of firms to trade through FDI or outsourcing, see for instance, [3], [4]. Most of these models can not be quantitatively analyzed and proffer only a high level insight into the decision making process.

Tax has a significant impact on SCF, as the product material moves across boundaries. Certain parts of the world offer special economic zones - also known as free trade zones - where goods bound for export can be manufactured, assembled, and stored with attractive tax holidays. Therefore, substantial research has gone into integrating taxes and other regulatory factors in the global supply chain design ([5], [6]). Recently, a mixed integer non-linear programming model that incorporates the import and export tax liabilities at various stages of the global supply chain has been proposed [7]. Besides tax, there are certain other factors with positive (for example, acquaintance) or negative (e.g. economic and cultural heterogeneity) influence that have a marked influence on the overall supply chain formation. However, these factors have been overlooked thus far in the literature.

1.1   Motivation

The literature abounds in expository research on supply chain formation (SCF) and network planning. However, almost all of these techniques analyze the problem of selecting an alternative at a given stage using a deterministic cost model while neglecting altogether the uncertainty in the surrounding ecosystem, which encompasses all the factors that might influence the supply chain formation. For instance, there are certain factors in most supply chain ecosystems, such as infrastructure, local demand and proximity to key markets, availability of skilled labor, inventory handling facilities, government regulations and incentives, financial costs (e.g. in acquisition of land), transportation, and tax and freight considerations, etc. While some of these factors, notably tax considerations and inventory handling costs, have been incorporated in the existing models, a vast majority of these factors still remains unaccounted. Furthermore, most of the sub-factors that determine these factors may change over a period of time, thereby triggering a change in the impact of these factors. Therefore, we believe there is a need for a generic probabilistic framework that seamlessly incorporates and integrates these factors for understanding the dynamics of the supply chains. This adaptive modeling of supply chains is fundamental to explaining the replacement of an extant end-to-end supply chain with a new one, as the different factors governing the SCF change over time. In this work, we explicate this dynamic aspect of supply chains using a statistical model EcoSupply.

2   Problem Definition and Notation

Consider a multi-stage global supply chain network, where each stage represents an activity such as production or assembly. We assume that the Supply chain has

N stages: $S_1, S_2, \ldots, S_N$. There are k_i alternatives, $S_{i1}, S_{i2}, \ldots, S_{ik_i}$, at any stage S_i to accomplish the activity of that stage. Each alternative q at a stage p is expressed as a d-dimensional observation or feature vector of factors x_{pq}, with the data observation corresponding to a factor r denoted by x_{pqr}, whereby the probability of a factor r being favorable is given by $\theta_{pqr}, 0 \leq \theta_{pqr} \leq 1, r \in \{1, 2, \ldots, d\}$. Let D_{pq} represent a set of n d-dimensional data vectors corresponding to the qth alternative in stage S_p, and D_{pqr} represent a set of n samples corresponding to factor r, assumed to be independent and identically distributed (i.i.d), $\{x_{pqr}^1, x_{pqr}^2, \ldots, x_{pqr}^n, x'_{pqr}\}$. Further, let τ_{pqr} denote the threshold above which a factor r is perceived favorable, and w_{pqr} and $C_{pqr}^{D_{pqr}}$ denote respectively the weight or perceived importance of r, and the estimated cost associated with r based on D_{pqr}, at the alternative q in the stage p. Finally, let C_{pq}^{init} and $C_{pq}^{D_{pq}}$ denote the initial unaccounted cost (which disregards the impact of factors), and the estimated cost based on D_{pq}, taking into consideration the ecosystem, if alternative q is chosen in stage p.

Then, the problem of probabilistic modeling of SCF is formulated as follows: find the probability of any supply chain, $SC = A_1, A_2, \ldots, A_N$, formed by choosing an alternative A_i from each stage $1 \leq i \leq N$. Intuitively, the greater this probability, the more likely the formation of SC, compared to any other supply chain. Furthermore, this probability might change over time, as more data is accumulated or the impact of various factors varies.

3   The EcoSupply Model

Two types of factors need to be considered: a) the factors local to an alternative, and b) the factors governed by a pair of alternatives at successive stages in the supply chain network. Below, we describe how these factors are modeled using the EcoSupply framework.

3.1   Modeling the Impact of Factors Specific to an Alternative

At any instant of time, each of the underlying factors in the ecosystem can be considered as being favorable or unfavorable towards selection of a particular alternative at a particular instant of time, e.g. there might be a fear of shortage in supply of raw materials at a particular alternative deeming a high cost for that alternative. Our aim is to continually learn the favorable probabilities as more data is accumulated over time. Each of the factors in the ecosystem can be perceived as Bernoulli variables representing unknown probability distributions. Then, the estimate for an observation x_{pqr} (which takes one of the two values: 1(favorable) or 0(unfavorable)) conditioned on the parameter θ_{pqr} is given by,

$$P(x_{pqr}|\theta_{pqr}) = \theta_{pqr}^{x_{pqr}} (1 - \theta_{pqr})^{1-x_{pqr}} \tag{1}$$

Then, we have the following result.

Lemma 1. *Let $D_{pqr} = \{x_{pqr}^1, x_{pqr}^2, \ldots, x_{pqr}^n\}$ be a set of n i.i.d samples drawn according to a probability distribution characterized by θ_{pqr}. If θ_{pqr} has a uniform prior distribution, then*

$$P(x_{pqr}|D_{pqr}) = \left(\frac{s_{pqr}^{D_{pqr}} + 1}{n+2}\right)^{x_{pqr}} \left(1 - \frac{s_{pqr}^{D_{pqr}} + 1}{n+2}\right)^{1 - x_{pqr}}$$

where $s_{pqr}^{D_{pqr}} = \sum_{j=1}^n x_{pqr}^j$

Proof.

$$P(D_{pqr}|\theta_{pqr}) = P(x_{pqr}^1, x_{pqr}^2, \ldots, x_{pqr}^n|\theta_{pqr})$$

$$= P(x_{pqr}^1|\theta_{pqr})P(x_{pqr}^2|\theta_{pqr})\ldots P(x_{pqr}^n|\theta_{pqr})$$

$$= \theta_{pqr}^{\sum_{j=1}^n x_{pqr}^j} + (1 - \theta_{pqr})^{\sum_{j=1}^n (1 - x_{pqr}^j)} \qquad [using\ (1)]$$

$$= \theta_{pqr}^{s_{pqr}^{D_{pqr}}} (1 - \theta_{pqr})^{n - s_{pqr}^{D_{pqr}}} \qquad (2)$$

Now,

$$p(\theta_{pqr}|D_{pqr}) = \frac{p(D_{pqr}|\theta_{pqr})p(\theta_{pqr})}{\displaystyle\int_{\theta_{pqr}} p(D_{pqr}|\theta_{pqr})p(\theta_{pqr})\, d\theta_{pqr}}$$

In the absence of any prior knowledge about θ_{pqr}, assuming a uniform distribution[1] in the interval $[0, 1]$, we obtain,

$$p(\theta_{pqr}|D_{pqr}) = \frac{p(D_{pqr}|\theta_{pqr})}{\displaystyle\int_0^1 p(D_{pqr}|\theta_{pqr})\, d\theta_{pqr}}$$

$$\Rightarrow P(x_{pqr}|D_{pqr}) = \int_{\theta_{pqr}} P(x_{pqr}|\theta_{pqr})p(\theta_{pqr}|D_{pqr})\, d\theta_{pqr}$$

$$= \int_0^1 P(x_{pqr}|\theta_{pqr}) \frac{p(D_{pqr}|\theta_{pqr})}{\displaystyle\int_0^1 p(D_{pqr}|\theta_{pqr})\, d\theta_{pqr}}\, d\theta_{pqr} \qquad (3)$$

Using (2),

$$\int_0^1 p(D_{pqr}|\theta_{pqr})\, d\theta_{pqr} = \int_0^1 \theta_{pqr}^{s_{pqr}^{D_{pqr}}} (1 - \theta_{pqr})^{n - s_{pqr}^{D_{pqr}}}\, d\theta_{pqr}$$

---

[1] In general, each of the factors is dependent on several sub-factors, and may follow an arbitrary distribution, e.g. the supply of raw materials may not be uniform and may vary from time-to-time, depending on a change in the capability of the source or trade restrictions. This is not a very stringent assumption, for example, refer [9] for modeling a Gaussian prior on θ_{pqr}.

From the definition of beta function, for $a, b > 0$,

$$\beta(a,b) = \int_0^1 t^{a-1}(1-t)^{b-1}dt = \frac{\Gamma(a)\Gamma(b)}{\Gamma(a+b)}$$

where $\Gamma(.)$ denotes the gamma function. Then, evaluating the gamma function on integral arguments, we get,

$$\int_0^1 p(D_{pqr}|\theta_{pqr})\,d\theta_{pqr} = \frac{\Gamma(s_{pqr}^{D_{pqr}}+1)\Gamma(n-s_{pqr}^{D_{pqr}}+1)}{\Gamma(n+2)}$$

$$= \frac{s_{pqr}^{D_{pqr}}!(n-s_{pqr}^{D_{pqr}})!}{(n+1)!}$$

which in the light of (3) yields,

$$P(x_{pqr}|D_{pqr}) = \frac{(n+1)!}{s_{pqr}^{D_{pqr}}!(n-s_{pqr}^{D_{pqr}})!}\int_0^1 P(x_{pqr}|\theta_{pqr})p(D_{pqr}|\theta_{pqr})\,d\theta_{pqr}$$

$$= \frac{(n+1)!}{s_{pqr}^{D_{pqr}}!(n-s_{pqr}^{D_{pqr}})!}\int_0^1 \theta_{pqr}^{s_{pqr}^{D_{pqr}}+x_{pqr}}(1-\theta_{pqr})^{n-s_{pqr}^{D_{pqr}}+1-x_{pqr}}\,d\theta_{pqr} \quad [using\ (1)\ and\ (2)]$$

$$= \frac{(s_{pqr}^{D_{pqr}}+x_{pqr})!(n-s_{pqr}^{D_{pqr}}+1-x_{pqr})!}{s_{pqr}^{D_{pqr}}!(n-s_{pqr}^{D_{pqr}})!(n+2)}$$

$$\Rightarrow P(x_{pqr}=1|D_{pqr}) = \frac{s_{pqr}^{D_{pqr}}+1}{n+2},$$

and,

$$P(x_{pqr}=0|D_{pqr}) = 1 - \frac{s_{pqr}^{D_{pqr}}+1}{n+2}$$

$$\Rightarrow P(x_{pqr}|D_{pqr}) = \left(\frac{s_{pqr}^{D_{pqr}}+1}{n+2}\right)^{x_{pqr}}\left(1-\frac{s_{pqr}^{D_{pqr}}+1}{n+2}\right)^{1-x_{pqr}}$$

In the next lemma, we show how the conditional density estimate can be incrementally updated on arrival of a new observation.

Lemma 2. *Let a new observation, x'_{pqr}, is recorded that results in an enhanced data set, $D'_{pqr} = D_{pqr}\bigcup\{x'_{pqr}\}$. Then, assuming the mutual independence of the d factors, the ratio of conditional probabilities,*

$$\frac{P(x_{pq}|D'_{pq})}{P(x_{pq}|D_{pq})} = \prod_{r=1}^{d}\left(\frac{n+2}{n+3}\right)\left[\frac{s_{pqr}^{D_{pqr}}+x'_{pqr}+1}{s_{pqr}^{D_{pqr}}+1}\right]^{x_{pqr}}\left[\frac{n-s_{pqr}^{D_{pqr}}+2}{n-s_{pqr}^{D_{pqr}}+1}\right]^{1-x_{pqr}}$$

Proof. It follows from Lemma 1 that

$$P(x_{pqr}|D'_{pqr}) = \left(\frac{s_{pqr}^{D'_{pqr}}+1}{n+3}\right)^{x_{pqr}}\left(1-\frac{s_{pqr}^{D'_{pqr}}+1}{n+3}\right)^{1-x_{pqr}}$$

$$= \left(\frac{s_{pqr}^{D_{pqr}} + x'_{pqr} + 1}{n+3} \right)^{x_{pqr}} \left(1 - \frac{s_{pqr}^{D_{pqr}} + x'_{pqr} + 1}{n+3} \right)^{1-x_{pqr}}$$

$$\Rightarrow \frac{P(x_{pqr}|D'_{pqr})}{P(x_{pqr}|D_{pqr})} = \left(\frac{n+2}{n+3} \right) \left[\frac{s_{pqr}^{D_{pqr}} + x'_{pqr} + 1}{s_{pqr}^{D_{pqr}} + 1} \right]^{x_{pqr}} \left[\frac{n - s_{pqr}^{D_{pqr}} + 2}{n - s_{pqr}^{D_{pqr}} + 1} \right]^{1-x_{pqr}}$$

Generalizing to the d-dimensional multivariate case by assuming that these d factors are mutually independent, we obtain,

$$P(x_{pq}|D'_{pq}) = P(x_{pq}|D_{pq}) \prod_{r=1}^{d} \left(\frac{n+2}{n+3} \right) \left[\frac{s_{pqr}^{D_{pqr}} + x'_{pqr} + 1}{s_{pqr}^{D_{pqr}} + 1} \right]^{x_{pqr}} \left[\frac{n - s_{pqr}^{D_{pqr}} + 2}{n - s_{pqr}^{D_{pqr}} + 1} \right]^{1-x_{pqr}}$$
$$(4)$$

Therefore, using Lemma 2, we can incrementally update the conditional density estimate on arrival of x'_{pqr}. Let τ_{pqr} be the threshold that determines if the factor r is favorable at alternative q in stage p. Then, one of the ways to compute the effective cost is given by,

$$C_{pq}^{D_{pq}} = C_{pq}^{init} \left[\sum_{j \in J_{pq}^{D_{pq}}} w_{pqj} e^{\frac{\tau_{pqj} - P(x_{pqj}=1|D_{pqj})}{m_{pqj}}} - |J_{pq}^{D_{pq}}| + 1 \right], \ where \quad (5)$$

$$J_{pq}^{D_{pq}} = \{ j : \tau_{pqj} > P(x_{pqj} = 1|D_{pqj}) \}$$

The weights w signify the importance of the different factors; the values τ can be adjusted to reflect the penalty in case of factors not meeting the desired threshold levels; and the scaling parameters m control the non-linearity of the model. Note that if a factor is deemed favorable with respect to the corresponding threshold, given the available data, then it does not add to the initial cost estimate, which disregards the ecosystem.

Theorem 1. *The overall estimated cost taking into account all the factors in the ecosystem at an alternative q in stage p based on D'_{pq}, for the cost model proposed in (5), is given by,*

$$C_{pq}^{D'_{pq}} = C_{pq}^{init} [\sum_{r \in J_{pq}^{D'_{pq}}} \left(I_{\tau_{pqr}}^{D_{pqr}} C_{pqr}^{D_{pqr}} + (1 - I_{\tau_{pqr}}^{D_{pqr}}) w_{pqr} \right)$$

$$e^{\frac{\tau_{pqr} - P(x_{pqr}=1|D_{pqr})}{m_{pqr}}} \left\{ \frac{x'_{pqr}}{P(x_{pqr}=1|D_{pqr})} - I_{\tau_{pqr}}^{D_{pqr}} \right\} - |J_{pq}^{D'_{pq}}| + 1]$$
$$(6)$$

Proof. Two cases are possible:

Case 1: $\tau_{pqr} > P(x_{pqr} = 1|D_{pqr})$ and $\tau_{pqr} > P(x_{pqruwc} = 1|D'_{pqr})$
Then, using

$$\frac{P(x_{pqr} = 1|D'_{pqr})}{P(x_{pqr} = 1|D_{pqr})} = \frac{(n+2)(s_{pqr}^{D_{pqr}} + x'_{pqr} + 1)}{(n+3)(s_{pqr}^{D_{pqr}} + 1)}, \quad we \ obtain,$$

$$C_{pqr}^{D'_{pqr}} = C_{pqr}^{D_{pqr}} e^{\frac{\tau_{pqr} - P(x_{pqr}=1|D_{pqr})}{m_{pqr}}\left\{\frac{(n+2)(s_{pqr}^{D_{pqr}} + x'_{pqr} + 1)}{(n+3)(s_{pqr}^{D_{pqr}} + 1)} - 1\right\}}$$

$$= C_{pqr}^{D_{pqr}} e^{\frac{\tau_{pqr} - P(x_{pqr}=1|D_{pqr})}{m_{pqr}}\left\{\frac{(n+2)x'_{pqr}}{s_{pqr}^{D_{pqr}} + 1} - 1\right\}}$$

$$= C_{pqr}^{D_{pqr}} e^{\frac{\tau_{pqr} - P(x_{pqr}=1|D_{pqr})}{m_{pqr}}\left\{\frac{x'_{pqr}}{P(x_{pqr}=1|D_{pqr})} - 1\right\}} \quad \left[since \ P(x_{pqr} = 1|D_{pqr}) = \frac{s_{pqr}^{D_{pqr}} + 1}{n+2}\right]$$

Case 2: $\tau_{pqr} < P(x_{pqr} = 1|D_{pqr})$ and $\tau_{pqr} > P(x_{pqr} = 1|D'_{pqr})$
It is straightforward to see,

$$C_{pqr}^{D'_{pqr}} = w_{pqr} e^{\frac{\tau_{pqr} - P(x_{pqr}=1|D_{pqr})}{m_{pqr}}\left\{\frac{x'_{pqr}}{P(x_{pqr}=1|D_{pqr})}\right\}}$$

These two cases can be expressed together as,

$$C_{pqr}^{D'_{pqr}} = \left(I_{\tau_{pqr}}^{D_{pqr}} C_{pqr}^{D_{pqr}} + (1 - I_{\tau_{pqr}}^{D_{pqr}})w_{pqr}\right) e^{\frac{\tau_{pqr} - P(x_{pqr}=1|D_{pqr})}{m_{pqr}}\left\{\frac{x'_{pqr}}{P(x_{pqr}=1|D_{pqr})} - I_{\tau_{pqr}}^{D_{pqr}}\right\}}$$

where $I_{\tau_{pqr}}^{D_{pqr}}$ is an indicator variable which takes value 1 if $\tau_{pqr} > P(x_{pqr} = 1|D_{pqr})$, else 0. Then, the overall cost considering all the factors, in accordance with (5), is given by (6).

We note that C_{pq}^{init} takes into consideration the influence of factors prevalent at the different alternatives on the effective costs. Similarly, costs involved among alternatives at successive stages (for instance, due to transport, tax, and handling of inventory in transaction etc.) can also be incorporated, taking into account uncertainty as gathered from historical data. The value of weights assigned to the different categories may be suitably adjusted for analyzing the overall cost across disparate supply chain application domains.

3.2   Modeling the Impact of Acquaintances and Distances

The impact of previous experience as a result of relationships among the different entities (e.g. suppliers/consumers at successive stages) is another important factor that has been overlooked thus far in the literature: if the experience is fruitful, the entities are likely to transact together as a part of supply chain again. In fact, this behavior is even more pronounced in case of global supply chains as the experience, between entities at successive alternatives, percolates down the supply chain. Furthermore, the distance dimensions also play a crucial role in the formation of global supply chains. In [8], these distances have been characterized into the following categories: cultural (e.g. religion, race, social norms, language), administrative and political (e.g. colony-colonizer links, currencies, trading arrangements), geographic (e.g. climate, waterway access, transportation and communication links, physical remoteness), and economic (e.g. information/knowledge, costs and quality of natural, financial and human resources, different consumer incomes)

An important observation is in order. These distances are a function of a pair of disparate alternatives at successive stages rather than being dependent on a single alternative. Thus, the whole process of the supply chain formation can be analyzed by using the following model:

1. Each of the alternatives is represented by a node.
2. For each node S_{pq}, a probability value $P_{pq} = \mathrm{P}(S_{pq})$ is calculated using (6) from Theorem 1 as

$$P_{pq} = \frac{\displaystyle\sum_{j=1}^{|S_p|} C_{pj}^{D''_{pj}} - C_{pq}^{D''_{pq}}}{(|S_p| - 1)\displaystyle\sum_{j=1}^{|S_p|} C_{pj}^{D''_{pj}}} \qquad (7)$$

3. The impact of acquaintance between alternatives, S_{ij} and $S_{l,k}$, $i \in \{1, 2, \ldots, N-1\}$, $l = i+1$, $j \in \{1, 2, \ldots, |S_i|\}$, $k \in \{1, 2, \ldots, |S_{i+1}|\}$, on SCF is reflected by the corresponding acquaintance edge having probability,

$$P_{ijlk}^{AB} = \frac{AB_{ijlk}}{\displaystyle\sum_{j=1}^{|E_{ij}|} AB_{ijlk}}$$

where, AB_{ijlk} denotes acquaintance benefit of alternative k at stage l; and E_{ij}^{AB} is the set of acquaintance edges that are outbound from alternative j at stage i.

4. The impact of distance between alternatives, S_{ij} and S_{lk}, $i \in \{1, 2, \ldots, N-1\}$, $l = i+1$, $j \in \{1, 2, \ldots, |S_i|\}$, $k \in \{1, 2, \ldots, |S_l|\}$, on SCF is reflected by a corresponding distance edge having probability,

$$P_{ijlk}^{DC} = \frac{\displaystyle\sum_{j=1}^{|E_{ij}^{DC}|} DC_{ijlk} - DC_{ijlk}}{(|E_{ij}^{DC}| - 1)\displaystyle\sum_{j=1}^{|E_{ij}^{DC}|} DC_{ijlk}}$$

with $DC_{ijlk} = W_C f_C(DC_{ijlk}^{C}) + W_A f_A(DC_{ijlk}^{A}) + W_G f_G(DC_{ijlk}^{G}) + W_E f_E(DC_{ijlk}^{E})$; where W_C, W_A, W_G, W_E are non-negative weights indicating the importance of the different dimensions corresponding to cultural, administrative, geographic, and economic distance respectively; $f_C, f_A, f_G, f_E : \Re^+ \to \Re^+$ are monotonically increasing functions that map the respective distance values to their equivalent perceived costs; and E_{ij}^{DC} is the set of distance edges that are outbound from alternative j at stage i.

5. The *acquaintance edge* and the *distance edge* between every pair of nodes in the underlying model are replaced by a single edge called the *influence edge* (with same orientation as the acquaintance edge) between the same nodes. The probability on this edge is given by,

$$P_{ijlk} = W * P_{ijlk}^{AB} + (1 - W) * P_{ijlk}^{DC}, \ 0 \le W \le 1$$

(where W indicates a relative preference for acquaintance over distance.)

$$= W \frac{AB_{ijlk}}{\displaystyle\sum_{j=1}^{|E_{ij}|} AB_{ijlk}} + (1 - W) \frac{\displaystyle\sum_{j=1}^{|E_{ij}^{DC}|} DC_{ijlk} - DC_{ijlk}}{(|E_{ij}^{DC}| - 1) \displaystyle\sum_{j=1}^{|E_{ij}^{DC}|} DC_{ijlk}} \tag{8}$$

Note that (8) is a valid probability measure since the sum of probabilities on all influence edges equals 1. Additionally, defining the probabilities this way is intuitive since the greater the acquaintance and the lesser the distance between two particular alternatives at successive stages is, the more likely the possibility of these alternatives being aligned again in a supply chain is.

4 Explaining the Dynamics of Supply Chain Formation

The dynamics of supply chain formation can be elegantly enunciated by using the following algorithm, based on the EcoSupply Model:

1. For each alternative q in stage p, draw a node with a probability value P_{pq} computed using (7).
2. Define the edge probabilities, for every pair of nodes representing alternatives at successive stages, using (8).
3. Add a dummy node, *Start*, which represents the stage S_0, and outbound edges to every node in S_1, with probability on each edge set to $\frac{1}{|S_1|}$. Further, set the probability value at *Start* to 1. (Note that this node serves the purpose of modeling multiple sources in the supply chain, which is another issue that has not been addressed in the literature thus far.)
4. Add a dummy node, *End*, which represents the stage S_{N+1}, and inbound edges from every node in S_N, with probability on each edge set to 1. Further, set the probability value at *End* to 1.
5. The probability of formation of a particular supply chain, $SC = A_1 A_2 \ldots A_N$, with $A_i \ i \in \{1, 2, \ldots, N\}$ denoting the alternative chosen at the stage i, is given by,

$$P_{SC} = P_{StartA_1} \prod_{i=1}^{N} P_{A_i} P_{A_i A_{i+1}} \tag{9}$$

The algorithm considers all the factors, depending on a particular alternative or a pair of alternatives at adjacent stages, that define the ecosystem: the probability on the nodes indicates the influence of factors restricted to a location whereas the probability on the edges indicates the influence of factors governing more than a single location. A change in any of these factors results in change in the probability values, given by (7) and (8), and a corresponding change in the probability of formation of an end to end supply chain, as indicated by (9). A relatively favorable ecosystem at an alternative, with respect to other alternatives, results in an increase in the corresponding probability of that alternative being a preferred choice for its stage, in the end-to-end supply chain.

5   Summary and Future Work

Modeling the impact of ecosystem on the supply chain formation is a topic of immense significance and has wide practical implications. Factors such as tax constraints and inventory handling costs have been well studied in the literature; however, several other important considerations such as the economic and cultural heterogeneity that constitute the entire ecosystem have been conspicuously ignored. In this paper, we proposed a generic Bayesian framework, EcoSupply, to model the dynamics of the supply chain formation. Specifically, we have illustrated how a change in the ecosystem accompanies a change in the local business alignments, and thereby the global supply chain dynamics. We have also showed how the acquaintances among the stakeholders greatly influence the future decisions regarding their collaboration. An important future direction would be to apply the EcoSupply model in different domains, for instance, food industry, automobile industry, financial sector, etc.

References

1. Walsh, W.E., Wellman, M.P., Ygge, F.: Combinatorial auctions for supply chain formation. In: Proceedings of the 2nd ACM Conference on Electronic Commerce (EC), pp. 260–269 (2000)
2. Parkes, D.C.: iBundle: An efficient ascending price bundle auction. In: ACM Conference on Electronic Commerce, pp. 148–157 (1999)
3. Antras, P., Helpman, E.: Global sourcing. J. Polit. Econ. 112, 552–580 (2004)
4. Grossman, G.M., Helpman, E.: Outsourcing in a global economy. Rev. Econ. Stud. 72(1), 135–159 (2005)
5. Goetschalckx, M., Vidal, C.J., Dogan, K.: Modeling and design of global logistics systems: A review of integrated strategic and tactical models and design algorithms. Eur. J. Oper. Res. 143(1), 1–18 (2002)
6. Oh, H.C., Karimi, I.A.: Regulatory factors and capacity-expansion planning in global chemical supply chains. Ind. Eng. Chem. Res. 43, 3364–3380 (2004)
7. Balaji, K., Viswanadham, N.: A tax integrated approach for global supply chain network planning. IEEE Trans. Auto. Sc. and Engg. (T-ASE) 5(4), 587–596 (2008)
8. Ghemawat, P.: Distance Still Matters. Harvard Business Review (2001)
9. Duda, R.O., Hart, P.E., Stork, D.G.: Pattern Classification, 2nd edn. Wiley-Interscience, Hoboken (2007)

A Data-Mining Method for Detection of Complex Nonlinear Relations Applied to a Model of Apoptosis in Cell Populations

Henrik Saxén and Frank Pettersson

Thermal and Flow Engineering Lab., Dept. Chemical Engineering, Åbo Akademi University, Biskopsgatan 8, FI-20500 Åbo, Finland
{Henrik.Saxen,Frank.Pettersson}@abo.fi

Abstract. In studying data sets for complex nonlinear relations, neural networks can be used as modeling tools. Trained fully connected networks cannot, however, reveal the relevant inputs among a large set of potential ones, so a pruning of the connections must be undertaken to reveal the underlying relations. The paper presents a general method for detecting nonlinear relations between a set of potential inputs and an output variable. The method is based on a neural network pruning algorithm, which is run repetitively to finally yield Pareto fronts of solutions with respect to the approximation error and network complexity. The occurrence of an input on these fronts is taken to reflect its relevance for describing the output variable. The method is illustrated on a simulated cell population sensitized to death-inducing ligands resulting in programmed cell death (apoptosis).

Keywords: Nonlinear modeling, data mining, pruning, apoptosis model.

1 Introduction

Neural networks have during the last decades established their position as tools for nonlinear modeling due to their universal approximation capabilities. A central problem in many practical modeling (e.g., data mining) problems is to detect relevant inputs among a set of potential ones. In neural network modeling, this problem is not easily solved, since a large number of model parameters (weights) inherent in this modeling technique often results in over-parameterization [1] and the weights cannot in themselves tell whether an input is important or not. To address this problem, both constructive and destructive algorithms, with growing or shrinking networks, have been proposed in the literature [2-4]. Many of these methods are, however, rather sensitive to noise in the signals and some of them include time-consuming retraining steps. The work of the present paper is based on an efficient pruning algorithm [5] that is run repetitively to yield statistical information about the relevance of the potential inputs. The technique is applied to detect the most central ones among a large number of parameters in a model of programmed cell death, apoptosis. Monte-Carlo simulations with the complex biological model, described by a large set of differential equations, are applied to obtain a distribution of the time to apoptosis after sensitizing the

K. Deb et al. (Eds.): SEAL 2010, LNCS 6457, pp. 687–695, 2010.

cells to death-inducing ligands. The parameters with strongest effect on the time to apoptosis are determined, and the results are analyzed and discussed on the basis of the underlying problem.

2 Detection Method

The algorithm to be outlined searches for a nonlinear model of a set of potential inputs, $\mathbf{x}$, dim $(\mathbf{x}) = N$, approximating an output, y. The model is implemented as a feedforward neural network of multi-layer type with a single layer of hidden nonlinear units and a single linear output node. The algorithm is based on the practical observation that for such a network with an arbitrary choice of weights in its lower layer of connections, $\mathbf{W}$, there is usually a weight vector, $\mathbf{v}$, to the output node that will lead to a relatively good solution, $\hat{y}$, of the approximation problem. The vector $\mathbf{v}$ can be determined by a simple matrix inversion. With this as the starting point, the pruning algorithm can be condensed into [5]:

1. Choose a sufficient number of hidden nodes, m, and generate a random weight matrix, $\mathbf{W}^{(0)}$, for the lower part of the network. Set the iteration index to $k = 1$.

2. Equate to zero, in turn, each non-zero weight, $w_{ij}^{(k-1)}$, of $\mathbf{W}^{(k-1)}$, and determine the optimal upper-layer weight vector, $\mathbf{v}$, by minimizing $F = \sum(\hat{y} - y)^2$ with linear least squares. Save the corresponding value of the objective function, $F_{ij}^{(k)}$.

3. Find the minimum of the objective function values, $\mathbf{F}^{(k)} = \min_{ij}\{F_{ij}^{(k)}\}$. Set $\mathbf{W}^{(k)} = \mathbf{W}^{(k-1)}$ and equate to zero the weight corresponding to the minimum objective function value, $w_{\tilde{ij}}^{(k)} = 0$ with $\tilde{ij} = \arg\min_{ij}\{F_{ij}^{(k)}\}$.

4. Set $k = k + 1$. If $k < m \cdot N$, go to 2. Else, end.

The algorithm gives rise to a vector expressing how the approximation error evolves along with the elimination of weights. Typically, the errors decrease initially as directly detrimental weights are removed (note that the lower-layer weights are not retrained!), but finally the error increases as useful connections (an inputs) are removed. Naturally, due to the random initial weight matrix, not all runs will produce useful information. Therefore, it is necessary to run the algorithm from a large number of weight matrices, and to treat the arising information, e.g., statistically. It should be stressed that the method avoids laborious non-linear retraining of the networks after each pruning step, which would often be associated with a risk of convergence problem, in particular for large-scale problems. The success of the method has been demonstrated on simulated problems with known solutions [5,6], and on different challenging real-world problems (e.g., [7,8]). Here, the following approach is proposed to extend the original method to systematically refine the information gained in the pruning analysis:

i.   Run the algorithm from a large number of initial matrices.
ii.   Note the performance of networks with a number of lower-layer connections of K or less.
iii. Create Pareto-frontier with non-dominated solutions among these, and consider the Pareto fronts ranked $1...r$.
iv.   Among all these, register the number of connections to the individual input variables. This number can be considered to reflect the importance of a specific input.

3 Application

3.1 Model of Apoptosis

Apoptosis, also called programmed cell death, is a mechanism of active, controlled cellular elimination, which together with cell growth and division regulates key biological processes including embryonic development, immune responses, and tissue homeostasis. The mechanism of apoptosis is complex (cf. Fig. 1, with abbreviations in the following text given in parentheses): Stimulated by ligands (L), rerceptors (R) at the cell exterior activate death receptors (e.g., Fas/CD95/Apo-1) in the cell interior, which recruit cytoplasmic proteins, forming a death-inducing signaling complex (DISC). In the DISC, the initiator caspase-8 is activated by cleavage of procaspase-8 (pcas8), resulting in downstream activation of effector caspases (e.g., cas3, cas6) which propagate the apoptotic signal, with positive feedback. Caspase-8 activation is regulated by inhibitory proteins c-FLIP, which may prevent or delay apoptosis.

Mathematical modeling of biological systems has become an important tool for simulation-based hypothesis testing [9,10]. Since the apoptotic signaling pathway has a complex dynamic behavior involving several reactions and feedback loops, it is difficult to infer its overall response by examining individual parts of the system separately. Instead, a systems view is required, considering the main factors affecting the different steps of apoptosis. A detailed mathematical model of apoptosis was presented by Bentele et al. [11], using ordinary differential equations for 41 chemical components in 32 reactions, modeling the reaction rates by mass action and Michaelis-Menten kinetics. The model is schematically illustrated in Fig. 1.

Toivonen et al. [12] extended the model by including generation and degradation reactions of c-FLIP, and also introduced stochastics. By assuming individual cells to show differences in both the concentrations of various molecules and in the chemical reaction rates, these variables (X) were assumed to be independent stochastic lognormally distributed variables

$$X_i(0) = X_i^0 e^{\varepsilon_i}, \varepsilon_i \sim N(0, \sigma_{x_i}),$$
(1)

where X_i^0 are the nominal initial concentrations and rate factors, here taken as those given in the study by Bentele et al. [11] and $N(0,\sigma)$ is a normal distribution with zero mean and variance σ.

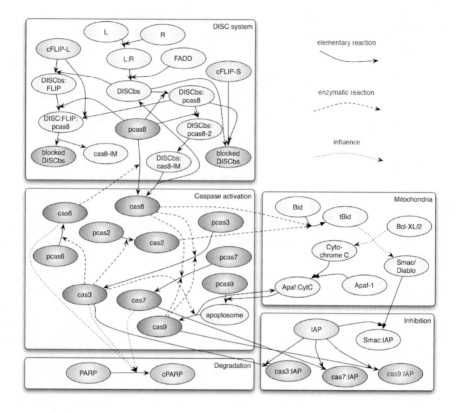

Fig. 1. Interaction between the components in the apoptosis model by Bentele et al. [11]

In order to determine the occurrence of apoptosis, the time to apoptosis, t_{ap}, was taken as the time elapsed from exposure to ligands to the instant when the concentration of activated caspase-3 (cas3) had decreased to half of its maximum

$$[cas3](t_{ap}) = \frac{1}{2}\max_t [cas3](t) \,, \tag{2}$$

where t_{ap} is larger than the time at which the maximum caspase-3 concentration is achieved. This definition gives times to apoptosis in general agreement with experimental results from cell lines.

In the present study the relatively complex apoptosis model of Toivonen et al. [12] is analyzed with the aim to detect the primary factors that influence apoptosis. This is done by applying the method outlined in section 2 by modeling the time to apoptosis (from Eq. (2)) using the stochastically varied initial concentrations and rate parameters of the chemical reactions as inputs.

3.2 Data Sets

The apoptosis model was run 1000 times with $\sigma = 0.3$ in Eq. (1), where the number of parameters subjected to the stochastic variations was totally 59, using a nominal ligand concentration of $[L]^0 = 50$ ng/ml. In accordance with earlier observations [11,12], this ligand concentration is sufficient to give rise to apoptosis in practically all runs, with a distribution of the time to apoptosis reported in the upper panel of Fig. 2. A corresponding set of runs with a lower ligand concentration, approaching the threshold below which apoptosis does not occur, resulted in the distribution in the lower panel of the figure, where only 66% of the runs resulted in apoptotic behavior. The reason for studying these two different data sets was that it was expected that different model parameters may play key roles for predicting the apoptotic behavior far from and close to the threshold value. For instance, the concentrations of c-FLIP would be expected to be variables of interest in the latter case, where high levels of these proteins are known to be able to prevent apoptosis [11,12].

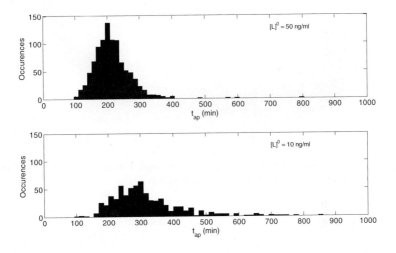

Fig. 2. istribution of the time to apoptosis for 1000 runs each of the model with log-normally distributed parameters ($\sigma = 0.3$ in Eq. (1)) for two different ligand concentrations, $[L]^0$

3.3 Results

The method of Section 2 was applied on the data, after first scaling all 59 inputs and the output (t_{ap}) to the range (0,1). In the 100 runs of each data set with $N = 59$ input variables and $m = 10$ hidden nodes, the final networks with a maximum complexity (expressed in terms of lower-layer connections) of $K = 30$ were considered. These were used to assemble the $r = 10$ best Pareto frontiers and the occurrences of the different inputs on these were noted.

Figure 3 illustrates the results of the pruning algorithm (steps 1-5 in Section 2) for the case with $[L]^0 = 10$ ng/ml. The densely connected networks (in the rightmost part of the figure) are seen to give rise to an average prediction error of about 13%, but as connections are being removed the error decreases below 8%, to finally rise to 12% when all but one connection is left. Focusing on the networks with a lower-layer complexity of 30 or less, i.e., those in the leftmost part of the graph, the 10 best Pareto frontiers are established and the inputs corresponding to these networks are registered. It should be noted that the corresponding errors for the first data (with $[L]^0 = 50$ ng/ml) set are considerably smaller, occasionally as low as 3.5%. This reflects the fact that in the case of strongly apoptotic conditions, the time to apoptosis shows smaller variance and is, therefore, more easily predicted by the models. Figure 4 shows the performance of the best models on the Pareto frontiers for the two data sets with a lower-layer complexity of 1…30 in the networks.

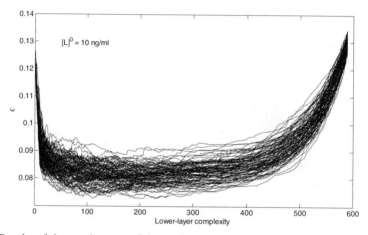

Fig. 3. Results of the pruning part of the method (cf. steps 1-5 in Section 2) for 100 runs on apoptosis data generated with $\sigma = 0.3$ and $[L]^0 = 10$ ng/ml

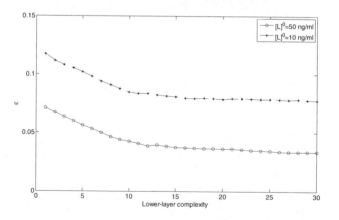

Fig. 4. Approximation provided by best networks on the Pareto frontiers with a lower-layer complexity ranging between one and thirty for the two data sets

The upper panel of Fig. 5 illustrates by bar graphs the occurrence of the 59 input variables in the neural models on the first ten Pareto frontiers for the data with $[L]^0 =$ 50 ng/ml, while the lower panel shows the corresponding results for the case with $[L]^0$ = 10 ng/ml. Some general observations can be made: For both data sets studied, twelve of the inputs appear more than 100 times and much more frequently than the others, and the differences in the occurrences between the two data sets are relatively small (Table 1). This indicates that the identified variables have a general impact on the time to apoptosis in the runs that correspond to conditions where apoptosis occurs.

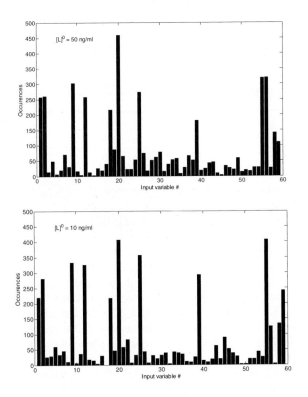

Fig. 5. Occurrence of input variables in the models on the first ten Pareto frontiers for the data set with $[L]^0$ = 50 ng/ml (upper panel) and $[L]^0$ = 10 ng/ml (lower panel)

The most frequently occurring variable (#20) in both data sets is the rate constant for the general degradation of all components after the apoptotic machinery has come to the "point of no return". Considering the definition of the time to apoptosis used in this study (cf. Eq. (2)), it is natural that this variable plays an important role. The second most important variable is the initial concentration of the ligand (#55). Also this finding is easily understood, since the ligand-to-receptor ratio is a triggering factor for apoptosis [11] (cf. uppermost box in Fig. 1). Of equal importance in theu upper panel of Fig. 5, but of considerably less importance in the lower panel, is the 56th

694   H. Saxén and F. Pettersson

Table 1. Inputs occurring more than 100 times on the 10 best Pareto fronts among the networks with a lower-layer complexity of 30 or less

Input	Variable	Occurrence	
		Data set 1	Data set 2
1	K_{LR}	257	220
2	$K_{DISC,pc\text{-}8}$	260	280
9	[procaspase-8]	302	333
12	$K_{DISC,FLIP}$	257	326
18	[R]	216	217
20	$K_{degrad,st}$	459	407
25	$K_{c\text{-}8,c\text{-}3}$	273	357
39	[IAP]	181	293
55	[L]	319	406
56	$C_{APOP,eff.act}$	320	124
58	[c-FLIP$_L$]	141	134
59	[c-FLIP$_S$]	110	241

variable, i.e., a constant in the degradation equation after apoptosis has occurred (cf. variable #20). The lower level of occurrence of this, and also of the 20th variables in the second data set, can be understood by noting that as the ligand concentration is closer to the threshold for apoptosis, other variables will play a more central role for the time to, or occurrence of, apoptosis. Another important variable, with occurrences around 300, is the initial concentration of procaspase-8 (variable #9). This variable is known to be central, because it is the origin of the initiator caspase for the caspase cascade, which eventually leads to apoptosis (uppermost box). A practically equally important variable is the rate constant (variable #25) for the enzymatic effect of caspase-8 on the reaction rate of procaspase-3 to caspase-3. As seen in the middle box of Fig. 1, this reaction is an important step in the initial phase of the caspase cascade.

Furthermore, parameters related to the formation of the death-inducing signaling complex (variable #2 and #12), are, as expected, found important, because here the competition between recruitment of procaspase-8 and c-FLIP affects the outcome of the signaling (uppermost box). The final ones among the important variables detected by the method are #18, #39, #58 and #59, which correspond to the initial concentration of receptors, of IAP, as well as of the long and short c-FLIP variants (c-FLIP$_L$ and c-FLIP$_S$). Noting that the ligand-to-receptor ratio is central for triggering apoptosis, as well as the inhibiting role played by the c-FLIP proteins at the DISC formation (uppermost box) and by IAP in reacting with the effector caspases (lowermost right box), the importance of these variables for the apoptotic behavior is noted. Interestingly, at close-to-threshold ligand-to-receptor ratios, the importance of the short variant of c-FLIP (i.e, c-FLIP$_S$) seems to be accentuated.

In summary, all of the major inputs of the neural models that have been detected by the method must be considered key parameters affecting the outcome of the apoptosis machinery. The method can thus be considered efficient for finding the salient factors for approximating a dependent variable on the basis of a data set with a large number of potential factors.

4 Conclusions

The paper has presented a method by which complex nonlinear relations can be detected in data sets. The core of the method is a pruning algorithm for feedforward neural networks, which gradually removes the least important of the remaining connections in the lower layer of the network. Pareto frontiers with the approximation error and the number of lower-layer weights for the models of lowest complexity obtained in repetitive pruning runs are established, and the occurrences of the input variables in the models on the fronts are taken to reflect their relevance. The methodology has been illustrated by applying it on two data sets generated by a model of a biological system, considering programmed cell death (apoptosis) in cell populations with stochastically varying model parameters. It was found successful in detecting the most central parameters initiating, preventing or propagating the apoptotic signal in the cell.

References

1. Principe, J.C., Euliano, N.R., Lefebvre, W.C.: Neural and adaptive systems: Fundamentals through simulations. John Wiley & Sons, New York (1999)
2. Frean, M.: The Upstart Algorithm: Method for Constructing and training Feedforward neural networks. Neural Comput. 2, 198–209 (1991)
3. García-Pedrajas, N., Ortiz-Boyer, D.: A cooperative constructive method for neural networks for pattern recognition. Pattern Recog. 40, 80–98 (2007)
4. Hassibi, B., Stork, D.G., Wolff, G.J.: Optimal Brain Surgeon and General Network Pruning. In: Int. Conf. Neural Networks, pp. 293–299. IEEE Press, New York (1993)
5. Saxén, H., Pettersson, F.: Method for the selection of inputs and structure of feedforward neural networks. Comput. Chem. Eng. 30, 1038–1045 (2006)
6. Saxén, H., Pettersson, F.: A data mining method applied to a metallurgical process. In: IEEE Symp. Comput. Intell. Data Mining, Hawaii, pp. 368–375 (2007)
7. Saxén, H., Pettersson, F.: Nonlinear prediction of the hot metal silicon content in the blast furnace. ISIJ Int. 47, 1732–1737 (2007)
8. Pettersson, F., Suh, C., Saxén, H., Rajan, K., Chakraborti, N.: Analyzing Sparse Data for Nitride Spinels Using Data Mining, Neural Networks, and Multiobjective Genetic Algorithms. Mater. Manuf. Proc. 24, 2–9 (2009)
9. Bhalla, U.S., Iyengar, R.: Emergent properties of networks of biological signaling pathways. Science 283, 381–387 (1999)
10. Kitano, H.: Systems biology: A brief overview. Science 295, 1662–1664 (2002)
11. Bentele, M., Lavrik, I., Ulrich, M., Stösser, S., Heermann, D., Kalthoff, H., Krammer, P., Eils, R.: Mathematical modeling reveals threshold mechanism in CD95-induced apoptosis. J. Cell Biol. 166, 839–851 (2004)
12. Toivonen, H., Meinander, A., Westerlund, M., Pettersson, F., Mikhailov, A., Asaoka, T., Eriksson, J.E., Saxén, H.: Modeling reveals that dynamic regulation of c-FLIP levels determines cell-to-cell distribution of CD95-mediated apoptosis (2010) (submitted manuscript)

An Implementation of Pareto Set Pursuing Technique for Concept Vehicle Design

Dhanesh Padmanabhan and Rajkumar Vaidyanathan[*]

India Science Lab, GM R&D
Bangalore, 560066
dhanesh.padmanabhan@gm.com, rajvaidyanathan@gmail.com

Abstract. In this paper, couple of multi-objective optimization (MOO) techniques is implemented on a concept vehicle design problem. The Pareto Set Pursuing (PSP) method, proposed by Shan and Wang [1] is compared with a commercial version of the NSGA-II algorithm. PSP uses a sequential surrogate model generation and progressive importance sampling approach as compared to the NSGA-II, which uses exact function evaluations. This comparative study is initially carried out with the aid of a continuous analytical test problem followed by a mixed discrete continuous, vehicle design problem. Based on these studies, it was found that PSP performed reasonably well as compared to the NSGA-II and offered considerable savings on the number of function evaluations. Additionally, it also provided an evenly spread Pareto frontier. For the concept vehicle design problem, the performance of PSP was comparable with NSGA-II in terms of the accuracy of Pareto frontier and better in terms of the spread of the Pareto frontier.

Keywords: Multi-Objective Optimization, Pareto Set Pursuing, Pareto Frontiers, Importance Sampling, Concept Vehicle Design.

1 Introduction

Early stages of vehicle design involve complex interplay of corporate goals (for e.g. profit, market share) and engineering performance objectives (for e.g. best-in-segment fuel economy and roominess). The performance objectives vary depending on the vehicles being designed and corporate goals behind the introduction of these vehicles in the market. The problem of achieving these performance objectives can be formulated as a Multi-Objective Optimization (MOO) problem and solved to obtain Pareto frontiers. These Pareto frontiers can be used for repetitive and fast evaluation of design alternatives such as in Gurnani et al [2] to arrive at a final design alternative [3]. Several MOO methods [4] exist that can be used to solve such problems. A typical MOO problem formulation is given by:

$$"\min_{x \in \Omega}" \ [f_1(x), f_2(x), .., f_m(x)]^T. \tag{1}$$

---

[*] Currently working at Oracle India Pvt Ltd.

K. Deb et al. (Eds.): SEAL 2010, LNCS 6457, pp. 696–705, 2010.

$f_1(x)$, $f_2(x)$, .., $f_m(x)$ are m objective functions that need to be minimized, x is the design variable vector comprising of a continuous design variable vector, x^c of length n^c and a discrete design variable vector, x^d of length n^d. Ω is the feasible set in the design variable space and can be represented as

$$\Omega = \left\{ x : g(x) \leq 0, x_i^{c,l} \leq x_i^c \leq x_i^{c,u} \; \forall i \in 1,..,n^c, x_j^d \in D_j \; \forall j \in 1,..,n^d \right\}. \qquad (2)$$

$g(x)$ is a vector of inequality constraint functions, $x_i^{c,l}$ and $x_i^{c,u}$ are the lower and upper bounds respectively of x_i^c. D_j is the set of all the discrete values or choices that x_j^d can take, for example, tire sizes, material types etc. There is no single feasible solution that minimizes all objectives simultaneously but there are potentially many possible "optimal" solutions to the MOO problem given by Equation 1. Such solutions are called Pareto optimal or non-dominated solutions.

One of the categories of MOO methods generates non-dominated solutions through population-based techniques such as Multi-Objective Genetic Algorithms (MOGA) [2] and other techniques based on statistical simulations such as the one investigated in this work. These methods are of interest for this work, since the main goal is to obtain a representation of the Pareto frontier for the concept vehicle design problem that tends to be a mixed discrete continuous problem involving several objectives. Some of the criteria that are essential for an approach to be used for the MOO problems in this work are:

- the ability to handle discontinuities in the Pareto Frontier,
- the ability to generate even spread of Pareto points, and
- the ability to converge to global Pareto optimal solutions with minimal number of exact (costly) function evaluations.

Methods like MOGA require lot of function evaluations, which in turn requires either costly exact function evaluations or cheap but accurate surrogate models which reduces computational cost. For problems with large number of design variables and highly nonlinear responses, the estimation of an accurate surrogate model would require a large number of costly exact function evaluations. Recently, Shan and Wang [1] have proposed a method called Pareto Set Pursuing (PSP) and they have shown that PSP performs well without demanding much on the accuracy of the surrogate model for certain test problems. Additionally, the costly exact function evaluations are done only for points near to the Pareto frontier unlike MOGA where all the designs are evaluated using exact function evaluations or highly accurate surrogate models. In this work, we have implemented a modified PSP for couple of MOO problems, one of them being a mixed discrete continuous problem. In the next section, PSP is explained along with some modifications.

2   Pareto Set Pursuing Method

The flow chart of the modified PSP method [1] is shown in Figure 1. PSP uses a progressive importance sampling approach and, in principle, iteratively generates an

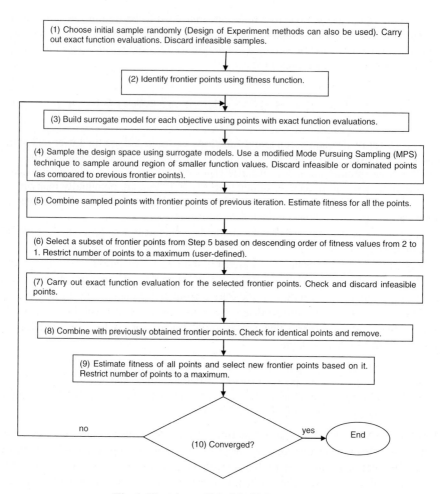

Fig. 1. Flowchart of Modified PSP Algorithm

updated and more accurate surrogate model for every objective. This surrogate model is used to sample progressively towards the Pareto front.

Some of the important terminologies used in the flowchart are explained below.

2.1 Fitness Function

The fitness function is given by the following equation:

$$G_i = \left[1 - \max_{j \neq i}\left(\min\left(f_{s1}^i - f_{s1}^j, f_{s2}^i - f_{s2}^j, ..., f_{sm}^i - f_{sm}^j\right)\right)\right], \qquad (3)$$

where G_i denotes the fitness value of the ith design and f_{sk}^i is the kth objective scaled between 0 and 1 based on the maximum and minimum values of all the objectives used for fitness estimation. Based on this definition, it can be shown that the Pareto set points have a fitness value in the range [1,2] and non-Pareto set points have a fitness value in the range [0,1).

2.2 Surrogate Model

In this work, both quadratic polynomial and linear spline functions are used as surrogate models. Shan and Wang have suggested the use of linear spline functions in addition to the quadratic polynomials since they prevent the fitting of non-existent curvatures. The quadratic polynomial model for k^{th} objective function is given as:

$$\hat{f}_k(x) = \beta_0 + \sum_{i=1}^{n} \beta_i x_i + \sum_{i=1}^{n} \beta_{ii} x_i^2 + \sum_{i<j}^{n} \sum_{j=1}^{n} \beta_{ij} x_i x_j , \tag{4}$$

where βs are the regression coefficients, x_i, $(i=1,...,n)$ are design variables and $\hat{f}_k(x)$ is the response. The minimum number of design points needed to estimate the regression coefficients are $(n+1)(n+2)/2$ where n is the number of design variables and an equivalent or greater number of designs are drawn in Step 1 of PSP. The linear spline function is given as

$$\hat{f}_k(x) = \sum_{i=1}^{l} \alpha_i \|x - x^{(i)}\| . \tag{5}$$

Here α_i are the coefficients that are estimated using least squares approach and $x^{(i)}$ represents the ith design. In the present work, the transition from quadratic polynomial to linear spline function is controlled by the average value of G. The choice of the average value of G, when the transition happens, is left for the user to select ($\sim$ 1.03 - 1.04 in this study). It is also ensured that a minimum number of iterations (5 in this study) are carried out with quadratic polynomial based surrogate model before the transition to linear spline function occurs. This is to make sure that the properties of the whole design space are captured adequately in the initial phase. Once the near Pareto frontier points are obtained, linear spline is used which performs well locally.

2.3 Mode Pursuing Sampling

Fu and Wang [5] constructed a sampling guidance function for each objective function given by the following equation:

$$z_k(x) = c_0 - \hat{f}_k(x), \qquad c_0 \geq \hat{f}_k(x) . \tag{6}$$

In the work presented here, the following sampling guidance function has been used instead:

$$z_k(x) = \frac{1}{c_0 + \hat{f}_k(x)}, \qquad c_0 > -\hat{f}_k(x) . \tag{7}$$

$z_k(x)$ is nonnegative over the sample space and once normalized can be interpreted as the probability density function (PDF). The sampling is then carried out based on this

PDF using a Mode Pursuing Sampling (MPS) technique [5]. In this technique, N number of uniformly distributed samples are generated in **x**. $z_k(\mathbf{x})$ is evaluated for these samples and the samples are segregated in K contours or bins such that N/K samples with highest values of $z_k(\mathbf{x})$ are contained in bin 1, N/K samples with next highest values of z are contained in bin 2 and so on. Then discrete probabilities are assigned to each bin. The discrete probability is proportional to the average value of z in the particular bin. For drawing a sample in Step 4 of PSP (see algorithm), one first picks a bin according to the discrete probabilities. Then one randomly picks a sample from each bin (equal probability for each sample in a bin). This technique causes the newly sampled design points to lie around the maximum of $z_k(\mathbf{x})$ and thereby the minimum of $\hat{f}_k(\mathbf{x})$. The choice of Fu and Wang's sampling guidance function (Equation 6) will give samples that follow a PDF similar to $-\hat{f}_k(\mathbf{x})$, whereas the sampling guidance function in Equation 7 will guarantee greater chance of sampling in the regions closer to minimum of $\hat{f}_k(\mathbf{x})$. This choice of sampling guidance function was observed to generate the desired number of non-dominated points more quickly than the Fu and Wang's sampling guidance function.

2.4 Frontier Point Selection

The fitness evaluation in Step 5 gives a set of points with fitness value greater than or equal to 1. In Step 6, Shan and Wang have used a heuristic approach to choose the number of frontier points from this set. This is based on the ratio of the frontier points in Step 5 (say N) to the number of frontier points in the previous iteration (say M).

- If N/M < 2 ⇒ N points are selected.
- If 2 ≤ N/M < 4 ⇒ M points are selected.
- If N/M ≥ 4 ⇒ 2M points are selected.

To select these points, Shan and Wang have used a sampling guidance function based on the fitness of the frontier points (all the N points from Step 5). This function is used to sample in the regions of high fitness value. In this work, the frontier points are selected in a descending order of fitness values between 2 and 1 instead of any guiding function. Additionally, it is also made sure that the number of frontier points chosen in Step 6 does not exceed a user specified number. The limit on the number of frontier points is imposed because it is more important to obtain a diverse set of Pareto solutions than a set of clustered solutions.

2.5 Convergence Criteria

The percentage of identical frontier points retained in consecutive iterations of the algorithm should be greater than a user-specified value (in this work 90-95% is used). Further research needs to be done to determine the optimum choice of this value since the ease of attaining a certain target percentage was seen to depend largely on the dimensionality of the problem.

2.6 Summary of Modifications to PSP

Following are the essential differences between Shan and Wang's implementation of PSP as compared to the current implementation:

1. A different sampling guidance function has been used in Step 4.

2. Shan and Wang computed exact constraint functions to check for infeasible solutions in Step 4. In this work, bounds on objective functions have also been imposed for the concept vehicle design problem. In the modified PSP, the approximate objective functions (obtained using surrogate models) are checked for infeasibilities in Step 4. Since these surrogate models can have inaccuracies because of which the infeasibilities maybe incorrectly determined; the bound constraints on the objectives are used to re-evaluate feasibility in Step 7.

3. A different approach is used for the selection of frontier points in Step 6.

4. The total number of candidate frontier points is restricted to a maximum user-defined number in Steps 6 and 9.

3 Implementation Studies

The modified PSP technique was compared with an implementation of NSGA-II [6] algorithm within the commercial package ISIGHT[1]. Both the approaches were implemented on a concept vehicle design problem. Prior to studying the impact of these two approaches on the vehicle design case, they were tested on an analytical problem, which has been solved by Shan and Wang in their study.

3.1 Analytical Test Problem

This is a nonlinear problem with three objectives, three continuous design variables and one constraint in addition to bounds on design variables. They are as follows:

$$f_1(x) = 25 - (x_1^3 + x_1^2(1 + x_2 + x_3) + x_2^3 + x_3^3)/10 \,,$$
$$f_2(x) = 35 - (x_1^3 + 2x_2^3 + x_2^2(2 + x_1 + x_3) + x_3^3)/10 \,,$$
$$f_3(x) = 50 - (x_1^3 + x_2^3 + 3x_3^3 + x_3^2(3 + x_1 + x_2))/10 \,,$$
$$g_1(x) = 12 - x_1^2 - x_2^2 - x_3^2 \geq 0 \,,$$
$$0 \leq x_1 \leq 5, 0 \leq x_2 \leq 5, 0 \leq x_3 \leq 5.$$

3.2 Concept Vehicle Design

The concept vehicle design process involves a multidisciplinary analysis framework as described in Fenyes et al [7] and Padmanabhan [8]. In this study, a sedan is designed based on 9 objectives, namely, minimum time taken to accelerate from 50 to 70 mph, maximum cargo volume, maximum diagonal head clearance for front driver or passenger, maximum head room for front driver or passenger, maximum fuel economy, maximum head room for rear passenger, maximum shoulder room, minimum curb weight and maximum knee clearance for the rear passenger. The design variables consisted of continuous design variables such as wheelbase (lengthwise distance between front and rear wheels), vertical center position of driver, vertical center position of rear passenger, vehicle height and front track (distance between front left and right wheels) and one discrete design variable involving five different configurations of engines and tires. The discrete design variable is treated as a continuous variable in the problem and configurations are ordered based on increasing order of the acceleration ratings of the engines. This was observed to give good R^2

---

[1] Trademark of Dassault Systèmes.

values for the surrogate models used in the PSP implementation. Additionally, in this problem, constraints were enforced on occupant roominess parameters such as head clearance and rear knee clearance such that minimum values of these parameters were guaranteed. There were also bounds on the continuous design variables determined by the manufacturing flexibility. The bounds on the wheelbase were also set different for each engine and tire configuration. Each of the objectives was dependent only on a selected set of design variables. Hence, during the surrogate model development phase of the PSP algorithm, the knowledge of dependencies between objectives and design variables were used.

3.3 Results and Discussion

Several metrics exist for assessing the quality of frontier or non-dominated solutions obtained from different MOO techniques [9]. In this work, PSP and NSGA-II available within ISIGHT (henceforth referred to as NSGA-II only) were compared on two criteria: (i) nearness to actual Pareto frontier and (ii) evenness of the distribution of non-dominated points.

For the nearness to Pareto frontier, the points from PSP and NSGA-II are combined and the overall non-dominated points were found. For each MOO technique, the percentage of non-dominated points present in the overall non-dominated points is computed. This gives a measure of quality of the solution in terms of closeness to the Pareto frontier. The MOO technique with the highest percentage is the best in terms of nearness to exact Pareto frontier.

For the evenness of the distribution, the nearest neighbor distance for each non-dominated solution is found for each MOO technique. For each MOO technique, the minimum, maximum, mean and standard deviations of the nearest neighbor distances were found. The technique with the lowest standard deviation is considered to be the best in terms of the evenness of distribution. For the analytical test problem, visual comparison of the non-dominated solution was also performed.

The Pareto frontiers obtained for the analytical test problem using PSP and NSGA-II are shown in Figure 2. It can be seen that the Pareto points obtained using PSP are seen to dominate those obtained using NSGA-II and the spread is more uniform for the PSP results. Looking at Table 1, it can be seen that the number of exact function evaluation is less for PSP. About 92% of non-dominated (ND) points obtained using PSP contributed to the overall non-dominated (OND) points as compared to the 40.5% of ND points from NSGA-II. This clearly resulted in PSP contributing to a larger percentage of OND points. The distribution of the nearest neighbor distances was more even (smaller standard deviations) for non-dominated solutions obtained from PSP as compared to those obtained using NSGA-II.

The results obtained using PSP and NSGA-II for the vehicle design problem are compared in Table 2. Since the initial sampling in the case of PSP is random, 3 runs were carried out with all the other parameters fixed and the results from all the three runs were used for comparison. For this problem, NSGA-II was found to be slightly better than PSP in terms of the percentage contributions to the overall non-dominated points. PSP out-performed NSGA-II in the number of function evaluations and the evenness of spread (lower standard deviation of the nearest neighbor distance).

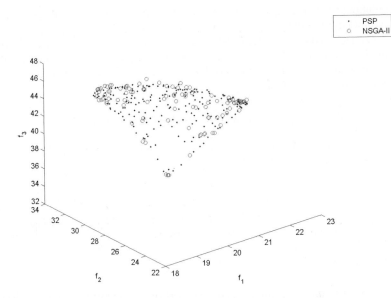

Fig. 2. Pareto Frontier obtained using PSP and NSGA-II for the Analytical Test Problem

Table 1. Comparison of results obtained from PSP and NSGA-II

	PSP	NSGA-II
# of exact function calls	564	5000
% of ND points in OND points	92.00	40.50
Minimum nearest neighbor distance	0.161	0.023
Maximum nearest neighbor distance	1.192	1.945
Mean nearest neighbor distance	0.363	0.351
Std. Dev. of nearest neighbor distance	0.131	0.286

Table 2. Comparison of results obtained from PSP and NSGA-II

	Run1		Run2		Run3	
	PSP	NSGA-II	PSP	NSGA-II	PSP	NSGA-II
# of exact function calls	1503	25000	2427	25000	2055	25000
% of ND points in OND points	95.80	98.60	94.20	98.00	93.60	99.20
Minimum nearest neighbor distance	0.064	0.001	0.065	0.001	0.065	0.001
Maximum nearest neighbor distance	0.554	0.445	0.479	0.445	0.467	0.445
Mean nearest neighbor distance	0.185	0.128	0.188	0.128	0.191	0.128
Std. Dev. of nearest neighbor distance	0.064	0.076	0.068	0.076	0.064	0.076

To highlight the spread in the Pareto set, the number of Pareto points in each discrete combination of the vehicle design problem is shown in Table 3. It was seen that PSP generated almost a uniform distribution between the different combinations as compared to NSGA-II. This, along with the large decrease in the number of exact function evaluations, highlighted the advantage of using PSP over the commercially implementation of NSGA-II.

Table 3. Comparison of results obtained from PSP and NSGA-II.

Discrete combination	PSP (average of 3 runs)	NSGA-II
1	133	195
2	105	49
3	84	74
4	93	31
5	84	151

4 Conclusion

The Pareto Set Pursuing technique and NSGA-II implementation within ISIGHT were implemented on an analytical problem with continuous design variables and a concept vehicle design problem with mixed continuous and discrete design variables. It was noticed that PSP outperformed NSGA-II for both the problems in terms of the required number of exact function evaluations and the evenness of the spread of Pareto frontier. The non-dominated solutions from PSP outnumbered the solutions from NSGA-II for the analytical problem. In the case of the vehicle design problem, this difference became blurry. Additionally, it has to be noted that there is considerable overhead in PSP at various stages, especially in building splines and MPS sampling. Overall, PSP is promising in terms of reducing the exact functional evaluations and also in obtaining a well-defined Pareto frontier. The selection of the parameter values used for the convergence criteria and controlling the transition of surrogate models requires further study. The treatment of discrete variables in the PSP approach also needs further study. In this work, the discrete variables were transformed to 'pseudo' continuous variables guided by prior engineering knowledge. Techniques such as NSGA-II are known to handle discrete variables. With regards to PSP, there has been some recent work done by Khokhar et al. [10] focusing on mixed discrete-continuous problems.

Acknowledgments. The authors would like to acknowledge Anil Maddulapalli and Joseph Donndelinger for their contributions.

References

1. Shan, S., Wang, G.G.: An Efficient Pareto Set Identification Approach for Multiobjective Optimization on Black-Box Functions. ASME J. Mech. Des. 127(5), 866–874 (2005)
2. Gurnani, A., Ferguson, S., Lewis, K., Donndelinger, J.: A constraint-based approach to feasibility assessment in preliminary design. AI EDAM 20, 351–367 (2006)
3. Olson, D.L.: Decision Aids for Selection Problems. Springer, New York (1996)

4. Deb, K.: Multi-Objective Optimization using Evolutionary Algorithms. John Wiley and Sons Ltd, Chichester (2001)
5. Fu, J.C., Wang, L.: A random-discretization based Monte Carlo sampling method and its applications. Meth. Comp. App. Prob. 4, 5–25 (2002)
6. Deb, K., Pratap, A., Agarwal, S., Meyarivan, T.: A fast and elitist multi-objective genetic algorithm: NSGA-II. IEEE Trans. Evol. Comp. 6(2), 181–197 (2002)
7. Fenyes, P. A., Donndelinger, J. A., Bourassa, J-F. A new system for multidisciplinary design and optimization of vehicle architectures. In: 9th AIAA Symposium on Multidisciplinary Analysis and Optimization, Paper No. AIAA-2002-5509 (2002)
8. Padmanabhan, D.: Techniques for Construction of Pareto Frontier Approximations. In: 46th AIAA/ASME/ASCE/AHS/ASC Structures, Structural Dynamics & Materials Conference. Paper No. AIAA-2005-1817 (2005)
9. Wu, J., Azarm, S.: Metrics for Quality Assessment of a Multiobjective Design Optimization Solution Set. ASME J. Mech. Des. 123(8), 18–25 (2001)
10. Khokhar, Z.O., Vahabzadeh, H., Ziai, A., Wang, G.G., Menon, C.: Mixed Variable Pareto Set Pursuing (PSP) Method and its Performance for Multi-Objective Design Optimization. ASME J. Mech. Des. (March 2010) (accepted)

EPIC: Efficient Integration of Partitional Clustering Algorithms for Classification

Vikas K. Garg[1] and M.N. Murty[2]

[1] IBM Research - India
vikaskga@in.ibm.com
[2] Department of Computer Science and Automation (CSA)
Indian Institute of Science (IISc), Bangalore, India
mnm@csa.iisc.ernet.in

Abstract. Partitional algorithms form an extremely popular class of clustering algorithms. Primarily, these algorithms can be classified into two sub-categories: a) k-means based algorithms that presume the knowledge of a suitable k, and b) algorithms such as Leader, which take a distance threshold value, τ, as an input. In this work, we make the following contributions. We 1) propose a novel technique, EPIC, which is based on both the number of clusters, k and the distance threshold, τ, 2) demonstrate that the proposed algorithm achieves better performance than the standard k-means algorithm, and 3) present a generic scheme for integrating EPIC into different classification algorithms to reduce their training time complexity.

1 Introduction

Clustering or unsupervised classification of patterns into groups based on similarity is a very well studied problem in machine learning and related disciplines [1]. Partitional clustering algorithms assign the data points into a pre-defined number of clusters. These algorithms can be broadly classified into two categories, based on how the number of clusters is specified. The k-means algorithm [5] is an immensely popular clustering algorithm that takes k, the number of clusters, as an input explicitly. There are many partitional clustering algorithms, such as BIRCH [2], DBSCAN [3], and Leader [4], which take as input a distance threshold value τ instead. This threshold value, indirectly, determines the number of clusters obtained using these techniques. We believe that a hybrid technique, which uses both k and τ in the clustering process, would be more useful since more domain knowledge can be easily incorporated. In our work, we propose a variant of the k-means algorithm, EPIC, to accomplish exactly the same goal. EPIC, an anagram of the initials of "**E**fficient **I**ntegration of **P**artitional **C**lustering for **C**lassification", initially assigns the data points to k_1 clusters, where $k_1 < k$, k being the tentative number of desired clusters. Then, an iterative process is followed to refine the clusters using the specified threshold distance, τ. We demonstrate that the proposed algorithm performs fewer distance computations than the k-means algorithm and thus provides better time performance, without making any assumptions about the distribution of the input data.

K. Deb et al. (Eds.): SEAL 2010, LNCS 6457, pp. 706–710, 2010.

2   The EPIC Algorithm

Inputs: A dataset to be clustered: $X = \{x_i, y_i\}_{i=1}^{N}$, where $x_i \in \Re^d$; a radius threshold parameter: τ; an approximate number of clusters: k; the maximum number of iterations allowed for the conventional k-means algorithm to converge: m (If m is not provided, take m to be 100, as is the common practice).

1. Let n be the maximum number of levels. Set n to some value $\leq \lfloor \frac{mk}{2} \rfloor + 1$. Initialize $count = 1$.
2. Cluster X into $k_1 = \left\lfloor \frac{1}{k^{n-2}} \left\{ \frac{2(n-1)}{m} \right\}^{n-1} \right\rfloor$ clusters using MacQueen's 2-pass k-means algorithm.
3. Compute the radius $\{r_i^1\}_{i=1}^{k_1}$ of each cluster $\{c_i^1\}_{i=1}^{k_1}$ and determine D, the maximum radius of any cluster.
4. Set $\tau_1 = \min(\tau, D - \epsilon)$, where $\epsilon \to 0$ is an extremely small positive quantity. Set $\tau = \tau_1$.
5. Set the level, $t = 1$.
6. For every cluster $\{c_i^t\}_{i=1}^{k_t}$, if $r_i^t > \tau_t$
   - split c_i^t using k-means into $(\frac{r_i^t}{\tau_t})^d$ clusters.
7. Let k_{t+1} be the total number of clusters. If $k_{t+1} < k$,
   - set $\tau_{t+1} = \tau_t \sqrt{\frac{k_t}{k_{t+1}}}$
   - set $t = t + 1$
   - Set $count = count + 1$
   - If $count < n$ and $\tau_{count-1} \geq D \left\{ \frac{2(count - 1)}{mk} \right\}^{\frac{1}{d}}$
     • Compute the radius $\{r_i^t\}_{i=1}^{k_t}$ of each cluster $\{c_i^t\}_{i=1}^{k_t}$
     • go to step 6.
8. Return the clusters with their centers.

2.1   Bound on Number of Distance Computations, Relation between τ and k, and Maximum Permissible Levels

Consider the given dataset $X = \{x_1, x_2, \ldots, x_N\}$, where $x_i \in \Re^d$ are independent samples drawn from an identical distribution. The number of distance computations using k-means on X for $m \geq 2$ iterations is, $ND_1 = mNk - k^2$. Also, the number of distance computations in first step of EPIC, using MacQueen's 2-pass algorithm [6], is $ND_{L_1} = 2Nk_1 - (k_1)^2$. Let C_i^t denote the ith cluster at level t, with center c_i^t. Then, after the first level of clustering, we have k_1 clusters: $C_1^1, C_2^1, \ldots, C_{k_1}^1$ with centers $c_1^1, c_2^1, \ldots, c_{k_1}^1$ respectively. Now, let us define the radius of cluster C_i^t, $r_i^t = max_{x_j \in C_i^t} d(x_j, c_i^t)$, where $d(x, y)$ is the distance between x and y. In the EPIC algorithm, the ith cluster at level t is partitioned at level $t + 1$ if $r_i^t \geq \tau_t$. Let k_t denote the number of clusters at level t. Clearly, $k_1 = k_1$. For each cluster i, $1 \leq i \leq k_t$, define an indicator variable $Z_i^t = 1_{\{r_i^t > \tau_t\}}$, and the probability $p_i^t = P(Z_i^t = 1)$. Further, let $|C_i^t|$ denote the number of data points assigned to C_i^t. If a cluster C_i^t is partitioned at $t + 1$, the next level, then the expected number of distance computations at level $t + 1$,

$$NDL_{t+1} = \sum_{i=1}^{k_t} p_i^t \left[2|C_i^t| \left(\frac{r_i^t}{\tau_t} \right)^d - \left(\frac{r_i^t}{\tau_t} \right)^{2d} \right].$$ The expected nummber of distance

computations, if EPIC proceeds till the nth level,

$$ND_2 = NDL_1 + \sum_{t=1}^{n-1} NDL_{t+1} = 2Nk_1 - (k_1)^2 + \sum_{t=1}^{n-1} \sum_{i=1}^{k_t} p_i^t \left[2|C_i^t| \left(\frac{r_i^t}{\tau_t} \right)^d - \left(\frac{r_i^t}{\tau_t} \right)^{2d} \right]$$

Now,

$$ND_1 - ND_2 = mNk - k^2 - 2Nk_1 + (k_1)^2 - \sum_{t=1}^{n-1} \sum_{i=1}^{k_t} p_i^t \left[2|C_i^t| \left(\frac{r_i^t}{\tau_t} \right)^d - \left(\frac{r_i^t}{\tau_t} \right)^{2d} \right]$$

$$= mNk - k^2 - 2Nk_1 + (k_1)^2 - \sum_{t=1}^{n-1} \sum_{i=1}^{k_t} 2p_i^t |C_i^t| \left(\frac{r_i^t}{\tau_t} \right)^d + \sum_{t=1}^{n-1} \sum_{i=1}^{k_t} p_i^t \left(\frac{r_i^t}{\tau_t} \right)^{2d}$$

Taking $A = mNk - k^2 - 2Nk_1 + (k_1)^2$, we get

$$A - \sum_{t=1}^{n-1} \sum_{i=1}^{k_t} 2p_i^t |C_i^t| \left(\frac{r_i^t}{\tau_t} \right)^d \le ND_1 - ND_2 \le A + \sum_{t=1}^{n-1} \sum_{i=1}^{k_t} p_i^t \left(\frac{r_i^t}{\tau_t} \right)^{2d} \qquad (1)$$

Let $P_t = max_i \; p_i^t$. Further, define $\alpha_t = max(1, max_i \; (\frac{r_i^t}{\tau_t}))$ so that the following bound holds for all $r_i^t : 0 \le r_i^t \le \alpha_t \tau_t$. Then, (1) implies

$$\Rightarrow A - \sum_{t=1}^{n-1} 2P_t \alpha_t^d \sum_{i=1}^{k_t} |C_i^t| \le ND_1 - ND_2 \le A + \sum_{t=1}^{n-1} k_t P_t \alpha_t^{2d}$$

$$\Rightarrow A - 2N \sum_{t=1}^{n-1} P_t \alpha_t^d \le ND_1 - ND_2 \le A + \sum_{t=1}^{n-1} k_t P_t \alpha_t^{2d} \; [since \; \sum_{i=1}^{k_t} |C_i^t| = N]$$

$$(2)$$

Further, since the radius of any cluster may never exceed D, the maximum distance between any pair of points in X, we must have for all t, $\alpha_t \tau_t \le D$.

$$\Rightarrow \alpha_t^d \le \left(\frac{D}{\tau_t} \right)^d \le \left(\frac{D}{\tau_{n-1}} \right)^d = M(say) \qquad (3)$$

since for all t, $\tau_t \ge \tau_{n-1}$. It follows from (2) that

$$A - 2MN(n-1) \le ND_1 - ND_2 \le A + \sum_{t=1}^{n-1} k_t M^2 \qquad (4)$$

The number of clusters at a level $t + 1$ is maximum when all the clusters at the previous level t are partitioned. Then, the number of clusters at level $t + 1$,

$$k_{t+1} \le k_t \left(\frac{r_i^t}{\tau_t} \right)^d \le k_t \alpha_t^d \le M k_t \qquad (5)$$

Recursively simplifying (5) till t equals 1, we get,

$$k_{t+1} \leq Mk_t \leq M^2 k_{t-1} \leq M^t k_1$$

$$\Rightarrow A - 2MN(n-1) \leq ND_1 - ND_2 \leq A + k_1 \left\{ \frac{M^{n+1} - M^2}{M-1} \right\} \quad [using\ (4)]$$

This gives a bound on difference in number of distance computations. Now, we want $ND_1 \geq ND_2$. Then,

$$A - 2MN(n-1) \geq 0 \Rightarrow M \leq \frac{A}{2N(n-1)}$$

Plugging in the values of A and M,

$$\left(\frac{D}{\tau_{n-1}}\right)^d \leq \frac{mNk - k^2 - 2Nk_1 + (k_1)^2}{2N(n-1)} = \frac{mk - 2k_1}{2(n-1)} - \frac{k^2 - (k_1)^2}{2N(n-1)} \leq \frac{mk}{2(n-1)} \tag{6}$$

Then, the relation between τ and k is given by,

$$D \geq \tau \geq \tau_{n-1} \geq D\left\{\frac{2(n-1)}{mk}\right\}^{\frac{1}{d}} \geq D\left\{\frac{2}{mk}\right\}^{\frac{1}{d}}$$

Also, we can find a bound on the maximum number of permissible levels:

$$n_{max} = \lfloor \frac{mk}{2} \rfloor + 1$$

where n_{max} is the maximum number of levels, which ensures that EPIC is computationally more efficient than the k-means algorithm. We want to bound the value of M, since it is directly involved in the expression for difference in number of computations. Using (3) and (6), we obtain

$$\left(\frac{D}{\tau}\right)^d \leq M \leq \frac{mk}{2(n-1)} \tag{7}$$

Finally, to complete the unification of τ and k, we must ensure that the number of clusters at the termination of EPIC algorithm is bounded by k, irrespective of the value of k_1. Then, using (3), we must have for any value of M given by (7),

$$k_1 M^{n-1} \leq k$$

$$\Rightarrow k_1 \left\{\frac{mk}{2(n-1)}\right\}^{n-1} \leq k$$

which in the wake of (7) yields

$$k_1 = \left\lfloor \frac{1}{k^{n-2}} \left\{\frac{2(n-1)}{m}\right\}^{n-1} \right\rfloor$$

3   Application of EPIC to Classification

A two-level implementation of EPIC can be employed to reduce the time complexity of various classification algorithms. We present below a generic technique for the integration of EPIC into classification algorithms to improve their performance.

Inputs: A set of training examples and corresponding class labels, $X = \{x_i, y_i\}_{i=1}^{N}$, where $x_i \in \Re^d$, and $y_i \in \Gamma$, the set of labels; the number of clusters, k.

1. Cluster X into k clusters and determine the radius of each cluster.
2. Set τ to some value in the range indicated by (7).
3. Train the classifier using the centroids of those clusters that have their radius greater than τ.
4. Determine the clusters which form a part of the classification model. Sub-cluster these clusters.
5. Train the classifier using the centroids of the clusters (obtained in the previous step), which have their radius greater than τ.
6. Again determine the clusters in the classification model and train the classifier with the patterns in these clusters.

The training time complexity of a classifier integrated with two-level EPIC can be shown to be linear[1].

References

1. Jain, A.K., Murty, M.N., Flynn, P.J.: Data Clustering: A Review. ACM Computing Surveys 31(3) (1999)
2. Zhang, T., Ramakrishnan, R., Livny, M.: BIRCH: An Efficient Data Clustering Method for Very large Databases. In: Proceedings of the 1996 ACM SIGMOD International Conference on Management of Data, pp. 103–114 (1996)
3. Ester, M., Kriegel, H.-P., Sander, J., Xu, X.: A density-based algorithm for discovering clusters in large spatial databases with noise. In: Proceedings of the Second International Conference on Knowledge Discovery and Data Mining (KDD), pp. 226–231 (1996)
4. Spath, H.: Cluster Analysis Algorithms for Data Reduction and Classification. Ellis Horwood, Chichester
5. Kanungo, T., Mount, D.M., Netanyahu, N.S., Piatko, C.D., Silverman, R., Wu, A.Y.: An Efficient k-means Clustering Algorithm: Analysis and Implementation. IEEE Transactions on Pattern Analysis and Machine Intelligence (PAMI), 881–892 (2002)
6. MacQueen, J.: Some methods for classification and analysis of multivariate observations. In: Proceedings of the Fifth Berkeley Symposium on Mathematical Statistics and Probability, vol. 1, pp. 281–297 (1967)
7. Garg, V.K., Murty, M.N.: Pragmatic Data Mining: Novel Paradigms for Tackling Key Challenges. Technical Report TR/2009/11, CSA, IISc (2009), http://csa.iisc.ernet.in/TR/2009/11/

---

[1] We omit this analysis and other integral sections like the experimental results due to space constraints. A comprehensive treatment is provided in our technical report [7].

Toward Optimal Disk Layout of Genome Scale Suffix Trees

Vikas K. Garg

IBM Research - India
vikaskga@in.ibm.com

Abstract. Suffix trees provide for efficient indexing of numerous sequence processing problems in biological databases. We address the pivotal issue of improving the search efficiency of disk-resident suffix trees by improving the storage layout from a statistical learning viewpoint. In particular, we make the following contributions: we (a) introduce the Q-Optimal Disk Layout(Q-OptDL) problem in the context of suffix trees and prove it to be NP-Hard, and (b) propose an algorithm for improving the layout of suffix trees that is guaranteed to perform asymptotically no worse than twice the optimal disk layout.

Keywords: Suffix Trees, 0/1 Knapsack, Statistical Learning.

1 Introduction

The suffix tree is an immensely popular data structure for indexing colossal scale biological repositories [1]. However, owing to substantially increased storage space requirements, for most practical bio-informatics applications, the suffix tree needs to be disk-resident. Consequently, searching for a pattern requires random traversal of suffix links, which results in increased I/O activity.

Layout strategies have been proposed in the context of suffix trees. In [2], a layout strategy, Stellar, was experimentally shown to improve search performance on a representative set of real genomic sequences. Our work is most closely related to [3], wherein the authors provide a self-adjusting layout that optimizes the total number of disk accesses over a sequence of unknown queries. However, the authors study the layout problem for suffix trees without considering the suffix links. Dispensing away the suffix links not only affects the construction time but also renders impractical several search algorithms that require traversing both edge and suffix links. In our work, we do a statistical analysis of the layout problem without ignoring the presence of suffix links. Furthermore, linear time algorithms for suffix array construction, such as [4], can be used to efficiently build suffix trees via suffix arrays. In [5], the authors show how to use suffix arrays for executing various suffix tree algorithms. The overhead of accessing disk in context of suffix trees is higher than the logarithmic penalty of suffix arrays [6]. Thus, design of better disk layout algorithms, to speed up search in suffix trees, remains an important practical consideration for developers of genomic tools. Our work is a significant step toward accomplishing this goal.

K. Deb et al. (Eds.): SEAL 2010, LNCS 6457, pp. 711–715, 2010.

2   Hardness of the Disk Layout Problem

Suppose we need to access a node x in the process of determining a potential match of the pattern. Let $P_1(x)$ and $P_2(x)$ respectively denote the probability that the parent or suffix parent of node x is present in the memory. Further, let $C_1(x)$ and $C_2(x)$ be the costs of accessing x when at least one of its parent or suffix parent is present in the memory and when none is present respectively. We note that, $C_2(x) \geq C_1(x)$ since $C_2(x)$ involves an additional I/O operation in order to bring a parent or suffix parent of x into the memory. Now expected cost of accessing x is given by $C(x)$, where
$C(x) = $ [Probability that parent(x)/suffix parent(x) is present in memory] * $C_1(x) + $ [Probability that none of the parent(x) and suffix parent(x) is present in memory] * $C_2(x)$.

$$\Rightarrow C(x) = (1 - (1 - P_1(x))(1 - P_2(x))) * C_1(x) + (1 - P_1(x))(1 - P_2(x)) * C_2(x)$$
$$= C_1(x) + (C_2(x) - C_1(x))(1 - P_1(x))(1 - P_2(x))$$

2.1   The Q-Optimal Disk Layout Problem

Given a large scale suffix tree S and a set of patterns (possibly infinite) Q to be matched with S, the Q-Optimal Disk Layout (Q-OptDL) problem is to find an arrangement L of nodes belonging to S on disk such that the overall cost of accessing the nodes of S on patterns in Q is minimum for L.

Theorem 1

The Q-OptDL problem is NP- Hard.

Proof. By definition, an optimal layout minimizes overall sum of costs over patterns in Q. So our objective function can be stated as,
$$minimize \sum_Q \sum_x [C_1(x) + (1 - P_1(x))(1 - P_2(x)) * (C_2(x) - C_1(x))]$$
Now, we relax the problem setting by assuming C_1 and C_2 as average memory and disk access costs respectively. Then, the objective function is given by
$$minimize \sum_Q \sum_x (1 - P_1(x))(1 - P_2(x)) * (C_2 - C_1)$$
Let $X_j \in \{0, 1\}$ be an indicator variable to represent whether node x is accessed. A node j lying on the disk that is not accessed does not contribute to the cost. Then, since $C_2 \geq C_1$, the objective function becomes
$$maximize \sum_Q \sum_{j=1}^n P(x) * X_j \ [where \ P(x) = [1 - (1 - P_1(x))(1 - P_2(x))]$$

Let the capacity of the main memory be M. The reduction algorithm takes a knapsack of capacity M and a singleton set Q, and tries to put some l nodes one by one into it, out of a total n potential candidates, based on the probability given by $P(x)$. Hence, the problem degenerates to the same formulation as the 0/1 Knapsack problem, and the rest of the proof follows along exactly the same lines as proving the NP-Hardness of the 0/1 Knapsack.

3   Improving the Disk Layout

We note that in genome databases, the consecutive base nucleotides have relatively different proportions. Our post-construction algorithm Approx. Q-OptDL exploits this fact by bringing, in a probabilistic fashion, a node's more frequent successors to the same disk page. The inputs to the algorithm are: (a) $\mathbf{r}$: root of the subtree to be traversed, (b) $\mathbf{B}$: capacity of the disk-page in terms of the number of nodes, and (c) $\mathbf{Q}$: set of patterns to be matched[1]. Algorithm 3.1 can be invoked periodically to improve the layout.

3.1   Approx. Q-OptDL(r,B,Q)

```
queue ← r
nodecount ← 0;
while queue not empty, do
        r ← queue;          //remove from the queue
        if r not visited then
                mark r as visited and increment nodecount
        while there is an unmarked child c of r, do
                P_{rc}^{Q} ← Relative proportion of base at c among
                        all the unmarked base child nodes of r in Q
                if c not marked visited AND nodecount < B then
                        mark c as visited with a probability P_{rc}^{Q}
                        if c is marked visited then
                        increment nodecount;
                        queue ← c;          //insert into the queue
                        s ← suffix-link(c);
                        if s not visited AND nodecount < B then
                                mark s as visited
                                increment nodecount
                                queue ← s
        if nodecount ≥ B then
                while queue not empty do
                        m ← queue
                        Approx. Q-OptDL(m,B,Q)
```

3.2   Performance Bound on Approx. Q-OptDL

In the following discussion, cost refers to the I/O activity caused due to limitations of the underlying disk layout.

---

[1] Note that Q represents some sort of prior knowledge about the search queries. In the absence of any such information, the sequence corresponding to the suffix tree could be used to initialize the probability values. When a new query comes, Q and the probability values in Q can be updated incrementally.

Theorem 2

The suffix tree disk layout obtained using *Approx. Q-OptDL* (Algorithm 3.1) has an asymptotic performance within twice that of the optimal disk layout.

Proof. Let P_{opt} and P denote the cost associated with the optimal layout and the layout L obtained using Algorithm 3.1 respectively, over an infinite number of patterns. Further, let $P_k(e)$ denote the cost of layout L while accessing Q, a set of k patterns. Then, $P = \lim_{k \to \infty} P_k(e)$.

Now, when we access a particular node x with the closest child node x', an I/O operation may be required if the next base in the pattern being matched is not present in the memory. Then, the conditional cost admitted due to this mismatch is given by $P_k(e|x, x')$. Suppose that during the matching process, at a particular selection step, the optimal layout chooses a node x with base θ, while the layout L using Algorithm 3.1 chooses node x'_k with a base θ'_k, then since base θ and θ'_k are conditionally independent of the nodes x and x', we have

$$P(\theta, \theta'_k | x, x'_k) = P(\theta|x)P(\theta'_k|x'_k)$$

A mismatch between the two layouts happens if $\theta \neq \theta'_k$ results in an I/O. Then, the conditional cost of this mismatch is given by

$$P_k(e|x, x'_k) = 1 - \sum_{i=1}^{m} P(\theta = t_i, \theta'_k = t_i | x, x'_k)$$

where m denotes the number of bases($m = 4$ for DNA)

$$\Rightarrow P_k(e|x, x'_k) = 1 - \sum_{i=1}^{m} P(t_i|x)P(t_i|x'_k) \tag{1}$$

We also note that instead of only Q, if different sets of patterns are used, then different layouts would be chosen by Algorithm 3.1. So, we take an *average layout*, under which conditional cost $P(e|x)$ is given by

$$P(e|x) = \int P(e|x, x'_k)p(x'_k|x)dx'_k \tag{2}$$

where $p(x'_k|x)$ represents the conditional density of x'_k on x. Using (1) and (2),

$$\lim_{k \to \infty} P_k(e|x) = \int [1 - \sum_{i=1}^{m} P(t_i|x)P(t_i|x'_k)]\delta(x'_k - x)dx'_k = 1 - \sum_{i=1}^{m} P^2(t_i|x)$$

Thereby, the asymptotic cost under layout L is given by

$$P = \lim_{k \to \infty} P_k(e) = \lim_{k \to \infty} \int P_k(e|x)p(x)dx$$

$$\Rightarrow P = \int [1 - \sum_{i=1}^{m} P^2(t_i|x)]p(x)dx \simeq \int [1 - P^2(t_{max}|x)]p(x)dx \qquad (3)$$

$$\Rightarrow P \simeq \int [2(1 - P(t_{max}|x))]p(x)dx$$

where t_{max} refers to the base with greatest probability, that is put into a disk page accordingly by Algorithm 3.1. Now,

$$\sum_{i=1}^{m} P^2(t_i|x) = P^2(t_{max}|x) + \sum_{i \neq max} P^2(t_i|x)$$

We seek to bound this sum by minimizing the second term subject to:
(a) $P(t_i|x) \geq 0$, and (b) $\sum_{i \neq max} P^2(t_i|x) = 1 - P(t_{max}|x) = P_{opt}(e|x)$
Note that (b) holds, since the optimal layout would tend to have least probability of incurring an I/O. Also, $\sum_{i=1}^{m} P^2(t_i|x)$ is minimized if all of the *a posteriori* conditional costs except that pertaining to t_{max} are equal. In the light of foregoing discussion, $P(t_i|x) = 1 - P_{opt}(e|x)$ *if* $i = max$; else $P(t_i|x) = \frac{P_{opt}(e|x)}{m-1}$.
We arrive at the following inequalities,

$$\sum_{i=1}^{m} P^2(t_i|x) \geq (1 - P_{opt}(e|x))^2 + \frac{P_{opt}^2(e|x)}{m - 1}$$

$$1 - \sum_{i=1}^{m} P^2(t_i|x) \leq 2P_{opt}(e|x) - \frac{m}{m-1}P_{opt}^2(e|x) \qquad (4)$$

Noting that the conditional variance $Var[P_{opt}(e|x)] \geq 0$, we get

$$\int P_{opt}^2(e|x)p(x)dx \geq P_{opt}^2 \qquad (5)$$

Then, using (3), (4), and (5), we obtain the following asymptotic bound:

$$P_{opt} \leq P \leq P_{opt}(2 - \frac{m}{m-1}P_{opt}) \qquad (6)$$

References

1. Gusfield, D.: Algorithms on Strings, Trees and Sequences: Computer Science and Computational Biology. Cambridge University Press, Cambridge (1997)
2. Bedathur, S.J., Haritsa, J.R.: Search- Optimized Suffix- Tree Storage for Biological Applications. In: Bader, D.A., Parashar, M., Sridhar, V., Prasanna, V.K. (eds.) HiPC 2005. LNCS, vol. 3769, pp. 29–39. Springer, Heidelberg (2005)
3. Ko, P., Aluru, S.: Optimal self-adjusting trees for dynamic string data in secondary storage. In: Ziviani, N., Baeza-Yates, R. (eds.) SPIRE 2007. LNCS, vol. 4726, pp. 184–194. Springer, Heidelberg (2007)
4. Kärkkäinen, J., Sanders, P., Burkhardt, S.: Linear work suffix array construction. JACM 53(6), 918–936 (2006)
5. Abouelhoda, M.I., Kurtz, S., Ohlebusch, E.: Replacing suffix trees with enhanced suffix arrays. J. Discrete Algorithms 2(1), 53–86 (2004)
6. Dementiev, R., Kärkkäinen, J., Mehnert, J., Sanders, P.: Better external memory suffix array construction. ACM Journal of Experimental Algorithmics 3.4, 12 (2008)

Author Index

Printing: Mercedes-Druck, Berlin
Binding: Stein+Lehmann, Berlin